BIOGEOCHEMISTRY OF GLOBAL CHANGE

BIOGEOCHEMISTRY OF GLOBAL CHANGE

Radiatively Active Trace Gases

SELECTED PAPERS FROM THE TENTH INTERNATIONAL SYMPOSIUM ON ENVIRONMENTAL BIOGEOCHEMISTRY, SAN FRANCISCO, AUGUST 19-24, 1991

SPRINGER-SCIENCE+BUSINESS MEDIA, B.V.

Papers presented at the 10th International Symposium on Environmental Biogeochemistry, San Francisco, California, 19–23 August 1991

Originally published by Chapman & Hall, Inc. in 1993

Library of Congress Cataloging-in-Publication Data

Biogeochemistry of global change : radiatively active trace gases / editor, Ronald S. Oremland.
p. cm.
Includes bibliographical references and index.
ISBN 978-0-412-04141-9 ISBN 978-1-4615-2812-8 (eBook)
DOI 10.1007/978-1-4615-2812-8
1. Biogeochemistry—Congresses. 2. Greenhouse gases—Congresses. 3. Bioclimatology—Congresses. 4. Climatic changes—Congresses.
I. Oremland, Ronald S.
QH343.7.B54 1993
574.5'222—dc20 93-19052
CIP

Contents

Terrestrial Systems

Preface

Public awareness has increased about the prospect of global climatic change occurring in the coming century. Whereas broad changes in the Earth's climate have taken place in the geologic past, they have, with some significant exceptions, been of a gradual nature. Gradual change allows time for plants and animals to evolve via natural selection and to produce species capable of thriving in the new climate. Additionally, the geographic range of species already present will either expand or contract with an altered climate, depending on their ability to establish themselves in new environment.

It is now possible that anthropogenic activity may induce such a change and that this may occur at a rate too rapid for present ecosystems to adapt to or be replaced by systems suited for the new climate. Of even greater concern is whether agricultural production can be sustained to meet human needs in the face of such broad environmental change. Although there is a great disparity of views among climatologists with regard to the extent, significance, and potential disruptive effects of climate change, there is general agreement that some degree of change will occur.

The basis for this type of climatic change is directly related to the increased abundance of radiatively active trace gases in the atmosphere. "Radiatively active" refers to the capacity of certain gases, such as CO_2, CH_4, N_2O, and chlorofluorocarbons (CFCs), to absorb infrared energy reflected from sunlight incident on Earth's surface. The increased abundance of these gases derives from expanded human industrial and agricultural activities occurring over the past two centuries. Because many of these heat-trapping gases have demonstrated continuously increasing patterns of atmospheric mixing ratios over time periods ranging from decades to centuries, a "greenhouse effect" has occurred as a consequence of their buildup. In addition, some of these gases, such as CFCs and N_2O, pose a threat to the integrity of the stratospheric ozone layer. Still others, such as dimethylsulfide, although not "radiatively active," nonetheless influence climate because their oxidation products induce cloud condensation. Hence, these gases act in complex, interdependent patterns that affect atmospheric chemistry and physics. The sum of their influences on these processes, as well as on global change, is difficult to predict. In addition, their sources and fates are not well-known.

With the exception of only a few cases (e.g., CFCs), most of these gases arise from mixed sources, that is, from industry as well as agriculture, biomass-burning, and emissions from unperturbed ecosystems. Similarly, recognized sinks are diverse and include atmospheric photochemical reactions, dissolution into the oceans, and biological uptake by soil and plant communities. Understanding the relative importance of these sources and sinks to the past, present, and future global balance of these gases, as well as their influence on the biota and global habitability, poses a difficult and interdisciplinary challenge to environmental scientists and managers. Clearly, this topic encompasses significant aspects of numerous disciplines, including atmospheric chemistry, oceanography, paleoclimatology, geochemistry, soil and plant sciences, and microbiology, to name but a few. In short, this falls under the general discipline of "environmental biogeochemistry," with the scope of the problem being beyond the expertise of any individual investigator. It follows that no single conference or book can hope to cover adequately the breadth of this rapidly moving field. Nonetheless, because of the diverse and disparate perspectives of the disciplines listed above, it is essential for individual researchers to gain an appreciation of the breadth of the field. One mechanism to achieve this is by holding interdisciplinary conferences that allow a free exchange of experimental results on the topic of the biogeochemistry of radiatively active trace gases. This volume consists of invited and contributed papers presented at the Tenth International Symposium on Environmental Biogeochemistry (ISEB), held in San Francisco during the week of 19–25 August 1991. There are 45 chapters in this volume, with diverse topics ranging from geologic temporal scales of mass transfer to aspects of microbial biochemistry related to the formation of CH_4. Most of the chapters deal with experimental assessments of the fluxes of trace gases or the biological formation (or consumption) of trace gases in various ecosystems.

I conceived the idea of hosting such a meeting when I attended the Ninth ISEB in Moscow in 1989. I previously organized a meeting devoted to atmospheric methane at the 1987 American Chemical Society meeting in Denver, but I was interested in conducting a meeting having a broader theme. Inasmuch as the ISEB is an international scientific organization and because "radiatively active" trace gases are truly a problem of global concern, the ISEB would be an appropriate forum for an expanded conference. My offer was accepted, and as the ISEB meets on alternate, odd-numbered years, I returned home thinking that two years was more than sufficient to put it all together. I soon realized that time was already quite short and that I would be working on this project for the next three years, if you include editing this volume. My life was occupied with worrying about where to hold the conference, what agencies were willing to support it, mailing lists and flyers, bank accounts and bills, session topics and chairpersons, reservations and cancellations, abstracts, manuscripts, publishers, receptions, tours, banquets, and, of course, whether the San Francisco Giants would play a home game that week. (They did!) So much for my naiveté!

There are many individuals and institutions to thank for bringing this conference and book to fruition, but as the result of space restrictions, I can acknowledge only some of them. First of all, the scientific agencies of the United States Government that provided generous financial support for the event: The Geological Survey, the National Aeronautics and Space Administration, the National Oceanic and Atmospheric Administration, the Department of Energy, the Environmental Protection Agency, the Department of Agriculture, and the National Science Foundation. In particular, the National Science Foundation acted as a funding/clearinghouse for several of the above agencies, in addition to providing its own share of funds. I am indebted to the efforts of Dr. Jarvis Moyers, who successfully guided the various ISEB funds through the bureaucratic maze, and did so in a timely fashion. If for nothing else, I wish to thank him for being able to achieve all this, even in the face of my many annoying, panic-stricken phone calls! I wish to thank Jo Ann Glennon and her staff at the Cathedral Hill Hotel for making the meeting go so smoothly, and the members of the ISEB 10 National Committee for the programmatic work, and the ISEB 10 Local Committee for numerous arrangements. As all the manuscripts in this volume received two anonymous peer reviews, I thank these nameless souls for their efforts. Most important, I am deeply grateful to my associate, Ms. Francis Strohmaier, who was involved in every facet of this undertaking, and whose attention to detail and tireless secretarial efforts not only helped me to retain my sanity, but made both the conference and the publication of the book possible.

Ronald S. Oremland
Chairman, ISEB
Menlo Park, California
September 22, 1992

Paleoatmospheres and Climates

Biogeochemical Cycles of Carbon on a Hierarchy of Time Scales

James C.G. Walker

Space Physics Research Laboratory, Department of Atmospheric, Oceanic and Space Science, and Department of Geological Sciences, The University of Michigan, Ann Arbor, Michigan 48109

ABSTRACT

The carbon dioxide content of the atmosphere affects life, climate, and the chemistry of oceans and sedimentary rocks. The processes that control atmospheric carbon dioxide can be analyzed in terms of biogeochemical cycles that transfer carbon between reservoirs. The small atmospheric reservoir is closely linked to other small reservoirs, shallow sea and biota. Transfer of carbon between these small reservoirs is rapid; residence times are short. The small reservoirs are connected by slower transfers to a much larger deep sea reservoir, and the coupled reservoirs of ocean, atmosphere, and biota are coupled, in turn, to very much larger reservoirs in sedimentary rocks. The exogenic reservoirs of sedimentary rocks, ocean, atmosphere, and life are coupled, in turn, by the very slow exchange of carbon with a large mantle reservoir. This hierarchy of interacting reservoirs controls atmospheric carbon dioxide on all time scales. Particular problems with restricted time scales can be analyzed in terms of a subset of the total system. Observations that illustrate interactions on different levels of the hierarchy include the seasonal change of carbon dioxide partial pressure, the glacial to interglacial change, and the evolving distribution of carbonate sediments.

Introduction

What controls the amount of carbon dioxide in the atmosphere? How has this amount changed in the past? How might it change in the future? To answer these questions, we must examine the biogeochemical cycles of carbon on various time scales. In terms of the evolution of environmental composition, there is probably no other constituent as interesting and important as carbon. Carbon dioxide is the substrate for photosynthesis and, thus, for all life. As a greenhouse gas, it has a great effect on climate. It leaves a record of change in sedimentary rocks because dissolved carbon affects marine chemistry, the pH of sea water, and the formation of carbonate sediments. In addition, there is a close relationship between the

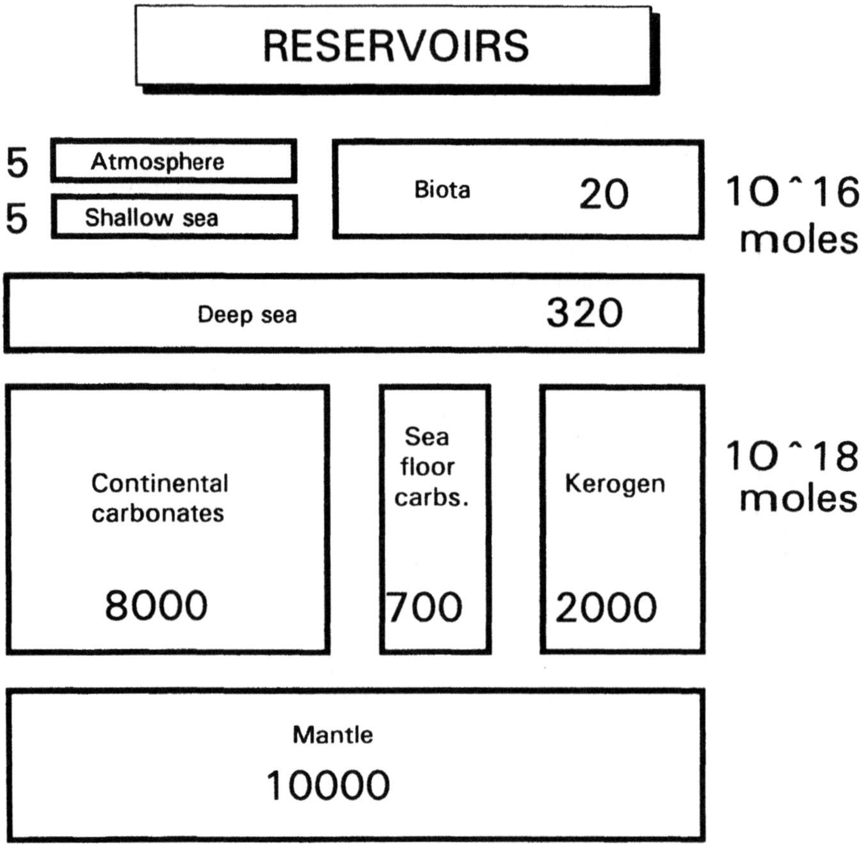

Figure 2. Global carbon reservoirs. Amounts are expressed in 10^{16} moles of carbon for the atmosphere, ocean, and biota reservoirs and in 10^{18} moles for the sedimentary rock and mantle reservoirs. In this and other figures, 10^18 means 10^{18}.

The Carbon System

For the carbon system there are a number of reservoirs that are important on different time scales and for different problems. A summary is presented in Figure 2.[20,32,39–41,62,81,82,91,100] Reservoir amounts are expressed in units of 10^{16} moles for atmosphere, shallow sea, biota, and deep sea at the top of the figure and in units of 10^{18} moles for sedimentary reservoirs and the mantle at the bottom of the figure. The atmosphere, the shallow sea, and the biota reservoirs are relatively small. The deep sea contains much more carbon than any of these. The sedimentary reservoirs, continental carbonates, sea floor carbonates, and kerogen, are very much larger still. The size of the mantle reservoir is not well-known, but it is probably at least as large as the combined surface (or exogenic) reservoirs.[24,33,45,103] In principle, to understand what controls the carbon dioxide content of the atmosphere, we must consider exchanges of carbon with all of these other reservoirs. In practice, how much to consider depends on the problem of interest and its time scale. But, do not forget that the atmosphere, even the atmosphere and the ocean, are a very small part of Earth's total carbon complement.

In the sedimentary level of Figure 2, a distinction was made among reduced carbon in the kerogen reservoir, oxidized carbon in the form of continental carbonates, and oxidized carbon in the form of sea floor carbonates. The distinction between reduced and oxidized carbon is important to the coupled biogeochemical cycles of oxygen.[11,12,14,16,38,59,99,101,104] The distinction between carbonate reservoirs is important because of their very different residence times. Sea floor carbonates have a short residence time, geologically speaking, and continental carbonates have a long residence time.[93–95] This difference may have influenced the geological history of carbon dioxide.[22]

To provide some information on rates of exchange, the residence times of carbon in these same reservoirs are shown in Figure 3. The quoted residence times correspond to the most rapid process removing carbon from the reservoir in question. Carbon survives for about 5 years in the atmosphere, 20 years in the biota, 10 years in the shallow sea, and much longer, 1500 years, in the deep sea. The time scale is different in the lower half of the figure. Carbon resides in sedimentary rocks for tens to hundreds of millions of years and in the mantle for billions of years. The time scale of the problem under study determines which of these reservoirs must be included in a theoretical analysis. In general, exchange between small reservoirs is rapid, and exchange between large reservoirs is slow.[92] There exists a natural hierarchy of reservoirs, residence times, and exchange rates. I shall now work my way up this hierarchy, starting with the small and fast. I shall illustrate the interactions of the carbon cycle by means of selective examples of change, eschewing changes that have been deduced theoretically in favor of changes that have been directly observed. This is not the place to review all of the suggestions that have been made about change in the carbon cycle during the course of Earth's history.

Seasons to Centuries

A familiar illustration of carbon cycle changes on a short time scale is the carbon dioxide concentration measured by Charles Keeling at Mauna Loa, shown in Figure 4.[28] As is known, the secular change in this record is a consequence in part of the burning of fossil fuels and in part of changes in land

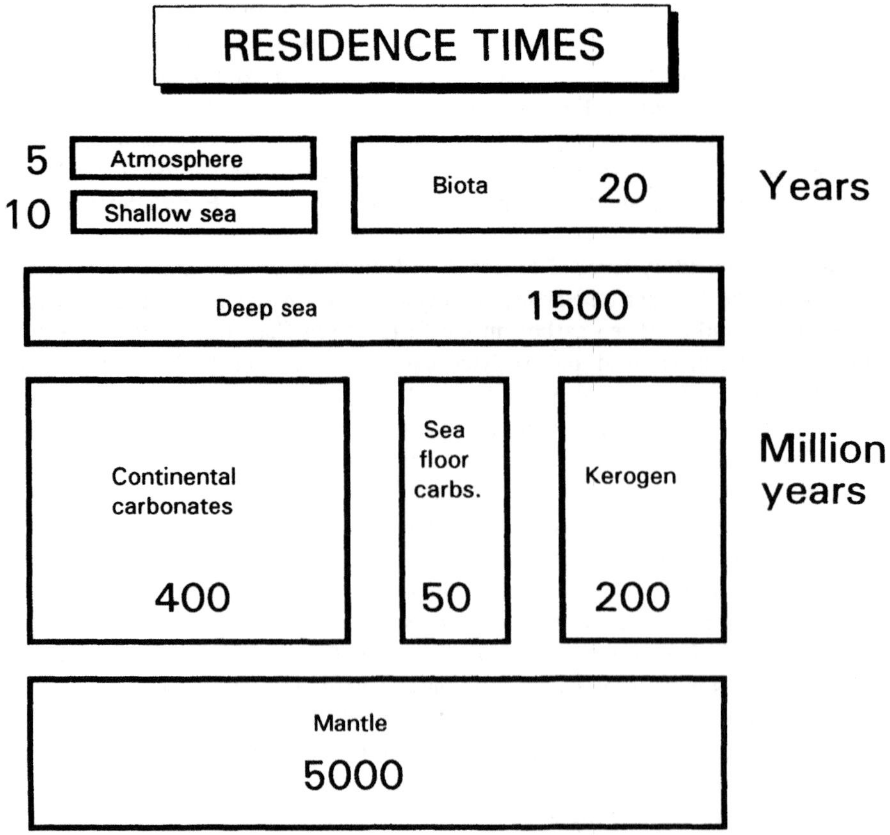

Figure 3. Residence times of carbon in global reservoirs. Times are expressed in years for the atmosphere, ocean, and biota reservoirs and in millions of years for the sedimentary rock and mantle reservoirs.

use.[37,41,43,74,88,116,117] I will not discuss these phenomena further, but want to consider, instead, the seasonal changes that are also evident in Figure 4. Carbon dioxide rises during the winter and decreases during the summer. There is nothing mysterious about these changes. They are a consequence of the exchange of carbon dioxide between the atmosphere and the biota with some modulation arising from exchange with the shallow sea. Figure 5 shows the reservoirs and the rates of exchange that influence the seasonal change of carbon dioxide. Reservoir amounts are expressed in units of 10^{16} moles and rates of exchange in units of 10^{16} moles/year. About 10^{16} moles of carbon are exchanged each year between the atmosphere and biota in the processes of photosynthesis and decay. Exchange with the shallow sea occurs at a comparable rate.

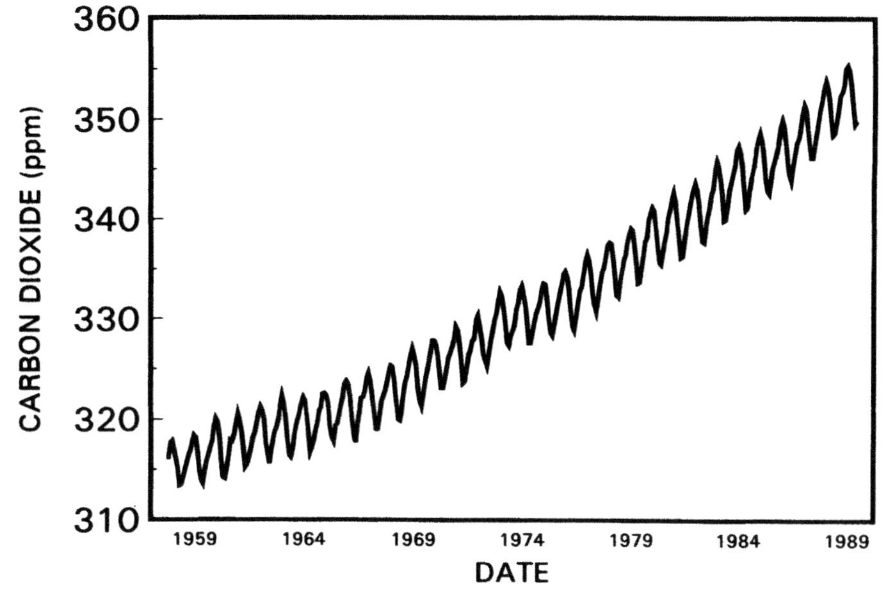

Figure 4. Carbon dioxide concentration in parts per million by volume measured by Keeling at Mauna Loa (from Ref. 28).

These are the reservoirs and the processes that one needs to consider if analyzing a problem on a time scale of years, a problem such as the seasonal change of carbon dioxide. To simulate the seasonal change of carbon dioxide, the cyclist does not have to consider all of the other reservoirs. Their roles are negligible. The associated rates of exchange are too slow. Obviously, carbon dioxide goes down during the summer because the rate of photosynthesis exceeds the rate of decay and it goes up during the winter because the reverse is the case. The response of the atmosphere to changing rates of photosynthesis and decay is modulated by the movement of carbon dioxide into and out of the shallow sea.[4,35,36,42,52,53,66,73,83,89]

Figure 5, however, introduces several other questions for which the answers are less obvious. For example, what determines the proportion of carbon dioxide in the atmosphere as compared to the biota? Why are there not 20 moles in the atmosphere and 5 in biomass? What sets these limits? Frankly, we are not really sure. There are various possibilities. Some people argue that the rate of photosynthesis increases with the carbon dioxide concentration in what is called the carbon dioxide fertilization effect.[1,25,37,53,60,114] In a system sense, such a feedback would work, yielding a stable partitioning of carbon between atmosphere and biota. On the other hand, some people attribute more importance to a

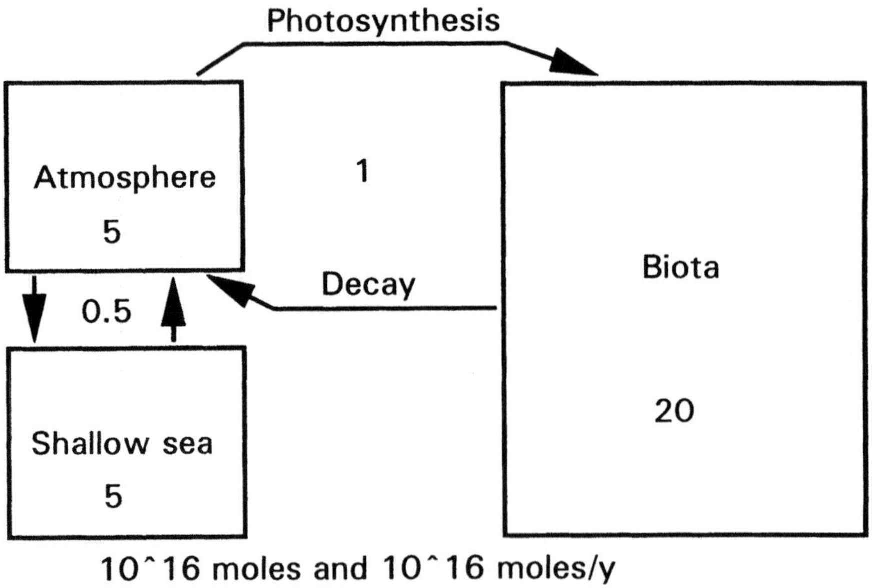

Figure 5. Reservoirs and transfer processes that affect carbon dioxide on time scales of about a year. Reservoir amounts are expressed in 10^{16} moles and rates of exchange in 10^{16} moles/year. The biota reservoir includes soil organic matter.

presumed increase in the rate of decay with increasing temperature,[44,46,77] temperature being an increasing function of atmospheric carbon dioxide through the greenhouse effect. Such a feedback, by itself, would be destabilizing, with increasing temperature driving ever more carbon dioxide from the biota into the atmosphere. Many biogeochemical cyclists suspect that ecological interactions result in almost a first-order rate law for conversion of biomass into carbon dioxide. More biomass supports more animals and more grazing and, therefore, more decay. According to this analysis, then, the mass of carbon in the biota is controlled by the rate of photosynthesis and the ecological interactions of the food web. The kind of question discussed in this paragraph would be easy to answer if we knew the global rate laws, but answers remain speculative today because we do not know the rate laws.

Quite another question is what sets the total amount of carbon in the closely coupled reservoirs shown in Figure 5. Why are there 30×10^{16} moles of carbon in atmosphere, biota, and shallow sea rather than 60×10^{16} or 15×10^{16} moles? This is a question that we must consider if we want to know what sets the level of carbon dioxide in the atmosphere. But this question can only be examined in terms of the next level of the reservoir hierarchy. Figure 6 shows the

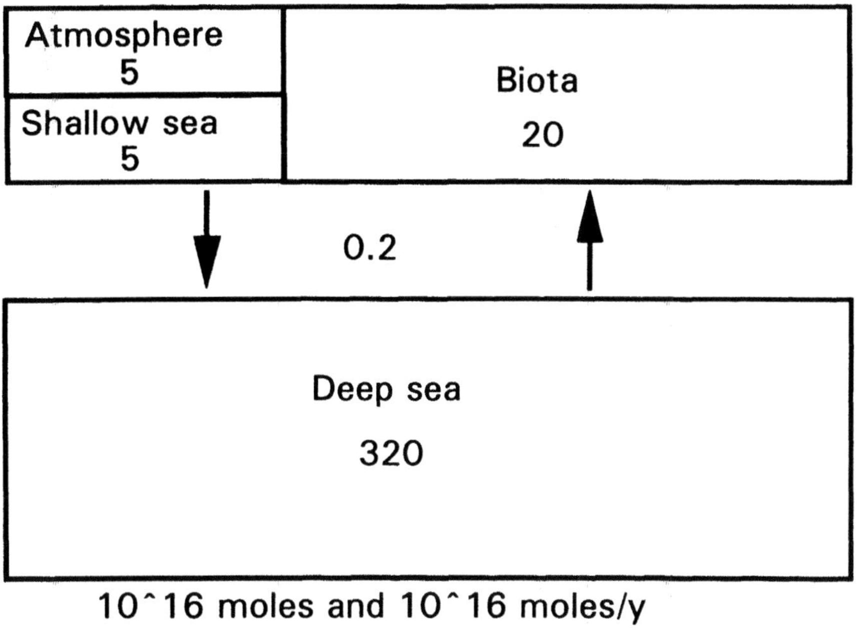

Figure 6. Reservoirs and transfer processes that affect carbon dioxide on time scales of about a thousand years. Reservoir amounts are expressed in 10^{16} moles and rates of exchange in 10^{16} moles/year.

system with the next level included. In this figure, reservoirs are still measured in 10^{16} moles and rates of exchange in 10^{16} moles per year. The residence time of carbon in the deep sea is about 1600 years. Because the atmosphere, biota, and shallow sea equilibrate among themselves on a time scale much shorter than this, the figure shows these reservoirs lumped together.

The carbon content of atmosphere, biota, and shallow sea combined is controlled by the exchange of carbon, on a time scale of thousands of years, with the much larger deep sea reservoir. The balance between carbon in the surface reservoirs and carbon in the deep sea depends on marine chemistry and the pH of sea water. This is one of the better-understood areas of carbon cycle modeling.[20,70] The pertinent reactions are summarized in Figure 7. Carbon dioxide dissolves and dissociates so that the partial pressure of carbon dioxide is proportional to the hydrogen ion concentration. Most of the carbon dissolved in sea water is in the form of the bicarbonate ion. So carbon dioxide partial pressure, proportional to the mass of carbon dioxide in the atmosphere, depends on the total dissolved carbon content of the ocean and the pH of sea water.

SPECIATION DEPENDS ON pH

$$CO_2 + H_2O = H_2CO_3 = H^+ + HCO_3^-$$

$$P(CO_2) = a[H^+][HCO_3^-]$$

$$HCO_3^- = H^+ + CO_3^=$$

$$[CO_3^=] = b[HCO_3^-]/[H^+]$$

$$CO_3^= + Ca^{++} = CaCO_3$$

Figure 7. The balance between dissolved and gaseous carbon depends on the pH of sea water. Average sea water is approximately saturated with respect to calcium carbonate.

Thousands of Years

Figure 8 presents an example of change on a time scale of tens of thousands of years. It is a record of the carbon dioxide amount in the atmosphere over the last 150,000 years, deduced from measurements of the properties of bubbles of air in Antarctic ice.[5] The record exhibits a remarkable change in carbon dioxide partial pressure between glacial periods with their maxima at 150,000 and 20,000 years ago and intervening interglacial periods. Carbon dioxide concentration is small during glacial times and large during interglacials. The climate feedback is, therefore, positive. At the end of an ice age, the greenhouse effect of increasing carbon dioxide makes it easier for the ice to melt. The causes of this change are not yet established beyond disagreement, but change on this time scale is almost certainly driven by the transfer of carbon in and out of the deep ocean reservoir. Such transfer, in turn, results from the change in the pH of sea water caused by the change in total dissolved carbon or alkalinity or both, or

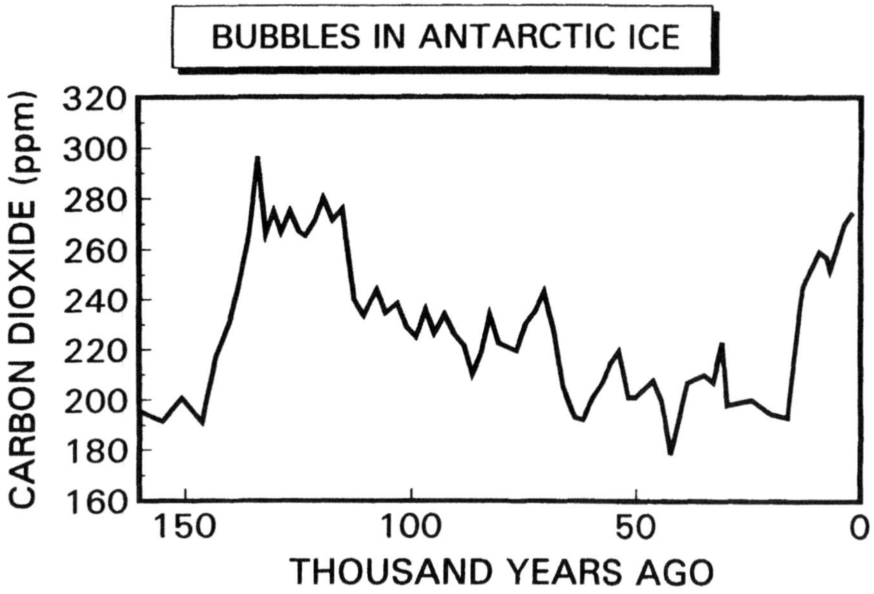

Figure 8. Change in atmospheric carbon dioxide during the last 150,000 years deduced from measurements of bubbles of air in Antarctic ice (from Ref. 5). In the earth sciences, some people plot time going forward to the right, as done here, and some plot age increasing to the right. To avoid confusion, be sure to check the axes.

possibly redistribution of these properties between surface ocean and deep ocean.[6–9,18,21,26,54,55,64,67,71,79,84] The rock cycle may be involved, which introduces the next level in the carbon cycle hierarchy.

Millions of Years

Figure 9 shows features of the biogeochemical cycles of carbon that are important on time scales of about a million years. On this time scale, ocean, atmosphere, and biota are closely coupled and effectively equilibrated among themselves. The combined reservoir of surface carbon interacts slowly with the very much larger sedimentary reservoirs. Because the sedimentary reservoirs are so large, I use different units in Figure 9. Reservoir sizes are expressed in 10^{18} moles and rates of exchange in 10^{18} moles/million years. Residence times, calculated by dividing rates into reservoirs, are 100 million years or so for the sedimentary reservoirs.

I have shown the exchange in and out of the kerogen reservoir as being in

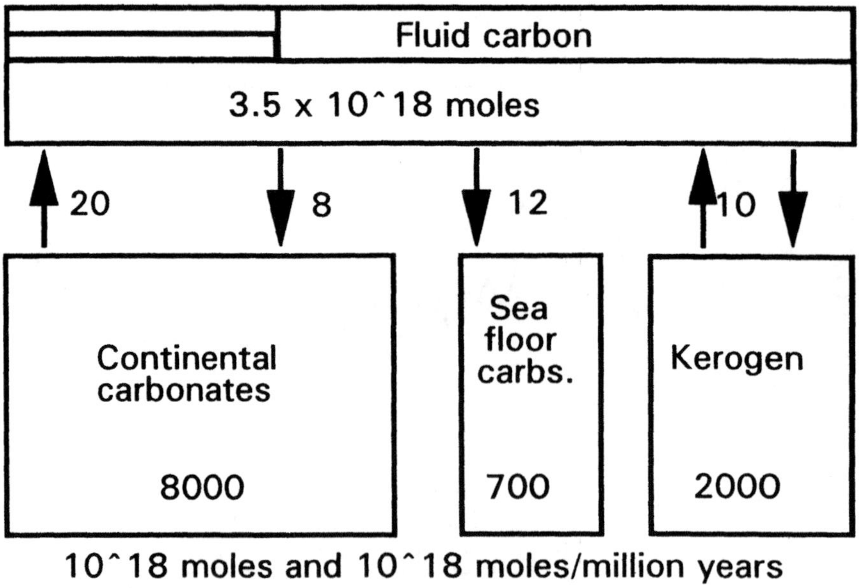

Figure 9. Reservoirs and transfer processes effective on the million year time scale. Reservoir amounts are expressed in 10^{18} moles and rates of exchange in 10^{18} moles/million years.

balance, but the data are by no means precise enough to confirm that such a balance exists at present. The kerogen budget, of course, is where the geochemical cycles of carbon connect with the geochemical cycles of oxygen[10,38–40,59,99–101] because the organic matter of kerogen is produced by photosynthesis, with a corresponding production of oxygen. Imbalances in the kerogen budget cause the concentration of atmospheric oxygen to change. The residence time of oxygen in the atmosphere against rock cycle interactions is a few million years. Exchange with the kerogen reservoir is also where the biogeochemical cycles of carbon affect the record of carbon isotopic composition preserved in sedimentary rocks[58,61,80,106] because organic matter is relatively deficient in the heavy carbon isotope. So there is an isotopic record of changes in the kerogen budget which contains information concerning both the carbon budget and the oxygen budget. Much effort in the reconstruction of ancient oceanic and atmospheric composition is devoted to this record.[11,12,14,16]

The present state of balance of the kerogen reservoir may not be known, but the budgets of the sedimentary carbonate reservoirs are clearly not balanced. Figure 9 makes an important distinction between long-lived continental carbonates and short-lived sea floor carbonates. Carbonate in these reservoirs recycles

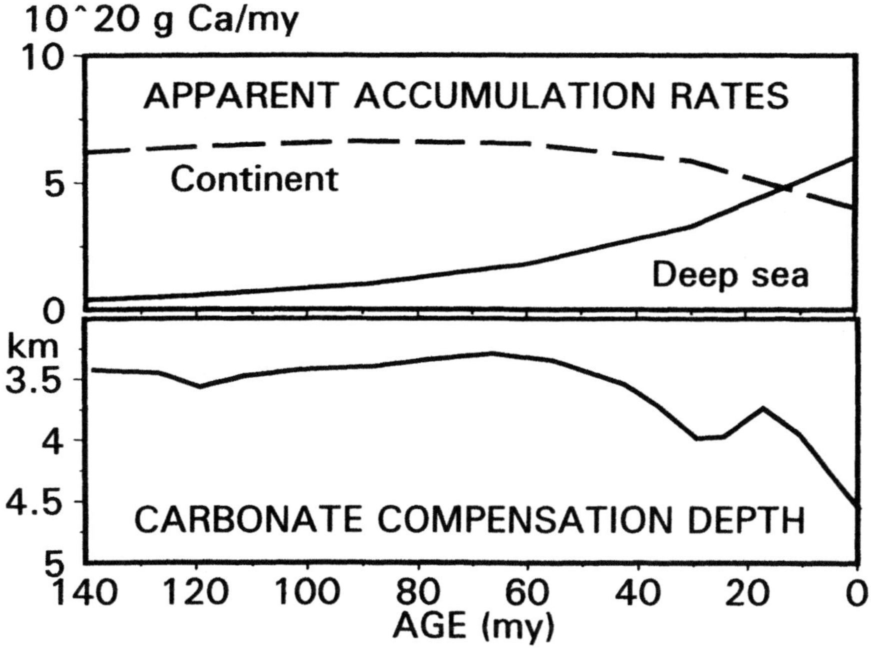

Figure 10. Carbonate accumulation rates during the last 150 million years and the depth at which sea water is significantly undersaturated with respect to calcite (from Ref. 115).

to atmosphere and ocean partly by exposure to the atmosphere followed by dissolution of the carbonate minerals and partly by metamorphic decarbonation. It is most interesting that the sea floor carbonate reservoir is presently growing at the expense of the continental carbonate reservoir. In this discussion, the present should be understood to mean this period of Earth's history, not this instant of human history. I am talking about Earth system evolution, not the impact of human activities.

Imbalance in the Budget of Carbonate Sediments

The data on carbonate accumulation rates over the last 140 million years are shown in Figure 10. The figure shows a smooth version of observations that have been reviewed by Wilkinson and Walker.[115] The decline in apparent accumulation rate going back in time is a consequence of the disappearance of old rocks caused by erosion. The point to note is that the pelagic carbonate reservoir in the geologically recent past has been growing while the continental carbonate res-

GROWTH OF DEEP SEA CARBONATE RESERVOIR

☐ CAUSES

- Evolution of carbonate-secreting plankton
- Decrease of shelf area

☐ CONSEQUENCES

- Transfer from long-lived continental reservoir to short-lived sea floor
- Transfer from exogenic system to mantle

Figure 11. Causes and consequences of the growth of the deep sea carbonate reservoir.

ervoir has been declining. At the same time, as Figure 10 shows, there has been an increase in the depth of the calcium carbonate compensation level in the global ocean. The ocean has become more saturated with respect to calcium carbonate. This secular change in the carbon cycle has interesting consequences and interesting possible causes.

There are two mechanisms that may be causing this change in the global carbon cycle. One is the origin and evolutionary expansion of the plankton that precipitate calcium carbonate. Their shells rain into the deep ocean to accumulate as deep sea carbonates, but they evolved just 100 or 150 million years ago. Deep sea carbonate sediments could not accumulate in abundance before the carbonate-secreting plankton arose.[17] Do we have, here, an unambiguous interpretation of change in Earth's history? Unfortunately not. At the same time as the plankton have been evolving and diversifying, there has been a marked decrease in the area of shallow water continental shelves, the shelves on which continental carbonates accumulate. The ocean must precipitate carbonate minerals. If shelves are not available, the deep sea will have to take their place.[34,75] We just do not know whether the changes shown in Figure 10 in the locus of accumulation of carbonates and in the saturation state of the ocean are caused by biological change or by tectonic change. As summarized in Figure 11, either mechanism could explain the observations.[107,111]

Although the debate about causes continues, there is little argument about important consequences of these changes. The system is transferring carbon from a long-lived continental reservoir to a short-lived sea floor reservoir; carbon on the sea floor recycles ten times faster than carbonate on the continents. Faster recycling will have a marked impact on the carbon cycles.[22,98] At the same time, these evolving processes are transferring carbon from the exogenic system of sediments, ocean, and atmosphere to the mantle because the potential for subduction of sea floor sediments is much greater than for continental sediments.[69,76] If these trends continue, we can expect to end up on a world without continental limestones. It will be a very different world. All of the carbon will have gone back into the deep sea, and much of it will have gone back into the mantle. Although it is not clear precisely which mechanism has driven the change, it is clear that the change was not possible until carbonate-secreting plankton had evolved. The underlying cause, therefore, is biological evolution. This phenomenon is not the only example of change in the global chemical environment caused by biological evolution, but it is one of the best documented.

Rock Cycle

I turn now to the question of what controls the proportion of carbon in ocean, atmosphere, and biota as opposed to the proportion of carbon in the sedimentary reservoirs? What determines the partitioning? Once again, this is a question that has to be answered before we can claim to understand what controls the amount of carbon dioxide in the atmosphere. Once again, the answer is not known, but it involves the rock cycle.

Calcium carbonate sediments are recycled in part by dissolution. The corresponding circulation of carbon is understood reasonably well, at least in principle. The addition of dissolved calcium and carbonate ions to the ocean raises the pH of sea water and increases the degree of saturation with respect to calcium carbonate by reactions summarized in Figure 7. The ocean system can be expected to respond by precipitating calcium carbonate minerals fast enough to balance input and output. The fate of carbon recycled by metamorphic decarbonation is less clear. This carbon is released to the atmosphere as carbon dioxide gas. The supply is augmented by fluxes of carbon dioxide from the mantle.[2,24,29–31,45,68,109] On geological time scales, the added carbon dioxide must be neutralized by dissolved cations released from noncarbonate (silicate) minerals and removed from the fluid reservoirs by precipitation of new carbonate minerals.[19]

The traditional view is that the release of cations occurs mainly by weathering processes taking place on land, as shown in Figure 12. The balance between fluid carbon and sedimentary carbon is set by the silicate weathering process. In principle, the rate of weathering depends on the amount of carbon dioxide in the atmosphere, not necessarily as a first-order reaction, but still as an increasing

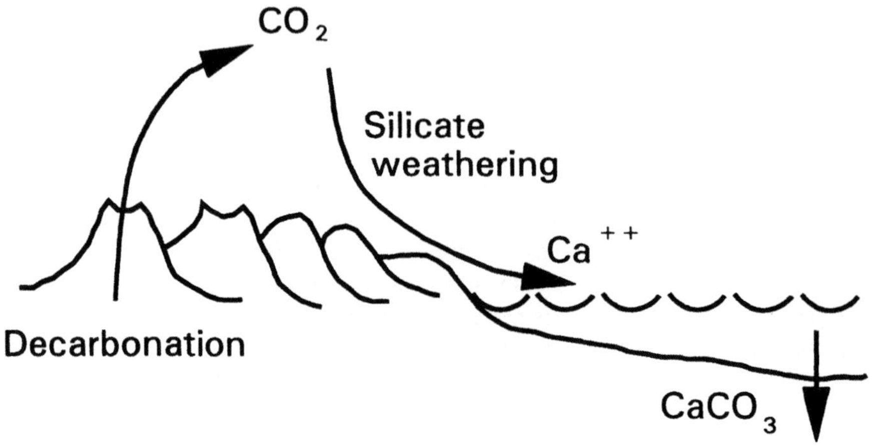

Figure 12. Silicate weathering neutralizes volcanic carbon dioxide.

function of carbon dioxide amount. The interaction constitutes a negative feedback. If volcanism and metamorphism release more carbon dioxide, the atmospheric content increases, the rate of silicate weathering increases, and the cations released by weathering fix more carbon into the reservoir of sedimentary carbonate rocks. This mechanism has been explored by a number of biogeochemical cyclists.[13,15,47,49–51,57,62,65,96,113]

Although the terrestrial weathering mechanism is popular, I am not convinced that it is correct. There is, as yet, no direct evidence that the global rate of silicate weathering is an increasing function of carbon dioxide partial pressure. In common with a number of others,[23,27,63,72,78,90,96] I suspect that the weathering rate is much more sensitive to such factors as geography, topography, lithology, and vegetation. Terrestrial weathering is certainly an important source of neutralizing cations, but it may not be the feedback process that controls the relative sizes of fluid and sedimentary carbon reservoirs.

Another possibility is sea floor weathering, in which sea water attacks sea floor basalt, releasing calcium ions and precipitating calcium carbonate. This calcium carbonate, which forms as veins and dispersed crystals within the sea floor basalt,[3,85–87] is carried down into the mantle by subduction or released as volcanic carbon dioxide when the sea floor is heated tectonically or metamorphically. The process is illustrated schematically in Figure 13. The distinction between sea floor weathering and continental weathering as the controlling source of neutralizing cations is not trivial. In the continental weathering mechanism, it is the partial pressure of atmospheric carbon dioxide that is buffered. In the sea floor weathering mechanism, it is the pH of sea water that is buffered. The continental weathering mechanism involves climate in the control of carbon

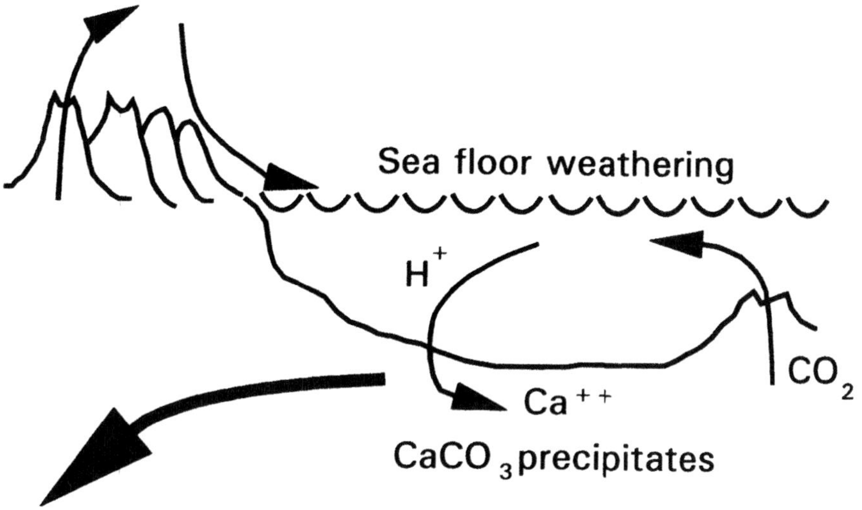

Figure 13. Sea floor weathering neutralizes volcanic carbon dioxide.

cycling[113] and suggests that important changes in carbon dioxide would have resulted from the growth of continents[48,56,108] and the evolution of land plants.[63,96,97,112] On the other hand, the continental weathering mechanism cannot have stabilized the carbon cycle on an Archean world with little continental land area.[108] Some other mechanism must have exercised control, possibly the sea floor weathering mechanism. In short, the two mechanisms imply quite different histories for atmospheric carbon dioxide.[105] At present, their relative importance is not known.

Billions of Years

So, finally, I come to the question of what determines how much carbon there is in the exogenic system of ocean, atmosphere, biota, and sedimentary rocks. Once again, if we cannot answer this question, we cannot really claim to understand what controls atmospheric carbon dioxide. The mass of exogenic carbon is determined by coupling to the mantle by way of degasing of carbon dioxide at mid-ocean ridges, and subduction of carbon, partly in the form of sea floor sediments and partly in the form of carbonate deposited within the sea floor by submarine weathering reactions. The estimated rate of exchange with the mantle, shown in Figure 14,[24,45,68,109] yields a residence time of exogenic carbon of about 5 billion years. The mantle reservoir is uncertain but probably

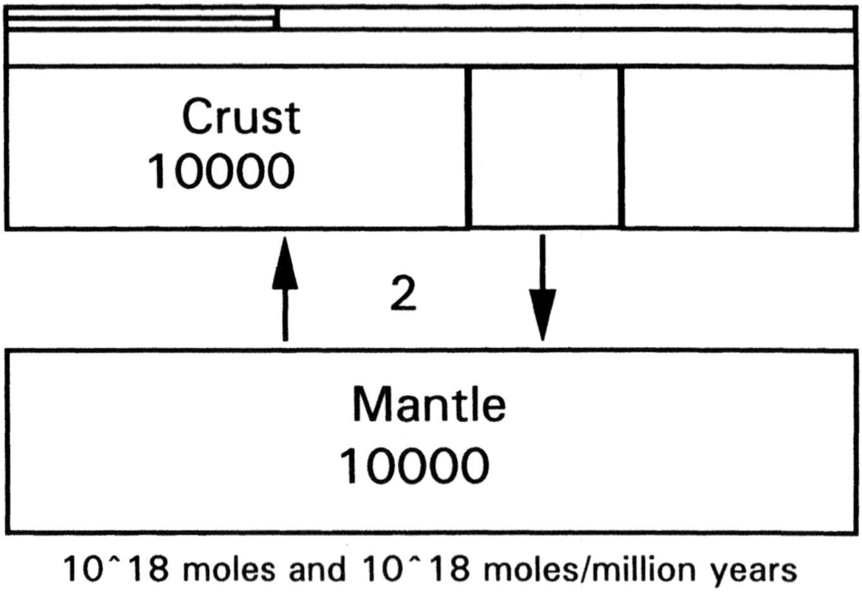

Figure 14. Coupling to the mantle on times of billions of years. Reservoir amounts are expressed in 10^{18} moles and rates of exchange in 10^{18} moles/million years.

comparable in size to the exogenic reservoir. With the inclusion of mantle carbon, I have accounted for all of Earth's carbon, or at least all of it that matters to the atmosphere and to we creatures who live and evolve in the atmosphere.

Conclusions

The biogeochemical cycles of carbon can be analyzed in terms of the hierarchy of interactions shown in Figure 15. Small reservoirs of carbon interact rapidly, on a time scale of years. Large reservoirs react more slowly, with residence times increasing toward the bottom of the figure. To understand, interpret, or predict the carbon dioxide content of the atmosphere, the biogeochemical cyclist must consider the rapid exchange of carbon between the small reservoirs of atmosphere, shallow sea, and biota, slow exchange with the deep sea and sedimentary reservoirs, and very slow exchange with the mantle reservoir. What needs to be considered within this hierarchy of interactions depends on the particular problem under study and its characteristic time scale.

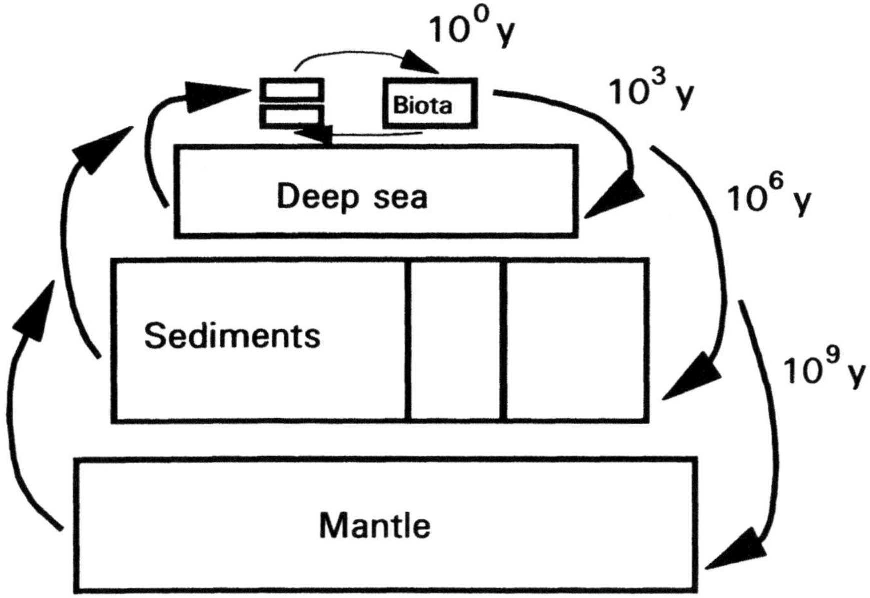

Figure 15. Hierarchy of interactions in the biogeochemical cycles of carbon.

Acknowledgments

My ideas about the biogeochemical cycles of carbon have benefited from discussions with Bruce Wilkinson, Bradley Opdyke, Jim Kasting, and John Hayes. This research was supported in part by the National Aeronautics and Space Administration under Grant NAGW-176.

References

1. Adams, R.M., C. Rosenzweig, R.M. Peart, J.T. Ritchie, B.A. McCarl, J.D. Glyer, R.B. Curry, J.W. Jones, K.J. Boote, and L.H. Allen. 1990. Global climate change and US agriculture. Nature **345**:219–224.
2. Allard, P., J. Carbonnelle, D. Dajlevic, J. Le Bronec, P. Morel, M.C. Robe, J.M. Maurenas, R. Faivre-Pierret, D. Martin, J.C. Sabroux, and P. Zettwoog. 1991. Eruptive and diffuse emissions of CO_2 from Mount Etna. Nature **351**:387–391.
3. Alt, J.C., J. Honnorez, C. Laverne, and R. Emmermann. 1986. Hydrothermal alteration of a 1 km section through the upper crust, Deep Sea Drilling Project Hole 504B: Mineralogy, chemistry, and evolution of seawater-basalt interactions. J. Geophys. Res. **91**:10309–10335.

4. Bacastow, R.B., C.D. Keeling, and T.P. Whorf. 1985. Seasonal amplitude increase in atmospheric CO_2 concentration at Mauna Loa, Hawaii, 1959–1982. J. Geophys. Res. **90**:10529–10540.

5. Barnola, J.M., D. Raynaud, Y.S. Korotkevich, and C. Lorius. 1987. Vostok ice core provides 160,000 year record of atmospheric CO_2. Nature **329**:408–414.

6. Berger, W.H. 1982. Increase of carbon dioxide in the atmosphere during deglaciation: The coral reef hypothesis. Naturwiss. **69**:87–88.

7. Berger, W.H. 1985. CO_2 increase and climate prediction: Clues from deep-sea carbonates. Episodes **8**:163–168.

8. Berger, W.H., and R.S. Keir. 1984. Glacial-Holocene changes in atmospheric CO_2 and the deep-sea record. In J.E. Hansen and T.S. Takahashi (eds.), Climate Processes and Climate Sensitivity, American Geophysical Union, Washington, DC, pp. 337–351.

9. Berger, W.H., and A. Spitzy. 1988. History of atmospheric CO_2: constraints from the deep-sea record. Paleoceanography **3**:401–411.

10. Berner, R.A. 1982. Burial of organic carbon and pyrite in the modern ocean: Its geochemical and environmental significance. American J. Sci. **282**:451–473.

11. Berner, R.A. 1987. Models for carbon and sulfur cycles and atmospheric oxygen: Application to Paleozoic history. American J. Sci. **287**:177–196.

12. Berner, R.A. 1989. Biogeochemical cycles of carbon and sulfur and their effect on atmospheric oxygen over Phanerozoic time. Palaeogeogr. Palaeoclimatol. Palaeoecol. **75**:97–122.

13. Berner, R.A. 1990. Atmospheric carbon dioxide levels over Phanerozoic time. Science **249**:1382–1386.

14. Berner, R.A., and D.E. Canfield. 1989. A new model for atmospheric oxygen over Phanerozoic time. American J. Sci. **289**:333–361.

15. Berner, R.A., A.C. Lasaga, and R.M. Garrels. 1983. The carbonate–silicate geochemical cycle and its effect on atmospheric carbon dioxide over the past 100 million years. American J. Sci. **283**:641–683.

16. Berner, R.A., and R. Raiswell. 1983. Burial of organic carbon and pyrite sulfur in sediments over Phanerozoic time: a new theory. Geochim. Cosmochim. Acta **47**:855–862.

17. Boss, S.K., and B.H. Wilkinson. 1991. Planktogenic/eustatic control on cratonic/oceanic carbonate accumulation. J. Geology **99**:497–513.

18. Boyle, E.A. 1988. The role of vertical chemical fractionation in controlling late Quaternary atmospheric carbon dioxide. J. Geophys. Res. **93**:15701–15715.

19. Broecker, W.S. 1971. A kinetic model for the chemical composition of sea water. Quaternary Res. **1**:188–207.

20. Broecker, W.S., and T.H. Peng. 1982. Tracers in the Sea. Lamont–Doherty Geological Observatory, Columbia University, Palisades, NY.

21. Broecker, W.S., and T.H. Peng. 1987. The role of $CaCO_3$ compensation in the

glacial to interglacial atmospheric CO_2 change. Global Biogeochem. Cycles **1**:15–29.

22. Caldeira, K. 1991. Continental–pelagic carbonate partitioning and the global carbonate–silicate cycle. Geology **19**:204–206.
23. Capo, R.C., and D.J. DePaolo. 1990. Seawater strontium isotopic variations from 2.5 million years ago to the present. Science **249**:51–55.
24. Des Marais, D.J. 1985. Carbon exchange between the mantle and the crust and its effect upon the atmosphere: Today compared to Archean time. In E.T. Sundquist and W.S. Broecker (eds.), The Carbon Cycle and Atmospheric CO_2: Natural Variations Archean to Present, American Geophysical Union, Washington, DC, pp. 602–611.
25. Esser, G. 1987. Sensitivity of global carbon pools and fluxes to human and potential climatic impacts. Tellus **39B**:245–260.
26. Farrell, J.W., and W.L. Prell. 1989. Climatic change and $CaCO_3$ preservation: An 800,000 year bathymetric reconstruction from the Central Equatorial Pacific Ocean. Paleoceanography **4**:447–466.
27. Francois, L.M., and J.C.G. Walker. 1992. Modelling the Phanerozoic carbon cycle and climate: Constraints from the $^{87}Sr/^{86}Sr$ isotopic ratio of seawater. American J. Sci. **292**:81–135.
28. Freyer, H.D. 1986. Interpretation of the Northern Hemispheric record of $^{13}C/^{12}C$ trends of atmospheric CO_2 in tree rings. In J.R. Trabalka and D.E. Reichle (eds.), The Changing Carbon Cycle: A Global Analysis, Springer-Verlag, NY, pp. 125–150.
29. Gerlach, T. 1991. Etna's greenhouse pump. Nature **351**:352–353.
30. Gerlach, T.M. 1989. CO_2 from magma-chamber degassing. Nature **337**:124.
31. Gerlach, T.M. 1991. Present-day CO_2 emissions from volcanos. EOS **72**:249–254.
32. Gregor, C.B., R.M. Garrels, F.T. Mackenzie, and J.B. Maynard (eds.). 1988. Chemical Cycles in the Evolution of the Earth, John Wiley and Sons, New York.
33. Harris, N. 1989. Carbon dioxide in the deep crust. Nature **340**:347–348.
34. Hay, W.W., and J.R. Southam. 1977. Modulation of marine sedimentation by the continental shelves. In N.R. Anderson and A. Malahoff (eds.), The Fate of Fossil Fuel CO_2 in the Oceans, Plenum Press, New York, pp. 569–604.
35. Heimann, M., and C.D. Keeling. 1989. A three-dimensional model of atmospheric CO_2 transport based on observed winds: 2. Model description and simulated tracer experiments. In D.H. Peterson (ed.), Aspects of Climate Variability in the Pacific and the Western Americas, American Geophysical Union, Washington, DC, pp. 237–275.
36. Heimann, M., C.D. Keeling, and C.J. Tucker. 1989. A three-dimensional model of atmospheric CO_2 transport based on observed winds: 3. Seasonal cycle and synoptic time scale variations. In D.H. Peterson (ed.), Aspects of Climate Variability in the Pacific and the Western Americas, American Geophysical Union, Washington, DC, pp. 277–303.

37. Hobbie, J., J. Cole, J. Dungan, R.A. Houghton, and B. Peterson. 1984. Role of biota in global CO_2 balance: The controversy. BioScience **34**:492–498.

38. Holland, H.D. 1973. Ocean water, nutrients, and atmospheric oxygen. In Proceedings of Symposium on Hydrogeochemistry and Biogeochemistry, The Clarke Co., Washington, DC, Vol. 1, pp. 68–81.

39. Holland, H.D. 1978. The Chemistry of the Atmosphere and Oceans, Wiley-Interscience, New York.

40. Holland, H.D. 1984. The Chemical Evolution of the Atmosphere and Oceans, Princeton University Press, Princeton, NJ.

41. Houghton, R.A. 1986. Estimating changes in the carbon content of terrestrial ecosystems from historical data. In J.R. Trabalka and D.E. Reichle (eds.), The Changing Carbon Cycle: A Global Analysis, Springer-Verlag, New York, pp. 173–193.

42. Houghton, R.A. 1987. Biotic changes consistent with the increased seasonal amplitude of atmospheric CO_2 concentrations. J. Geophys. Res. **92**:4223–4230.

43. Houghton, R.A., J.E. Hobbie, J.M. Melillo, B. Moore, B.J. Peterson, G.R. Shaver, and G.M. Woodwell. 1983. Changes in the carbon content of terrestrial biota and soils between 1860 and 1980: A net release of CO_2 to the atmosphere. Ecol. Monogr. **53**:235–262.

44. Houghton, R.A., and G.M. Woodwell. 1989. Global climatic change. Sci. American **260**(4):36–44.

45. Javoy, M., F. Pineau, and C.J. Allegre. 1982. Carbon geodynamic cycle. Nature **300**:171–173.

46. Jenkinson, D.S., D.E. Adams, and A. Wild. 1991. Model estimates of CO_2 emissions from soil in response to global warming. Nature **351**:304–306.

47. Kasting, J.F. 1984. The carbonate–silicate geochemical cycle and its effect on atmospheric carbon dioxide over the past 100 million years. American J. Sci. **284**:1175–1182.

48. Kasting, J.F. 1987. Theoretical constraints on oxygen and carbon dioxide concentrations in the Precambrian atmosphere. Precambrian Res. **34**:205–229.

49. Kasting, J.F., and S.M. Richardson. 1985. Seafloor hydrothermal activity and spreading rates: the Eocene carbon dioxide greenhouse revisited. Geochim. Cosmochim. Acta **49**:2541–2544.

50. Kasting, J.F., S.M. Richardson, J.B. Pollack, and O.B. Toon. 1986. A hybrid model of the CO_2 geochemical cycle and its application to large impact events. American J. Sci. **286**:361–389.

51. Kasting, J.F., O.B. Toon, and J.B. Pollack. 1988. How climate evolved on the terrestrial planets. Sci. American **256**(2):90–97.

52. Keeling, C.D., R.B. Bacastow, A.F. Carter, S.C. Piper, T.P. Whorf, M. Heimann, W.G. Mook, and H. Roeloffzen. 1989. A three-dimensional model of atmospheric CO_2 transport based on observed winds: 1. Analysis of observational data. In D.H. Peterson (ed.), Aspects of Climate Variability in the Pacific and the Western Americas, American Geophysical Union, Washington, DC, pp. 165–236.

53. Keeling, C.D., S.C. Piper, and M. Heimann. 1989. A three-dimensional model of atmospheric CO_2 transport based on observed winds: 4. Mean annual gradients and interannual variations. In D.H. Peterson (ed.), Aspects of Climate Variability in the Pacific and the Western Americas, American Geophysical Union, Washington, DC, pp. 305–363.

54. Keir, R.S., and W.H. Berger. 1985. Late Holocene carbonate dissolution in the equatorial Pacific: Reef growth or neoglaciation? In E.T. Sundquist and W.S. Broecker (eds.), The Carbon Cycle and Atmospheric CO_2: Natural Variations Archean to Present, American Geophysical Union, Washington, DC, pp. 208–220.

55. Knox, F., and M.B. McElroy. 1984. Changes in atmospheric CO_2: Influence of the marine biota at high latitude. J. Geophys. Res. **89**:4629–4637.

56. Koster van Groos, A.F. 1988. Weathering, the carbon cycle, and the differentiation of the continental crust and mantle. J. Geophys. Res. **93**:8952–8958.

57. Kuhn, W.R., J.C.G. Walker, and H.G. Marshall. 1989. The effect on Earth's surface temperature from variations in rotation rate, continent formation, solar luminosity, and carbon dioxide. J. Geophys. Res. **94**:11129–11136.

58. Kump, L.R. 1991. Interpreting carbon-isotope excursions: Strangelove oceans. Geology **19**:299–302.

59. Kump, L.R., and R.M. Garrels. 1986. Modeling atmospheric O_2 in the global sedimentary redox cycle. American J. Sci. **286**:337–360.

60. LaMarche, V.C., D.A. Graybill, H.C. Fritts, and M.R. Rose. 1984. Increasing atmospheric carbon dioxide: Tree ring evidence for growth enhancement in natural vegetation. Science **225**:1019–1021.

61. Lasaga, A.C. 1989. A new approach to isotopic modeling of the variation of atmospheric oxygen through the Phanerozoic. American J. Sci. **289**:411–435.

62. Lasaga, A.C., R.A. Berner, and R.M. Garrels. 1985. An improved model of atmospheric CO_2 fluctuations over the past 100 million years. In E.T. Sundquist and W.S. Broecker (eds.), The Carbon Cycle and Atmospheric CO_2: Natural Variations Archean to Present, American Geophysical Union, Washington, DC, pp. 397–411.

63. Lovelock, J.E., and M. Whitfield. 1982. Life span of the biosphere. Nature **296**:561–563.

64. Lyle, M., and N. Pisias. 1990. Ocean circulation and atmospheric CO_2 changes: Coupled use of models and paleoceanographic data. Paleoceanography **5**:15–41.

65. Marshall, H.G., J.C.G. Walker, and W.R. Kuhn. 1988. Long-term climate change and the geochemical cycle of carbon. J. Geophys. Res. **93**:791–801.

66. Marston, J.B., M. Oppenheimer, R.M. Fujita, and S.R. Gaffin. 1991. Carbon dioxide and temperature. Nature **349**:573–574.

67. Martin, J.H. 1990. Glacial-Interglacial CO_2 change: The iron hypothesis. Paleoceanography **5**:1–14.

68. Marty, B., and A. Jambon. 1987. $C/^3He$ in volatile fluxes from the solid Earth: implications for carbon geodynamics. Earth Planetary Sci. Letters **83**:16–26.

69. Meade, C., and R. Jeanloz. 1991. Deep-focus earthquakes and recycling of water into the Earth's mantle. Science **252**:68–72.

70. Millero, F.J. 1979. The thermodynamics of the carbonate system in seawater. Geochim. Cosmochim. Acta **43**:1651–1661.

71. Mix, A.C. 1989. Influence of productivity variations on long-term atmospheric CO_2. Nature **337**:541–544.

72. Molnar, P., and P. England. 1990. Late Cenozoic uplift of mountain ranges and global climate change: chicken or egg? Nature **346**:29–34.

73. Mooney, H.A., P.M. Vitousek, and P.A. Matson. 1987. Exchange of materials between terrestrial ecosystems and the atmosphere. Science **238**:926–932.

74. Neftel, A., E. Moor, H. Oeschger, and B. Stauffer. 1985. Evidence from polar ice cores for the increase in atmospheric CO_2 in the past two centuries. Nature **315**: 45–47.

75. Opdyke, B.N., and B.H. Wilkinson. 1988. Surface area control of shallow cratonic to deep marine carbonate accumulation. Paleoceanography **3**:685–703.

76. Peacock, S.M. 1990. Fluid processes in subduction zones. Science **248**:329–337.

77. Post, W.M., T.-H. Peng, W.R. Emanuel, A.W. King, V.H. Dale, and D.L. DeAngelis. 1990. The global carbon cycle. American Sci. **78**:310–326.

78. Raymo, M.E., W.F. Ruddiiman, and P.N. Froelich. 1988. Influence of late Cenozoic mountain building on ocean geochemical cycles. Geology **16**:649–653.

79. Sarmiento, J.L., and R. Toggweiler. 1984. A new model for the role of the oceans in determining atmospheric PCO_2. Nature **308**:621–624.

80. Schidlowski, M., J.M. Hayes, and I.R. Kaplan. 1983. Isotopic inferences of ancient biochemistries: carbon, sulfur, hydrogen, and nitrogen. In J.W. Schopf (ed.), Earth's Earliest Biosphere, Princeton University Press, Princeton, NJ, pp. 149–186.

81. Schlesinger, W.H. 1986. Changes in soil carbon storage and associated properties with disturbance and recovery. In J.R. Trabalka and D.E. Reichle (eds.), The Changing Carbon Cycle: A Global Analysis, Springer-Verlag, New York, pp. 194–220.

82. Schlesinger, W.H. 1991. Biogeochemistry: An Analysis of Global Change. Academic Press, San Diego, CA.

83. Siegenthaler, U. 1990. El Nino and atmospheric CO_2. Nature **345**:295–296.

84. Siegenthaler, U., and T. Wenk. 1984. Rapid atmospheric CO_2 variations and ocean circulation. Nature **308**:624–626.

85. Staudigel, H., and S.R. Hart. 1983. Alteration of basaltic glass: Mechanism and significance for the oceanic crust–seawater budget. Geochim. Cosmochim. Acta **47**:337–350.

86. Staudigel, H., S.R. Hart, and S. Richardson. 1981. Alteration of the oceanic crust: Processes and timing. Earth Planetary Sci. Letters **52**:311–327.

87. Staudigel, H., S.R. Hart, H.-U. Schmincke, and B.M. Smith. 1989. Cretaceous

ocean crust at DSDP Site 417 and 418: Carbon uptake from weathering versus loss by magmatic outgassing. Geochim. Cosmochim. Acta **53**:3091–3094.

88. Stuiver, M. 1978. Atmospheric carbon dioxide and carbon reservoir changes. Science **199**:253–258.

89. Tans, P.P., I.Y. Fung, and T. Takahashi. 1990. Observational constraints on the global atmospheric CO_2 budget. Science **247**:1431–1438.

90. Tardy, Y., R. N'Kounkou, and J.L. Probst. 1989. The global water cycle and continental erosion during Phanerozoic time (570 my). American J. Sci. **289**: 455–483.

91. Tingle, T.N., and H.W. Green. 1987. Carbon solubility in olivine: Implications for upper mantle evolution. Geology **15**:324–326.

92. Veizer, J. 1988. The Earth and its life: Systems perspective. Origins of Life **18**:13–39.

93. Veizer, J., and S.L. Jansen. 1979. Basement and sedimentary recycling and continental evolution. J. Geology **87**:341–370.

94. Veizer, J., and S.L. Jansen. 1985. Basement and sedimentary recycling 2: Time dimension to global tectonics. J. Geology **93**:625–643.

95. Veizer, J., P. Laznicka, and S.L. Jansen. 1989. Mineralization through geologic time: Recycling perspective. American J. Sci. **289**:484–524.

96. Volk, T. 1987. Feedbacks between weathering and atmospheric CO_2 over the last 100 million years. American J. Sci. **287**:763–779.

97. Volk, T. 1989. Rise of angiosperms as a factor in long-term cooling. Geology **17**:107–110.

98. Volk, T. 1989. Sensitivity of climate and atmospheric CO_2 to deep-ocean and shallow-ocean carbonate burial. Nature **337**:637–640.

99. Walker, J.C.G. 1974. Stability of atmospheric oxygen. American J. Sci. **274**: 193–214.

100. Walker, J.C.G. 1977. Evolution of the Atmosphere, Macmillan Publishing Company, New York.

101. Walker, J.C.G. 1980. The oxygen cycle. In O. Hutzinger (ed.), The Natural Environment and the Biogeochemical Cycles, Springer-Verlag, Berlin, pp. 87–104.

102. Walker, J.C.G. 1982. Evolution of the atmosphere's chemical composition. Impact **32**:261–269.

103. Walker, J.C.G. 1983. Carbon geodynamic cycle. Nature **303**:730–731.

104. Walker, J.C.G. 1984. How life affects the atmosphere. BioScience **34**:486–491.

105. Walker, J.C.G. 1986. Earth History: The Several Ages of the Earth, Jones and Bartlett Publishers, Inc., Boston.

106. Walker, J.C.G. 1986. Global geochemical cycles of carbon, sulfur, and oxygen. Marine Geol. **70**:159–174.

107. Walker, J.C.G. 1989. The geological record of atmospheric evolution. Indian J. Radio Space Res. **17**:147–154.

108. Walker, J.C.G. 1990. Precambrian evolution of the climate system. Global and Planetary Change **82**:261–289.

109. Walker, J.C.G. 1991. Degassing. In T.M. Donahue, K.K. Trivers, and D.M. Abramson (eds.), Planetary Sciences, American and Soviet Research, National Academy Press, Washington, DC, pp. 191–202.

110. Walker, J.C.G. 1991. Numerical Adventures With Geochemical Cycles, Oxford University Press, New York.

111. Walker, J.C.G. 1991. Feedback processes in the biogeochemical cycles of carbon. In S.H. Schneider and P.J. Boston (eds.). Scientists on Gaia, MIT Press, Cambridge, MA, 183–190.

112. Walker, J.C.G., and J.I. Drever. 1988. Geochemical cycles of atmospheric gases. In B. Gregor and J.B. Maynard (eds.), Chemical Cycles in the Evolution of the Earth, John Wiley and Sons, New York, pp. 55–76.

113. Walker, J.C.G., P.B. Hays, and J.F. Kasting. 1981. A negative feedback mechanism for the long-term stabilization of Earth's surface temperature. J. Geophys. Res. **86**:9776–9782.

114. Walker, J.C.G., and J.F. Kasting. 1992. Effects of fuel and forest conservation on future levels of atmospheric carbon dioxide. Palaeogeography, Palaeoclimatology, Palaeoecology (Global and Planetary Change Section) **97**:151–189.

115. Wilkinson, B.H., and J.C.G. Walker. 1989. Phanerozoic cycling of sedimentary carbonate. American J. Sci. **289**:525–548.

116. Woodwell, G.M., J.E. Hobbie, R.A. Houghton, J.M. Melillo, B. Moore, B.J. Peterson, and G.R. Shaver. 1983. Global deforestation: Contribution to atmospheric carbon dioxide. Science **222**:1081–1086.

117. Woodwell, G.M., R.H. Whittaker, W.A. Reiners, G.E. Likens, C.C. Delwiche, and D.B. Botkin. 1978. The biota and the world carbon budget. Science **199**:141–146.

Ice Core Records as a Key to Understanding the History of Atmospheric Trace Gases

D. Raynaud

Laboratoire de Glaciologie et Géophysique de l'Environnement, B.P. 96, 38402 St. Martin d'Hères Cedex, France

ABSTRACT

Ice cores have been revealed as a powerful indicator of global environmental conditions of the preindustrial time and the glacial–interglacial cycle (the last 150,000 years). Of special interest is the most direct evidence of past changes in radiatively active trace gases found in the ice record. We present here a review of the results concerning three of the main radiatively active gases: CO_2, CH_4, and N_2O.

The record indicates that the three gases had significantly lower concentrations during the preindustrial time and that we cannot disregard possible global fluctuations of the preindustrial background over the last millennium. The large increase (by up to a factor of 2 in the case of CH_4) between preindustrial time and today is largely attributed to anthropogenic activities.

The glacial–interglacial cycle has been documented for CO_2 and CH_4. Their variations are generally well-correlated with global climatic changes showing strongly depleted concentrations (by up to about a factor of 2 in the case of CH_4) during the full glacial conditions relative to interglacial periods. The gas records obtained from ice cores can be interpreted, at least qualitatively, in terms of flux changes between ocean, continental biosphere, and atmosphere, and, in the case of CH_4, of changes in tropospheric destruction rate. These changes are assumed to be induced by direct or indirect climatic forcings. In turn, the changes in CO_2 and CH_4 are participating on the paleoclimatic variability and information can be obtained from the record on the climate sensitivity of greenhouse gases.

Introduction

Certain atmospheric trace gases (TG) are, climatically speaking, important because they are radiatively active and their atmospheric lifetime is long enough to influence the climate at a global scale. Such gases include CO_2, CH_4, N_2O, and CFCs. The evolution of these trace gases has been documented with precision by direct atmospheric measurements, only for the most recent period (for

Table 1. Current atmospheric concentrations and rates of increase of key radiatively active gases (from Ref. 14).

	CO_2	CH_4	CFC-11	CFC-12	N_2O
Mean atmospheric concentrations for 1990	353 ppmv	1.72 ppmv	280 pptv	484 pptv	310 ppbv
Current rate of annual atmospheric increase	1.8 ppmv (0.5%)	0.015 ppmv (0.9%)	9.5 pptv (4%)	17 pptv (4%)	0.8 ppbv (0.25%)

instance, for CO_2 since 1958 and for CH_4 since 1978). The atmospheric data clearly indicate increasing trends (Table 1) which are largely, if not totally, due to various human activities.

Analysis of gases trapped in ice opens now the possibility to document widely the anthropogenic influence back in time and to extend the TG record over time periods (Little Ice Age, glacial–interglacial cycle) encompassing climatic change with little or no human disturbance. In this chapter, we will first discuss how ice traps air and which parameters could affect the atmospheric TG record frozen in the ice. We will then review the existing ice records from CO_2, CH_4, and N_2O and what we could learn in terms of environmental biogeochemistry and climatic impact from these records. We will, also, stress the potential for other measurements to be performed on trace gases extracted from ice cores.

Method

How Air Bubbles Become Trapped in Polar Ice?

The air trapping process is diagrammatically illustrated in Figure 1. The snow deposited in the area free of summer melting at the surface of the ice sheets is compressed and sintered as a result of water vapor diffusion and plastic deformation of the solid H_2O grains under the weight of subsequently fallen snow.[32] During this stage, called firn, of the transformation of snow into ice, the air found in the open pores becomes trapped in bubbles which close off during the last step of the sintering process, mainly in the bottom part of the firn layer.[33] When all of the pores are finally closed off, the resulting material is called ice.

The trapping process complicates the determination of the age of the air

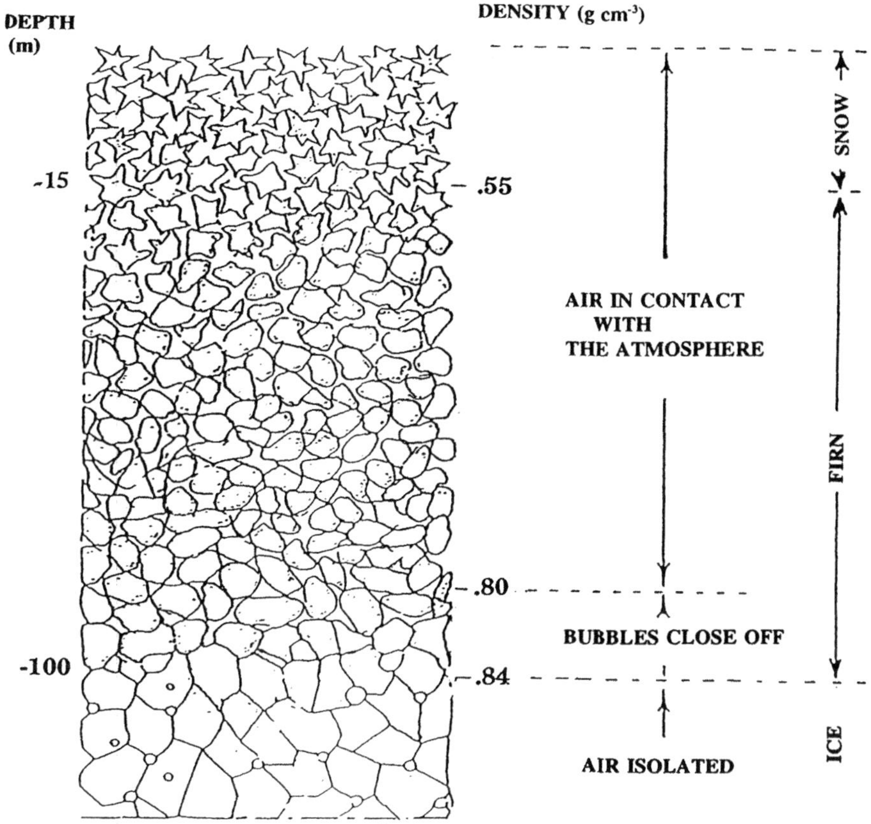

Figure 1. Diagram of the snow–ice transformation illustrating the trapping process of air bubbles. The indicated depths correspond to the site of Vostok.

enclosed in an ice sample. First, because the bubbles close off after the snow deposition, the age of the air is younger than the age of the precipitation (see Table 2). Second, because all the pores found in a layer corresponding to the same deposition time do not close at the same time, the age of the air trapped in an ice sample analyzed for TG, which typically results from the close-off of several thousands of bubbles, will integrate a time interval ranging between a few years and several hundred years (see Table 2).

Parameters Which Could Affect the Atmospheric Record of the Trace Gases Trapped in Ice

The question is to know to what extent the composition of the air extracted from the ice is the same as the original atmospheric composition prevailing at the

Table 2. Antarctica—Range of the chronological parameters used for evaluating the age of the air trapped in ice under current conditions.

	Law Dome DEO 8	Vostok
Age difference between snow deposition and air enclosure	35 years	2500 years
Time interval integrated by the air extracted from an ice sample	14 years	300 years

NOTE: At station DEO 8 on the Law Dome, the snow transforms very quickly in ice, mainly because of the particularly high snow accumulation rate (11). In contrast, the station Vostok in the central part of East Antarctica experiences a very low accumulation rate of snow and extremely cold temperatures, leading to a very slow firn densification (data from Ref. 3).

time of the trapping. As illustrated in Figure 2, there are several parameters which could make the gas record measured on ice samples different from the real atmospheric evolution. A recent thorough discussion of the physical mechanisms involved can be found in Ref. 32. Although there is still fundamental work to be carried out to understand processes like air mixing, the current knowledge of the different mechanisms involved suggest that the CO_2 record and probably also the CH_4 and N_2O records can be well preserved (within a few percent) in ice.

In fact, the main reasons for confidence arise from the ice core records of CO_2 and CH_4 themselves. The CO_2 record obtained from ice core measurements connects within a few ppmv with the atmospheric record around 1958 (see the section on CO_2) and more recent results show that the fit between ice record and atmospheric record of CH_4 is also achieved.[11] Furthermore, the good agreement obtained for the last glacial–interglacial change between the records of CO_2 and CH_4 measured on ice cores taken at different places on both the Antarctic and Greenland ice sheets support the significance of the long-term TG records from ice cores.

The CO_2 Record

The CO_2 Increase Starting from "Preindustrial" Time

This CO_2 increase has been mainly documented on two antarctic cores.[13,22,24] The precise reconstruction from the Siple core (see Figure 3 taken from Ref. 13)

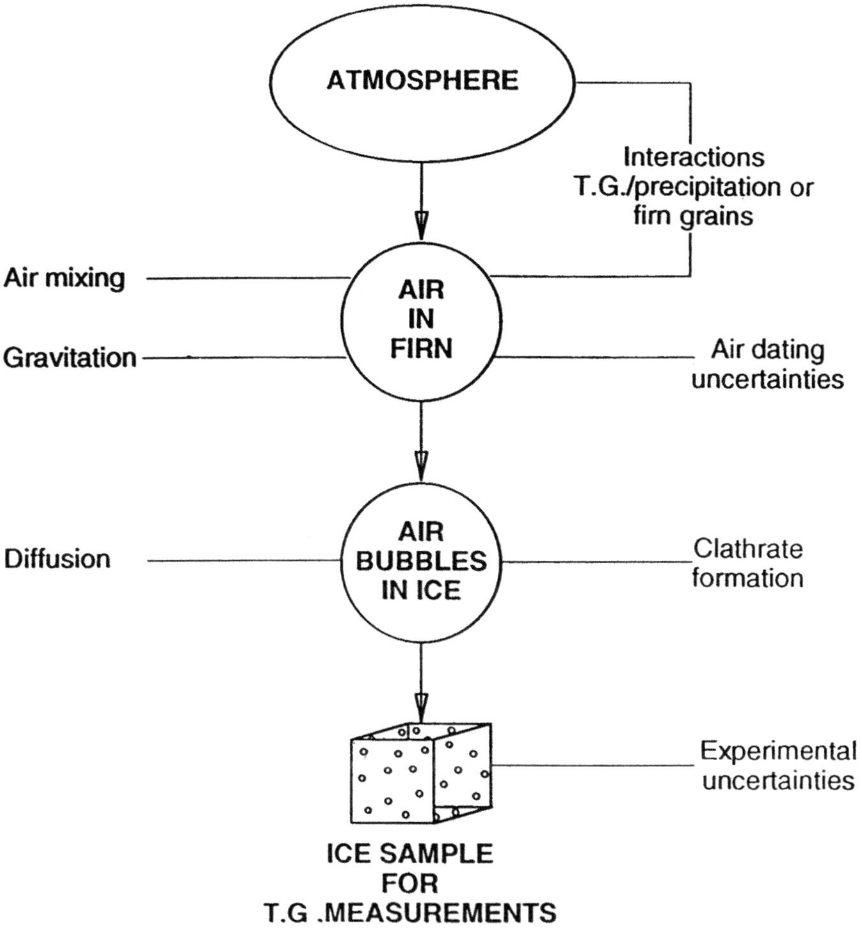

Figure 2. Diagram illustrating the physical processes involved in the air trapping by ice and the sources of uncertainty when measuring the atmospheric record from ice core analysis. Only in the upper part of the firn, where convection occurs, can the air be considered well mixed and homogeneous with the surrounding atmosphere. Deeper, the air column tends to reach a state of diffusive equilibrium in which the heaviest components become enriched by gravitational effect at the bottom of the column. In the sites where the snow transforms very quickly in ice, diffusive equilibrium may not be reached before the pores close off. Measurements of the gas diffusion constants and of various tracers in the firn are progressing and now both incomplete air mixing and gravitational effects can be better taken into account. After the bubbles close off, the bubble size decreases and the air pressure progressively increases in the bubbles. Several hundred meters below the surface, the pressure conditions lead to the formation of air clathrates.

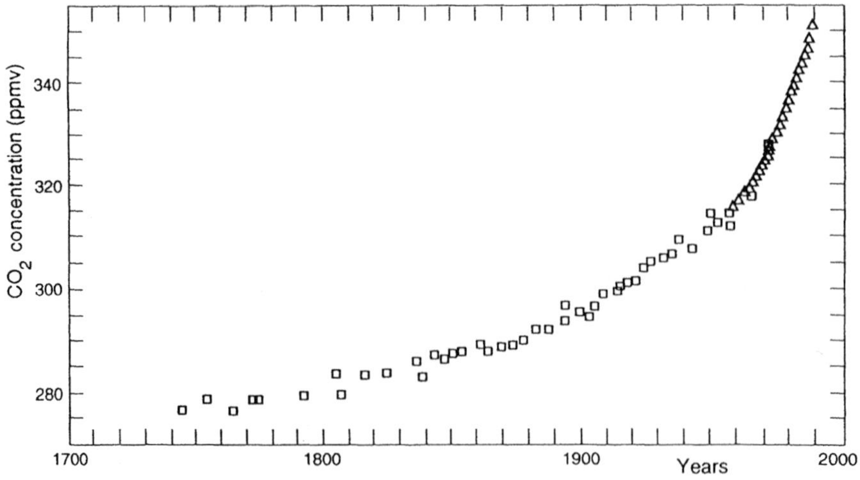

Figure 3. Precise reconstruction of the CO_2 increase following the preindustrial time from measurements performed on the Siple ice core (after Ref. 13). Note the good agreement between the ice core data (squares) and the CO_2 record directly measured in the atmosphere starting from 1958 (triangles).

stresses the good connection, within a few ppmv, between ice and atmospheric records, the main uncertainties to evaluate the fitting of the two records being due to the air dating and the CO_2 measurements of ice samples. We should note that the atmospheric CO_2 concentrations at Siple (Antarctica) may slightly deviate from the Mauna Loa record which is the atmospheric record shown in Figure 3.

The Antarctic ice core data also indicate that the significant increase started around 1800 from a mean CO_2 concentration of about 280 ppmv, as also shown by recent measurements performed on the GISP 2 core retrieved in the central part of Greenland (38).

In terms of carbon cycle, the release of C in the atmosphere due to anthropogenic activities during the period 1850–1986 is estimated as follows[14]:

- 195 ± 20 GtC due to fossil fuel burning and other minor industrial sources,
- 117 ± 35 GtC due to deforestation and changes in land use.

This means that the increase of 60 ppmv in atmospheric CO_2 concentrations for the same period, as deduced for the combined ice core–atmosphere record, represents about 40% of the total anthropogenic input. The rest of the human input must have been absorbed by the ocean and the continental biosphere, although we cannot totally disregard a minor and parallel decrease of CO_2 concentrations due to nonanthropogenic causes over the same period.

The Last 1000 Years

There are a few measurements from ice cores suggesting that the "preindustrial" atmospheric level of CO_2 has not been constant over the last 1000 years. Raynaud and Barnola[28] suggest that the CO_2 concentrations were lower by about 10 ppmv around the Little Ice Age (Figure 4). Further back in time, measurements performed by Siegenthaler et al.[34] indicate that the atmospheric CO_2 concentrations may have increased rapidly by about 10 ppmv around the thirteenth century and then decreased more slowly (Figure 4).

These results stress the possibility of atmospheric CO_2 variations of the order of 10 ppmv during the "preindustrial" era. The new ice cores obtained recently in Central Greenland (Eurocore, GISP 2, GRIP) are promising for investigating in more detail such variations and, thus, better understanding the role of the ocean and the biosphere on the carbon cycle in relation with the climatic fluctuations over the last millennium.

The Glacial–Interglacial Cycle

The first significant results about the long-term CO_2 record from ice cores showed that CO_2 concentrations (190–200 ppmv) during the Last Glacial Maximum (around 18,000 years ago) were lower than Holocene concentrations (270–280 ppmv) by 25–30% (see, for instance, Refs. 9 and 21). The discovery of this large increase of CO_2 parallel with the last glacial–interglacial warming has been documented from several ice cores taken both in Antarctica and Greenland and has focused attention on the role played by greenhouse gases in the worldwide warming associated with the deglaciation.

The CO_2 record was later extended to the full glacial–interglacial cycle over about the last 150,000 years[2,3] by analyzing the Vostok (East Antarctica) ice core. The Vostok record shows, in a general way, a remarkable correlation between changes in ice isotopic composition, a proxy for antarctic climate, and changes in CO_2 (Figure 5).

The Vostok core data confirm the CO_2 increase linked with the last deglaciation (around 15 kyr BP in Figure 5) and highlight the global character of this feature showing that the previous glacial–interglacial warming (around 140 kyr BP in Figure 5) was accompanied by a very similar atmospheric CO_2 increase. Beside the large glacial–interglacial signal, a dominant ~20 kyr periodic component, close to the orbital precession frequency, is shown by spectral analysis.

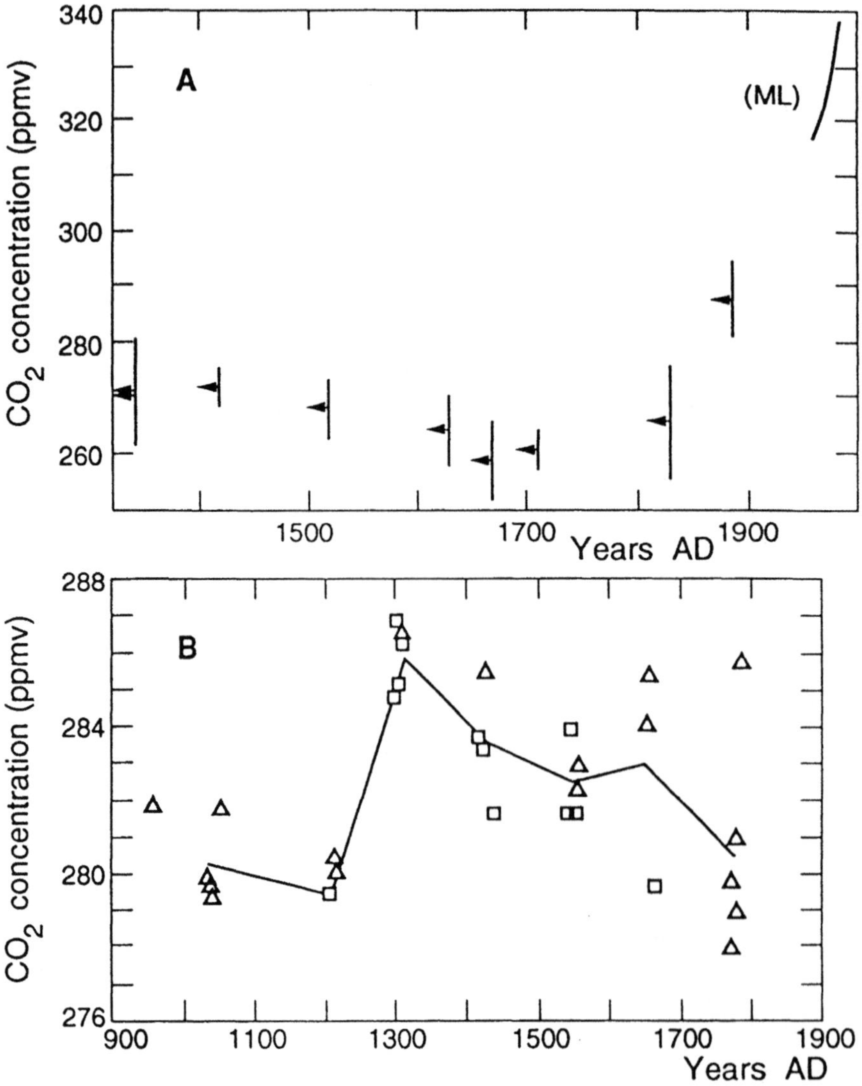

Figure 4. Ice core measurements suggesting fluctuations of the CO_2 "preindustrial" level over the last millennium. (A) Measurements on the D 57 (Antarctica) ice core indicate a possible decrease of about 10 ppmv linked with the Little Ice Age.[28] Note that the absolute values of the CO_2 concentrations probably have to be increased by about 10 ppmv as a result of a systematic experimental error which has been detected after publication of the results. ML corresponds to the atmospheric Mauna Loa record. (B) Measurements on an ice core taken at South Pole Station indicate an increase in the atmospheric CO_2 concentrations around the thirteenth century. The figure (after Ref. 34) shows an increase of about 6 ppmv. But, because of the long time interval required for air enclosure at South Pole, the real atmospheric change should have been larger (of the order of 10 ppmv).

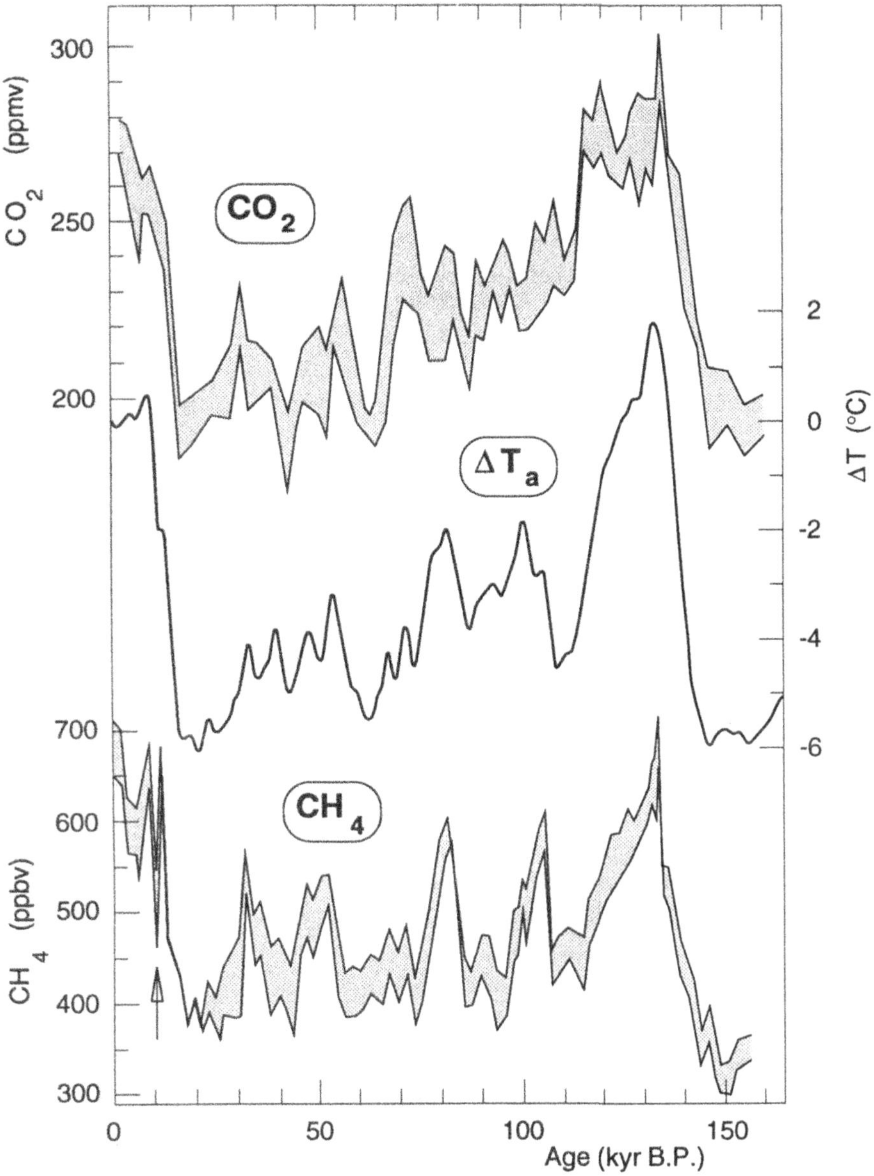

Figure 5. The Vostok ice core record of atmospheric CO_2 and CH_4 (after Refs. 2 and 7). The temperature curve represents the change of the temperature at the top of the inversion layer over Antarctica as deduced from the isotopic composition of the Vostok ice core.[15]

These features strongly suggest that a coupling exists between atmospheric CO_2 and climate over the last glacial cycle.

During the penultimate glacial–interglacial transition, the Vostok core reveals that the CO_2 increase is in phase or slightly lags (by less than 1000 years) the Antarctic climate warming. CO_2 results obtained with a high temporal resolution from the Byrd (West Antarctica) ice core indicate for the last glacial–interglacial transition that CO_2 lags the Antarctic temperature by less than 1200 years.[23] In contrast, as documented by the Vostok core, the CO_2 decrease lags by several thousands of years the major cooling in Antarctica (cooling starting around 130 kyr BP in Figure 5) corresponding to the onset of the glacial period.

The mechanisms suggested for explaining the glacial–interglacial increase of CO_2 involve mainly oceanic modifications through a combination of biological productivity and oceanic circulation changes (see, for instance, Ref. 5 and references therein). Possible contributions of changes in sea ice extent[6,12] and terrestrial carbon storage[1,26] changes should also be considered. The difference in the CO_2–climate phase relationship observed between glacial–interglacial and interglacial–glacial transitions suggests that different mechanisms have been operating to force the CO_2 changes observed in the climate cycle.

Finally, the discovery[35] in certain Greenland ice cores of "abrupt" changes in CO_2 concentrations of around 30–40 kyr BP (during the ice age period) opens the question of the occurrence of rapid (within less than 100 years) shifts of the order of 50 ppmv of atmospheric CO_2 concentrations parallel with abrupt climatic events. So far, the numerous measurements performed on the Antarctic Byrd ice core to search for these "abrupt changes" have been unsuccessful and it is, therefore, unclear if they are real or represent artifacts in the Greenland ice core records.[14] Several ice cores have been investigated for the Younger Dryas (for a review, see Ref. 16), a period of climatic reversal which occurred during the last glacial–interglacial warming. No strong CO_2 signal has been detected, but the glacial–interglacial CO_2 increase seems to stall with CO_2 concentrations remaining pretty constant around this time.

The CH_4 Record

Analysis of CH_4 from ice samples was initiated by Robbins et al.[31] and has subsequently been continued during the 1980s.

The "Preindustrial" and "Industrial" Eras

Ice core data for CH_4 are available for the last few thousands years.[8,10,17,25,29,27,36] The results indicate a mean "preindustrial" level around 700 ppbv and document the important increase over the most recent period, especially over the last century, which is highly correlated with the increase of the human population. The results obtained on the ice core which has trapped the most recent air show a good connection between the ice core record and the direct CH_4 measurements in the atmosphere.[11] A decrease of the CH_4 concentrations might be linked with the Little Ice Age[19] but more precise measurements are required.

The Last Climatic Cycle

A large glacial–interglacial increase in CH_4 concentrations has been first observed both from Greenland[37] and Antarctic[29] ice, showing that, during the last and the penultimate climatic transitions, atmospheric concentrations in CH_4 nearly double (from about 0.35 to 0.65 ppmv). The CH_4 record has subsequently been documented over all of the last climatic cycle by analyzing the Vostok core.[7] It shows strong variations of past atmospheric concentrations in the 0.35–0.65-ppmv range, i.e., well below the present mean atmospheric concentration (> 1.7 ppmv). As with CO_2, the overall correlation is remarkable ($r^2 = 0.78$) between the CH_4 levels and the climate at Vostok. Nevertheless, as shown in Figure 5, there are marked differences between the CH_4 and CO_2 evolution curves:

- The decrease of CH_4 centered around 120 kyr BP in Figure 5 is roughly in phase with cooling toward glacial conditions as depicted by the climatic record of the Vostok core.
- The CH_4 fluctuations found during the glacial period between 110 and 20 kyr BP in Figure 5 are well-defined as four peaks with highest values only slightly less than the concentrations observed during the Holocene and the previous interglacial. This part of the record shows much better similarities than for CO_2 with the temporal variations of antarctic temperature.
- The CH_4 record shows, at the end of the last glacial–interglacial transition, a large oscillation (marked with an arrow in Figure 5), whose amplitude is ~60% of the full glacial–interglacial CH_4 variation. This oscillation is suspected to be associated with the abrupt, large and short in duration cooling experienced in the Northern Hemisphere between ~11 and 10 kyr BP and known as the Younger Dryas event.

Prior the anthropogenic perturbation, the flux of methane evolved from bacterial processes in freshwater wetland areas was most likely the dominant permanent source[7] and the CH_4 destruction by OH oxidation in the troposphere the main operating sink.

Although a quantitative evaluation of the changes in CH_4 sources and sinks

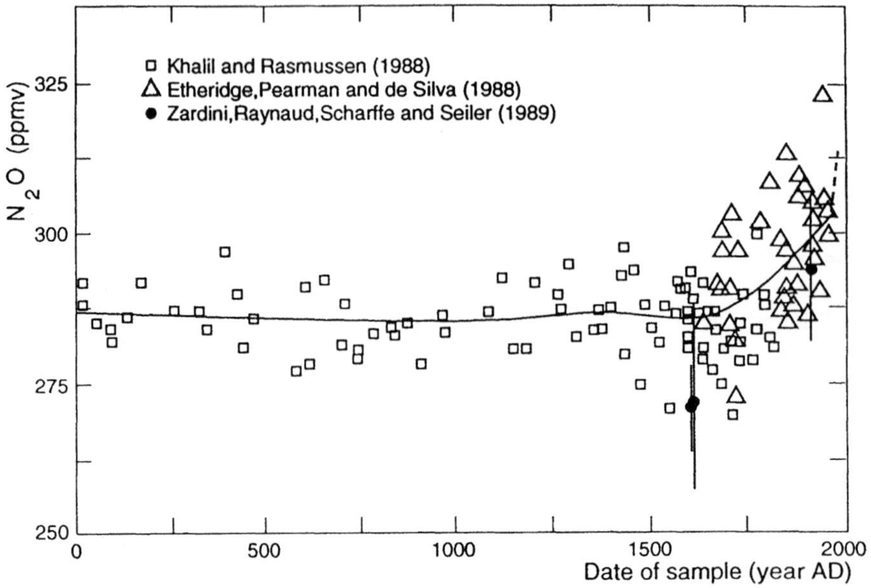

Figure 6. N_2O measurements from ice core samples for the last 2000 years (after a compilation from Ref. 14).

requires a large effort in modeling and the acquisition of new paleodata, the climatically induced changes operating on the wetland areas at low latitudes and northern high latitudes appear to be critical when explaining the atmospheric CH_4 record over the last climatic cycle.[7,29] The differences outlined between the CO_2 and CH_4 records most likely reflect the predominant sources which are suspected to be mainly oceanic in the case of CO_2 and from continental areas in the case of CH_4. Thus, the time lag between CO_2 and CH_4, when the climate is cooling toward glacial conditions, may reflect a delayed response and a longer time constant of the mechanisms driving changes in the chemical circulation of the ocean and, consequently, the atmospheric CO_2 changes recorded for this period.

The N_2O Record

The record of N_2O from ice cores is less documented than those of CO_2 and CH_4 and concerns mainly the last 2000 years. Figure 6 shows a compilation of the available data[10,18,39] for this period. The record indicates a mean "preindustrial'' atmospheric concentration of about 285 ppbv, compared with 310 ppbv for 1990. The scattering of the data is important and likely reflects uncertainties in experimental procedures and dating. The highest concentrations are found in the youngest part of the record, reflecting the increasing trend due to the anthropic influence. The lowest values around 1600–1700 AD in Figure 6 could be linked with the Little Ice Age.

There are only a few sparse measurements for earlier periods. The values observed from the Dome C (Antarctica) core by Zardini et al.[36] for levels corresponding to the Holocene period and the last glacial–interglacial transition suggest an increase of at least 20 ppbv when the climate was warming up.[40]

Climatic Impact of the Trace Gas Changes over the Climatic Cycle

The changes recorded for atmospheric CO_2 or CH_4 levels appear to have been driven by consequences of climatic changes, but the contribution of the greenhouse gas changes to the variability of past climate must be evaluated. These changes can affect Earth's temperature in three ways: the direct radiative effect, chemical feedbacks (in the case of CH_4), and climatic feedbacks. A recent review article[20] deals with this subject.

The direct radiative effect arises from the trapping of infrared energy by the radiatively active gases. The glacial–interglacial increases of CO_2 and CH_4 result in global surface equilibrium warmings of about 0.5°C for CO_2 and 0.08°C for CH_4.

The CH_4 chemical feedbacks are due to possible CH_4-induced modifications through atmospheric chemistry of tropospheric ozone and stratospheric water vapor, two other radiatively active gases. We estimate that in the case of the glacial–interglacial CH_4 increase, the chemical feedback would contribute by an additional warming of about 0.07°C.[7] There are uncertainties in this evaluation, but it does emphasize a potentially important amplification of the direct radiative effect of CH_4 as a result of its chemical activity. Both CH_4 and CO_2 may affect stratospheric ozone and, thus, induce an additional but likely small impact on radiative forcing.

So the addition of the direct radiative effects of CO_2 and CH_4 with the effect of the CH_4 feedback leads to a mean global warming of about 0.65°C in the case of the glacial–interglacial transitions. In fact, because of the various fast climatic feedbacks essentially related to tropospheric water vapor, sea ice, and clouds, the climate warming due to the increase of greenhouse gases could have been different. For instance, the climate warming linked with the CO_2 and CH_4 glacial–interglacial increases (assuming we can add the effect of the two gases) would be, with boundary conditions corresponding to the current climate, about 1.5 to 4.5 times the value of 0.65°C as shown from $2 \times CO_2$ General Circulation

Model experiments.[14] This range illustrates in fact the major uncertainty linked with the evaluation of the climatic feedbacks when dealing with the greenhouse effect.

An effort has been made for assessing how paleoclimate data can be used to specify the role of greenhouse gases in explaining past global climate changes. These changes include not only the above-mentioned fast feedback processes but also longer-term processes such as those linked with slow changes in several boundary conditions. One approach[20] has been to evaluate the relative role of the different climatic forcings which may have contributed to the climatic signal as recorded in the Vostok core over the last climatic cycle. A linear multivariate analysis is used in which the climatic output (the Vostok temperature record) is decomposed into a component that can be accounted for by climatic forcings plus a residual which is minimized in a least-square sense. The analysis accounts for five climatic forcings: the ice volume taken as a proxy of the Northern Hemisphere forcing; a Southern Hemisphere component represented by the local annual insolation; the greenhouse contribution (CO_2 and CH_4); the aerosol loading; and the amount in cloud condensation nuclei (with the Vostok dust and non-sea-salt sulphate records taken as proxy of these two last forcings). As far as greenhouse gases are concerned, the interesting result is that a reasonable estimate is 50 ± 10% for the overall contribution of the greenhouse gases to the Vostok temperature change over the last climatic cycle. Most of the other half of the glacial–interglacial change may be associated with the Northern Hemisphere forcing.

It is interesting to compare these results with temperature changes deduced from GCM simulations for the last deglaciation. Broccoli and Manabe[4] and recently Rind et al.[30] showed that, as inferred from the Vostok results, a significant part (i.e., ~40–50% from these models) of the glacial–interglacial global scale warming may, indeed, be attributed to the greenhouse effect (i.e., direct radiative forcing and associated fast feedbacks). Thus, Vostok results and comparison with GCM results support the conclusion that greenhouse gases along with the Northern Hemisphere ice sheets have significantly participated to the glacial–interglacial changes by amplifying the orbital forcing. This analysis could be improved when a completed N_2O record becomes available.

References

1. Adams, J.M., H. Faure, L. Faure-Denard, J.M. McGlade, and F.I. Woodward. 1990. Increases in terrestrial carbon storage from the Last Glacial Maximum to the present. Nature **348**:711–714.

2. Barnola, J.M., D. Raynaud, Y.S. Korotkevich, and C. Lorius. 1987. Vostok ice core: A 160,000 year record of atmospheric CO_2. Nature **329**:408–414.

3. Barnola, J.M., P. Pimienta, D. Raynaud, and Y.S. Korotkevich. 1991. CO_2–climate relationship as deduced from the Vostok ice core: a re-examination based on new measurements and on a re-evaluation of the air dating. Tellus **43B**:83–90.

4. Broccoli, A.J., and S. Manabe. 1987. The influence of continental ice, atmospheric CO_2, land albedo on the climate of the Last Glacial Maximum. Climate Dynamics **1**:87–99.

5. Broecker, W.S., and T.C. Peng. 1986. Glacial to interglacial changes in the operation of the global carbon cycle. Radiocarbon **28**:309–327.

6. Budd, W.F. 1980. The importance of the polar regions for the atmospheric carbon dioxide concentrations. In G.I. Pearman, (ed.) Carbon Dioxide and Climate: Australian Research, Australian Academy of Science, pp. 115–128.

7. Chappellaz, J., J.M. Barnola, D. Raynaud, Y.S. Korotkevich, and C. Lorius. 1990. Ice-core record of atmospheric methane over the past 160,000 years. Nature **345**:14.

8. Craig, J., and C.C. Chou. 1982. CH_4: the record in polar ice cores. Geophys. Res. Lett. **9**:1221–1224.

9. Delmas, R.J., J.M. Ascencio, and M. Legrand. 1980. Polar ice evidence that atmospheric CO_2 20,000 years B.P. was 50 % of present. Nature **284**:155–157.

10. Etheridge, D.M., G.I. Pearman, and F. de Silva. 1988. Atmospheric trace-gas variations as revealed by air trapped in an ice core from Law Dome, Antarctica, Annals Glaciology **10**:28–33.

11. Etheridge, D.M., G.I. Pearman, and P.J. Fraser. 1992. Changes in tropospheric methane between 1841 and 1978 from a high accumulation-rate Antarctic ice core. Tellus. **44B**:282–294.

12. Faure, F. 1987. Mécanisme d'amplification du cycle climatique global: l'effet de couvercle de la glace de mer contrôle le CO_2 atmosphérique, C.R. Acad. Sci. Paris, série II **305**:523–527.

13. Friedli, H., K. Loetscher, H. Oeschger, U. Siegenthaler, and B. Stauffer. 1986. Ice core record of the $^{13}C/^{12}C$ record of atmospheric CO_2 in the past two centuries. Nature **324**:237–238.

14. Intergovernmental Panel on Climate Change. 1990. In J.T. Houghton, G.J. Jenkins, and J.J. Ephraums (ed.), Climate Change, Cambridge University Press, Cambridge.

15. Jouzel, J., C. Lorius, J.R. Petit, C. Genthon, N.I. Barkov, V.M. Kotlyakov, and V.N. Petrov. 1987. Vostok ice core: a continuous isotope temperature record over the last climatic cycle (160,000 years). Nature **329**:403–408.

16. Jouzel, J., J.R. Petit, N.I. Barkov, J.M. Barnola, J. Chappellaz, P. Ciais, V.M. Kotlyakov, C. Lorius, V.N. Petrov, D. Raynaud, and C. Ritz. 1992. The last deglaciation in Antarctica: further evidence of a "Younger Dryas" type climatic event. In E. Bard and W.S. Broecker (eds.), The Last Deglaciation, Absolute and Relative Dating, NATO ASI series, Springer-Verlag, Berlin, Heidelberg.

17. Khalil, M.A.K., and R.A. Rasmussen. 1982. Secular trends of atmospheric methane (CH_4). Chemosphere **11**:877–883.

18. Khalil, M.A.K., and R.A. Ramussen. 1988. Nitrous oxide: trends and global mass balance over the last 3000 years. Annals Glaciology **10**:73–79.

19. Khalil, M.A.K., and R.A. Rasmussen, 1989. Climate-induced feedbacks for the global cycles of methane and nitrous oxide. Tellus **41B**:554–559.

20. Lorius, C., J. Jouzel, D. Raynaud, J. Hansen, and H. Le Treut. 1990. The ice-core record: climate sensitivity and future greenhouse warming. Nature **347**:139–145.

21. Neftel, A., H. Oeschger, J. Schwander, B. Stauffer, and R. Zumbrunn. 1982. Ice core measurements give atmospheric CO_2 content during the past 40,000 years. Nature **295**:222–223.

22. Neftel, A., E. Moor, H. Oeschger, and B. Stauffer. 1985. Evidence from polar ice cores for the increase in atmospheric CO_2 in the past two centuries. Nature **315**: 45–47.

23. Neftel, A., H. Oeschger, T. Staffelbach, and B. Stauffer. 1988. CO_2 record in the Byrd ice core 50,000–5,000 years BP. Nature **331**:609–611.

24. Pearman, G.I., D. Etheridge, F. de Silva, and P.J. Fraser. 1986. Evidence of changing concentrations of atmospheric CO_2, N_2O and CH_4 from air bubbles in Antarctic ice. Nature **330**:248–250.

25. Pearman, G.I., and P.J. Fraser. 1988. Sources of increased methane. Nature **332**: 489–490.

26. Prentice, K.C., and I.Y. Fung. 1990. The sensitivity of terrestrial carbon storage to climate change. Nature **346**:48–51.

27. Rasmussen, R.A., and M.A.K. Khalil. 1984. Atmospheric methane in the recent and ancient atmospheres: concentrations, trends, and interhemispheric gradient. J. Geophys. Res. **89**:11599–11605.

28. Raynaud, D., and J.M. Barnola. 1985. An Antarctic ice core reveals atmospheric CO_2 variations over the past few centuries. Nature **315**:309–311.

29. Raynaud, D., J. Chappellaz, J.M. Barnola, Y.S. Korotkevich, and C. Lorius. 1988. Climatic and CH_4 cycle implications of glacial–interglacial CH_4 change in the Vostok ice core. Nature **333**:655–657.

30. Rind, D., D. Peteet, and G. Kukla. 1989. Can Milankovitch orbital variation initiate the growth of ice sheets in a General Circulation Model? J. Geophys. Res. **41**:12851–12871.

31. Robbins, R.C., L.A. Cavanagh, L.J. Salas, and E. Robinson. 1973. Analysis of ancient atmospheres; J. Geophys. Res. **78**:5341–5344.

32. Schwander, J. 1989. The transformation of snow to ice and the occlusion of gases, In H. Oeschger and C.C. Langway, Jr. (eds.), The Environmental Record in Glaciers and Ice Sheets, Report of the Dahlem Workshop held in Berlin 1988, March 13–18, John Wiley and Sons, Chichester, pp. 51–67.

33. Schwander, J., and B. Stauffer. 1984. Age difference between polar ice and the air trapped in its bubbles. Nature **311**:45–47.

34. Siegenthaler, U., H. Friedli, H. Loetscher, E. Moor, A. Neftel, H. Oeschger, and

B. Stauffer. 1988. Stable isotope ratios and concentrations of CO_2 in air from polar ice cores. Annals Glaciology **10**:151–156.

35. Stauffer, B., H. Hofer, H. Oeschger, J. Schwander, and U. Siegenthaler. 1984. Atmospheric CO_2 concentrations during the last glaciation. Annals Glaciology **5**:760–764.

36. Stauffer, B., G. Fischer, A. Neftel, and H. Oeschger. 1985. Increases in atmospheric methane recorded in Antarctic ice core. Science **229**:1386–1388.

37. Stauffer, B., E. Lochbronner, H. Oeschger, and J. Schwander. 1988. Methane concentration in the glacial atmosphere was only half that of the preindustrial Holocene. Nature **332**:812–814.

38. Wahlen, M., D. Allen, and B. Deck. 1991. Initial measurements of CO_2 concentrations (1530 to 1940 AD) in air occluded in the GISP 2 ice core from central Greenland. Geophys. Res. Lett. **18**:1457–1460.

39. Zardini, D., D. Raynaud, D. Scharffe, and W. Seiler. 1989. N_2O measurements of air extracted from Antarctic ice cores: implications on atmospheric N_2O back to the last glacial–interglacial transition. J. Atmos. Chem. **8**:189–201.

40. *Note added in proof.* Very recent data covering the full glacial–interglacial transition (Leuenberger, M., and U. Siegenthaler. 1992. Ice-age atmospheric concentration of nitrous oxide from an Antarctic ice core. Nature **360**:449–451) indicate that the N_2O increase associated with the last glacial–interglacial warming was about 70 ppbv.

Ancient Ice Air Content of the Vostok Ice Core

I.P. Semiletov

Pacific Oceanological Institute, Far Eastern Branch, Academy of Sciences, Russia, Vladivostok 690032, Russia

ABSTRACT

A deep ice core drilled at Vostok Station, Antarctica, provides a record of atmospheric climate and CO_2 that is representative of global changes over the last glacial–interglacial–penultimate glacial cycle (about 200 kyr). The final Vostok core in 1989 approached 2540 m. Here we present and discuss the initial CO_2 record obtained to 2340 m. A combination of static head-space extraction and stripping the remaining dissolved gas was followed by gas-chromatographic analysis. The obtained data show the "warm" maxima indicated by CO_2 ice core content at 2280 m. The link between the process of air capture and release during the snowflake formation, snow accumulation, metamorphism, and sintering was considered. It was shown that the carbon ionic species, which were trapped during the snowflake formation, might be the cause for the discrepancy between the CO_2 measurements by "wet" and "dry" techniques. The possible mechanism at the Younger Dryas signal is also considered.

Introduction

A prominent characteristic of glacier ice is the inclusion in the ice of gas bubbles, cavities, or gas-hydrates. Gas trapped in bubbles of glacier ice may be of different origins, depending on the processes responsible for the formation of the ice. In inland glaciers with very cold climates where the surface temperature is always below the freezing point, ice is formed by snow compacting and sintering of ice crystals. The gas enclosed in such ice is expected to be uniform and similar to that of the atmosphere. Since diffusion may be considered to be insignificant,[14] the composition and quantity of the trapped gas remain constant. This, of course, is only true if the snow grains do not contain significant amounts of dissolved gases and if there is no selective adsorption of gas components at the snow surfaces.[39]

In searching for ancient air, many glaciologists have analyzed gas in ice samples from Greenland and Antarctica.[6,20,22,30,34] A component of special interest is the CO_2. According to present knowledge based on the Milankovich hypothesis, the optimum in the Holocene, the last interglacial period, was passed about 9400 years ago. However, the climatic effect caused by anthropogenic emission of CO_2 and other greenhouse trace gases into the atmosphere favors a correction in the scenario of future natural climatic change. Here, we present and discuss the initial CO_2 record obtained to 2340 m in the Vostok ice core. Vostok Station is located in the central part of East Antarctica (78°27′S and 106°51′E) at an elevation of 3490 m. The present mean annual temperature is −55.5°C, and the snow accumulation rate is about 2.3 g cm^{-2} yr^{-1}. A deep ice core extracted at Vostok Station provides a record of atmospheric climate and CO_2 that is representative of global changes over the last glacial–interglacial cycle and the penultimate glacial period. The borehole provides a record that is essentially undisturbed by ice flow conditions.

Materials and Methods

The study of CO_2 ice core content was carried out during in the 34th Soviet Antarctic Expedition during February–March 1989. The ice core samples were extracted from Vostok Station (the borehole, 4G-2) and from a hole drilled at site "105-km" of the transsect Mirny–Vostok. The samples were then transferred to RV "Professor Vize," where the study of CO_2 paleovariations was performed. Only representative samples were used for the study.[35,36] There are several methods available for extracting the gases from an ice sample. These include:

1. Melting the ice sample in a vacuum-tight container and pumping off the gases during or after the melting process[7]
2. Extraction of the air of a few single air bubbles by a micro syringe[7]
3. Grinding an ice sample at low temperature in a vacuum container, so that a proportion of the air bubbles are opened and the gases may be pumped off[7]
4. Sublimation of an ice sample at low temperature. The water molecules are absorbed on an absorber, and the gas molecules extracted[27]
5. Gas extraction by purging the melt water with He after the surface of the ice has been purged clean of adsorbed gas. The extracted gas is then absorbed in a cold trap followed later by release and volumetric analysis[11]
6. Melting of the ice in the He flushed container with subsequent head-space gas-chromatographic analysis.[32]

At present, it is supposed that the most reliable gas measurements of the air inclusions in ice cores are provided by the modification of the melting ice sample technique or by the improved dry extraction technique with fast crushing of small ice sample.[1,23,30] However, there are differences in the CO_2 values obtained by wet extraction and dry extraction. The comparison of data on the total inorganic carbon content (ΣCO_2) and the CO_2 concentration of the escaped gas shows discrepancies for samples from Byrd Station and Camp Century which exceed the analytical errors.[22]

The ΣCO_2 values normally exceed the corresponded values of CO_2 concentration in the air bubbles by 15–30%. This discrepancy might be caused by impurities in the ice core. First, the alkaline ice matrix absorbs a minor fraction of the CO_2 in the bubbles. Second, the dissolution of carbonate dust particles increases the ΣCO_2 in the melted water. However, the analysis of the foregoing factors shows that this effect should be negligible. Thus, the carbonate content is significantly higher in Camp Century ice than in the Byrd core, but the ΣCO_2 values are quantitatively comparable.[22] The upper limit of CO_2 originating from dissolution of dust particles is approximately 2 μl CO_2 kg ice.[39] This value is approximately 5% of the mean ΣCO_2 value, i.e., within analytical error. Also, the consideration of the CO_2 record in the Byrd ice core means that alkaline dust is not responsible for the drop in the CO_2 level.[23]

Part of the CO_2 in ice is enclosed within the crystals.[39] Therefore, the relatively low values of CO_2 concentration obtained by dry extraction might be caused by the incomplete extraction of the air inclusion due to the presence of gas-hydrate. Gas-hydrate inclusions generally range from <10 μm to >300 μm.[18,38] Sizes of ice grains exceed or are similar to sizes of gas-hydrates. This effect might be significant for deep ice layers where ice grains are largest and the existence of gas-hydrate predominates.

The foregoing uncertainties are not expected in the wet extraction technique chosen for low cost and simplicity. All ice samples were kept frozen prior to analysis. The ice core sample was removed from the optimal conditions for gas extraction and gas-chromatographic (GC) analysis, with the following techniques being tested:

1. Stripping of small melted samples, 5–10 g, after the flushing of ice at cold conditions
2. Stripping of melted ice after dissolution of air bubbles in the water solution under mechanical pressing (repeated injection of a small part of the "squeezed" sample in a glass stripping cylinder)
3. Static head-space extraction during the melting of the ice in hydrogen atmosphere followed by pneumatic injection of a small part of the equilibrated gas phase in the GC line
4. Static headspace extraction during the melting of the ice in a hydrogen atmosphere followed by complete injection of head-space gas by carrier gas (hydrogen) and by the stripping of remained dissolved gases.

The last method was used for maximum sensitivity of the analysis and convenience of the calibration by the CO_2 gas standard. It should be noted that the first and second methods were not used due to some technical difficulties. The third method is acceptable for the measurement of nitrogen-oxygen content, but not for CO_2 measurement.

The ice core sample was kept in a plastic storage bag. The outer crust of the core was shaved off with a knife. The ice core sample was cut into two or three pieces by heated nichrome fiber for replication of measurements on each sample and to determine the reproducibility of the measured concentrations. Afterward, the inner part of the ice core was removed using a laboratory thermodrill. The optimal diameter of ice sample to avoid contamination by drilling solution (kerosene and freon) is approximately 45 mm. The height of samples did not exceed 35 mm. The scheme of the gas analysis is shown in Figure 1. When the core was of a suitable size, it was sealed in a plexiglass cylinder with a glass porous filter installed near the bottom. The other container, which is acceptable for the analysis, contains two valves on the top (see Figure 1, right upper side). This container was flushed 10 times with hydrogen and pressurized to 1.2 atm under cold conditions. The same results were obtained at flushing temperatures of −40°C to −10°C. To check the flushing efficiency, concentrations of carbon dioxide, nitrogen, and oxygen were measured before the core was melted. After the container returned to room temperature, the hydrogen in the head space contained gases which were equilibrated with the meltwater. The head-space gas was then injected into the GC line. The remaining dissolved CO_2 was stripped by the hydrogen stream. Stripping is achieved by means of flushing through the porous glass filter in the cylinder bottom or by using the capillary tube which is pulled down to the bottom through a hermetic stuffing-box in the top (Figure 1, right upper side). Complete extraction was confirmed by record showing the absence of CO_2 in the hydrogen stream. The ratio of meltwater/head space ranged from 0.12 to 0.3. The change in the sample's size from 20 to 50 g did not affect the results.

Carbon dioxide is measured by a gas chromatograph LChM-80MD with a Poropak-T,Q (4 m, 80–120 mesh) and a flame-ionization detector run isothermally at 40°C with hydrogen carrier gas.[35,36] The calibrations of the absolute concentrations of CO_2 were based on two standards of CO_2 in nitrogen (manufactured at West Germany). Conversion of CO_2 to CH_4 after a Porapac column is done in a Ni-catalyst column (14 cm, Chromaton, 80–100 mesh, coated by Ni) at 380°C. Almost all measurements were found to be reproducible to within ±5–10%. In general, results depend on ice core fractures. A total of 91 measurements on 49 ice samples were performed. The Vostok data of total gas content which were obtained in the 34th Soviet Antarctic Expedition were used to calculate the CO_2 concentrations in air inclusions. These data are the result of the Soviet–France collaboration at the Vostok Station (V. Lipenkov, personal communication).

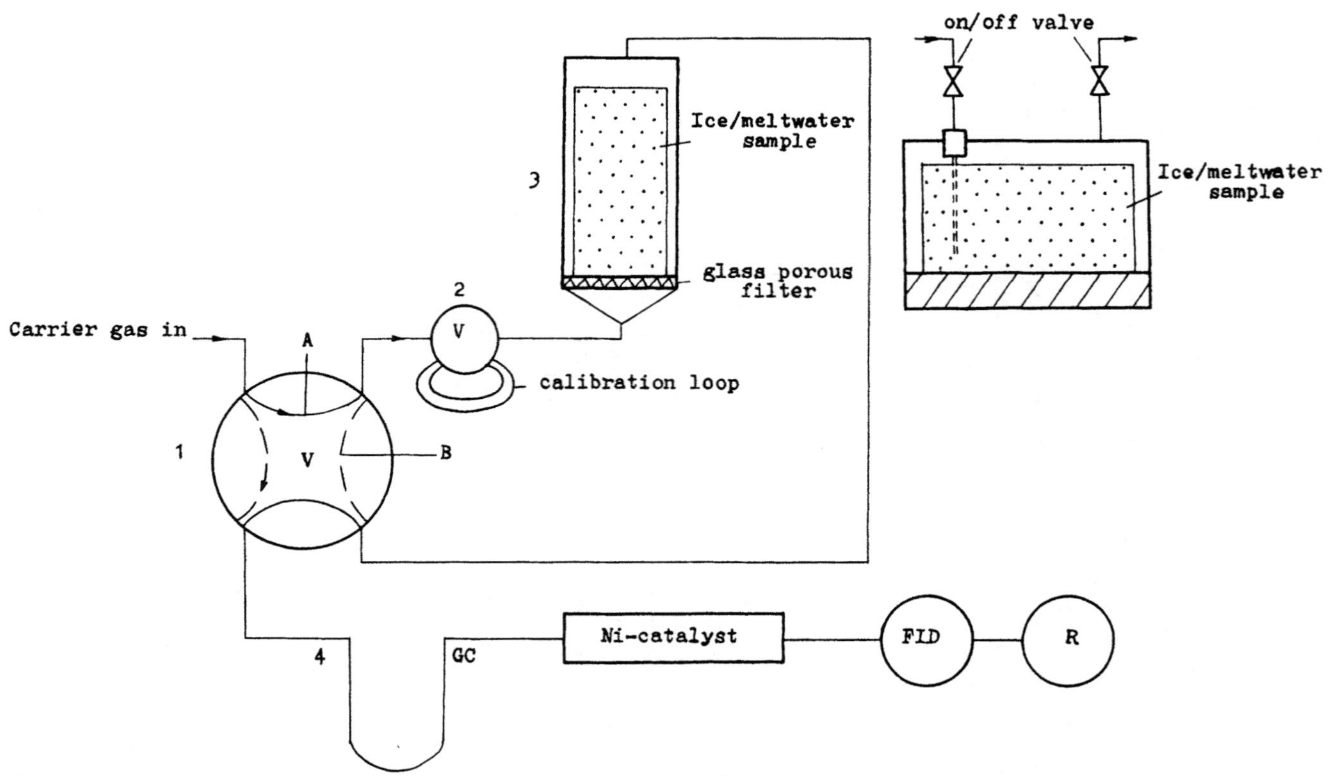

Figure 1. The scheme of the GC analysis for the study of ice air content: 1—the valve distributor (A—flushing position, B—nonflushing position); 2—the calibrating valve; 3—the cell (container) for the gas extraction; 4—GC column with Porapak-T; 5—Ni catalyst; 6—the flame-ionization detector (FID); 7—the recorder. Right upper side: The other container which is acceptable for the analysis.

Results

The concentrations of CO_2 deduced from the analyses of ice cores are shown in Figure 2 (A—"105-km" record, B—Vostok record). Presented in the upper part of Figure 3.2B are the ages corresponding to the Vostok ice core depth, following the Lorius chronology.[19]

The record presented in Figure 2A shows the Holocene—the last ice age. The Holocene is characterized by five minima separated by slightly warmer episodes. The maxima of CO_2 concentration of 317 ppm at 431 m horizon correlates with the presence of the CO_2 concentration maxima which was obtained for Camp Century at the 400–500-m layer.[22] However, this maxima is significantly greater than the corresponding values for Byrd Station and for Camp Century. It should be noted that the CO_2 concentrations for the "105-km" site were calculated by using the theoretical value of the total gas content. This value is based on the data of Raynaud.[29,31] He suggests that the mean total gas content (*V*) obtained for each formation site is proportional to the elevation of the site. We use the mean *V* value, 0.11 $cm^3\ g^{-1}$, for the 105-km site elevation of 1500 m.[26] This value agrees well with the mean *V* value measured on ice formed in Antarctica [New Byrd, depth interval 100–900 m below surface (b.s.), elevation 1650 m] and Greenland (Camp Century, 200–800m b.s., 1840 m). The comparison of the "105-km" CO_2 record with the previous study for the Byrd station agrees with data of the Institute of Physics in Bern.[22] These values are lower by about 10% compared with values obtained by the laboratory of Glaciology, Grenoble.[2,19] It should be noted that the background *V* variations may approach 5–7% which are caused by seasonal and other changes.[29]

The Byrd results indicate an increase of the CO_2 concentration in the depth interval 600–500 m. Also, a large increase in CO_2 is observed the depth interval 400–300 m at Vostok Station[2] and in the depth interval 1200–1000 m at Byrd Station and Camp Century.[22] This large CO_2 increase is associated with the change from the last glacial maximum (18 kyr BP) to the Holocene. We suggest that these synchronous changes might be used for the solution of the question on the palaeo-rate of snow accumulation.

According to Barnola et al.,[2] the 2083-m-long ice core at Vostok Station provides a unique opportunity to extend the ice record of atmospheric CO_2 over the last glacial–interglacial cycle back to the penultimate ice age about 160 kyr ago. Consequently, the CO_2 record of the 2340-m ice core presented in Figure 3.2B reflects the penultimate ice age. For comparison, the CO_2 record obtained at the Laboratory of Glaciology and Geophysics, Grenoble[19] in the interval 1650–2000 m is included in Figure 2B. The differences between the graph of presented CO_2 and the Grenoble values do not exceed the analytical error of 5–10%. However, at the horizon 1621 m and in the interval 2000–2083 m, the corresponding difference is about 60–80 ppmv (Table 1). It is clear that the

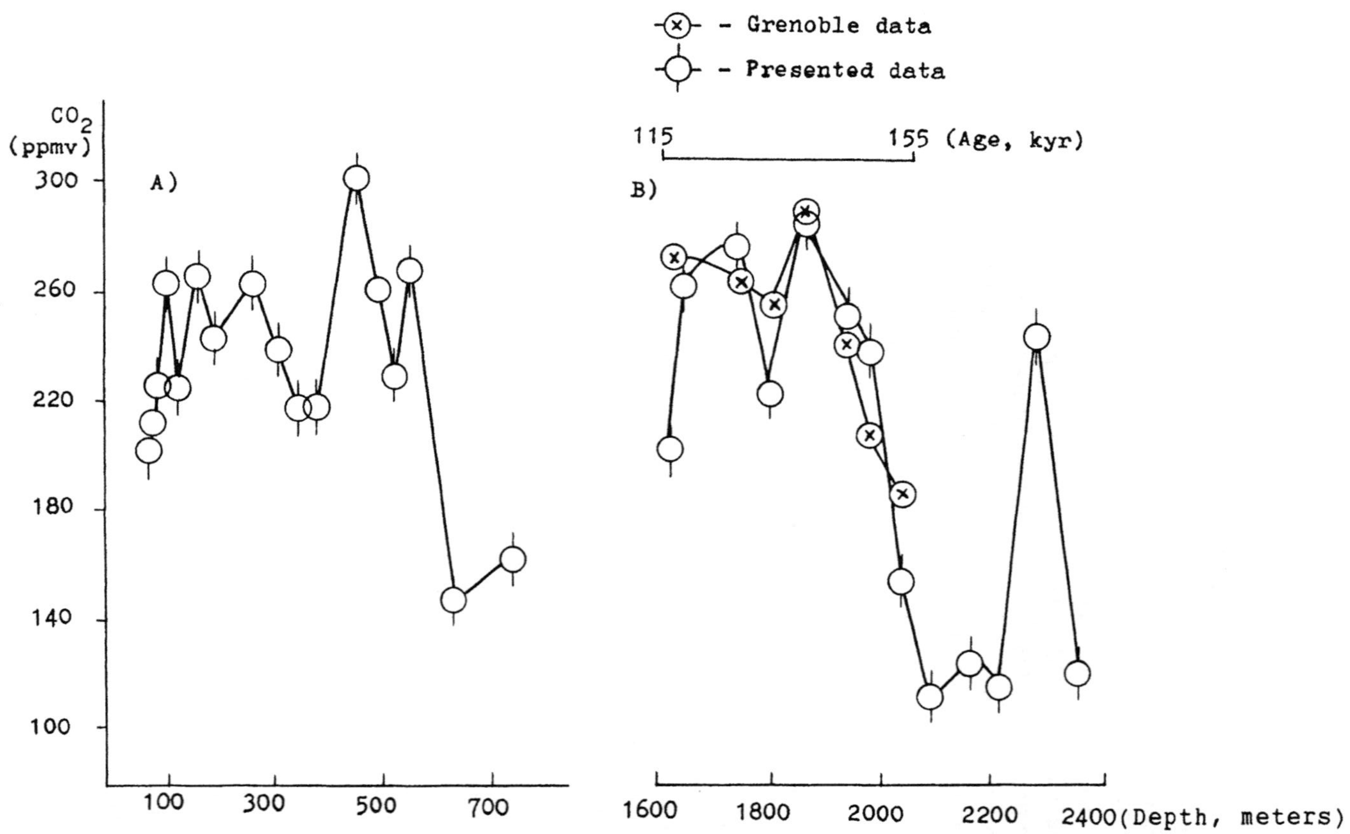

Figure 2. Atmospheric CO_2 records (ppmv), depth (m): (A) at ''105-km'' site of the transsect Mirniy–Vostok; (B) at Vostok Station. In the upper part (B), the age of the ice core depth, following the Lorius chronology, is presented (Ref. 19). For comparison, the CO_2 data obtained by the Grenoble group are shown (Ref. 2).

Table 1. Atmospheric CO_2 data from Vostok ice core (see Figure 2B).

	CO_2 Concentrations (ppmv)	
Depth (m)	Grenoble data (Refs. 2 and 19)	Presented data
1621	276	207
1649	271	266
1741	267	283
1800	268	222
1860	291	296
1921	249	260
1982	216	240
2040	192	156
2101	—	115
2160	—	125
2220	—	117
2280	—	251
2340	—	126

foregoing difference is not caused by the gas-hydrate factor. The underestimating of our data with respect to Grenoble values might be explained in part by incomplete water vapor trapping before GC analysis using the thermal conductivity detector.[35] Experiments indicate that the presence of water vapor in the analyzed air is not systematic but, when it occurs, leads to errors around the "best estimates" ranging from +3 to +22 ppmv on the positive side and from −3 to −16 ppmv on the negative side.[2] However, this cause cannot lead to the selective decrease of CO_2 recorded in the glacial episodes only. We can see the same CO_2 decreasing for the last ice age (Figure 2A) and for the penultimate ice age (Figure 2B). We suggest that this effect might be connected partially with the changes in pH values and other species of the atmospheric water carbonate system before the transformation of water drop to snowflake (see Discussion). The ice record of atmospheric CO_2 shows a sharp increase at depth 2280 m. This oscillation is based on two values of CO_2 concentration which were measured at different times. The foregoing positive oscillation might be associated with a relatively abrupt "short-time" warming.

Both records show rapid and large increases of CO_2 concentrations associated with the two deglaciations and exhibit the highest CO_2 values during the interglacial periods. Correlations between our records and Grenoble data occur in the corresponding depth intervals. The records also show a relatively abrupt CO_2 decrease associated with the last glaciation and penultimate glaciation. During the two glaciations, our data show a low constant CO_2 level in comparison with previous values.

Discussion

One potential forcing of global climate change may be changes in atmospheric CO_2. It has been shown that changes in the concentration of CO_2 in the Pleistocene atmosphere were of sufficient magnitude to have had a significant effect on global climate.[16,20,22,37] The ice core CO_2 records offer a unique opportunity to examine the link that exists between a climatic atmospheric time series and astronomical and CO_2 forcings. However, as stated above, the study of the ice core CO_2 record is not only a very important component but the record is also difficult to measure. It is known that air inclusions formed by dry sintering reflect the composition of the ancient atmosphere. There is good reason to believe that air inclusions are reliable for the study of air macrocomponents by wet extraction and dry extraction which are constant within a limit of about $\pm 1\%$ for nitrogen and oxygen concentration.[10,15] In contrast, the examination of the CO_2 records for some ice samples can show either great differences or good agreement between wet and dry techniques.[1,22,30] What is the nature of the measured CO_2 values? What values can be obtained from natural ice? For the initial understanding of main causes and effects, we will consider briefly the link of processes of air trapping and release during snowflake formation–precipitation–metamorphism.

In general, the gas content in the water drop–snowflake system is determined by the altitude of air trapping, temperature humidity regime, and aerosol concentrations. It is known that particles of clay mineral are more effective nuclei for crystallization than sea salt.[17] The air trapping altitude above Greenland, therefore, might be lower than in Antarctica. As a consequence, the amount of primary air trapped during the formation of snowflakes in the Northern Hemisphere might be higher in comparison with the Antarctic. The upper limit of this effect might provide for 10–15% of the mean total gas content if the equilibrated water drop freezes at the 600-mb level in the atmosphere.[36] Natural snow crystals contain a number of tiny air bubbles, and snowflake clusters are very significant.[3] However, Maeno and Kuroiwa[21] observed air bubbles in snow crystals and found that bubbles decreased in size over the period of a few days; and small bubbles, a few micrometers in diameter, disappeared completely. They attributed the disappearance and shrinkage of small gas bubbles to dissolution into defects in the ice lattice. The dissolution into ice seems quite likely because the existence of air components has been confirmed by many chemical analyses. However, during metamorphism, the process of gas leakage from primary crystals would take place. Stauffer and Berner[39] have found that firn and snow grains of 1 mm diameter stored at a constant CO_2 partial pressure and a constant temperature would reach equilibrium in CO_2 concentration in approximately one year. In a cold dry-snow zone, such as at the Byrd Station or Vostok Station, small firn grains are in contact with atmospheric air for centuries before they sinter to ice. Obviously, for samples from the lower part of the firn or from the ice at such a location, we can expect a uniform CO_2, O_2, N_2 content, even if the O_2 diffusion constant is smaller in comparison with the CO_2 diffusion constant as was stated by Hemmingsen.[13]

In spite of the foregoing consideration, we suppose that the effect of primary air trapping might explain the differences between wet and dry extraction techniques. This supposition is based on the simple calculation of the carbonate system in the atmospheric water droplets equilibrated with air CO_2 level at about 300 ppmv before freezing. We used Lyman's ionization constants for carbonic acid and coefficients of CO_2 solubility at temperature 0°C.[5] We found that the equilibrated carbonate-ion content is neglible. The main ionic component is bicarbonate-ion which is changed at the upper limit of 0.6–6.0 μM per kg water for pH values of 5–6. These values were calculated for ground CO_2 level at 300 ppm. The total CO_2 contained in the air bubbles with a CO_2 concentration at 300 ppm corresponds to 1.3–1.5 μm C per kg ice. Because of the altitude's trapping of carbonate species by snowflake, the contribution of the bicarbonate ion may vary significantly. For example, the measured bicarbonate concentration in present rain water is changed from 0.3 to 1.5 μM C per kg water.[11] As a result, the remaining bicarbonate ice content might provide all the observed differences between wet and dry extraction techniques.

Consider briefly the possible causes of the observed lowest CO_2 level in ice core during glaciations. We assume that this effect might be connected to low accumulation rates during cold periods. During these periods of time delay before sintering and complete isolating of primary absorbed air, the carbon ionic species and air bubbles are great in comparison with interglacial periods with relatively high accumulation rates. As a consequence, the leakage of components of primary absorbed air would be more significant than during "warm" time. But this question is open to the possibility of leakage of carbon ionic species during metamorphism. We believe that the ionic components produced by the ionization of polar gases such as CO_2 and H_2S in the rain drops are fixed partly or completely in the solid phase of water.

Second, the slightly incomplete CO_2 extraction from the meltwater might take place due to a slight increase in alkalinity in the glacial intervals. The increase might be connected with the enhanced aerosol transport toward the Antarctic continent.[25] Concentrations of alkali and alkaline earth elements in Antarctic glacial ice as in the Greenland material increase during the glacial stages, but to a much smaller extent than those in Greenland ice.[7,8] Carbon dioxide in the ice core might interact with impurities in the ice, particularly alkaline dust. However, it was shown that this effect is of minor importance for the Antarctic ice core (Byrd Station) because the Antarctic ice is slightly acidic along the whole core.[23] Thus, this effect is neglected for the direct absorption of CO_2 in the ice, but the effect may act as a buffer mechanism with a slight increase of pH values in meltwater from the glacial-stage ice core. In principle, this mechanism would shift the carbonate equilibrium to carbonate-ion formation. This mechanism might change the kinetics of the extraction of the free CO_2 from the meltwater. For example, we have examined isothermal CO_2 "stripping" from a small sample (0.412 cm^3) of strong alkaline solution (sea water, pH25 = 8.12). It was found that analytical zero in the "stripper"–carrier gas (hydrogen) is reached after "stripping" for 40 min (carrier gas stream 17 cm^3 min^{-1}). Acid was then added to the solution and about 30% of the total inorganic carbon was obtained during the complete extraction. The foregoing consideration is speculative, but it might explain the slight effect in relatively alkaline ice samples.

Third, during glacial stages, the carbonate equilibrium might be shifted toward free CO_2 in comparison with relatively warm stages. This shift is explained by a significant temperature decrease in the carbonate system of water drops before snowflake formation. As a consequence, the concentration of carbon-ionic species in deposited snow is lower in comparison with interglacial stages.

The observed CO_2 maxima at horizon 2280 m would indicate a "sudden" warming before 160 kyr ago. However, this question requires additional study, as the assumption of a "short-term" warming is based on limited data. Nevertheless, this phenomenon appears to be realistic and is consistent with $\delta^{18}O$ data obtained earlier. Thus, according to Imbrie et al.,[16] the observed decrease in $\delta^{18}O$ of approximately 1.4 times during the interval 150 to 170 kyr should indicate glacier melting. Similar trends of $\delta^{18}O$ were observed during the previous interglacial optimum of about 122 to 125 kyr ago. The same relationship, based on two measurements, was observed for the Camp Century ice at a depth of about 1200 m.[22] CO_2 and $\delta^{18}O$ parameters measured on ice samples from Dye 3 show the same strongly fluctuating values.[24]

Consider briefly a possible mechanism to explain the short-term CO_2–$\delta^{18}O$ signal. The $\delta^{18}O$ values of an ice core are determined originally by the process of precipitation. The main bulk of precipitation comes from clouds in polar ice sheets.[33] Thus, the isotopic composition of oxygen in snowfalls reflects the range of temperature in the corresponded cloud sheet. The existence of cloud sheets is determined mainly by wet-temperature regimes and aerosol concentrations. Therefore, for a constant temperature and humidity regime, the changes in aerosol concentrations might determine the altitude for the appearance of rain–snow clouds. The strong injection of aerosols as nuclei of crystallization in the relatively high and cold levels might cause snow precipitation with relatively "cold" $\delta^{18}O$ values. Then, during gravitation precipitation, aerosols might cause cloud formation with relatively "warm" $\delta^{18}O$ values. It should be noted that this process is valid for paleotemperature reconstructions that are based also on the ratio of D/H. Likewise, the corresponding values of bicarbonate content in snowflakes should be lower or higher in comparison with the mean ambient atmospheric conditions. Bicarbonate content is ruled by the value of CO_2 partial pressure which is determined by the altitude of the snowflake formation. We suggest that the most probable source of strong short-term aerosol's injection is connected with extremely high volcanic activity. This idea is supported by the correlation between the Allerod–Younger Dryas $\delta^{18}O$ signal and the Eiffel eruption of 10.0–12.0 kyr BP, which were the last European eruptions north of the Alps. Also, numerous large $SO^{2-}{}_4$ peaks in the Dye 3 core suggest that volcanic activity was very high during this time interval.[14] In the Northeast Atlantic, the Vedde ash layer dated at 10.6 kyr BP is found.[9] This ash layer is associated with a Younger Dryas cold event in the Northeast Atlantic, although it apparently coincides with a low meltwater discharge rate.[4] This hypothesis might explain why real meltwater events coincides with cold Younger Dryas oscillation. It might explain also the great Younger Dryas oscillations in the Northern Hemisphere where there was great volcanic activity in comparison with the Southern Hemisphere.

Conclusion

Consideration that the process of air capture release during the snowflake formation precipitation and sublimation are linked shows that polar gas CO_2 should be fixed mainly as the bicarbonate ion in the ice structure. The difference between the ΣCO_2 value which is obtained by the "wet" technique and "dry" CO_2 air content might be explained in this way.

The discussed antarctic CO_2 paleovariations show a correlation with previous CO_2 records in interglacial studies, but some values are significantly lower in the glacial intervals. A significant maxima of CO_2 concentration was found at a depth of 2280 m.

A speculative mechanism for the imitation of the Younger Dryas signal is presented. The mechanism is based on the relations between strong aerosol volcanic injection to the atmosphere and the altitude of snow cloud formation. The snow bicarbonate concentration and $\delta^{18}O$ value are presented as records which reflect conditions of the cloud sheet before snowflake formation. The known volcanic events in the Northern Hemisphere are in good agreement with the Younger Dryas CO_2–$\delta^{18}O$ oscillations. These events might explain the spatial dilution of the Younger Dryas signal from regions with strong volcanic activity.

Acknowledgments

I thank N. Barkov, A. Gusev, and V. Lipenkov for helpful discussions. I should like to express my thanks to Professor O. Bordovsky for his interest in this work. I also thank

Dr. Wm. K. Johnson and two anonymous reviewers for their comments on the manuscript and some editing. The work was supported by the Arctic and Antarctic Research Institute, by Project "Hydrochemistry" of the World Ocean Program and by the Institute of Geography, Academy of Sciences, Russia. I am also grateful to the Division of Ice Drilling, Gorniy Institute, which provided a drilling program of the Soviet Antarctic Expedition.

References

1. Barnola, I.M., D. Raynaud, A. Neftel, and H. Oeschger. 1983. Comparasion of CO_2 measurements by two laboratories on air from bubbles in polar ice. Nature **303**:410–413.
2. Barnola, I.M., D. Raynaud, Y.S. Korotkevich, and C. Lorius. 1987. Vostok ice core provides 160,000 years record of atmospheric CO_2. Nature **329**:408–414.
3. Blanchard, D.C., and A.H. Woodcock. 1957. Bubble formation and modification in the Sea and its Meteorological Significance. Tellus **9**:145–158.
4. Broecker, W.S., D.M. Peteet, and D. Rind. 1985. Does the ocean–atmosphere system have more than one stable mode of operation. Nature **315**:21–26.
5. Broecker, W.S., D.M. Peteet, and D. Rind. 1983. Carbon dioxide sub-group of the Joint panel on oceanographic tables and standards, Report of Meeting (Miami, Florida, 21–23 September 1983), UNESCO.
6. Coachman, L.K., E. Hemmingsen, and P.F. Scholander. 1956. Gas enclosures in a temperate glacier. Tellus **8**:415–423.
7. Cragin, J.H., M.M. Herron, C.C. Langway, Jr., and G. Klouda. 1977. Interhemispheric comparison of changes in the composition of atmospheric precipitation during the late Cenozoic Era. In M.J. Dunbar (ed.), Polar Oceans, Arctic Institute of North America, pp. 617–631.
8. De Angelis, M., N.I. Barkov, and V.N. Petrov. 1987. Aerosol concentrations over the last climatic cycle (160 kyr) from an Antarctic ice core. Nature **325**:318–320.
9. Fairbanks, R.G. 1989. A 17,000-year glacio-eustatic sea level record: influence of glacial melting rates on the Younger Dryas event and deep-ocean circulation. Nature **342**:637–642.
10. Fireman, E.L., and T.L. Norris. 1982. Ages and composition of gas trapped in Allen Hills and Byrd core ice. Earth Planetary Sci. Lett. **60**:339–350.
11. Graedel, T.E. 1984. Effect of below-cloud gas scavenging on raindrop chemistry over remote ocean regions. Atmosph. Environ. **18**:1835–1842.
12. Hammer, C.U., H.B. Clausen, and W. Dansgaard. 1980. Greenland ice sheet evidence of past-glacial volcanism and its climatic impacts. Nature **288**:230–235.
13. Hemmingsen, E. 1959. Permiation of gases through ice. Tellus **11**:354–359.
14. Herron, M.M., and C.C. Langway, Jr. 1985. Chloride, nitrate, and sulfate in the Dye 3 and Camp Century, Greenland Ice Cores. In C.C. Langway, Jr., H.

Oeschger, and W. Dasgaard (eds.), Greenland Ice Core: Geophysics, Geochemistry, and the Environment. American Geophysical Union, Washington, DC, pp. 71–76.

15. Horibe, Y., K. Shigehara, and C.C. Langway, Jr. 1985. Chemical, and isotopic composition of air inclusions in a Greenland, ice core. Earth Planetary Sci. Lett. **73**:207–210.
16. Imbrie, J., I.D. Hayes, D.C. Martinson, A. McIntyre, A.C. Mix, I.I. Morley, N.G. Pisias, W.L. Prell, and N.J. Shackleton. 1984. The orbital theory of Pleistocene Climate. In A.L. Berger (ed.), Milancovich and Climate, D. Reidel, London, pp. 269–305.
17. Rasool, S.I. (ed.). 1973. Chemistry of the Lower Atmosphere, Plenum Press, New York, London.
18. Lipenkov, V. Ya., and A.N. Salamatin. 1986. Volume relaxation of the ice core from the bore hole at Vostok Station. Antarctica **30**:59–71 (in Russian).
19. Lorius, C., N.I. Barkov, I. Jouzel, Y.S. Korotkevich, V.M. Kotlyakov, and C. Raynaud. 1988. Antarctic Ice Core: CO_2 and Climatic Change over the last Climatic Cycle. EOS **69**:681–684.
20. Lorius, C., I. Jouzel, D. Raynaud, J. Hansen, and H. Le Treut. 1990. The ice-core record: climate sensitivity and future greenhouse warming. Nature **347**:139–145.
21. Maeno, N., and D. Kuroiwa. 1967. Methamorphism of air bubbles in a snow crystal. J. Glaciology **6**:561–564.
22. Neftel, A., H. Oeschger, J. Schwander, B. Stauffer, and R. Zumbrunn. 1982. Ice core sample measurements give atmospheric CO_2 content during the past 40,000 years. Nature **295**:220–223.
23. Neftel, A., H. Oeschger, T. Staffelbach, and B. Stauffer. 1988. CO_2 record in the Byrd ice core 50,000–5,000 years BP. Nature **331**:609–611.
24. Oeschger, H. 1985. The contribution of ice core studies to the understanding of environmental processes. In C.C. Langway, Jr., H. Oeschger, and W. Dansgaard (eds.), Greenland Ice Core: Geophysics, Geochemistry, and the Environment. American Geophysical Union, Washington, DC, pp. 9–17.
25. Petit, J. R., M. Briat, and A. Royer. 1981. Ice age aerosol content from East Antarctic ice core samples and past wind strength. Nature **293**:391–394.
26. Petrov, V.N. 1975. The Atmospheric Supply of Antarctic Ice Covering, Gidrometeoizdat Press, Leningrad (in Russian).
27. Raynaud, D., and R. Delmas. 1977. Composition des gaz contenus, dans la glace polaire. IAHS-AISH Publ. **118**:377–381.
28. Raynaud, D., and B. Lebel. 1979. A total gas content and surface elevation of polar ice sheets. Nature **281**:289–291.
29. Raynaud, D., and I.M. Whillans. 1982. Air content of the Byrd core and past changes in the West Antarctic ice sheet. Annals Glaciology **3**:269–273.
30. Raynaud, D., R. Delmas, I.M. Auscencio, and M. Legrand. 1982. Gas extraction from polar ice cores: a critical issue for studying the evolution of atmospheric CO_2 and ice sheet surface elevation. Annals Glaciology **3**:265–268.

31. Raynaud, D. 1983. The total gas content in polar ice core. In G. de Q. Robin (ed.), The climatic Record in Polar Ice Sheets. Cambridge University Press, Cambridge, pp. 00–00.

32. Rasmussen, R.A., and M.A.K. Khalil. 1984. Atmospheric methane in the recent and ancient atmospheres: concentrations, trends, and interhemispheric gradient. J. Geophys. Res. **89**:11599–11605.

33. Robin, G. de Q. 1983. Ice sheets: isotopes and temperatures. In G. de Q. Robin (ed.), The Climatic Record in Polar Ice Sheets Cambridge University Press, Cambridge, pp. 1–18.

34. Scholander, P.F., E.A. Hemmingsen, L.K. Coachman, and D.C. Nutt. 1961. Composition of gas bubbles in Greenland icebergs. J. Glaciology **3**:813–822.

35. Semiletov, I.P., A.M. Gusev, N.I. Barkov, N.V. Pozdnyakov, and V. Ya Lipenkov. 1989. Carbon dioxide palaeovariations in the antarctic ice cores. Reports of the Academy of Sciences, USSR (Doklady Akademii Nauk SSSR) **309**:196–199.

36. Semiletov, I.P. 1992. On palaeovariations of atmosphere composition in the ice core air inclusions. Antarctica, issue 31 (in Russian).

37. Shackleton, N.J., M.A. Hall, J. Line, and C. Shuxi. 1983. Carbon isotope data in core V19-30 confirm reduced carbon dioxide concentration in the ice age atmosphere. Nature **306**:319–322.

38. Shoji, H., and C.C. Langway, Jr. 1982. Air hydrate inclusions in fresh ice core. Nature **298**:548–550.

39. Stauffer, B., and W Berner. 1978. CO_2 in natural ice. J. Glaciology **21**:291–300.

Late Jurassic Paleoclimate of Pangea Based on Results from a General Circulation Model

George T. Moore

Earth System Science Center, Pennsylvania State University, University Park, Pennsylvania 16802

Darryl N. Hayashida

Chevron Petroleum Technology Company, La Habra, California 90631

Charles A. Ross

GEOSTRAT Consultants
600 Highland Dr., Bellingham, Washington 98225

Stephen R. Jacobson

Chevron Petroleum Technology Company, La Habra, California 90631

ABSTRACT

The progressive disintegration of Pangea occurred during the Jurassic period. The Late Jurassic Epoch represented a time when rift systems divided Gondwana in two, and separated northern Gondwana from North America. Rising eustatic sea level throughout the Jurassic flooded large parts of the continents particularly in the Northern Hemisphere. These conditions created a series of zonally oriented continents and contributed significantly to climate amelioration in the mid-Mesozoic. This chapter focuses on the changed paleoclimate caused by Pangea's disintegration.

General circulation models (GCM) that run on today's supercomputers can be valuable tools in recreating the paleoclimate of brief and important geologic time intervals. We report results from a seasonal GCM simulation for the Late Jurassic. The geological and geochemical record favors an elevated greenhouse effect.

Model results from a simulation using 1120 ppm CO_2, four times the preindustrial level, produce a paleoclimate that fits the geologic record. Sea ice is restricted to the high latitudes, making landfall only in restricted areas. The trade winds bring heavy seasonal rainfall to eastern Gondwana and to the Tethys Sea margins. A strong summer monsoon occurs over southeast Asia. The distribution of coal-forming environments correlates with precipitation sufficient to maintain gymnosperm forests and soil moisture distribution. Evaporites are localized to areas of negative precipitation-minus-evaporation. These simulations predict the distribution of large-scale zonal phenomena (e.g., evaporites, carbonates, coals) and more complex, environmentally sensitive deposits (e.g., corals). The GCM results are impressive considering the model's simplistic nature, coarse grid, and many parameterizations.

Introduction

On a first-order scale, present-day climate is driven by the size, location, and topography of the continental land areas. It is reasonable to believe that this collective forcing system has prevailed throughout geologic time. Because the planet and atmosphere are dynamic and respond to varying influences with time, the study of paleoclimate is further complicated by factors exterior to just land–sea relationships.

Paleoclimate is forced by paleogeography, together with thermally and tectonically induced mountains, Milankovitch-type orbital variations, sea level fluctuations, and variations in the concentration of radiatively active trace gases within the paleoatmosphere. In turn, paleoclimate created the environments in which sediments were deposited. We use the record of climatically sensitive sedimentary rocks with their contained faunas and floras to test the sensitivity of the model results from multiple simulations. We conclude that the simulation used in this paper with 1120 ppm CO_2, four times the preindustrial level, best fits the geologic record.[29]

We report here results from a Late Jurassic (150 Ma) (Figure 1) seasonal paleoclimate simulation generated using a general circulation model (GCM). When compared with biostratigraphic and lithostratigraphic compilations, the model results show the simulation fits the geologic record very well. The Late Jurassic represents a time when seaway connections became established between northern and southern Gondwana as well as between Gondwana and North America (Figure 2). The rising eustatic sea level throughout the Jurassic flooded large parts of the continents and expanded the previously existing seaway between North America and Eur-Asia. These flooding conditions contributed significantly to climate amelioration in the mid-Mesozoic. However, as observed by others for different time frames, the paleoclimate certainly was strongly seasonal in the continental interiors.[17,27,39]

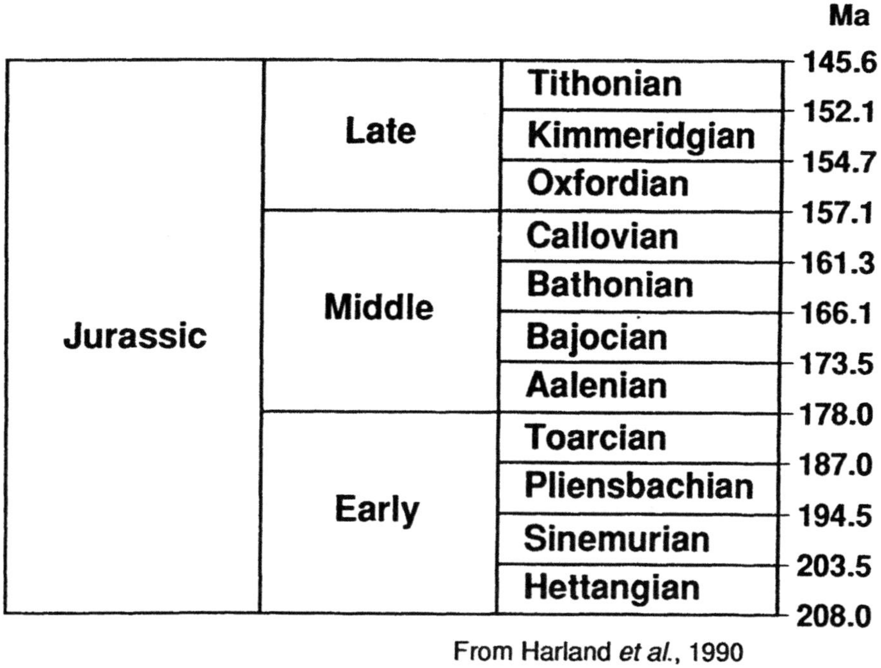

Figure 1. Subdivision of the Jurassic period showing the time interval of the Kimmeridgian/Tithonian used in this paleoclimate simulation (from Ref. 25).

Model Description

The GCM used for this study is the Community Climate Model (CCM) developed at the National Center for Atmospheric Research (NCAR), Boulder, Colorado. The CCM ran on a Cray X-MP at Chevron Oil Field Research Company (presently: Chevron Petroleum Technology Company), La Habra, California. It was derived from the three-dimensional Australian spectral model developed and described by Bourke,[7] Bourke et al.,[8] and McAvaney et al.[28] The CCM uses thermodynamic and hydrodynamic laws applied to the atmosphere. The vertical dimension consists of nine atmospheric levels extending to 33 km in seasonal simulations. The modeled atmosphere extends through the troposphere and into the middle stratosphere. The atmosphere is coupled thermally but not dynamically to a mixed-layer ocean (50 m deep) with a 5° × 5° surface grid.[42] The atmospheric grid consists of 40 (latitude) by 48 (longitude) grid points with each grid cell having a resolution of approximately 4.5° latitude by 7.5° longi-

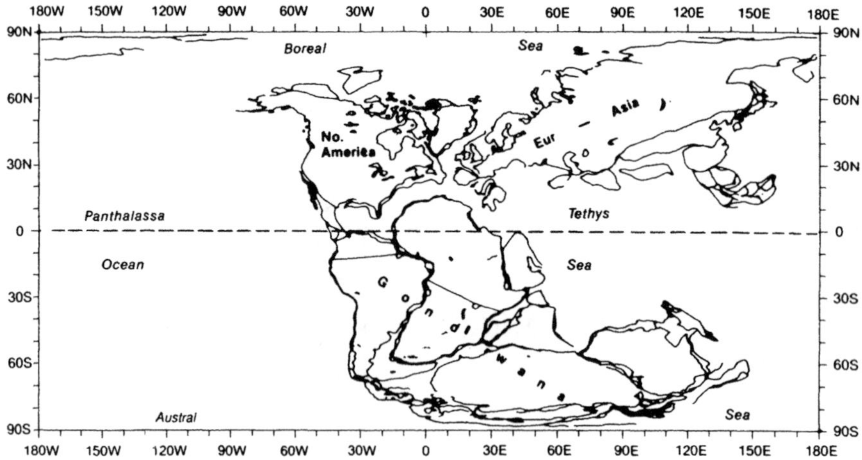

Figure 2. Late Jurassic paleogeography showing modern shorelines. Land masses and bodies of water referred to in text are identified. Reconstruction based on Rowley (36).

tude. The CCM has been shown to model present-day climate reasonably well.[8,28,32]

Bourke et al.[8] and McAvaney et al.[28] describe the parametrizations of physical processes in the CCM. Ramanathan et al.[33] added a radiation-cloudiness formulation which included an absorptance formulation for CO_2. An improved hydrologic cycle treatment was added by Washington and Williamson[44] and Washington and Meehl.[43] It utilizes the model-computed precipitation and evaporation rates to simulate the accumulation of soil moisture. The grid cell capacity, the equivalent of soil saturation, is assumed to be 15 cm. Any excess is considered to be runoff, a parameter not used in this chapter.

The Earth's orbital eccentricity, axial obliquity, and precession, as well as its length of day and year, have changed throughout geologic time. Further, the Sun's irradiation changes on a long-term geological trend,[19,20,45] as well as periodic short-term cycles.[1,19,] Reid[34] has shown that solar irradiance may have varied between 0.1% and 1% over the last few centuries, and possibly by 0.6–0.8% between recent sunspot cycles. These variables are poorly constrained beyond the Pleistocene.

Stellar evolutionary theory predicts that solar luminosity has increased from about 70–75% of the current value over geologic time (4.5 Ga).[19,20] However, questions about values, rates of change, and even the theory remain. Using the standard or linear model of increasing irradiance, the Late Jurassic Sun should have had about 98% of the present Sun's luminosity. The short-term variation in solar irradiation of up to 1% is in the same magnitude and general range as the 2.3% calculated for the Late Jurassic using stellar evolutionary theory. We acknowledge that variations in solar irradiance have occurred over geologic time and that the changes undoubtedly altered the paleoclimate. However, these variables are poorly constrained. Therefore, in this study, we used the present orbital characteristics of the Earth and solar constant of 1.370×10^3 W m^{-2}.

This simulation employs a seasonal model in which solar insolation varies throughout the year. The seasonal version of the model requires about 12 model years to come to steady state, after which climate statistics can be taken confidently. The maps in this report are from 90-day seasonal averages of June/July/August (JJA) for model year 15.5 and December/January/February (DJF) for model year 16.0.

Experiments using elevated CO_2 concentrations to alter the greenhouse effect have been performed using the CCM for the present,[42,43] and Cretaceous.[5] The CCM, therefore, has been shown to be a reasonable tool for modeling atmospheres with different CO_2 concentrations and, therefore, greenhouse effects. For the simulation, we used 1120 ppm CO_2, four times the preindustrial level of 280 ppm CO_2,[2] an approximate average of various workers estimates.[29]

Boundary Conditions

Paleogeography

Two rift systems propagating westward out of the Tethys Sea as shown by paleogeographic reconstructions,[36,37,47] and a third persistent rift system connecting the Boreal Sea with the Tethys Sea[48] fragmented the Late Permian–early Mesozoic megacontinent by Late Jurassic time. These major plate tectonic events produced oceanic seaways that separated Pangea into three major continental segments: North America, Eur-Asia, and Gondwana (Figure 2). Gondwana can be subdivided further into north and south members. The paleogeography used for these simulations shows the relationships at the beginning of the Kimmeridgian [middle Late Jurassic (see Figure 1)] and were used as the ocean–land input boundary conditions for the paleoclimate simulation.

Enhancing the new oceanic pathways were the effects of the Jurassic eustatic sea level rise. In the Kimmeridgian (middle Late Jurassic), the eustatic sea level reached a Jurassic maximum,[24,26] probably in response to increased sea floor spreading and rifting of the Pangean megacontinent. This widespread flooding of continental margins and development of epicontinental seaways is shown in the difference between the present continental outlines and the land outline in model resolution format (Figure 3).

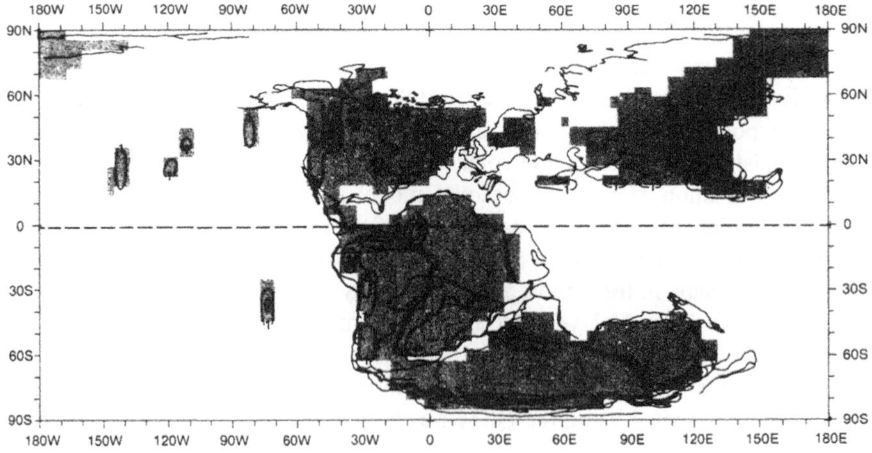

Figure 3. Late Jurassic paleogeography with geologically interpreted paleotopography (contours in kilometers). Shaded areas represent land grid cells in model resolution. Reconstruction based on Rowley[36] with islands in the Panthalassa Ocean derived from Ross and Ross.[35]

Paleotopography

Little qualitative or quantitative information is available for the elevations of Late Jurassic mountains. Mountains are geologically self-destructive environments. To develop this input boundary condition, we used the "mountain" distribution in the latest Jurassic from Ziegler et al.[48]; interpreted them into the plate tectonic framework of Scotese and Golonka[37]; and assigned values in 1 km intervals (up to 3 km) using the present topographic analogues occurring in equivalent plate tectonic settings as a model (Figure 3).

Previous Paleoclimate Studies Using GCMs

GCMs have been used by various workers to model paleoclimate of the Eocene (Sloan[38]; Sloan and Barron[39,41]); mid-Cretaceous (Barron and Washington[4]; Barron[3]); Late Jurassic (Moore et al.[29]; Valdes and Sellwood[49]); mid-Jurassic (Chandler[11]); Late Permian–Triassic (Kutzbach and Gallimore[27]); and Late Permian (Moore et al.[30]). Only some of these simulations (Eocene, mid-Cretaceous, Late Jurassic, and Late Permian) have been run using the same GCM. The sources of the paleogeography differ (Eocene and mid-Cretaceous versus Late Jurassic versus Late Permian) and one uses idealized hemispherically symetrical continents

(Late Permian–Triassic). Various authors developed their own paleotopography; some idealized, some interpreted from plate tectonics, all of which is smoothed over the coarse grids (e.g., 4.5° × 7.5° latitude × longitude) of the GCMs. So, although it may appear from just the references that this new subdiscipline of the earth sciences has made great strides in modeling the last 255 Ma, there are monumental inconsistencies permeating the tools and results. In addition, when recent studies compared 14 GCMs for cloud-climate feedback[9] and 17 GCMs for snow-climate feedback,[10] the results varied widely. The comparisons emphasized the need for model improvements based on additional data and coordination among the institutions developing the models. Whereas model inconsistencies should be kept in mind when viewing the details of sequential simulations, important general and regional results have been obtained from this kaleidoscope of GCMs and approaches.

Crowley et al. ran sensitivity experiments simulating the seasonal cycle of temperatures on Gondwana[15] and Pangea[16] for the Kazanian using a two-dimensional energy balance climate model. They concluded that large temperature cycles are a feature of giant continents and that the temperatures rose so high as to create areas which were uninhabitable.[16]

Kutzbach and Gallimore[27] modeled the paleoclimate of Pangea using a low-resolution GCM with idealized hemispherically symmetrical paleogeography and flat continents. They made a series of sensitivity experiments including elevated CO_2. They showed that widespread continental aridity, precipitation on eastern coasts, western tropical coasts, and areas poleward of 40° latitude, strong monsoonal circulation, and extreme temperatures with hot summers and cold winters in the mid- to high latitudes characterized the Pangean model world.

Moore et al.[30] used the high-resolution CCM with the ocean–land grid cells reflecting a Late Permian paleogeography from reconstructions by Ziegler[46] and Ross and Ross[35] and variable paleotopography inferred from plate tectonic settings to model paleoclimate. Our conclusions [30] were, in part, similar to those of Kutzbach and Gallimore[27] where they related to the Southern Hemisphere. In the Northern Hemisphere, however, the results differed. Moore et al.[30] results showed orographically forced precipitation, no monsoons, and evaporitic conditions in three areas where evaporites appear in the geologic record.

Model Results and Discussion

Moore et al.[29] made a comprehensive analysis of the modeled Late Jurassic paleoclimate. This study utilizes five parameters that are important for evaluating the conditions under which climatically sensitive sediments are deposited. They include surface temperature, precipitation, wind, precipitation-minus-evaporation

(P−E), and soil moisture. For this study, we used the seasonal extremes, December/January/February (DJF) and June/July/August (JJA).

Surface Temperature

The surface temperature maps show the seasonal variation in the oceans and seaways. On land, the maps show the seasonal moderation of tropical to mid-latitude coastal temperatures by marine conditions, temperature gradients, and the mid- and high latitude temperature extremes of the continental interiors. Sea surface temperatures (SST) during both seasons in the tropical Panthalassa Ocean and Tethys Sea range from 25°C to slightly above 30°C (Figure 4). The temperature range (20–30°C) in the tropical–subtropical sea fits the distribution of corals compiled from the literature.[29]

The high-latitude seas are distinctly different. The Northern Hemisphere (N.H.) Boreal Sea is covered by extensive DJF winter sea ice to almost 50°N.[29] (The model computes sea ice when the SST drops below −1.8°C, so the 0° isotherm approximates the sea ice front.) In the JJA season, the sea ice front retreats poleward by 10° latitude or more. The Austral Sea is considerably warmer and only limited sea ice forms in the highest latitudes in JJA.

The biofacies and lithofacies record of the Late Jurassic indicates that the world was warm[6,18,21,23] with zonally distributed seasons, an expanded tropical zone,[23] and largely ice-free continents.[12–14,18]

The interior DJF temperatures of Gondwana reach ~50°C in the subtropics. The margins of the continent are temperate. Similar JJA thermal highs reaching ~45°C occur in the subtropical latitudes of both North America and Eur-Asia. Due to the positions of both continents, the highs are in their southern regions. Temperatures along the continental margins are far milder. Southern Gondwana JJA interior temperatures reach −35°C in a concentric pattern strongly influenced by the continental outline. In the N.H., the DJF temperature pattern is distinctly zonal, whereas in both hemispheres, the summer season's temperature patterns are zonal over the oceans and strongly gradient across the continents.

The summer subtropical temperature gradients across the continents and JJA winter temperatures on southern Gondwana parallel observations of others that the continents have wide temperature ranges depending on locations and size. Only the continental margins moderated by the ocean may be described as equable. The interiors experience high summer and low winter temperatures. As established by others, the term equable, referring to temperatures, is a misnomer for large continents and should be used with caution, if at all.[17,40] The summer and winter temperature maxima are milder than the Late Permian extremes reported by Moore et al.[30] using the CCM and elevated CO_2, but similar to Crowley et al.[16] using an energy balance model.

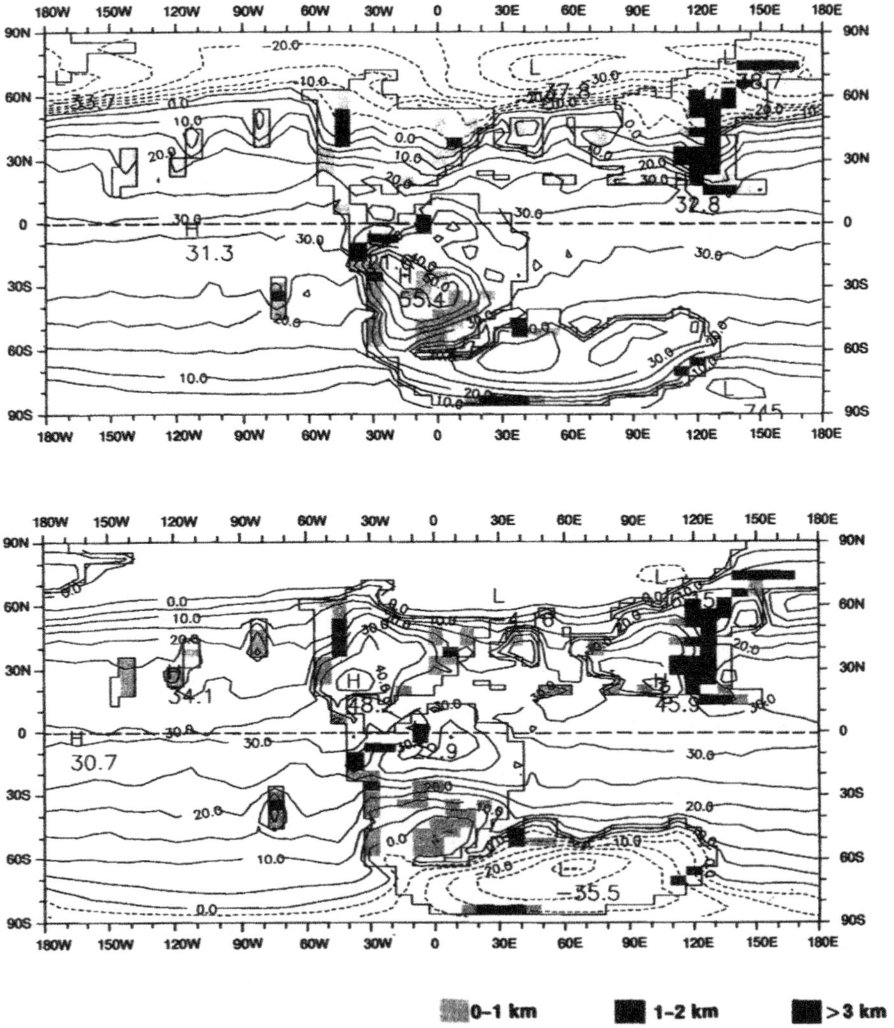

Figure 4. (Top) December/January/February and (bottom) June/July/August Late Jurassic surface temperature (°C) distribution. Contour interval. CI = 5°C. Continents and islands shown as outlines; elevations in patterns of grey.

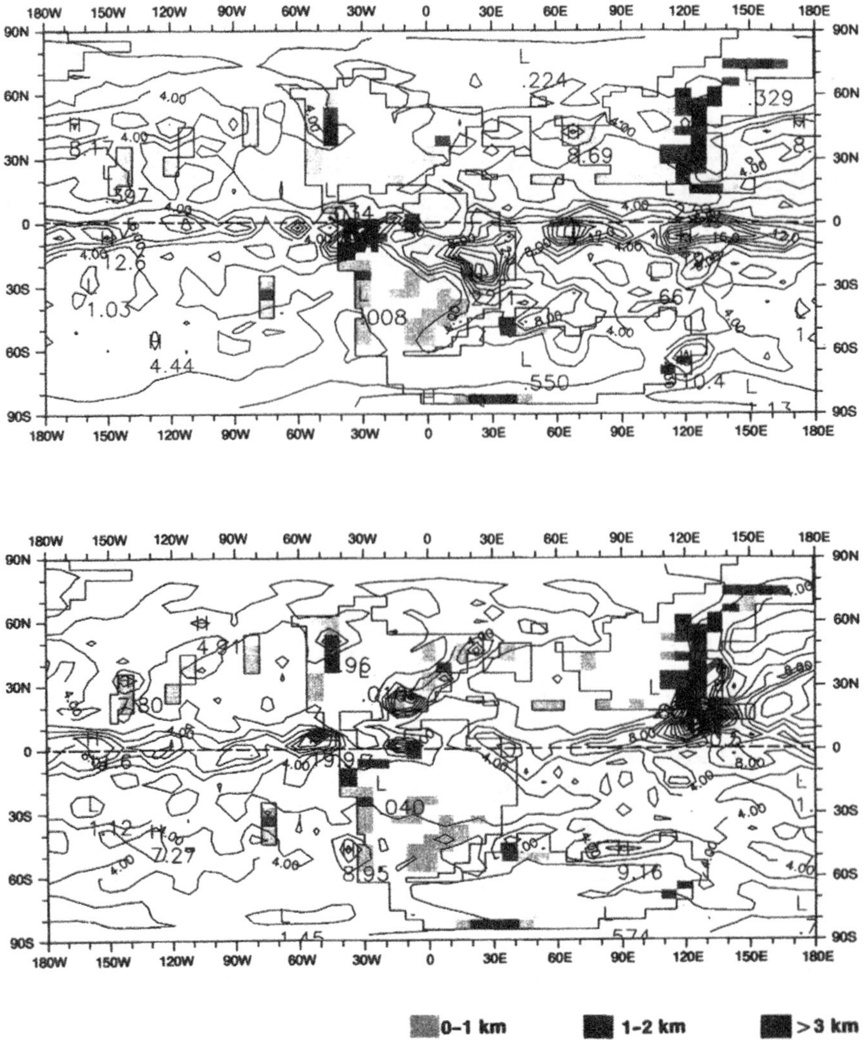

Figure 5. (Top) December/January/February and (bottom) June/July/August maps of Late Jurassic precipitation (mm d^{-1}). M = monsoon related precipitation. CI = 2 mm d^{-1}. Continents and islands shown as outlines; elevations in patterns of grey.

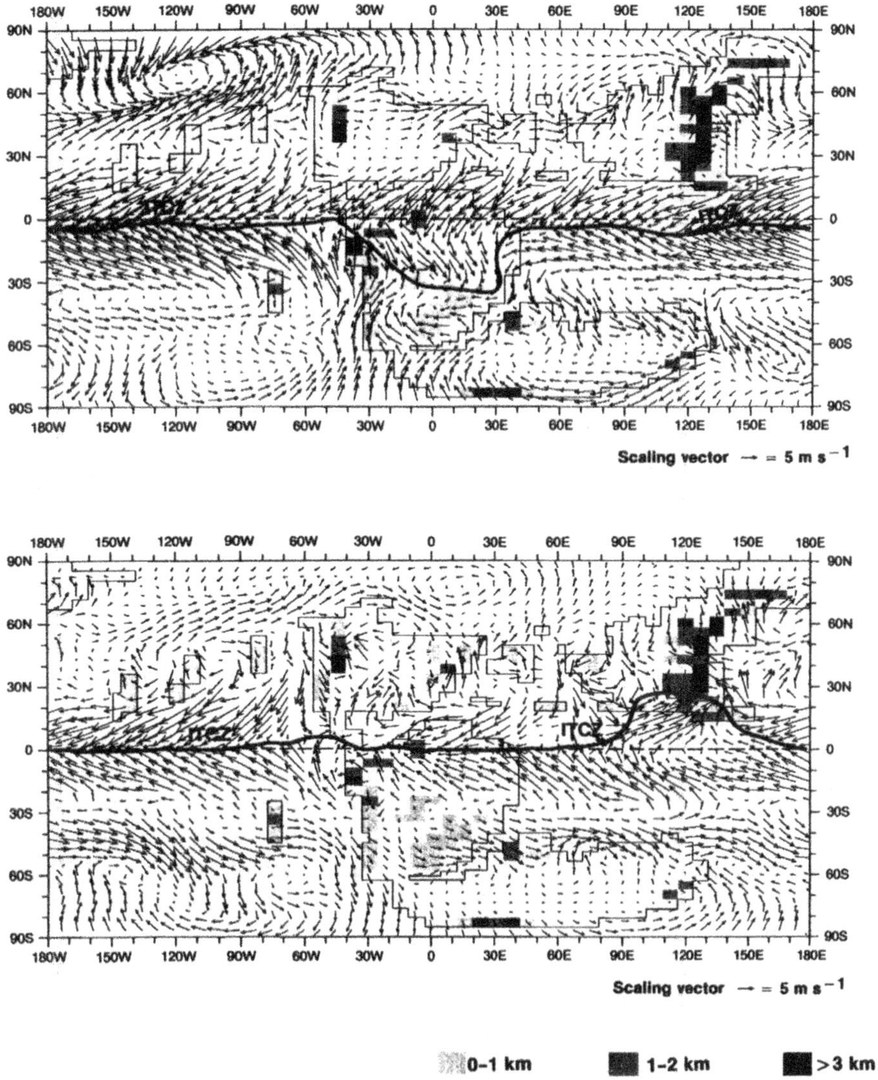

Figure 6. (Top) December/January/February and (bottom) June/July/August maps of Late Jurassic surface wind vectors. ITCZ = Intertropical Convergence Zone. Continents and islands shown as outlines; elevations in patterns of grey.

Precipitation

The modeled Late Jurassic DJF rainfall possesses a strong zonal, equatorial orientation (Figure 5). Heavy precipitation centers in the Tethys Sea and the monsoon in northwestern Gondwana are focused between 5°N and 20°S. The extensive precipitation center in east-central Gondwana in this season results from landfall of the strong northeast and weaker southeast trade winds with a vast, uninterrupted fetch. In the alternate JJA season, the southeast trade winds strengthen and bring moderate precipitation to Gondwana's northeastern margin. Minor DJF centers of rainfall also occur in extreme southeastern Gondwana and northwestern North America. The western North American DJF precipitation results from northward circulation of marine air on the fringe of the strong polar high that develops over the Boreal Sea. The JJA rainfall, in contrast, is forced by the orographic effects of the ancestral Sierra Nevada and the southward movement of air associated with the subtropical high over the northeastern Panthalassa Ocean.

The JJA precipitation patterns have a more complex distribution. A nearly continuous band of moderate precipitation with two intense centers parallels the equator. One of the centers, an intense monsoon, occurs in the N.H. tropics on the southeastern margin of Asia in the northeastern Tethys Sea (Figures 5 and 6). The heavy rainfall results from the combined effects of the landfall of warm tropical marine air and the orographic effects of the south end of the eastern Asian mountain chain. The tropical N.H. monsoonal circulation is duplicated in a diminutive way over southeastern North America. The interior of the continents are arid (Figure 5).

The map of total annual precipitation shows a lack of zonation (Figure 7). Rather each continent forces its own pattern with moderate to high rainfall patterns on the eastern sides of continents and arid regions in the centers and on the mid-latitude western margins. Further, the higher paleolatitudes above 60° generally are arid. These results support the conclusions of Parrish et al.[31] and Hallam[21–23] from the biofacies and lithofacies record that the interiors of the continents were arid.

Precipitation-Minus-Evaporation

One of the most compelling correlations between the geological record and the simulation is the distribution of evaporites[31] and the P − E net moisture map (Figure 8). The zero line divides where precipitation exceeds evaporation ($P - E > 0$; positive values and solid lines) or vice versa ($P - E < 0$; negative values and dashed lines). When the annual P − E results are compared to Tithonian evaporite distribution, 20 localities (71%) fall within net evaporative areas, 5 (18%) are on the zero line, and the remaining 3 (11%) are in net precipitation

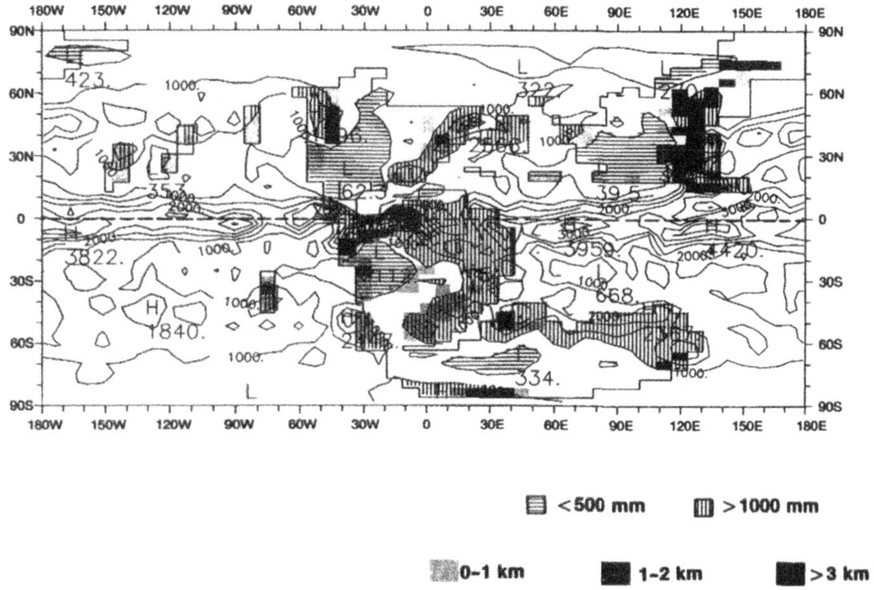

Figure 7. Total annual precipitation (mm) for the Late Jurassic simulation. CI = 500 mm. Continents and islands shown as outlines; elevations in patterns of grey.

areas (Figure 8). The N.H. and S.H bands of evaporites are divided by an equatorial zone of positive P − E (Figure 8). This positive region in the Tethys Sea coincides with the belt of high annual equatorial rainfall (Figure 7). Correlation of the results with known evaporites in the geologic record is remarkably consistent.

Soil Moisture

The DJF and JJA soil moisture maps (Figure 9) show where the surface temperature permits seasonal continental precipitation to exceed evaporation (Figure 8). Accumulation of soil moisture usually is associated with centers of heavy precipitation. To investigate how soil moisture may regulate the size and pattern of the continental biomass, the distribution of Tithonian coals from Parrish et al.[31] is added to the seasonal maps. Of the 20 Tithonian coal localities cited by Parrish et al.,[31] only 5 fall in areas lacking DJF or JJA seasonal soil moisture. Three localities were on the northeast, southeast, and southwest margins of Gondwana, and the other two were in northeastern Asia. The 15 coal localities represent a 75% correlation with seasonal soil moisture. This represents

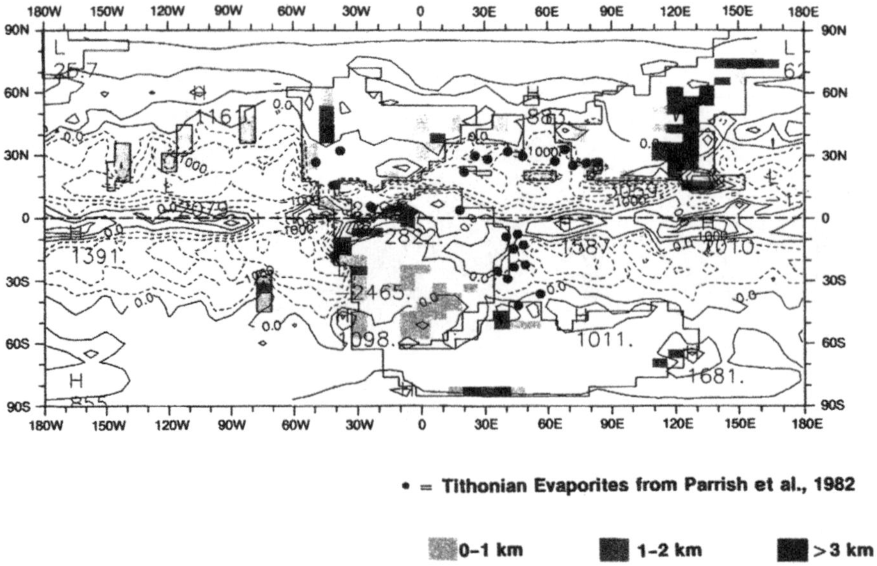

Figure 8. Map of annual precipitation-minus-evaporation (mm) for the Late Jurassic simulation with Tithonian evaporite distribution. Solid lines = net precipitation; dashed lines = net evaporation. CI = 500 mm. Continents and islands shown as outlines; elevations in patterns of grey.

a good, positive correlation. The points not falling into seasonal moist zones may be victims of model resolution and/or model coastal definition.

Conclusions

The Late Jurassic represented the final phase of disintegration for the mega-continent Pangea. The crustal breakup of Pangea by two westward migrating spreading centers out of the western Tethys Sea and a rising sea level that began flooding large parts of Eur-Asia profoundly altered the paleoclimate. Even so, the major continental blocks were extensive, which resulted in the interiors being harsh and arid.

Environmentally sensitive lithotopes and biotopes indicate that the Late Jurassic world was warm. We found that a world with an enhanced greenhouse effect (1120 ppm CO_2) fits the geology quite well. The distribution of sea ice that is largely offshore of the continents fits the record of no reported glacial deposits.

DJF and JJA oceanic rainbelts are in the tropics within 15° of the equator. The

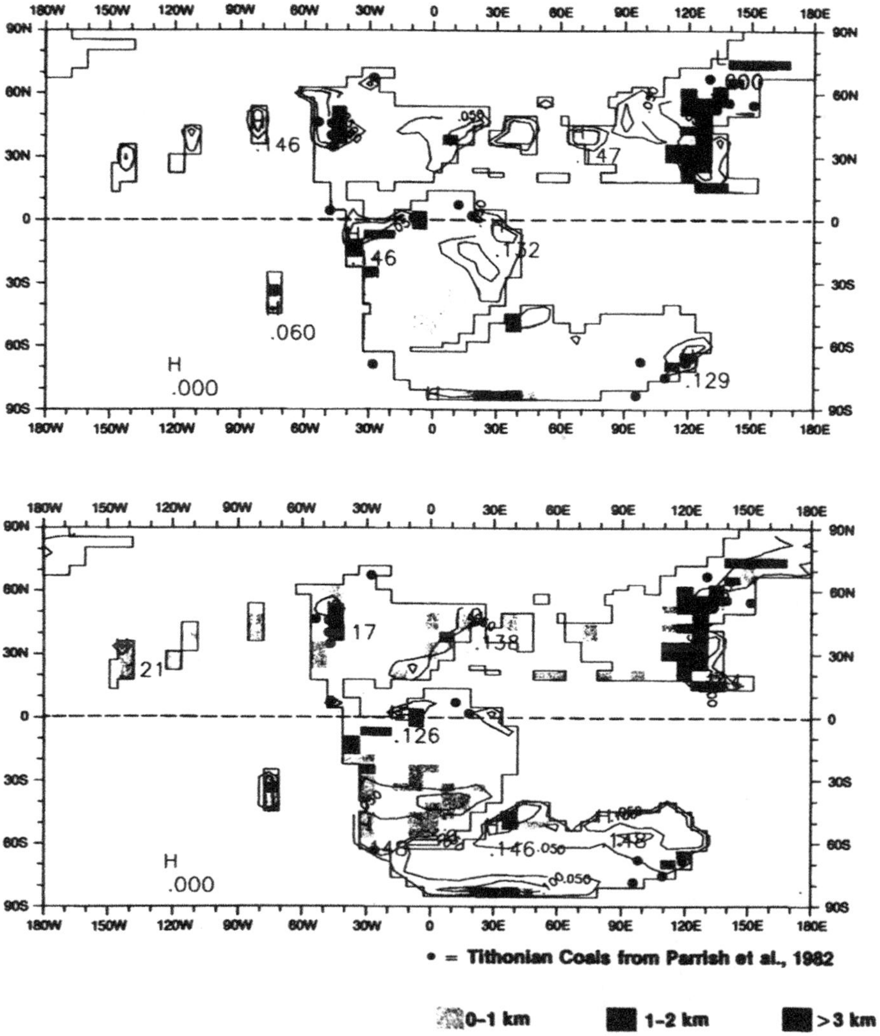

Figure 9. (Top) December/January/February and (bottom) June/July/August maps of Late Jurassic soil moisture (m). CI = 0.050 m. Continents and islands shown as outlines; elevations in patterns of grey.

trade winds bring heavy DJF precipitation to eastern Gondwana and in JJA to the margins of the Tethys Sea. A massive JJA monsoonal circulation focuses heavy precipitation on southeastern Asia. Weaker monsoonal conditions develop near the isthmus connecting North America with Gondwana. Total annual precipitation shows that the eastern parts of the continents receive precipitation as well as the tropical western margins. The interiors and mid-latitude western margins are arid.

Evaporites precipitate where evaporation (E) exceeds precipitation (P). Areas of negative annual P − E shows an 89% correlation with localities containing Tithonian evaporites.

The GCM used in this study, the CCM developed at NCAR, which runs on state-of-the-art supercomputers can be a valuable tool for simulating the paleoclimate of brief and important geologic time intervals. These results are impressive considering the fragmentary nature of the geologic record, the imprecision of plate location, the poorly constrained paleoatmospheric greenhouse gas concentrations, the large grid cell size of the CCM, and the many parameterizations in the code.

Acknowledgments

We thank D. G. Rowley for a reconstruction on which the ocean–land boundary conditions are based. We slightly modified the version he provided us. C. A. Meyer, L. C. Sloan, and two anonymous reviewers provided many constructive comments which materially improved the manuscript. We thank them. J. D. Koishor and J. L. Bube prepared the figures. J. J. McLellan and B. A. Villegas typed the manuscript. We appreciate their efforts. We thank Chevron Oil Field Research Company for permission to publish this chapter.

References

1. Baliunas, S., and R. Jastrow. 1990. Evidence for long-term brightness changes of solar-type stars. Nature **348**:520–522.
2. Barnola, J.M., D. Raynaud, Y.S. Korotkevich, and C. Lorius. 1987. Vostok ice core provides 160,000-year record of atmospheric CO_2. Nature **329**:408–414.
3. Barron, E.J. 1985. Numerical climate modeling; a frontier in petroleum source rock prediction: results based on Cretaceous simulations: AAPG Bull. **69**:448–459.
4. Barron, E.J., and W.M. Washington. 1982. Cretaceous climate: A comparison of atmospheric simulations with the geologic record: Palaeogeog. Palaeoclim. Palaeoecol. **40**:103–133.
5. Barron, E.J., and W.M. Washington. 1985. Warm Cretaceous climates: High at-

mospheric CO_2 as a plausible mechanism: The carbon cycle and atmospheric CO_2: Natural variations Archean to Present. A.G.U. Geophy. Mon. **32**:546–553.

6. Berlin, T.S., D.P. Naydin, V.N. Saks, R.V. Teis, and A.V. Khabakov. 1966. Jurassic and Cretaceous climate in northern USSR, from paleotemperature determinations. Internat. Geol. **9**:1080–1092.
7. Bourke, W. 1974. A multi-level spectral model. I. Formulations and hemispheric integrations: Monthly Weather Rev. **102**:687–701.
8. Bourke, W., B. McAvaney, K. Puri, and R. Thurling. 1977. Global modeling of atmospheric flow by spectral methods. In J. Chang (ed.), Methods in Computational Physics, Academic Press, New York, vol. 17, pp. 267–324.
9. Cess, R.D., et al. 1989. Interpretation of cloud-climate feedback as produced by 14 atmospheric general circulation models. Science **245**:513–516.
10. Cess, R.D., et al. 1991. Interpretation of snow-climate feedback as produced by 17 general circulation models. Science **253**:888–892.
11. Chandler, M.A. 1989. Aridity at mid and low latitudes in warmer climates: A simulation of the Jurassic climate using the GISS GCM (Abs.). EOS **70**:1010.
12. Christie-Blick, N. 1982. Pre-Pleistocene glaciation on Earth: Implications for climatic history of Mars. Icarus **50**:423–443.
13. Crowell, J.C., and L.A. Frakes. 1970. Phanerozoic glaciation and the causes of ice ages. Amer. J. Sci. **268**:193–224.
14. Crowell, J.C., and L.A. Frakes. 1975. The late Paleozoic glaciation. In K.S.W. Campbell (ed.), Gondwana Geology, National University Press, Canberra, Australia, pp. 313–331.
15. Crowley, T.J., J.G. Mengel, and D.A. Short. 1987. Gondwana's seasonal cycle. Nature **329**:803–807.
16. Crowley, T.J., W.T. Hyde, and D.A. Short. 1989. Seasonal cycle variations on the supercontinent of Pangaea. Geology **17**:457–460.
17. Crowley, T.J., and G.R. North. 1991. Paleoclimatology, Oxford University Press, Oxford.
18. Frakes, L.A. 1979. Climates Throughout Geologic Time, Elsevier Scientific Publishing Co., New York.
19. Gérard, J.-C. 1990. Modelling the climate response to solar variability. Phil. Trans. Royal Soc. London, Ser. A **330**:561–574.
20. Gilliland, R.L. 1989. Solar evolution. Palaeogeog. Palaeoclim. Palaeoecol. **75**:35–55.
21. Hallam, A. 1982. The Jurassic climate. In Climate in Earth History, National Academy Press, Washington, DC, pp. 159–162.
22. Hallam, A. 1984. Continental humid and arid zones during the Jurassic and Cretaceous. Palaeogeog. Palaeoclim. Palaeoecol. **47**:195–223.
23. Hallam, A. 1985. A review of Mesozoic climates. J. Geol. Soc. London **142**:433–445.

24. Hallam, A. 1988. A reevaluation of Jurassic eustasy in the light of new data and the revised Exxon curve. In C.K. Wilgus et al. (eds.), Sea Level Changes—An Integrated Approach, SEPM Spec. Pub., no. 42, Society Economic Paleontologists and Mineralogists, Tulsa, OK, pp. 261–273.
25. Harland, W.B., R.L. Armstrong, A.V. Cox, L.E. Craig, A.G. Smith, and D.G. Smith. 1990. Geological Time Scale 1989, Cambridge University Press, Cambridge.
26. Haq, B.U., J. Hardenbol, and P.R. Vail. 1988. Mesozoic and Cenozoic chronostratigraphy and cycles of sea-level change. In C.K. Wilgus et al. (eds.), Sea Level Changes—An Integrated Approach, SEPM Spec. Pub. 42, Society of Economic Paleontologists and Mineralogists, Tulsa, OK, pp. 71–108.
27. Kutzbach, J.E., and R.G. Gallimore. 1989. Pangaean climates: Megamonsoons of the megacontinent. J. Geophys. Res. **94**:3341–3357.
28. McAvaney, B.J., W. Bourke, and K. Puri. 1978. A global spectral model for simulation of the general circulation. J. Atmos. Sci. **35**:1557–1583.
29. Moore, G.T., D.N. Hayashida, C.A. Ross, and S.R. Jacobson. 1992. The paleoclimate of the Late Jurassic world: I. Results from a general circulation model. Palaeogeog. Palaeoclim. Palaeoecol., **93**:113–150.
30. Moore, G.T., C.J. Peoples, and S.C. Luhn. 1989. The paleoclimate of the early Late Permian World. EOS **70**:1010.
31. Parrish, A.M. Ziegler, and C.R. Scotese. 1982. Rainfall patterns and the distribution of coals and evaporites in the Mesozoic and Cenozoic. Palaeogeog. Palaeoclim. Palaeoecol. **40**:67–101.
32. Pitcher, E.J., R.C. Malone, V. Ramanathan, M.L. Blackmon, K. Puri, and W. Bourke. 1983. January and July simulations with a spectral general circulation model. J. Atmos. Sci., **40**:580–604.
33. Ramanathan, V., E.J. Pitcher, R.C. Malone, and M.L. Blackmon. 1983. The response of a spectral general circulation model to refinements in radiative processes. Amer. Meteor. Soc. **40**:605–630.
34. Reid, G.C. 1991. Solar total irradiance variations and the global sea surface temperature record. J. Geophys. Res. **96**:2835–2844.
35. Ross, C.A., and J.R.P. Ross. 1983. Late Paleozoic accreted terranes of western North America. In C.H. Stevens (ed.), Pre-Jurassic rocks in western North American suspect terranes, Pacific Sec., Society Economic Paleontologists and Mineralogists, Los Angeles, CA, pp. 7–22.
36. Rowley, D.B. 1992. Preliminary Jurassic reconstructions of the Circum-Pacific region. In G.E.G. Westermann (ed.), The Jurassic of the Circum-Pacific, IGCT Project 171, Cambridge University Press, Cambridge, pp. 15–27.
37. Scotese, C.R., and J. Golonka. 1992. Paleogeographic Atlas: PALEOMAP Project, University of Texas at Arlington, Arlington, TX.
38. Sloan, L.C. 1990. Determination of critical factors in the simulation of Eocene global climate, with special reference to North America. Ph.D. dissertation, Penn State University.

39. Sloan, L.C., and E.J. Barron. 1989. Eocene climate based on results from atmospheric circulation model experiments. EOS **70**:374–375.

40. Sloan, L.C., and E.J. Barron. 1990. "Equable" climates in Earth history? Geology **18**:489–492.

41. Sloan, L.C., and E.J. Barron. 1992. A comparison of Eocene climate model results to quantified paleoclimatic interpretations. Palaeogeog. Palaeoclim. Palaeoecol. **93**:183–202.

42. Washington, W.M., and G.A. Meehl. 1983. General circulation model experiments on the climatic effects due to a doubling and quadrupling of carbon dioxide concentrations. J. Geophys. Res. **88**:6600–6610.

43. Washington, W.M., and G.A. Meehl. 1984. Seasonal cycle experiment on the climate sensitivity due to a doubling of CO_2 with an atmospheric general circulation model coupled to a simple mixed-layer ocean. J. Geophys. Res. **89**:9475–9503.

44. Washington, W.M., and D.L. Williamson. 1977. A description of the NCAR global circulation models. In J. Chang (ed.), Methods in Computational Physics, Academic Press, New York, Vol. 17, pp. 111–169.

45. White, O.R. 1977. The Solar Output and Its Variation, Colorado Associated University Press, Boulder, CO.

46. Ziegler, A.M. 1981. Paleozoic paleogeography. In M.W. McElhinny and D.A. Valencio (eds.), Paleoreconstruction of the Continents, American Geophysical Union, Washington, DC, and Geological Society of America, Boulder, CO, pp. 31–37.

47. Ziegler, A.M., C.R. Scotese, and S.F. Barrett. 1983. Mesozoic and Cenozoic paleogeographic maps. In P. Brosche and J. Sündermann (eds.), Tidal Friction and the Earth's Rotation II, Springer-Verlag, Berlin, pp. 240–252.

48. Ziegler, P.A. 1988. Evolution of the Arctic–North Atlantic and the western Tethys: AAPG Mem. 43, Tulsa, OK, 198 p., 30 plates.

49. Valdes, P.J., and B.W. Sellwood, 1992. A palaeoclimate model for the Kimmeridgian. Palaeogeog. Palaeoclim. Palaeoecol. **95**:47–72.

Distributions, Reactions, Sources, and Sinks

The Sensitivity of the Global Atmosphere to Anthropogenic Methane Sources

Natalia G. Andronova and Igor L. Karol

Voeikov Main Geophysical Observatory, 194018, Leningrad, Russia

ABSTRACT

In this chapter, we consider a change in anthropogenic methane sources to be the external forcing for the global atmosphere. The global atmosphere is represented simply by a linearized system of interactions near their equilibrium values of annually averaged temperature and methane, carbon monoxide, tropospheric ozone, stratospheric ozone, total ozone, hydroxyl, and water vapor.

The properties of the system are studied by the Cause-and-Effect Analysis (CEA) technique. The CEA technique enables the identification and visualization of the pathways from the system's external parameters to its internal variables and all of the system's feedbacks and allows the quantitative estimation of the sensitivity of any internal variable to a change in any external parameter.

It is shown that the system's feedbacks yield a net loop effect that is much less than unity, which gives a large sensitivity to any external forcing; that an increase in the anthropogenic methane source results in increases in methane, ozone, and temperature in the troposphere (positive sensitivities), and a decrease in tropospheric hydroxyl (negative sensitivity); and that the "temperature–wetlands" link decreases significantly the negative sensitivity of the methane and carbon monoxide concentrations to the temperature changes.

Introduction

Photochemical and radiative processes in the global atmosphere, which is a part of the earth's geophysical system, define a complex system of interactions which complicates investigation and prognosis of the system's behavior under external forcing. However, such a complex system can be linearized near its equilibrium position and then analyzed qualitatively by the Cause-and-Effect Analysis (CEA) technique. This technique allows identification and visualization of the pathways from the system's external parameters to its internal variables, as well as all of the system's feedbacks and permits the quantitative estimation of the sensitivity of any internal variable to a change in any external parameter. The rules of the CEA technique have been presented by Heise,[10] together with illustrative applications for social and economical problems, whereas the application of the CEA technique to photochemical problems has been presented by Andronova[2] and Andronova and Schlesinger.[4]

In this chapter, the CEA technique is used to investigate the role of a change in anthropogenic methane sources for the global atmosphere, represented simply as a linearized system of photochemical and radiative interactions near their equilibrium annually averaged values of temperature and the concentrations of methane, carbon monoxide, tropospheric ozone, stratospheric ozone, total ozone, hydroxyl, and water vapor.

The second section of this chapter presents the simplified and linearized system of atmospheric photochemical and radiative processes and also the analysis method. This method is used in the third section to estimate the system's net loop effect, and in the fourth section to investigate the sensitivity of the system to an increase in anthropogenic methane sources. The last section presents the main conclusions.

Photochemical System and Analysis Method

Generally, the set of atmospheric photochemical interactions can be expressed by the set of differential mass-balanced equations

$$\frac{d(Y)}{dt} + q \cdot (Y) = Q, \tag{1}$$

where (Y) is the concentration of gas Y, $q\cdot(Y)$ is the total photochemical sink of Y, and Q is the total source of Y. In general, the photochemical sinks and sources of gas Y depend on the system's internal variables $Y = \{Y_k\}$ and/or external variables $E = \{E_j\}$. Linearizing Eq. (1) near its equilibrium point $x_0 = \{(Y_k)_0, (E_j)_0\}$ gives

$$[Y_i] = \sum_k l_{ik}\,[Y_k] + \sum_j \alpha_j[E_j], \tag{2}$$

where $[x] = \Delta x/x_0$, $\Delta x = x - x_0$, l_{ik} represents the intensity of the influence of internal variable $[Y_k]$ on internal variable $[Y_i]$, and $[E_j]$ can be interpreted as the intensity of the influence of external forcing j on the internal variable $[Y_i]$ with the relative input α_j. The values of l_{ik} and α_j, which are constants and depend on the equilibrium position x_0, define the relative input of variables Y_k and E_j to the total source and/or to the total sink of variable Y_i. Thus, if the variable Y_k is the single sink (source) of variable Y_i, then $l_{ik} = -1$ ($l_{ik} = 1$);

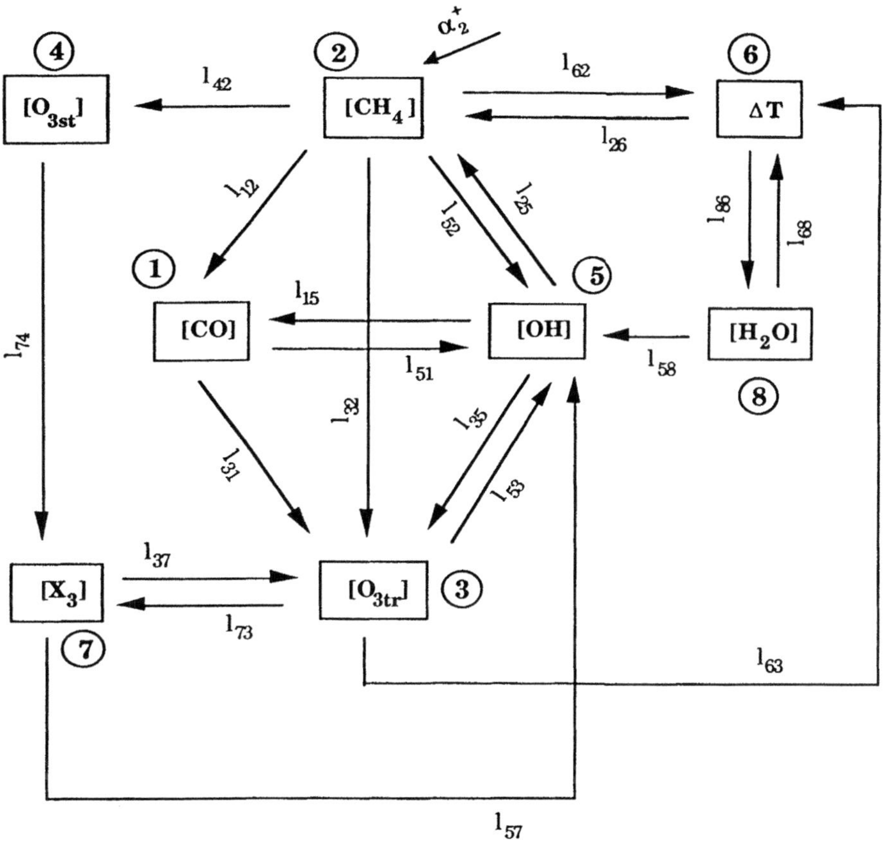

Figure 1. The system of interconnections of the greenhouse atmosphere.

and if the variable E_j is the single sink (source) of variable Y_i, then $\alpha_j = -1$ ($\alpha_j = 1$).

Equation (2) is an algebraic equation which has a graph analogue that can be analyzed by the CEA technique.[4]

The atmospheric photochemical system connected with methane is presented in Figure 1, wherein the chemical symbols in square brackets represent deviations of the global, annually averaged gas concentrations (Y) from their corresponding equilibrium values (Y_0) in relative units, $[Y] = \Delta Y/Y_0$, $\Delta Y = Y - Y_0$, the causes of which are small disturbances which drive the system to a new equilibrium position. Thus, $[CH_4]$ represents the relative deviation of the tropospheric concentration of methane, and [CO], [OH], and $[H_2O]$ represent the relative deviation of the concentrations of hydroxyl radical (as the main methane sink), carbon monoxide (as one of the main hydroxyl sinks), and water vapor (as the main hydroxyl source). The relative deviation of total ozone in an atmospheric column and the deviation in absolute units (Kelvin) of the tropospheric temperature, represented by $[X_3]$ and ΔT, are the variables which define the conditions for the atmospheric photochemical interactions. The relative deviation of the concentrations of tropospheric ozone and the averaged ozone concentration above 30 km, represented by $[O_{3tr}]$ and $[O_{3st}]$, are the main parts of the total ozone. We will consider the external forcing of our system to be the increase in the total anthropogenic methane source (E_2) with relative input α_2^+.

In Figure 1, each variable is represented by $[Y_l]$, and an arrow with the symbol l_{ij} defines the direction and intensity of the influence of variable $[Y_j]$ on variable $[Y_i]$. The set of system links $\{l_{ij}\}$ creates a set of loops, the net loop effect of which defines the system's reaction to the external forcing. According to Heise,[10] the system's net loop effect (D) is

$$D = [(1 - L')(1 - L'') \cdots (1 - L^{`})]^* = [1 - L]^*, \tag{3}$$

where L′, L″, . . . , L` are the effects of each loop of the system given by the product of the numerical values of the links l_{ij} that comprise it (see Figure 1); the superscript * means that products which contain common indices, thereby defining the effects of "touching ways," are removed from Eq. (3); and $(1 - L)$ is a shorthand notation.

To define the parameters on which our system's net loop effect depends, we consider the following subsystems of increasing complexity:

a. a "nongreenhouse" troposphere (NGT), consisting of the four elements: $[CH_4]$, $[CO]$, $[OH]$, and $[O_{3tr}]$;
b. a greenhouse troposphere (GT), consisting of the six elements: $[CH_4]$, $[CO]$, $[OH]$, $[O_{3tr}]$, $[H_2O]$, and ΔT;
c. the full system.

According to Heise,[10] the value of a system's sensitivity coefficient (ψ_{ik}) of variable $[Y_i]$ to external forcing (E_k) is

$$\psi_{ik} = \frac{[(e'_{ik} + e''_{ik} + \cdots)(1 - L)]^*}{D}, \qquad [Y_i] = \psi_{ik}\,\alpha_k\,[E_k], \tag{4}$$

where $(1 - L)$, D, and the superscript * have the same meaning as in Eq. (3); e'_{ik}, e''_{ik}, . . . are the effects of all the possible pathways by which the disturbance E_k, with relative input α_k into variable Y_k, passes through the system until it reaches the output variable Y_i. The numerical values of the pathways e'_{ik}, e''_{ik}, . . . are defined in the same manner as that for L', L'', . . . , that is, as the product of all the l_{ij} along the pathway. From Eq. (4), the role of D can be seen, namely, the set of system loops amplifies the external forcing if $|D| < 1$, and damps or "controls" the external forcing if $|D| > 1$; also, if D is close to zero, then the system can be very sensitive to an external disturbance.

The System's Loops and Net Loop Effect

In this section, we consider three models of increasing complexity to investigate the system's loops and to estimate the system's net loop effect.

Loops of the "Nongreenhouse" Troposphere

Consider the system for the four variables, $[Y_1] = [CO]$, $[Y_2] = [CH_4]$, $[Y_3] = [O_{3tr}]$, and $[Y_5] = [OH]$, presented in Figure 2a. The interconnections between these variables, defined in principle by photochemical reactions between them, create the following seven loops:

$$L_{151} = l_{51}l_{15}; \quad L_{1521} = l_{51}l_{25}l_{12}; \quad L_{252} = l_{52}l_{25}; \quad L_{353} = l_{53}l_{35};$$
$$L_{2352} = l_{32}l_{53}l_{25}; \quad L_{1351} = l_{31}l_{53}l_{15}; \quad L_{21352} = l_{12}l_{31}l_{53}l_{25},$$

with the net loop effect, D_{NGT}, given by

$$D_{NGT} = [1 - L]^* = 1 - L_{151} - L_{1521} - L_{252} - L_{353} - L_{2352} - L_{1351} - L_{21352}, \tag{5}$$

where the effects of the "touching ways" (the products of the loops having common indexes) have been excluded.

To estimate the value of D_{NGT}, we assume that

$$l_{25} = -1; \quad l_{15} = -1 + l_{12}; \quad l_{35} = l_{31} + l_{32},$$

which means that

1. hydroxyl provides the dominant methane sink in troposphere by reaction (A.3) of the Appendix;
2. on one side, hydroxyl provides the dominant carbon monoxide sink in the troposphere by reaction (A.1) and, on the other side, hydroxyl provides the same input to the carbon monoxide source in the troposphere as methane by reaction (A.2);

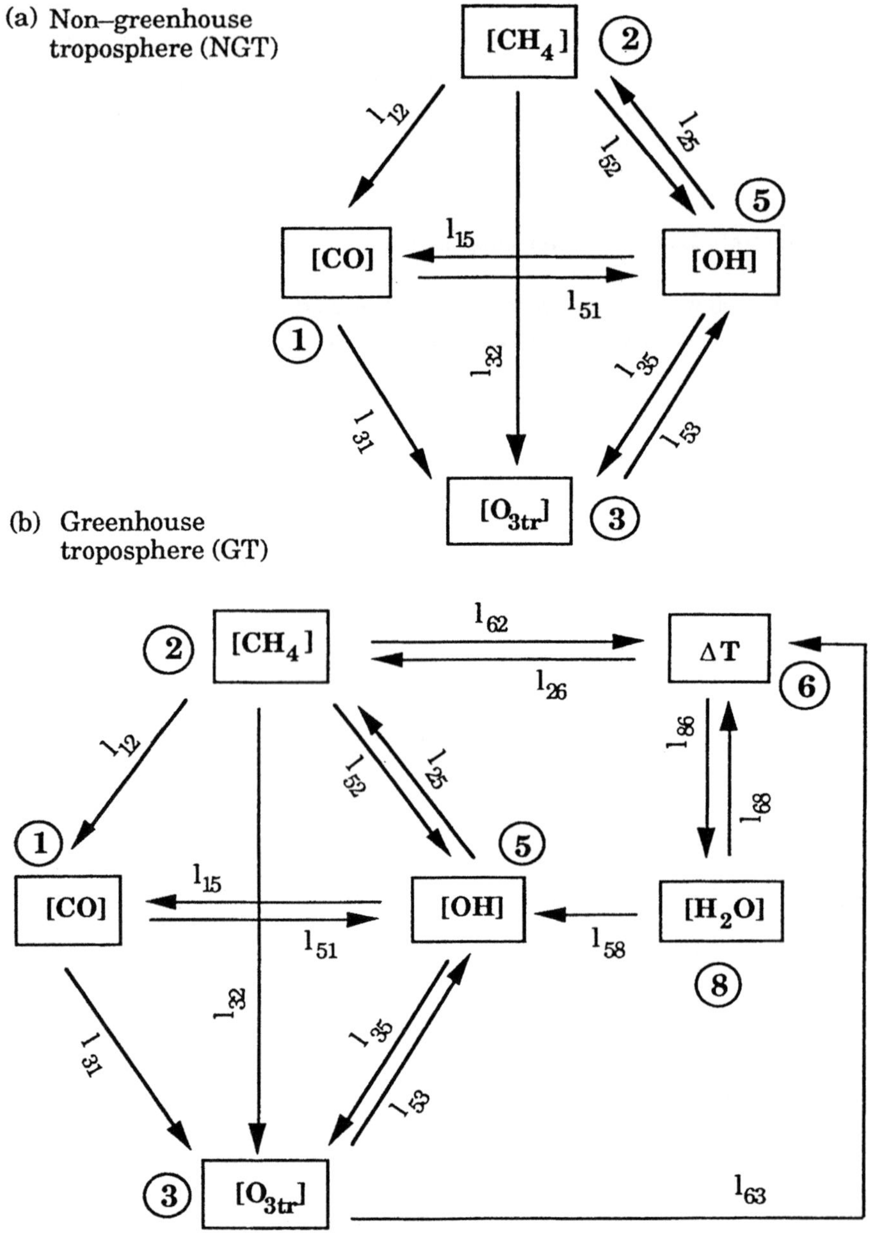

Figure 2. The interactions of system: (a) ''nongreenhouse'' troposphere; (b) greenhouse troposphere.

Table 1. Values of the links l_{ij} of the system.

l_{ij}	Min/Max Value	Mean Value	References and Comments
l_{12}	0.2/0.4	0.3	12
l_{15}		−0.7	$l_{15} - l_{12} = -1$
l_{25}	−1.0	−1.0	The single sink
l_{26}	0.02 (1/K)	0.02 (1/K)	3
l_{31}	0.06/0.11	0.09	11,12
l_{32}	0.01/0.03	0.02	12
l_{35}		0.11	$l_{35} - l_{31} - l_{32} = 0$
l_{37}	−0.4/−0.6	−0.5	1,5
l_{42}	0/0.06	0.03	12
l_{51}	−0.22/−0.08	−0.15	12
l_{52}	−0.3/−0.1	−0.2	12
l_{53}	0.1/0.4	0.25	12
l_{57}	0.06/0.48	0.27	1,5
l_{58}		0.25	$l_{53} = l_{58}$
l_{62}	0.3/0.4 (K)	0.35 (K)	12
l_{63}	0.6/0.7 (K)	0.65 (K)	12
l_{68}	6.0/10.0 (K)	7.5 (K)	7,14
l_{73}	0.1/0.2	0.15	1,5
l_{74}	0.3/0.4	0.35	1,5
l_{86}	0.02/0.07 (1/K)	0.045 (1/K)	15

3. hydroxyl provides the same input to the ozone source as methane and carbon monoxide together by reactions (A.4) and (A.5) in the NO_x-enriched troposphere.

These assumptions allow presentation of Eq. (5.5) as

$$D_{NGT} = 1 + l_{51} + l_{52}, \tag{6}$$

which means that D_{NGT} is less than unity for all values of parameters because l_{51} and l_{52} are negative.

We get the estimates of the l_{51} and l_{52} values from the results of the one-dimensional photochemical model experiments presented by Karol[12] as shown in Table 1. According to the values of l_{ij}, the mean value of the net loop effect for the "nongreenhouse" troposphere, $\overline{D}_{NGT}$, is

$$\overline{D}_{NGT} = 0.65. \tag{7}$$

From Eq. (7) we conclude that, because the value of the net loop effect is less than unity, the "nongreenhouse" troposphere is very sensitive to external forcing.

Loops of the Greenhouse Troposphere

The addition of two more elements to the four-element "nongreenhouse" troposphere, ΔT and $[H_2O]$, forms six new links (see Figure 2b), among which l_{62}, l_{63}, and l_{68} reflect the radiative properties of methane, ozone, and water vapor, respectively; l_{58} shows that the water vapor is a photochemical source of hydroxyl; and l_{86} and l_{26} give the influences of temperature on the surface sources of methane and water vapor, respectively. These links form 12 new loops, which create the net loop effect of the greenhouse troposphere as

$$
\begin{aligned}
D_{GT} = {} & D_{NGT}(1 - L_{686}) - L_{262}(1 - L_{151} - L_{353} - L_{1521} - L_{1351}) \\
& + L_{151}(L_{2362} + L_{13621}) - (L_{2362} + L_{13621} + L_{26852} + L_{25362} \\
& + L_{36853} + L_{153621} + L_{136251} + L_{136851} + L_{236852} + L_{1368521}).
\end{aligned} \tag{8}
$$

It is qualitatively clear that, because the loop input to the net loop effect decreases as the number of loop indexes increases because the relative values of the links are less than unity, the value of D_{GT} is dominated by the first term and the first part of the second term of Eq. (8). To estimate the value of D_{GT}, we use the values of l_{ij} from Table 1, obtained from the different model estimates indicated in the last column of Table 1. Indeed, according to these estimates, the net loop effect of the greenhouse troposphere can be represented approximately by

$$D_{GT} \cong D_{NGT}(1 - L_{686}), \tag{9}$$

and we conclude that loop L_{686} for the "temperature–water vapor" connection dominates the net loop effect. For the mean values of the links l_{68} and l_{86} from Table 5.1, we obtain

$$\overline{D}_{GT} \cong 0.65\,\overline{D}_{NGT} = 0.43. \tag{10}$$

The inclusion of loop L_{686} into the "nongreenhouse" troposphere decreases the net loop effect, thereby making the troposphere more sensitive to external forcing.

Loops of the Greenhouse Atmosphere

The complete system of interactions having all eight elements is presented in Figure 1 and contains 30 loops. Among the new links, l_{73} and l_{74} show that tropospheric ozone and ozone above 30 km, respectively, are parts of the total ozone; l_{37} and l_{57} represent the influences of the total ozone on the concentrations of the tropospheric ozone and hydroxyl through the influence on the photodissociation rates; and l_{42} reflects the influence of methane on the stratospheric ozone through the chemical reaction methane with the chlorine radical, which is one of the considerable ozone sinks in stratosphere. The net loop effect of this complete system of interconnections, after exclusion of terms with small absolute values, can be written as

$$(11) \qquad D_{GA} \cong (1 - L_{686})D_{NGT}(1 - L_{373}) = D_{GT}(1 - L_{373}),$$

from which one can obtain

$$D_{GT} < D_{NGT} < D_{GA},$$

where conditions $L_{686} > 0$ and $L_{373} < 0$ have been used. This leads to the conclusion that, whereas the interactions between temperature and water vapor decreased the net loop effect of the "nongreenhouse" troposphere, the interconnection between stratospheric ozone and total ozone increases the net loop effect. For the average values of the parameters in Table 5.1, we obtain

$$(12) \qquad \overline{D}_{GA} = 0.46.$$

Sensitivity Analysis

We will consider the external forcing of the complete system to be the increase in the total anthropogenic methane source (E_2) with relative input $\alpha_2^+ = 0.6$.[8]

According to Eq. (4), we have obtained the sensitivity coefficients ψ_{i2}, for $i = 1, \ldots, 8$, to the increase in anthropogenic methane source. These calculations reveal that the sensitivities of the tropospheric concentrations of methane, ozone, carbon monoxide, and water vapor, and the tropospheric temperature, to an increase in the methane source intensity (i.e., due to anthropogenic origin) are positive: $\psi_{22} = 0.8$, $\psi_{32} = 0.06$, $\psi_{12} = 0.4$, $\psi_{82} = 0.02$, $\psi_{62} = 0.4$, but $\psi_{52} = -0.2$. This means that a 1% increase in the anthropogenic methane source ($E_2 = 0.01$) leads to $[CH_4] = 0.8\%$ (the system reduces the introduced disturbance by 0.2%), $[O_{3tr}] = 0.1\%$, $[CO] = 0.4\%$, $[H_2O] = 0.02\%$, $\Delta T = 0.004$ K, and $[OH] = -0.2\%$. As one can see, the largest sensitivities to an increase of the anthropogenic methane source are for the atmospheric concentrations of methane, carbon monoxide, and hydroxyl.

Table 2. The growth (% per year) of the anthropogenic methane sources (A_i) and the total methane source ($E_2 = [Q_2]$) during 1970–1980, $[x] = (x-x_0)/x_0$, $x_0 = x$ (1970).

Methane Sources	Share of Total Source[a]	$[A_i]$[b]	E_{2i}[c]
Rice paddies	0.2	0.76	0.15
Cattle breeding	0.2	1.4	0.28
Landfills	0.08	3.7	0.30
Fossil fuel mining			
Coal	0.05	2.6–3.2[d]	0.13–0.16
Oil	0.02	3.3	0.066
Gas	0.05	2.3–3.6[d]	0.12–0.18
Total	0.6		1.04–1.14

[a]Reference 8.
[b]Reference 16.
[c]$E_{2i} = (A_i/Q_2)\cdot[A_i]$, $E_2 = \Sigma E_{2i}$.
[d]Reference 6.

Based on the above sensitivity of the atmospheric methane concentration and the approximately 1% per year recent increase of the atmospheric methane concentration,[17] the required rate of increase of the methane source is

$$(13) \qquad [E_2] = \frac{[CH_4]}{\alpha_2^+ \psi_{22}} = \frac{0.01}{0.5} = 0.02,$$

that is, about 2% per year. Table 2 presents the recent rate of increase of the intensity of different anthropogenic methane sources and their total growth. As one can see, the recent growth rate of the total methane source due to its anthropogenic part is 1.04–1.14% per year. From this result, we conclude that either the atmospheric methane concentration is somewhat more sensitive to its source than is estimated here, or the anthropogenic sources play a more important role in the total source than previously estimated. For example, we have the conviction that fossil fuel mining should play a more important role in the atmospheric methane budget than reported by Cicerone and Oremland[8] because this human activity provides the most rapid methane emission to the atmosphere from the Earth.

Table 3 demonstrates the role of link l_{26} in the atmospheric sensitivity. Here we consider link l_{26} as the influence of climate change on the natural sources of methane, particularly on the methane flux from wetlands. The second column in Table 3 presents the main pathways for the influence of global warming on the

Table 3. The sensitivity of troposphere composition (%) to a doubling of CO_2.

Output	The Main Pathways	$[Y_j]$	Other Studies
$[CH_4]$	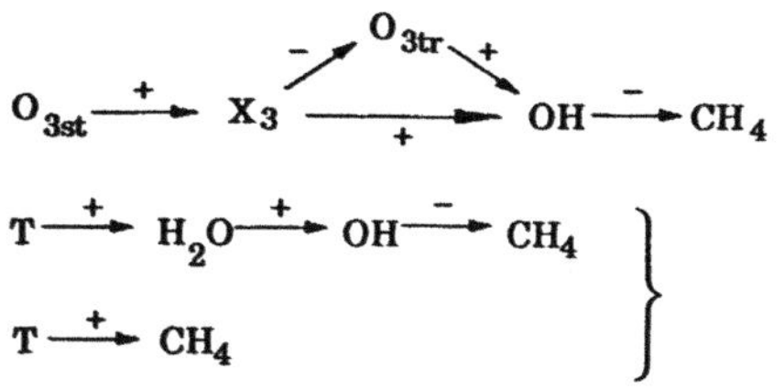	–1 (–10)	(–9, –17) [a,b]
$[CO]$	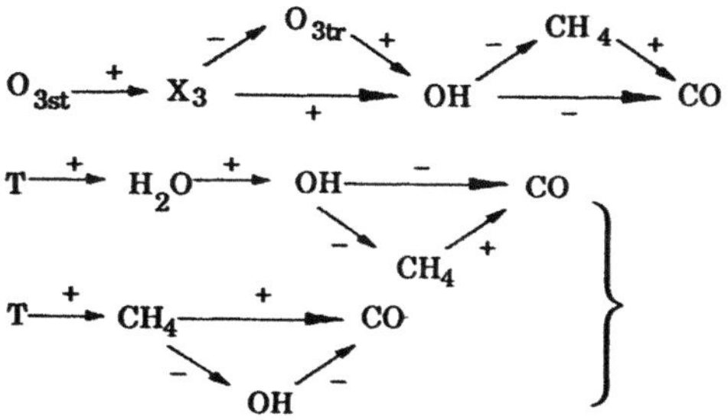	–3 (–10)	(–7) [a]
$[OH]$	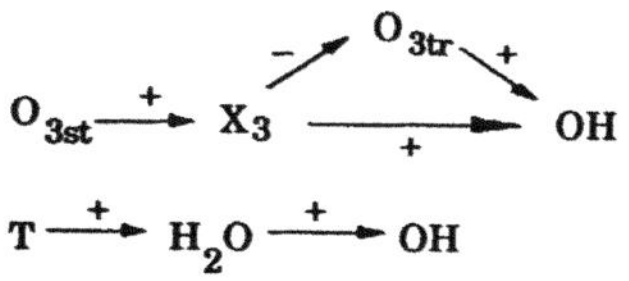	+10 (+10)	(+3, +8) [a]
$[O_{3tr}]$	$O_{3st} \xrightarrow{+} X3 \xrightarrow{-} O_{3tr}$	–5 (–5)	(–5, –11) [a,b]
$[H_2O]$	$T \xrightarrow{+} H_2O$	+25 (+25)	(+10, +36) [c]
$[X_3]$	$O_{3st} \xrightarrow{+} X3$	+7 (+7)	(+2, +6) [d]

a) Callis (1983); b) Hameed and Pinto (1979); c) Atm. Ozone (1985); d) Owens (1985)

[a]Reference 7; [b]Reference 9; [c]Reference 5; [d]Reference 13.

concentrations of water vapor, methane, carbon monoxide, hydroxyl, tropospheric ozone, and total ozone. For water vapor, this influence comes directly through the change in the tropospheric temperature. For methane, carbon monoxide, and hydroxyl, this influence comes both through the changes in the stratospheric ozone concentration and the tropospheric temperature.[5] For the tropospheric ozone and total ozone, this influence comes through the change in the concentration of stratospheric ozone. The third column in Table 3 presents our estimate of the sensitivity of the atmospheric composition to the global warming induced by a doubling of the carbon dioxide concentration, and in the parentheses this sensitivity is presented for the case $l_{26} = 0$. The role of the link l_{26} comes through the creation of additional pathways for the temperature influence on the concentrations of methane and carbon monoxide. These pathways significantly reduce the negative sensitivity of the methane and carbon monoxide to temperature change, making them more sensitive to changes in their sources. The last column in Table 3 presents the antecedent model estimations of the sensitivity of the atmospheric composition to global warming induced by a doubling of the carbon dioxide concentration. These estimates did not include the possible influence of temperature on the methane flux from wetlands. One can see that there is good agreement between our estimates for the case $l_{26} = 0$ and these earlier estimates presented in the last column of Table 3.

Conclusion

In this chapter, we have used the CEA technique to investigate the sensitivity of the linearized and simplified system of photochemical and radiative interactions in the global atmosphere to a change in the anthropogenic methane sources for annually averaged conditions.

Consideration of three systems with increasing complexity showed that:

1. the sensitivity of the "nongreenhouse" troposphere system is predominantly determined by interactions between hydroxyl and carbon monoxide, and between hydroxyl and methane;
2. the sensitivity of the greenhouse troposphere system is also determined by the interactions between temperature and water vapor;
3. the sensitivity of the complete system is, furthermore, determined by the interactions between total ozone and tropospheric ozone.

As a result, the "nongreenhouse" troposphere is less sensitive to external forcing than the "greenhouse" atmosphere.

The system of atmospheric photochemical interactions yields a large sensitivity for the atmospheric methane concentration to changes in the intensity of the anthropogenic methane sources, with a 1% increase of the atmospheric methane concentration in response to a 2% increase of the intensity of the total methane source. This sensitivity is somewhat less than that inferred from the present observed increase in atmospheric methane concentration and the total methane source due to anthropogenic activity.

At the present intensity of anthropogenic methane sources, the role of wetlands (link l_{26}) is negligible. But, under a future global warming, this link can significantly increase the sensitivity of the atmospheric concentrations of methane and carbon monoxide to changes in anthropogenic methane sources.

Acknowledgment

The authors express their gratitude to Professor M. E. Schlesinger for editing their English and for useful discussions.

Appendix

(1) Hydroxyl is the CO sink, and carbon monoxide is the OH sink:

$$OH + CO \rightarrow H + CO_2 \tag{A.1}$$

(2) Methane and hydroxyl are the CO sources ($NO_x = \{NO_2, NO\}$):

$$\begin{aligned} CH_4 + OH &\rightarrow CH_3 + H_2O \\ CH_3 + O_2 + M &\rightarrow CH_3O_2 + M \\ CH_3O_2 + NO_x &\rightarrow NO_x + CH_3O \\ CH_3O + O_2 &\rightarrow H_2CO + HO_2 \\ H_2CO + h\nu &\rightarrow CO + H_2 \end{aligned}$$

$$CH_4 + OH + h\nu \rightarrow CO + \text{products} \tag{A.2}$$

(3) Methane is the OH sink and hydroxyl is the CH_4 sink:

$$CH_4 + OH \rightarrow CH_3 + H_2O \tag{A.3}$$

(4) Carbon monoxide and hydroxyl are the O_{3tr} sources in the NO_x-enriched troposphere:

$$\begin{aligned} OH + CO &\rightarrow H + CO_2 \\ H + O_2 + M &\rightarrow HO_2 + M \\ HO_2 + NO_x &\rightarrow NO_x + OH \\ NO_x + h\nu &\rightarrow NO_x + O \\ O + O_2 + M &\rightarrow O_3 + M \end{aligned}$$

$$OH + CO + h\nu \rightarrow O_3 + \text{products} \tag{A.4}$$

(5) Methane and hydroxyl are the O_{3tr} sources in the NO_x-enriched troposphere:

$$\begin{aligned} CH_4 + OH &\rightarrow CH_3 + H_2O \\ CH_3 + O_2 + M &\rightarrow CH_3O_2 + M \\ CH_3O_2 + NO_x &\rightarrow NO_x + CH_3O \\ NO_x + h\nu &\rightarrow NO_x + O \\ O + O_2 + M &\rightarrow O_3 + M \end{aligned}$$

$$CH_4 + OH + h\nu \rightarrow O_3 + \text{products} \tag{A.5}$$

(6) Water vapor and ozone are the OH sources:

$$\begin{aligned} O_3 + h\nu &\rightarrow O_2 + O(^1D) \\ H_2O + O(^1D) &\rightarrow OH + OH \end{aligned}$$

$$O_3 + h\nu + H_2O \rightarrow OH + OH + \text{products} \tag{A.6}$$

References

1. Aleksandrov E.L., et al. 1982. Atmospheric Ozone and its Variability, Hydrometizdat, Leningrad, (in Russian).
2. Andronova, N.G. 1990. The causal analysis of the photochemical interactions. Izvestia Acad. Nauk, Fizika Atmosfery Okeana **26**:150–157 (in Russian).
3. Andronova, N.G. 1990. The role of natural wetlands in the evolution of land sources of methane. EOS **71**:A21B-3.
4. Andronova, N.G., and M.E. Schlesinger. 1991. The application of cause-and-effect analysis to mathematical models of geophysical phenomena: 1. Formulation and sensitivity analysis. J. Geophys. Res. **96**:941–946.

5. Atmospheric Ozone. 1985. WMO, Report No. 16, Geneva.
6. Barns, D.W., and J.A. Edmonds. 1991. An Evaluation of the Relationship Between the Production and Use of Energy and Atmospheric Methane Emission. Report under Contract No. DE-AC06-76OR-1830.
7. Callis, L.B., et al. 1983. On the relationship between the greenhouse effect, atmospheric photochemistry and species distribution. J. Geophys. Res. **88**:1401–1426.
8. Cicerone, R.J., and R.S. Oremland. 1988. Biogeochemical aspects of atmospheric methane. Glob. Biogeochem. Cycles **2**:299–327.
9. Hammed, S., and J. Pinto. 1979. Sensitivity of the predicted CO–OH–CH_4 perturbation to tropospheric NO_x concentration. J. Geophys. Res. **84**:763–768.
10. Heise, D.R. 1981. Causal Analysis, Wiley, New York.
11. Hough A.M., and R.G. Derwent. 1990. Changes in the global concentration of tropospheric ozone due to human activities. Nature **344**:645–648.
12. Karol, I.L. (ed.). 1986. Radiative-Photochemical Atmospheric Models, Hydrometizdat, Leningrad (in Russian).
13. Owens, A.J., et al. 1985. A coupled one-dimensional radiative-convective, chemistry-transport model of the atmosphere: 1. Model structure and steady-state perturbations. J. Geophys. Res. **90**:2283–2311.
14. Ring D., et al. 1991. Positive water vapour feedback in climate models confirmed by satellite data. Nature **349**:500–503.
15. Schlesinger, M.E. (ed.). 1991. Greenhouse-gas-induced Climatic Change: A Critical Appraisal of Simulations and Observations, Elsevier, Amsterdam.
16. Statistical Yearbook 1979/1980. 1981. Thirty-first issue. Department of International Economic Social Affairs, New York.
17. Steele, L.P., et al. 1987. The global distribution of methane in the troposphere. J. Atmos. Chem. **5**:125–171.

A Time-Dependent Two-Dimensional-Model Study of the Trend in Atmospheric Methane

S.-M. Fan, J.M. Rodriguez, M.K.W. Ko, and N.D. Sze*

Atmospheric and Environmental Research, Inc., Cambridge, Massachusetts

ABSTRACT

The atmospheric concentration of methane (CH_4) has increased from about 0.7 to 1.7 ppm in the last two centuries. Causes of the increase are uncertain. Possibilities include increased emissions from the use of fossil fuel, landfills, biomass burning, domestic cattle, and rice paddies. Another possibility is a reduction in the photochemical removal rate of CH_4 in the atmosphere due to decreasing concentrations of OH. Such reductions could result from increasing CO emissions and the strong coupling between CO, OH, and CH_4. We use a time-dependent two-dimensional photochemical model of the atmosphere to calculate the trend in CH_4 emission required to maintain the observed increasing concentrations of CH_4 and CO over the past two decades, and illustrate the potential impact of changes in tropospheric OH. The results show that about 40% of the methane trend could be due to decreasing concentrations of tropospheric OH, caused mostly by increasing abundances of CO, with the remaining 60% due to increasing emissions of CH_4. However, our calculations only include the direct coupling between CO, CH_4, and OH. Uncertainties due to the impact of other factors, such as increases in tropospheric ozone and NO_x ($NO + NO_2$), or decreases in stratospheric ozone, could result in different calculated trends for tropospheric OH. This study illustrates that strategies aimed at controlling increases in methane must consider not only methane emissions, but also the anthropogenic emissions for other tropospheric species, particularly CO.

Introduction

Atmospheric methane plays important roles in both the climate and chemistry of the Earth's atmosphere. Absorption and reemission of planetary radiation by methane molecules contribute to increasing the surface temperature of the planet,

---

*Now at Department of Earth and Planetary Sciences, Harvard University, Cambridge, MA 02138.

making methane an important greenhouse gas. Oxidation of CH_4 is a source of stratospheric H_2O, and its reaction with chlorine atoms serves as the principal sink for stratospheric reactive chlorine. In the troposphere, CH_4 oxidation constitutes a source of tropospheric ozone and CO.

The concentrations of atmospheric CH_4 have been increasing at about 1% per year since systematic measurements of the gas began in the late 1970s.[1,18,42] Analyses of air extracted from polar ice reveal that the methane concentration two hundred years ago was a factor of 2.5 lower than at present,[5,35,41] indicating that the increasing trend has lasted for quite some time. The coincidence of the rise in atmospheric methane and that of industrial and agricultural activities suggests that anthropogenic sources of methane have been increasing and could be partly responsible for the observed trend. Methane emissions have been observed from coal mining, natural gas drilling, landfills, biomass burning, domestic cattle, and rice paddies. Natural methane emissions may also have changed due to man's perturbation of ecosystems.

Since both CO and CH_4 constitute sinks for tropospheric OH, increases in the atmospheric burdens of these species could result in a possible long-term decrease in the tropospheric OH concentration. Such a decrease would, in turn, reduce the rate of removal of CH_4 in the atmosphere and further increase the methane concentrations. It is, therefore, possible that a considerable fraction of the observed methane increase could also be due simply to decreases in tropospheric OH rather than directly to increases in the CH_4 emission. Early model studies[3,44] predicted an increase of similar magnitude to observations due to suppression of OH by anthropogenic CO emissions when the interactions between CO, CH_4, and OH cycles were taken into account. Wofsy[49] also recognized, from his studies of spatial distributions of CO and CH_4, that increased CO and CH_4 emissions could suppress global tropospheric OH. The temporal trends of OH, CO, and methane have been investigated in more recent model studies (see, for example, Refs. 16, 24, and 45). These studies have utilized the best available estimates of the budgets and emission trends of methane (e.g., Refs. 4 and 14,) and CO (e.g., Refs. 29 and 50) to calculate corresponding trends in the atmospheric burdens of CH_4, CO, and OH.

Our approach in this chapter is slightly different. We adopt the *observed* increases in CH_4 and CO concentrations as boundary conditions to a two-dimensional time-dependent photochemical model to calculate self-consistently both the resulting trend in tropospheric OH and the trends in methane emissions required to maintain the observed increases in CH_4 burden. In particular, we address the question of how much of the increase in atmospheric CH_4 could be accounted for by (a) increases in emissions and (b) decrease in tropospheric OH. Observed ozone trends in the stratosphere[43] and in the troposphere of the Northern Hemisphere[2] (N.H.) as well as possible trends in NO_x, could also impact tropospheric OH and lead to increasing trends. Uncertainties in these processes preclude a complete resolution of this issue from a modeling point of view. Thus, the results presented here should be regarded as an illustration of the possible contribution of different mechanisms to the observed methane trends and, in particular, of the fact that observed increases in methane burden are not easily translated into corresponding increases in methane emissions.

Model Description

The AER two-dimensional zonal-mean model has been described in detail by Ko et al.[20,21] and Rodriguez et al.[38] The model has 19 latitude bands of equal width (9.5°) from pole to pole and 17 vertical intervals, each at ~3 km, from the surface to the stratopause. The tropopause is specified at top of level 5 (~15 km altitude) in the tropics and at top of level 3 (~9 km) at high latitudes. The first layer from the surface includes but does not resolve the planetary boundary layer since exchange between the PBL and the free troposphere is not explicitly represented. Concentrations calculated and/or adopted at the lowest layer (midpoint of about 1.5 km) thus roughly correspond to those of the free troposphere above the boundary layer.

The model includes complete photochemical processes of reactive species of hydrogen, nitrogen, oxygen, chlorine, and bromine. The model also includes chemistry of CO, chlorofluorocarbons (CFCs, HCFCs), N_2O, and, in particular, the complete oxidation processes of CH_4 and C_2H_6. We adopt the more recently reported reaction rate of CH_4 with OH,[46] which is 25% slower than previously accepted.

The model uses specified distribution of temperature, transport by advection and diffusion, and climatology of H_2O. The radiation field is calculated assuming multiple scattering under clear sky conditions. Distributions of trace species are calculated with appropriate boundary conditions which are discussed below for CH_4, CO, and NO_x. A globally averaged surface deposition velocity of 0.1 cm s^{-1} was adopted for O_3, based on measurements of deposition velocities over both oceans and continents[9,22,23,48] and weighing the measurements by the relative land and ocean coverage. Surface concentrations of CFCs are fixed at present-day levels; the calculations, thus, do not simulate the trend in stratospheric O_3 and its impact on tropospheric chemistry. The adopted transport and boundary conditions for other gases have been discussed in a previous study[38] using the same model. The transport parametrization has been constrained to yield latitudinal gradients of CFC-11 in agreement with ALE/GAGE data.[34,50]

Nitrogen oxides (NO_x) play an important part in tropospheric chemistry by influencing production of O_3 and OH radicals. Measurements of NO_x in the remote troposphere are sparse and concentrations variable.[10,36] We consider in this study two cases for NO_x concentrations. The "low NO_x" calculations adopt a fixed mixing ratio of NO_y (total reactive nitrogen = $NO_x + 2 \times N_2O_5 + HO_2NO_2$ + PAN + HNO_3) of 100 pptv at the lowest altitude grid point at all latitudes and seasons. (This boundary condition implies a surface flux of about 10 TgN/yr.) An additional source of NO_x from lightning is included in the tropics, with magnitude equal to 2 TgN/yr and distributed between 4 and 15 km (see Ref. 20 for a discussion of lightning sources). The "high NO_x" calculations assumes a flux of 20 TgN/yr uniformly distributed at the surface and a source of NO_x from tropical lightning at 4 TgN/yr. The "low NO_x" case yields NO_x concentrations of about 6 to 12 pptv in the lower free troposphere, a factor of 2–3 smaller than those adopted in the study of Spivakovsky et al.[40] and represent a lower bound for acceptable NO_x values. The concentrations calculated in the "high NO_x" case are consistent with those of Spivakovsky et al.

Time-dependent simulations are carried out for a 25-year period between 1962 and 1987. The boundary conditions for CH_4 and CO for 1987 are specified as follows. For CH_4, the observed tropospheric mixing ratios in September 1987 [1] with latitudinal variations of 1.65 ppm in the SH to 1.77 ppm in the mid-high latitude N.H. are adopted. The seasonal variations of CH_4 are relatively small and ignored. The lower boundary conditions for CO are taken from the climatology compiled by Spivakovsky et al.,[40] which include both latitudinal and seasonal variations. The CO mixing ratios of Spivakovsky et al.[40] are based on measurements made in the early 1980s, but generally higher than reported for 1987 by Khalil and Rasmussen[17] by 20 ppbv or so. We did not attempt to modify the CO climatology of Spivakovsky et al. due to the large spatial and temporal variabilities of measurements.

Blake and Rowland[1] reported an average trend of CH_4 at 16 ppb/yr from 10 years (1978–1987) of global measurements. This is confirmed by independent measurements (1980–1988) of Khalil and Rasmussen.[18] The trend for CO concentrations is less well-defined. Rinsland and Levine[37] deduced a trend for free tropospheric CO concentrations at ~2 ppb/yr from 1951 to 1985 from infrared total column amount measurements in central Europe. Global measurements of CO by Khalil and Rasmussen[17] show the rates of increase are slower, ~0.8 ppb per year from 1980 to 1987. However, CO increases in the Southern Hemisphere have not been confirmed.[13]

The model was initially run with 1962 boundary conditions for the first year until the solution was annually periodic, i.e., agreed with the previous year within 0.2%. Time-dependent model calculations were then carried out for 25 years with concentrations of CH_4 and CO increasing each year at specified rates. In two multiyear time-dependent calculations, concentrations of CH_4 and CO are assumed to have been increasing at rates of 16 ppb/yr and 0.8 ppb/yr, respectively, for 25 years to the 1987 level, for the two different concentrations of NO_x discussed above. No temporal changes in NO_x emissions were included in any of the cases.

Results

Our model derived atmospheric lifetimes for CH_3CCl_3 are, respectively, 6.4 years for the "low NO_x" calculation and 5.4 years for the "high NO_x" calculation. The corresponding lifetimes calculated for CH_4 are 14 years and 11 years, respectively. The globally and annually averaged number densities of tropospheric OH are 6.3×10^5 and 7.4×10^5 radicals cm^{-3} in the last year of the model calculations for "low NO_x" and "high NO_x" concentrations, respectively. Analysis of the ALE/GAGE measurements by Prinn et al.,[32] using a multibox model for the troposphere and stratosphere, yielded a value of 6.4 (+1.2, −0.9) years for the lifetime of CH_3CCl_3 and an average tropospheric OH concentration of 7.7×10^5 radicals cm^{-3}. More recent estimates by Prinn et al.[33] indicate a lifetime of 5.7 (+0.7, −0.6) years for methylchloroform and an average tropospheric OH of $8.7(\pm 1) \times 10^5$ radicals cm^{-3}, based on longer records of methylchloroform concentrations and emissions.

Figure 1 shows the calculated trends of annually and globally averaged tropospheric OH concentrations. The model-derived concentrations of tropospheric OH decrease at an average rate ~0.4% per year between 1962 and 1987, when concentrations of CO and CH_4 increase at 0.8 ppbv per year (~1%/year) and 16 ppbv/year (~1.1% per year), respectively. The OH trends show only slight dependence on the NO_x concentrations, −0.38% per year at low NO_x vs. −0.41% per year at high NO_x concentrations. In a third time dependent simulation (not shown) with nitrogen oxides calculated from the "high NO_x" boundary conditions, methane increases of 16 ppb/yr were adopted, but CO was assumed to remain constant at the 1987 level. The calculated OH trend is then ~−0.10% per year. The calculated trend of −0.4%/year in Figure 1 is thus due mostly to the adopted increase in CO.

Figure 1 also suggests that the calculated OH trend could be quite different, and even reverse sign, if concentrations of NO_x were not held fixed in the simulation. For example, comparison of the "low NO_x" and "high NO_x" results indicate that the globally-averaged OH concentration could have *increased* by 10% if the global abundances of tropospheric NO_x were to increase by factors of 2–3 between 1962 and 1987. The issue of NO_x trends and their impact on tropospheric OH and CH_4 will be discussed further in the next section.

Figure 2 shows the calculated *net* emissions (actual emission—soil uptake) of CH_4, which are required to maintain the trends in CH_4 burden adopted in the model. The assumed concentration of CH_4 implies that the burden has changed by 1.1%/year over the 25 years period. The calculated OH shows an average decrease of 0.4%/yr in the same period. The calculated net emissions increase at an average rate of 0.7%/yr over the 25-year period. Therefore, in the scenario adopted in these calculations, only 0.6 of the observed trend in CH_4 is directly attributable to an increase in net methane emissions; the remaining 0.4 of the

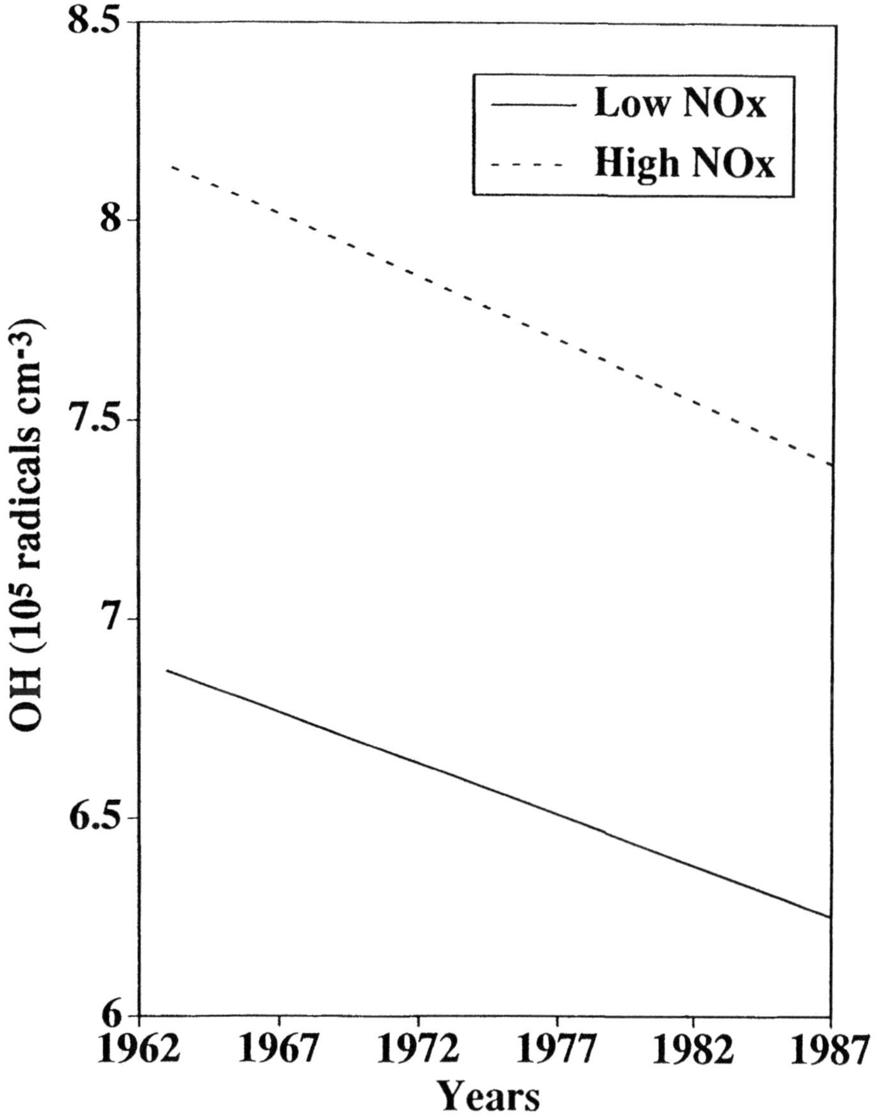

Figure 1. Model-derived tropospheric concentration of OH radicals from 1962 to 1987. Curves represent global and annual averages. See text for details of calculations.

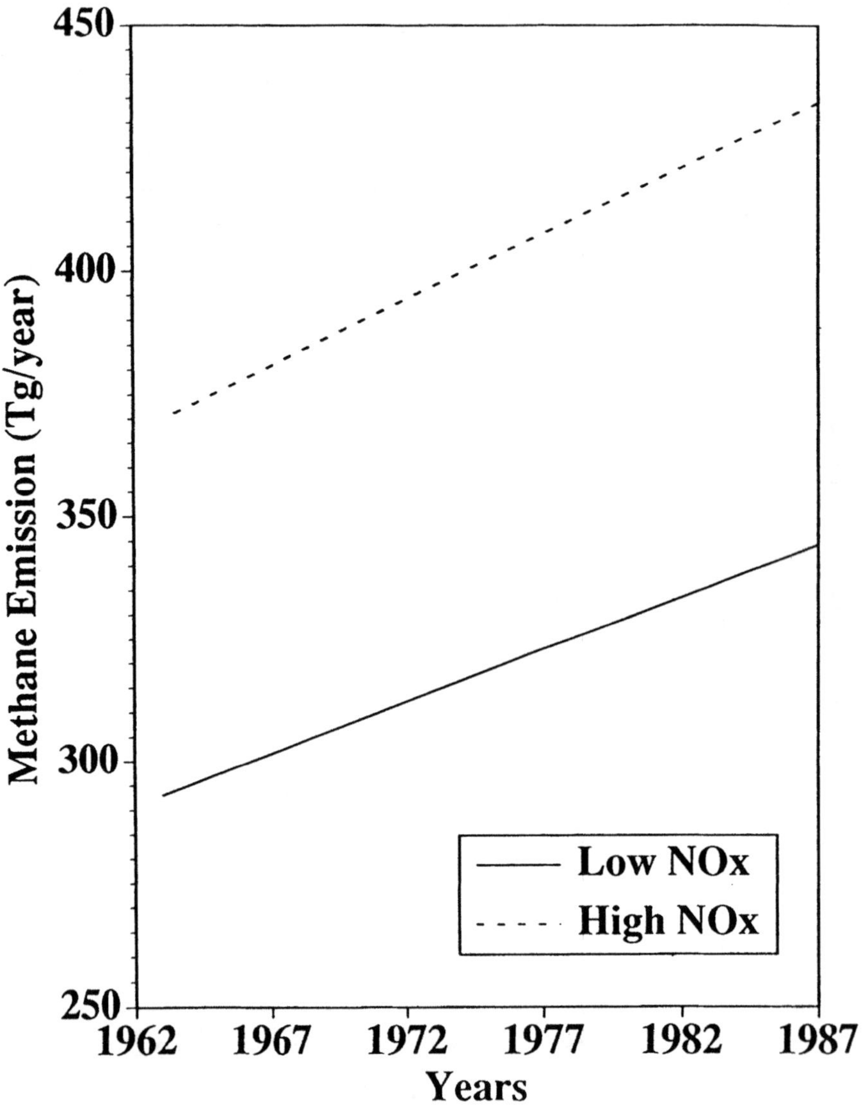

Figure 2. Model-derived annual CH_4 emission in the atmosphere from 1962 to 1987 needed to maintain observed trends in CH_4 burden.

trend is due to a reduction in the methane atmospheric sink as a result of reductions in tropospheric OH.

The absolute increases in CH_4 net emissions correspond to a total of 50 and 65 $TgCH_4$/yr between 1962 and 1987, for the "low NO_x" and "high NO_x" cases, respectively. We note that these increases could be due to a combination of changes in methane sources and soil uptake. The soil uptake term for late 1980s conditions is estimated to be about 30 $TgCH_4$/yr.[4] Since it is unlikely that this sink could have changed by over a factor of 2 in the past 25 years, we conclude that the bulk of the change in *net* emission arises from changes in CH_4 sources.

Discussion

The model results presented above illustrate possible perturbations to tropospheric OH, CH_4, CO, and their interrelations. For simultaneous increases of CO at 0.8 ppb per year and of CH_4 at 16 ppb per year, the globally averaged concentration of tropospheric OH may decrease by 0.4% per year. The decrease in the methane sink due to this OH trend accounts for about 0.4 of the trend in atmospheric CH_4 concentrations. These results are similar to those of Thompson and Cicerone[45] and Isaksen and Hov[16] from their respective one- or two-dimensional model calculations with similar boundary conditions. We stress, however, that both the absolute magnitude and trends of tropospheric OH are uncertain at present. These uncertainties are discussed below.

Uncertainties in the Present-Day OH Concentration

Owing to the lack of direct observations of tropospheric OH, we must use indirect derivations of its tropospheric abundances to validate the model-calculated OH. Using observed concentrations of CH_3CCl_3 and its estimated release rate, Prinn et al.[32,33] have derived the atmospheric lifetime for CH_3CCl_3. If we assume that the atmospheric lifetime is determined primarily by reaction with OH in the troposphere, one obtains a value for the global mean concentration of OH. Measurements of ^{14}CO also provide a constraint on the tropospheric OH where ^{14}C is almost all produced in the atmosphere by cosmic ray neutrons.[8,47] Uncertainties in the ocean and stratospheric sinks and emission inventory of CH_3CCl_3 and in the production and distribution of ^{14}CO indicate that estimates for the present average concentrations of tropospheric OH, based on the above measurements, could be uncertain by at least 30%.

There are many factors that affect the tropospheric OH concentrations calculated by the model. For example, our calculations adopt a clear sky approximation in the radiation. More important, two-dimensional models cannot account very well for the longitudinal asymmetries in OH fields due to variations in cloudiness, albedo, and concentrations of O_3, CH_4, NMHCs, CO, NO_x, and water vapor. These longitudinal asymmetries could be at least a factor of 2.[40] Their impact on the globally averaged OH is difficult to estimate without a full three-dimensional model. In addition, there are insufficient data to adequately constrain many of these variables on a global scale.

The choice of input parameters and boundary conditions are partly constrained by available observations, but also by the calculated lifetimes of methyl chloroform. Any changes/refinements in the input parameters, such as background tropospheric composition, input emissions, or radiation field, would need to satisfy not only observational constraints but yield methyl chloroform lifetimes consistent with those derived from ALE/GAGE. Our experience shows that there is enough uncertainty in the available observations of tropospheric composition and other parameters that agreement can be obtained.

Uncertainties in OH Trends

The recent reanalysis of the ALE/GAGE data for CH_3CCl_3 by Prinn et al.[33] suggests an *increase* in global tropospheric OH, at a rate of 1 ± 0.8%/yr. Inherent uncertainties in the analysis, however, preclude a definitive answer as to the sign and magnitude of OH trends.

Model predictions are hampered by, among other things, a lack of detailed data of NO_x and O_3 in the remote atmosphere. Our understanding of the global distribution and trends of NO_x at present requires resolution of several uncertainties. These include the role of anthropogenic activity in determining NO_x in remote environments,[15,25,26,31] the contribution from biomass burning in the tropics,[7] and the role of PAN and other long-lived reservoirs in the long-range transport of NO_x species.[6,39]

Bojkov[2] reviewed surface and balloon measurements of tropospheric ozone and concluded that ozone concentration was increasing at about 1% per year in large parts of the Northern Hemisphere. This trend can be partly (~0.2% per year) explained by the increases in CH_4 and CO based on our model results. It could also be attributed to ozone export to the free troposphere from polluted regions where ozone is produced mostly in the planetary boundary layer due to surface emissions of NO_x and CO and reactive hydrocarbons. The global distribution of ozone in the troposphere derived from satellite observations,[11,12] indeed, shows ozone plumes spreading from continent to ocean in the Northern Hemisphere as the year proceeds from spring to summer. Ozone plumes are also observed spreading from Africa and South America to the southern oceans due to biomass burning. Increased ozone concentrations could also lead to increased OH production.

Satellite observations of column ozone averaged between 65N and 65S show a decrease of ~3% over some 11 years (0.26% per year) from November 1978 to May 1990.[43] It is also found that the trend is near zero at the equator and increases toward both poles, being ~5%/decade at mid latitudes. Radiative calculations by Madronich and Granier[30] suggest that the increase in tropospheric UV radiation due to the decrease in stratospheric ozone could lead to global OH trends of +0.4%/year. These results are consistent with the analysis of Prinn et al.[33] However, tropospheric UV increases could also be affected by possible increases in aerosols.[28] This effect is difficult to quantify.

Impact on Derived CH_4 Trends

Given all the uncertainties, it is possible that concentrations of OH in the troposphere could have remained constant or even increased in the last decade. In particular, we note that if OH had been increasing in the past 20 years, the increase in net CH_4 emissions is actually faster than the increase in atmospheric concentrations. If global tropospheric OH trends were available from observations, modifications to our results can be estimated as follows.

The observed trend of atmospheric CH_4 burden can be represented by

(1) $$\frac{1}{B}\frac{dB}{dt} \approx \frac{1}{\phi}\frac{d\phi}{dt} + \frac{1}{\tau}\frac{d\tau}{dt},$$

where B, ϕ, and τ are, respectively, the global burden, the net emission (emission—uptake), and the photochemical lifetime of CH_4. Our detailed model calculations indicate that Eq. (1), derived from a steady-state expression (i.e., $B \approx \phi\tau$), is appropriate when the annual increase of methane burden is much less than the annual emission (or loss). Our model results also show that the relative change of methane lifetime is related to that of the global mean concentration of hydroxyl radical [OH] by

(2) $$\frac{1}{\tau}\frac{d\tau}{dt} \approx \frac{1}{[\mathrm{OH}]}\frac{d[\mathrm{OH}]}{dt}.$$

Combining Eqs. (1) and (2), we have

(3) $$\frac{1}{\phi}\frac{d\phi}{dt} \approx \frac{1}{B}\frac{dB}{dt}\frac{1}{[\mathrm{OH}]}\frac{d[\mathrm{OH}]}{dt}.$$

We can then use Eq. (3) to estimate the required change in net emission, for given observed changes in both CH_4 burden and tropospheric OH. Table 1 shows the trends in methane emissions that are needed to maintain the observed CH_4

Table 1. Methane trends (%/yr).

Observed Trend in Concentration	Assumed Trend in Lifetime	Derived Trend in Emission
1.1	0.4[a]	0.7[a]
1.1	0.0	1.1[b]
1.1	−1.0	2.1[c]

[a]Derived from model results, assuming OH trend is due to coupling to observed increases in CH_4 and CO only.
[b]This is derived from Eq. (6.1) and corresponds to constant OH concentration.
[c]This is derived from Eq. (6.1) and corresponds to 1%/yr increase in OH concentration.

trend for various assumed OH trends. We note that, in general, we should not expect the magnitude of the trend in methane emissions to match the observed trend in methane burden.

Conclusions

It is shown, using our two-dimensional photochemical model, that the increasing concentrations of CH_4 and CO could suppress the concentrations of OH in the troposphere. A decrease in tropospheric OH may be partly responsible for the observed increase of methane. In the absence of trends in NO_x or tropospheric UV radiation, a trend in global OH concentrations of about −0.4%/yr is predicted. Model results show that methane emissions may have increased only 0.7%/yr while the concentrations of CH_4 have increased 16 ppb/yr (1.1%/yr averaged over the modeled period, 25 years) and CO 0.8 ppb/yr. Methane emissions, thus, contribute about 0.6 of the trend directly. The global trends in tropospheric OH and methane emissions, however, are not predicted with confidence due to uncertainties about the extent to which anthropogenic NO_x may have impacted on tropospheric chemistry, or the behavior of tropospheric UV radiation in response to stratospheric ozone changes and/or tropospheric aerosols.

Calculations in which only methane increased at 16 ppbv/yr yielded decreases in global tropospheric OH of only 0.1%/yr. Comparison of these results to the scenarios in which both methane and CO increased illustrates the potentially large impact that emissions of CO could have on tropospheric OH and, thus, on

CH_4 trends. Our results point out that strategies aimed at reducing the increase in atmospheric concentrations of CH_4 must take into consideration not only methane emissions but also other processes which may affect the global concentration of OH. Reduction of global CO emission can be accomplished by exercising improved biomass management and reduced consumption of fossil fuels.

Acknowledgments

This work was supported by contracts with the Alternative Fluorocarbon Environmental Acceptability Study (AFEAS) and by NASA's Atmospheric Chemistry, Modeling and Analysis Program.

References

1. Blake, D.R., and F.S. Rowland. 1988. Continuing worldwide increase in tropospheric methane, 1978 to 1987. Science **239**:1129–1131.
2. Bojkov, R.D. 1988. Ozone changes at the surface and in the free troposphere, In I.S.A. Isaksen (ed.), Tropospheric Ozone: Regional and Global Scale Interactions. D. Reidel, Boston, pp. 83–96.
3. Chameides, W.L., S.C. Liu, and R.J. Cicerone. 1977. Possible variations in atmospheric methane. J. Geophys. Res. **82**:1795–1798.
4. Cicerone, R.J., and R.S. Oremland. 1988. Biogeochemical aspects of atmospheric methane. Global Biogeochem. Cycles **2**:299–327.
5. Craig, H., and C.C. Chou. 1982. Methane: The record in polar ice cores. Geophys. Res. Lett. **9**:1221–1224.
6. Crutzen, P.J. 1979. The role of NO and NO_2 in the chemistry of the troposphere and stratosphere. Ann. Rev. Earth Planet. Sci. **7**:443–472.
7. Crutzen, P.J., and M.O. Andreae. 1990. Biomass burning in the tropics: Impact on atmospheric chemistry and biogeochemical cycles. Science **250**:1669–1678.
8. Derwent, R.G., and A. Volz-Thomas. 1989. The tropospheric lifetimes of halocarbons and their reactions with OH radicals: An assessment based on the concentration of ^{14}CO. In Scientific Assessment of Stratospheric Ozone: 1989, Volume II, Appendix: AFEAS Report. Global Ozone Research and Monitoring Project, World Meteorological Organization, Geneva, pp. 125–146.
9. Fan, S.-M., S.C. Wofsy, P.S. Bakwin, D.J. Jacob, and D.R. Fitzjarrald. 1990. Atmosphere-biosphere exchange of CO_2 and O_3 in the central Amazon forest. J. Geophys. Res. **95**:16851–16864.
10. Fehsenfeld, F.C., D.D. Parrish, and D.W. Fahey. 1988. The measurement of NO_x

in the non-urban troposphere. In I.S.A. Isaksen (ed.), Tropospheric Ozone: Regional and Global Scale Interactions. D. Reidel, Boston, pp. 185–215.

11. Fishman, J., K. Fakhruzzaman, B. Cros, and D. Nganga. 1989. Identification of widespread pollution in the Southern Hemisphere deduced from satellite analyses. Science **252**:1693–1696.

12. Fishman, J., C.E. Watson, J.C. Larsen, and J.A. Logan. 1990. Distribution of tropospheric ozone determined from satellite data. J. Geophys. Res. **95**:3599–3617.

13. Fraser, P.J. 1989. Atmospheric trends and distributions of carbon monoxide. Report to IPCC Working Group 1. Panel on Methane, Nonmethane Hydrocarbons, and Carbon Monoxide, Cambridge, MA.

14. Fung, I., J. John, J. Lerner, E. Matthews, M. Prather, L.P. Steele, and P.J. Fraser. 1991. Three-dimensional model synthesis of the global methane cycle. J. Geophys. Res. **96**:13033–13065.

15. Galloway, J.N., and D.M. Whelpdale. 1987. WATOX-86 overview and western North Atlantic Ocean S and N atmospheric budgets. Global Biogeochem. Cycles. **1**:261–281.

16. Isaksen, I.S.A. and O. Hov. 1987. Calculation of trends in the tropospheric concentration of O_3, OH, CO, CH_4 and NO_x. Tellus **39B**:271–285.

17. Khalil, M.A.K., and R.A. Rasmussen. 1988. Carbon monoxide in the Earth's atmosphere: Indications of a global increase. Nature **332**:242–245.

18. Khalil, M.A.K., and R.A. Rasmussen. 1990. Atmospheric methane: Recent global trends. Environ. Sci. Technol. **24**:549–553.

19. Ko, M.K.W., M.B. McElroy, D. Weisenstein, and N.D. Sze. 1989. Lightning: A possible source of stratospheric odd nitrogen. J. Geophys. Res. **91**:5395–5404.

20. Ko, M.K.W., K.K. Tung, D. Weisenstein, and N.D. Sze. 1985. A zonal mean model of stratospheric tracer transport in isentropic coordinates: Numerical simulations for nitrous oxide and nitric acid. J. Geophys. Res. **90**:2313–2329.

21. Ko, M.K.W., N.D. Sze, and D. Weisenstein. 1989. The role of dynamical and chemical processes in determining the stratospheric concentration of ozone in one-dimensional and two-dimensional models. J. Geophys. Res. **94**:9889–9896.

22. Lenschow, D.H., R. Pearson, Jr., and B.B. Stankov. 1981. Estimating the ozone budget in the boundary layer by use of aircraft measurements of ozone eddy flux and mean concentration. J. Geophys. Res. **86**:7291–7297.

23. Lenschow, D.H., R. Pearson, Jr., and B.B. Stankov. 1982. Measurements of ozone vertical flux to ocean and forest. J. Geophys. Res. **87**:8833–8837.

24. Levine, J.S., C.P. Rinsland, and G.M. Tennile. 1985. The photochemistry of methane and carbon monoxide in the troposphere in 1950 and 1985. Nature **318**: 254–257.

25. Levy, II, H., and W.J. Moxim. 1987. Fate of US and Canadian combustion nitrogen emissions. Nature **328**:414–416.

26. Levy, II, H., and W.J. Moxim. 1989. Simulated global distribution and deposition of reactive nitrogen emitted by fossil fuel combustion. Tellus **41B**:256–271.

27. Liu, S.C., and M. Trainer. 1988. Responses of the tropospheric ozone and odd hydrogen radicals to column ozone change. J. Atmos. Chem. **6**:221–234.

28. Liu, S.C., S.A. McKeen, and S. Madronich. 1991. Effect of anthropogenic aerosols on biologically active ultraviolet radiation. Geophys. Res. Lett. **18**:2265–2268.

29. Logan, J.A., M.J. Prather, S.C. Wofsy, and M.B. McElroy. 1981. Tropospheric Chemistry: A global perspective. J. Geophys. Res. **86**:7210–7254.

30. Madronich, S., and C. Granier. 1992. Impact of recent total ozone changes on tropospheric ozone photodissociation, hydroxyl radicals, and methane trends. Geophys. Res. Lett. **19**:465–467.

31. Penner, J.E., C.S. Atherton, J. Dignon, S.J. Ghan, J.J. Walton, and S. Hameed. 1991. Tropospheric nitrogen: A three-dimensional study of sources, distributions and deposition. J. Geophys. Res. **96**:959–990.

32. Prinn, R., D. Cunnold, R. Rasmussen, P. Simmonds, F. Alyea, A. Crawford, P. Fraser, and R. Rosen. 1987. Atmospheric trends in methylchloroform and the global average for the hydroxyl radical. Science **238**:945–950.

33. Prinn, R., D. Cunnold, P. Simmonds, F. Alyea, R. Boldi, A. Crawford, P. Fraser, D. Gutzler, D. Hartley, R. Rosen, and R. Rasmussen. 1992. Global average concentration and trend for hydroxyl radicals deduced from ALE/GAGE trichloroethane (methylchloroform) data for 1978–1990. J. Geophys. Res. **97**:2445–2461.

34. Prinn, R.G., P.G. Simmonds, R.A. Rasmussen, R.D. Rosen, F.N. Alyea, C.A. Cardelino, A.J. Crawford, D.M. Cunnold, P.J. Fraser, and J.E. Lovelock. 1983. The atmospheric lifetime experiment, 1. Introduction, instrumentation and overview. J. Geophys. Res. **88**:8353–8367.

35. Rasmussen, R.A., and M.A.K. Khalil. 1984. Atmospheric methane in the recent and ancient atmospheres: Concentrations, trends, and interhemispheric gradient. J. Geophys. Res. **89**:11599–11605.

36. Ridley, B.A. 1991. Recent measurements of oxidized nitrogen compounds in the troposphere. Atmos. Environ. **25A**:1905–1926.

37. Rinsland, C.P., and J.S. Levine. 1985. Free tropospheric carbon monoxide concentrations in 1950 and 1951 deduced from infrared total column amount measurements. Nature **318**:250–254.

38. Rodriguez, J.M., M.K.W. Ko, N.D. Sze, and C.W. Heisey. 1991. Impact of biomass burning on tropospheric CO and OH: A two-dimensional study. In J.S. Levine (ed.), Global Biomass Burning: Atmospheric, Climatic, and Biospheric Implications. The MIT Press, Cambridge, MA, pp. 351–359.

39. Singh, H.B. 1987. Reactive nitrogen in the troposphere. Environ. Sci. Technol. **21**:320–327.

40. Spivakovsky, C.M., R. Yevich, J.A. Logan, S.C. Wofsy, M.B. McElroy, and M.J. Prather. 1990. Tropospheric OH in a three-dimensional chemical tracer model: An assessment based on observations of CH_3CCl_3. J. Geophys. Res. **95**:18441–18472.

41. Stauffer, B., G. Fischer, A. Neftel, and H. Oeschger. 1985. Increases in atmospheric methane recorded in Antarctic ice core. Science **229**:1386–1388.

42. Steele, L.P., P.J. Fraser, R.A. Rasmussen, M.A.K. Khalil, T.J. Conway, A.J. Crawford, R.H. Gammon, K.A. Masarie, and K.W. Thoning. 1987. The global distribution of methane in the troposphere. J. Atmos. Chem. **5**:125–172.

43. Stolarski, R.S., P. Bloomfield, R.D. McPeters, and J.R. Herman. 1991. Total ozone trends deduced from Nimbus 7 TOMS data. Geophys. Res. Lett. **18**:1015–1018.

44. Sze, N.D. 1977. Anthropogenic CO emissions: Implications for atmospheric CO–OH–CH_4 cycle. Science **195**:673–675.

45. Thompson, A.M., and R.J. Cicerone. 1986. Atmospheric CH_4, CO and OH from 1860 to 1985. Nature **321**:148–150.

46. Vaghjiani, G.L., and A.R. Ravishankara. 1991. New measurement of the rate coefficient for the reaction of OH with methane. Nature **350**:406–409.

47. Volz, A., D.H. Ehhalt, and R.G. Derwent. 1981. Seasonal and latitudinal variation in ^{14}CO and the tropospheric concentration of OH radicals. J. Geophys. Res. **86**:5163–5171.

48. Wesely, M.L., D.R. Cook, and R.M. Williams. 1981. Field measurement of small ozone fluxes to snow, wet bare soil, and lake water. Boundary Layer Meteorol. **20**:459–471.

49. Wofsy, S.C. 1976. Interactions of CH_4 and CO in the Earth's atmosphere. Ann. Rev. Earth Planet. Sci. **4**:441–469.

50. WMO. 1985. Atmospheric ozone, global ozone research and monitoring project. Report No. 16, World Meteorological Organization, Geneva.

The Influence of Tropical Biomass Burning on Climate and the Atmospheric Environment

Meinrat O. Andreae

Biogeochemistry Department, Max Planck Institute for Chemistry, D-6500 Mainz, Germany

ABSTRACT

This chapter presents a review of the historical, economic, and environmental aspects of biomass burning. The use and extent of biomass burning in agriculture and various economic activities and its geographical distribution are outlined. The resulting emissions of gases and aerosols to the atmosphere are estimated based on field and laboratory measurements. Chemical processes in the emission plumes and their atmospheric transport on the regional to global scale are evaluated. The impact of the emissions from biomass burning on climate and the environment is shown to be, in many respects, of a magnitude comparable to the effects of the emissions from fossil fuel combustion.

Introduction

Air pollution became part of the human environment as soon as our early ancestors learned to tame and use fire; they must have been exposed to large amounts of carcinogenic compounds from the smoke of open fires in caves, huts, and tents. Even today, pollution from biomass burning is a serious health concern in countries like India, where biomass is the principal domestic fuel, and a dense pall hangs over villages and towns during the evening and morning hours. Even in North America, many communities have had to issue ordinances regulating domestic burning in fireplaces and wood-burning stoves to control atmospheric pollution in winter. In this chapter, I address the larger-scale effects of biomass burning on the regional and global atmospheric environment. This impact is most dramatically evident in the tropics, where airports like Santarém, in the middle of the Amazon Basin, must frequently be closed during the burning season because of poor visibility resulting from the smoke from gigantic fires hundreds to thousands of kilometers away. Such large fires and the resulting smoke plumes are a common sight to air travelers in the tropics and are even frequently observed by the Space Shuttle crews.[175]

The History of Fire

Fires have occurred on Earth ever since the evolution of land plants some 350 to 400 million years ago. Once the production of organic matter by plants had supplied the fuel for wildfires, climatic and ecological parameters (length and intensity of dry seasons, lightning frequency, etc.) established a natural fire frequency. With the advent of herbivorous animals, the accumulation rate of natural fire potential was controlled by the interplay of phyto-production on one hand and fire frequency and herbivory on the other.[147] For example, a high rate of grazing by large reptiles (dinosaurs) may have contributed to the predominance of savanna-like vegetation on Earth during the late Cretaceous era (ca. 70 million years ago), which would have been accompanied by a high fire frequency and low fire intensity. High concentrations of black carbon (soot and charcoal) in Cretaceous–Tertiary boundary sediments suggest that the end of the age of dinosaurs was accompanied by large fires which may have perturbed the Earth's climate and thereby contributed to the dinosaurs' demise.[173,174] Early humans eventually overcame the fear of fire shared by all other primates and learned about 1.5 to 2 million years ago to make use of fire as a tool for food preparation, hunting, and landscape control.[84] They developed quite sophisticated fire-management strategies, examples of which survived up to this century among the Aborigines of Australia,[86,127] where *Homo sapiens* arrived about 40,000 years ago. At about the same time, pollen records show a shift from pyrophobic vegetation (e.g., Araucaria spp.) to pyrotolerant and pyrophilic species (e.g., Eucalyptus spp.), and the amount of charcoal particles in sediment cores increases by three orders of magnitude. More information on the relationship between fire and human evolution can be found in the review papers by Pyne,[126] James,[84] and Schüle.[147]

On a quantitative level, the history of biomass burning through geological and historical time is difficult to establish since little physical evidence remains other than charcoal particles buried in sediments. Measurements of charcoal in dated sediment cores have shown some clear correlations between the rate of burning and human activity, e.g., the increase of charcoal in Australian sediments coinciding with human settlement as mentioned earlier. Charcoal layers in Amazonian soils also document widespread catastrophic fires during periods of drought in Amazonia in prehistoric times.[141] In temperate forests, the combination of tree-ring dating and fire scar identification can provide a high-resolution record of fire activity on century to millenial time scales.[78,91,158,160] This approach cannot be used in tropical forests, however, where annual tree-rings are usually absent. Clearly, much research remains to be done to clarify the relationship between changing human population densities and agricultural/industrial activity on one hand and biomass burning rate on the other.

Biomass Burning in Agriculture and Economy

Biomass burning has a variety of objectives: (1) clearing of forest and brushland for agricultural use, (2) control of brush, weeds, and litter accumulation on grazing and crop lands, (3) nutrient regeneration on grazing and crop lands, (4) control of fuel accumulation in forests, (5) clearing of vegetation to obtain access for hunting and gathering, and (6) energy production for cooking and heating. Finally, there is the category of wildfires, which are not started and controlled by people, but which often have large economic and ecological consequences.

Clearing of Forest and Brushland for Agricultural Use

Two types of forest clearing for agricultural use may be differentiated: shifting agriculture, where the land is allowed to return to forest vegetation after a relatively short period of use, and permanent removal of forest to be replaced by grazing or crop lands. In both instances, the clearing and burning initially follows the same pattern: Trees are felled and the vegetation is left for some time to dry out to obtain better burning efficiency. The material is then set on fire, often after bulldozing it into large piles. The efficiency of this first burn is variable; it often does not exceed 10–30%.[61,62] This low initial efficiency is due to the large fraction of rainforest biomass residing in the tree trunks, only a small portion of which tends to be consumed during the first burn. The remaining material is then moved into piles and left to rot or dry. It is usually set on fire again when it is judged to be dry enough. As a result of this practice, some 40% of the above-ground carbon from forest-clearing enters the atmosphere through combustion, whereas the rest is released through rotting, i.e., microbial decomposition.

In shifting agriculture, which in 1980 was practiced by some 500 million people worldwide[103] the cleared areas are used for agriculture for a few years until yields decline, and then are abandoned and new areas cleared. The land area under shifting agriculture is on the order of 300–500 million hectares and is expanding annually by ~1.3%.[103,121] In 1980 alone, an area of ~5.1 million hectares was cleared for this purpose according to FAO estimates,[60] and it is safe to assume that the current rate of clearing is well above this figure.

In traditional shifting agriculture, the use of small parcels of land and long fallow periods (some 20 years) allowed the regrowth of forest vegetation. In contrast to this sustainable mode of shifting agriculture, most current practice entails deforestation along a front at the forest edge ("pioneer front"), rapid exploitation of the land, and little chance of forest regrowth on the degraded zone behind the deforestation front. Sometimes, this land can subsequently be used for permanent agriculture. Usually, however, this attempt fails due to poor soil conditions or ineffective land management, and the areas turn into degraded land not suitable for productive agricultural use. This destructive type of shifting agriculture is especially characteristic of displaced rural populations in developing countries with high rates of population growth. It is currently one of the leading causes of deforestation, especially in Africa, where it accounts for over 70% of rain forest destruction (see Ref. 52 and references therein).

The permanent removal of rainforest, driven not only by expanding human populations requiring additional food and living space, but also by large-scale resettlement programs and land speculation tactics, is progressing at a rapid rate. Still, the global rate of deforestation is highly uncertain. Based on the work by Lanly,[102] Houghton et al.,[83] and Detwiler et al.,[51] a global deforestation rate of about 22 million ha/yr during the 1970s was estimated by Hao et al.[74] Based on more recent assessments, a Study Commission of the German Bundestag estimated that the rate of deforestation for the year 1989 was 14.2 million ha/yr in primary closed tropical forests (i.e., rainforests) and 9.3 million ha/yr in primary open forests (i.e., wooded savannas), giving a total destruction rate of 23.5 million ha/yr.[52] The area of secondary forests destroyed each year is highly uncertain, but may be of the same magnitude.

Biomass Burning in Tropical Savannas and Brushland

Tropical savannas cover an area (1530 million ha) roughly similar to that of the tropical forests (1440 million ha[102]). Savannas typically consist of a more or less continuous layer of grass with interspersed trees and shrubs. They are burned periodically at intervals which may range from one to four years.[100,117,118,145] This frequency may be increasing in some regions as a result of increasing population and more intensive use of rangeland.[117] Although lightning may start some fires in savannas, most investigators are convinced that almost all savanna fires are set by humans.

Savanna-burning is carried out for several reasons: Probably foremost is the control of weeds, shrubs, tree seedlings, and litter accumulation. It is also perceived as a form of pest control, reducing the population of insects, snakes, etc. Another reason for burning is the belief that it is an effective way to return the nutrients stored in the dry vegetation, which accumulates during the dry season, to the soils and thereby promote regrowth of fresh grass during the following rainy season. In the long run, however, this belief may be fallacious, as we will discuss in the section on atmospheric emissions from fires. Many of the tree savannas and open forest formations, particularly dry deciduous and semideciduous forests are burned annually to facilitate access and harvest of nonwood forest products.[68,154] Finally, fires are also set to drive game for hunting, a practice which may go back to the earliest use of fire by man.

The savanna area burned each year is very large: Hao et al.[74] estimate ~750 million ha, much more than the tropical rainforest area burned annually. As a

Table 1. Biomass exposed to fire and burned as domestic fuel. Charcoal use is indicated separately.

Region	Population[a] (1985)(millions)	Biomass fuel Exposed (Tg dm/yr)	Burned (Tg dm/yr)	Charcoal[b] (Tg dm/yr)
Tropical America	430	220	190	7.5
Africa	560	280	250	9.3
Asia	2820	1410	1270	3.3
Oceania	25	13	11	0.0
Total tropics	3830	1910	1720	20.1
USA and Canada	260	110	100	0.5
Western Europe	380	54	49	0.2
USSR and Eastern Europe	390	78	70	0.2
Total extratropical	1030	250	220	0.9
World total	4860	2160	1940	21.0

[a]Reference 59.
[b]Reference 60.

result, savanna-burning releases about three times as much CO_2 and other trace gases to the atmosphere as deforestation burning, in spite of the much higher biomass density of the rainforest. In Africa, savanna fires account for almost 90% of the emissions from savanna- and forest-burning combined, and for almost one-third of the global emissions from biomass burning. The great importance of savanna fires in Africa is due to the very large area covered by this landscape type in Africa and to the very high frequency at which African savannas are burned (~75% per year).

Biomass Fuel: Wood, Charcoal, and Agricultural Waste

Biomass fuel, in the form of wood, charcoal, and agricultural waste, is a major source of energy in developing countries, and its gathering often represents a large part of the workload of local populations. Due to the highly distributed nature of fuel-wood use, the annual amount of wood-burning is difficult to estimate. The number given by FAO[60] for 1987, 1050 Tg dm/yr (teragram dry matter per year; one teragram = one million tons = 10^{12} grams), is almost certainly an underestimate because it does not cover the use of wood which is not marketed. In Table 1, I present an estimate of biomass fuel use based on a combination of the FAO estimate with population and per capita fuel-use statistics. For the extratropical regions, I have multiplied the FAO estimates by a factor of 1.5 to account for fuel-wood that is not marketed. For the tropical regions, I have multiplied the 1985 population numbers with an estimated annual biomass fuel consumption of 500 kg per person. According to Scurlock and Hall,[148] urban populations consume about 500 kg biomass fuel per person and year, whereas about twice this rate applies to rural populations. They estimate that perhaps two-thirds of the rural energy use in China comes from agricultural wastes. In Yunnan province, China, the fuel-wood consumption is estimated to be about 1 m^3 (720 kg) per person per year.[107] Bilsborrow and DeLargy[19] give per capita annual fuel-wood consumption rates of 2.0 m^3 (1400 kg) for the Sudan, with consumption in urban areas up to 2.8 m^3 (2000 kg). On the other hand, Joshi[87] estimates that in India only 160 and 350 kg biomass fuel per person and year are consumed by urban and rural populations, respectively. Overall, the average value of 500 kg per year for the tropics worldwide appears to be a reasonable, and probably somewhat conservative estimate.

Charcoal production for domestic and industrial use has become an attractive alternative to the direct use of wood as fuel since its higher energy density makes it easier and cheaper to transport and market. During charcoal production, which consists of the controlled partial combustion (pyrolysis) of dry wood, a mixture of water, methane, carbon monoxide, methanol, tar, and other volatile products is distilled off, that, in principle, can be captured and processed into marketable by-products. However, with most of the charcoal kilns in the developing countries, these volatile compounds are released to the atmosphere. This makes the production of charcoal a potent source of trace gases, especially methane.[49] Depending on the efficiency of the charcoal kiln, the ratio of carbon input (in the form of wood) to carbon output (in the form of charcoal) varies between about 2.0 and 1.2 in industrial kilns.[14] In the primitive kilns used by small operators in developing countries, this ratio may be much higher, on the order of 10 (R.A. Delmas, personal communication, 1990). For the purpose of estimating the amount of carbon released to the atmosphere due to charcoal production and use, I have, therefore, multiplied the amount of charcoal burned in the tropics by a factor of 4 to account for losses during production (Table 2).

Burning of Agricultural Wastes

A substantial amount of agricultural waste, e.g., straw and stalks, is used as domestic fuel and has already been included in our estimates of biomass fuel combustion. Burning of agricultural waste in the fields is extremely difficult to quantify because of its distributed nature. No statistics are available, mostly because material of direct economic value is not involved. A highly conspicuous example is the burning of sugar cane fields before harvesting, which is a source of highly visible air pollution in many tropical regions.[93] Barnard[16] estimates

Table 2. Global estimates of annual amounts of biomass burning and of the resulting release of carbon to the atmosphere.

Burning Source	Biomass Burned Tropical	Extratropical (Tg dm/yr)	Total	Carbon Released[a] (Tg/a)
Forest	1260	1150	2410	1080
Savanna	3690	—	3690	1660
Biomass fuel	1720	220	1940	880
Charcoal	20	1	21	82
Agric. waste in fields	420	420	850	380
World Total	7120	1800	8910	4080

[a]Based on a carbon content of 45% in dry biomass. In the case of charcoal, the rate of burning (based on FAO production statistics) has been multiplied by 4 to account for losses in the production process (see text).

that the amount of crop residue produced in the developing world equals about 1700 Tg dm/yr and that a similar amount is produced in the developed countries. Following the suggestion of Crutzen and Andreae,[43] I estimate that 25% of agricultural waste, i.e., 850 Tg dm/yr, is burned in the fields. It may be important to attempt a careful reexamination of this source since it appears to be one of the major contributors to atmospheric emissions from biomass-burning and since it has the potential of being reduced substantially through alternative agricultural practices.

Prescribed Burning and Wildfires in Forests

Wildfires are most common in temperate and boreal forests since undisturbed tropical forests are usually too moist to allow the propagation of wildfires. According to Brown and Davis,[26] most wildfires are the result of human activities (including, e.g., sparks from railroad engines), with only a small fraction (10–30%) initiated by lightning. Although individual wildfires may be very large and lead to smoke plumes visible over great distances, the area burned per year is relatively low compared to the areas burned in tropical forests and savannas. Stocks (personal communication, 1991) estimates that about 8 million ha of temperate and boreal forest are subject to wildfire each year. It should be noted, however, that wildfire activity is subject to large year-to-year variations and that the largest wildfires, such as the Siberian fire of 1915 which consumed over 14 million ha, make a significant contribution to global emissions at the time of their occurrence. The role of wildfires may increase in the future, as climate change will promote droughts and forest diebacks, and, therefore, more frequent wildfires.[18]

Prescribed fires are a commonly used tool of forest management, particularly in North America and Australia. They serve to reduce the accumulation of dry, combustible plant debris, which is a major cause of destructive wildfires. They are also used to eliminate shrubby vegetation which competes with tree crops for nutrients. In North America, only about 2 million ha are burned annually in prescribed fires. The latest figures from Australia, on the other hand, show 40–130 million ha burned annually (B. Stocks, personal communication, 1991). Owing to the lighter fuel loads associated with understory burning, the amount of carbon burned per hectare is relatively small. Still, some Australian authors estimate that 250–800 Tg C are released annually through this practice.[32] In Table 2, I have included an amount of 900 Tg dm/yr, corresponding to ~400 Tg carbon, for prescribed burning in Australia.

Seasonality of Biomass-Burning

Frequent mention is made in the literature of "burning seasons" and it is often assumed that practically no burning takes place outside of these periods. Indeed, a large fraction of the burning of forests and savannas takes place during the dry season and, therefore, burning is most intensive in the Northern Hemisphere from November to March and in the Southern Hemisphere from June to September. However, experience in tropical countries shows that burning can be observed almost whenever and wherever there is plant material dry enough to burn. This applies to cooking and heating fires, burning of garden and agricultural wastes, burning of the areas surrounding living quarters to control insects, weeds, snakes, etc., and often burning just for the pleasure of watching the flames and smoke. By the end of the rainy season, many farmers are waiting for cleared areas to become dry enough to burn, leading to an outbreak of burning in late May and early June in the southern hemispheric part of Brazil. The large-scale forest-clearing burns are usually started later in the dry season to ensure driest fuel conditions and, thus, better burning efficiency.

Emissions to the Atmosphere from Biomass-Burning

Carbon Dioxide

The burning of organic material produces water vapor and carbon dioxide as primary products. In this sense, burning can be looked on as the abiotic equivalent of the respiratory catabolism of biological material, returning the carbon fixed by plant photosynthesis back to the atmosphere as CO_2. Thus, as long as the same amount of vegetation is allowed to grow back on the burnt area, as, for example, in periodic savanna fires, biomass-burning does not influence the atmospheric CO_2 budget in the long term. In contrast, the clearing of tropical forests contributes directly to the CO_2 greenhouse effect since the CO_2 emitted from the burning and decay of the large amount of biomass stored in a rainforest (as much as 600 to/ha) cannot be taken up again by the vegetation which will eventually grow on the same site, usually grass or an agricultural crop. In most cases, the emission of CO_2 from the decay of above- and below-ground organic matter left behind after the clearing fires is actually greater than that from the burning itself.

Earlier estimates of the net amount of CO_2 added to the atmosphere from deforestation fell into a relatively narrow range: Houghton et al.[83] gave an estimate of 900–2500 Tg C/yr, Detwiler and Hall[50] suggested a range of 400–1600 Tg C/yr, and Hao et al.[74] calculated 700–2000 Tg C/yr. However, this agreement mostly reflects the fact that all authors based their estimates on the same data: The estimates on land-use conversion from FAO statistics[102,120] and the biomass densities given by Whittaker and Likens,[170] Brown and Lugo,[28] and Detwiler et al.[51] It now appears that the FAO statistics have significantly underestimated deforestation rates and that deforestation rates have been strongly accelerating over the past decade.[82,121] Houghton[82] estimates that the net CO_2 release to the atmosphere due to deforestation falls in the range 1100–3600 Tg C/yr. This wide range reflects the great prevailing uncertainty regarding the areal extent of deforestation and the original and successional biomass loading.[27] In spite of this uncertainty, it is evident that net CO_2 release from deforestation is significant compared with the present annual emission of CO_2 from fossil-fuel-burning (~5200 Tg C/yr). One can, therefore, conclude that the destruction of tropical forests contributes about 18–40% of the global CO_2 greenhouse effect.

Other Trace Gases

Together with CO_2, a large variety of gases and particles are emitted from fires (Tables 3 and 4). Among them are the products of incomplete combustion of carbon compounds [e.g., carbon monoxide (CO), methane (CH_4), and other hydrocarbons] and of compounds containing other nutrient elements [e.g., nitric oxide (NO) and sulfur dioxide (SO_2) from the nitrogen and sulfur in amino acids and proteins]. Particulate matter (aerosol) in the smoke consists of organic substances, black (soot) carbon, and inorganic materials, e.g., potassium carbonate and silica. The emission ratios in Table 3 are the molar ratio between the species of interest and CO_2 in the smoke, with the exception of the aerosol emission ratios which are expressed as g/kg C(CO_2). (The ratios have been multiplied by 1000 to make the numbers easier to read.)

Table 3. Emission ratios for trace gases and aerosols from biomass-burning, based on field and laboratory studies.

Species	Field Studies[a] Range	Laboratory Studies[b] Range	Laboratory Studies[b] Avg.	Best Guess[c]
GASES[d]				
CO	22–175	32–160	73	80
CH_4	0.9–16.0	1.7–11.4	5.1	8
NMHC[e]	3.0–11.0	1.7–38.6	14.3	10
C_2H_6	0.1–1.7	0.1–1.7	0.7	0.7
C_2H_4	0.3–2.7	0.3–5.1	1.5	1.5
C_2H_2	0.6–0.6	0.1–2.9	0.7	0.7
C_3H_8	0.0–1.2	0.02–0.5	0.2	0.2
C_3H_6	0.1–1.4	0.1–1.9	0.8	0.8
Aromatic	1.5–1.5	0.3–4.3	1.4	1.4
N_2	—	0.8–9.9	4.7	4.7
N_2O	0.09–2.2	0.01–0.3	0.1	0.1
NO_x	1.5–8.0	0.9–7.7	2.0	2.3
NH_3	0.9–5.3	0.1–2.5	0.6	1.3
RCN[f]	—	0.1–1.4	0.6	0.6
SO_x[g]	0.1–0.34	0.08–0.4	0.19	0.25
COS	0.005–0.016	—	—	0.011
CH_3Cl	0.023–0.124	0.013–0.320	0.117	0.12
H_2	10.3–33	—	—	27
AEROSOLS[h]				
TPM[i]	11–82	—	—	25
POC[j]	6.3–54	—	—	17
EC[k]	0.4–16	—	—	2
Potassium	0.24–0.58	—	—	0.4
OZONE[l]	4.8–40	—	—	30

[a]Based on data in Refs. 6, 10, 29, 34–37, 45–46, 48, 58, 70, 72, 75–77, 119, 128, 131, 157, 164–167. The range shown is a range of means from these studies, not a range of individual measurements.
[b]Based on data from Refs. 96, 109, 110, and Bingemer (personal communication, 1990). The range shown is from individual burns of different types of biomass (savanna grasses, pine straw, etc.).
[c]Based on a more or less subjective evaluation of the information in footnotes a and b.
[d]Results are expressed as moles of substance X emitted per 1000 moles of CO_2 emitted.
[e]Nonmethane hydrocarbons (C_2–C_{10}).
[f]$(CN)_2 + HCN + CH_3CN + C_2H_5CN + C_2H_4CN$.
[g]SO_2 + aerosol sulfate.
[h]All data in g/kg $C(CO_2)$.
[i]Total particulate matter.
[j]Particulate organic carbon (including elemental carbon).
[k]Elemental (black, "soot") carbon.
[l]Photochemically produced ozone in the burning plumes.

In Table 4, I combine the estimates of biomass-burning rates in the tropics and worldwide (Table 2) with my best estimate of the emission ratios (Table 3) to derive regional and global rates of pyrogenic emissions. The uncertainty of these estimates is about 50% in the case of CO_2, where the emission amount is relatively well-constrained, and about a factor of 2 for most of the other gases, where additional uncertainty is contributed by the limited data base on emission ratios.

In spite of these uncertainties in our quantitative estimates, it is quite evident from this compilation that biomass-burning results in globally important contributions to the atmospheric budget of several of these pollutant gases. When one considers that much of the burning is concentrated in relatively distinct regions, e.g., the African savannas, and occurs over a limited time period, it is not surprising that it results in levels of atmospheric pollution which rival and sometimes exceed those in the industrialized regions of the developed nations.

Dry plant biomass consists of about 45% carbon (by weight), most of the remainder being hydrogen and oxygen. On a mass basis, the nutrient element contents are relatively low: about 0.3–3.8% nitrogen, 0.1–0.9% sulfur, 0.01–0.3% phosphorus, and 0.5–3.4% potassium.[24] Consequently, the emissions from biomass combustion are dominated by the oxides of carbon, CO_2 and CO. The fraction of CO emitted depends on the characteristics of the fire: hot, fast fires with a good supply of oxygen produce relatively little CO, whereas smoldering fires emit a large fraction of this gas.[110,165] Pyrolysis and incomplete oxidation in oxygen-deficient fires are also responsible for the emission of CH_4, nonmethane hydrocarbons (NMHC), hydrogen gas (H_2), and various partially oxidized organic compounds, e.g., alcohols, aldehydes, ketones, and organic acids.[13,36,70,72,76,109,110] Consequently, these compounds are also released preferentially during the smoldering stages. Figure 1 shows the sequence of the emission of CO_2 (maximum in the flaming stage), CO (maximum in the early smoldering stage), and the organic acids (maxima in the late smoldering stage) from an experimental fire (D. Scharffe, W. Hartmann, and M.O. Andreae, unpublished data).

Nitrogen is present in plant biomass mostly as amino groups ($R\text{-}NH_2$) in the amino acids of the proteins. During combustion, it is released by pyrolytic decomposition of the organic matter and then partially or completely oxidized to various volatile nitrogen compounds. NO represents some 10–30% of the nitrogen initially contained in the fuel, and other nitrogen compounds (NO_2, N_2O, NH_3, HCN, organic nitriles, and nitrates) account for another 20–30% of the fuel nitrogen.[72,109,110] The remainder, up to 50% of the fuel nitrogen, is released in the form of molecular nitrogen.[96] Almost none of the fuel nitrogen is left in the ash.

Table 4 shows that biomass-burning releases large amounts of the gases which are the main ingredients of "classical" smog chemistry in the atmosphere:

Table 4. Comparison of global emissions from biomass-burning with emissions from all sources (including biomass-burning).

Species	Emission Ratio (10^3)	Emissions[a] Tropics	Emissions[a] World	All Sources	Percent	Reference for "All Sources" Estimate
CO_2 (net from deforestation)	—	2400	2400	7600[b]	32	21
Carbon burned		3200	4090	—	—	
CO_2 (gross from combustion)	880	2820	3590	8790[c]	41	21
GASES						
CO	80	220	290	1100	26	171
CH_4	8	23	29	380	8	33
NMHC[d]	10	28	36	100	36	57
N_2	4.7	31	39	150	26	138
N_2O	0.1	0.7	0.8	≤6	≥14	163
NO_x	2.3	7.6	9.6	40	24	111
NH_3	1.3	4.3	5.5	44	12	10
RCN	0.6	2.0	2.5	—	—	
SO_x	0.25	1.9	2.4	200	1.2	4
COS	0.011	0.08	0.11	1.4	8	90
CH_4Cl	0.12	1.00	1.28	2.3	55	171
H_2	27	13	16	36	45	146
AEOROSOLS[e]						
TPM	25	70	90	1500	6	125
POC[f]	17	47	59	180	33	53
EC[g]	2.5	7	9	22	40	161
Potassium	0.4	1.1	1.4	—	—	
OZONE	30	338	431	1100	39	42

[a]Emissions are given as Tg C/yr for CO_2, CO, CH_4, and NMHC, as Tg N, S, and Cl for N species, S species, and CH, Cl, respectively.

[b]Deforestation plus fossil-fuel-burning.

[c]Biomass burning plus fossil-fuel-burning.

[d]Nonmethane hydrocarbons (excluding isoprene and terpenes).

[e]Total particulate matter. Emission ratios for aerosols in g/kg C (CO_2)

[f]Particulate organic carbon (including elemental carbon).

[g]Elemental (black, soot) carbon.

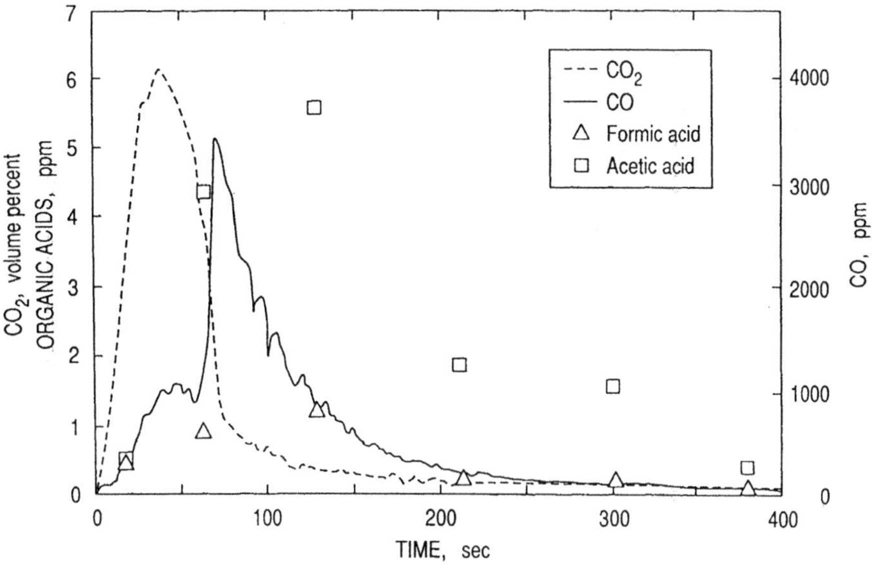

Figure 1. Concentrations of CO_2, CO, formic acid, and acetic acid in the smoke from an experimental biomass fire as a function of time. The flaming stage in this fire lasted for about 60 seconds. Concentrations are in volume percent for CO_2 and in volume mixing ratios (ppm) for the other species (1% = 10,000 ppm).

hydrocarbons, carbon monoxide, and nitrogen oxides (NO_x). About one-quarter to one-third of the global emissions of these compounds is attributable to biomass-burning. (I have ignored in this budget the natural emissions of the very short-lived terpenes and isoprene.) Since only about one-half of the total source flux of these species is from anthropogenic emissions (the rest from natural sources), I find that biomass-burning accounts for roughly about one-half of the anthropogenic source of atmospheric hydrocarbons, CO, NO_x, and tropospheric ozone.

Methane (CH_4) and nitrous oxide (N_2O) contribute to the atmospheric greenhouse effect and are long-lived enough to enter the stratosphere and participate in the ozone cycle there. The pyrogenic emission of methane accounts for as much as 10% of the global source flux. In the case of N_2O, recent evidence[37,110] shows that the emission ratios previously published were too high due to a storage artifact. N_2O emission factors determined in laboratory burning experiments (average 0.09×10^{-3} mol N_2O/mol CO_2) are still somewhat lower than the results from field measurements conducted without sample storage (average 0.14×10^{-3}; 37). A recent reevaluation of the global N_2O budget[163] suggests that the total source strength for this gas is less than 6 Tg N/yr and that the emission flux required to explain the observed increase of N_2O in the atmosphere is about 3–4.5 Tg N/yr. Assuming an emission factor of 0.1×10^{-3}, biomass-burning can account for over 15% of the total source strength and for 20–30% of the atmospheric increase.

The global atmospheric budget of ammonia (NH_3) is not well-known. Based on the discussion in Andreae et al.,[10] I estimate global emissions to be in the range of 20–80 Tg N/yr, with a "best guess" estimate of 44 Tg N/yr. Microbial release from animal excreta and soils makes up the largest fraction of these emissions, whereas biomass burning contributes only some 11%.

Whereas nitrogen is almost totally volatilized during combustion, about 20–50% of the fuel sulfur remains in the ash.[48,139,166] This is due to the difference in the speciation of sulfur and nitrogen in plants. In contrast to nitrogen, only part of the sulfur in plants is organically bound in the form of sulfur-containing amino acids in proteins (50–90%: 162). Only this organic sulfur fraction is likely to be volatilized during combustion, mostly to SO_2, but also to a small extent to carbonyl sulfide (COS). The rest of the plant sulfur is in the form of sulfate and sulfate esters which are pyrolyzed to sulfate during combustion. Sulfate is not volatile at the temperatures prevailing in biomass fires and is either retained in the ash or, to some extent, incorporated into smoke particles.[139] The same applies to phosphorus, which is also present largely as phosphate esters that are broken down to involatile phosphate during combustion.

Biomass-burning contributes only about 1% to the total atmospheric sulfur budget and represents only about 3% of the anthropogenic part of the emissions. Still, since most of the natural sulfur emissions are from the oceans and most of the anthropogenic emissions are concentrated in the industrialized regions of the temperate latitudes, biomass-burning does make a pronounced impact on the sulfur budget over remote continental regions, e.g., the Amazon and Congo Basins.[6,8,20] Here, sulfur fluxes to the vegetation may be enhanced by as much as a factor of 5 due to tropical biomass-burning.

Methyl chloride, CH_3Cl, is also released by biomass-burning and has actually been proposed as a tracer for pyrogenic emissions.[131] This gas is of great significance for the stratospheric ozone budget; it is the second largest source of stratospheric chlorine after the man-made chlorofluoromethanes. The pyrogenic source corresponds to about 1.3 Tg Cl/yr (1.8 Tg CH_3Cl/yr), more than 50% of the global emission.

Emission of Aerosol Particles (Smoke)

Even though the smoke from biomass fires is one of the most obvious emissions, often visible in the tropics as a continuous pall hanging in the air for days or weeks, quantitative estimates on the amounts of particulate matter released are still quite uncertain. This is mostly due to the wide range of emission ratios observed for aerosols in field studies. The production of smoke particles from fires depends strongly on the type of combustion, with flaming fires having relatively low emission ratios [on the order of 5–15 g/kg C(CO_2)] and smoldering fires having much higher emissions [15–50 g/kg C(CO_2)[58,157]]. Consequently, savanna fires, which burn relatively hot and efficiently, tend to have much lower smoke emissions relative to the amount of fuel consumed than the less efficiently burning deforestation fires in the tropics.[165]

Our measurements in Amazonia showed that the particulate organic carbon (POC) content of the smoke aerosol is about 66%, which is consistent with the aerosol particles being mostly composed of partially oxygenated organic matter.[8] Subsequent authors have reached similar conclusions.[29,58,157,165] The content of black elemental carbon (EC) in smoke particles from biomass-burning is highly variable. In smoldering fires, it is as low as 4% [weight percent carbon in total particulate matter (TPM)], whereas in intensively flaming fires, it can reach 40%.[29,58,123,164] Consistent with this observed increase in the relative amount of black carbon emissions with combustion efficiency, Cachier et al.[29] found the highest ratios of black carbon to total carbon in aerosol samples which they attributed to savanna fires. The "best guess" aerosol emission factors in Table 3 (25 g TPM, 17 g POC, and 2.5 g EC per kg CO_2-carbon emitted) are based on an assessment of literature data from savanna and forest fires in the tropics[8,29,165] and from forest fires in temperate regions[58,128,157,164] and reflect an attempt to integrate the different types of emission regimes.

In Table 4, the emission of total particulate matter (TPM) from biomass-burning is estimated at 90 Tg/yr, only about 6% of the global aerosol production. However, much of this production consists of large dust particles, which reside only briefly in the atmosphere. The smaller smoke particles are much more long-lived and more active in scattering solar radiation. In fact, the total number of smoke particles produced by biomass fires is on the same order as the number of aerosol particles formed from the oxidation of SO_2 from fossil-fuel-burning. Of the particulate organic carbon (POC) released (60 Tg/yr), about 15% (9 Tg/yr) is in the form of black elemental carbon. Based on these estimates, POC and EC from biomass-burning account for about one-third each of the organic and elemental carbon aerosol released globally.[53,161]

Charcoal Formation and Carbon Storage

Seiler and Crutzen[149] pointed out that, once formed, charcoal and soot carbon are not readily reoxidized back to CO_2, even over geological time scales, and that the burial of this refractory elemental carbon provides a long-term sink for atmospheric CO_2, and, incidentally, a net source of oxygen to the atmosphere. Although this mechanism appears very plausible on a qualitative basis, its quantitative importance remains controversial.

The production of aerosol elemental carbon plays only a small role in this context. The emission ratio of 2.5×10^{-3} g EC per kg CO_2-carbon corresponds to about 1.1 g EC per kg of burned biomass. Since only a fraction of the above-ground biomass is burned (20–80%, depending on vegetation type), the fraction of biomass which becomes airborne as charcoal or soot particles is very small, probably less than 0.1% (see also Refs. 155 and 156). The main mechanism for charcoal storage must, therefore, be its direct incorporation into terrestrial soils and sediments. Seiler and Crutzen[149] present a review of the literature and show that charcoal layers do, indeed, occur frequently in tropical and temperate soils. Fearnside et al.[62,63] measured the amount of charcoal produced in clearing fires in Amazonia and found a relatively low rate of charcoal production, only about 3.6% of the above-ground carbon exposed to fire. From observations during a prescribed burn in a Florida pine forest,[38] an elemental carbon yield of 5.4% of the carbon exposed to fire can be derived. From these very limited data and the amounts of biomass exposed to fire and burned annually, a range of 0.3–0.7 Pg C/yr for the global rate of charcoal formation and the resulting sequestering of CO_2 can be estimated. Thus, carbon sequestration by charcoal formation and burial could reduce by as much as 10% the ~7600 Tg C released per year to the world atmosphere by human activity.

Transport and Photochemistry in the Smoke Plumes

Long-Range Transport of Smoke Plumes

The smoke from biomass fires rises buoyantly in the atmosphere, entraining ambient air. Frequently, clouds form on the smoke plume, but usually reevaporate without rainfall. In cases where they do produce rain, some particulate matter and soluble gases may already be scavenged and removed at this stage. The height to which the smoke plumes can rise is limited in the tropics by the trade wind inversion at an altitude of some 3–5 km. When the rising smoke plumes have become trapped below an inversion layer and start spreading horizontally, they tend to become stretched into relatively thin layers of great horizontal extent. Such thin haze layers can travel very far from their point of origin without losing their identity.

The further fate of the smoke-laden air masses depends on the regional large-scale atmospheric circulation. Over tropical Africa, the plumes usually travel in a westerly direction and toward the equator. As they approach the Intertropical Convergence Zone (ITCZ), vertical convection in the troposphere intensifies, destroying the layered structure and resulting in the distribution of the pyrogenic emissions throughout the lower troposphere. Finally, in the ITCZ region, smoke and gases from biomass burning may be injected into the middle and upper troposphere, possibly even into the stratosphere. Air masses from the biomass-burning regions in South America usually move toward the south and southeast due to the effect of the Andes barrier on the large-scale circulation. Here again, they may become entrained into a convergence zone, the seasonal South Atlantic Convergence Zone (SACZ). This convergence zone becomes established in austral spring, when biomass-burning is abundant. Indeed, the data from the space-borne MAPS instrument shows high levels of CO in the middle to upper troposphere near the ITCZ and the SACZ.[132,133]

Evidence for the efficiency of these transport mechanisms is abundant. The smoke plumes from the fires in South America and Africa are easily seen from space and have been the subject of investigations from the Space Shuttle.[175] They have also been mapped using various space-borne sensors, in particular the Advanced-Very-High-Resolution-Radiometer (AVHRR) instruments on the NOAA series satellites.[88,151] At ground level, haze and smoke from fires some thousand kilometers away is still dense enough that airports in the middle of the Amazon Basin must frequently be closed during the burning season.

Less obvious than the visible smoke, but even more indicative of the large extent to which the emissions from burning are transported around the globe, are the results of chemical measurements from satellites, aircraft, and research vessels. Soot carbon and other pyrogenic aerosol constituents have been measured at sea level during research cruises over the remote Atlantic and Pacific.[3,7] Carbon monoxide and ozone originating from fires in Africa were consistently observed at altitudes of 1.5–5 km during the NASA/CITE-3 research flights off the coast of Brazil.[5] On an even larger scale, high levels of ozone and CO are seen from satellites over the tropical regions of Africa and South America, and large areas of the surrounding oceans.[66,132,133,169]

Smog Chemistry in the Smoke Plumes

Biomass fires emit essentially the same constituents as fossil-fuel-burning, namely, CO, hydrocarbons, and NO_x, the starting materials for the production of O_3 and photochemical smog. Once such a mixture is exposed to sunlight, hydrocarbons are oxidized photochemically first to various peroxides, aldehydes, etc., then to CO. This CO is added to the amount directly emitted from the fires and is finally oxidized to CO_2. In the presence of elevated levels of NO_x, as is the case in the smoke plumes, the oxidation of CO and hydrocarbons is accompanied by the formation of ozone.[41] The efficiency of ozone formation, i.e., the amount of ozone formed per molecule of hydrocarbon oxidized, depends on the proportions of hydrocarbons, NO_x, and O_3 present in the reaction mixture and, thus, on the history of transport and mixing of the air mass involved.[31] Plumes from smoldering fires, which have a very low NO_x/hydrocarbon ratio appear to produce relatively little O_3,[172] whereas the highest amounts of O_3 production have been observed in plumes from the hot savanna fires of Africa.[5,9]

Environmental Effects

Ozone Pollution of the Troposphere

The ozone levels produced in the plumes often reach values typical of industrialized regions.[5,8,9,39,40,44,47,94,113] The highest concentrations, in the range of 50 to 100 ppb, are usually found in discrete layers at altitudes of 1–5 km (Figure 2). At ground level, concentrations are normally lower and show a pronounced daily cycle, with a minimum at night and a maximum near midday. This cycle is controlled by the balance of O_3 sources and sinks: At night, O_3 consumption by deposition to vegetation and by reaction with NO emitted from soils reduces the concentration of ozone near the ground; during the day, it is replenished by photochemical ozone formation and downward mixing of ozone-rich air from higher altitudes. Over an unpolluted tropical rainforest, e.g., in the central Amazon Basin during the wet season, O_3 levels are very low: below 10 ppb at midday and below 3 ppb at night.[71,92] Under somewhat more polluted conditions, e.g., in central Amazonia during the dry season, nighttime values are still very low, but daytime O_3 concentrations reach 10–30 ppb at the height of the tree crowns. Finally, in the savanna regions, where biomass-burning currently has the strongest impact, surface ozone concentrations in excess of 40 ppb are frequently measured,[40] and even values as high as 90 ppb have been reported.[95] Ozone episodes with ground level concentrations of 80–120 ppb must be expected to be common in the tropics, particularly during the dry season when photochemically reactive air becomes trapped under a subsiding inversion layer. These levels are similar to those observed in the highly polluted regions of the eastern United States and of Central Europe (Ref. 112 and references therein). Studies in temperate forest regions have linked such levels of ozone pollution to the tree and vegetation damage which has become widespread in Europe and North America. In view of the sharp increase of O_3 with altitude frequently observed in the tropics, the likelihood of vegetation damage by ozone is especially high in mountainous regions, where the ground surface intersects the levels where the highest ozone concentrations are encountered.

Perturbation of Oxidant Cycles in the Troposphere

The large-scale increase of tropospheric ozone in the tropics, expected to persist in the future, indicates a fundamental change in the chemical behavior of the troposphere. Many gases, particularly hydrocarbons, are continuously emit-

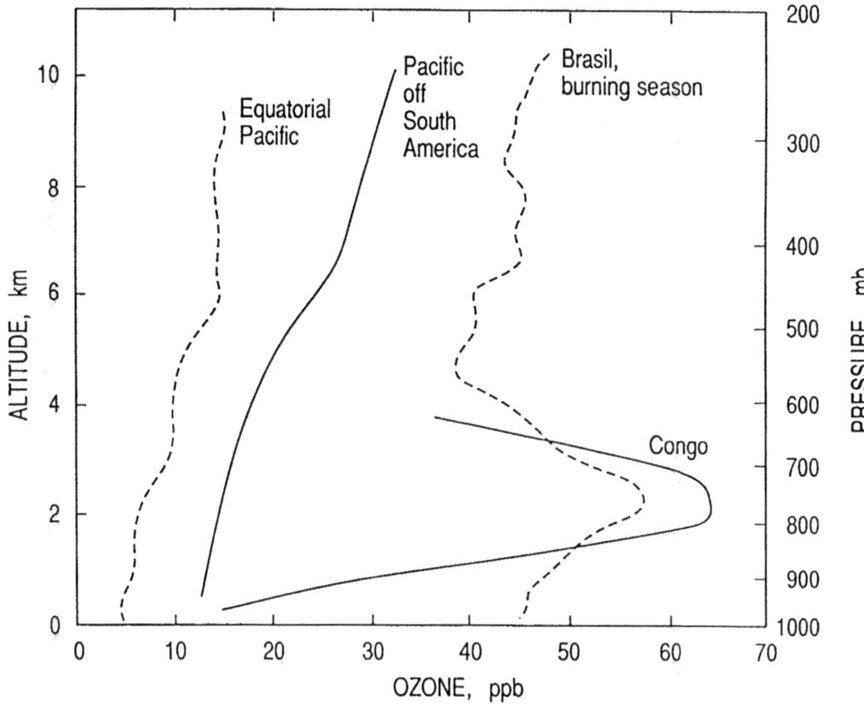

Figure 2. Vertical profiles of ozone in the tropical troposphere. The profile over the equatorial Pacific shows no influence from biomass-burning,[65] whereas the profile over the Pacific off South America[150] suggests ozone enhancement due to long-range transport from the tropical continents. The ozone profiles over Brazil[44] and the Congo[9] show high ozone concentrations between 1 and 4 km altitude due to photochemical production in biomass-burning plumes (Figure modified from Ref. 64).

ted into the atmosphere from natural and anthropogenic sources. A buildup of these gases in the atmosphere is prevented by a self-cleaning mechanism, whereby these substances are slowly ''combusted'' photochemically to CO_2. The key molecule responsible for this oxidation process is the hydroxyl radical (OH). The reaction chains involved are such that ozone and OH are consumed when the concentration of NO_x is low, which is normally the case in the unpolluted troposphere. Based on the observed increase of methane and CO in the atmosphere, Crutzen[41] suggested the possibility of a global decrease in OH and O_3, leading through a feedback mechanism to a further increase in CH_4 and CO, and a possibly unstable condition of the atmosphere. Injection of large amounts of NO_x from biomass-burning may counteract this feedback since hydrocarbon oxidation in the presence of elevated amounts of NO_x *produces* additional amounts of O_3 and OH. The model calculations of Keller et al.[89] predict a tenfold increase in regional OH concentrations to result from deforestation and biomass-burning in the tropics. This effect may be further enhanced by the pyrogenic emission of CH_2O [$\sim$(2–3) $\times 10^{-3}$ relative to CO_2], which is photolyzed in the tropical atmosphere within a few hours; this process partially leads to production of HO_2 radicals via the formation of H and CHO. Whereas considerable uncertainty persists over the net effect of these changes, it now appears likely that biomass-burning results in pronounced regional enhancements of O_3 and OH levels while, on a global scale, it contributes to decreasing average OH concentrations. This would lead to a buildup of the many gases that are removed from the atmosphere by OH. This change may be an extremely important perturbation of global atmospheric chemistry.[43]

Climate Change

Increasing concentrations of a number of trace gases in the atmosphere (CO_2, the chlorofluorocarbons, CH_4, N_2O, and O_3) are thought to cause an increase in global temperatures through the "greenhouse effect" (Figure 3; Ref. 130). I have already discussed the fact that the CO_2 release from the clearing of tropical rainforests is responsible for about 30% of the increase of CO_2 in the atmosphere. Ozone is another efficient absorber of infrared radiation, and its increase in the troposphere contributes about 15% to the overall greenhouse warming. Photochemical ozone production from pyrogenic emissions accounts for about one-half of the increase of ozone into the troposphere (Figure 3), roughly the same amount as produced from fossil-fuel emissions in the industrialized regions of the Northern Hemisphere.[42] Ozone soundings[47,94] and the MAPS data[133] show that the effect of ozone production from biomass-burning is quite pronounced even in the middle and upper troposphere, where it contributes most significantly to greenhouse warming. The third major source of tropospheric ozone, downward mixing from the stratosphere, is of the same magnitude as each of the photochemical sources, but does not contribute to the increase in ozone since it has remained constant from preindustrial times.

While biomass-burning accounts for only about 8% of the global emission of methane (another "greenhouse gas"), it is responsible for a much greater fraction of the *increase* in global emissions since many of the other sources, e.g., natural wetlands, remain constant or are actually shrinking. From estimates of the temporal trends of methane source strengths from 1940 to 1980 given by Bolle et al.,[22] one can conclude that biomass-burning has contributed about 20% to the increase in methane emissions over that time period. Much the same argument applies to N_2O, where biomass-burning may account for about one-quarter of the observed increase in the tropospheric burden.

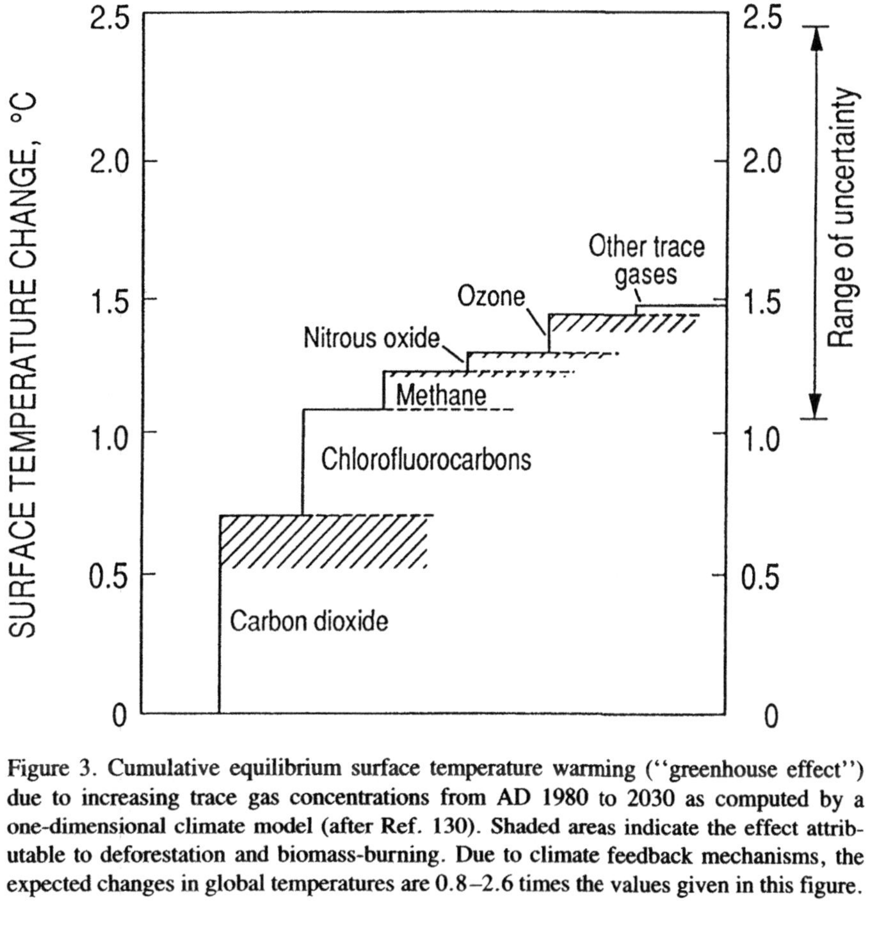

Figure 3. Cumulative equilibrium surface temperature warming (''greenhouse effect'') due to increasing trace gas concentrations from AD 1980 to 2030 as computed by a one-dimensional climate model (after Ref. 130). Shaded areas indicate the effect attributable to deforestation and biomass-burning. Due to climate feedback mechanisms, the expected changes in global temperatures are 0.8–2.6 times the values given in this figure.

The climatic effect of smoke aerosols is beyond current understanding because of the complexity of the interactions involved. Owing to the limited lifetime of aerosol particles (on the order of days) compared to the lifetime of most of the ''greenhouse gases,'' their climatic effects act initially more on a regional than a global scale. However, since the equatorial regions, particularly the Amazon Basin, the Congo Basin, and the area around Borneo, are extremely important in receiving solar energy and in redistributing this heat vertically through the atmosphere, any change affecting the operation of these ''heat engines of the atmosphere'' must be viewed with great concern.

Aerosols can influence climate directly through changing the Earth's radiation balance. They reflect sunlight back into space, and thus reduce the amount of heat the Earth receives. Black soot aerosol, on the other hand, may absorb sunlight and result in heating of the atmosphere. Robock[135,136] has shown that large daytime temperature drops can occur below smoke plumes from forest fires. With the strong expansion of tropical biomass-burning, this effect may well have masked the expected greenhouse temperature rise during the dry season on the tropical continents.[124]

A great source of concern is the influence of pyrogenic particles on the behavior and properties of clouds. Cloud droplets form on aerosol particles, called cloud condensation nuclei (CCN). The properties of the cloud depend on the number of available CCN: The more CCN, the more droplets form, which, given a constant amount of available water, results in a smaller droplet size. Clouds made up of smaller droplets are whiter and reflect more sunlight back into space, but are less likely to produce rain. Since clouds are one of the most important controls on the heat balance of the Earth, any large-scale modification of cloud properties is likely to have a strong impact on global climate. The limited body of data that is available to date on the production of CCN by biomass fires suggests strongly that pyrogenic aerosol particles are very effective as CCN, and that large numbers of such particles are released during burning.[54–56,73,129,137] Aging for 20 hours increased the percentage of pyrogenic aerosol particles which were effective as CCN to almost 100%.[73]

The changes in precipitation efficiency which may be caused by increased abundance of CCN add to the perturbation of the hydrological cycle in the tropics caused by changes in the characteristics of the land surface brought about by deforestation and desertification. Tropical forests are extremely efficient in returning water that has fallen on them as rain back to the atmosphere in the form of water vapor. There it can form clouds and rain again, and the cycle can repeat itself many times. A region like the Amazon Basin can thus retain water for a long time and maintain a large "standing stock" of water. If the forest is replaced by grassland or, as often is the case, converted into a nearly unvegetated surface by erosion and loss of top soil, water runs off quickly and returns through streams and rivers to the ocean without much chance of recycling. Beyond the unfavorable consequences that such large-scale changes in the availability of water would have on human activities, this modification of the hydrological cycle may itself perturb climate.[104,152] Evaporation of water at the ground results in the transfer of heat energy from the Earth's surface to the atmosphere, where it is stored, transported by atmospheric convection, and finally released when the vapor condenses into cloud droplets. This mechanism serves to redistribute heat energy vertically in the atmosphere. If less water is available for this purpose, the heat absorbed at the ground must be removed by radiation and dry convection, leading to higher surface temperatures and a change in the vertical distribution of heat. Such perturbations of the heat and water balance of the atmosphere must be viewed with great concern.

Table 5. Rainwater pH and acid deposition at some continental tropical sites and in the Eastern United States.

	pH		Rainfall	Deposition	
	Mean[a]	Range	(cm)	(kg H^+/ha/yr)	Ref.
Venezuela					
San Eusebio	4.6	3.8–6.2	158	0.39	153
San Carlos	4.8	4.4–5.2	—	—	67
La Paragua	4.7	4.0–5.0	—	—	143
Brazil					
Manaus, dry season	4.6	3.8–5.0	240[b]	0.29[b]	12
Manaus, wet season	5.2	4.3–6.1			
Australia					
Groote Eylandt	4.3		—	—	101
Katherine	4.8	4.2–5.4	—	—	67
Jabiru	4.3		—	—	13
Ivory Coast					
Ayame	4.6	4.0–6.5	179	0.41	98
Congo					
Boyele	4.4		185	0.74	99
Eastern USA	4.3	3.0–5.9	130	0.67	17

[a]Volume weighted.
[b]Annual average.

Acid Deposition

After acid rain had become a notorious environmental problem in Europe and North America, it came as somewhat of a surprise to scientists to learn that it was widespread in the tropics as well (Table 5): Acid rain was reported from Venezuela,[67,143,153] from Brazil,[6,11] from Africa,[97–99] and from Australia.[13] In all instances, organic acids (especially formic and acetic acids) and nitric acid were shown to account for a large part of the acidity, in contrast to the situation in the industrialized temperate regions, where sulfuric acid and nitric acid predominate. Recent evidence shows that the organic acids can to a large extent be attributed to direct emission from biomass-burning and to the photochemical formation of formic and acetic acid in the plumes.[159,79] The presence of elevated levels of organic and black carbon in tropical rain during the dry season has also been presented as evidence for the importance of pyrogenic emissions for rainwater chemistry in the tropics.[30]

In regions impacted by biomass-burning, pyrogenic NO_x emissions overwhelm, by far, the natural sources of NO_x, e.g., soil emissions and NO formation in lightning bolts. Nitric acid is formed photochemically from the NO_x emitted in the fires. A modeling study simulating the effects of biomass-burning and a moderate amount of additional pollution from logging, etc., suggests that rainwater pH values of ~4.2 can be expected in the tropics due to the formation of nitric acid alone.[89] The addition of organic acids to precipitation can be expected to drive pH down to values below 4. Such values have indeed been frequently observed in Africa (J.-P. Lacaux, personal communication, 1989). For comparison, the mean pH measured during the period 1963–1982 at the Hubbard Brook site in New Hampshire, a classical site for the study of acid rain, was 4.2,[108] and the mean pH in rain sampled throughout the eastern United States in 1980 was 4.3.[17]

Acidic substances in the atmosphere can be deposited to plants and soils either by rain and fog (wet deposition) or by the direct removal of aerosols and gases at the surface (dry deposition). In the humid tropics, wet deposition accounts for most of the deposition flux, whereas in the savanna regions, especially during the dry season, dry deposition dominates. Acid deposition has been linked to forest damage in Europe and the eastern United States. It is thought that such damage is usually caused by a combination of factors, including acid deposition, ozone, and the concurrent exposure to other sources of stress, e.g., drought or insect infestation.[23] Acid deposition can act on an ecosystem following two major pathways: direct damage through the deposition of acidic aerosols and gases on leaves, or soil acidification. The danger of leaf injury is serious only at pH levels below 3.5, which is rarely encountered in the tropics.[116] However, in view of the presence of the highly soluble gaseous nitric and organic acids from biomass-burning, it is quite likely that fog and dew can reach such pH levels during the dry season in many regions. Tropical forests may be inherently more sensitive to foliar damage than temperate forests due to the longer average leaf life in the tropics, where the leaves of many trees are shed only after two or more years. The resulting longer exposure of individual leaves may make cumulative damage more pronounced.[116]

Many species of tropical animals undergo stages in their life cycle during which they live in rainwater which has collected in bromeliads or similar plants. Further species, like frogs and salamanders, depend on water accumulated between dead leaves, mosses, etc., to keep their skin moist. Acid deposition poses a serious risk to these animals, which are essential components of the forest ecosystem. This danger is most acute for insects, which are essential for the pollination of many rainforest plant species.[15]

Tropical soils were thought to be relatively resistant to rapid acidification due to their high sulfate adsorption capacity.[116] This view may, however, not be justified since acid deposition in the tropics is dominated by nitric rather than sulfuric acid. Furthermore, there are large areas where highly weathered soils predominate, which are highly susceptible to acidification, e.g., in Venezuela.[142] Even in relatively resistant soils, chronic exposure to acid deposition will eventually lead to soil acidification and associated problems, like leaching of aluminum, manganese, and other cations. Microbial processes in the soils are also at risk of being perturbed. Nitrogen cycling is likely to be influenced, both by decreasing pH and by the addition of nitrate and ammonium ions from wet and dry deposition. Uptake of ammonium by plants and microbial oxidation of ammonium in the soils further contribute to soil acidification. This is particularly relevant here since ammonium is a major constituent of pyrogenic emissions.

Disruption of Nutrient Cycles and Soil Degradation

Tropical ecosystems, both natural and agricultural, are frequently deficient in nitrogen, sulfur, or both.[140] When such an area is burned, a substantial part of the nitrogen and sulfur present in the ecosystem is volatilized. If these elements were redeposited relatively nearby, there would be no net gain or loss of nutrients. However, some 50% of fuel nitrogen is emitted as molecular nitrogen, resulting in a significant overall loss of fixed (nutrient) nitrogen.[43] The interactions between fire and grazing in grasslands add a further complication.[80] The nitrogen loss due to burning is much greater from ungrazed than from grazed areas since the latter have less above-ground biomass exposed to fire. This effect is large enough that grazing may control whether a burned grassland gains or loses nitrogen.

On the other hand, ecosystems that are not burned, e.g., remaining areas of intact rainforest, will receive an increased nutrient input. Studies of rainwater chemistry in the central Amazon Basin suggest that as much as 90% of the sulfur and nitrogen deposited there is from sources outside the forest ecosystem, with long-range transport of emissions from biomass-burning playing a major role.[12] The effects of such increasing inputs of nutrients to the rainforests are not known. It must be remembered, however, that the increased nutrient inputs are accompanied by an increase in acid deposition and ozone concentration. The impact of these complex environmental perturbations on tropical forest ecosystems will be difficult to analyze.

Besides the immediate volatilization of nitrogen during the burns, long-term changes in the microbial cycling of nitrogen in the soils appear to result from the use of burning as an agricultural tool.[1,106] NO and N_2O emissions from soils at experimental sites that were burned were about twice as high as from soils at unburned sites, and this effect persisted for at least 6 months following the burn.[2] Postburn measurements at a Venezuelan savanna site showed enhanced NO emissions by a factor of 10 for the 4 days during which the measurements were made.[85] Levine et al.[105] found an increase in biogenic CH_4 and NO emissions from wetlands following vegetation burn-off. Other studies have also shown enhanced fluxes of nitrogen oxides from soils following forest clearing and conversion to grazing land,[115,134] but in these studies the effect of burning was not isolated explicitly. In more extensive studies, Luizão et al.[114] did not observe enhanced N_2O fluxes from sites that were only burned and cleared, but observed a threefold enhancement of emissions on 3–4-year-old pasture sites. According to these researchers, the enhanced emissions may be caused by increased input of oxidizable carbon from grass roots or rhizomes, or compaction of the soil surface by the cattle.

On the other hand, studies by Sanhueza et al.[144] in a Venezuelan savanna and by Goreau and de Mello[69] on a cleared forest site in Amazonia showed that forested areas emit about 10 times more N_2O than secondary grassland ecosystems derived from deforestation. Thus, although a disturbed tropical forest ecosystem may for some time emit more N_2O, this may only be temporary, and, in the long run, less N_2O may be emitted. It is clear that further research is needed to elucidate the effects of biomass-burning on nutrient cycling and especially on nitrogen volatilization. The few studies available do suggest, however, that deforestation and biomass-burning contribute to the inputs of nitrogen oxides and the resulting perturbations of air quality and climate well beyond the direct emissions from the fires.

Perturbation of Stratospheric Chemistry and the Ozone Layer

Many of the trace gases released by biomass-burning are involved in the chemical reaction cycles which control the stratospheric ozone layer. The fact that some biomass-burning occurs in close vicinity to the Intertropical Convergence Zone is particularly relevant: In this region of intensive vertical convection, even relatively short-lived gases, e.g., nonmethane hydrocarbons, may be injected into the stratosphere. The most important pyrogenic gas entering the stratosphere is probably methyl chloride, about 50% of which now comes from biomass-burning and which is the second largest source of stratospheric chlorine. COS is the only sulfur gas which is able to escape tropospheric oxidation and diffuse to the stratosphere. Here it is oxidized to form sulfuric acid aerosol particles, which scatter sunlight, and participate in the reaction cycles which control the abundance of stratospheric ozone.[25,81,122] It is impossible at this time to predict what the overall effect of biomass-burning on the fluxes of all of these species into the stratosphere and on the chemical reactions there will be. However, the potential for disruption of stratospheric chemical cycles is great enough to warrant future investigation by modeling and experiment.

References

1. Anderson, I.C., J.S. Levine, M.A. Poth, and P.J. Riggan. 1988. Enhanced biogenic emissions of nitric oxide and nitrous oxide following surface biomass burning. J. Geophys. Res. **93**:3893–3898.

2. Anderson, I.C., and M.A. Poth. 1989. Semiannual losses of nitrogen as NO and N_2O from unburned and burned chaparral. Global Biogeochem. Cycles **3**:121–135.

3. Andreae, M.O. 1983. Soot carbon and excess fine potassium: Long-range transport of combustion-derived aerosols. Science **220**:1148–1151.

4. Andreae, M.O. 1990. Ocean-atmosphere interactions in the global biogeochemical sulfur cycle. Marine Chem. **30**:1–29.

5. Andreae, M.O., B.E. Anderson, J.E. Collins, G.L. Gregory, G.W. Sachse, J.D. Bradshaw, D.D. Davis, V.W.J.H. Kirchhoff, and R.B. Chatfield. 1992. Influence of plumes from biomass burning on atmospheric chemistry over the equatorial Atlantic during CITE-3. J. Geophys. Res. (in preparation).

6. Andreae, M.O., and T.W. Andreae. 1988. The cycle of biogenic sulfur compounds over the Amazon Basin. I. Dry season. J. Geophys. Res. **93**:1487–1497.

7. Andreae, M.O., T.W. Andreae, R.J. Ferek, and H. Raemdonck. 1984. Long-range transport of soot carbon in the marine atmosphere. Science of the Total Environment **36**:73–80.

8. Andreae, M.O., E.V. Browell, M. Garstang, G.L. Gregory, R.C. Harriss, G.F. Hill, D.J. Jacob, M.C. Pereira, G.W. Sachse, A.W. Setzer, P.L. Silva Dias, R.W. Talbot, A.L. Torres, and S.C. Wofsy. 1988a. Biomass-burning emissions and associated haze layers over Amazonia. J. Geophys. Res. **93**:1509–1527.

9. Andreae, M.O., A. Chapuis, B. Cros, J. Fontan, G. Helas, C. Justice, Y.J. Kaufman, A. Minga, and D. Nganga. 1992. Ozone and Aitken nuclei over equatorial Africa: Airborne observations during DECAFE 88. J. Geophys. Res. **97**: 6137–6148.

10. Andreae, M.O., A.C. Delany, S. Liu, J. Logan, L.P. Steele, H. Westberg, and R. Zika. 1989. Key aspects of species related to global biogeochemical cycles. In D.H. Lenschow and B.B. Hicks (eds.), Global Tropospheric Chemistry: Chemical Fluxes in the Global Atmosphere, NCAR, Boulder, CO, pp. 9–29.

11. Andreae, M.O., R.W. Talbot, T.W. Andreae, and R.C. Harriss. 1988b. Formic and acetic acid over the central Amazon region, Brazil. I. Dry Season. J. Geophys. Res. **93**:1616–1624.

12. Andreae, M.O., R.W. Talbot, H. Berresheim, and K.M. Beecher. 1990. Precipitation chemistry in central Amazonia. J. Geophys. Res. **95**:16987–16999.

13. Ayers, G.P., and R.W. Gillett. 1988. Acidification in Australia. In H. Rodhe and R. Herrera (eds.), Acidification in Tropical Countries (SCOPE 36), J. Wiley and Sons, Chichester, pp. 347–400.

14. Ayres, R.U., L.W. Ayres, and J.A. Tarr. 1987. An Historical Reconstruction of

Major Atmospheric Gaseous Emissions in the U.S., 1880–1980. Final Report prepared for Resources of the Future, Inc., by Variflex Corp., Pittsburgh, PA.

15. Baker, H.G., K.S. Bawa, G.W. Frankie, and P.A. Opler. 1983. Reproductive biology of plants in tropical forests. In F.B. Golley (ed.), Ecosystems of the World 14A, Tropical Rain Forest Ecosystems, Elsevier, Amsterdam, pp. 183–215.

16. Barnard, G.W. 1990. Use of agricultural residues as fuel. In J. Pasztor and L.A. Kristoferson (eds.), Bioenergy and the Environment, Westview, Boulder, CO, 410 p.

17. Barrie, L.A., and J.M. Hales. 1984. The spatial distributions of precipitation acidity and major ion wet deposition in North America during 1980. Tellus **36B**: 333–355.

18. Betancourt, J.L., T.W. Swetnam, S.W. Leavitt, and D.R. Cayan. 1992. Interannual variation in global wildfire activity and its relevance to the global carbon budget. In R.S. Oremland (ed.), The Biogeochemistry of Global Change: Radiative Trace Gases, Chapman & Hall, New York (in press).

19. Bilsborrow, R.E., and P.F. DeLargy. 1991. Land use, migration and natural resource deterioration: The experience of Guatemala and the Sudan. In K. Davis and M.S. Bernstam (eds.), Resources, Environment, and Population: Present Knowledge, Future Options, Oxford University Press, New York, pp. 125–147.

20. Bingemer, H.G., M.O. Andreae, T.W. Andreae, P. Artaxo, G. Helas, D.J. Jacob, N. Mihalopoulos, and B.C. Nguyen. 1992. Sulfur gases and aerosols in and above the equatorial African rainforest. J. Geophys. Res. **97**:6207–6217.

21. Bolin, B., B.R. Döös, J. Jäger, and R.A. Warrick. 1986. The Greenhouse Effect, Climatic Change, and Ecosystems (SCOPE 29), J. Wiley and Sons, Chichester.

22. Bolle, H.-J., W. Seiler, and B. Bolin. 1986. Other greenhouse gases and aerosols: Assessing their role for atmospheric radiative transfer. In B. Bolin, B.R. Döös, J. Jäger, and R.A. Warrick (eds.), The Greenhouse Effect, Climatic Change, and Ecosystems (SCOPE 29), J. Wiley and Sons, Chichester, pp. 157–203.

23. Bormann, F.H. 1985. Air pollution and forests: An ecosystem perspective. Bioscience **35**:434–441.

24. Bowen, H.J.M. 1979. Environmental Chemistry of the Elements, Academic Press, London.

25. Brasseur, G.P., C. Granier, and S. Walters. 1990. Future changes in stratospheric ozone and the role of heterogeneous chemistry. Nature **348**:626–628.

26. Brown, A.P., and K.P. Davis. 1973. Forest Fire, Control and Use, McGraw-Hill, New York.

27. Brown, S., A.J.R. Gillespie, and A. Lugo. 1989. Biomass estimation methods for tropical forests with applications to forest inventory data. Forest Science **35**:881.

28. Brown, S., and A.E. Lugo. 1984. Biomass of tropical forests: A new estimate based on forest volumes. Science **223**:1290–1293.

29. Cachier, H., M.-P. Brémond, and P. Buat-Ménard. 1989. Carbonaceous aerosols from different tropical biomass burning sources. Nature **340**:371–373.

30. Cachier, H., and J. Ducret. 1991. Influence of biomass burning on equatorial African rains. Nature **352**:228–230.

31. Chatfield, R.B., and A.C. Delany. 1990. Convection links biomass burning to increased tropical ozone; However, models will tend to overpredict O_3. J. Geophys. Res. **95**:18473–18488.

32. Cheney, N.P., R.J. Raison, and P.H. Khanna. 1980. Release of carbon to the atmosphere in Australian vegetation fires. In G.I. Pearman (ed.), Carbon Dioxide and Climate: Australian Research, Australian Academy of Science, Canberra, pp. 153–158.

33. Cicerone, R.J., and R.S. Oremland. 1988. Biogeochemical aspects of atmospheric methane. Global Biogeochem. Cycles **2**:299–327.

34. Cofer, W.R., III, J.S. Levine, D.I. Sebacher, E.L. Winstead, P.J. Riggan, J.A. Brass, and V.G. Ambrosia. 1988. Particulate emissions from a mid-latitude prescribed chaparral fire. J. Geophys. Res. **93**:5207–5212.

35. Cofer, W.R., III, J.S. Levine, D.I. Sebacher, E.L. Winstead, P.J. Riggan, B.J. Stocks, J.A. Brass, V.G. Ambrosia, and P.J. Boston. 1989. Trace gas emissions from chaparral and boreal forest fires. J. Geophys. Res. **94**:2255–2259.

36. Cofer, W.R., III, J.S. Levine, E.L. Winstead, P.J. LeBel, A.M. Koller, Jr., and C.R. Hinkle. 1990. Trace gas emissions from burning Florida wetlands. J. Geophys. Res. **95**:1865–1870.

37. Cofer, W.R., III, J.S. Levine, E.L. Winstead, and B.J. Stocks. 1991. New estimates of nitrous oxide emissions from biomass burning. Nature **349**:689–691.

38. Comery, J.A. 1981. Ph.D. thesis. University of Washington, Seattle, Washington.

39. Cros, B., R. Delmas, B. Clairac, J. Loemba-Ndembi, and J. Fontan. 1987. Survey of ozone concentrations in an equatorial region during the rainy season. J. Geophys. Res. **92**:9772–9778.

40. Cros, B., R. Delmas, D. Nganga, B. Clairac, and J. Fontan. 1988. Seasonal trends of ozone in equatorial Africa: Experimental evidence of photochemical formation. J. Geophys. Res. **93**:8355–8366.

41. Crutzen, P.J. 1987. The role of the tropics in atmospheric chemistry. In R.E. Dickinson (ed.), The Geophysiology of Amazonia, J. Wiley, New York, pp. 107–130.

42. Crutzen, P.J. 1988. Tropospheric Ozone: An Overview. In I.S.A. Isaksen (ed.), Tropospheric Ozone, Reidel, Dordrecht, pp. 3–32.

43. Crutzen, P.J., and M.O. Andreae. 1990. Biomass burning in the tropics: Impact on atmospheric chemistry and biogeochemical cycles. Science **250**:1669–1678.

44. Crutzen, P.J., A.C. Delany, J.P. Greenberg, P. Haagenson, L. Heidt, R. Lueb, W. Pollock, W. Seiler, A.F. Wartburg, and P.R. Zimmerman. 1985. Tropospheric chemical composition measurements in Brazil during the dry season. J. Atmos. Chem. **2**:233–256.

45. Crutzen, P.J., L.E. Heidt, J.P. Krasnec, W.H. Pollock, and W. Seiler. 1979. Biomass burning as a source of atmospheric gases CO, H_2, N_2O, NO, CH_3Cl, and COS. Nature **282**:253–256.

46. De Angelis, D.G., D.S. Ruffin, J.A. Peters, and R.B. Reznik. 1980. Source Assessment: Residential Combustion of Wood, EPA-600/2-80-042b. Environmental Protection Agency, Washington, DC.

47. Delany, A.C., P. Haagensen, S. Walters, A.F. Wartburg, and P.J. Crutzen. 1985. Photochemically produced ozone in the emission from large-scale tropical vegetation fires. J. Geophys. Res. **90**:2425–2429.

48. Delmas, R. 1982. On the emission of carbon, nitrogen and sulfur in the atmosphere during bushfires in intertropical savannah zones. Geophys. Res. Lett. **9**: 761–764.

49. Delmas, R.A., A. Marenco, J.P. Tathy, B. Cros, and J.G.R. Baudet. 1991. Sources and sinks of methane in the African savanna. CH_4 emissions from biomass burning. J. Geophys. Res. **96**:7287–7299.

50. Detwiler, R.P., and C.A.S. Hall. 1988. Tropical forests and the global carbon cycle. Science **239**:42–47.

51. Detwiler, R.P., C.A.S. Hall, and P. Bogdonoff. 1985. Land use change and carbon exchange in the tropics: II. Estimates for the entire region. Environmental Management **9**:335–344.

52. Deutscher Bundestag. 1990. Protecting the Tropical Forests: A High-Priority International Task. Deutscher Bundestag, Bonn.

53. Duce, R.A. 1978. Speculations on the budget of particulate and vapor phase nonmethane organic carbon in the global troposphere. Pure Appl. Geophys. **116**: 244–273.

54. Désalmand, F. 1987. Observations of CCN concentrations south of the Sahara during a dust haze. Atmos. Res. **21**:13–28.

55. Désalmand, F., J. Baudet, and R. Serpolay. 1982. Influence of rainfall on the seasonal variations of cloud condensation nuclei concentrations in a sub-equatorial climate. J. Atmos. Sci. **39**:2076–2082.

56. Eagan, R.C., P.V. Hobbs, and L.F. Radke. 1974. Measurements of cloud condensation nuclei and cloud droplet size distributions in the vicinity of forest fires. J. Appl. Meteorol. **13**:553–557.

57. Ehhalt, D.H., J. Rudolph, and U. Schmidt. 1986. On the importance of light hydrocarbons in multiphase atmospheric systems. In W. Jaeschke (ed.), Chemistry of Multiphase Atmospheric Systems, Springer-Verlag, Berlin, pp. 321–350.

58. Einfeld, W., D.E. Ward, and C.C. Hardy. 1991. Effects of fire behavior on prescribed fire smoke characteristics: A case study. In J.S. Levine (ed.), Global Biomass Burning: Atmospheric, Climatic, and Biospheric Implications, MIT Press, Cambridge, MA, pp. 412–419.

59. FAO. 1986. FAO Production Yearbook 1985, Food and Agriculture Organization, Rome.

60. FAO. 1989. Yearbook of Forest Products 1987 (1976–1987), Food and Agriculture Organization, Rome.

61. Fearnside, P.M. 1985. Burn quality prediction for simulation of the agricultural

system of Brazil's Transamazon Highway colonists for estimating human carrying capacity. In G.V. Govil (ed.), Ecology and Resource Management in the Tropics, International Society for Tropical Ecology, Varanasi, India.

62. Fearnside, P.M. 1990. Fire in the tropical rain forest of the Amazon Basin. In J. Goldammer (ed.), Fire in the Tropical Biota: Ecosystem Processes and Global Challenges, Springer-Verlag, Berlin, pp. 106–116.

63. Fearnside, P.M., M. Keller, N.L. Filho, and F.M. Fernandes. 1989. Rainforest burning and the global carbon budget: Biomass, combustion efficiency and charcoal formation in the Brazilian Amazon. Unpublished manuscript.

64. Fishman, J. 1988. Tropospheric ozone from satellite total ozone measurements. In I.S.A. Isaksen (ed.), Tropospheric Ozone: Regional and Global Scale Interactions, Reidel, Dordrecht, pp. 111–123.

65. Fishman, J., and J.C. Larsen. 1987. The distribution of total ozone and stratospheric ozone in the tropics: Implications for the distribution of tropospheric ozone. J. Geophys. Res. **92**:6627–6634.

66. Fishman, J., C.E. Watson, J.C. Larsen, and J.A. Logan. 1990. Distribution of tropospheric ozone determined from satellite data. J. Geophys. Res. **95**:3599–3617.

67. Galloway, J.N., G.E. Likens, W.C. Keene, and J.M. Miller. 1982. The composition of precipitation in remote areas of the world. J. Geophys. Res. **87**:8771–8786.

68. Goldammer, J.G. 1988. Rural land-use and wildland fires in the tropics. Agroforestry Systems **6**:235–252.

69. Goreau, T.J., and W.Z. de Mello. 1988. Tropical deforestation: Some effects on atmospheric chemistry. Ambio **17**:275–281.

70. Greenberg, J.P., P.R. Zimmerman, L. Heidt, and W. Pollock. 1984. Hydrocarbon and carbon monoxide emissions from biomass burning in Brazil. J. Geophys. Res. **89**:1350–1354.

71. Gregory, G.L., E.V. Browell, L.S. Warren, and C.H. Hudgins. 1990. Amazon Basin ozone and aerosol: Wet-season observations. J. Geophys. Res. **95**:16903–16912.

72. Griffith, D.W.T., W.G. Mankin, M.T. Coffey, D.E. Ward, and A. Riebau. 1991. FTIR remote sensing of biomass burning emissions of CO_2, CO, CH_4, CH_2O, NO, NO_2, NH_3, and N_2O. In J.S. Levine (ed.), Global Biomass Burning: Atmospheric, Climatic and Biospheric Implications, MIT Press, Cambridge, MA, pp. 230–239.

73. Hallett, J., J.G. Hudson, and C.F. Rogers. 1989. Characterization of combustion aerosols for haze and cloud formation. Aerosol Sci. Technol. **10**:70–83.

74. Hao, W.-M., M.-H. Liu, and P.J. Crutzen. 1990. Estimates of annual and regional releases of CO_2 and other trace gases to the atmosphere from fires in the tropics, based on the FAO statistics for the period 1975–1980. In J.G. Goldammer (ed.), Fire in the Tropical Biota: Ecosystem Processes and Global Challenges, Springer-Verlag, Berlin, pp. 440–462.

75. Hegg, D.A., L.F. Radke, P.V. Hobbs, C.A. Brock, and P.J. Riggan. 1987. Nitrogen and sulfur emissions from the burning of forest products near large urban areas. J. Geophys. Res. **92**:14701–14709.

76. Hegg, D.A., L.F. Radke, P.V. Hobbs, R.A. Rasmussen, and P.J. Riggan. 1990. Emissions of some trace gases from biomass fires. J. Geophys. Res. **95**:5669–5675.

77. Hegg, D.A., L.F. Radke, P.V. Hobbs, and P.J. Riggan. 1988. Ammonia emissions from biomass burning. Geophys. Res. Lett. **15**:335–337.

78. Heinselmann, M.L. 1973. Fire in the virgin forests of the Boundary Waters Canoe Area, Minnesota. Quaternary Res. **3**:329–382.

79. Helas, G., H. Bingemer, and M.O. Andreae. 1992. Organic acids over equatorial Africa: Results from DECAFE 88. J. Geophys. Res. **97**:6187–6193.

80. Hobbs, N.T., D.S. Schimel, C.E. Owensby, and D.S. Ojima. 1991. Fire and grazing in the tallgrass prairie: Contingent effects on nitrogen budgets. Ecology **72**:1374–1382.

81. Hofmann, D.J., and S. Solomon. 1989. Ozone destruction through heterogeneous chemistry following the eruption of El Chichón. J. Geophys. Res. **94**:5029–5041.

82. Houghton, R.A. 1991. Tropical deforestation and atmospheric carbon dioxide. Climatic Change **19**:99–118.

83. Houghton, R.A., R.D. Boone, J.M. Melillo, C.A. Palm, G.M. Woodwell, N. Myers, B. Moore, III, and D.L. Skole. 1985. Net flux of carbon dioxide from tropical forests in 1980. Nature **316**:617–620.

84. James, S.R. 1989. Hominid use of fire in the Lower and Middle Pleistocene: A review of the evidence. Current Anthropology **30**:1–26.

85. Johansson, C., H. Rodhe, and E. Sanhueza. 1988. Emission of NO in a tropical savannah and a cloud forest during the dry season. J. Geophys. Res. **93**:7180–7192.

86. Jones, R. 1979. The fifth continent: problems concerning human colonisation of Australia. Ann. Rev. Anthropology **8**:445–466.

87. Joshi, V. 1991. Biomass burning in India. In J.S. Levine (ed.), Global Biomass Burning: Atmospheric, Climatic, and Biospheric Implications, MIT Press, Cambridge, MA, pp. 185–193.

88. Kaufman, Y.J., C.J. Tucker, and I. Fung. 1990. Remote sensing of biomass burning in the tropics. J. Geophys. Res. **95**:9927–9939.

89. Keller, M., D.J. Jacob, S.C. Wofsy, and R.C. Harriss. 1991. Effects of tropical deforestation on global and regional atmospheric chemistry. Climatic Change **19**: 139–158.

90. Khalil, M.A.K., and R.A. Rasmussen. 1984. Global sources, lifetimes and mass balances of carbonyl sulfide (OCS) and carbon disulfide (CS_2) in the earth's atmosphere. Atmos. Environ. **18**:1805–1813.

91. Kilgore, B.M., and D. Taylor. 1979. Fire history of a sequoia-mixed conifer forest. Ecology **60**:129–142.

92. Kirchhoff, V.W.J.H., I.M.O. daSilva, and E.V. Browell. 1990. Ozone measurements in Amazonia: Dry season versus wet season. J. Geophys. Res. **95**:16913–16926.

93. Kirchhoff, V.W.J.H., E.V.A. Marinho, P.L.S. Dias, R. Calheiros, R. Andre, and C. Volpe. 1989. 03 and CO from burning sugar cane. Nature **339**:264.

94. Kirchhoff, V.W.J.H., and C.A. Nobre. 1986. Atmospheric chemistry research in Brazil: Ozone measurements at Natal, Manaus, and Cuiabá. Revista Geofisica **24**:95–108.

95. Kirchhoff, V.W.J.H., and R.A. Rasmussen. 1990. Time variations of CO and O_3 concentrations in a region subject to biomass burning. J. Geophys. Res. **95**: 7521–7532.

96. Kuhlbusch, T.A., J.M. Lobert, P.J. Crutzen, and P. Warneck. 1991. Molecular nitrogen emissions from denitrification during biomass burning. Nature **351**:135–137.

97. Lacaux, J.P., R. Delmas, G. Kouadio, B. Cros, and M.O. Andreae. 1992. Precipitation chemistry in the Mayombe Forest of equatorial Africa. J. Geophys. Res. **97**:6195–6206.

98. Lacaux, J.P., J. Servant, and J.G.R. Baudet. 1987. Acid rain in the tropical forests of the Ivory Coast. Atmos. Environ. **21**:2643–2647.

99. Lacaux, J.P., J. Servant, M.L. Huertas, B. Cros, R. Delmas, J. Loemba-Ndembi, and M.O. Andreae. 1988. Precipitation chemistry from remote sites in the African equatorial forest. Eos Trans. AGU **69**:1069.

100. Lacey, C.J., J. Walker, and I.R. Noble. 1982. Fire in Australian tropical savannas. In B.J. Huntley and B.H. Walker (eds.), Ecology of Tropical Savannas, Springer-Verlag, Berlin, pp. 246–272.

101. Langkamp, P.J., and M.J. Dalling. 1983. Nutrient cycling in a stand of Acacia holoserica. Aust. J. Bot. **31**:141–149.

102. Lanly, J.P. 1982. Tropical forest resources. FAO Forestry Paper No. 30. Food and Agriculture Organization, Rome.

103. Lanly, J.P. 1985. Defining and measuring shifting cultivation. Unasylva **37**:17–21.

104. Lean, J., and D.A. Warrilow. 1989. Simulation of the regional climatic impact of Amazon deforestation. Nature **342**:411–413.

105. Levine, J.S., W.R. Cofer, III, D.I. Sebacher, R.P. Rhinehart, E.L. Winstead, S. Sebacher, C.R. Hinkle, P.A. Schmalzer, and A.M. Koller, Jr. 1990. The effect of fire on biogenic emissions of methane and nitric oxide from wetlands. J. Geophys. Res. **95**:1853–1864.

106. Levine, J.S., W.R. Cofer, III, D.I. Sebacher, E.L. Winstead, S. Sebacher, and P.J. Boston. 1988. The effects of fire on biogenic soil emissions of nitric oxide and nitrous oxide. Global Biogeochem. Cycles **2**:445–449.

107. Li, J.-N. 1991. Population effects on deforestation and soil erosion in China. In K. Davis and M.S. Bernstam (eds.), Resources, Environment, and Population:

Present Knowledge, Future Options, Oxford University Press, New York, pp. 254–258.

108. Likens, G.E., F.H. Bormann, R.S. Pierce, J.S. Eaton, and R.E. Munn. 1984. Long-term trends in precipitation chemistry at Hubbard Brook, New Hampshire. Atmos. Environ. **18**:2641–2647.

109. Lobert, J.M. 1989. Verbrennung pflanzlicher Biomasse als Quelle atmosphärischer Spurengase: Cyanoverbindungen, CO, CO_2 und NO_x. Ph.D. thesis, Mainz, Germany.

110. Lobert, J.M., D.H. Scharffe, W.M. Hao, T.A. Kuhlbusch, R. Seuwen, P. Warneck, and P.J. Crutzen. 1991. Experimental evaluation of biomass burning emissions: Nitrogen and carbon containing compounds. In J.S. Levine (ed.), Global Biomass Burning, MIT Press, Cambridge, MA, pp. 289–304.

111. Logan, J.A. 1983. Nitrogen oxides in the troposphere: Global and regional budgets. J. Geophys. Res. **88**:10785–10807.

112. Logan, J.A. 1985. Tropospheric ozone: Seasonal behavior, trends, and anthropogenic influence. J. Geophys. Res. **90**:10463–10482.

113. Logan, J.A., and V.W.J.H. Kirchhoff. 1986. Seasonal variations of tropospheric ozone at Natal, Brazil. J. Geophys. Res. **91**:7875–7881.

114. Luizão, F., P.A. Matson, G.P. Livingston, R. Luizão, and P.M. Vitousek. 1989. Nitrous oxide flux following tropical land clearing. Global Biogeochem. Cycles **3**:281–285.

115. Matson, P.A., P.M. Vitousek, G.P. Livingston, and N.A. Swanberg. 1988. Nitrous oxide flux from Amazon ecosystems: Fertility and disturbance effects. Eos Trans. AGU **69**:320.

116. McDowell, W.H. 1988. Potential effects of acid deposition on tropical terrestrial ecosystems. In H. Rodhe and R. Herrera (eds.), Acidification in tropical countries, J. Wiley and Sons, Chichester, pp. 117–139.

117. Menaut, J.-C. 1983. The vegetation of African savannas. In F. Bourliere (ed.), Ecosystems of the World 13: Tropical Savannas, Elsevier, Amsterdam, pp. 109–149.

118. Menaut, J.-C., L. Abbadie, F. Lavenu, P. Loudjani, and A. Podaire. 1991. Biomass burning in West African savannas. In J.S. Levine (ed.), Global Biomass Burning: Atmospheric, Climatic, and Biospheric Implications, MIT Press, Cambridge, MA, pp. 133–142.

119. Miner, S. 1969. Preliminary air pollution survey of ammonia, Report APTD-69-25. National Air Pollution Control Administration, Raleigh, NC.

120. Myers, N. 1980. Conversion of Tropical Moist Forests, National Academy of Sciences, Washington, DC.

121. Myers, N. 1989. Deforestation Rates in Tropical Forests and their Climatic Implications, Friends of the Earth, London.

122. Oppenheimer, M. 1987. Stratospheric sulphate production and the photochemistry of the Antarctic circumpolar vortex. Nature **328**:702–704.

123. Patterson, E.M., and C.K. McMahon. 1984. Absorption characteristics of forest fire particulate matter. Atmos. Environ. **18**:2541–2551.

124. Penner, J.E., S.J. Ghan, and J.J. Walton. 1991. The role of biomass burning in the budget and cycle of carbonaceous soot aerosols and their climate impact. In J.S. Levine (ed.), Global Biomass Burning: Atmospheric, Climatic and Biospheric Implications, MIT Press, Cambridge, MA, pp. 387–393.

125. Peterson, J.T., and C.E. Junge. 1971. Sources of particulate matter in the atmosphere. In W.H. Matthews, W.W. Kellogg, and G.D. Robinson (eds.), Man's Impact on the Climate, MIT Press, Cambridge, MA, pp. 310–320.

126. Pyne, S.J. 1982. Fire in America: A Culture History of Wildland and Rural Fire, Princeton University Press, Princeton, NI.

127. Pyne, S.J. 1991. Burning Bush: A Fire History of Australia, Henry Holt and Co., New York.

128. Radke, L.F., D.A. Hegg, J.H. Lyons, C.A. Brock, P.V. Hobbs, R. Weiss, and R. Rasmussen. 1988. Airborne measurements on smokes from biomass burning. In P.V. Hobbs and M.P. McCormick (eds.), Aerosols and Climate, A. Deepak Publishing, pp. 411–422.

129. Radke, L.F., J.L. Stith, D.A. Hegg, and P.V. Hobbs. 1978. Airborne studies of particles and gases from forest fires. J. Air Pollut. Contr. Assoc. **28**:30–34.

130. Ramanathan, V., R.J. Cicerone, H.B. Singh, and J.T. Kiehl. 1985. Trace gas trends and their potential role in climate change. J. Geophys. Res. **90**:5547–5566.

131. Rasmussen, R.A., L.E. Rasmussen, M.A.K. Khalil, and R.W. Dalluge. 1980. Concentration distribution of methyl chloride in the atmosphere. J. Geophys. Res. **85**:7350–7356.

132. Reichle, H.G., Jr., V.S. Conners, J.A. Holland, W.D. Hypes, H.A. Wallio, J.C. Casas, B.B. Gormsen, M.S. Saylor, and W.D. Hasketh. 1986. Middle and upper tropospheric carbon monoxide mixing ratios as measured by a satellite borne remote sensor during November 1981. J. Geophys. Res. **91**:10865–10888.

133. Reichle, H.G., Jr., V.S. Connors, J.A. Holland, R.T. Sherrill, H.A. Wallio, J.C. Casas, E.P. Condon, B.B. Gormsen, and W. Seiler. 1990. The distribution of middle tropospheric carbon monoxide during early October 1984. J. Geophys. Res. **95**:9845–9856.

134. Robertson, G.P., and J.M. Tiedje. 1988. Deforestation alters denitrification in a lowland tropical rain forest. Nature **336**:756–759.

135. Robock, A. 1988. Enhancement of surface cooling due to forest fire smoke. Science **242**:911–913.

136. Robock, A. 1991. Surface cooling due to smoke from biomass burning. In J.S. Levine (ed.), Global Biomass Burning: Atmospheric, Climatic and Biospheric Implications, MIT Press, Cambridge, MA, pp. 463–476.

137. Rogers, C.F., J.G. Hudson, B. Zielinska, R.L. Tanner, J. Hallett, and J.G. Watson. 1991. Cloud condensation nuclei from biomass burning. In J.S. Levine (ed.), Global Biomass Burning: Atmospheric, Climatic and Biospheric Implications, MIT Press, Cambridge, MA, pp. 431–438.

138. Rosswall, T. 1981. The biogeochemical nitrogen cycle. In G.E. Likens (ed.), Some Perspectives of the Major Biogeochemical Cycles (SCOPE 17), J. Wiley and Sons, Chichester, pp. 25–50.

139. Sanborn, P.T., and T.M. Ballard. 1991. Combustion losses of sulphur from conifer foliage: Implications of chemical form and soil nitrogen status. Biogeochemistry **12**:129–134.

140. Sanchez, P.A. 1976. Properties and Management of Soils in the Tropics, J. Wiley, New York.

141. Sanford, R.L., Jr., J. Saldarriaga, K.E. Clark, C. Uhl, and R. Herrera. 1985. Amazon rain-forest fires. Science **227**:53–55.

142. Sanhueza, E., G. Cuenca, M.J. Gomez, R. Herrera, C. Ishizaki, I. Marti, and J. Paolini. 1988. Characterization of the Venezuelan environment and its potential for acidification. In H. Rodhe and R. Herrera (eds.), Acidification in Tropical Countries (SCOPE 36), J. Wiley and Sons, Chichester, pp. 197–225.

143. Sanhueza, E., W. Elbert, A. Rondon, M. Corina Arias, and M. Hermoso. 1989. Organic and inorganic acids in rain from a remote site of the Venezuelan savannah. Tellus **41B**:170–176.

144. Sanhueza, E., W.M. Hao, D. Scharffe, L. Donoso, and P.J. Crutzen. 1990. N_2O and NO emissions from soils of the northern part of the Guayana Shield, Venezuela. J. Geophys. Res. **95**:22481–22488.

145. Sarmiento, G., and M. Monasterio. 1975. A critical consideration of the environmental conditions associated with the occurrence of savanna ecosystems in tropical America. In F.B. Golley and E. Medina (eds.), Tropical Ecological Systems, Springer-Verlag, Berlin, pp. 223–250.

146. Schmidt, U. 1975. Molecular hydrogen in the atmosphere. Tellus **27**:93.

147. Schüle, W. 1990. Landscapes and climate in prehistory: Interactions of wildlife, man and fire. In J. Goldammer (ed.), Fire in the Tropical Biota: Ecosystem Processes and Global Challenges, Springer-Verlag, Berlin, pp. 273–318.

148. Scurlock, J.M.O., and D.O. Hall. 1990. The contribution of biomass to global energy use (1987). Biomass **21**:75–81.

149. Seiler, W., and P.J. Crutzen. 1980. Estimates of gross and net fluxes of carbon between the biosphere and the atmosphere from biomass burning. Climatic Change **2**:207–247.

150. Seiler, W., and J. Fishman. 1981. The distribution of carbon monoxide and ozone in the free troposphere. J. Geophys. Res. **86**:7255–7266.

151. Setzer, A.W., and M.C. Pereira. 1991. Amazon biomass burnings in 1987 and their tropospheric emissions. Ambio **20**:19–22.

152. Shukla, J., C. Nobre, and P. Sellers. 1990. Amazon deforestation and climate change. Science **247**:1322–1325.

153. Steinhardt, U., and H.W. Fassbender. 1979. Características y composición química de las lluvias de los Andes occidentales de Venezuela. Turrialba **29**: 175–182.

154. Stott, P.A., J.G. Goldammer, and W.L. Werner. 1990. The role of fire in the tropical lowland deciduous forests of Asia. In J.G. Goldammer (ed.), Fire in the Tropical Biota: Ecosystem Processes and Global Challenges, Springer-Verlag, Berlin, pp. 32–44.

155. Suman, D.O. 1984. The production and transport of charcoal formed during agricultural burning in central Panama. Interciencia **9**:311–313.

156. Suman, D.O. 1988. The flux of charcoal to the troposphere during the period of agricultural burning in Panama. J. Atmos. Chem. **6**:21–34.

157. Susott, R.A., D.E. Ward, R.E. Babbitt, D.J. Latham, L.G. Weger, and P.M. Boyd. 1990. Fire Dynamics and Chemistry of Large Fires (Final Report). USDA Forest Service, Missoula, MT.

158. Swetnam, T.W., and J.L. Betancourt. 1990. Fire-Southern Oscillation relations in the southwestern United States. Science **24**:1017–1020.

159. Talbot, R.W., K.M. Beecher, R.C. Harriss, and W.R. Cofer, III. 1988. Atmospheric geochemistry of formic and acetic acids at a mid-latitude temperate site. J. Geophys. Res. **93**:1638–1652.

160. Tande, G.F. 1979. Fire history and vegetation pattern of coniferous forests in Jasper National Park, Alberta, Canada. Canadian J. Botany **57**:1912–1931.

161. Turco, R.P., O.B. Toon, R.C. Whitten, J.B. Pollack, and P. Hamill. 1983. The global cycle of particulate elemental carbon: A theoretical assessment. In H.R. Pruppacher, R.G. Semonin, and W.G.N. Slinn (eds.), Precipitation Scavenging, Dry Deposition, and Resuspension, Elsevier, New York, pp. 1337–1351.

162. Turner, J., and M.J. Lambert. 1980. Sulfur nutrition in forests. In D.S. Shriner, C.R. Richmond, and S.E. Lindberg (eds.), Atmospheric Sulfur Deposition: Environmental Impact and Health Effects, Ann Arbor Science Publishers, Ann Arbor, MI, pp. 321–333.

163. Vitousek, P.M. 1991. Trace gas exchanges in agricultural ecosystems. Paper presented at the 10th International Symposium on Environmental Biogeochemistry, San Francisco, CA, 19–23 August 1991.

164. Ward, D.E. 1986. Field Scale Measurements of Emission from Open Fires. Technical Paper presented at the Defense Nuclear Agency Global Effects review. Defense Nuclear Agency, Washington, DC.

165. Ward, D.E., and C.C. Hardy. 1991. Smoke emissions from wildland fires. Environ. Internat. **17**:117–134.

166. Ward, D.E., C.K. McMahon, and D.F. Adams. 1982. Laboratory measurements of carbonyl sulfide and total sulfur emissions from open burning of forest biomass. Paper presented at the 75th Annual Meeting of the Air Pollution Control Association, New Orleans, LA, 20–25 June 1982, 1–16.

167. Ward, D.E., A.W. Setzer, Y.J. Kaufman, and R.A. Rasmussen. 1991. Characteristics of smoke emissions from biomass fires of the Amazon region—BASE-A experiment. In J.S. Levine (ed.), Global Biomass Burning: Atmospheric, Climatic, and Biospheric Implications, MIT Press, Cambridge, MA, pp. 394–402.

168. Ward, D.E., R.A. Susott, R.E. Babbitt, and C.C. Hardy. 1990. Properties and concentration of smoke near the ground from biomass field tests. Paper presented at the Symposium on Smoke/Obscurants XIV; 17–19 April 1990, Laurel, MD.

169. Watson, C.E., J. Fishman, and H.G. Reichle, Jr. 1990. The significance of biomass burning as a source of carbon monoxide and ozone in the Southern Hemisphere tropics: A satellite analysis. J. Geophys. Res. **95**:16443–14450.

170. Whittaker, R.H., and G.E. Likens. 1975. The biosphere and man. In H. Lieth and R. Whittaker (eds.), Primary Productivity of the Biosphere, Springer-Verlag, Berlin, pp. 325–328.

171. WMO. 1985. Atmospheric Ozone. Global Ozone Research and Monitoring Project—Report No. 16. World Meteorological Organization, Geneva.

172. Wofsy, S.C., G.W. Sachse, G.L. Gregory, D.R. Blake, J.D. Bradshaw, S.T. Sandholm, H.B. Singh, J.A. Barrick, R.C. Harriss, R.W. Talbot, M.A. Shipham, E.V. Browell, D.J. Jacob, and J.A. Logan. 1992. Atmospheric chemistry in the Arctic and Subarctic: Influence of natural fires, industrial emissions, and stratospheric inputs. J. Geophys. Res. **97**:16,731–16,746.

173. Wolbach, W.S., I. Gilmour, E. Anders, C.J. Orth, and R.R. Brooks. 1988. Global fire at the Cretaceous–Tertiary boundary. Nature **334**:665–669.

174. Wolbach, W.S., R.S. Lewis, and E. Anders. 1985. Cretaceous extinctions: Evidence for wildfires and search for meteoritic material. Science **230**:167–170.

175. Wood, C.A., and R. Nelson. 1991. Astronaut observations of global biomass burning. In J.S. Levine (ed.), Global Biomass Burning: Atmospheric, Climatic and biospheric Implications, MIT Press, Cambridge, MA, pp. 29–40.

Long-Term Effects of Fossil-Fuel-Burning and Deforestation on Levels of Atmospheric CO_2

James F.D. Kasting

Department of Geosciences, Penn State University, University Park, Pennsylvania 16802

James C.G. Walker

Space Physics Research Laboratory, Department of Atmospheric, Oceanic, and Space Science, and Department of Geological Sciences, University of Michigan, Ann Arbor, Michigan 48109

ABSTRACT

Atmospheric CO_2 levels are currently increasing as a consequence of the burning of fossil fuels and deforestation of tropical lands. Here, we present numerical simulations of atmospheric CO_2 concentrations over the next few hundreds to hundreds of thousands of years for various patterns of fossil-fuel consumption and land use. The computer model predicts that atmospheric CO_2 concentrations could exceed 2000 ppm (parts per million) within the next few centuries if we consume most of the available fossil fuel and if the current trend toward deforestation continues. This prediction is relatively insensitive to the future rate of fossil-fuel-burning unless the consumption rate is several times lower than the present rate. Conserving existing forests or planting new ones might lower the peak CO_2 level by a factor of 2 if CO_2 fertilization of plant growth is effective in natural settings and if soil carbon storage does not decrease as the global climate warms. The model also predicts that most of the anthropogenic CO_2 will be removed within the next 5000 to 10,000 years, but that more than a million years will pass before atmospheric CO_2 returns to its original, preindustrial value.

Introduction

The potential effect of fossil-fuel-burning and deforestation on the Earth's climate is a matter of both scientific and popular interest. Numerous theoretical studies have addressed the questions of how much the climate might warm if atmospheric CO_2 doubles and of how fast this change might occur. Most of these studies have focused on the next 50 to 100 years, as this is the time frame of greatest immediate concern. Several attempts have been made to project CO_2 levels farther into the future,[1,4,5,13,21,23] but these appear to have had little impact outside of the immediate, carbon-cycle-modeling community.

Limiting our thinking to the next century is dangerous because the time constant for CO_2 removal is much longer than 100 years and because the amount of CO_2 that might ultimately be generated from fossil-fuel-burning and deforestation is considerably larger than that needed to double the atmospheric concentration. At the current mixing ratio of 350 ppm, the atmosphere contains some 6×10^{16} moles of CO_2. Recoverable fossil-fuel reserves as of 1982[5] totaled 35×10^{16} moles C; 85% of this is coal and lignite, with the rest divided between petroleum, natural gas, oil shale, and tar sands. The living biosphere contains $\sim 5 \times 10^{16}$ moles C and humus in soils accounts for $\sim 13 \times 10^{16}$ moles C.[12] If all of this fossil and biospheric carbon were oxidized instantaneously, and if half the resulting CO_2 dissolved in the oceans, as is approximately true of past anthropogenic CO_2 emissions,[5] atmospheric CO_2 would increase by a factor of ~ 5.5.

The actual problem of projecting future atmospheric CO_2 levels is, of course, quite complex. The atmospheric concentration depends on the rate of fossil-fuel consumption, the rate of deforestation or reforestation, the effect of increased CO_2 concentrations and climate change on existing ecosystems, and the rate of CO_2 uptake by the oceans and sediments. Future rates of fossil-fuel consumption are difficult to predict, so the CO_2 production rate is uncertain. The current rate of fossil-fuel consumption[22] is 5×10^{14} moles C yr^{-1}; this would be sufficient to exhaust the fossil-fuel reserves in ~ 700 years. The rate at which CO_2 is being released from deforestation is thought to lie in the range of $(0.3–2) \times 10^{14}$ moles yr^{-1},[12] or 6–40% of the fossil-fuel combustion rate.

The loss rate for CO_2 is also uncertain because it is affected by a number of different processes that interact on a variety of time scales. These processes deserve some elaboration because they are in general poorly understood. The next section discusses these loss processes and shows how they fit into the natural, long-term, CO_2 geochemical cycle.

CO_2 Loss Processes

The major processes that can remove anthropogenic CO_2 from the atmosphere have been described by Broecker and Peng.[5] Atmospheric CO_2 equilibrates rapidly (in 10 years or less) with dissolved carbonate species [$(CO_2)_{aq}$, HCO_3^-, and $CO_3^=$] in the surface ocean. However, the capacity of the surface ocean to absorb CO_2 is rather small. Most of the uptake occurs by way of the equilibrium reaction

(1) $$CO_2 + CO_3^{=} + H_2O \rightleftharpoons 2HCO_3^-.$$

Given an average depth of 75 m and an average carbonate ion concentration of 200 μmoles/l,[5] the surface ocean can absorb only about 0.5×10^{16} moles of added CO_2, or ~1.5% of the estimated fossil-fuel inventory. The deep ocean, which has a lower $CO_3^{=}$ concentration but a larger volume, has a much greater capacity for CO_2 uptake[5]: 15×10^{16} moles, or roughly 40% of the fossil reservoir. The deep ocean mixes slowly, however; over a thousand years is required to fully utilize the buffering capacity provided by reaction (1).

Still greater amounts of injected CO_2 can be removed by weathering of carbonate rocks on land or by reaction with carbonate sediments on the ocean floor. Both processes can be represented chemically as

(2) $$CO_2 + CaCO_3 + H_2O \rightleftharpoons Ca^{++} + 2HCO_3^-.$$

Dolomite, $CaMg(CO_3)_2$, can be dissolved by an analogous reaction. It is generally believed that dissolution of sea-floor carbonates will provide the greatest amount of CO_2 buffering over the next few centuries. The uppermost 10 cm of marine sediments (the depth to which burrowing organisms penetrate) contains some 8×10^{16} moles of calcium carbonate.[5] This figure increases to 40×10^{16} moles if one includes the carbonate that could eventually be exposed by bioturbation.[5] (We use a somewhat higher figure, 70×10^{16} moles, in our model,[26] but only a portion of this carbonate ever dissolves.) Thus, in principle, dissolution of carbonate sediments could neutralize all of the CO_2 produced by fossil-fuel-burning. The minimum time required for this to happen is several thousand years since corrosive, CO_2-enriched water must circulate over the sediments several times before they are likely to dissolve.[5]

Reactions (1) and (2) remove CO_2 from the atmosphere by storing it in the oceans as bicarbonate ion. To balance the CO_2 cycle, some process must remove bicarbonate from the oceans; otherwise, volcanic outgassing would cause the total amount of CO_2 stored in the combined atmosphere/ocean system to increase with time. The rate of volcanic outgassing of CO_2 is estimated at $(5-11) \times 10^{12}$ moles yr^{-1},[2,3,10,26] or 50–100 times slower than the current rate of fossil-fuel-burning. The process that removes most of the volcanic CO_2 (and, hence, prevents bicarbonate ion from accumulating in the oceans) is the weathering of silicate rocks on the continents, followed by the deposition of carbonate sediments on the sea floor. If one represents all silicate rocks by the mineral wollastonite ($CaSiO_3$), the net result of these two processes can be written as

(3) $$CaSiO_3 + CO_2 \rightarrow CaCO_3 + SiO_2.$$

The time scale for silicate weathering, that is, the residence time of carbon in the combined atmosphere/ocean system, is approximately 400,000 years.[3]

CO_2 is also removed from the atmosphere/ocean system by the organic carbon cycle. Roughly 20% of the carbon in modern marine sediments is organic.[15] We assume that this loss of organic carbon from the oceans is exactly balanced by an influx of organic carbon from weathering of kerogen in rocks; thus, the influence of the marine organic carbon cycle is minimal in our model. Burial of marine organic carbon is unlikely to play a major role in removing anthropogenic CO_2 unless ocean fertilization techniques (e.g., Ref. 16) can increase the burial rate dramatically.

The terrestrial biosphere could also take up CO_2 if the total amount of carbon stored in forests or soils were to increase significantly. Indeed, a recent study of the global CO_2 budget by Tans et al.[24] suggests that such a phenomenon may be occurring at present (although obviously not in tropical regions that are being deforested). One reason that carbon storage in undisturbed forests might be increasing is that C3 plants, which include most trees, grow faster under enhanced CO_2 levels. For C3 plants grown under controlled, greenhouse conditions, the relative growth rate can be approximated by the function[8,26]

(4) $$F(CO_2) = 2.22(1 - \exp[-0.003(pCO_2 - 80)]).$$

This function has been adjusted to yield a relative growth rate of unity at the preindustrial CO_2 level of 280 ppm. Whether this same fertilization function applies to natural ecosystems, where factors other than CO_2 availability may limit the growth rate, is a matter of debate.

Even if CO_2 fertilization is the missing sink for carbon,[24] this does not necessarily imply that forests will continue to be a net sink for CO_2 in the future because other factors, e.g., the effect of climatic warming on soil carbon storage,[12] may counterbalance this effect. We show below that predicting how terrestrial ecosystems will respond to increased CO_2 concentrations is essential to understanding how climate may change over the next few centuries.

A Model Calculation of CO_2 Uptake

The CO_2 uptake processes described in the previous section have been incorporated into a computer model of the global carbon cycle.[26] This model is unique in that it is the only model to date that incorporates the long-term aspects of the geochemical carbon cycle (i.e., rock weathering); hence, it is the only model that is capable of tracking the fossil-fuel CO_2 pulse from beginning to end. It should be emphasized from the start that the detailed predictions of this, or any other, model are uncertain because the CO_2 uptake processes are difficult to parametrize and because the computer models are, at best, highly simplified approximations to the real atmosphere/ocean system. Nonetheless, the basic nature of the predictions should be reasonably robust because the results depend to a large extent on relative reservoir sizes and approximate transfer times, both of which are fairly well-understood.

The Walker and Kasting model[26] consists of a set of eight boxes representing the atmosphere, terrestrial biosphere (forests plus soils), warm and cold surface ocean, thermocline, and three deep ocean reservoirs (Figure 1). The independent variables tracked by the model include atmospheric CO_2 concentration, biospheric carbon content, and five quantities in each ocean box: total dissolved inorganic carbon, alkalinity, dissolved phosphorus, ^{13}C, and ^{14}C. Dissolved

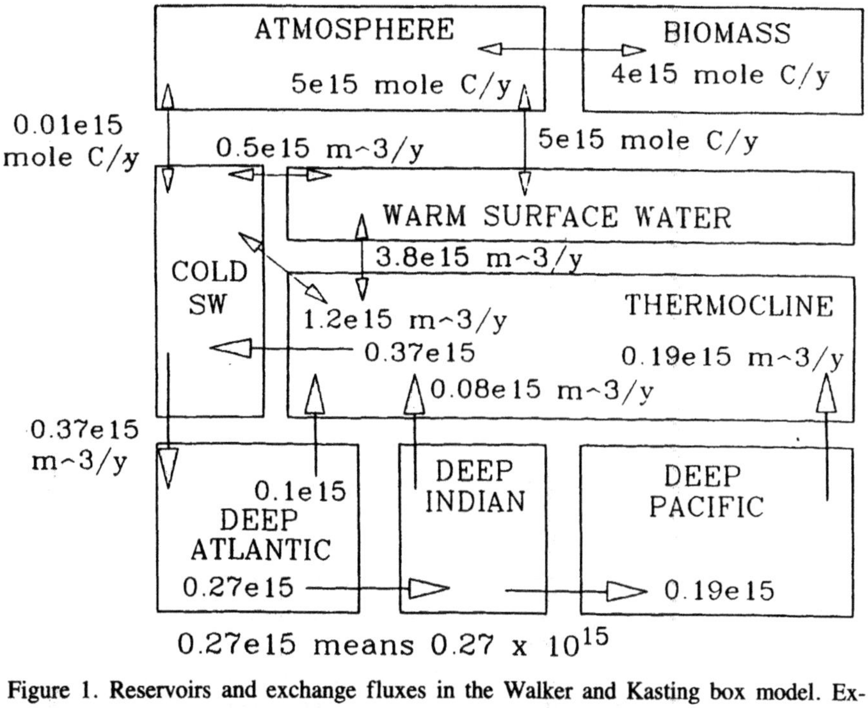

Figure 1. Reservoirs and exchange fluxes in the Walker and Kasting box model. Exchange fluxes between the various ocean boxes are given in cubic meters per year; fluxes between the atmosphere and other reservoirs are in moles of carbon per year. 1e10 means 1×10^{10}; m^3 means m^3.

phosphorus is assumed to control the rate of marine productivity in the model. The carbon isotopes are included as tracers of productivity and oceanic circulation.

In addition to the circulation fluxes shown in Figure 1, the computer model includes a flux of particulate matter from the warm surface ocean to the thermocline and deep ocean basins. The particles, which represent the remains of marine organisms, consist of a mixture of organic carbon and calcium carbonate. Most (>90%) of these particles dissolve (or are oxidized) as they descend through the water column. As such, they constitute a "biological pump" that tends to remove dissolved CO_2 from surface water and to concentrate it in the deep ocean. Some fraction of the calcium carbonate in the descending particles survives the passage through the ocean and is buried in sediments. The amount buried depends on the carbonate saturation depth which, in this model, represents the boundary above which calcite ($CaCO_3$) is stable and below which it dissolves. The area of ocean floor above the carbonate saturation depth is estimated from the calculated carbonate ion content of deep water, the solubility product of calcite, and the average hypsommetry of the ocean basins.[17] In the real ocean, a transition zone is observed over which the extent of dissolution of $CaCO_3$ increases with depth and the percentage of $CaCO_3$ in sediments decreases. That complication is ignored in the model.

Finally, the model also includes a simplified representation of the rock cycle (Figure 2). The magnitude of the present carbonate and silicate weathering rates [reactions (2) and (3)] is estimated from the data of Meybeck.[18] The rate of carbonate weathering is assumed to vary linearly with pCO_2; the silicate weathering is assumed to vary as $pCO_2^{0.3}$. The CO_2 dependence of the silicate weathering rate is intermediate between that of Berner et al.[3] and Walker et al.[25] The assumption of a linear CO_2 dependence for carbonate weathering is made so as to maximize the rate of uptake of fossil-fuel CO_2. These weathering rate laws incorporate the idea that both surface temperature and possibly global runoff rates increase with increasing atmospheric CO_2, and that the combination of these factors tends to accelerate rates of chemical weathering.

After adjusting the oceanic circulation rates and particle fluxes to try to match the present system as closely as possible, the model was used to simulate the observed rise in atmospheric CO_2 during the past two centuries (Figure 3). The assumed anthropogenic CO_2 sources were consumption of fossil fuels (rate taken from Ref. 14) and deforestation of parts of the Northern Hemisphere and tropics (rate estimated from Refs. 6 and 11). The symbols in Figure 3 show measurements of CO_2; the solid curve shows atmospheric CO_2 concentrations projected by the model. The model reproduces the observed CO_2 trend reasonably well, but only when CO_2 fertilization [Eq. (4)] is taken into account. When this effect is omitted, the rate of CO_2 increase predicted by the model is much faster than observed. This "missing carbon" problem is common to all current carbon cycle

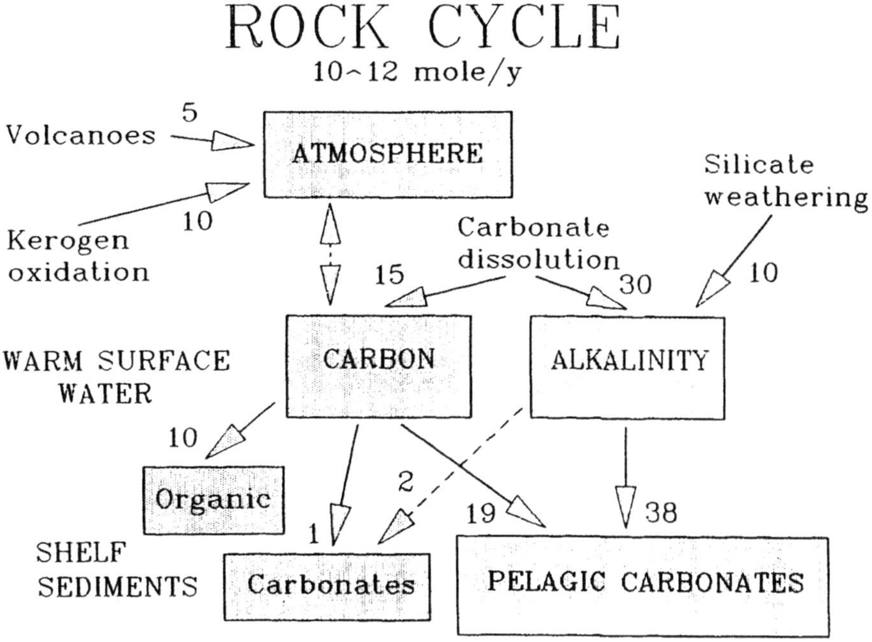

Figure 2. Schematic diagram of the long-term CO_2 geochemical cycle (i.e., the rock cycle). Arrows show fluxes of carbon in units of 10^{12} moles yr^{-1}.

models (e.g., Refs. 7 and 20); it mirrors the "missing sink" problem identified by Tans et al.[24] in their study of the global CO_2 budget.

It is admittedly arguable whether the missing carbon has been taken up by CO_2-fertilized plants. Let us assume for the moment, however, that the fertilization effect is real and operative. One can then use the model to predict future atmospheric CO_2 levels for different assumptions concerning rates of fossil-fuel consumption and patterns of deforestation/reforestation. Some possible scenarios are shown in Figure 4. In the "save fuel" scenario (bottom panel) the present rate of fossil-fuel-burning is assumed to continue for the next 700 years until the entire 35×10^{16} moles of carbon in the 1982 recoverable reserves are exhausted. The "burn fuel" scenario assumes that the same total amount of fossil fuel is consumed in less than 400 years. The "save forests" scenario (upper panel) assumes that deforestation ceases immediately and that the areal extent of forests remains constant into the indefinite future. The "burn forests" scenario assumes that the present trend of deforestation continues until only 30% of the original, preindustrial forest area remains. The quantity termed "biomass potential" shown in the upper panel is the area of forested land weighted by carbon storage, normalized to a value of unity for the preindustrial Earth.

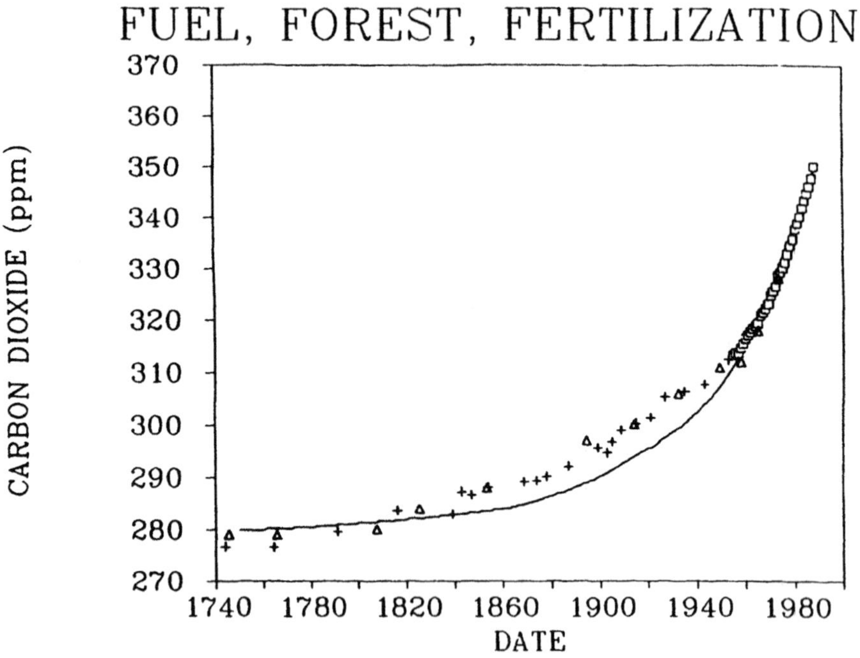

Figure 3. Comparison between observed CO_2 concentrations (represented by symbols) and CO_2 concentrations predicted by the Walker/Kasting box model. Squares represent CO_2 measurements from Mauna Loa (from Ref. 14); crosses and triangles represent ice core data (from Refs. 9, 19).

The various combinations of the above assumptions yield four different cases that were examined with the model (Figure 5). The most pessimistic (but, arguably, the most realistic) of these is represented by the scenario labeled "burn both," in which the fossil fuels are burned rapidly and the forests are assumed to disappear. Atmospheric CO_2 concentrations are predicted to reach ~2100 ppm in this case; the peak occurs in about the year 2350 AD when the fossil fuel reserves are nearly exhausted. Somewhat surprisingly, the "burn forests, save fuel" results are not tremendously different: atmospheric CO_2 peaks at ~2000 ppm in the year 2700. Burning the fossil fuels at a slower rate delays the CO_2 increase by 300–400 years but does little to diminish its eventual magnitude. Given a little additional thought, this result is not so surprising after all. In both cases, the time scale for CO_2 input (hundreds of years) is significantly shorter than the time scale for CO_2 removal (thousands to hundreds of thousands of years). An order-of-magnitude decrease in the present fossil-fuel consumption rate would be needed to significantly reduce the height of the atmospheric CO_2 peak.[26]

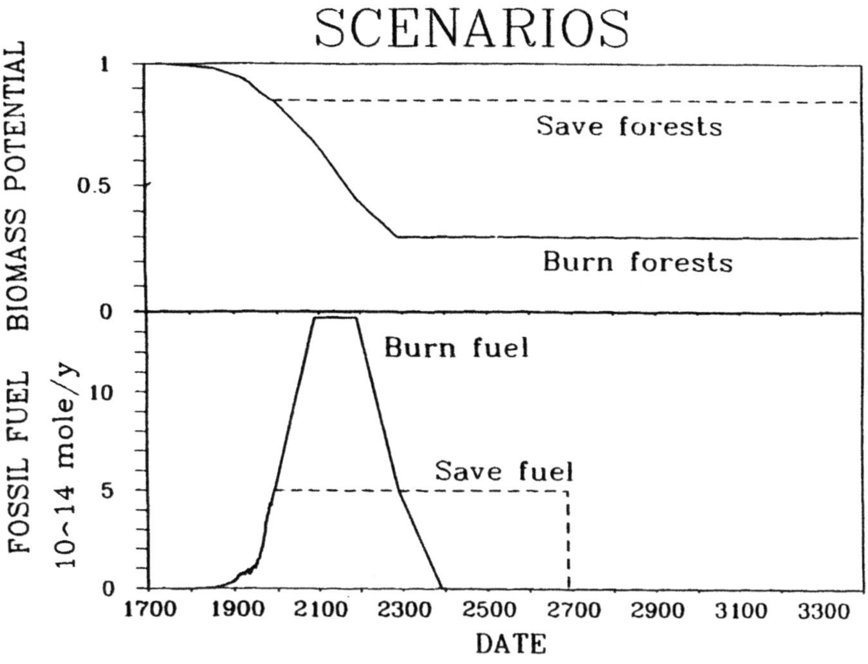

Figure 4. Hypothetical scenarios for future rates of fossil-fuel-burning and patterns of land use. The "save fuel" curve corresponds to burning the existing fossil-fuel reserves at the present rate. The total amount of fossil fuel consumed is the same in both cases.

What is remarkable in this model is the large reduction in future atmospheric CO_2 levels that results from preserving the forests. The two lower curves in Figure 5 indicate that the peak CO_2 level is reduced to ~1100 ppm in both the "burn fuel" and "save fuel" scenarios. Roughly 1000 ppm of CO_2, or three times the amount now present in the atmosphere, has been absorbed by the biomass reservoir as a consequence of CO_2 fertilization. This implies a doubling of the total amount of carbon stored in forests and soils. Whether this could actually occur without a significant increase in the area covered by forests is not clear. CO_2 fertilization is also assumed to apply in the "burn forests" scenarios, but its effect on atmospheric CO_2 is much smaller because of the smaller area of forested land.

Taken at face value, these results would seem to indicate that preserving the forests is more important in the long term than is conserving fossil fuel. This conclusion may be premature, however, for two reasons: First, as mentioned earlier, CO_2 fertilization may be less effective for natural ecosystems than it is in a controlled, greenhouse environment. Second, even if trees do grow faster in

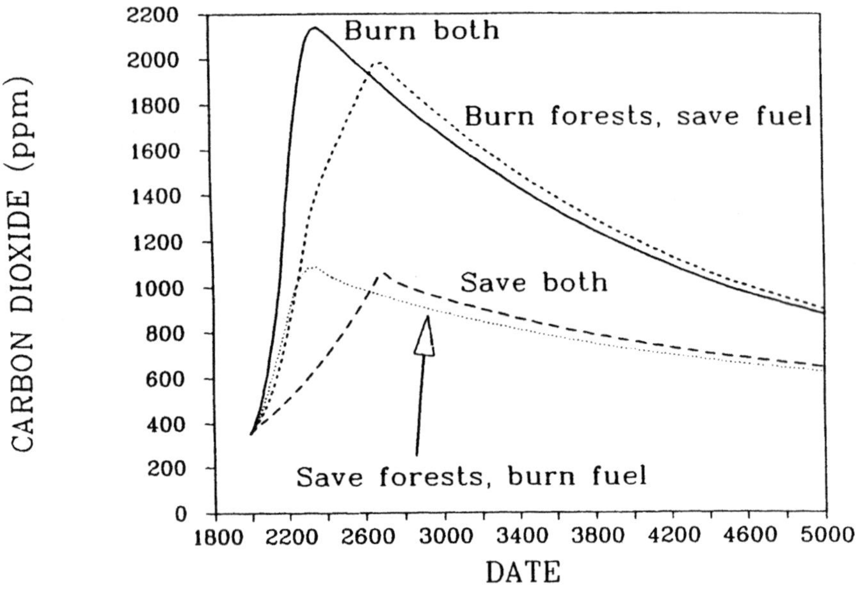

Figure 5. Atmospheric CO_2 concentrations predicted by the box model for the different scenarios shown in Figure 4. All of these calculations assume that CO_2 fertilization is real and will continue to affect growth rates in the future.

a high-CO_2 environment, it is not obvious that soil carbon storage will increase proportionately. Indeed, Houghton and Woodwell[12] have argued (convincingly, we believe) that soil carbon storage will likely decrease as the climate warms. Their prediction is based partly on the observation that tropical soils today contain much less carbon than do soils in temperate and boreal ecosystems. The reason is that organic material decays much faster in a warm climate. Thus, if global warming converts temperate ecosystems into subtropical ones, much of the organic carbon stored in these soils could be returned to the atmosphere.

This phenomenon is crudely simulated in the computer model by increasing the decay rate of biomass by a factor of 2 for a 10°C rise in surface temperature. The model predicts that global surface temperatures will rise about 2°C for each doubling of atmospheric CO_2; hence, the maximum temperature increase for the scenarios considered in Figure 5 is about 6°C. The corresponding increase in the decay rate is $2^{0.6}$, or a factor of roughly 1.5. Including this "enhanced decay" rate assumption in the model yields the results shown in Figure 6. The curves shown are for the "save forests, burn fuel" scenario; the terms "enhanced growth" and "constant growth" indicate whether or not CO_2 fertilization has been included in the calculation. It can be seen from Figure 6 that the beneficial effects of preserving the forests are greatly diminished when either the fertilization effect is eliminated or the decay rate is enhanced. We conclude that a better understanding of CO_2 fertilization and more detailed computer models are needed to determine whether terrestrial ecosystems can have a major impact on the uptake of fossil-fuel CO_2.

The long-term recovery of the system is shown in Figure 7. The solid curve in the bottom panel shows atmospheric CO_2 levels predicted for the pessimistic. "burn both" scenario of Figure 5. The other curves show the consequences of suppressing sea-floor carbonate dissolution (dashed curve) or assuming a weaker, square-root dependence of the carbonate weathering rate on atmospheric CO_2 (dotted curve). In each case, the bulk of the anthropogenic CO_2 is removed within a few thousand years, but the recovery to the preindustrial steady state requires over one million years. The relatively rapid decline of atmospheric CO_2 during the next few thousand years is a consequence of reactions (1) and (2); the much

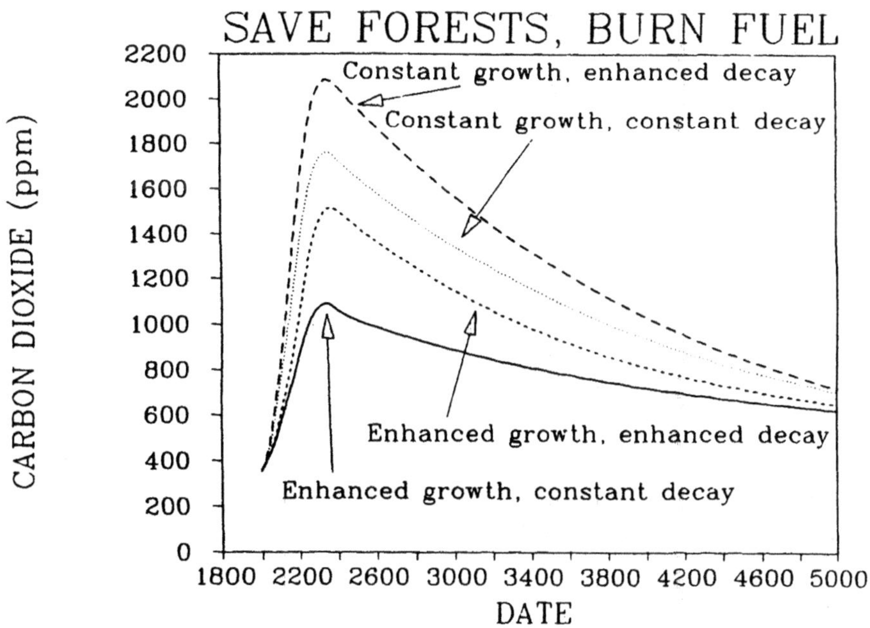

Figure 6. Predicted atmospheric CO_2 concentrations for the "save forests, burn fuel" scenario of Figure 5, but with various assumptions concerning CO_2 fertilization and soil carbon decay rates. The solid curve is the same as in Figure 5; the other curves show the effects of suppressing CO_2 fertilization and/or increasing the decay rate.

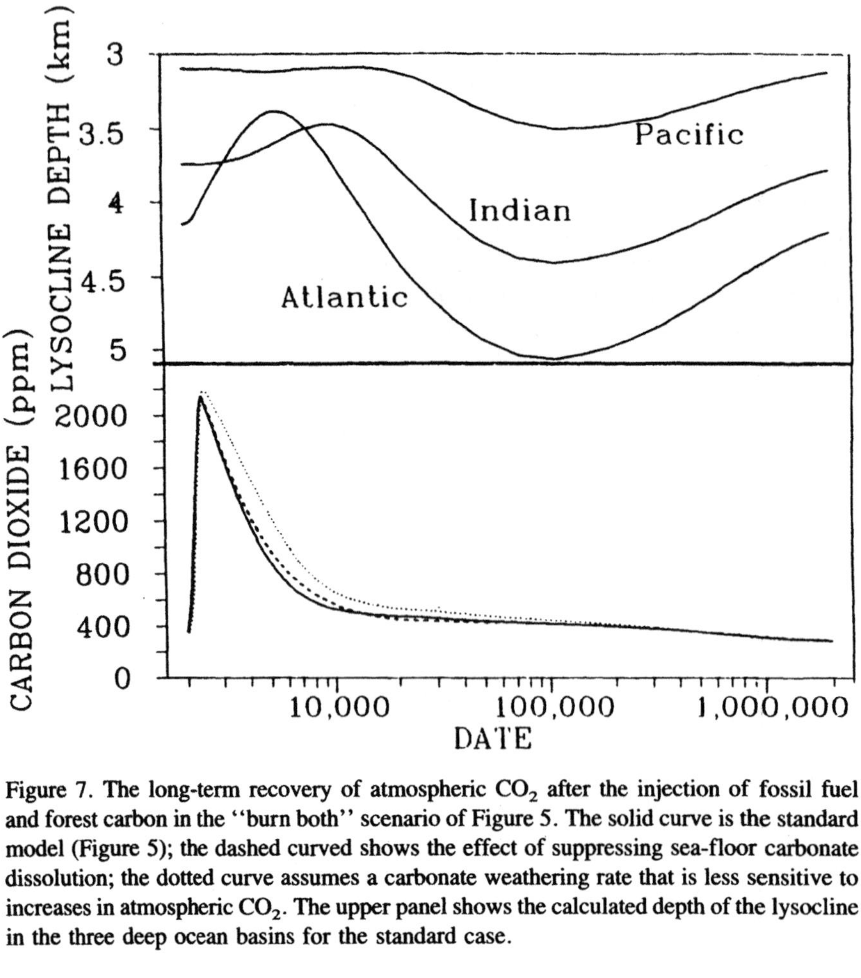

Figure 7. The long-term recovery of atmospheric CO_2 after the injection of fossil fuel and forest carbon in the "burn both" scenario of Figure 5. The solid curve is the standard model (Figure 5); the dashed curved shows the effect of suppressing sea-floor carbonate dissolution; the dotted curve assumes a carbonate weathering rate that is less sensitive to increases in atmospheric CO_2. The upper panel shows the calculated depth of the lysocline in the three deep ocean basins for the standard case.

slower decline beyond 10,000 AD is brought about by the weathering of silicate rocks on the continents [reaction (3)]. The calcium released by silicate weathering ultimately removes the excess fossil fuel CO_2 by sequestering it in carbonate sediments.

Conclusion

Atmospheric CO_2 concentrations are rising because humans are adding CO_2 to the atmosphere 50–100 times faster than the natural rate of input from volcanos.

The time scale over which the fossil-fuel reserves are likely to be consumed and forests chopped down is much shorter than the time required for CO_2 to be taken up by the oceans and sediments. Atmospheric CO_2 concentrations could increase by a factor of 6 or more over the next few centuries if current energy and land-use practices remain in place. Reducing fossil-fuel consumption by conserving energy and switching to alternative energy sources can reduce the rate of CO_2 buildup, but it will do little to diminish the eventual peak CO_2 concentration unless the CO_2 production rate is reduced to a small fraction of its present value. Preserving existing forests, or planting new ones, could reduce the peak CO_2 concentration by as much as a factor of 2, but only if CO_2 fertilization is effective and if soil carbon storage does not decrease as the climate warms. This suggests that reforestation efforts are worthwhile but that we should not count on reforestation to solve the global warming problem by itself. The CO_2 released by fossil-fuel-burning and deforestation will influence the Earth's climate for the next few hundreds of thousands of years. We should, therefore, be cautious about the size of the perturbation that we induce.

Acknowledgment

This work was supported in part by NASA Grant NAGW-176 to the University of Michigan.

References

1. Bacastrow, R.B., and A. Bjorkstrom. 1981. Comparison of ocean models for the carbon cycle. In B. Bolin (ed.), Carbon Cycle Modelling, SCOPE 16. J. Wiley, New York, pp. 29–79.
2. Berner, R.A. 1990. Atmospheric carbon dioxide levels over Phanerozoic time. Science **249**:1382–1386.
3. Berner, R.A., A.C. Lasaga, and R.M. Garrels. 1983. The carbonate–silicate geochemical cycle and its effect on atmospheric carbon dioxide over the past 100 million years. Amer. J. Sci. **283**:641–683.
4. Bolin, B., E.T. Degens, and P. Duvigneaud. 1979. Carbon Cycle Modelling. In B. Bolin, E.T. Degens, S. Kempe, et al. (eds.), The Global Carbon Cycle, SCOPE 13. J. Wiley, New York., pp. 1–28.
5. Broecker, W.S., and T.-H. Peng. 1982. Tracers in the Sea, Palisades, Lamont-Doherty Geological Observatory, New York.
6. Detwiler, R.P. and C.A.S. Hall. 1988. Tropical forests and the global carbon cycle. Science **239**:42–47.

7. Enting, I.G., and G.I. Pearman. 1986. The use of observations in calibrating and validating carbon cycle models. In J.R. Trabalka and D.E. Reichle (eds.), The Changing Carbon Cycle: A Global Analysis, Springer-Verlag, New York, pp. 425–458.
8. Esser, G. 1987. Sensitivity of global carbon cycle pools and fluxes to human and potential climatic impacts. Tellus **39B**:245–260.
9. Friedli, H., H. Lotscher, H. Oeschger, et al. 1986. Ice core record of the 13C/12C ratio of atmospheric CO_2 in the past two centuries. Nature **324**:237–238.
10. Holland, H.D. 1978. The Chemistry of the Atmosphere and Oceans, J. Wiley, New York.
11. Houghton, R.A., J.E. Hobbie, J.M. Melillo, et al. 1983. Changes in the carbon content of terrestrial biota and soils between 1860 and 1980: A net release of CO_2 to the atmosphere. Ecol. Monogr. **53**:235–262.
12. Houghton, R.A., and G.M. Woodwell. 1989. Global climatic change. Sci. Amer. **260**:36–44 (April).
13. Keeling, C.D., and R.B. Bacastrow. 1977. Impact of industrial gases on climate. In Energy and Climate: Studies in Geophysics. National Academy of Science, Washington, DC, pp. 72–95.
14. Keeling, C.D., R.B. Bacastrow, A.F. Carter, et al. 1989. A three-dimensional model of atmospheric CO2 transport based on observed winds. 1. Analysis of observational data. In D.H. Peterson (ed.), Aspects of Climate Variability in the Pacific and the Western Americas, American Geophysical Union, Washington, DC, pp. 165–236.
15. Lasaga, A.C., R.A. Berner, and R.M. Garrels. 1985. An improved model of atmospheric CO_2 fluctuations of the past 100 million years. In E.T. Sundquist and W.S. Broecker (eds.), The Carbon Cycle and Atmospheric CO2: Natural Variations Archean to Present, American Geophysical Union, Washington, DC, pp. 397–411.
16. Martin, J.H. 1990. Glacial-interglacial CO_2 change: the iron hypothesis. Paleoceanography **5**:1–13.
17. Menard, H.W., and S.M. Smith. 1966. Hypsommetry of ocean basin provinces. J. Geophys. Res. **71**:4305–4325.
18. Meybeck, M. 1987. Global chemical weathering of surficial rocks estimated from river dissolved loads. Amer. J. Sci. **287**:401–423.
19. Neftel, A., E. Moor, H. Oeschger, et al. 1985. Evidence from polar ice cores for the increase in atmospheric CO_2 in the past two centuries. Nature **315**:45–47.
20. Post, W.M., T.-H. Peng, W.R. Emanuel, et al. 1990. The global carbon cycle. Amer. Sci. **78**:310–326.
21. Revelle, R., and W. Munk. 1977. The carbon dioxide cycle and the biosphere. In Energy and Climate: Studies in Geophysics, National Academy of Science, Washington, DC, pp. 140–158.
22. Rotty, R.M., and G. Marland. 1986. Fossil fuel combustion: Recent amounts, patterns, and trends of CO_2. In J.R. Trabalka and D.E. Reichle (eds.), The Changing Carbon Cycle: A Global Analysis, Springer-Verlag, New York, pp. 474–490.

23. Sundquist, E.T. 1986. Geologic analogs: Their value and limitations in carbon dioxide research. In J.R. Trabalka and D.E. Reichle (eds.), The Changing Carbon Cycle: A Global Analysis, Springer-Verlag, New York, pp. 371–402.

24. Tans, P.P., I.Y. Fung, and T. Takahashi. 1990. Observational constraints on the global atmospheric CO_2 budget. Science **247**:1431–1438.

25. Walker, J.C.G., P.B. Hays, and J.F. Kasting. 1981. A negative feedback mechanism for the long-term stabilization of Earth's surface temperature. J. Geophys. Res. **86**:9776–9782.

26. Walker, J.C.G., and J.F. Kasting. 1992. Effects of fuel and forest conservation on predicted levels of atmospheric carbon dioxide. Global and Planet. Change. **97**: 151–189.

Models of Oceanic and Terrestrial Sinks of Anthropogenic CO_2: A Review of the Contemporary Carbon Cycle

Inez Fung

NASA Goddard Space Flight Center, Institute for Space Studies, 2880 Broadway, New York, New York 10025

ABSTRACT

The atmospheric increase of carbon dioxide in the 1980s represents only ~57% of the release from fossil-fuel combustion. A balanced budget thus requires the CO_2 sinks to remove 2.9–5.0 GtC/yr, an amount equal to 43% of the fossil fuel plus 100% of the deforestation inputs. Oceanic uptake estimated directly from data on air–sea differences in the partial pressures of CO_2 is ~1.6 GtC/yr; uncertainties in this estimate due to inadequate sampling both in time and space are difficult to quantify. Ocean models yield oceanic sink strengths ~30% of the fossil-fuel inputs alone, implying a significant net terrestrial sink. Interhemispheric gradient of CO_2 in the atmosphere points to a Northern Hemisphere mid-latitudes sink, whereas deconvolution of the historical records suggests that the sink has been operative for the past 50 years. Although isotopic data can potentially provide powerful constraints on the partitioning of the CO_2 sink between land and sea, the conclusions are highly dependent on the $\delta^{13}C$ value of the respired CO_2. Understanding ecosystem dynamics with turnover times of 10–100 years is, thus, tantamount.

Introduction

Carbon dioxide is the most abundant and single most important greenhouse gas (other than water vapor) in the atmosphere. Its role in the radiative balance and its potential for altering the climate of the Earth have been recognized for over a hundred years. Chemical measurements of atmospheric CO_2 were made in the nineteenth century at a few locations.[19,21] However, the modern high-precision record of CO_2 in the atmosphere did not begin until 1958, the International Geophysical Year (IGY), when C.D. Keeling of Scripps Institution of Oceanography pioneered measurements of CO_2 using an infrared gas analyzer at Mauna Loa Observatory in Hawaii and at the South Pole. Since 1974, background measurements of atmospheric CO_2 have been made continuously at four stations (Pt. Barrow, Alaska; Mauna Loa, Hawaii; American Samoa; and the

South Pole) by the Climate Monitoring and Diagnostics Laboratory (CMDL; previously known as the Geophysical Monitoring for Climatic Change program) of the National Oceanic and Atmospheric Administration (NOAA) of the U.S. Department of Commerce. In addition to the continuous monitoring stations, NOAA/CMDL also operates a cooperative sampling network. Flask samples of air are collected weekly from these sites and are shipped to the CMDL facility in Boulder, Colorado for analysis. The sampling network began before 1970 at a few initial sites, expanded to a network of 15 stations in 1979, and, as of 1986, consisted of ~26 stations.[12,25,36] Besides the U.S. program, surface measurements of atmospheric CO_2 around the globe are made by many countries including Australia, Canada, France, Italy, Japan, New Zealand, Spain, Germany, and Switzerland.

Contemporary Atmospheric CO_2 Concentrations

The Mauna Loa CO_2 record shows that CO_2 steadily increased from 315 parts per million by volume (ppm) in 1958 to 351 ppm in 1988. This corresponds to an increase at the rate of 0.36% per year, or a mean increase of 1.1 ± 0.2 ppm per year. From 1958 to 1988, CO_2 at Mauna Loa increased by 36 ppm; over the same period, fossil-fuel combustion was a source of 123 gigatons (1 Gt = 10^{12} kg) of carbon as CO_2 to the atmosphere, which is equivalent to 59 ppm of CO_2. The apparent fraction of the fossil-fuel sources of CO_2 that remained in the atmosphere during this period is, thus, 57%. As other net sources of CO_2, particularly land-use modification, have been important during this period, the actual fraction of anthropogenic carbon emissions remaining in the atmosphere is less than the apparent airborne fraction and is estimated to be 48 ± 8% for 1980–1989.[30]

The CO_2 concentrations from the ~26 global distributed sites in the NOAA/CMDL cooperative flask sampling network show large-scale coherent, temporal and spatial variations.[12,36] Concentrations of CO_2 at all the stations are increasing at the rate of ~1.5 ppm per year, similar to the rate of increase at Mauna Loa.

Historical CO_2 Concentration

Bubbles in natural ice contain samples of ancient air. Analysis by gas chromatography and laser infrared spectroscopy of gases occluded in gas bubbles in polar ice has provided a unique reconstruction of atmospheric CO_2 history prior to the modern high-precision instrumental record.[39] Deep ice cores have been drilled from many locations in both Greenland and Antarctica.

From the ice core data, it is deduced that in preindustrial times, ~1800, the CO_2 concentration was 285 ± 10 ppm and has increased at an accelerating rate since the industrial era.[38,44,48] The ice core data reveal the possible existence of

Table 1. Perturbation carbon balance for 1980–1989.
Unit is Gt C/yr.

Fossil fuel	5.5
Land use modif.	0.4–2.5
Total sources	5.9–8.0
Atmos. increase	3.0
Net sink	2.9–5.0

natural fluctuations of the order of ± 10 ppm occurring at decadal time scales during the last few thousand years.[13,39,48]

The Contemporary Carbon Budget

The atmosphere exchanges CO_2 with the terrestrial biosphere and with the oceans. Averaged over decades, sources must approximately equal sinks if the system is to remain in quasi-steady-state; however, the individual flux in each direction may be large (50–100 Gt C/yr). The fluxes of carbon to the atmosphere associated with anthropogenic activities are roughly ten times smaller than the natural fluxes of carbon. However, the net anthropogenic fluxes are unidirectional perturbations on the total carbon cycle. Thus, analysis of the *trends* of carbon inventory, or markers of CO_2 exchanges in the different carbon reservoirs, will provide information about the contemporary carbon cycle even though there are still large uncertainties in the mean cycle itself.

The contemporary carbon budget is summarized in Table 1. In the 1980s, fossil-fuel CO_2 source was relatively constant, at ~5.5 Gt C/yr. Land-use modification was estimated to range between 0.4–2.5 Gt C/yr,[14,29] recently updated to 1.1–3.6 Gt C/yr.[28] Here, we shall use the older estimates so as to compare with budget studies already published. Over this period, atmospheric increase, as determined from the NOAA/CMDL network, was 3.0 Gt C/yr. Mass balance, thus, requires a net sink 2.9–5.0 Gt C/yr.

It has generally been assumed that the oceans are the primary sink for the anthropogenic CO_2. In the following, we review the data and models that focus on this sink. Several perspectives are taken and will be discussed in the remainder of the chapter: the oceanic, historical, atmospheric, terrestrial, and isotopic.

Oceanic Perspective

The capacity of the ocean to take up excess CO_2 is illustrated by the following schematic model taken from Broecker et al.[6] Carbon chemistry in the ocean can be simplified to

$$CO_2 + H_2O \rightleftarrows HCO_3^- + H^+,$$

$$CO_3^= + H^+ \rightleftarrows HCO_3^-.$$

The reversible reactions are governed by kinematic dissociation constants $k_1(T,S)$ and $k_2(T,S)$, which are known functions of temperature T and salinity S of the water. The reactions also conserve carbon and charge:

(1) $$k_1(T,S) = \frac{[HCO_3^-][H^+]}{[CO_2]},$$

(2) $$k_2(T,S) = \frac{[HCO_3^-]}{[CO_3^=][H^+]},$$

(3) $$\Sigma C = [CO_2] + [HCO_3^-] + [CO_3^=],$$

and

(4) $$\text{ALK} = [HCO_3^-] + 2[CO_3^=],$$

where ΣC is the total concentration of inorganic carbon, and ALK is carbonate alkalinity, net charge of the species not involved in carbon chemistry. The partial pressure of CO_2 in water, pCO_2, is governed by Henry's Law:

(5) $$pCO_2 = \beta(T,P)\,[CO_2],$$

where β is the solubility of CO_2 at temperature T and pressure P.

Equations (3)–(5) summarize the inorganic carbon chemistry in the ocean. They can be combined into a single expression for pCO_2:

(6) $$pCO_2 = pCO_2\,(T,S,\Sigma C,\text{ALK},\ldots).$$

Changes in pCO_2 can then be written as

(7) $$d\,\ell n\,pCO_2 = \frac{\partial\,\ell n\,pCO_2}{\partial T}\,dT + \frac{\partial\,\ell n\,pCO_2}{\partial(\ell n\,\Sigma C)}\,d(\ell n\,\Sigma C) + \cdots.$$

The first coefficient on the right-hand side of Eq. (7) expresses the change of pCO_2 with temperature, approximately 4.3%/°C at present pCO_2 values. The second coefficient is the Revelle factor, or buffer factor, which ranges from 14 at 0°C to ~8 at 30°C, with a mean of 10 for surface waters. It states that a 1% change in the total carbon concentration results in a 10% change in the partial pressure of CO_2 in the surface waters. The implication of the Revelle factor for the anthropogenic carbon cycle is discussed below.

The net exchange of CO_2 between the atmosphere and the oceans depends on the difference in CO_2 partial pressures across the air–sea interface. Consider an initial equilibrium, denoted by the superscript 0, between the atmosphere and the ocean. Introduce to this equilibrium an amount ΔF of excess CO_2. At final equilibrium, denoted by the superscript 1, a fraction γ of the excess CO_2 remains in the atmosphere, while $(1-\gamma)$ enters the ocean. CO_2 is inert in the atmosphere, so that

$$\Delta CO_{2\text{air}} = [CO_2]^1_{\text{air}} - [CO_2]^0_{\text{air}} = \gamma\ \Delta F.$$

In the ocean, the excess carbon is partitioned among all the carbon species, so that

$$\Delta\Sigma C = \Sigma C^1 - \Sigma C^0 = (1 - \gamma)\ \Delta F.$$

At both the initial and final equilibria, atmospheric and oceanic CO_2 concentrations are equal, so that $\Delta[CO_2]_{\text{air}} = \Delta[CO_2]_{\text{water}}$. Using the Revelle factor, we obtain

$$\frac{\gamma \Delta F/[CO_2]^0}{(1-\gamma)\Delta F/\Sigma C^0} = 10. \tag{8}$$

If the entire water column is involved in the equilibrium, $\Sigma C^0 = 8740$ moles/m^2 and $[CO_2]^0 = 145$ moles/m^2. Solving Eq. (8), $\gamma = 0.14$. In other words, at maximum, the ocean can take up 86% of the anthropogenic input. This maximum estimate would be altered with consideration of the other processes of the oceanic carbon cycle, in particular, carbonate dissolution. The amount of anthropogenic CO_2 actually absorbed, as a function of time, depends, *inter alia,* on the rate of vertical mixing in the ocean column.

Because of the variability of the oceanic carbon system and the precision of ocean carbon measurements, the oceanic signature of anthropogenic CO_2 has not been demonstrated unambiguously from direct measurements. In the Atlantic, pCO_2 in the surface water increased by 8 ± 8 μatm from 1958 to the mid-1970s.[58] With available pCO_2 measurements, a global net flux of 2.6 Gt C/yr into the oceans in the 1980s is estimated.[8,59,60] An update results in a net flux of 1.6 Gt C/yr.[61–63,65] The large reduction in the recent estimate stems from improved sampling of the seasonal variations of pCO_2 in surface waters: pCO_2 values are higher in the summer in the gyres and higher in the winter in other regions. Hence, estimation of the annual mean based on measurements made in one season would be biased. Given that a large expanse of the oceans is still unsampled, the current estimate of the net flux into the ocean must not be viewed as definitive.

Another approach to estimate directly the amount of anthropogenic carbon in the ocean is by examining the cumulative changes of ΣC in oceanic reservoir. A difficulty in this approach is the lack of baseline or historical data of ΣC from which to estimate change. As water masses in the ocean interior move primarily along constant density (isopycnal) contours, concentration differences between the ocean surface and interior locations along the same isopycnal have been used to infer the anthropogenic CO_2 signal.[5,9,10] Considerable controversy exists about this procedure, as mixing and biological processes also alter CO_2 concentrations; correction schemes for these other processes remain problematic due to insufficient data.[7,11,52,53]

CO_2 Uptake as a Function of Time

Models of varying complexity have been used to estimate the oceanic uptake of anthropogenic CO_2. One-dimensional and two-dimensional models, and the data for calibrating the models, have been reviewed by Björkström.[1]

A commonly used one-dimensional model is the box-diffusion model.[40] Here, a well-mixed atmosphere is coupled to a horizontally averaged column of water. The water column is topped by a well-mixed layer, i.e., box, of thickness h, with a diffusive interior characterized by a diffusivity κ. The mixed layer thickness, typically $h = 75$ m, is taken from observations of the density structure in the ocean. The value of κ is estimated from the distribution of tracers in the ocean. There is no known tracer with the same invasion history as anthropogenic CO_2. Models are generally calibrated with natural C-14 and bomb C-14, yielding an oceanic uptake of 1.7 and 2.4 Gt C/yr for the period 1980–1989.[30,56] The difference between the two can be understood in terms of their residence time in the ocean. It is ~2000 years for natural C-14, so that κ represents diffusivity of a "clogged" pipe. The invasion history is ~20 years for bomb C-14, so that κ would represent unimpeded diffusion into a relatively empty reservoir. These two time scales bracket the history of fossil fuel, ~200 years, and hence the uptake of fossil-fuel CO_2.

A variation of the one-dimensional box-diffusion model is the outcrop-diffusion model,[54] where the mixed layer caps only a fraction of the ocean. For the remainder, ~10% of the area, the diffusive interior exchanges directly with the atmosphere. Uptake of the outcrop diffusion model is 3.6 Gt C/yr for 1980–1989. In this model, CO_2 entering at the outcrop is mixed immediately within an internal layer, thus underestimating the resistance to further atmospheric inva-

Table 2. Uptake of anthropogenic CO_2 for 1980–1989, as estimated by different oceanic models. The net sink required for carbon balance is 2.9–5.0 GtC/yr.

Ocean Model	Calculated Ocean Sink (Gt C/yr)	Inferred Land Sink (GtC/yr)	Ref.
Box-diffusion model			
Bomb ^{14}C calibr.	2.4	0.5–2.6	56
Natural ^{14}C calibr.	1.7	1.2–3.3	40
Outcrop-diffusion model	3.6	−0.7–1.4	54
Hamburg 3-D OGCM	1.5	1.4–3.5	37
GFDL 3-D OGCM	1.9	1.0–3.1	50

sion. The CO_2 uptake of the outcrop diffusion model is, thus, considered to be an upper bound.

Models that have geographic resolution include the multibox model of the world ocean[45] and the 84-box model of the North Atlantic.[2] Here, the sizes of the boxes are determined by geographic and hydrographic data, whereas a broader suite of tracer and dynamic constraints are brought in to calibrate the exchange coefficients among the boxes. The estimates of CO_2 uptake by the models are in the range of the one-dimensional box-diffusion models.

Three-dimensional ocean general circulation models (GCMs) are constructed based on physical laws of fluid dynamics. The flow fields are calculated for a given pattern of wind stress and buoyancy fluxes at the surface, and the models are validated by comparing the circulation statistics to those observed. Tracer distributions are, thus, used to evaluate, rather than calibrate, the simulated flow characteristics.[49,66,67] Ocean GCMs estimate CO_2 sinks of 1.2–1.9 Gt C/yr.[37,50]

Model estimates of the ocean uptake of anthropogenic CO_2 fall short of that required for carbon balance (Table 2). In the one-dimensional models, a terrestrial sink is generally implemented. The functional form of the terrestrial sink is made proportional to the atmospheric CO_2 concentration, and the sink is referred to as "CO_2 fertilization," based on mathematical rather than ecological considerations. By residual, the magnitude of the terrestrial sink is inferred to range between −0.7 and 3.5 Gt C/yr.

Historical Perspective

The history of the carbon balance since the preindustrial era provides a different perspective on the sink of anthropogenic CO_2. The global carbon balance can be written as

$$\frac{\partial C_a}{\partial t} - \text{FOSSIL} - \text{OCEAN} = \text{LAND} \tag{9}$$

The first term on the left-hand side of Eq. (9), changing atmospheric concentration C_a as a function of time, can be estimated from ice core data; the second, input from fossil-fuel-burning, extrapolated backwards from today's values; and the third, the net ocean uptake, calculated from a range of ocean models as described above. This deconvolution thus yields, as a residual, an indirect estimate of the net exchange with the land.

Siegenthaler and Oeschger[56] estimated a cumulative net source from the terrestrial biosphere of ~90–150 Gt C for the period ~1700 to 1980, with the peak emission around the 1900s. From 1959 to 1983, the biosphere was only a weak source, with a net release rate of 0–0.9 Gt C/yr, less than the direct estimate of the land-use source of CO_2.

Houghton[27] further separates the net terrestrial exchange into a land-use component and a missing component:

$$\text{LAND} = \text{DEFORESTATION} + \text{MISSING}. \tag{10}$$

The history of the missing carbon sink is shown in Figure 1. Dependent on the model used to estimate the oceanic uptake, there was either no missing sink or a small source up until the 1940s. Thereafter, all the oceanic models imply a missing sink that was largest in the 1970s. In the 1980s, a land sink is still implied, albeit at reduced strength.

The time history of the missing sink corresponds with that of several factors that affect the biosphere. For example, after the Second World War, there have been changes in agricultural practices, especially the abandonment of agricultural lands and increased fertilizer application that would affect terrestrial carbon cycling. Also, surface air temperature declined from the 1940s to 1970s, and then a decade of rapid warming ensued. Although these external changes are known theoretically to affect growth, decomposition, and other processes of the terrestrial biosphere, the actual extent of their impact is unknown.

Atmospheric Perspective

Information about the forcing of a system is contained in its response. Atmospheric CO_2 measurements from the NOAA/CMDL and other networks show that annually averaged CO_2 concentrations are higher in the Northern Hemisphere than in the Southern Hemisphere. The interhemispheric difference was ~1 ppm in the 1960s and is ~3.0 ppm now, reflecting the Northern Hemisphere

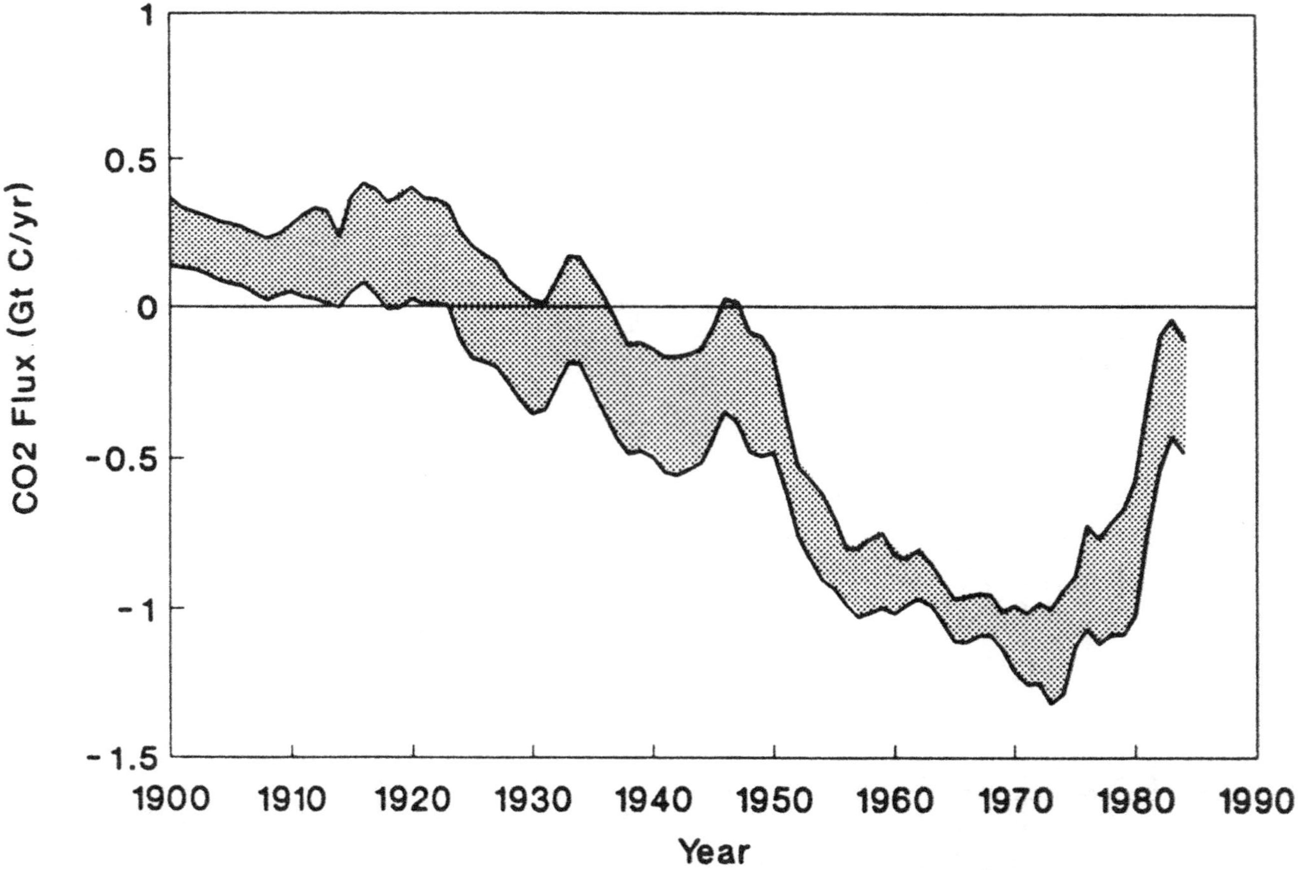

Figure 1. The ''missing'' biotic flux as determined from the deconvolution analysis. Shaded region spans the upper and lower bounds of the flux. Positive values denote fluxes to the atmosphere. (Adapted from Houghton[27] with the permission of the author.)

mid-latitude source of fossil-fuel CO_2. This gradient has remained approximately constant in the 1980s. Also evident in the north–south distribution of atmospheric CO_2 is the relative maximum of ~1 ppm in the equatorial region, caused mainly by the outgassing of CO_2 from supersaturated surface waters of the equatorial oceans. Although tropical deforestation may also contribute to the equatorial maximum in atmospheric CO_2, models of the global carbon cycle suggest that the observations are inconsistent with a net deforestation source greater than approximately 1.5 Gt C/yr.[33,43,64]

Two-dimensional (latitude–height) and three-dimensional models of atmospheric circulation and transport have used this information to constrain the budget of CO_2. These models essentially solve the equation of conservation of CO_2. The transport characteristics are determined independent of the CO_2 cycle and are validated directly using observed atmospheric circulation statistics or indirectly using distribution of tracers. In some of these models, spatial and temporal variations of the CO_2 sources and sinks are specified *a priori* and CO_2 is then distributed in the atmosphere according to the transport fields.[22–24,26,34,35,41,42,65] In others, the observed atmospheric CO_2 distributions are specified and the inverse of the two-dimensional conservation equations are solved for the zonally averaged net source/sink functions.[16,64] The inverse approach thus yields information about the latitudinal distribution of the net exchange, but cannot distinguish among the components that make up the net exchange.

In the 1980s, about 96% of the fossil-fuel source was in the Northern Hemisphere. If all the fossil-fuel CO_2 remained airborne, then the interhemispheric difference would be ~4.4 ppm, larger than the 3.0 ppm observed (Figure 2). The deforestation source is mainly in the tropics and contributes little to the interhemispheric gradient of CO_2 in the atmosphere. A large southern ocean sink would depress the Southern Hemisphere concentration relative to the north and further exaggerate the north–south gradient of CO_2 in the atmosphere. Hence, the sinks have to dominate in the Northern Hemisphere.[26,34,35,42,65]

The identity of the sink cannot be uniquely determined using the atmospheric CO_2 data alone. In Tans et al.,[65] the extensive pCO_2 data[59,60,61] constrained the net oceanic CO_2 exchange, at least in the Northern Hemisphere where there are relatively abundant good-quality oceanic pCO_2 observations. They thus obtained an oceanic sink that may range from 0.3 to 0.7 Gt C/yr, with a corresponding terrestrial sink that ranges from 2.0 Gt C/yr to 4.2 Gt C/yr. Because the Northern Hemisphere oceans are still undersampled, the partitioning between the land and ocean sinks obtained with the presently available oceanic observations is tentative. However, their model calculation showed that for the ocean to be a sole sink of anthropogenic CO_2, the annually averaged air–sea difference in pCO_2 must be ~70 μatm in the North Atlantic and Pacific, about double that observed. They argued that the discrepancy between the required and observed annual mean

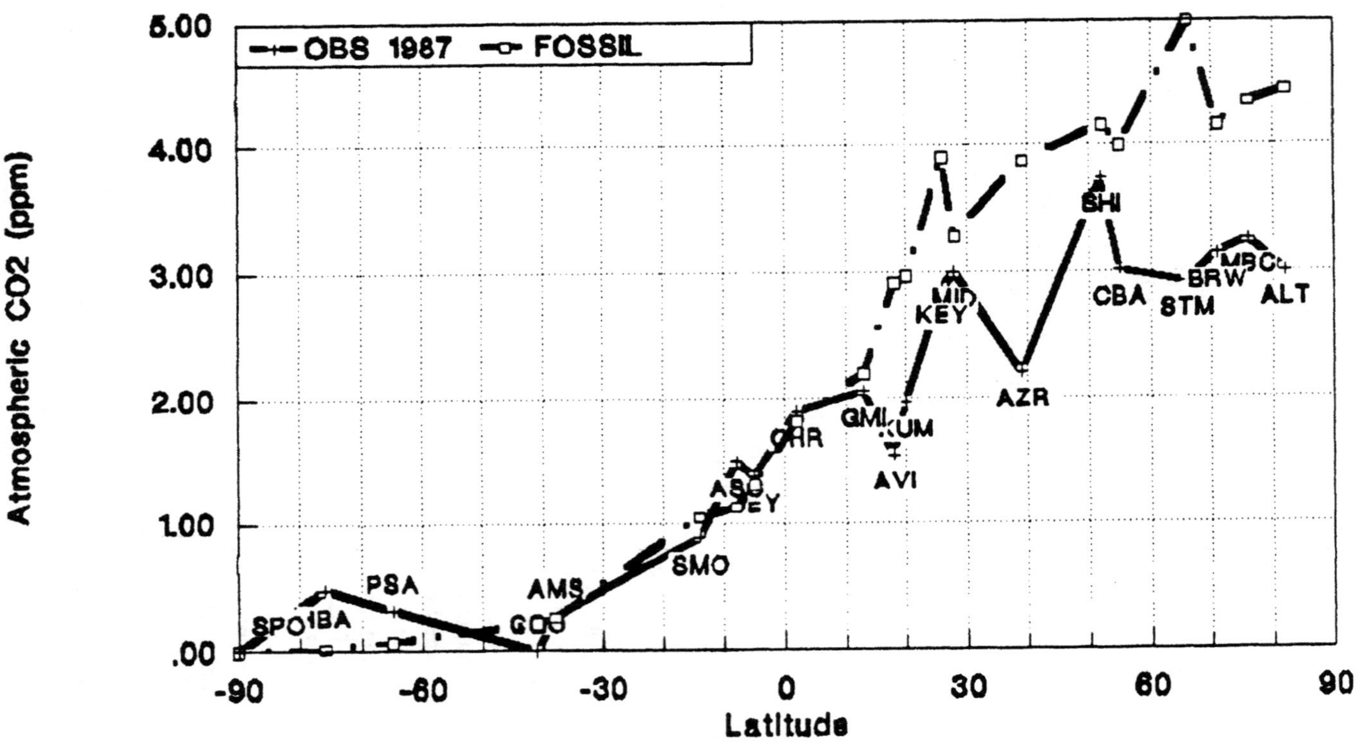

Figure 2. Latitudinal distribution of the annually averaged CO_2 concentration near the surface. The solid line joins the concentrations measured at the NOAA/CMDL network, whereas the dotted line joins the values simulated by the three-dimensional tracer model for fossil fuel release with an airborne fraction of 100%.

oceanic pCO_2 is too large to be explained by known measurement uncertainties. A land sink is thus required to balance the budget and match the north–south gradient in the atmosphere. In their model calculation, Keeling et al.[34,35] assumed that the air–sea pCO_2 difference was ~60 μatm in the northern oceans, and so the oceanic uptake was 2.2 Gt C/yr, and the terrestrial sink needed to balance the budget was 2.1 Gt C/yr. This sink is assumed to be proportional to the annual net primary productivity of the ecosystems, and so dominates in the tropics.

Terrestrial Perspective

Carbon cycling in the terrestrial biosphere can be represented schematically by exchanges among several well-mixed boxes with different turnover times (Figure 3). The boxes would comprise (1) photosynthetic compartment, with turnover times of seasons; (2) woody compartment, with turnover times of years to decades; (3) litter and surface soil pool, with turnover times of decades to centuries; and (4) long-term soil pool, with geologic turnover times.[15] Though a multibox terrestrial biosphere has been an integral part of many models of the contemporary carbon cycle, these studies have focused on deriving or constraining the magnitude of the net flux from the land (including land-use modication) and have not included sufficient biospheric dynamics to extract information about the mechanism of the missing carbon sink on land.[46,56]

The gross primary productivity of the terrestrial biosphere is ~80–100 Gt C/yr. In other words, 80–100 Gt C enters into the photosynthetic compartment annually. If the biosphere were at equilibrium, the same amount of carbon is returned via respiration and decomposition processes from all four compartments. Recently, net CO_2 fluxes into the biosphere, or biospheric disequilibrium, have been observed by direct flux measurements in secondary mid-latitude forests recovering from climatic or human disturbances.[32,68] Also, a net uptake of 0.85–1.1 Gt C/yr has been simulated for boreal forests using an ecosystem process model.[3,4]

The historical perspective suggests a cumulative biospheric accumulation of 80 Gt C over the past 50 or so years.[27] From the consideration of time scales alone, it is clear that the excess carbon cannot be stored in the photosynthetic compartment or in the long-term soil pool.[51] Increases in the woody compartment in tree-rings (e.g., Ref. 62) provide some support for this carbon accumulation. Litter pools and upper layers of the soil, with turnover times of ~50 years, are also likely candidate sites for carbon storage. Over this period, these pools have been subjected to a cooling followed by a warming trend as well as to increased fertilization and other changes in agricultural practice. In principle, these pools could have responded to these perturbations via retarded temperature-

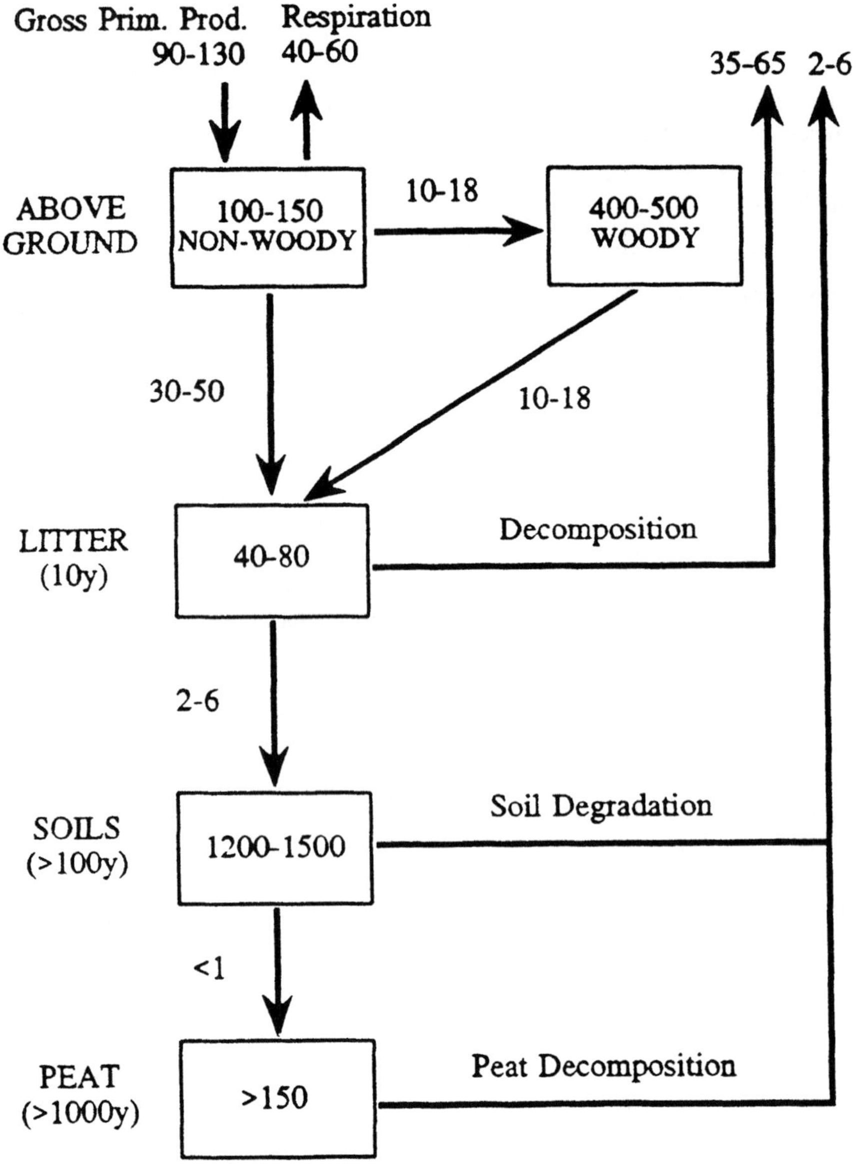

Figure 3. Schematic representation of the biosphere. Reservoir sizes are given in Gt C, whereas fluxes (arrows) are in Gt C/yr.

regulated decomposition rates, or via increased storage due to increased nitrogen content. There is little information to determine whether actual, albeit transient, storage of CO_2 has been realized from these potential feedbacks.

Isotopic Perspective

An important perspective on the contemporary carbon budget is contained in the variations of isotopic concentrations in the different carbon reservoirs. Isotopic concentrations are expressed as

$$\delta^{13}C = \left(\frac{R_{sample}}{R_{standard}} - 1\right) \times 1000, \tag{11}$$

where $R = {}^{13}C/{}^{12}C$. $R_{standard} = 0.011237$ is the ${}^{13}C/{}^{12}C$ ratio of the conventional marine carbonate standard, PeeDee Belemnite. $\delta^{13}C$ values are about -6 to $-8‰$ for the atmosphere, 1.5 to 2‰ for the ocean, and -25 to $-30‰$ for the terrestrial biosphere. Fossil-fuel CO_2, derived from plant carbon, has the same $\delta^{13}C$ value.

The exchange of CO_2 and ${}^{13}CO_2$ between two reservoirs (denoted by subscripts i and j) can be illustrated using the following simple model:

$$\frac{\partial C_i}{\partial t} = -F_{ij} + F_{ji}\ , \tag{12}$$

$$\frac{\partial R_i C_i}{\partial t} = -\alpha_{ij} R_i F_{ij} + \alpha_{ji} R_j F_{ji}. \tag{13}$$

Here, C_i is the mass of CO_2 in reservoir i; F_{ij} is the flux of CO_2 from reservoir i to reservoir j. R_i is the ${}^{13}C/{}^{12}C$ ratio in reservoir i, and α_{ij} is the fractionation factor associated with fluxes from reservoir i to reservoir j. We shall use the subscripts a, o, b to denote the atmosphere, ocean, and terrestrial biosphere, respectively. Fractionation factors associated with fluxes from the atmosphere to the ocean and the terrestrial biosphere, α_{ao}, and α_{ab}, are ~0.998 and ~0.982, respectively. A sink of atmospheric CO_2, be it oceanic or terrestrial, would preferentially remove ${}^{12}C$ relative to ${}^{13}C$ from the atmosphere, thus increasing the ${}^{13}C/{}^{12}C$ ratio in the atmosphere and decreasing the ${}^{13}C/{}^{12}C$ ratio in the recipient reservoir. Starting with an atmosphere with 700 Gt C and $\delta^{13}C$ of $-7‰$ and holding all fractionation factors constant, an additional invasion of

1 Gt C/yr into the oceans would increase the atmospheric $\delta^{13}C$ value by 0.0028‰ per year, whereas increasing terrestrial photosynthesis by 1 Gt C/yr would increase the atmospheric $\delta^{13}C$ by 0.026‰/yr. The factor-of-10 difference between the two atmospheric signatures thus allows the discrimination between oceanic and terrestrial sources/sinks of atmospheric CO_2. Changes in $\delta^{13}C$ of each of the reservoirs provide crucial constraints on the contribution of each reservoir to the global carbon budget.

Values of $\delta^{13}C$ of atmospheric CO_2 has decreased from −6.5 to −7.8‰ from 1800 to 1984.[34] If all the fossil-fuel CO_2 remained airborne, $\delta^{13}C$ of the atmospheric reservoir would have decreased by 5.3‰ over the same period, instead of the 1.3‰ observed. The partitioning between terrestrial and ocean sinks not only should yield the correct $\delta^{13}C$ in the atmosphere, it should, in turn, be reflected in the $\delta^{13}C$ changes in the respective reservoirs.

Air–sea exchange of CO_2 involves several fractionation steps: from HCO_3^- and $CO_3^=$ ions to CO_2, in the solubility of CO_2 in sea water, and in gas exchange across the air–sea interface. These have been reviewed by Siegenthaler and Munnich.[55] If the oceans were the sole sink for the anthropogenic CO_2, then the $\delta^{13}C$ of the oceanic mixed layer would have decreased by ~0.2‰ between 1970 and 1990, based on a one-dimensional box-diffusion model. The calculated decrease is less than the difference between that measured in 1990[47] and that previously measured in the 1970s during the GEOSECS experiment. Thus, the observed decrease is too small to support the hypothesis of the oceans being the sole sink.

There are many fractionation steps in carbon cycling in the terrestrial biosphere, with the rate-limiting step determining the resultant ^{13}C fraction in a particular carbon pool or compartment. The $\delta^{13}C$ of plant material thus depends on the $\delta^{13}C$ of the air, stomatal conductance, and carbon assimilation rate, as well as the allocation of carbon among the different compartments.[17,46,57] These environmental and physiological influences are generally acknowledged in carbon cycle models that attempt to use $\delta^{13}C$ of tree-rings to deconvolve the net CO_2 flux from the land.[46,56,58] The ^{13}C contents of the litter and soil pools are not well-known and are generally assumed to be −25‰ in these models.

Influx of CO_2 to the biosphere carries the ^{13}C signature of the atmosphere. As atmospheric $\delta^{13}C$ (and corresponding R_a) decreases with time with the addition of fossil-fuel and deforestation CO_2, it is expected that $\delta^{13}C$ of the biosphere would decrease even if there is no net change in the ^{12}C balance of the biosphere. Efflux of CO_2 from the biosphere is the sum of the CO_2 respired from leaves, woody materials, litter, and soils. It will be a weighted mean of the atmospheric $\delta^{13}C$ returned via autotrophic respiration on time scales of seasons, $\delta^{13}C$ of the woody components and litter compartments with turnover times of a year to decades, and $\delta^{13}C$ of the soil pools with turnover times of centuries or longer. At any moment in time, the $\delta^{13}C$, or R_b, value of the respired CO_2 depends on the mean age (relative to current atmospheric values) of the carbon in the biospheric pools and the time history of $\delta^{13}C$ of each of the pools involved. Thus, it is important to determine the $\delta^{13}C$ trends of the different biospheric compartments.

Observations of $\delta^{13}C$ variations in tree-rings show a wide range of variations. The extremes are bracketed by Western European trees who show a maximum $\delta^{13}C$ decline of ~1.5‰ over the past 50 years,[20] and by several Tasmanian trees, who showed little decline except for <0.5‰ in the past 20 years.[18] Although there are differing interpretations of the $\delta^{13}C$ records from the different sites, tree-ring $\delta^{13}C$ records do provide first estimates of biospheric $\delta^{13}C$ trends.

Equation (12) can be used to illustrate schematically the sensitivity of atmospheric $\delta^{13}C$ (R_a) to assumptions about the history of $\delta^{13}C$ of the biosphere. Here, we express every term as a steady-state value, denoted by superscript °, and a departure from steady state, denoted by the superscript ′. At steady state with respect to both CO_2 and ^{13}C exchanges, $F^\circ_{ab} = F^\circ_{ba}$, and $\alpha_{ab} R^\circ_a = \alpha_{ba} R^\circ_b$. Since $\alpha_{ba} = 1$, $\alpha_{ab} = R^\circ_b/R^\circ_a$. We assume steady state in the 1700s with $\delta C^\circ_a =$ −6.3‰ and $\delta C^\circ_b =$ −25‰. Equation (12) can thus be written

$$
\begin{aligned}
\frac{\partial R_a C_a}{\partial t} &= -\alpha_{ab}(R^\circ_a + R'_a)(F^\circ_{ab} + F'_{ab}) + \alpha_{ba}(R^\circ_b + R'_b)(F^\circ_{ba} + F'_{ba}) \\
&= -\alpha_{ab} R'_a F^\circ_{ab} + \alpha_{ba} R'_b F^\circ_{ba} - \alpha_{ab} R^\circ_a F'_{ab} + \alpha_{ba} R^\circ_b F'_{ba} \\
&\quad - \alpha_{ab} R'_a F'_{ab} + \alpha_{ba} R'_b F'_{ba}.
\end{aligned}
\tag{14}
$$

Now consider atmosphere–biosphere exchanges in the 1980s. We fix $F^\circ_{ab} = 60$ Gt C/yr, $F'_{ab} = 0$, and $\delta C_a =$ −7.5‰. Figure 4 shows the sensitivity of R'_a to $R'_b - R'_a$ for different values of F'_{ba}. It is clear that even when the exchange is in steady state with respect to ^{12}C ($F'_{ba} = 0$), δC_a would increase if $R'_b \neq R^1_a$, and the magnitude of the increase would depend on the degree of disequilibrium between atmospheric and biospheric $\delta^{13}C$.

The locations of the anthropogenic CO_2 sink can also be examined from the perspective of the $\delta^{13}C$ distribution in the atmosphere. Figure 5 shows the north–south gradient of atmospheric $\delta^{13}C$ observed and that caused by fossil fuel and equilibrium ocean. The combination of land and ocean sink has to account for a very small but detectable difference of 0.05–0.10‰ between the hemispheres. It is clear that a mid-latitude land sink > 2.0 Gt C/yr, as postulated by Tans et al.[65] is inconsistent with the interhemispheric $\delta^{13}C$ gradient. Keeling et al. assumed that 55% of F_{ba} returned with δC_b of −25.53‰ and the remainder with δC_b of −24.93‰ in 1980, so that effectively $\alpha_{ab} R_a$ and $\alpha_{ba} R_b$ corresponded to an effective $\delta^{13}C$ difference of ~0.3‰. There are little data to determine if such a large difference is biologically reasonable. If the trends of $\delta^{13}C$ in tree-rings can be applied to the litter and upper soil pools, the $\delta^{13}C$ difference would imply a long lag between uptake and release of CO_2 in the biosphere. As can be seen in Figure 4, this large difference would also favor a lower biospheric sink to fit the atmospheric north–south gradient.

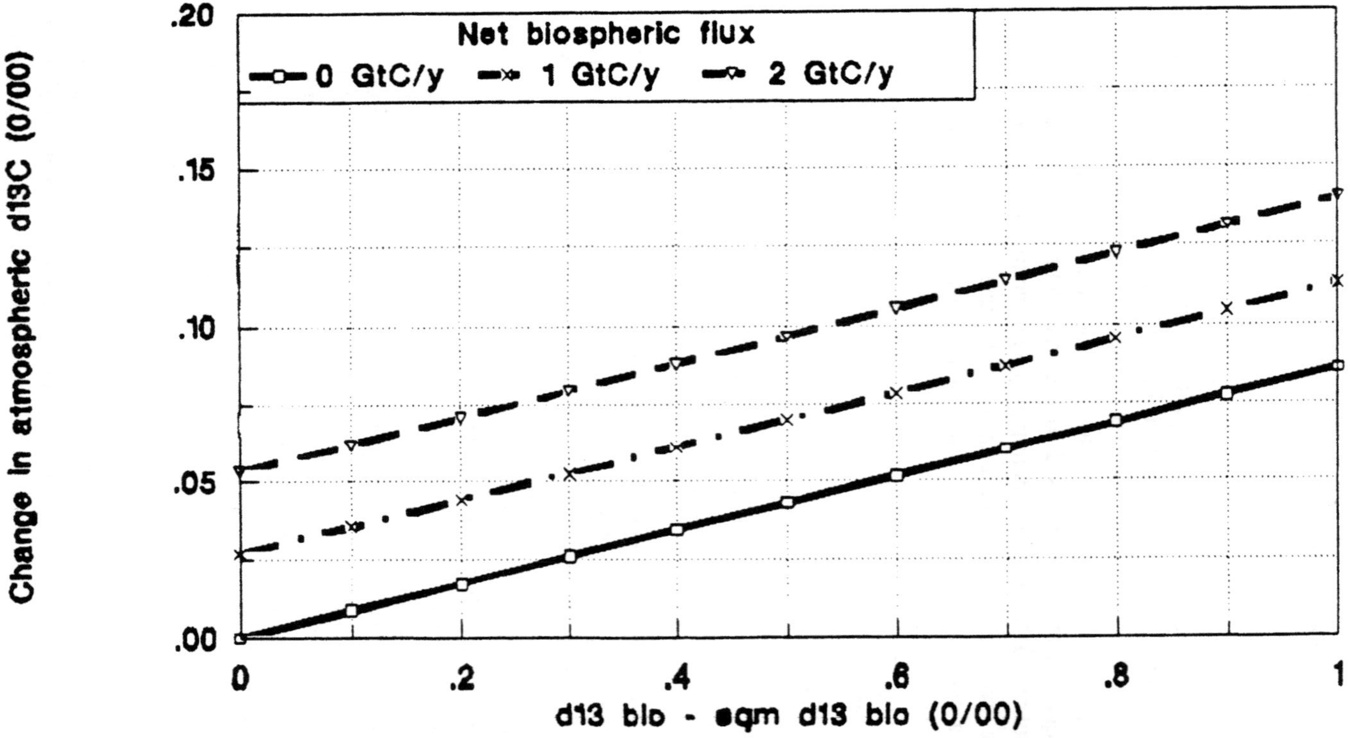

Figure 4. Sensitivity of atmospheric $\delta^{13}C$ to atmosphere exchange disequilibrium, $\alpha_{ba}R_b - \alpha_{ab}R_a$, expressed in ‰, for different strengths of the net biospheric sink.

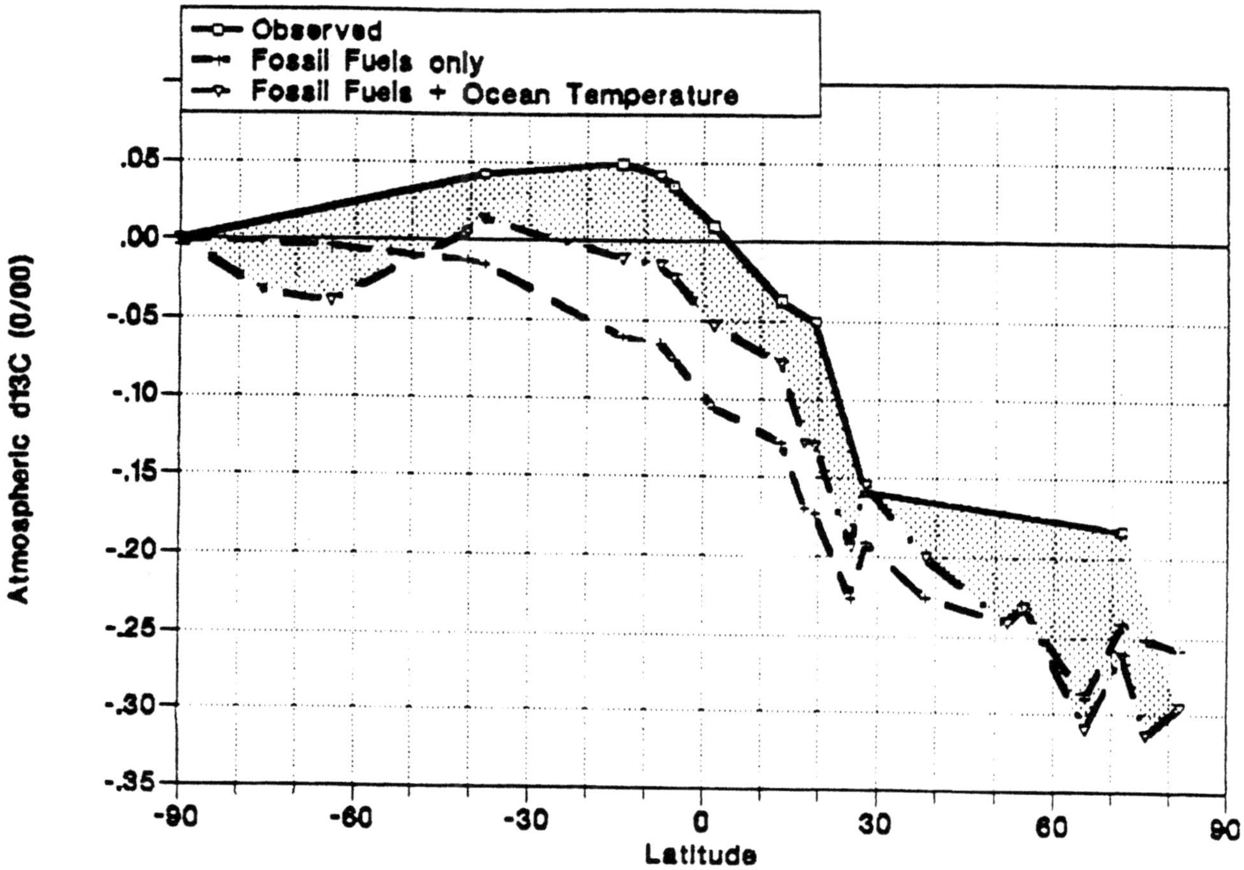

Figure 5. Observed and modeled latitudinal $\delta^{13}C$ in the atmosphere for 1980. The shaded region denotes the signal from the total carbon sink required to match the observations.

The simple model did not address possible changes in α_{ab} or α_{ba} due to CO_2 fertilization, changes in temperature, and other abiotic parameters. It is clear that using isotopic information for the interpretation of the sink of anthropogenic CO_2 is a difficult problem. Improved understanding of the control and dynamics of the influx of CO_2 from the atmosphere to storage in the woody compartments is only half the story. The other half resides in the return cycle through litter and soils. Thus, documentation of the history of ^{13}C and ^{14}C fractions in these pools, and an understanding of the controls and dynamics of the processes of decomposition and soil respiration, is crucial.

Conclusion

In this chapter, we have presented a series of models, with oceanic, atmospheric, and biospheric, as well as historical and isotopic perspectives, to investigate the partitioning of the anthropogenic CO_2 sink between land and sea. The ocean models suggest that the oceans have absorbed ~30% of the fossil-fuel carbon in the 1980s. The atmospheric models suggest that the CO_2 sink has to dominate in the Northern Hemisphere. This, when combined with the most comprehensive compilation of pCO_2 observations in the Northern Hemisphere, suggests a land sink that predominates in mid-latitudes in the Northern Hemisphere. Deconvolution of the CO_2 sink from the historical records of atmospheric CO_2, fossil fuel and land-use sources suggests that the missing sink was operative in the past 50 years. This suggests that the biosphere is acting as a transient carbon storage in the woody material, litter, and upper layers of the soils which turn over in time scales of 10–100 years. Isotopic data can potentially provide powerful constraints on the partitioning of the CO_2 sink between land and ocean. The biospheric model suggests that conclusions about the land–sea partitioning of the anthropogenic sink are highly dependent on the $\delta^{13}C$ value of the respired CO_2. Understanding the dynamics of the turnover on the time scales of 10–100 years is, thus, tantamount.

Acknowledgments

The author thanks Chris Field for a most stimulating discussion on carbon isotopic exchange. The research was supported by NASA Biogeochemistry and Geophysics Branch, NASA EOS Interdisciplinary Science Program, DOE Carbon Dioxide Research Program, and Global Climate Change Division, Office of Policy Analysis of the Environmental Protection Agency.

References

1. Björkström, A. 1986. One-dimension and two-dimensional ocean models fo predicting the distribution of CO_2 between the ocean and the atmosphere. In The

Changing Carbon Cycle, A Global Analysis, Springer-Verlag, New York, pp. 258–278.

2. Bolin, B., A. Björkström, and B. Moore. 1989. Uptake by the Atlantic Ocean of excess atmospheric carbon dioxide and radiocarbon. In Contribution of Geophysics to Climate Change Studies, Proceedings of IUGG Symposium, Vancouver, IUGG/AGU, Washington, pp. 57–78.
3. Bonan, G.B. 1991. Atmosphere–biosphere exchange of carbon dioxide in boreal forests. J. Geophys. Res., **96**:7301–7312.
4. Bonan, G.B. 1991. Seasonal and annual carbon fluxes in a boreal forest landscape. J. Geophys. Res., **96**:17329–17338.
5. Brewer, P.G. 1978. Direct observation of the oceanic CO_2 increase. Geophys. Res. Lett., **5**:997–1000.
6. Broecker, W.S., T.T. Takahashi, J.J. Simpson, and T.H. Peng. 1979. Fate of fossil fuel carbon dioxide and the global carbon budget. Science, **206**:409–418.
7. Broecker, W.S., T.H. Peng, P. Brewer, and T. Takahashi. 1982. Chenology challenged. Trans. of the Amer. Geophysical Union, **63**:979.
8. Broecker, W.S., J.R. Ledwell, T. Takahashi, R. Weiss, L. Melivat, L. Memery, T.-H. Peng, B. Hahne, and K.O. Munnich. 1986. Isotopic versus micrometeorologic ocean CO_2 fluxes: a serious conflict. J. Geophys.Res., **91**:10517–10527.
9. Chen, C.T. 1982. Oceanic penetration of excess CO_2 in a cross section between Alaska and Hawaii. Geophys. Res. Lett., **9**:117–119.
10. Chen, C.T. 1982. On the distribution of anthropogenic CO_2 in the Atlantic and southern oceans. Deep Sea Res., **29**:563–580.
11. Chen, C.T., F.J. Millero, and R.M. Pytkowics. 1982. Comment on calculating the oceanic CO_2 increase: A need for caution by A.M. Shiller. J. Geophys. Res., **87**:2083–2085.
12. Conway, T.J., P.P. Tans, L.S. Waterman, K.W. Thoning, K.A. Masarie, and R.H. Gammon. 1988. Atmospheric carbon dioxide measurements in the remote global troposphere, 1981–1984. Tellus, **40B**:81–115.
13. Delmas, R.J., J.M. Ascencio, and M. Legrand. 1980. Polar ice evidence that atmospheric CO_2 20,000 yr-BP was 50% of present. Nature, **284**:155–157.
14. Detweiler, R.P., and C.A. Hall. 1988. Tropical forests and the global carbon cycle. Science, **239**:42–47.
15. Emanuel, W.R., G.G. Killough, W.M. Post, and H.H. Shugart. 1984. Modeling terrestrial ecosystems in the global carbon cycle with shifts in carbon storage capacity by land-use change. Ecology, **65**:970–983.
16. Enting, I.G., and J.V. Mansbridge. 1991. Latitudinal distribution of sources and sinks of CO_2: results of an inversion study. Tellus, **43B**:156–170.
17. Farquhar, G.D., M.H. O'Leary, and J.A. Berry. 1982. On the relationship between carbon isotope discrimination and the intercellular carbon dioxide concentration in leaves. Aust. J. Plant Physiol., **9**:121–137.

18. Francey, R.J. 1981. Tasmanian tree rings belie suggested anthropogenic $^{13}C/^{12}C$ trends. Nature, **290**:232–235.

19. Fraser, P.J., W.P. Elliott, and L.S. Waterman. 1986. Atmospheric CO_2 record from direct chemical measurements during the 19th century. In The Changing Carbon Cycle, A Global Analysis, Springer-Verlag, New York, pp. 66–68.

20. Freyer, H.D., and N. Belacy. 1983. $^{13}C/^{12}C$ records in northern hemispheric trees during the past 500 years, Anthropogenic impact and climatic superstitions. J. Geophys. Res., **88**:6844–6852.

21. From, E., and C.D. Keeling. 1986. Reassessment of the late 19th century atmospheric carbon dioxide variations in the air of western Europe and the British Isles based on an unpublished analysis of contemporary air masses by G.S. Callendar. Tellus, **38B**:87–105.

22. Fung, I. 1986. Analysis of the seasonal and geographic patterns of atmospheric CO_2 distributions with a 3-D tracer model. In The Changing Carbon Cycle, A Global Analysis, Springer-Verlag, New York, pp. 459–473.

23. Fung, I., K. Prentice, E. Matthews, J. Lerner, and G. Russell. 1983. A 3-D tracer model study of atmospheric CO_2: Response to seasonal exchanges with the terrestrial biosphere. J. Geophys. Res., **88**:1282–1294.

24. Fung, I., C.J. Tucker, and K.C. Prentice. 1987. Application of advanced very high resolution radiometer vegetation index to study atmosphere–biosphere exchange of CO_2. J. Geophys. Res., **92**:2999–3015.

25. Gammon, R.H., W.D. Komhyr, and J.T. Peterson. 1986. The global atmospheric CO_2 distribution 1968–1983: Interpretation of the results of the NOAA/GMCC measurement program. In The Changing Carbon Cycle, A Global Analysis, Springer-Verlag, New York, pp. 1–15.

26. Higuchi, K. 1983. Effect of two different atmospheric models on the absorptive rate of excess atmospheric carbon by the sea. Geophys. Res. Lett., **10**:869–872.

27. Houghton, R.A. 1989. The long-term flux of carbon to the atmosphere from changes in land use. In Proceedings of the Third International Conference on Analysis and Evaluation of Atmospheric CO_2 Data Present and Past, Environmental Pollution Monitoring and Research Programme. No. 59, World Meteorological Organization, Geneva, pp. 80–85.

28. Houghton, R.A. 1991. Tropical deforestation and atmospheric carbon dioxide. Climatic Change, **19**:99–118.

29. Houghton, R.A., R.D. Boone, J.R. Fruci, J.E. Hobbie, J.M. Melillo, C.A. Palm, B.J. Peterson, G.R. Shaver, G.M. Woodwell, B. Moore, D.L. Skole, and N. Myers. 1987. The flux of carbon from terrestrial ecosystems to the atmosphere in 1980 due to changes in land use: Geographic distribution of the global flux. Tellus, **39B**:122–139.

30. Houghton, J.T., G.J. Genkins, and J.J.E. (eds.). 1990. Climate Change, the IPCC Scientific Assessment, Cambridge University Press, Cambridge.

31. Jacoby, G., and R. D'Arrigo. 1989. Reconstructed Northern Hemisphere annual

temperature since 1671 based on high-latitude tree-ring data from North America. Climatic Change, **14**:39–59.

32. Kauppi, P.E., K. Mielikainen, and K. Kuusela. 1992. Biomass and carbon budget of European forests, 1971 to 1990. Science, **256**:70–74.
33. Keeling, C.D., and M. Heimann. 1986. Meridional eddy diffusion model of the transport of atmospheric carbon dioxide, 2, Mean annual carbon cycle. J.Geophys. Res., **91**:7782–7796.
34. Keeling, C.D., R.B. Bacastow, A.F. Carter, S.C. Piper, T.P. Whorf, M. Heimann, W.G. Mook, and H. Roeloffzen. 1989. A three dimensional model of atmospheric CO_2 transport based on observed winds: 1. Analysis of observational data. In Aspects of climate variability in the Pacific and the Western America, Geophysical Monograph 55, American Geophysical Union, Washington, DC, pp. 165–236.
35. Keeling, C.D., S. Piper, and M. Heimann. 1989. A three-dimensional model of atmospheric CO_2 transport based on observed winds: 4. Mean annual gradients and interannual variations. In Aspects of climate variability in the Pacific and the Western America, Geophysical Monograph 55, American Geophysical Union, Washington, DC, pp. 305–363.
36. Komhyr, W.D., R.H. Gammon, T.B. Harris, L.S. Waterman, T.J. Conway, W.R. Taylor, and K.W. Thoning. 1985. Global atmospheric CO_2 distribution and variations from 1968–1982 NOAA/GMCC CO_2 flask sample data. J. Geophys. Res., **90**:5567–5596.
37. Maier-Reimer, E. 1991. Personal communication.
38. Neftel, A., E. Moore, H. Oeschger, and B. Stauffer. 1985. Evidence from polar ice cores for the increase in atmospheric CO_2 in the past two centuries. Nature, **315**: 45–47.
39. Oeschger, H., and B. Stauffer. 1986. Review of the history of atmospheric CO_2 recorded in ice cores. In The Changing Carbon Cycle, A Global Analysis, Springer-Verlag, New York, pp. 89–108.
40. Oeschger, H., U. Siegenthaler, U. Schotterer, and A. Gugelmann. 1975. A box diffusion model to study the carbon dioxide exchange in nature. Tellus, **27**:168–192.
41. Pearman, G.I., and P. Hyson. 1981. The annual variation of atmospheric CO_2 concentration observed in the northern hemisphere. J. Geophys. Res., **86**:9839–9843.
42. Pearman, G.I., and P. Hyson. 1986. Global transport and inter-reservoir exchange of carbon dioxide with particular reference to stable isotopic distributions. J. Atmos. Chem., **4**:81–124.
43. Pearman, G.I., P. Hyson, and P.J. Fraser. 1983. The global distribution of atmospheric carbon dioxide, 1, Aspects of observations and modeling. J. Geophys. Res., **88**:3581–3590.
44. Pearman, G.I., D. Etheridge, F. deSilva, and P.J. Fraser. 1986. Evidence of changing concentrations of atmospheric CO_2, N_2O and CH_4 from air bubbles in Antarctic ice. Nature, **320**:248–250.

45. Peng, T.-H., and W.S. Broecker. 1985. The utility of multiple tracer distributions in calibrating models for uptake of anthropogenic CO_2 by the ocean thermocline. J. Geophys. Res., **90**:7023–7035.

46. Peng, T.-H., W.S. Broecker, H.D. Freyer, and S. Trumbore. 1983. A deconvolution of the tree-ring based $\delta^{13}C$ record. J.Geophys. Res., **88**:3609–3620.

47. Quay, P.D., B. Tilbrook, and C.S. Wong. 1992. Oceanic uptake of fossil fuel CO_2: Carbon-13 evidence. Science, **256**:74–79.

48. Raynaud, D., and J.M. Barnola. 1985. An Antarctic ice core reveals atmospheric CO_2 variations over the past few centuries. Nature, **315**:309–311.

49. Sarmiento, J.L. 1983. A simulation of bomb tritium entry into the Atlantic Ocean. J. Phys. Oceanogr., **13**:1924–1939.

50. Sarmiento, J.L., J.C. Orr, and U. Siegenthaler. 1991. A perturbation simulation of CO_2 uptake in an ocean general circulation model. J.Geophys. Res., **97**:3621–3646.

51. Schlesinger, W.H. 1990. Evidence from chronosequence studies for a low carbon-storage potential of soils. Nature, **348**:232–234.

52. Shiller, A.M. 1981. Calculating the oceanic CO_2 increase: A need for caution. J. Geophys. Res., **86**:11083–11088.

53. Shiller, A.M. 1982. Reply to comments of Chen, Millero, and Pytkowica. J. Geophys. Res., **87**:2086.

54. Siegenthaler, U. 1983. Uptake of excess CO_2 by an outcrop-diffusion model of the ocean. J. Geophys. Res., **88C**:3599–3608.

55. Siegenthaler, U., and K.O. Munnich. 1981. $^{13}C/^{12}C$ fraction during CO_2 transfer from air to sea. In Carbon Cycle Modelling, SCOPE 16, J. Wiley and Sons, Chichester, pp. 249–257.

56. Siegenthaler, U., and H. Oeschger. 1987. Biospheric CO_2 emissions during the past 200 years reconstructed by deconvolution of ice core data. Tellus, **39B**:140–154.

57. Stuiver, M., R.L. Burk, and P.D. Quay. 1984. $^{13}C/^{12}C$ ratios in tree rings and the transfer of biospheric carbon to the atmosphere. J. Geophys. Res., **89D**:11731–11748.

58. Takahashi, T., D. Chipman, and T. Volk. 1983. Geographical, seasonal, and secular variations of the partial pressure of CO_2 in surface waters of the North Atlantic Ocean: the results of the North Atlantic TTO Program. In Proceedings: Carbon Dioxide Research Conference: Carbon Dioxide, Science and Consensus CONF-820970, U.S. Dept. of Energy, pp. II.123–II.143.

59. Takahashi, T., D. Chipman, W. Smethie, J. Goddard, S. Trumbore, G. Mathieu, and S. Sutherland. 1985. Assessment of carbon dioxide sink/source in the oceanic areas: The results of 1982–1984 investigation. Final Technical Report for Contract DE-AC02-81-ER600000 A & B to the Department of Energy, 53, Lamont-Doherty Geological Observatory, Palisades, NY.

60. Takahashi, T., J. Goddard, S. Sutherland, D. Chipman, and C. Breeze. 1986. Seasonal and geographic variability of carbon dioxide sink/source in the oceanic areas: Observations in the North and equatorial Pacific Ocean: 1984–1986 and

global summary. Final Technical Report for Contract MM 19X-89675C to the Department of Energy, 66, Lamont-Doherty Geological Observatory, Palisades, NY.

61. Takahashi, T., J. Goddard, S. Sutherland, G. Mathieu, and D. Chipman. 1988. Assessment of carbon dioxide sink/source in the North Pacific Ocean: Seasonal and geographic variability, 1986–1987. Final Technical Report for Contract 19X-89675C to the Department of Energy, 66p, Lamont-Doherty Geological Observatory, Palisades, NY.

62. Takahashi, T., G. Mathieu, D. Chipman, J. Goddard, and L.-M. Ma. 1991. Assessment of carbon dioxide sink/source in the North Pacific Ocean. Final Technical Report for Contract 19X-91319C to the Department of Energy, 38, Lamont-Doherty Geological Observatory, Palisades, NY.

63. Takahashi, T., J. Goddard, D. Chipman, S. Sutherland, and G. Mathieu. 1991. Assessment of carbon dioxide sink/source in the North Pacific Ocean: Seasonal and graphic variability, 1984–1989. Final Technical Report for Contract 19X-SC248C to the Department of Energy, 111, Lamont-Doherty Geological Observatory, Palisades, NY.

64. Tans, P.P., T.J. Conway, and T. Nakazawa. 1989. Latitudinal distribution of the sources and sinks of atmospheric carbon dioxide derived from surface observations and an atmospheric transport model. J. Geophys. Res., **94D**:5151–5172.

65. Tans, P.P., I.Y. Fung, and T. Takahashi. 1990. Observational constraints on the global atmospheric CO_2 budget. Science, **247**:1431–1438.

66. Toggweiler, J.R., K. Dixon, and K. Bryan. 1989. Simulations of radiocarbon in a coarse-resolution, world ocean model, I: Steady state, pre-bomb distribution. J. Geophys. Res., **94**:8217–8242.

67. Toggweiler, J.R., K. Dixon, and K. Bryan. 1989. Simulations of radiocarbon in a coarse-resolution, world ocean model, II: Distributions of bomb-produced ^{14}C. J. Geophys. Res., **94**:8243–8264.

68. Wofsy, S.C., J.E. Munger, P.S. Bakwin, B.C. Daube, and T.R. Moore. 1991. Net CO_2 uptake by northern woodlands. Trans. Am. Geophys. Union (abstract), **72**:76.

Terrestrial Systems

Agriculture, the Global Nitrogen Cycle, and Trace Gas Flux

Peter M. Vitousek

Department of Biological Sciences, Stanford University, Stanford, California 94305

Pamela A. Matson

Ecosystem Science and Technology Branch, NASA–Ames Research Center, Moffett Field, California 94035

ABSTRACT

The global budget of nitrous oxide is out of balance, with sources exceeding sinks by 30–40%, and our understanding of the causes of this imbalance is inadequate. Moreover, our knowledge of the budgets of other nitrogen-containing trace gases is worse than that for nitrous oxide. An analysis of the overall global nitrogen cycle and anthropogenic perturbations to it suggest that the global cycle has been altered to a greater extent than has nitrous oxide, in that fluxes of nitrogen across ecosystem boundaries have more than doubled as a consequence of human activity. The dominant sources of change to the overall cycle are agricultural intensification and land-use change, particularly in tropical regions. These processes should be evaluated carefully as causal factors for imbalances in global budgets of nitrogen-containing trace gases.

Introduction

The nitrogen cycle is second only to carbon in its potential to drive global change on the time scale of decades to centuries. Nitrogen-containing trace gases contribute significantly—and increasingly—to both radiative and chemical changes in Earth's atmosphere, and increased runoff of nitrogen to freshwater, estuarine, and coastal marine ecosystems can alter the dynamics of recipient ecosystems substantially. Moreover, nitrogen interacts strongly with the carbon cycle, regulating carbon flux in many ecosystems.[75] Finally, the nitrogen cycle is altered more by human activity than are other element cycles, in the sense that fluxes of nitrogen across ecosystem boundaries (biosphere–atmosphere, biosphere–hydrosphere) have been altered more for nitrogen than for any other element.

Despite the importance of the nitrogen cycle and the global significance of its alteration, the cycle remains poorly understood in important ways. For example, the radiatively active gas nitrous oxide is increasing in concentration at about 0.25% per year—a small change that, nevertheless, represents at least a 30% imbalance between sources and sinks.[56] The major source of nitrous oxide globally is tropical soils.[33,42] Until fairly recently, the major cause of the global increase was thought to be fossil-fuel combustion,[27] but it is now clear that fossil-fuel use makes only a small direct contribution.[37,54] Just what does cause the increase is uncertain[80]—although inverse modeling of sources based on atmospheric measurements demonstrates that much of the global increase is due to tropical sources.[56] However, flux measurements of nitrous oxide in the tropics are sparse, particularly in the human-modified areas that potentially could be sources of global change. For example, a recent review of nitrous oxide emissions from fertilized land summarized results from 104 sites, but none of them were south of California.[21]

Fluxes of other nitrogen-containing trace gases are still less well-defined, and tropical systems may be equally important in their dynamics. Nitric oxide contributes (often in a rate-limiting way[28]) to the formation of tropospheric ozone. Nitrous oxide is produced in large quantities by biomass-burning,[14] but there is little information on soil emissions from intensive agricultural sites in the tropics, or their consequences for atmospheric chemistry. Finally, there is increasing evidence that methane emission/absorption in nonflooded soil interacts strongly with the nitrogen cycle,[53,72] but these interactions have not been examined fully in human-modified tropical areas.

In this chapter, we review the global budget of nitrous oxide, showing that the imbalance between inputs to and losses from the atmosphere cannot be explained easily on the basis of available information. We then evaluate sources of human-caused global change in the nitrogen cycle, emphasizing changes in fluxes across ecosystem boundaries and demonstrating that agriculture is the dominant source of global change. Finally, we discuss the implications of these changes in the nitrogen cycle for fluxes of radiatively active trace gases.

Background—Sources of Increase in Nitrous Oxide

Interest in the source of tropospheric nitrous oxide dates to at least the early 1970s; at that time, the potential for stratospheric ozone depletion as a consequence of increased nitrous oxide was the motivating concern.[14,46] The fact that nitrous oxide concentrations were increasing globally was demonstrated clearly by Weiss,[81] and those observations were confirmed and extended by a number of subsequent studies.[11,58] Fertilizer nitrogen was one of the first potential sources of the increase to be examined; at that time, field studies evaluated temperate agricultural soils, where the vast majority of fertilizer nitrogen was then applied.[2,19,21,51,63,68,70] It was concluded that although nitrogen fertilizer could be a significant source of nitrous oxide, accounting for perhaps 0.5–1.0 Tg of the 3.5 Tg annual increment in atmospheric N_2O, it was not the dominant factor causing the observed increases in tropospheric concentrations.

The first more or less complete and balanced global budget for nitrous oxide was prepared by McElroy and Wofsy.[47,83] They used extrapolations of an initial set of flux and process measurements to conclude that the major background source of nitrous oxide (prior to widespread human activity) was tropical forest soils[33,34] and that the most important cause of the recent increase in its concentrations was production of nitrous oxide during fossil-fuel combustion in large industrial boilers.[27]

McElroy and Wofsy's conclusion that tropical forests are the major natural source of nitrous oxide has held up well through the collection of many more flux and process measurements[26,35,38,40,41,45,64,79] and the development of alternative regional budgets,[42,64] although estimates of the absolute magnitude of that natural tropical source have been reduced. Moreover, measurements of changes in atmospheric nitrous oxide concentrations at a number of stations, and calculations of source distributions based on them,[11,56] suggest that the majority of nitrous oxide is emitted in the tropics. The importance of the tropical forest source of nitrous oxide is consistent with the biogeochemistry of most tropical forests; on average, nitrogen circulation through biomass and rates of soil nitrogen transformations are much greater in both absolute terms and relative to other elements in tropical as compared with temperate or boreal forest ecosystems.[76,78] Nitrogen appears to function as an excess nutrient in the majority of lowland tropical forests.

In contrast, the suggestion that the increase in atmospheric concentrations of nitrous oxide was due to fossil-fuel combustion[27,47] is in error. A wholly unanticipated set of reactions in stored samples produces nitrous oxide in the presence of high concentrations of oxides of nitrogen, sulfur dioxide, and water vapor[37,54]; these reactions produce most of the nitrous oxide previously attributed to fossil-fuel combustion. Fossil fuels probably account for less than 10% of the annual increase in tropospheric nitrous oxide.

A number of other possible sources for increased nitrous oxide have been identified. These include tropical land-clearing,[40] biomass burning,[13] degassing of irrigation water,[49,62] nitrogen saturation of natural ecosystems caused by enhanced deposition of atmospheric nitrogen, which itself results from combustion or agricultural activity upwind,[1,48] and specialized industrial processes.[74] Each of these contributes at least a little nitrous oxide to the atmosphere, but it is difficult to find enough to account for the measured global increase even by pushing all of them toward their upper limits.[80]

Overall, it seems fair to suggest that our understanding of the global source budget for nitrous oxide is less certain now than we thought it was a few years ago. One way out of this impasse is to examine the magnitude of recent global changes in the nitrogen cycle as a whole–particularly in those components that involve human-caused alterations in fluxes across ecosystem boundaries. We can then examine changes in trace gas budgets in the context of changes in the overall cycle.

The Global Nitrogen Cycle

The global nitrogen cycle as it was understood as of the early 1980s is summarized in Table 1. Many of the fluxes in Table 1 are uncertain and/or based on decade-old information; it would be rewarding to update and reevaluate our understanding of the global cycle. Nevertheless, this summary provides a useful background against which to examine human-caused changes to the nitrogen cycle.

From our point of view, the most important changes in the nitrogen cycle are those that involve movement of nitrogen across the boundaries between land and atmosphere, land and water, and water and atmosphere; these changes have the potential to alter the flux of nitrogen through the processes that produce trace gases. Human-caused changes of this kind include those that alter the overall rate of fixation of atmospheric nitrogen, and those that increase the mobilization and loss of plant and soil nitrogen at greater than background rates. Such changes increase the amount of nitrogen cycling through dynamic forms (ammonium, nitrate, and labile organic nitrogen) within soils, water columns, or sediments. A small and variable fraction of this dynamic nitrogen can then "leak" as nitrous

Table 1. The major pools and fluxes of nitrogen in terrestrial ecosystems under background (unperturbed) conditions. Information summarized from Soderlund and Rosswall[71] and Schlesinger,[66] supplemented by sources described in the text.

Pools (in Pg)	
Soil	95
Plants	10
Annual Fluxes (in Tg)	
Through plants	1200
Biological N fixation in terrestrial ecosystems[a]	100
Fixation by lightning	10
River runoff	20
Denitrification	90

[a]An additional 5–20 Tg of N is fixed within marine ecosystems.[10]

Table 2. Magnitudes of global changes in nitrogen fixation or nitrogen mobilization and loss as a consequence of human activity. All values are means, or in the middle of reported ranges. Sources described in the text.

Process	N Flux (Tg)
New fixation	
Industrial N fixation (fertilizer N)	80
N fixation by legume crops	30
N fixation by fossil-fuel combustion	25
Mobilized by human activity	
Biomass-burning	40
Flux from cleared tropical lands	22

oxide and other nitrogen-containing trace gases each time it cycles through biologically available forms.[23]

The nitrogen cycle as it existed prior to substantial human alteration included fluxes of approximately 110 Tg of N from the atmosphere into biologically available forms in terrestrial ecosystems, with ~90% of this flux due to biological nitrogen fixation and the remainder due to lightning. (It must be acknowledged that the magnitude of natural biological nitrogen fixation globally is not at all well-known.) Losses of nitrogen from terrestrial ecosystems were approximately equivalent to inputs, with 20 Tg going to marine ecosystems and the remainder denitrified on land or in fresh water. Human alterations to the nitrogen cycle are vast relative to this background. They include deliberate industrial fixation of nitrogen for use in fertilizer, the widespread use of leguminous crops that fix atmospheric nitrogen, and incidental fixation of atmospheric nitrogen during fossil-fuel combustion (most dramatically in internal combustion engines). In addition, processes whereby human activity mobilizes stored soil and plant nitrogen at accelerated rates include biomass-burning and land-clearing/conversion to agricultural use. These changes are summarized in Table 2; we discuss each briefly below.

Nitrogen Fertilizer

In the late 1970s, global nitrogen fertilizer use was estimated at 60 Tg/yr, most of it in the temperate zone. Since then, however, the rate of increase in nitrogen fertilization has been slow in the developed countries of the temperate zone and extraordinarily rapid in the tropics. By the late 1980s, global nitrogen fertilizer use was approximately 80 Tg/yr, with more than half of that in the developing world (including China).[22] Moreover, the momentum of population growth ensures that this pattern will continue (or we will face worse problems than nitrous oxide); current projections suggest that 100 Tg/yr of nitrogen fertilizer will be applied in the developing world in 2025, versus <50 Tg in the now-developed countries.[20] Overall, industrial fixation of nitrogen for fertilizer represents the most substantial human-caused change in the global nitrogen cycle; it is increasingly concentrated in the tropics, and its significance as a source for increased fluxes of nitrogen trace gases appears worth a reexamination.

Leguminous Crops

The nitrogen fixed by leguminous crops (over and above that which would have been fixed by native vegetation occupying these sites) represents a human-caused net addition of nitrogen to terrestrial ecosystems. The magnitude of this flux was estimated at 30 Tg/yr in the early 1980s[71]; the expansion of leguminous crop area since that time suggests that it has increased, but no direct information is available. There are few published data on the magnitude of nitrogen trace gas fluxes from leguminous crops anywhere[21]; those few show a significant increase in flux compared to similarly managed nonlegumes. None of the measurements were carried out in tropical regions.

Incidental Fixation During Fossil-Fuel Combustion

Approximately 25 Tg of atmospheric nitrogen are fixed by fossil-fuel combustion, particularly in internal combustion engines. This nitrogen is mostly reactive NO_x; it can be deposited in that form downwind of where it is produced, or it can be oxidized to nitric acid and contribute to acid rain in downwind areas. Where it is deposited, recipient systems are fertilized with nitrogen, potentially leading to elevated fluxes of nitrogen-containing trace gases in those areas.[1,48] This process could be responsible for the elevated nitrous oxide fluxes observed in German forests subject to high rates of atmospheric nitrogen loading,[67] although similarly high fluxes have not been observed in acid rain-affected areas of eastern North America.[8,33] (Not all of the enhanced nitrogen deposition downwind of industrial areas is derived from combustion; much of it comes from volatilization of ammonia from nitrogen fertilizer and animal manures in upwind agricultural areas.)

Biomass-Burning

Combustion mobilizes a large fraction of the nitrogen in the burned biomass into the atmosphere[57,69]; accelerated mobilization of soil nitrogen also occurs for a period of time following fire.[3,44] Burning itself produces relatively little nitrous oxide[13,15] (from 0.2 to, at most, 1 Tg/yr), but it does mobilize between 24 and 57 Tg of nitrogen from plants to the atmosphere.[15] Perhaps half of that is lost as dinitrogen, and so plays no immediate additional biogeochemical role[39]; the rest significantly alters tropospheric chemistry before being deposited in downwind ecosystems, where it may alter nitrogen-cycling processes in the direction of greater production of nitrogen-containing trace gases. Just how much of the flux from biomass-burning represents a human-caused alteration is not certain—the frequency and effectiveness of natural fires is poorly known in many biomes. Nevertheless, it is clear that humans have increased this flux substantially, particularly in the tropics, and continue to do so.

Land-Clearing and Conversion

Most sites lose substantial quantities of carbon and nitrogen from soils when they are converted to agricultural use.[7,29,50,65] These changes are particularly large and rapid in tropical regions, where up to half of the carbon and nitrogen in surface soil can disappear in less than five years following land conversion. Matson et al.[44] estimated that 20–25 Tg of nitrogen are mobilized and lost annually by tropical land-clearing—in addition to the nitrogen lost during biomass-burning. Knowledge of the fate of this mobilized nitrogen is crucial to determining its consequences, but we have relatively little information on pathways of nitrogen loss in cleared tropical sites. A number of studies have reported increased nitrous oxide emissions from disturbed sites,[9,34,45] but such observations are not universal.[25,26] Luizão et al.[40] reported substantially elevated nitrous oxide losses from several tropical pastures ranging in age from 3 to 10 years, but no increase in fluxes from a recently cleared and burned tropical forest site. They suggested that the increase in fluxes from pastures could be due to altered soil aeration caused by soil compaction by ungulates, but alternative possibilities (i.e., increased oxidizable carbon from grass roots, increased soil temperature due to the effects of clearing) could also contribute.

Other Sources of Change

A number of other processes could contribute to changes in the overall global nitrogen budget or in the fluxes of nitrogen-containing trace gases. Drainage of organic soils for agriculture or forestry leads to rapid losses of soil carbon and nitrogen; these have been associated with very high local nitrous oxide fluxes.[19] Their global significance, however, is uncertain. Assuming a carbon : nitrogen ratio in peat of 20 and Armentano's[4] estimate of 30–370 Tg/yr of soil carbon oxidized in drained peatlands, from 1.5 to 18 Tg/yr of nitrogen could be mobilized annually by this pathway.

On the other hand, whereas cleared and converted sites mobilize and lose soil nitrogen, successional sites accumulate nitrogen in plants and soil. This accumulation is associated with reduced losses of nitrogen to streamwater and the atmosphere.[59–61,77] Consequently, where succession predominates over land-clearing (as in much of the temperate zone), current fluxes of nitrogen from land to the atmosphere are likely to be reduced as a consequence of past human activity. The implications of this accumulation in biomass are analogous to those in the carbon cycle, where it is necessary to compare losses of carbon caused by deforestation with carbon accumulation in successional vegetation and soils.[18] This problem is perhaps less severe for nitrogen than carbon, in that the tropical systems undergoing net land-clearing are much richer in nitrogen (and, hence, have higher potential trace gas fluxes) than the temperate systems undergoing net recovery.

Implications for Trace Gas Flux

Nitrous Oxide

Overall, human activity causes greatly increased movement of nitrogen between the atmosphere and terrestrial ecosystems. Nitrogen fixation, in particular, has more than doubled in comparison with background rates, a human-caused perturbation much greater in relative terms than any to the carbon cycle. In this sense, nitrous oxide fluxes have been altered less than has the overall nitrogen cycle, in that global sources and sinks of the gas are now "only" 30–40% out of balance—but that may simply reflect a lag before the increased rates of nitrogen fixation globally are matched by similarly increased rates of nitrous oxide emission.

One striking aspect of human modifications to the nitrogen cycle (Table 2) is the predominance of agriculture and land use among the causes of change. The production of nitrogen fertilizer and the planting of legume crops together fix more nitrogen annually than is thought to have been fixed in all natural ecosystems worldwide. The effects of this added nitrogen on nitrous oxide fluxes from temperate agricultural systems have received some attention, as discussed above; direct emissions from fertilized land probably represent a significant (0.5–1.0 Tg) but not controlling flux in a global context. There has been relatively little work in tropical agricultural systems, however, where approximately half of the fertilizer nitrogen used globally is now applied. The major exception is some research on nitrous oxide fluxes from rice paddies shortly after fertilization[52]; as would be expected, little nitrous oxide is produced in these permanently flooded soils. However, measurements have not been extended through a complete cropping cycle for paddy rice, or to other tropical crops.

There is reason to think that nitrous oxide emissions per unit of fertilizer nitrogen may prove to be greater in the tropics than the temperate zone. Nitrous oxide emissions from natural ecosystems are much greater on an areal basis from tropical than temperate forests[8,42], as are emissions per unit of nitrogen cycled between vegetation and soils.[41,78] Every time nitrogen passes through inorganic forms in tropical forest soil, the chance that it will be lost is greater than that in temperate sites. The reasons for this difference are not well-known; they may be related to the fact that nitrogen functions as an excess nutrient in most lowland tropical forests (which generally appear to be phosphorus or calcium limited) in comparison to higher latitude systems (which are generally nitrogen limited).[78] In any case, *if* the same comparison applies to tropical agricultural systems (now wholly unknown), the fraction of fertilizer N lost as nitrous oxide could be higher than that from temperate agricultural ecosystems. An understanding of fluxes of nitrous oxide from tropical agriculture would, therefore, be useful; there is a possibility that they represent the missing source that could explain the currently imbalanced global nitrous oxide budget.

Nitric Oxide

Nitric oxide is produced by the same processes that produce nitrous oxide, although generally in different ratios. Consequently, the same alterations to the nitrogen cycle that cause increased nitrous oxide emission could increase nitric oxide as well. Indeed, nitric oxide fluxes tend to be elevated in the same sites as nitrous oxide fluxes, although often at different times during the year or at different phases of wetting–drying cycles.[2,16,17,43] The global cycle of nitric oxide is more difficult to analyze than is nitrous oxide; NO is reactive in the troposphere, and so its global budget is not as well constrained by global measurements (although progress is being made; see Ref. 55). Moreover, emissions from the soil do not necessarily represent inputs into the troposphere; a large fraction of NO can react and be deposited in plant canopies.[5,6] Nevertheless, tropical forest and grassland soils appear to represent a greater source of nitric oxide than do temperate soils,[30–32] as is the case for nitrous oxide.

The dominant process that alters tropospheric nitric oxide concentrations in tropical regions is biomass-burning.[15,39] However, background fluxes against which to compare emissions of nitric oxide in fires are not well-known: There is limited information on intact forests and savannas in the tropics,[5,16,31,32,64] and next to nothing on agricultural systems (save agricultural-burning). Emissions from temperate agricultural systems generally exceed those from temperate forests; they increase with increasing nitrogen availability in the source ecosystem.[82] Moreover, the absence of a plant canopy at times during the cropping cycle allows a larger fraction of the emitted nitrogen to reach the atmosphere in annual crops than in forest ecosystems. It seems likely, therefore, that agricultural intensification increases nitric oxide fluxes even in the absence of biomass-burning—and it would be useful to know if any such increase could be sufficient to trigger the formation of elevated regional concentrations of tropospheric ozone.

Methane

The rapid global increase in methane concentrations is well-known to be driven in large part by tropical agriculture and land-use change; rice agriculture, livestock, biomass-burning, and increased termite populations have received substantial attention.[12,24] In addition, there is evidence that changes in the nitrogen cycle could also contribute changing methane fluxes. Steudler et al.[72] and Mosier et al.[53] recently demonstrated an association between methane uptake by nonflooded temperate soils and the nitrogen status of those soils. Both found that nitrogen fertilization caused an increase in nitrous oxide flux and a decrease in methane uptake by soils; the effect was very clear, although small enough that its global significance remains uncertain. Keller et al.[36] observed that land-clearing and conversion to agriculture decreased methane uptake in Panamanian soils, an observation consistent a nitrogen effect (although nitrogen was neither measured nor manipulated directly). If land-clearing and agricultural intensification (nitrogen fertilization, legume monocultures) generally have this effect in tropical regions, they could represent a small additional contribution to the global increase in tropospheric methane.

Conclusions

Overall, global alteration of the nitrogen cycle through land-use change and especially agricultural intensification is so substantial that we consider the relative *lack* of response in trace nitrogen species in the atmosphere to be somewhat surprising. Many of the potential sources of nitrogen trace gas flux (direct emissions from cleared land, increasing use of nitrogen fertilizer, emissions from drainage water and contaminated aquifers, emissions from "nitrogen saturated" downwind ecosystems, biomass-burning) can be related to agriculture; a reexamination of agricultural practices (particularly in the tropics) as a source for nitrogen trace gases, therefore, appears potentially rewarding.

Acknowledgments

This research was supported by NASA Biospherics Research Grant NAGW-1068. We thank Eric Davidson and Philip Robertson for helpful discussions of this topic, and Cheryl Nakashima for preparing the manuscript.

References

1. Aber, J.D., J.K. Nadelhoffer, P.A. Steudler, and J.M. Melillo. 1989. Nitrogen saturation in northern forest ecosystems: hypotheses and implications. Bioscience, **39**:378–386.
2. Anderson, I.C., and J.S. Levine. 1987. Simultaneous field measurements of biogenic emissions of nitric oxide and nitrous oxide. J. Geophys. Res. **92**:965–957.
3. Anderson, I.C., J.S. Levine, M.A. Poth, and P.J. Riggan. 1988. Enhanced biogenic emissions of nitric oxide and nitrous oxide following surface biomass burning. J. Geophys. Res. **93**:3893–3898.
4. Armentano, T.V. 1980. Drainage of organic soils as a factor in the world carbon cycle. BioScience **30**:825–830.
5. Bakwin, P.S., S.C. Wofsy, S.-M. Fan, M. Keller, S.E. Trumbore, and J.M. da Costa. 1990. Emission of nitric oxide from tropical forest soils and exchange of N_2O between the forest canopy and atmospheric boundary layers. J. Geophys. Res. **95**:755–16764.
6. Bakwin, P.S., S.C. Wofsy, and S.-M. Fan. 1990. Measurements of reactive nitrogen oxides (NO_y) within and above a tropical forest canopy in the wet season. J. Geophs. Res. **95**:16765–16772.
7. Bolin, B., P.J. Crutzen, P.M. Vitousek, R.G. Woodmansee, E.D. Goldberg, and R.B. Cook. 1983. Interactions of biogeochemical cycles. In B. Bolin and R.B. Cook (eds.), The Biogeochemical Cycles and Their Interactions, J. Wiley and Sons, Chichester, pp. 8–40.
8. Bowden, R.D., P.A. Steudler, J.M. Melillo, and J.D. Aber. 1990. Annual nitrous oxide fluxes from temperate forest soils in the northeastern United States. J. Geophys. Res. **95**:13997–14005.
9. Bowden, W.B., and F.H. Bormann. 1986. Transport and loss of nitrous oxide in soil water after forest clear-cutting. Science **233**:867–869.
10. Carpenter, E.J., and D.G. Capone. 1983. Nitrogen fixation by marine *Oscillatoria (Trichodesmium)* in the world's oceans. In E.J. Carpenter and D.G. Capone (eds.), Nitrogen in the Marine Environment, Academic Press, New York, pp. 65–103.
11. Cicerone, R. 1989. Analysis of sources and sinks of atmospheric nitrous oxide (N_2O). J. Geophys. Res. **94**:18265–19272.
12. Cicerone, R.J., and R.S. Oremland. 1988. Biogeochemical aspects of methane cycling. Global Biogeochem. Cycles **2**:299–327.
13. Cofer, W.R., J.S. Levine, E.L. Winstead, and B.J. Stocks. 1991. New estimates of nitrous oxide emissions from biomass burning. Nature **349**:689–691.
14. Crutzen, P. 1970. The influence of nitrogen oxides on the atmospheric ozone content. Quart. J. Roy. Met. Soc. **96**:320–325.
15. Crutzen, P.J., and M.O. Andreae. 1990. Biomass burning in the tropics: impact on atmospheric chemistry and biogeochemical cycles. Science **250**:1669–1678.

16. Davidson, E.A., P.M. Vitousek, P.A. Matson, R. Riley, G. García-Méndez, and J.M. Maass. 1991. Soil emissions of nitric oxide in a seasonally dry tropical forest of Mexico. J. Geophys. Res. **96**:15439–15445.

17. Davidson, E.A., P.A. Matson, P.M. Vitousek, R. Riley, K. Dunkin, G. García-Méndez, and J.M. Maass. 1993. Process regulation of soil emissions of NO and N_2O in a seasonally dry tropical forest. Ecology **74**:130–139.

18. Detwiler, R.P., and C.A.S. Hall. 1988. Tropical forests and the global carbon cycle. Science **239**:43–47.

19. Duxbury, J.M., D.R. Bouldin, R.E. Terry, and R.L. Tait III. 1982. Emissions of nitrous oxide from soils. Nature **298**:462–464.

20. EPA. 1990. Greenhouse gas emissions from agricultural ecosystems, IPCC Report, Washington, D.C.

21. Eichner, M.J. 1990. Nitrous oxide emissions from fertilized soils: summary of the available data. J. Env. Qual. **19**:272–280.

22. FAO. 1985. FAO Fertilizer Yearbook. United Nations Food and Agricultural Organization, Rome.

23. Firestone, M.K., and E.A. Davidson. 1989. Microbiological basis of NO and N_2O production and consumption in soil. In M.O. Andreae and D.S. Schimel (eds.), Exchange of Trace Gases Between Terrestrial Ecosystems and the Atmosphere, J. Wiley & Sons, New York.

24. Fung, I., J. John, J. Lerner, E. Matthews, M. Prather, L.P. Steele, and P.J. Fraser. 1991. Three-dimensional model synthesis of the global methane cycle. J. Geophys. Res. **96**:13033–13065.

25. García-Méndez, G., J.M. Maass, P.A. Matson, and P.M. Vitousek. 1991. Nitrogen transformations and nitrous oxide flux in a tropical deciduous forest in Mexico. Oecologia **88**:362–366.

26. Goreau, T.J., and W.Z. de Mello. 1987. Effects of deforestation on source and sinks of atmospheric carbon dioxide, nitrous oxide, and methane from central Amazonian soils and biota during the dry season: a preliminary study. In E. Salati, P.B. Vose, and T. Lovejoy (eds.), Biogeochemistry of Tropical Rainforests: Problems for Research, CENA, Piracicaba, Brazil.

27. Hao, W.M., S.C. Wofsy, M.B. McElroy, J.M. Beer, and M.A. Togan. 1987. Sources of atmospheric nitrous oxide from combustion. J. Geophys. Res. **92**:3098–3104.

28. Jacob, D.J., and S.C. Wofsy. 1990. Budgets of reactive nitrogen, hydrocarbons, and ozone over the Amazon forest during the wet season. J. Geophys. Res. **95**: 16737–16754.

29. Jenny, H. 1980. Soil Genesis with Ecological Perspectives, Springer-Verlag, New York.

30. Johansson, C., H. Rodhe, and E. Sanhueza. 1988. Emission of NO in a tropical savanna and cloud forest during the dry season. J. Geophys. Res. **93**:7180–7192.

31. Johansson, C., and E. Sanhueza. 1988. Emission of NO from savanna soils during the rainy season. J. Geophys. Res. **93**:14193–14198.

32. Kaplan, W.A., S.C. Wofsy, M. Keller, and J.M. da Costa. 1988. Emission of NO and deposition of O_3 in a tropical forest system. J. Geophys. Res. **93**:1389–1395.

33. Keller, M., T. Goreau, S. Wofsy, W. Kaplan, and M. McElroy. 1983. Production of nitrous oxide and consumption of methane by forest soils. Geophys. Res. Lett. **10**:1156–1159.

34. Keller, M., W.A. Kaplan, and S.C. Wofsy. 1986. Emissions of N_2O, CH_4, and CO_2 from tropical soils. J. Geophys. Res. **91**:11791–11802.

35. Keller, M., W. Kaplan, S. Wofsy, and J.M. da Costa. 1988. Emissions of N_2O from tropical soils: Response to fertilization with NH_4^+, NO_3^-, and PO_4^{3-}. J. Geophys. Res. **93**:1600–1604.

36. Keller, M., M.E. Mitre, and R.F. Stallard. 1990. Consumption of atmospheric methane in soils of central Panama: effects of agricultural development. Global Biogeochem. Cycles **4**:21–27.

37. Linak, W.P., J.A. McSorley, R.E. Hall, J.V. Ryan, R.K. Srivastava, J.O.L. Wendt, and J.B. Mareb. 1990. Nitrous oxide emissions from fossil fuel combustion. J. Geophys. Res. **95**:7533–7541.

38. Livingston, G., P. Vitousek, and P. Matson. 1988. Nitrous oxide flux and nitrogen transformations across a landscape gradient in Amazonia. J. Geophys. Res. **93**: 1593–1599.

39. Lobert, J.M., D.H. Scharffe, W.M. Hao, and P.J. Crutzen. 1990. Importance of biomass burning in the atmospheric budget of nitrogen-containing trace gases. Nature **346**:552–554.

40. Luizão, F., P. Matson, G. Livingston, R. Luizão, and P. Vitousek. 1989. Nitrous oxide flux following tropical land clearing. Global Biogeochem. Cycles **3**:281–285.

41. Matson, P.A., and P.M. Vitousek. 1987. Cross-system comparisons of soil nitrogen transformations and nitrous oxide flux in tropical forest ecosystems. Global Biogeochem. Cycles **1**:163–170.

42. Matson, P.A., and P.M. Vitousek. 1990. Ecosystem approach to a global nitrous oxide budget. Bioscience **40**:667–672.

43. Matson, P.A., and P.M. Vitousek. Biosphere-atmosphere interactions in a tropical deciduous forest ecosystem. In H.A. Mooney, E. Medina, and S.H. Bullock (eds.), The Tropical Deciduous Forest Ecosystem (in press).

44. Matson, P.A., P.M. Vitousek, J. Ewel, M. Mazzarino, and G. Robertson. 1987. Nitrogen transformations following tropical forest felling and burning on a volcanic soil. Ecology **68**:491–502.

45. Matson, P.A., P.M. Vitousek, G.P. Livingston, and N. Swanberg. 1990. Nitrous oxide flux from Amazonian forests: effects of soil fertility and disturbance. J. Geophys. Res. **95**:16789–16798.

46. McElroy, M., and J. McConnell. 1971. Nitrous oxide: A natural source of stratospheric NO. J. Atmos. Sci. **28**:1095–1098.

47. McElroy, M.B., and S.C. Wofsy. 1986. Tropical forests: Interactions with the

atmosphere. In G.T. Prance (ed.), Tropical Rain Forests and the World Atmosphere. AAAS Selected symposium 101, Westview Press, Boulder, CO.

48. Melillo, J.M., P.A. Steudler, J.D. Aber, and R.D. Bowden. 1989. Atmospheric deposition and nutrient cycling. In M.O. Andreae and D.S. Schimel (eds.), Exchange of Gases Between Terrestrial Ecosystems and the Atmosphere, J. Wiley and Sons, Chichester, pp. 263–280.

49. Minami, K., and A. Ohsawa. 1990. Emission of nitrous oxide dissolved in drainage water from agricultural land. In A.F. Bouwman (ed.), Soils and the Greenhouse Effect, J. Wiley and Sons, Chichester, pp. 502–509.

50. Moore, B., R.D. Boone, J.E. Hobbie, R.A. Houghton, J.M. Melillo, B.J. Peterson, G.I. Shaver, C.J. Vorosmarty, and G.M. Woodwell. 1981. A simple model for analysis of the role of terrestrial ecosystems in the global carbon budget. In B. Bolin (ed.), Carbon Cycle Modelling. J. Wiley and Sons, New York, pp. 365–385.

51. Mosier, A.R., W.D. Guenzi, and E.E. Schweizer. 1986. Soil losses of dinitrogen and nitrous oxide from irrigated crops in northeastern Colorado. Soil Sci. Soc. Amer. J. **50**:344–348.

52. Mosier, A.R., S.K. Mohanty, A. Bhadrachalam, and S.F. Chakravorti. 1990. Evolution of dinitrogen and nitrous oxide from the soil to the atmosphere through rice plants. Biol. Fert. Soils **9**:61–67.

53. Mosier, A., D. Schimel, D. Valentine, K. Bronson, and W. Parton. 1991. Methane and nitrous oxide fluxes in native, fertilized, and cultivated grasslands. Nature **350**:330–332.

54. Muzio, L.J. and J.C. Kramlich. 1988. An artifact in the measurement of N_2O from combustion sources. Geophys. Res. Lett. **15**(12):1369–1372.

55. Penner, J.E., C.S. Atherton, J. Dignon, S.J. Ghan, and J.J. Walton. 1991. Tropospheric nitrogen: a three-dimensional study of sources, distributions, and deposition. J. Geophys. Res. **96**:959–990.

56. Prinn, R., D. Cunnold, R. Rasmussen, P. Simmonds, F. Alyea., A. Crawford, P. Fraser, and R. Rosen. 1990. Atmospheric trends and emissions of nitrous oxide deduced from ten years of ALE–GAGE data. J. Geophys. Res. **95**:18369–18385.

57. Raison, R.J. 1979. Modification of the soil environment by vegetation fires, with particular reference to nitrogen transformations: a review. Plant and Soil **51**:73–108.

58. Rasmussen, R.A., and M.A.K. Khalil. 1986. Atmospheric trace gases: Trends and distributions over the last decade. Science **232**:1623–1624.

59. Robertson, G.P., and T.H. Rosswall. 1986. Nitrogen in West Africa: the regional cycle. Ecol. Monogr. **56**:43–72.

60. Robertson, G.P., and J.M. Tiedje. 1984. Denitrification and nitrous oxide production in successional and old-growth Michigan forests. Soil Sci. Soc. Amer. J. **48**:383–389.

61. Robertson, G.P., and J.M. Tiedje. 1988. Deforestation alters denitrification in a lowland tropical rain forest. Nature **336**:756–759.

62. Ronan, D., M. Mordeckai, and E. Almon. 1988. Contaminated aquifers are forgotten component of the global N_2O budget. Nature **335**:57–59.

63. Sahrawat, K.L., and D.R. Keeney. 1986. Nitrous oxide emissions from soils. Adv. Soil Sci. **4**:103–148.

64. Sanhueza, E., W.M. Hao, D. Scharffe, L. Donoso, and P.J. Crutzen. 1990. N_2O and NO emissions from soils of the Guyana Shield. J. Geophys. Res. **95**:22481–22488.

65. Schlesinger, W.H. 1986. Changes in soil carbon storage and associated properties with disturbance and recovery. In J.R. Trabalka, and D.E. Reichle (eds.), The Changing Carbon Cycle: A Global Analysis, Springer-Verlag, New York, pp. 194–220.

66. Schlesinger, W.H. 1991. Biogeochemistry: An Analysis of Global Change, Academic Press, San Diego.

67. Schmidt, J., W. Seiler, and R. Conrad. 1988. Emission of nitrous oxide from temperate forest soils into the atmosphere. J. Atmos. Chem. **6**:95–115.

68. Seiler, W., and R. Conrad. 1981. Field measurements of natural and fertilizer induced N_2O release rates from soils. J. Air Poll. Contr. Assn. **31**:767–772.

69. Seiler, W., and P. Crutzen. 1980. Estimates of gross and net fluxes of carbon between the biosphere and atmosphere from biomass burning. Climatic Change **2**:207–248.

70. Slemr, F., R. Conrad, and W. Seiler. 1984. Nitrous oxide emissions from fertilized and unfertilized soils in a subtropical region (Andalusia, Spain). J. Atmos. Chem. **1**:159–169.

71. Soderlund, R., and T.H. Rosswall. 1982. The nitrogen cycles. In O. Hutzinger (ed.), Handbook of Environmental Chemistry, Springer-Verlag, Berlin, pp. 62–81.

72. Steudler, P.P., R.D. Bowden, J.M. Melillo, and J.D. Aber. 1989. Influence of nitrogen fertilization on methane uptake in temperate forest soils. Nature **341**: 314–316.

73. Tans, P.O., I.Y. Fung, and T. Takahashi. 1990. Observational constraints on the global atmospheric CO_2 budget. Science **247**:1431–1439.

74. Thiemens, M.H., and W.C. Trogler. 1991. Nylon production: an unknown source of atmospheric nitrous oxide. Science **251**:932–934.

75. Vitousek, P.M., and R.W. Howarth. 1991. Nitrogen limitation on land and in the sea: how can it occur? Biogeochemistry **13**:87–115.

76. Vitousek, P.M., and P.A. Matson. 1988. Nitrogen transformations in tropical forest soils. Soil Biol. Biochem. **20**:361–367.

77. Vitousek, P.M., and W.A. Reiners. 1975. Ecosystem succession and nutrient retention: a hypothesis. BioScience **25**:376–381.

78. Vitousek, P.M., and R.L. Sanford. 1986. Nutrient cycling in moist tropical forest. Ann. Rev. Ecol. Syst. **17**:137–167.

79. Vitousek, P.M., P.A. Matson, C. Volkmann, J.M. Maass, and G. García-Mendéz.

1989. Nitrous oxide flux from seasonally-dry tropical forests: a survey. Global Biogeochem. Cycles **3**:375–382.

80. Watson, R.T., H. Rodhe, H. Oeschger, and U. Siegenthaler. 1990. Greenhouse gases and aerosols. In J.T. Houghton, G.J. Jenkins, and J.J. Ephraums (eds.), Climate Change: The IPCC Scientific Assessment, Cambridge University Press, Cambridge, pp. 1–40.

81. Weiss, R.F. 1981. The temporal and spatial distribution of tropospheric nitrous oxide. J. Geophys. Res. **86**(C8):7185–7195.

82. Williams, E.J., and F.C. Fehsenfeld. 1991. Measurement of soil nitrogen oxide emissions in three north American ecosystems. J. Geophys. Res. **96**:1033–1042.

83. WMO. 1986. Atmospheric ozone 1985: Assessment of our understanding of the processes controlling its present distribution and change, Global Ozone and Monitoring Project, Report 16, World Meteorological Organization, Geneva.

Sulfur Gas Emissions from African Savanna-Burning

B.C. Nguyen, N. Mihalopoulos, B. Bonsang, and J.P. Putaud

Centre des Faibles Radioactivités, Laboratoire mixte CNRS–CEA, F-91198, Gif-sur-Yvette Cedex, France

J.P. Lacaux

Laboratoire d'Aérologie, C.R.A., 65300 Lannemezan, France

ABSTRACT

To study fluxes and factors controlling the emission of sulfur gases from biomass-burning, field experiments have been conducted in an African savanna area during the dry season (January 1989 and 1991). During these experiments, COS, SO_2, CO_2, and CO have been simultaneously monitored. SO_2 and COS values range from 0.1 and 0.6 ppbv out of the fires to 115 and 42 ppbv in the plume, respectively. A significant correlation has been observed between COS and CO concentrations ($R = 0.96$), indicating a COS production during the smoldering combustion. SO_2 correlates with CO_2 ($R = 0.92$), indicating production during the flaming combustion. The increase in COS/CO_2 ratios ($\Delta COS/\Delta CO_2$) during field experiments ranged from 3×10^{-6} to 61×10^{-6} (12×10^{-6} mean value). COS emission from biomass-burning is estimated to be up to 0.02 Tg S/yr in African savanna areas and up to 0.07 Tg S/yr on a global basis, thus contributing about 10% to the global COS flux. Concerning SO_2, emission factor ($\Delta SO_2/\Delta CO_2$) ratios ranged from 27×10^{-6} to 65×10^{-6} (mean value 47×10^{-6}) during field experiments. Biomass-burning is estimated to release up to 0.3 Tg S/yr, which do not represent a significant SO_2 source on a global scale.

Introduction

Extensive anthropogenic biomass-burning that occurs in the tropics during the dry season is likely to produce large amounts of gases and particles with potential impact on the radiative and chemical balance of the troposphere.[12] In tropical Africa, savannas are burned almost continuously, from December to March in the Northern Hemisphere, and from June to October in the Southern Hemisphere. According to Hao et al.,[15] African savanna-burning could thus account for about 50% of the tropical biomass-burning emissions.

Biomass-burning emissions represent only a few percent of the global atmospheric sulfur budget. However, they could contribute significantly to the regional S budget over remote continental regions such as the Amazon and Congo basins.[1,5] Among the emitted sulfur species, sulfur dioxide (SO_2) and carbonyl sulfide (COS) are the predominant precursors of sulfate aerosols in the troposphere and the stratosphere during volcanic quiescent periods, respectively. Emissions of these compounds could, therefore, have a negative radiative effect on local and/or regional scales. Up to now, very limited data regarding sulfur gas emissions from biomass-burning have been reported. COS emission factors (n = 7) have only been measured from forest fires in Colorado by Crutzen et al.[11] During laboratory experiments, Delmas[13] estimated the total sulfur emissions (n = 2) from the African savanna-burning by evaluating the sulfur content of the biomass before and after burning. Andreae et al.[1] measured SO_2 emission ratios ($\Delta SO_2/\Delta CO_2$) in plumes (n = 3) during flights over Amazonia. In this chapter, we present results from field experiments conducted to study fluxes and factors controlling the emission of SO_2 and COS from African savanna-burning. During the experiments, CO_2 and CO concentrations were also monitored to be used as tracers of combustion processes.

Experimental

Field experiments were conducted during the dry season (January 1989 and January 1991) at the Lamto station (6°13′ N, 5°01′ W) in Ivory Coast (Figure 1). Lamto is located at the border of the tropical rainforest and savanna, within a large reserve of 20 km^2. In the savanna region, the vegetation consists mainly of graminea (*Hyparrhenia displandra* and *Loudetia simplex*). The dry season is characterized by a east to north–east dry wind coming from the sahalian regions (Harmattan). It is associated with an ~2-km-thick haze layer resulting from the advection of dust particles and smoke combustion aerosols. During the experiments, the temperature was between 30 and 32°C and the relative humidity was 80% in the morning and 40% in the afternoon. During a flight over the northern part of the country in January 1989 (Table 1), two samples (1 and 2) were collected at altitudes of 2400 m in the free troposphere and 200 m in the haze layer, respectively. Samples representing the ground-level evolution of the local background concentrations were collected at the station at an altitude of 10 m (samples 3 to 7). Combustion gas samples (samples 8 to 24) were collected during 5 intensive experiments about 5–10 m from the fires on January 16 and 18 in 1989 and on January 9–15 in 1991.

For COS, CO, and CO_2 analyses, samples were collected in 2-L electropol-

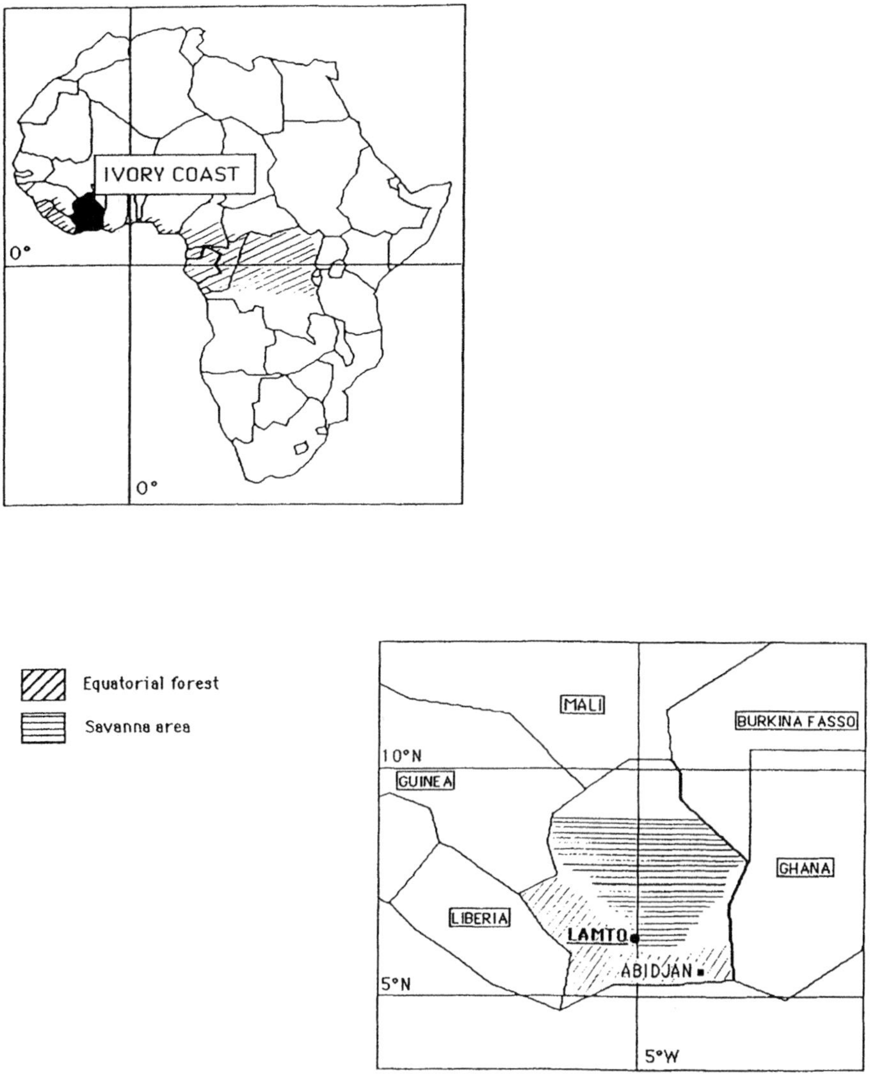

Figure 1. Location of the experimental station of Lamto in Ivory Coast.

ished stainless-steel canisters fitted with metal bellow valves. These canisters were previously conditioned by heating under vacuum. Several tests were conducted in our laboratory to assess the stability of hydrocarbons and COS inside the canisters. No significant concentration change was observed for COS over 1 month[21] and for hydrocarbons over 6 months.[8]

COS was cryogenically trapped at −90°C on conditioned 60–80 mesh Tenax GC loaded into a FEP/Teflon U column. The quantitative analysis of the samples was performed with a Varian 3400 gas chromatograph, equipped with a double-flame photometric detector operating in the sulfur mode. Precision was about 7%. Further details of the analytical system have been described by Belviso et al.[3] CO_2 was converted to CH_4 using a methanizer and measured by an Intersmat gas chromatograph. The GC was equipped with an 1/8-in. 2-m-long 60/80 mesh Carbosphere S packed column held at 105°C. The detector for CH_4 was a flame ionization detection (FID) and for CO was a reduction gas detector (RGD2). Precision for the CO_2 and CO determinations was better than 0.1%. SO_2 concentrations were measured using a modified West-Gaeke method.[26] Ambient air was bubbled through a bubler containing 10 ml of Na_2HgCl_4 solution (0.1 M) at a flow rate of 4 L/min. The collected SO_2 forms a complex with the Na_2HgCl_4, which is subsequently quantified with a spectrophotometer. The sampling time ranged from 5 to 15 min during burning experiments and from 1 to 6 h during measurements of background concentration. SO_2 samples were kept at 4°C during transport between the sampling site and the laboratory. Analysis were performed in the laboratory within two weeks after collection. More details about sampling, storage, and analysis of samples, and a comparison of West-Gaeke method with other analytical methods can be found in refs. 6 and 26. During the experiment, CO_2 concentrations were measured simultaneously with SO_2 sampling with a CO_2 nondispersive infrared analyzer (BINOS 100) equipped with two cells and having two ranges (0–500 ppm and 0–3000 ppm). CO_2 and SO_2 was determined with a precision of 1% and 10%, respectively.

Results and Discussion

The results of Lamto experiment are reported in Table 1 (COS, CO, CO_2) and Table 2 (SO_2 and CO_2). Samples collected during the aircraft flight are distinguished from the samples collected at the station and the samples collected in the fire plumes. The CO_2 values listed in Table 2 correspond to the integrated CO_2 concentrations measured during the SO_2 sampling.

Background Conditions

At ground level, background COS, SO_2, CO, and CO_2 concentrations are in the range of 0.43–0.6 ppbv, 0.1–0.7 ppbv, 0.18–0.47 ppbv, and 343–354

Table 1: Carbon dioxide, carbon monoxide, and carbonyl sulfide concentrations at Lamto during background conditions and fire experiments.

Sample No.	Date (MM/DD/19YY)	Sampling Conditions	CO_2 (ppmv)	CO (ppmv)	COS (ppbv)
Altitude					
1	01/13/1989	Aircraft 2400 m	350	0.10[a]	0.43
2	01/13/1989	Aircraft 200 m	352	N.M.[b]	0.70
Ground level					
Background					
3	01/16/1989	Lamto station	352	0.18	0.47
4	01/14/1991	Lamto station	354	0.47	0.60
5	01/15/1991	Lamto station	343	0.30	0.55
Tropospheric background at 10°N			350	0.10[b]	0.50[c]
Fire samples					
6	01/16/1989	Savanna fire	392	5.80	1.23
7	01/16/1989	Savanna fire	731	35.2	1.63
8	01/18/1989	Savanna fire	394	6.58	0.99
9	01/18/1989	Savanna fire	556	36.6	1.55
10	01/09/1991	Savanna fire	3190	360	41.9
11	01/09/1991	Savanna fire	1039	46.4	5.10
12	01/09/1991	Savanna fire	3769	299	16.9
13	01/09/1991	Savanna fire	4434	503	40.2
14	01/10/1991	Savanna fire	924	29.0	3.12
15	01/10/1991	Savanna fire	981	45.2	6.15
16	01/10/1991	Savanna fire	5333	274	18.1
17	01/11/1991	Savanna fire	2932	130	8.85
18	01/11/1991	Savanna fire	369	0.84	1.50
19	01/12/1991	Savanna fire	4648	242	16.3
20	01/12/1991	Savanna fire	4235	362	32.0
21	01/12/1991	Savanna fire	4410	338	37.7
22	01/15/1991	Savanna fire	414	3.24	2.59

[a]N.M.: not measured.
[b]Reference 20.
[c]Reference 5.

Table 2. Integrated ground-level concentrations of carbon dioxide and sulfur dioxide at Lamto during background conditions and fires experiments.

Sample No.	Date (MM/DD/19YY)	Sampling Conditions	SO_2 (ppbv)	CO_2 (ppmv)
Background				
1	01/14/1991	Lamto station	0.1	360
2	01/14/1991	Lamto station	0.7	360
3	01/14/1991	Lamto station	0.3	360
4	01/15/1991	Lamto station	0.2	360
Tropospheric background at 10°N			0.04[a]	
Fire samples				
5	01/09/1991	Savanna fire	48.8	1108
6	01/09/1991	Savanna fire	24.1	831
7	01/10/1991	Savanna fire	15.1	728
8	01/11/1991	Savanna fire	46.4	1411
9	01/12/1991	Savanna fire	115.1	2226
10	01/14/1991	Savanna fire	18.2	726
11	01/15/1991	Savanna fire	36.9	1559
12	01/15/1991	Savanna fire	15.4	931
13	01/15/1991	Savanna fire	36.2	1123

[a]Reference 25.

ppmv, respectively. For a few of these species, the measured background concentrations at ground level were higher than the reported tropospheric background values at the same latitude. Particularly, CO concentration was four times higher than the tropospheric background value, which denotes the influence of savanna burning on the budget of this compound. Considering the aircraft flight experiment, it can be seen that concentrations of all the species measured in the free troposphere are very close to the tropospheric background level at this latitude (No. 1, Table 1), increasing sharply in the fire plume sample (No. 2, Table 1) to reach values significantly higher than the background levels measured at Lamto (Nos. 3, 4, 5, Table 1). The concentrations measured in the fire plume (No. 2, Table 1) agree very well with those previously observed during the DECAFE experiment in the boundary layer of the equatorial forest of the Congo.[5,7,26] These observations are, therefore, consistent with an intense local production of carbon- and sulfur-containing gases in the boundary layer where the combustion products are localized.

Fire Plume Samples

In the fire plume samples (Nos. 6–22 in Table 1, and Nos. 5–13 in Table 2), COS and SO_2 concentrations reached values as high as 42 and 115 ppbv, re-

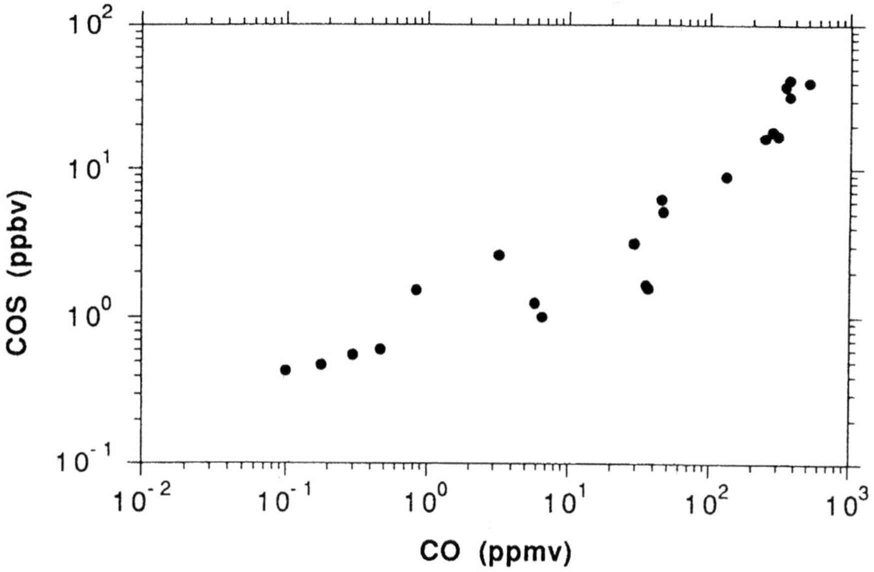

Figure 2. Correlation of COS with CO.

spectively, exceeding by a factor of 100 the measured background concentrations at Lamto station. Simultaneously, CO_2 concentrations ranged from 369 to 5333 ppmv and CO from 0.8 to 503 ppmv. As in the case of the sulfur compounds, it appears that CO_2 and CO concentrations in the fire plume samples far exceed the background values. However, the ratio of fire plume concentration to background concentration is significantly higher for CO (2.5×10^3) than for CO_2 (about 20), denoting a significant enhancement of CO production during combustion process.[9] Figure 2 shows the regression between COS and CO based on a compilation of data obtained in bushfire plumes during the two Lamto experiments. A significant correlation is observed between COS and CO, clearly indicating the role of the fires as a COS source.

Crutzen and Andreae[12] have shown from laboratory experiments that hydrocarbons, CO and CO_2, are not produced during the same stage of the combustion process: CO_2 is mainly produced during the flaming stage, whereas CO and hydrocarbons are mainly emitted during the smoldering stage. The significant correlations between COS and CO can, therefore, indicate an enhanced COS production during the smoldering stage. On the other hand, SO_2 correlates significantly with CO_2 ($R = 0.92$), suggesting a SO_2 production during the flaming stage, in agreement with the results obtained by Crutzen and Andreae[12] from laboratory experiments.

Emission Factor Estimates

Mixing ratios of sulfur gases above the background level measured at Lamto are used to derive emissions factors relative to CO_2 [Δ(SO_2 or COS)/ΔCO_2]. From the data presented in Tables 1 and 2, mean emission factors of $12 \pm 14 \times 10^{-6}$ and $47 \pm 13 \times 10^{-6}$ (on molar ratio) for COS and SO_2, respectively, can be estimated. The large uncertainty on the COS emission factor (more than 100%) is in line with our conclusion that COS and CO_2 are produced during different stages of the combustion. The mean COS emission factor of 12×10^{-6} agrees well with the values of 15.8×10^{-6} reported by Crutzen et al.[11] who obtained a range of $(5.4\text{–}28.6) \times 10^{-6}$ during forest fires in Colorado. It is also consistent with the two values of 6×10^{-6} and 42×10^{-6} reported by Bingemer et al.[5] during a flight above the equatorial forest of Congo.

Concerning SO_2, we are not aware of any direct previous determination of the emission factor ($\Delta SO_2/\Delta CO_2$) during field measurements from biomass-burning. Andreae et al.[1] have measured SO_2, sulfate (SO_4), and methanesulfonate (MSA) concentrations in the haze layer and in the boundary layer during flights over Amazonia. These authors have deduced a mean value of $\Delta SO_x/\Delta CO_2$ ($SO_x = SO_2 + SO_4^{2-}$ + MSA) of the order of 0.32×10^{-3} and a mean $\Delta SO_2/\Delta SO_x$ ratio of 0.072. Taking this ratio into account, we have calculated a mean $\Delta SO_2/\Delta CO_2$ ratio of the order of 23×10^{-6}, in reasonable agreement with our estimation of $47 \pm 13 \times 10^{-6}$.

Estimates of Regional and Global Emissions from Biomass-Burning

On the basis of the emission ratios relative to CO_2 calculated in the previous section, we can estimate the fluxes of SO_2 and COS from biomass-burning in the African savanna area and over the world. We have used Crutzen and Andreae's[12] CO_2 emission flux from biomass-burning (mean 240 Tmol/yr; range 130–340 Tmol/yr) and the fraction of biomass-burning allocated to the African savanna area which is estimated to 17%.[12,15] Based on our measured $\Delta SO_2/\Delta CO_2$ ratio of $47 \pm 13 \times 10^{-6}$, we can, thus, estimate the mean global SO_2 flux from biomass-burning to 0.3 Tg S/yr (0.06 from the African savanna area). This estimate is more than 2 orders of magnitude lower than the anthropogenic SO_2 flux[16,23] and about 40 times lower than the oceanic SO_2 production resulting from the atmospheric oxidation of reduced sulfur compounds emitted from the oceans (12 Tg S/yr).[2]

The mean $\Delta COS/\Delta CO_2$ emission factor of $12 \pm 13 \times 10^{-6}$ measured during this experiment has a great deal of uncertainty (more than 100%). As mentioned before, this is due to the fact that COS and CO_2 are not produced during the same stage of the combustion process. To estimate the global COS flux, we, therefore, preferred to evaluate the COS flux based on the relationship between COS and

Table 3. Global sources of COS.

Sources	Estimate (Tg S/yr)	Range	Ref.
Oceans	0.23	0.16–0.32	24
		0.11–0.21	18
Soils	0.07	0.04–0.11	14
			19
Volcanoes	0.005	0.001–0.02	4
Marshes	<0.001		10
Power plants, motor vehicles, chemical industry.	0.07	0.04–0.21	18
Biomass-burning		0.05–0.27	11
	0.07	0.03–0.11	This work
Subtotal	0.45	0.11–0.93	
$CS_2 \rightarrow COS$	0.32	0–1.1	18
	0.16	0.11–0.32	21
TOTAL	0.61–0.77	0.11–2.04	

CO (slope $= 0.085 \times 10^{-3}$, $R = 0.96$, Figure 2). We have used Crutzen and Andreae's[12] emission flux from biomass-burning (mean 26 Tmol/yr; range 10–42 Tmol/yr) based on the emission flux from biomass-burning and a CO/CO_2 emission factor of 10%, consistent with mean CO/CO_2 emission factor of 11.2% ($n = 41$) estimated from the results of our experiments. Thus, we calculated a global COS flux from biomass-burning of 0.07 Tg S/yr (range 0.03–0.11). This result is not so far from the COS flux of 0.09 Tg S/yr which could have been calculated from the $\Delta COS/\Delta CO_2$ emission factor of $12 \pm 13 \times 10^{-6}$ and the CO_2 flux from biomass-burning of 240 Tmol/yr.[12] Contrary to the SO_2 flux, COS emissions from biomass-burning represent an important contribution to the global COS budget. As shown in Table 3 which compiles the recent studies regarding the emissions of COS from various sources, biomass-burning contributes about 10% to the global COS flux and represents a COS source comparable to human activities and soil emissions.

Conclusion

During field biomass-burning experiments conducted in an African savanna area, SO_2, COS, CO_2, and CO have been simultaneously monitored. Significant correlations have been observed between COS and CO concentrations (R = 0.96), indicating a COS production during smoldering combustion. On the other hand, SO_2 correlates significantly with CO_2 ($R = 0.92$) indicating a production during flaming combustion. COS production by biomass-burning represents 0.02 Tg S/yr from African savanna areas and up to 0.07 Tg S/yr on a global basis, i.e., about 10% to the global COS flux. On the other hand, SO_2 production by biomass-burning contributes less than 1% to the global SO_2 flux.

Acknowledgments

This work was supported by the Centre National de la Recherche Scientifique, Ministère de la Coopération, and Commissariat à l'Energie Atomique. The authors thank the authorities of the Ivory Coast. Professor J. Baudet from Département de Physique, Université d'Abidjan, Professor J. Fontan, and Dr. R. Delmas from Université Paul Sabatier, and Mr. J.L. Tireford from the station of Lamto are particularly acknowledged for the organization of this campaign. This is CFR contribution No. 1275.

References

1. Andreae, M.O., et al. 1988. Biomass-burning emissions and associated haze layers over Amazonia. J. Geophys. Res. **93**:1509–1527.
2. Bates, T.S., J.D. Cline, R.H. Gammon, and S.R. Kelly-Hansen. 1987. Regional and seasonal variations in the flux of oceanic dimethylsulfide to the atmosphere. J. Geophys. Res. **92**:2930–2938.
3. Belviso, S., N. Mihalopoulos, and B.C. Nguyen. 1987. The supersaturation of carbonyl sulfide (OCS) in rain waters. Atmos. Environ. **21**:1363–1367.
4. Belviso, S., B.C. Nguyen, and P. Allard. 1986. Estimate of carbonyl sulfide (OCS) volcanic source strength deduced from OCS/CO2 ratios in volcanic gases. Geophys. Res. Lett. **13**:133–136.
5. Bingemer, H.G., M.O. Andreae, T.M. Andreae, P. Artaxo, G. Helas, D.J. Jacob, N. Mihalopoulos, and B.C. Nguyen. 1992. Sulfur gases and aerosols in and above the equatorial African rainforest. J. Geophys. Res. (in press).
6. Bonsang, B. 1980. Cycle atmospherique du soufre d'origine marine. Ph. D. thesis, Université de Picardie.
7. Bonsang, B., M. Kanakidou, G. Lambert, and J. Rudolph. 1988. Vertical profiles and fluxes of NMHC in the equatorial forest of Congo. AGU Fall Meeting. 5–9 December 1988, San Francisco.
8. Bonsang, B., M. Kanakidou, G. Lambert, and P. Monfray. 1988. The marine source of C2–C6 aliphatic hydrocarbons. J. Atmos. Chem. **6**:3–20.
9. Bonsang, B., G. Lambert, and C. Boissard. 1991. Light hydrocarbons emissions

from African savanna-burnings. In J.S. Levine (ed.), Global Biomass Burning, MIT Press, Cambridge, MA, pp. 155–161.

10. Carroll, M.A., L.E. Heidt, R.J. Cicerone, and R.G. Prinn. 1986. OCS, H_2S, and CS_2 fluxes from a salt water marsh. J. Atmos. Chem. **4**:375–395.

11. Crutzen, P., L.E. Heidt, J.P. Krasnec, W.H. Pollock, and W. Seiler. 1979. Biomass burning as a source of atmospheric trace gases. Nature **282**:253–256.

12. Crutzen, P., and M.O. Andreae. 1990. Biomass burning in the tropics: Impact on atmospheric chemistry and biogeochemical cycles. Science **250**:1669–1678.

13. Delmas, R. 1982. On the emission of carbon, nitrogen and sulfur in the atmosphere during bush fires in intertropical savannah zones. Geophys. Res. Lett. **9**:761–764.

14. Goldan, P.D., W.C. Kuster, D.L. Albritton, and Fehsenfeld. 1987. The measurement of biogenic sulfur emissions from soils and vegetation: three sites in the eastern United States revised. J. Atmos. Chem. **5**:439–467.

15. Hao, W.M., M.H. Liu, and P.J. Crutzen. 1990. Estimates of annual and regional releases of CO_2 and other trace gases to the atmosphere from fires in the tropics, based on the FAO statistics for the period 1975–1980. In Fire in the Tropical Biota, J.C. Goldammer (ed.), Springer-Verlag, New York, pp. 440–462.

16. Hameed, S., and J. Dignon. 1988. Changes in the geographical distributions of global emissions of NO_x and SO_x from fossil fuel combustion between 1966 and 1980. Atmos. Environ. **22**:441–449.

17. Johnson, J.E., and H. Harrison. 1986. Carbonyl sulfide concentrations in the surface waters and above the Pacific ocean. J. Geophys. Res. **91**:7883–7888.

18. Khalil, M.A.K., and R.A. Rasmussen. 1984. Global sources, lifetimes and mass balances of carbonyl sulfide (OCS) and carbon disulfide (CS2) in the earth's atmosphere. Atmos. Environ. **18**:1805–1813.

19. Lamb, B., H. Westberg, G. Allwine, L. Bamesberger, and A. Guenther. 1987. Measurement of biogenic sulfur emissions from soils, crops and trees using dynamic enclosure methods. J. Atmos. Chem. **5**:469–491.

20. Marenco, A., M. Macaigne, and S. Prieur. 1989. Meridional and vertical CO and CH_4 distributions in the background troposphere (70°N–60°S; 0–12 Km altitude) from scientifique aircraft measurements during the STRATOZ III experiment (June 1984). Atmos. Environ. **23**:185–200.

21. Mihalopoulos, N. 1989. Contribution à l'étude du cycle biogéochimique de l'oxysulfure de carbone dans l'atmosphère. Ph.D. thesis, University Paris 7.

22. Mihalopoulos, N., et al. 1991. The oceanic source of carbonyl sulfide. Atmos. Environ. **26**:1383–1394.

23. Moller, D. 1984, Estimations of the global man-made sulphur emission. Atmos. Environ. **18**:19–27.

24. Nguyen, B.C., B. Bonsang, and G. Lambert, 1974. The atmospheric concentration of sulfur dioxide and sulfate aerosol over antarctic, subantarctic areas and ocean. Tellus **26**:241–247.

25. Nguyen, B.C., B. Bonsang, and A. Gaudry. 1983. The role of the ocean in the global atmospheric sulfur cycle. J. Geophys. Res. **88**:10903–10914.

26. Rudolph, J., R. Koppman, and B. Bonsang. 1988. Measurements of light hydrocarbons and carbon monoxide over an equatorial forest in Africa. AGU Fall Meeting. 5–9 December 1988, San Francisco.

Methane Production and Uptake in Some Terrestrial Ecosystems of the Former USSR

N.S. Panikov, A.S. Belyaev, A.M. Semenov, and V.V. Zelenev

Institute of Microbiology, Russian Academy of Sciences, 117811, Moscow, Russian Federation

ABSTRACT

The production, emission, and uptake of methane was studied in different terrestrial ecosystems of the European plain of the USSR: hypersaline marshes of the Crimea, arable soils of the Moscow region, wetlands and dry forest soils of the Tver (Kalinin) and Syktyvkar regions, as well as tundra soils of the Polar Ural. Two main approaches were used: (1) CH_4-exchange rates between the soil and atmosphere were used by chamber techniques and (2) the estimation of CH_4 production and uptake in soil cores incubated under aerobic as well as anaerobic conditions. The highest rates of both CH_4 production and utilization occurred in the acid ombrotrophic bogs of the boreal forests. The wetlands were both sources and sinks for CH_4. The methane cycle in most studied soils was found to be a closed one: The processes of methanogenesis in the lower horizons were balanced by methane oxidation in the upper horizons. The only exceptions were hypersaline marshes and the waterlogged depressions of the ombrotrophic bogs where the activity of methanotrophs was absent or poorly expressed. In this case, significant CH_4 emission from soils to the atmosphere occurred (up to 50–120 mg CH_4–C m^{-2} h^{-1}). Kinetic approaches have been used for estimation of biomass and *in situ* activity of both methanogenic and methanotrophic bacteria.

Introduction

Methane exchange between soils and the atmosphere is an important problem in biogeochemistry.[1,2] Methane is involved in many chemical reactions with atmospheric gases and, through its infrared properties, has an influence on Earth's heat balance. The reason(s) for the increase in atmospheric CH_4 concentration of 1% per year is not clear.[3] The total annual flux of methane to atmosphere is estimated to be 374–714 Tg, from which 110–330 Tg has been attributed to different kinds of wetlands.[10] However, the data on nonanthropogenic sources and sinks of methane are unreliable. A large gap is evident with respect to methane production and consumption in the soils within the former USSR, which includes about 62% of the total wetland area of the Earth.

Our objective was to conduct a survey of the spatial and temporal distribution of sources and sinks of methane in some soils of the European part of the former USSR.

Material and Methods

Site Locations

The field observations were carried out in five geographical points of the European plain of the former USSR. Some physical/chemical characteristics are given in Table 1.

I. The Crimean region, Arabatskaya strelka (45°22′N, 35°21′E).

1. A hypersaline lagoon 50–100 m from the Sivash shore with a salt concentration 30% (area ~100 m^2; depth was up to 1.2 m).
2. A lagoon with a salt concentration of 5.5%, separated from Sivash by a long narrow strip of land 1–3 m wide (area of the lagoon ~300 m^2; depth was up to 1 m).
3. The dry automorphic soil on a watershed of the peninsula (about 1 km from Sivash) under xerophytic grass cover with domination of *Carex pachystilis*.

II. Moscow region, Serpukhov district (55°32′N, 37°39′E).

4. Arable gray forest soil under winter wheat.

III. Tver (Kalinin) region (56°10′N, 32°12′E).

5. An ombrotrophic bog (oligotrophic wetland with atmospheric type of nutrition) about 1.7 km,^{2} drained in 1972–1973.
6. The Big Rogovskoye ombrotrophic bog (7.3 km^2).
7. The undrained ombrotrophic bog, Lamtevsky Moch (4.1 km^2).

The following three sites were associated with Grustinskoe Lake:

8. A floating bog on the lake (~1–3 m in depth) consisting of partly decomposed plant material colonized by the *Carex-Sphagnum* community.

9. A minerotrophic groundwater fen located in a black alder geum–fern forest (area 0.1 km^2).
10. A mixed linden–spruce forest on sod–podzolic loamy–sandy soil.
11. A cowberry–moss pine forest Geras' kin Bor on podzolic sandy–loamy soil. The observations were made at the watershed area and in a 300-m^2 depression occupied by pine forest.

IV. Syktyvkar region (61°41′N, 50°45′E).

12. Kuz'-Nur bog, which consisted of a hummock-depression complex under dwarf shrub–sedge–sphagnum vegetation.
13. Kiya-Nyur bog which represented the transitional type of wetland with a hummock-depression complex under dead forest stand.
14. Field forest station Lyali which was a sod–podzolic soil under mixed forest. The measurements were made on two sites: within the watershed area and in the valley of a small forest brook.

V. Vorkuta region (67°59′N, 64°42′E).

15. A plane–hilly swamp of about 5 km^2 with small lakes, 10–20 m in diameter; permafrost was located at 40–50 cm.

Determination of CH_4 Emission

The static chambers described by Whalen and Reeburgh[12] were employed to quantify flux. The chambers consisted of two parts: (1) a permanent square stainless-steel collar base and (2) a removable Plexiglas cover. The first part of the chamber was inserted into the soil (depth = 10 cm) in advance (0.5–1.0 h). To avoid artifacts of CH_4 "squeezing," each site was provided with light bridges to accommodate the operator during all manipulations. The Plexiglas cover was connected with the base via a water seal only during measurements (usually for 30 min); the rest of the time the chambers were open to air. The air samples were taken from the isolator by hypodermic syringe through a rubber septum three times: immediately after cover installation and then after 15 and 30 min. The samples were stored in glass flasks under sealwater (10% NaCl). Methane emission was calculated by approximating the dynamics of CH_4 concentration by linear regression.

Determination of CO_2 Emission

These measurements were made in the same chambers, the only difference was that air samples were drawn into 300–500-cc PCV bags. The CO_2 exchange

Table 1. Methane emission from the soil to the atmosphere.

N*	SITE	GROUND WATER LEVEL (CM)	TEMPERATURE °C	E_h (mV)	pH	CH_4 FLUX (mg CH_4-C/ h * m^2)	CH_4/CO_2
1	2	3	4	5	6	7	8
	Crimean Region. Arabatskaya strelka						
1.	The lagoon center	+ (3 - 4)**	20 - 35	-	7.0	0.05 - 0.4	-
2.	1 m from shore (fresh deposits of *Zostera*	+ (5 - 30)	20 - 24	-	7.2	0.10 - 35.0	-
	1.5 m from inner shore (old deposits of *Zostera*	+(5 - 20)	20 - 24	-	7.2	0.05 - 0.32	-
3.	Hummock	< -100	20 - 35	-	-	0	-
	Between hummocks	< -100	20 - 35	-	-	0	
	Moscow region. Poushchino						
4.	Field experiment	< - 200	13 - 15	200-	5.5	0 - 0.04	0.001 - 0.005
	Tver (Kalinin) region. West-Dvinsky district						
5.	The central part between trenches	-30	9	+10 - (-160)	3.0	0.00 - 0.08	0.00 - 0.004
	Trench	+10	9	-200	4.0	3.9	0.05
6.	Mound	-30	12	+15	4.1	0.05 - 0.25	0.002 - 0.03
	Depression	+1 - 2	13	-15	4.0	5.2	0.24
	Lake shore	+10	13.5	-100	-	0.52	1.24
7.	Hillock	-10	15	-	-	115.2	-
	Between them	+1	15	-	-	6.3 - 84.7	-
8.	20 m from water	-1 - 2	8	-160	3.9	0.00 - 1.10	0 - 0.023
	100 m from water, *Sphagnum* spot	-5	9	-180	3.5	4.4	0.41
	Sphagnum + Carex sp.	-5	9	-180	4.0	36.6	1.74
9	Hillock	-15	8	-246	7.2	0.03	0.0004
	Waterlogged depression	+5	8	-246	7.2	0.07	0.009
10.	Under the trees, under herbage	< -50	9	+170	6.5	0.00	0.00
11.	Watershed	< -50	10	+160	4.0	0.00	0.00
	Depression	-20	10	+160	3.0	0.00	0.00

N*	SITE	GROUND WATER LEVEL (CM)	TEMPERATURE °C	E_h (mV)	pH	CH_4 FLUX (mg CH_4-C/ h * m^2)	CH_4/CO_2
1	2	3	4	5	6	7	8
	Syktyvkar region, Komi SSR						
12.	Under the pine forest mound	-70	15	-	4.2	0.00	0.00
	Depression	-10	15	-	4.2	0.00	0.00
13.	Without trees	-8	15	-	4.2	1.11 - 1.22	-
	Hillock	-50	14	-	-	-.52	-
	Depression	+1 - 3	14	-	4.2	0.20 - 4.23	-
14.	Watershed	< -50	12	+160	5.0	0.00	0.00
	Brook shore	-2	11	-	4.5	0.15	-
	Vorkuta region, Halmer-Yu						
15.	Mound	< -70	15	-	-	0.00	0.00
	Mound's base	-20	14	-	6.4	0.28 - 0.38	-
	Waterlogged part	0-	11	-	6.1	0.79 - 7.85	-

*Site number, see section Materials and Methods.

**"+" indicates groundwater level above the soil surface, and "−" indicates ground water level below the soil surface.

—Not determined.

rates were measured in the dark, the Plexiglas cover being shaded by nontransparent screen (aluminum foil or black cotton fabric).

Measurement of CH_4 Transformation in Soil (Peat) Cores

Serum glass flasks (500 cc, the outer diameter 77 mm), severed at their bottoms, were inserted into mineral soils to a depth of 8–10 cm. In the case of easily deformable peat soils, preliminary cutting by a special cylindrical knife (of the same diameter as the sampling flask) was performed. Then the flask was introduced into the cutting, and the soil core was carefully taken away. The core height was usually 95–100 cm, and the head-space volume was about 100 cc. The core was sealed at the bottom by a tight butyl rubber cap, and at the top by a silicone stopper. When core samples were taken below the level of the groundwater table, a plot area of 40 $\times$ 40 cm^2 was isolated from the water by a stainless-steel frame. Water was pumped out and soil cores were taken as described above. The flask with waterlogged soil did not contain any gas-filled head space, and all its free volume (from the peat surface to the stopper) was occupied by the water from the same horizons.

Four replicate cores from each horizon were brought to the field laboratory. The upper water layer ($\sim$100 cc) was carefully removed from the flooded samples and replaced by an equal volume of air. All cores were incubated outdoors at a temperature close to that *in situ* (the differences did not exceed 2°C). Gas samples (0.5 cc) were periodically withdrawn and analyzed for CH_4. After 1–2 days, methane was added into two of the four replicate cores (final concentration 50–10,000 ppm), and incubation was continued under aerobic conditions. The head space in another two replicates were replaced by a 1 : 1 mixture of H_2 : CO_2 (vol : vol), and incubation was carried out under anaerobic conditions.

Determination of CH_4 Concentration in the Soil Air

These measurements were made by a membrane probe technique.[5] The probe was made from a hollow L-shaped stainless-steel tube 10–20 mm in diameter. The lower perforated part was covered by a thin ($\sim$100 μm) polyethylene membrane and inserted horizontally into the soil at a chosen depth. The vertical part of the tube was sealed by a rubber septum. Steady-state concentrations of soil gases were usually established during 1–3 days.

Gas Determinations

CH_4 concentrations were analyzed by means of a portable gas chromatograph (XPM-4, USSR) equipped with a flame ionization detector (1 m column with aluminum oxide, H_2 as carrier, oven at 100°C). The detection limit was 0.2 ppm.

CO_2 was determined with a portable infrared gas analyzer. The detection limit was around 5 ppm.

Other Methods

pH in soil extracts was determined with glass electrodes on a portable pH meter (pH-150, USSR). In wetlands, measurements were made *in situ* at a depth of 10 cm. In aerobic soils, the determinations were made in 1*N* KCl extracts (soil : solution = 1 : 1). The redox potentials E_h were measured with platinum electrodes at the depths of 5–70 cm. The temperature was estimated by thermistors and mercury thermometers at depths of 2–10 cm. Atmospheric pressure was determined with aneroid barometers, and salinity of solutions was measured by refractometer.

Results and Discussion

Methane Emission from Different Soils

The results of single-point determinations made in June–July 1990 are summarized in Table 1. In the inner hypersaline lagoon (site 1), the CH_4 emission was rather low ($\leq$ 1 mg C m^{-2} h^{-1}). It rose to 35 mg C m^{-2} h^{-1} at Sivash shore (site 2), where the salinity was low (5.5%) and there was a large amount of available organic matter. Similar high rates of CH_4 emission were observed in ombrotrophic bogs. The highest absolute values were found at Lamtevsky Moch (up to 115 mg C m^{-2} h^{-1}), at the floating bog of the Grustinskoye Lake (4–37 mg C m^{-2} h^{-1}), as well as in waterlogged depressions of the forest wetlands—B. Rogovskoye (5.2 mg C m^{-2} h^{-1}), Halmer-Yu ($\leq$ 7.9 mg C m^{-2} h^{-1}), and Kiya-Nyur (4.2 mg C m^{-2} h^{-1}). In the drained Sosvyatakoe bog, the CH_4 emission was insignificant; during repetitive determinations, it has been detected only occasionally. Intensive methane emission (~4 mg C m^{-2} h^{-1}) was steadily recorded from drainage trenches. Methane emission was seldom observed from the elevated elements of microrelief and mesorelief of the ombrotrophic bogs. Moreover, on the hillocks and mounds, we sometimes detected consumption of CH_4 (e.g., Table 1, site 13).

In watershed aerobic soils (sites 3, 4, and 14), CH_4 emission was either absent or extremely low: $<$0.01 mg C m^{-2} h^{-1}. The detectable methane emission from control plots of arable gray forest soil (0.03–0.04 mg C m^{-2} h^{-1}) was detected only twice in more than 10 attempts. In the forest soils of Lyali station, significant CH_4 emission was noticed only in a waterlogged depression.

The contribution of methane to total emission of C compounds from soil was evaluated by the CH_4 : CO_2 ratio (Table 1) because CO_2 is the main volatile

C compound. This ratio was usually less than 0.001. However, in waterlogged depressions in wetlands, it increased to 0.2–1.7. Hence, the contribution of methane to total carbon flux in some ecosystems can be even higher than that of CO_2.

It follows from our preliminary results that CH_4 emission from soils varies broadly in time and space. Even within one particular location, the emission rates were not the same during several successive measurements and from several adjacent points. It was impossible to find one or several parameters (pH, E_h, temperature, vegetation type, etc.) which would have a high correlation with CH_4 emission. Also, it was not absolutely compulsory that high emission values are attributed to wetlands. Sometimes, more CH_4 was emitted from aerobic soils (e.g., from arable forest soil in Pushchino) than from peatlands. To elucidate this problem, we carried out study of the dynamics and spatial distribution of CH_4 sources and sinks for one particular ecosystem, using the ombrotrophic bog Sosvyatskoe of the Tver region.

Dynamics of Gas Emission

We followed the temporal variations in CH_4 and CO_2 fluxes from soil that occur on (1) daily, (2) weekly (every 5–10 days), and (3) seasonal cycles. All measurements were made at the same points, thereby overcoming the problem of spatial variations in CH_4 and CO_2 emission. Since the upper removable part of the chamber was connected to the base only for the short periods of time (30 min), it did not change the state of covered soil. Some of the results are shown in Figures 1–5. The diurnal cycles of CO_2 and CH_4 emission were dependent on the season. In the cold season, the rate of emission was practically constant during the entire day span or varied randomly (Figures 1 and 2). From April to September, statistically significant increases of CO_2 emission were observed. Some positive relationship was also recorded between the rates of CH_4 emission and the soil temperature. This relationship could be interpreted in two ways: as a stimulating temperature effect on the activity of methanogenic bacteria and as a positive influence of temperature on the rate of gas diffusion (or ebullition) from the soil to the atmosphere.

Examples of day-to-day variations in CO_2 and CH_4 fluxes are presented in Figures 3–5, all these measurements were made at a fixed time of the day—from 10:11 A.M. to 12:00 noon. CO_2 evolution rates exhibited generally regular rises and falls every 4–7 days: 1 maximum in April, 3 maxima in June, and 2 maxima in July. The oscillations of CH_4 fluxes were less regular. The absence of a close relationship between CO_2 and CH_4 fluxes would lead to the assumption that microbial populations involved in CH_4 transformation may not have a direct connection with plants as do aerobic soil microbes.

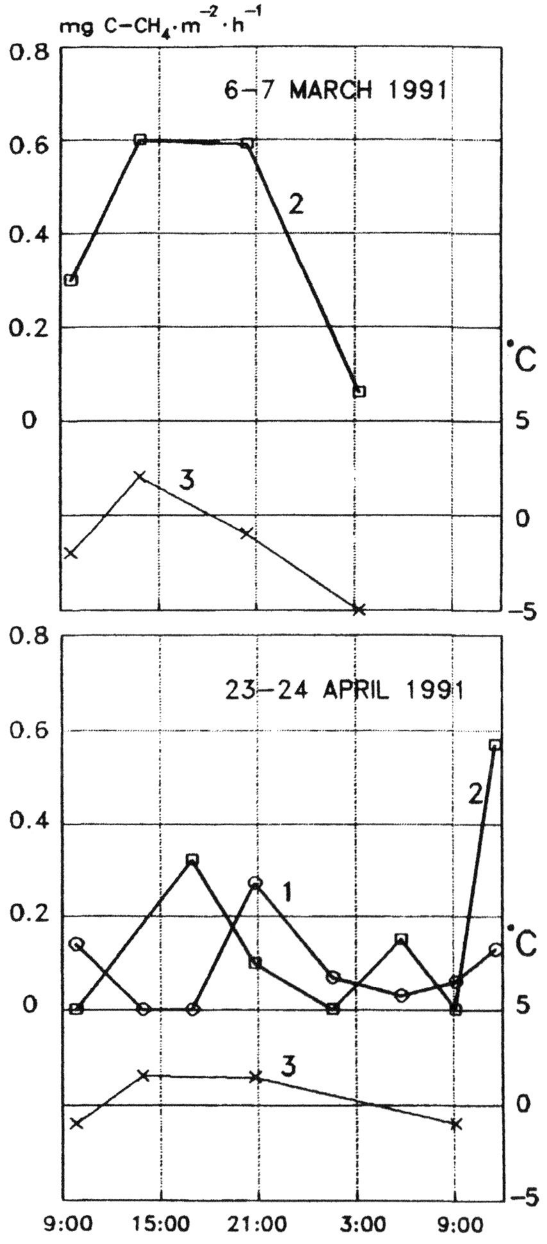

Figure 1. The diel variation of CH_4 fluxes from the Sosvyatskoe ombrotrophic bog. Curve 1: Chamber is located in the undrained central part of the bog. Curve 2: chamber is located in the trenches. Curve 3: the temperature of the peat surface.

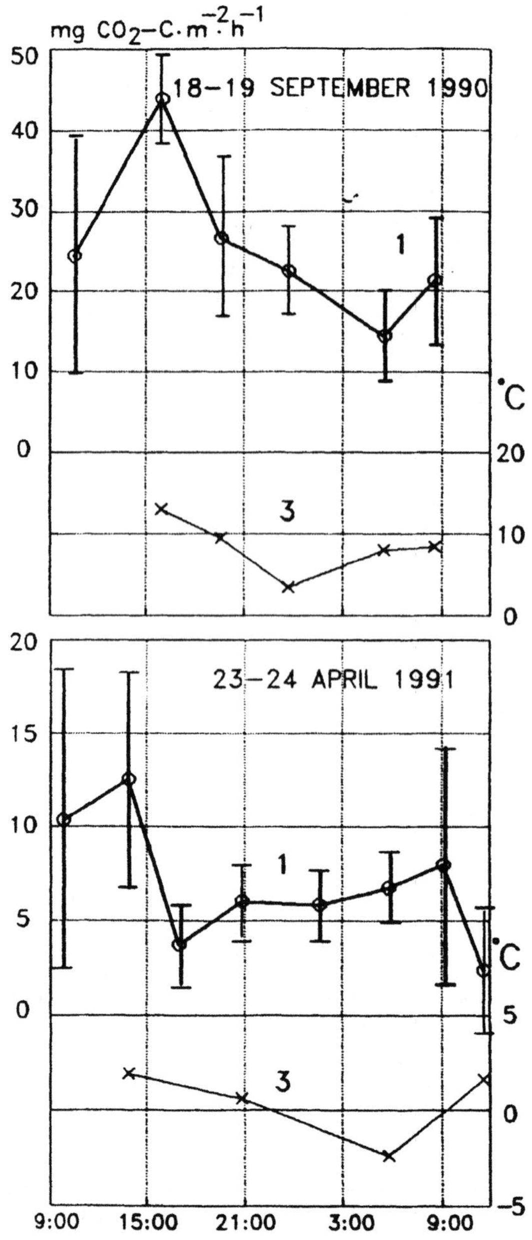

Figure 2. The diel variation in CO_2 fluxes from the Sosvyatskoe ombrotrophic bog. Curves 1–3 as in Figure 1. Points correspond to the means of measurements with four individual chambers. Bars represent standard errors.

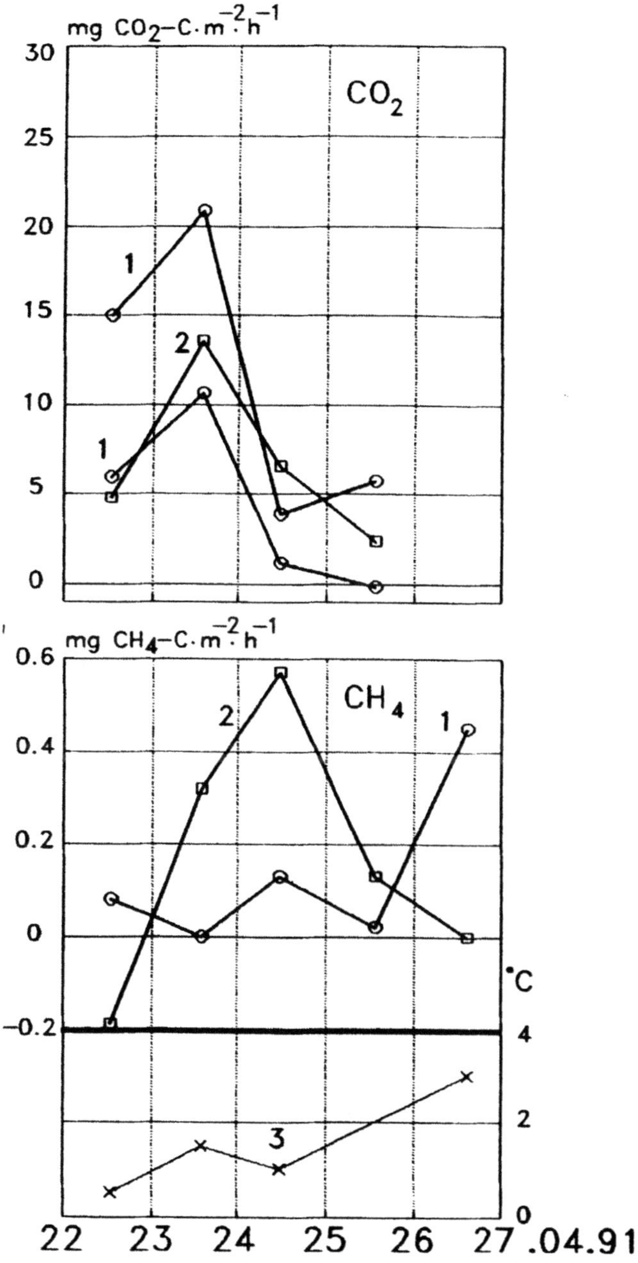

Figure 3. The day-to-day variations in CO_2 and CH_4 fluxes from the Sosvyatskoe ombrotrophic bog in April. Curves 1–3 as in Figure 1.

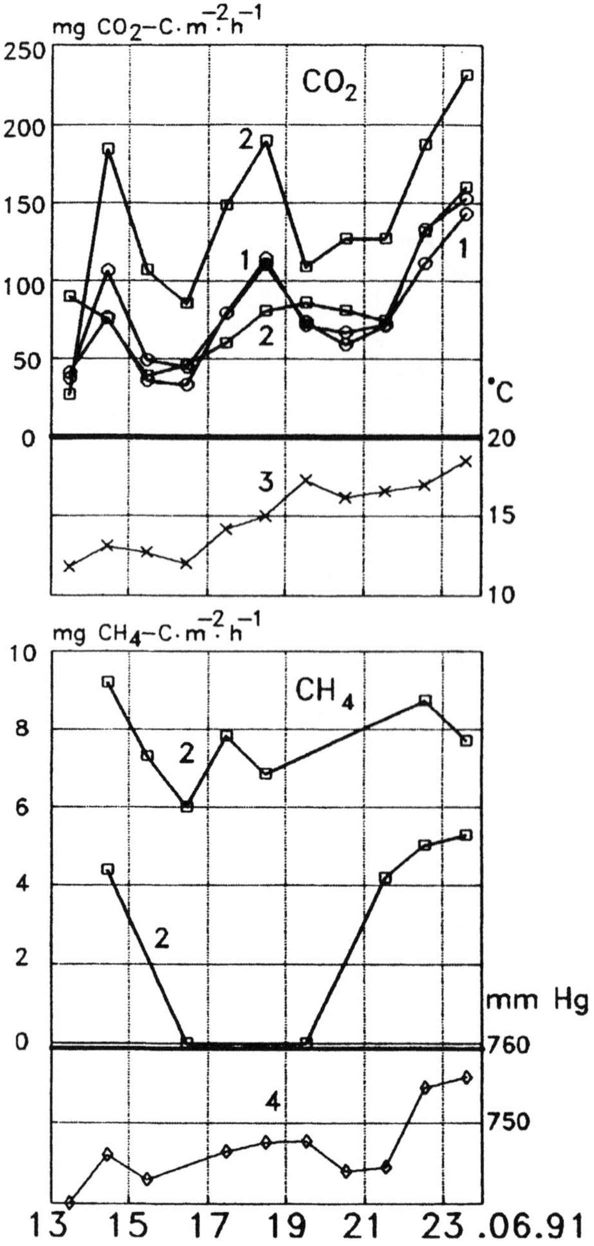

Figure 4. The day-to-day variations in CO_2 and CH_4 fluxes from the Sosvyatskoe ombrotrophic bog in June. Curves 1–3 as in Figures 1 and 4—the atmospheric pressure.

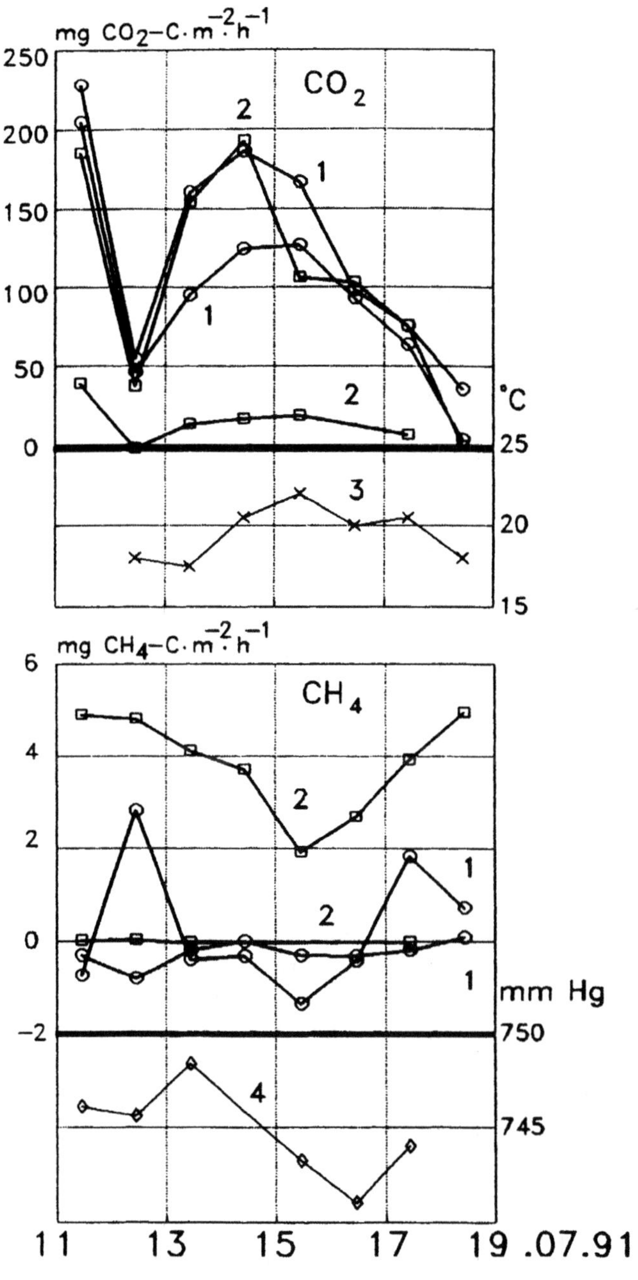

Figure 5. The day-to-day variations in CO_2 and CH_4 fluxes from the Sosvyatskoe ombrotrophic bog in July. Curves 1–4 as in Figure 1.

It is interesting that CH_4 emission from the bog center in July was negative, i.e., atmospheric methane was consumed (scavenged) on the interface between air and bog surface. In April, September, and November, net methane flux was steadily positive not only from the trenches but also from the bog center. In June, the CH_4 emission from the bog center was zero. The trenches always were sources of CH_4.

The influence of the two recorded environment factors—temperature and atmospheric pressure—explained only part of the observed dynamic patterns of fluxes because there were no complete synchronization of corresponding time series. A positive relationship was found between the CO_2 evolution rate and the temperature, as well as between CH_4 emission from trenches and atmospheric pressure.

Figure 6 shows the seasonal dynamics of CH_4 and CO_2 emission from the bog center and trenches. To make a reasonable approximation of the size of gas emission on an annual basis, we took into account all available data on daily and weekly variations. Numeric integration gave the following average values of annual emission:

methane from the bog center: 1.05 g of carbon m^{-2};
methane from the trenches: 14.7 g of carbon m^{-2};
carbon dioxide from the bog center: 314.3 g of carbon m^{-2};
carbon dioxide from the trenches: 261.0 g of carbon m^{-2}.

Hence, the dynamic studies demonstrate the significant contribution of methane to the gaseous C emission from the ombrotrophic bog: On average, it accounts for 0.33% of the total CO_2 + CH_4 flux from the bog center and 5.33% of that from the trenches.

Sharp differences between trenches and central parts of the ombrotrophic bog were found seasonally (Figure 6). Methane emission from trenches was positive the entire year, including winter. CH_4 fluxes positively correlated with soil and air temperature: The highest rates were observed in July, and the lowest were recorded during winter. In the central intertrench site, CH_4 evolution rates were inversely related to temperature: During summer, the flux was negative (i.e., the bog acted as a methane sink), and the highest positive evolution rate took place in winter (April and the end of November). The absolute values of CH_4 emission from central part of the bog was maximal in November: 0.8 mg C m^{-2} h^{-1}. This period corresponded to the beginning of bog freezing when the peat temperature was above zero and the air temperature became negative. Obviously, such a high CH_4 emission in November was not due to gas formation *de novo*, but due to transfer from low peat horizons across the temperature gradient.

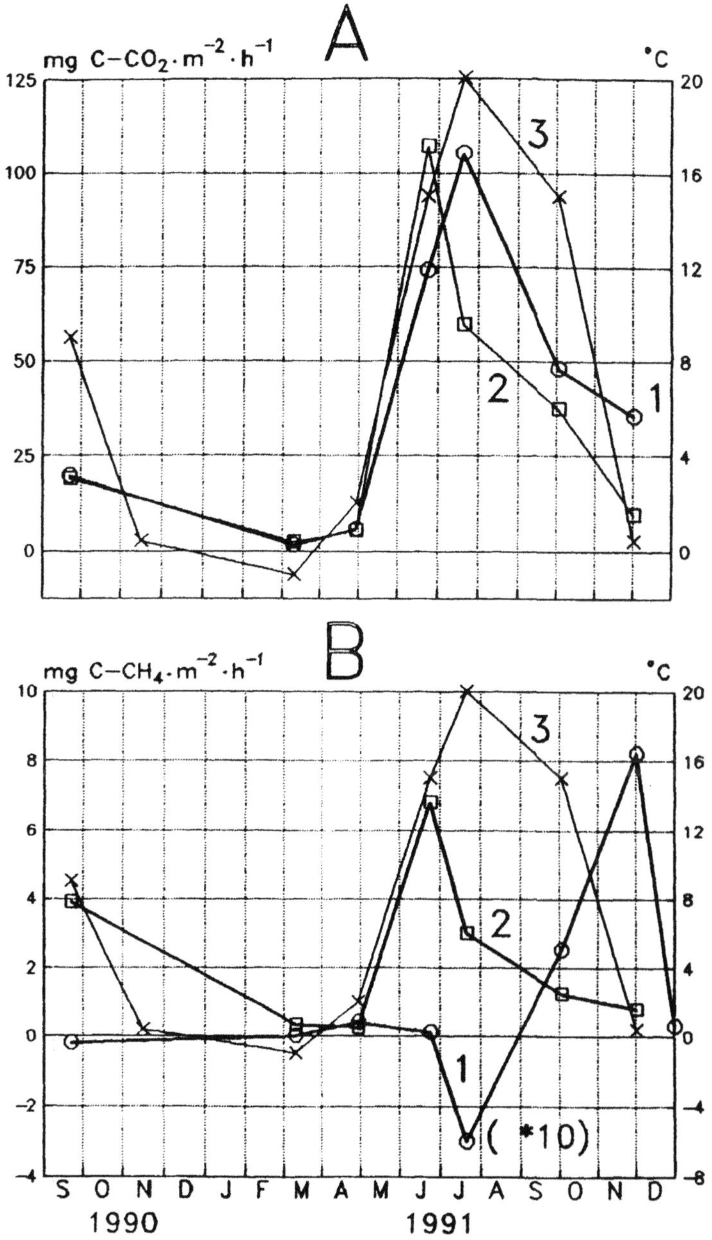

Figure 6. The seasonal dynamics of (A) CO_2 and (B) CH_4 emission from the Sosvyatskoe ombrotrophic bog in 1990/1991. Curves 1–3 as in Figure 1. Notice the difference in scales for curves 1 and 2 in B: The values for curve 1 are multiplied by 10.

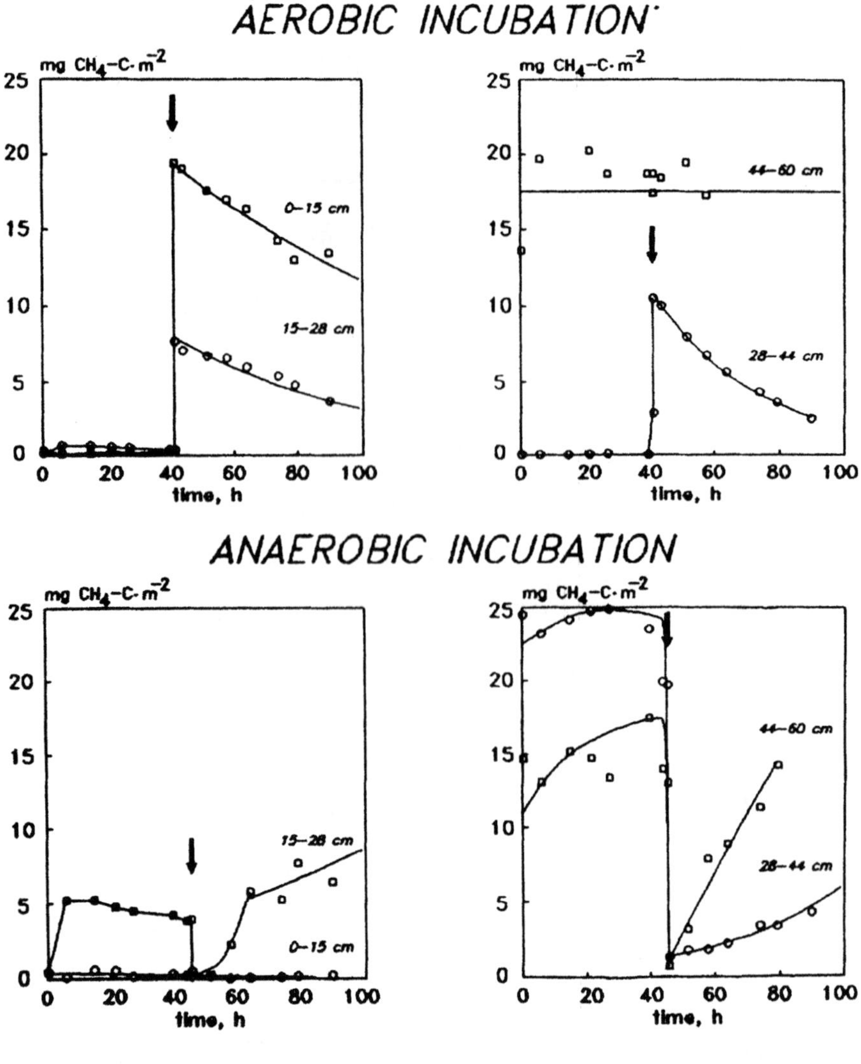

Figure 7. An example of data on methane uptake and formation in peat core incubated under aerobic and anaerobic conditions. Dynamics of CH_4 concentration in the head space of the cores was recorded before and after the replacement (↓) of the gas phase. The depth of the peat layer is indicated above each curve. Sampling time 9/18/1990.

Methane Sources and Sinks in the Soil Profile

An example of CH_4 production and consumption in soil cores from different depths of ombrotrophic bogs are shown in Figure 7. During the first day, in all core samples except for one from an upper horizon, CH_4 emission was observed. On the second day after replacement of the gas phases, conditions were created for either methanogenesis or aerobic CH_4 oxidation. The reaction of the microbial community to head-space replacement was immediate: In the upper horizons, oxidation occurred, whereas in the lower ones (below 44 cm), CH_4 emission occurred. Methanogenesis was never observed in anaerobically incubated surface soils, and oxidation was never observed in aerobically incubated soils from the lower horizons. In the intermediate horizons (15–44 cm), the existence of both sources and sinks of CH_4 was found.

The curves of CH_4 consumption were of a hyperbolic type and could be approximated by the Michaelis–Menten kinetic equation:

$$\frac{ds}{dt} = -V\frac{s}{K_S+s} \tag{1}$$

or after integration

$$K_S \ln\left(\frac{s_o}{s}\right) + s_o - s = Vt, \tag{2}$$

where s is the CH_4 concentration, s_o is the initial s value, V and K_S are the parameters of the Michaelis–Menten equation, and t is time.

To find the parameter values, we used a best-fit computer procedure.[7] With rare exceptions, the experimental curves did obey Eq. (2) as could be visually estimated from Figure 8. The numerical values of K_S and V estimated in this way are shown in Table 2 for all incubation experiments. K_S values varied from 36 to 2251 ppm of methane in air, which are equivalent to 0.2–8.4 μM methane. They fell within the interval (K_S 2–4 μM) reported for methane oxidation observed in wetland sediments.[6] Maximal activity of methane oxidation V varied in the range 0.00–11.65 mg CH_4–C h^{-1} m^{-2}. A dramatic increase of V took place in the central part of the bog in midsummer: from 0.3–0.4 in June to 11–12 mg CH_4–C h^{-1} m^{-2} in July. Therefore, we can conclude that intensive growth of methanotrophic bacteria occurs and only during a short seasonal period. High values of V were always observed in trenches, where both the potential and actual activity of methane oxidation was kept high.

Equation (1) gives the possibility to estimate not only potential but also the

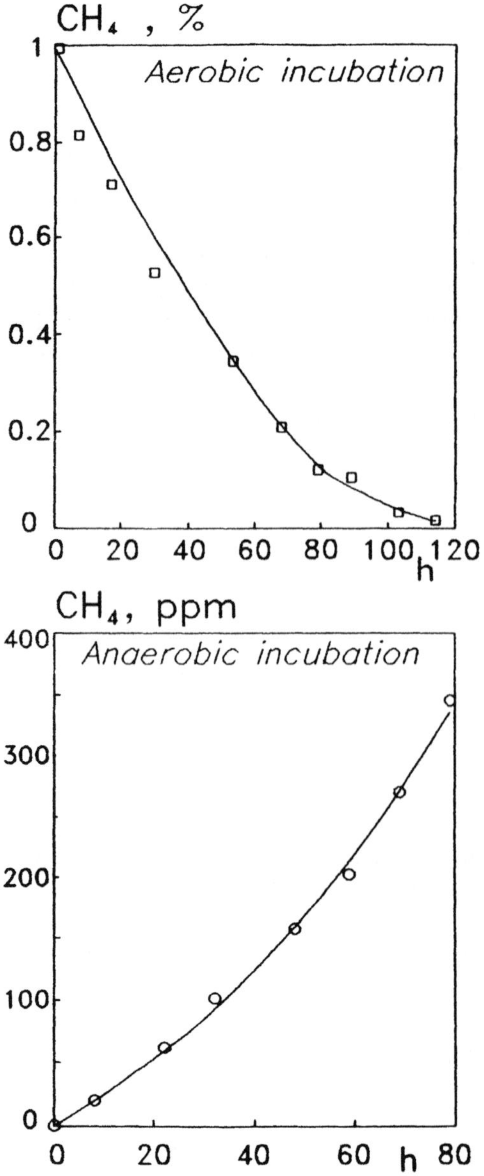

Figure 8. Illustration of the kinetic analysis of dynamic data on methane uptake and formation in the cores of the Sosvyatskoe ombrotrophic bog. Top: aerobic incubation, the curves of CH_4 uptake are calculated according to Eq. (2), peat cores are sampled from the horizons 0–10 cm in July 1991. Bottom: anaerobic incubation, the curves of CH_4 formation are calculated according to Eq. (4), peat cores are sampled from the horizons 30–40 cm.

Table 2. Seasonal dynamics of spatial distribution of CH_4 sources and sinks in ombrotrophic Sosvyatskoe bog (central undrained part of the bog).

Sampling Time	Horizon (cm)	CH_4 conc. in soil air (ppm)	Kinetic Parameters of CH_4 Uptake		Calculated* rate of *in situ* CH_4 uptake	Rate of Methane Production	The Balance of CH_4 as evaluated... (mg C h^{-1} m^{-2})	
			V (mg C h^{-1} m^{-2} of soil core)	K_m (ppm)	(mg C h^{-1} m^{-2} per horizon depth)		from soil core experiments	from CH_4 emission by chamber method
9/18/1990	0 - 15	4.0	0.147	1200	0.0005	0.000	-0.681	0
	15 - 28	7000	0.92	1677	0.742	0.260		
	28 - 44	12750	0.71	347	0.691	0.062		
	44 - 60	38200	0.00	-	0.000	0.420		
3/6/1991	0 - 10	3.0	0.000	-	0.000	0.000	-0.175	0
	10 - 20	1415	0.280	36	0.270	0.029		
	20 - 30	35770	0.076	51	0.075	0.054		
	30 - 40	32884	0.075	475	0.074	0.161		
4/23/1991	0 - 13	40	0.072	303	0.008	0.000	+0.432	+0.098
	13 - 26	1415	0.361	1419	0.180	0.021		
	26 - 43	36000	0.000	-	0.000	0.198		
	43 - 56	33000	0.000	-	0.000	0.401		
06/12/1991	0 - 10	250	0.370	1200**	0.064	0.000	-1.783	+0.11
	10 - 20	5650	0.420	1200	0.346	0.015		
	20 - 30	49454	0.950	1200	0.765	0.107		
	30 - 40	98047	0.890	1200	0.850	0.120		
7/11/1991	0 - 10	5	11.29	603	0.094	0.000	-0.410	-0.046
	10 - 20	80	11.65	1891	0.601	0.000		
	20 - 30	270	0.620	1200	0.114	0.383		
	30 - 40	360	0.550	1200	0.116	0.132		

*Calculated according to equation of Michaelis–Menten with experimentally determined parameters V and K_m.

**Particular values of K_m measured for this horizon appeared to be statistically unreliable; instead of them, we used the average values of that parameter which have been found for all peat samples.

actual activity. For this purpose, we used the CH_4 concentration in soil profiles (Tables 2 and 3) as the variable s to Eq. (1). Results of these calculations are presented in Tables 2 and 3.

The curves of methanogenesis under anaerobic conditions were often concave, i.e., resembling growth curves in batch culture. Hence, as a first approximation the following kinetic model may be used for quantitative description of methane formation:

$$(3) \qquad \frac{dx}{dt} = \mu x, \quad \mu = \frac{\mu_m c}{(K_c + c)} \sim \mu_m$$

$$\frac{ds}{dt} = \frac{\mu x}{Y},$$

where c is the concentration of methanogenic substrates ($H_2 + CO_2$), μ_m and K_c are the parameters of Monod's equation, and Y is the yield of biomass per unit of methane formed. After the introduction of an anaerobic gas mixture into core, the specific growth rate μ increases up to its upper limit μ_m ($\mu \sim \mu_m$ = constant at $c >> K_c$). For this situation, the dynamics of CH_4 production can be described by the integrated form of Eq. (3):

$$(4) \qquad s = \frac{x_0}{Y}\left[\exp\left(\mu_m t\right) - 1\right],$$

where x_0 is the initial biomass of methanogens. Hence, the exponential curve of CH_4 emission can be interpreted as a consequence of the methanogenic bacteria growth. Best fit of Eq. (4) to experimental data (Figure 8) gave the following parameters values: $\mu_m = 0.015\ h^{-1} = 0.36\ d^{-1}$, and $x_0/Y = 1.0$ mg C m^{-2}. If we use a Y value of 0.09 g of biomass C per gram of produced CH_4–C,[8] then $x_0 = 92$ μg of biomass–C m^{-2}, which results in a population density of 10^4 cells per gram of dry peat. Both values μ_m and x_0 fell within ranges reported for pure cultures of methanogens or for enumeration of these bacteria in soils.[4,11] However, Eq. (3) was not valid for all obtained experimental curves. Some of them were biphasic exponents; some were linear, some were hyperbolic. We estimated the average rates of CH_4 formation during the first day of anaerobic incubation (Tables 2 and 3).

On the basis of core determinations of CH_4 exchange rates, we evaluated two main components of the CH_4 balance (generation and consumption) for the central part of the ombrotrophic bog (Table 2, column 8). It was found that the algebraic sum of CH_4 fluxes for all peat layers gave values close to the CH_4 emission measured by the chambers (Table 2, column 9). A small systematic

Table 3. Dynamics distribution of CH_4 sources and sinks in the drainage trenches of Sosvyatskoe ombrotrophic bog.

Sampling Time	Horizon (cm)	CH_4 conc. in soil air (ppm)	Kinetic Parameters of CH_4 Uptake		Calculated* rate of *in situ* CH_4 uptake	Rate of Methane Production	The Balance of CH_4 as evaluated... (mg C h^{-1} m^{-2})	
			V (mg C h^{-1} m^{-2} of soil core)	K_m (ppm)	(mg C h^{-1} m^{-2} per horizon depth)		from soil core experiments	from CH_4 emission by chamber method
4/23/1991	0 - 10*	2	0.102	63	0.003	0.000	-0.116	+15
	10 - 20	35	0.315	63	0.113	0.000		
6/12/1991	0 - 8*	13	6.690	710	0.120	0.000	-0.185	+4.87
	12 - 20	17	1.720	427	0.065	0.000		
	0 - 10**	2	1.300	1200	0.002	0.000	-0.081	-0.016
	11 - 20	94	3.050	2251	0.122	0.043		
7/11/1991	0 - 10*	2675	4.400	1469	2.830	0.000	-4.102	+3.87
	10 - 20	3382	1.865	1200	1.370	0.098		

**Eriophorum* hummock.

**Part of the trench without plants.

Table 4. Vertical distribution of sources and sinks of methane in wetlands fluxes are expressed in mg CH_4–C m^{-2} h^{-1}.

Horizon	Rate of Methane uptake	Rate of Methane Production	E_h* mV	The Balance of CH_4 as Evaluated from soil core experiments	from CH_4 emission by chamber method
Hypersaline lagoon, Arabatskaya strelka					
0 - 4	0.000	0.082	n.d.	+0.138	+0.05 - +0.4
4 - 12	0.000	0.028			
12 - 22	0.000	0.028			
Big Rogovskoye bog, waterlogged depression					
0 - 15	0.140	0.027	-15	+3.29	+5.2
15 - 25	0.000	3.400			
Big Rogovskoye bog, mound					
0 - 17	0.800	0.000	+15	+0.49	+0.25
17 - 25	0.004	0.060			
25 - 35	0.000	1.230			
Floating bog on Grustinskoye Lake, 100m from water					
0 - 15	2.490	0.000	-180	+0.84	4.4
15 - 30	0.000	3.333			
Lowland groundwater fen, Grustinka					
0 - 15	0.140	0.400	+246	+1.00	+0.07 - +1.60
15 - 25	0.310	1.050			
Mixed forest on the watershed site, 100m from lowland swamp					
0 - 10	0.290	0.000	+170	-0.74	0
10 - 20	0.450	0.000			
Watershed site (Geraskin Bor) adjacent to Sosviatskoe bog					
0 - 15	0.175	0.000	+160	-0.175	0
Geraskin Bor, local wetland under pine					
0 - 15	0.140	0.000	+160	-1.23	0
15 - 25	0.900	0.000			
25 - 35	0.190	0.000			

*Average value from four to five measurements.

error always had a negative sign, evidently due to the overestimation of the CH_4 uptake rate. Obviously Eq. (1) is too simple to get a complete fit to the experimental data. Probably the most important deviation came from the fact that we did not consider the dependence of uptake rate on dissolved oxygen concentration. The main qualitative features of net CH_4 emission were reproduced adequately: the negative emission of CH_4 in July and increased positive emission in April.

For the trenches, soil core experiments and the chamber technique gave different results even from a qualitative point of view: The first approach showed net consumption of CH_4, whereas the second showed net emission (Table 3). This discrepancy could be explained by the fact that emitted methane was produced not in the trenches, but elsewhere—in the surrounding areas of the bog.

Data on all studied soils are summarized in Table 4. The general regularity already mentioned for the ombrotrophic bog was reproduced for all soils: In the upper parts of soil profiles, the process of CH_4 consumption was prevalent, whereas in the lower horizons, it was methanogenesis. The rates of CH_4 exchange were maximal in wetlands: lagoons, bogs, and fens. In contrast to opinion,[9] CH_4 uptake was most intensive in wetlands as opposed to dry soils. The consumption of methane was completely absent in the hypersaline lagoons.

Conclusions

The methane cycle in most of the studied soils was found to be a closed one: The processes of methanogenesis in the lower horizons were equal or less than potential CH_4 consumption. The exceptions were hypersaline lagoons, where methanotrophic bacteria were apparently absent, as well as the trenches and natural depressions in ombrotrophic bogs, where CH_4-consuming activity was insufficient to compensate for the transfer and intensive CH_4 generation in the lower horizons. Special attention should be paid to the question of what are the limiting factors for methanotroph development in ombrotrophic bogs. Probably the conditions are not adequate (low pH and nutrients content, low temperature) for intensive growth of these bacteria, but perhaps are sufficient for methanogenic bacteria. As a result, the methanotrophs are able to grow only sporadically in midsummer. Transiently (in the spring and fall) methane formation becomes more intensive than methane oxidation.

References

1. Andreae, M.O., and D.S. Shimel (eds). 1989. In Exchange of Trace Gases Between Terrestrial Ecosystems and the Atmosphere, J. Wiley and Sons, New York, p. 347.

2. Bouwman, A.F., (ed). 1990. In Soils and the Greenhouse Effect, New York et al. J. Wiley and Sons, New York, p. 575.
3. Cicerone, R.J., and R.S. Oremland. 1988. Biogeochemical aspects of atmospheric methane. Global Biogeochem. Cycles **2**:299–327.
4. Kalyuzhny, S.V., A.N. Nozhevnikova, and S.D. Varfolomeyev. 1985. The kinetics of *Methanosarcina vacuolata* growth on methanol. Microbilogia **54**:257–262.
5. Kasparov, S.V., N.S. Panikov, and O.I. Min'ko. 1985. Membrane probe for analysis of gas composition of soil air. Pochvovedenije (Soviet Soil Science) **11**:145–151.
6. King, G.M. 1990. Dynamics and controls of methane oxidation in a Danish wetland sediments. FEMS Microbial Ecology **74**:309–324.
7. Krut'ko, P.D., A.I. Maximov, and L.M. Skvortsov. 1988. Algorithms and Programs for Automatic Systems Construction, Radyo y svias, Moscow, p. 306.
8. Robertson, A.M., and R.S. Wolfe. 1970. Adenosine triphosphate pools in *Methanobacterium*. J. Bacteriol. **102**:43–53.
9. Schutz, H., W. Seiler, and H. Rennenberg. 1990. Soil and land use related sources and sinks of methane (CH_4) in the context of the global methane budget. In Soils and the Greenhouse Effect, J. Wiley and Sons, New York, pp. 269–285.
10. Stewart, J.W.B., I. Aselmann, A.F. Bouwman, and R.L. Desjardins. 1989. Extrapolation of flux measurements to regional and global scales. In Exchange of Trace Gases Between Terrestrial Ecosystems and the Atmosphere, J. Wiley and Sons, New York, pp. 155–174.
11. Varfolomejev, S.D. 1987. Kinetic regularities of microbial populations development. In S.D. Varfolomejev (ed.), Contemporary Problems of Biokinetics, Moscow University Press, Moscow, pp. 6–77.
12. Whalen, S.C., and W.S. Reeburgh. 1988. A methane flux time series for tundra environments. Global Biogeochem. Cycles **2**:399–409.

Emission of N-Oxides from Acid Irrigated and Limed Soils of a Coniferous Forest in Bavaria

Hans Papen, Bettina Hellmann, Hannelore Papke, and Heinz Rennenberg

Fraunhofer-Institute for Atmospheric Environmental Research, D-8100 Garmisch-Partenkirchen, Germany

ABSTRACT

The effects of acid irrigation and liming on the emission rates of N_2O and NO from an old spruce stand were measured seasonally. Mean annual N_2O fluxes were 25 μg N_2O–N m^{-2} h^{-1}. If this coniferous forest can be considered representative, we estimate a contribution of temperate coniferous forest soils to the global N_2O budget of 1 Tg y^{-1}. For NO, mean annual fluxes from temperate coniferous forest soils were estimated to be in the range of 0.5–24 μg NO–N m^{-2} h^{-1}. Studies with intact soil cores indicated that 0–54% of the N_2O and 12–42% of the NO emitted could be accounted for by denitrification and that N_2O/NO emission rates were inhibited by C_2H_2. However, since chemolithotrophic nitrifiers could never be detected in the control, water-irrigated, and acid-irrigated plots, and since denitrification was already inhibited by O_2, other biological processes different from denitrification and chemolithotrophic nitrification are concluded to be responsible for these N_2O/NO emissions.

Introduction

Nitrous oxide (N_2O) is an important greenhouse gas[11] with an atmospheric residence time of 100–170 years[13,25]; it is also involved in the depletion of stratospheric ozone.[11,13] The atmospheric concentration of N_2O is estimated to increase at present with a rate of 0.2–0.3% per year.[25,40] Nitric oxide (NO) also plays an important role in the chemistry of the atmosphere and is believed to contribute to an increase of tropospheric O_3 concentrations.[14]

In the past, research identifying significant sources of N_2O focused on the contribution of agricultural N fertilizers (e.g., Refs. 13 and 49), fossil-fuel combustion, and biomass-burning.[14] Estimates of the contribution of natural sources to the atmospheric N_2O budget, especially from temperate forests, still have a high degree of uncertainty. Temperate forests may be important sources of N_2O and NO since they occupy 24% of the total forested land on the earth.[7]

The composition and the activity of the natural soil microbial community in forests are believed to be different from agricultural soils.[44] Therefore, the biological sources of N_2O and NO may also be more diverse in forested than in agricultural soils.[44] Nitrogen cycling in temperate coniferous forests[36] is substantially different from N cycling in temperate deciduous forests and, therefore, may also function differently as sources and sinks for N trace gases.

Nitrogen trace gas fluxes from soils are believed to be mediated by microbial processes such as denitrification and chemolithotrophic nitrification (e.g., Refs. 2, 5, 23, 24, 32, 33, 34, 35, 39, 41–43, 46, 50, and 51), though other biological processes like nitrate respiration by *Enterobacteriaceae*[1,6] and assimilatory nitrate reduction of bacteria and fungi[6] have been demonstrated to produce N_2O in pure culture experiments. There is increasing information about the mechanisms that control nitrogen trace gas fluxes from different ecosystems.[15–17,26–28] In a study using intact soil cores from temperate forest soils,[44] it was demonstrated that biological processes other than denitrification and chemolithotrophic nitrification were responsible for most of the observed N_2O emissions.

Furthermore, the effects of acid rain leading to increased soil acidification, of liming to reduce soil acidity, and the consequences of enhanced N deposition to forest soils from the atmosphere[36] may influence the emission of N trace gases from soils. The purposes of this study were (1) to quantify N_2O and NO emission rates from a temperate coniferous forest in southern Germany, (2) to elucidate the effect of acid irrigation and liming on trace gas emissions from these soils, and (3) to utilize inhibitors to estimate the contribution of different microbial processes to N gas emissions.

Materials and Methods

Site

Field measurements of soil N_2O and NO emissions were performed in an 80-year-old spruce stand, the "Höglwald," about 16 km southeast of Augsburg, Bavaria, Germany.[29] The forest is divided into several experimental 2500-m^2 plots: A1 (untreated), C1 (irrigated with "rain-like" waters, pH 5.0–5.5), B1 (acid-irrigated with water containing sulfuric acid, pH 2.7–2.8, 3–4 kmoles of protons ha^{-1} yr), A2 (limed once in 1984 with 4000 kg/ha dolomite), B2 (acid-irrigated plus limed). Irrigations have been performed 15–18 times/yr totaling 170 mm/yr since 1984. About 40 kg N ha^{-1} y^{-1} is attributable to atmospheric deposition. Among this, two-thirds is in the form of ammonia, one-third in the form of nitrate. About the same amount of N (40 kg ha^{-1} y^{-1}) leaves the forest ecosystem in the form of nitrate. Data on atmospheric NO, NO_2, and O_3 gas-mixing ratios are summarized in Ref. 30. Soils are generally acidic, even when limed, with the following mean pH values in the O horizons: A1: 3.7; B1: 3.2; C1: 3.5; A2: 4.5; B2: 4.4.

Chambers for the Determination of N_2O and NO Fluxes

Static (N_2O measurements) and dynamic (NO measurements) chambers (N_2O boxes: 30 cm × 30 cm × 30 cm; NO boxes: 30 cm × 30 cm × 28 cm) were built from Plexiglas, mounted on stainless-steel forms (30 cm × 30 cm × 15 cm), and inserted 12 cm into the soil immediately before measurements (total volume of N_2O boxes: 29.71; of NO boxes: 25.21). The tops of the static and dynamic chambers could be opened and closed automatically by pneumatic devices, thereby exposing the enclosed soil area to ambient air after each measurement. The gas inlet plane of the dynamic chambers, located opposite to the outlet of the chambers, possessed seven holes (2.0 cm in diameter), thus avoiding pressure differences between the inlet and outlet of the boxes. A fan was installed in each box to enable rapid mixing of the atmosphere. Each chamber was equipped with thermoelements which measured the soil temperatures at the surface and at 1 and 5 cm depth. Every second chamber contained a probe (Vaisala, Helsinki, Finland) for continuous registration of the air temperature and humidity inside the boxes. Two dynamic and two static chambers were placed on two different plots for measurements lasting 72 h. Analysis of spatial variability revealed differences in N trace gas emissions in the range 8–25%.

N_2O Measurements with Closed Chambers

Following closure, 180 ml of the atmosphere inside the boxes were transported by a metal bellows pump via Teflon tubings to a sample loop (total volume: 3.14 ml) over different time intervals (0, 10, 30, 50, 70, 90 min). Before the gas mixture entered the sample loop, it was passed through a nafion-dryer (Perma Pure, Toms River, NJ, USA). After relaxing the sample gas in the sample loop, it was flushed directly by the carrier gas stream through an ascarite precolumn (trap for CO_2) onto a stainless-steel column (3 m × 1/8 in.) packed with Porapak N (60–80 mesh) in a GC (Shimadzu GC Mini-2E, Shimadzu Europa, Duisburg, Germany) equipped with a ^{63}Ni electron capture detector. Gas-chromatographic conditions for N_2O analysis were: carrier gas, nitrogen grade 5.0 (Messer-Griesheim, München, Germany), 40 ml min^{-1}; make-up gas, 5% methane in argon (ECD-quality, Messer-Griesheim, München, Germany), 2 ml min^{-1}; column temperature, 60°C; detector temperature, 300°C. The retention time for N_2O was 2.20 ± 0.01 min. At regular intervals, calibration of the gas chromatograph (GC) was performed using a standard gas mixture with a defined mixing ratio for N_2O (N_2O in synthetic air: 630 ppb, Messer Griesheim, München, Germany). N_2O fluxes were calculated from the slopes of the initial linear increase of the N_2O mixing ratio inside the individual boxes as described.[12] The detection limit for N_2O flux was 0.5–1.0 μg N_2O–N m^{-2} h^{-1}.

NO Measurements with Dynamic Chambers

Measurement of the NO fluxes and calibration of the NO_x analyzer were performed as described[50] with the following exceptions: the flow rate of ambient air through the chambers was 45 L min^{-1} and a NO_x instrument (Tecan CLD 770 AL ppt, Zellweger-Tecan-Systeme GmbH, München, Germany) with a sensitivity limit of < 50 ppt was used. Calibration of the instrument was performed in regular intervals with a multigas calibrator (Tecan-Envial series 100 Multigas-calibrator) using certified standard gas (524 ppb NO in nitrogen grade 5.0, Messer-Griesheim, München, Germany).

Sampling of Intact Soil Cores

Intact soil cores were taken during August/September 1990 from all plots by the following device: Plexiglas cylinders (height 25 cm; inner diameter 12 cm) covered by a stainless-steel plate were driven carefully through the O layers into the mineral soil (A horizon). By placing crumpled paper on top of the forest floor inside the cylinders, destruction of the natural structure of the O horizons was avoided and contamination of deeper horizons with material from higher horizons was prevented. As in the field measurements, litter material was not removed. The cylinders containing the intact soil cores were closed at the bottom by gas-tight Plexiglas lids and at the top by a water-tight, gas-permeable foil, and transported in insulated coolers (4°C) to the laboratory.

Laboratory Studies with Intact Soil Cores

The mineral soil was carefully removed from the cores. The intact O horizons (LOf1, Of2, Oh) remained in the Plexiglas column and were used for experiments. Bottom and tops of the Plexiglas columns were sealed gas-tight with Plexiglas lids; directly below the top lid, a fan was installed for rapid equilibration. The volume of the atmosphere above the soil varied from 1.0 to 1.5 L depending on the thickness of the O horizon. Cylinders containing the soil cores were incubated at 20°C in a water bath, which was about 2°C higher than average soil temperature (17.6°C at 1–5 cm depth) during the field surveys of August/September 1990. After temperature equilibration, a stream of 100 kPa O_2 (5.0, Messer-Griesheim, München, Germany) was passed initially through the soil core from the bottom to the top (flow rate: 3 L min^{-1}) to achieve rapid equilibration and to facilitate penetration of this inhibitor of denitrification in the soil column.[44] Thereafter, the gas was passed again above the soil surface and NO fluxes were determined. Then the cylinders were closed again and N_2O emissions were determined under an atmosphere of 100 kPa O_2. The difference in N_2O and NO emission rates between ambient air and pure oxygen was assumed to be due to denitrification since chemolithotrophic nitrification was not inhibited under these conditions.[44]

Thereafter, N_2O/NO fluxes from the soil cores were measured in an atmosphere of 100 kPa O_2 containing 10 Pa C_2H_2 (0.01% acetylene in 99.99% oxygen, Linde Chemie, München, Germany). A further reduction in N trace gas emissions under this atmosphere, therefore, is assumed to reflect the contribution of autotrophic ammonia oxidizers. Any remaining NO/N_2O emissions which are insensitive to 100 kPa O_2 plus 10 Pa acetylene were considered as preliminary evidence for unidentified microbial processes involved in N trace gas production. Drying of the soil cores was avoided by adding water vapor to the gases: gas streams were passed through a flask containing sterilized water prior to passage over the soil cores. The weight of individual soil cores at the beginning and end of an experiment showed no significant differences, indicating that changes in water content of the soil columns had not occurred. To test any production of N_2O and NO by abiological process, e.g., chemodenitrification,[10,47] parallel soil cores were autoclaved and analyzed for N_2O/NO emissions under the same experimental regime as described above.

Quantification of Chemolithotrophic Nitrifiers

Numbers of chemolithotrophic ammonium and nitrite oxidizers in the O horizons from parallel soil cores (in triplicates) were quantified by the most probable number (MPN) technique[38] using five inoculation tubes for each dilution step. MPN tubes were checked weekly for nitrite/nitrate produced from ammonium or nitrate produced from nitrite over a 3-month period.

Results

Field Results

Mean N_2O emission rates from the various treatments are summarized in Figure 1. The data demonstrate a seasonal pattern in N_2O emission rates from differently treated soils with highest values during late spring, decreasing values in late summer, lowest values during late autumn, and intermediate values in late winter. Diel fluctuations in N_2O emission rates were not observed and emission rates during day and night hours were not significantly different.

Mean N_2O emission rates from soils of the untreated and acid-irrigated plots in May/June 1990 (in $\mu g\ N_2O$–N $m^{-2}\ h^{-1}$) were similar (A1: 71.8 ± 16.3; B1:

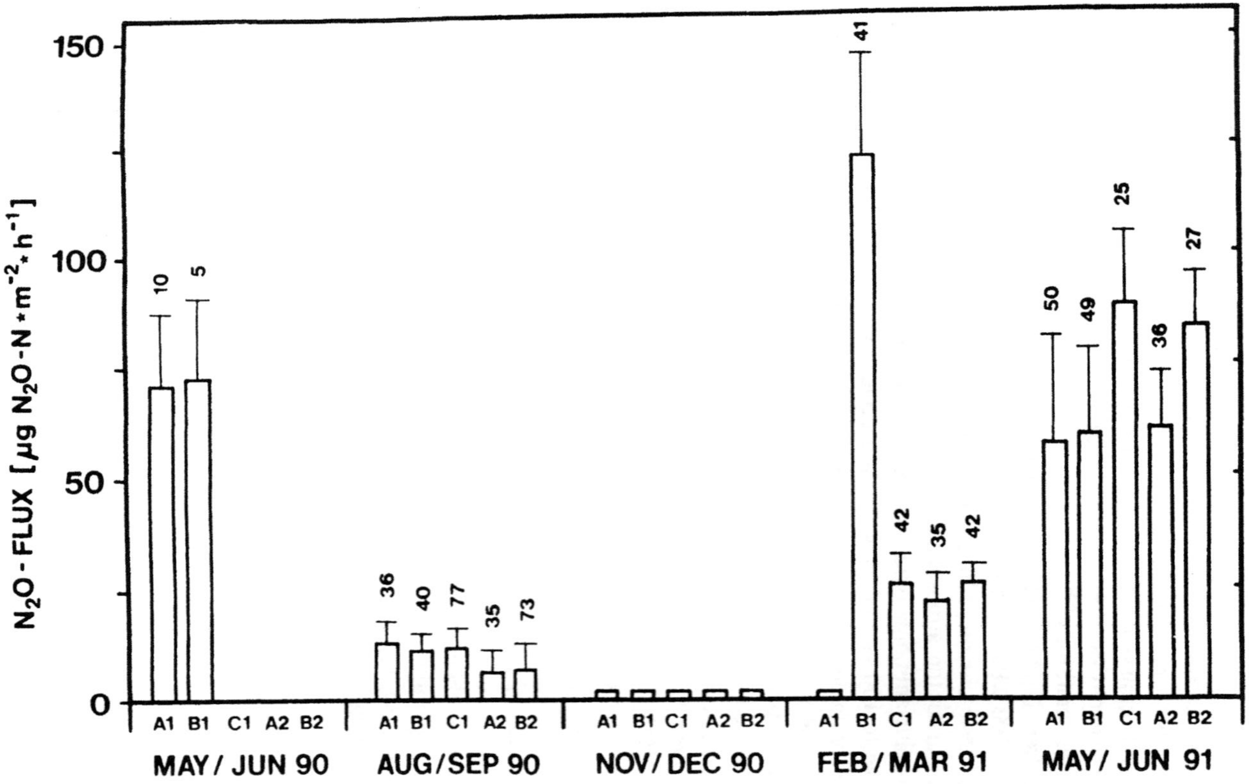

Figure 1. Mean N_2O fluxes during five field surveys in 1990/1991 from differently treated soils: A1: control plot; B1: acid-irrigated plot; C1: water-irrigated plot; A2: limed plot; B2: acid-irrigated plus limed plot. Data represents the means from two chambers over a 72-h measuring period. Numbers indicate the number of replicates (*n*).

74.3 ± 18.5). During August/September 1990, N_2O emission rates on A1/B1 were significantly ($p < 0.01$) lower than in May/June. Emission rates from these plots had decreased by about the same extent (A1: 15.1 ± 4.5 μg N_2O–N m^{-2} h^{-1}; B1: 11.1 ± 3.8 μg N_2O–N m^{-2} h^{-1}). N_2O emission rates from the water-irrigated plot C1 (12.3 ± 4.3 μg N_2O–N m^{-2} h^{-1}) were in the same range as from A1/B1. On the other hand, N_2O emission rates from the limed (A2: 6.2 ± 4.8 μg N_2O–N m^{-2} h^{-1}) and the acid-irrigated plus limed (B2: 6.7 ± 5.4 μg N_2O–N m^{-2} h^{-1}) plots were 50% lower than control (A1) and acid-irrigated plots (B1). During November/December 1990, when the soil was frozen and covered by snow, N_2O emission rates reached their lowest values (generally ≤ 1 μg N_2O–N m^{-2} h^{-1}). At the beginning of the February/March 1991 survey, the soil of A1 was still frozen and the N_2O emission rate was low (≤ 1 μg N_2O–N m^{-2} h^{-1}). Approximately 24 h later, when measurements were performed on B1, the snow had melted and the soil had begun to thaw. A burst of N_2O from the soil (B1) was observed over 48 h on this plot and was equivalent to 122.3 ± 23.2 μg N_2O–N m^{-2} h^{-1}. Mean N_2O emission rates on C1 (30.9 ± 6.1 μg N_2O m^{-2} h^{-1}), A2 (22.9 ± 5.9 μg N_2O–N m^{-2} h^{-1}), and B2 (26.0 ± 3.9 μg N_2O–N m^{-2} h^{-1}) had markedly increased compared to those in November/December 1990, but were not significantly different from each other. A further increase in mean N_2O emission rates was observed in May/June 1991 (in μg N_2O–N m^{-2} h^{-1}) on all plots: A1: 57.2 ± 13.8; B1: 59.1 ± 18.5; C1: 88.0 ± 16.3; A2: 59.7 ± 12.5; B2: 83.2 ± 12.0.

With regard to NO (Figure 2), only data for late autumn 1990 and late winter/spring 1991 are available. In contrast to N_2O, net emission as well as net deposition of NO was observed during all surveys. During November/December 1990, mean NO fluxes for A1 (12.4 ± 13.1 μg NO–N m^{-2} h^{-1}) and B1 (12.1 ± 11.45 μg NO–N m^{-2} h^{-1}) were nearly identical and increased to their maximum in the spring (May/June 1991: A1: 31.8 ± 25.3 μg NO–N m^{-2} h^{-1}; B1: 38.7 ± 18.5 μg NO–N m^{-2} h^{-1}). Mean NO emission rates (in μg NO–N m^{-2} h^{-1}) from the limed (A2 November/December 1990: 3.7 ± 5.7; February/March 1991: 7.5 ± 12.9; May/June 1991: 5.1 ± 15.3) and acid-irrigated plus limed plots (B2: November/December 1990: 3.6 ± 4.0; February/March 4.8 ± 19.77; May/June: 11.6 ± 6.7) were significantly ($p < 0.01$) lower than those of the control (A1) and acid-irrigated plots (B1) at all seasons. NO emission rates on the water-irrigated plot (C1) increased markedly in May/June 1991, but were lower or almost identical to values obtained on A2/B2.

Mechanism Studies

Results of the inhibition studies are summarized in Table 1. N_2O emission rates from soil cores held under pressurized air were similar to field rates. Only the B1 soil core showed markedly higher N_2O emission rates. About a 50%

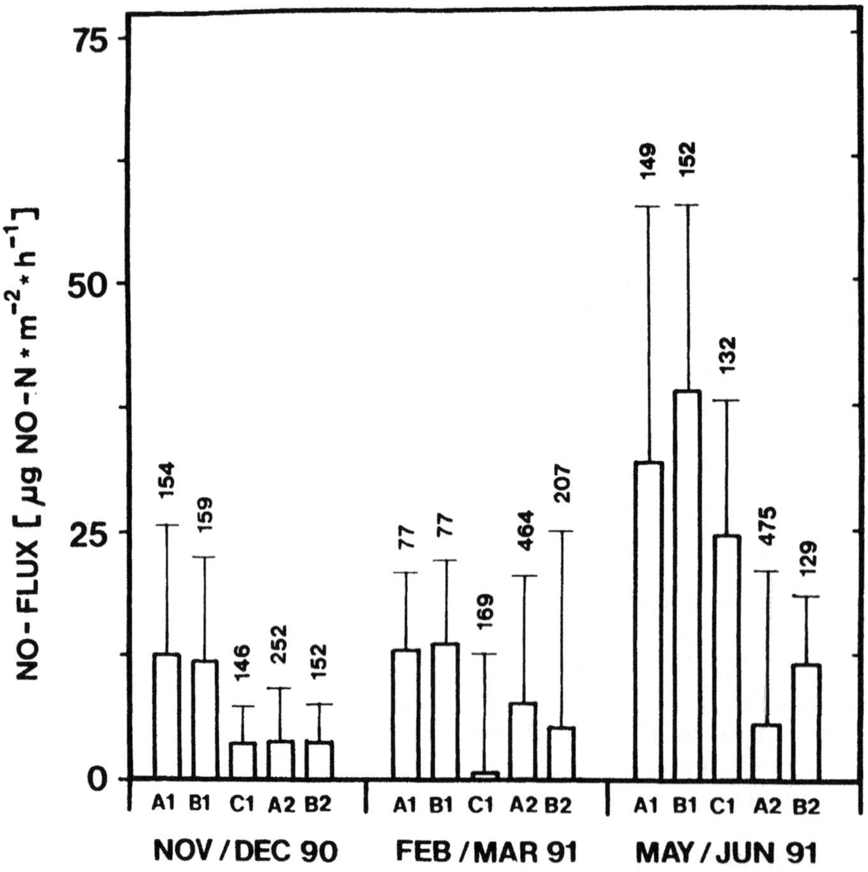

Figure 2. Mean NO fluxes during three field measuring surveys in 1990/1991 from differently treated soils.

reduction in N_2O emission from A2 and B2 plots compared to A1 and C1 plots was also observed.

By treatment with 100 kPa O_2, denitrification in the soil was inhibited,[44] whereas autotrophic ammonium oxidation and heterotrophic nitrification were not. Therefore, the decrease in N_2O/NO emissions in an atmosphere of 100 kPa O_2 most likely reflects the contribution of denitrification. As summarized in Table 1, depending on the soil treatment only between 0% (B1) and 54% (B2) of the N_2O emission rates observed could be accounted for by denitrification. Therefore, other microbial processes may have been responsible for a major part of the N_2O emissions from the soils.

Table 1. Differentiation between microbiological processes involved in N_2O/NO emissions from intact soil cores by use of gaseous inhibitors. Emission rate is in μg N m^{-2} h^{-1}.

Plot	Gas	Air	100 kPa O_2	100 kPa O_2 + 10 Pa C_2H_2
A1	N_2O	25.0 (100)	14.4 (58)	5.8 (23)
	NO	953.0 (100)	714.0 (75)	78.1 (8)
B1	N_2O	69.8 (100)	78.2 (112)	11.9 (17)
	NO	200.6 (100)	138.0 (69)	14.6 (7)
C1	N_2O	22.7 (100)	13.6 (60)	2.6 (11)
	NO	401.3 (100)	233.7 (58)	25.0 (6)
A2	N_2O	8.3 (100)	7.5 (90)	3.9 (47)
	NO	87.8 (100)	61.0 (70)	17.6 (20)
B2	N_2O	10.2 (100)	4.7 (46)	4.7 (46)
	NO	74.1 (100)	65.2 (88)	11.1 (15)

Note: Intact soil cores were taken in August/September 1990 and treated in the laboratory with different gas mixtures inhibiting different soil microbial processes to differentiate between different microbial processes (denitrification, chemolithotrophic nitrification, others) involved in N_2O/NO emissions as described in the text. Data in parentheses represent percentages of emission rates in air. Data given are means of emission rates of two soil cores each of one experiment in August/September 1990, but are representative of other independently performed experiments on soil cores taken at other seasons in the years 1990–1991.

Treatment with 10 Pa C_2H_2 in 100 kPa O_2 should inhibit chemolithotrophic ammonium oxidation in addition to denitrification since this concentration of C_2H_2 will inhibit the ammonium monooxygenase of autotrophic ammonium oxidizers[4,20,21]; the decrease in N_2O emission rates should, therefore, represent the contribution of autotrophic ammonium oxidation to the N_2O fluxes observed. As can be seen from Table 1, in most soil cores a further decrease in N_2O emission rates was observed. Since in soils of A1, B1, and C1 chemolithotrophic nitrifiers never could be detected,[38] this decrease may not be due to inhibition of chemolithotrophic ammonium oxidation. Other biological processes, also sensitive to 10 Pa C_2H_2, exist in these soils and contribute to the N_2O emissions observed. Regardless of season, chemolithotrophic nitrifiers were detected in the O horizons of limed (A2) and acid-irrigated plus limed plots (B2).[38] The numbers of chemolithotrophic ammonium/nitrite oxidizers, determined by MPN, during September/August 1990 were (in 10^3 g $soil^{-1}$): A2: 5.7/15.5 and B2: 3.7/11.5.

Results similar to those obtained for N_2O were obtained for NO in the soil core experiment. Between 12% and 42% of the NO emission rates observed could be accounted for by denitrification processes (Table 1). Treatment with 100 kPa O_2/10 Pa C_2H_2 resulted in a decrease (10-fold) in NO emission rates from soil cores of A1, B1, and C1. Since, in these soils, chemolithotrophic nitrifiers could not be detected, other microbiological processes, different from denitrification and chemolithotrophic ammonia oxidation, may have been responsible for these emissions. Soil cores which had been autoclaved either showed no N_2O/NO emissions or only marginal emissions, which were at the sensitivity limits of the instruments employed.

Discussion

Field Results

N_2O and NO emissions during five field surveys covering a 12-month period showed seasonal variations in N trace gas flux, whereas diel variations were not observed. Maximum seasonal emission rates were observed during late spring, which is comparable to the results of others.[7,18,48] In most cases, N_2O/NO emissions were highest on control and acid-irrigated plots. However, acid irrigation (B1) appeared not to have resulted in a further stimulation of N trace gas emissions. Furthermore, the water-irrigated plot (C1) showed no significantly enhanced N_2O/NO emission. This may reflect the observation that significant air deficits, which would result in stimulation of denitrification, never could be demonstrated in any soil layer of the "Höglwald" soils upon irrigation.[19] The reasons for the significant reduction ($p < 0.01$) in N_2O (August/September 1990) and NO emissions (observed in all field measurements) on the limed plots are not clear. It is speculated that the enhanced soil pH caused by liming allowed part of the denitrifier population in the soils to further reduce N_2O to N_2. Markedly reduced N_2O emission rates on limed plots were also described for a beech stand in northern Germany.[9] If liming more broadly reduces N trace gas emissions from acid soils, this approach could reduce N trace gas emissions to the atmosphere from ecosystems with acid soils. However, since chemolithotrophic nitrifiers are favored by liming,[38] other problems may arise, such as enhanced nitrate release from these soils into ground water.[29,38] Peak emissions of N_2O in late winter 1991 (B1) have also been described by others[7,48] when snow melt and soil thawing occurred. Whether these peak emissions (Figure 1) were due to a sudden release of entrapped N_2O during thawing or resulted from microbial activities in the soil remains unclear. However, the observation that high emission rates from B1 lasted for 48 h may be taken as an indication of the presence of microbial activity.

We have calculated mean N_2O fluxes from all experimental plots over a period of about one year (August 1990–June 1991) to be 25 μg N_2O–N m^{-2} h^{-1}. From this calculation, peak N_2O emission rates in February/March 1991 (B1) were omitted and emission rates in November/December 1990 were considered to be 1 μg N_2O–N m^{-2} h^{-1}. From this mean emission rate, a lower limit for N_2O emissions of temperate coniferous forests on a global scale of 1 Tg N/yr was calculated [using available global areal data for coniferous temperate forests of 4.88×10^6 km^2 (Ref. 53)]. This estimate is markedly higher than the recently published values of 0.012 Tg N/yr for temperate coniferous and deciduous forests.[7,8] Our estimate is, however, in good agreement with another estimate[48] of the total source strength of temperate forest 0.7–1.5 Tg N/yr. Emission rates for NO from a temperate coniferous forest in Sweden were reported to be in the range 0.36–2.88 μg NO–N m^{-2} h^{-1},[22] whereas a temperate mixed forest in the United States exhibited rates in the range 0.72–14.76 μg NO–N m^{-2} h^{-1}.[55] Mean NO emission rates from the "Höglwald" soils during three field measuring surveys were in the range 0.48–24.3 μg NO–N m^{-2} h^{-1} and, therefore, were closer to the range of NO emission reported by Williams et al.[55] Nevertheless, we stress that our estimates of the contribution of temperate coniferous forests to atmospheric N_2O and the emission ranges given for NO from the "Höglwald" soils still yield a high degree of uncertainty as do estimates of other authors. It cannot be excluded that the higher N trace gas fluxes from the Höglwald soils, as compared with values reported from other temperate forests, may reflect the influence of high anthropogenic inputs of N to this coniferous forest.

Mechanism Studies

From the experiments with intact soil cores (Table 1), it is concluded that regardless of treatment, denitrification contributed at most 50% to the N_2O and NO emissions. In most cases, the contribution of denitrification was markedly lower than 50%. Furthermore, it is very unlikely that chemolithotrophic nitrification was a source of N_2O and NO in soils of A1, B1, and C1 since these organisms never were detected in these experimental plots, whereas they were always detected on the limed plots (A2, B2). These results are in agreement with earlier findings demonstrating that organisms other than denitrifiers and nitrifiers accounted for 77–100% of the N_2O flux from intact soil cores.[44] We observed a major decrease in NO emission (one order of magnitude) with the acetylene treatment (Table 1), suggesting that nitrification processes were active and were a much more important mechanism for NO production/emission than for N_2O production, a finding earlier reported by others.[17] One possibility is that heterotrophic nitrification may be involved in N_2O/NO production.[1,3,37,45] Heterotrophic nitrification has been reported in many bacteria and fungi.[1,3,31,37,38,45,47] It is generally believed that chemolithotrophic ammonium oxidation is inhibited by low concentrations (10 Pa) of acetylene,[4,20,21] whereas heterotrophic nitrifiers are not. Lack of inhibition has only been described for a single *Anthrobacter* strain[21] and two fungi.[47,52] Recent results from experiments with pure cultures in our laboratory (Papen, unpublished results) using a highly heterotrophically nitrifying *Alcaligenes faecalis* strain[37] and other heterotrophic nitrifiers isolated from the Höglwald soils indicate that acetylene was also a strong inhibitor of heterotrophic nitrification activity in these strains. Therefore, it might be speculated that the dramatic reduction in NO emission from soil cores A1, B1, C1 after treatment with acetylene (0.01%) may reflect inhibition of heterotrophic nitrifiers present in high numbers in these soils (range: 10^6–10^7 g soil^{-1}; Papen, unpublished results). Recently, the contribution of heterotrophic nitrification to N trace gas fluxes from agricultural soils was estimated to be marginal at best.[54] In conclusion, more research has to be devoted to elucidate the contribution of heterotrophic nitrification to N_2O/NO fluxes from different ecosystems.

Acknowledgments

We thank Frau R. von Berg for the determination of the numbers of chemolithotrophic nitrifiers and expert technical assistance. This work was supported by the Bundesministerium fur Forschung und Technologie under contract No. BEO 0339175G3.

References

1. Anderson, I.C., and J.S. Levine. 1986. Relative rates of nitric oxide and nitrous oxide production by nitrifiers, denitrifiers, and nitrate respirers. Appl. Environ. Microbiol. **51**:938–945.
2. Anderson, I.C., and J.S. Levine. 1987. Simultaneous field measurements of biogenic emissions of nitric oxide and nitrous oxide. J. Geophy. Res. **92**:965–976.
3. Anderson, I., J. Homstead, D. Burdige, and M. Poth. 1991. Factors affecting emissions of NO and N_2O by autotrophic and heterotrophic nitrifiers. Abstracts of the General ASM Meeting—1991, Abstract N-1.
4. Bedard, C., and R. Knowles. 1989. Physiology, biochemistry, and specific inhibitors of CH_4, NH_4^+, and CO oxidation by methanotrophs and nitrifiers. Microbiol. Rev. **53**:68–84.
5. Blackmer, A.M., J.M. Bremner, and E.L. Schmidt. 1980. Production of nitrous oxide by ammonia-oxidizing chemoautotrophic microorganisms in soil. Appl. Environ. Microbiol. **40**:1060–1066.
6. Bleakley, B.H., and J.M. Tiedje. 1982. Nitrous oxide production by organisms other than nitrifiers or denitrifiers. Appl. Environ. Microbiol. **44**:1342–1348.
7. Bowden, R.D., P.A. Steudler, J.M. Melillo, and J.D. Aber. 1990. Annual nitrous oxide fluxes from temperate forest soils in the northeastern United States. J. Geophys. Res. **95**:13997–14005.

8. Bowden, R.D., J.M. Melillo, P.A. Steudler, and J.D. Aber. 1991. Effects of nitrogen additions on annual nitrous oxide fluxes from temperate forest soils in the northeastern United States. J. Geophys. Res. **96**:9321–9328.

9. Brumme, R., F. Beese, and N. Lotfield. 1989. Berichte des Forschungszentrums Waldökosysteme, Series B, **15**:83–89.

10. Chalk, P.M., and C.J. Smith. 1983. Chemodenitrification. In J.R. Freney and J.R. Simpson (eds.), Gaseous Losses of Nitrogen from Plant–Soil Systems, Martinus Nijhoff/Dr. W. Junk, Dordrecht, pp. 65–90.

11. Cicerone, R.J. 1987. Changes in stratospheric ozone. Science **237**:35–42.

12. Conrad, R., W. Seiler, and G. Bunse. 1983. Factors influencing the loss of fertilizer nitrogen into the atmosphere as N_2O. J. Geophys. Res. **88**:6709–6718.

13. Crutzen, P.J. 1976. Upper limits on atmospheric ozone reductions following increased application of fixed nitrogen to the soil. Geophys. Res. Lett. **3**:169–172.

14. Crutzen, P.J. 1979. The role of NO and NO_2 in the chemistry of the troposphere and stratosphere. Ann. Rev. Earth Planet. Sci. **7**:443–472.

15. Davidson, E.A., W.T. Swank, and T.O. Perry. 1986. Distinguishing between nitrification and denitrification as sources of gaseous nitrogen production in soil. Appl. Environ. Microbiol. **52**:1280–1286.

16. Davidson, E.A. 1991. Fluxes of nitrous oxide and nitric oxide from terrestrial ecosystems. In J.E. Rogers and W.B. Whitman (eds.), Microbial Production and Consumption of Greenhouse Gases. Methane, Nitrogen Oxides, and Halomethanes, American Society for Microbiology, Washington, DC, pp. 219–235.

17. Firestone, M.K., and E.A. Davidson. 1989. Microbiological basis of NO and N_2O production and consumption. In M.O. Andreae and D.S. Schimel (eds.), Dahlem Workshop Reports, Life Sciences Research Report 47, Exchange of Trace Gases Between Terrestrial Ecosystems and the Atmosphere, J. Wiley and Sons, Chichester, pp. 7–21.

18. Goodroad, L.L., and D.R. Keeney. 1984. Nitrous oxide emission from forest, marsh, and prairie ecosystems. J. Environ. Qual. **13**:448–452.

19. Grimmeisen, W., K. Kreutzer, and J. Bittersohl. 1986. Einfluss der Beregnung auf Matrixpotentiale und Bodendurchfeuchtung im Höglwald-Experiment. Forstw. Clb. **105**:295–299.

20. Hynes, R.H., and R. Knowles. 1978. Inhibition by acetylene of ammonia oxidation in *Nitrosomonas europaea*. FEMS Microbiol. Lett. **4**:319–321.

21. Hynes, R.K., and R. Knowles. 1982. Effect of acetylene on autotrophic and heterotrophic nitrification. Can. J. Microbiol. **28**:334–340.

22. Johansson, C.J. 1984. Field measurements of emission of nitric oxide from fertilized and unfertilized forest soils in Sweden. J. Atmos. Chem. **1**:429–442.

23. Johansson, C.J., and L. Granat. 1984. Emission of nitric oxide from arable land. Tellus **36B**:25–37.

24. Keller, M.T., J. Goreau, S.C. Wofsy, W.A. Kaplan, and M.B. McElroy. 1983. Production of nitrous oxide and consumption of methane by forest soils. Geophys. Res. Lett. **10**:1156–1159.

25. Khalil, M.A.K., and R.A. Rasmussen. 1983. Increase and seasonal cycles of nitrous oxide in the Earth's atmosphere. Tellus **35B**:161–169.

26. King, H.M. 1990. Dynamics and controls of methane oxidation in a Danish wetland sediment. FEMS Microbiol. Ecol. **74**:309–324.

27. Kramer, M., M. Baumgärtner, M. Bender, and R. Conrad. 1990. Consumption of NO by methanotrophic bacteria in pure culture and in soil. FEMS Microbiol. Ecol. **73**:345–350.

28. Kreitinger, J.P., T.M. Klein, N.J. Novick, and M. Alexander. 1985. Nitrification and characteristics of nitrifying microorganisms in an acid forest soil. Soil Sci. Soc. Amer. J. **49**:1407–1410.

29. Kreutzer, K., and J. Bittersohl. 1986. Untersuchungen über die Auswirkungen des sauren Regens und der kompensatorischen Kalkung im Wald. Forstw. Cbl. **105**: 273–282.

30. Kreuzig, R., and F. Korte. 1986. Luftchemische Charakterisierung des Standortes Höglwald. Forstw. Cbl. **105**:292–295.

31. Lang, E., and G. Jagnow. 1986. Fungi of a forest soil nitrifying at low pH values. FEMS Microbiol. Ecol. **38**:257–265.

32. Lipschultz, F., O.C. Zafiriou, S.C. Wofsy, M.B. McElroy, F.W. Valois, and S.W. Watson. 1981. Production of NO and N_2O by soil nitrifying bacteria. Nature **294**: 641–643.

33. Martikainen, P.J. 1985. Nitrous oxide emission associated with autotrophic ammonium oxidation in acid coniferous forest soil. Appl. Environ. Microbiol. **50**:1519–1525.

34. Matson, P.A., P.M. Vitousek, G.P. Livingston, and N.A. Swanberg. 1990. Sources of variation in nitrous oxide flux from amazonian ecosystems. J. Geophys. Res. **95**:16789–16798.

35. McKenney, D.J., K.F. Shuttleworth, J.R. Vriesacker, and W.I. Findlay. 1982. Production and loss of nitric oxide from denitrification in anaerobic Brookston clay. Appl. Environ. Microbiol. **43**:534–541.

36. Melillo, J.M., P.A. Steudler, J.D. Aber, and R.D. Bowden. 1989. Atmospheric deposition and nutrient cycling. In M.O. Andreae and D.S. Schimel (eds.), Dahlem Workshop Reports, Life Sciences Research Report 47, Exchange of Trace Gases Between Terrestrial Ecosystems and the Atmosphere, J. Wiley and Sons, Chichester, pp. 263–280.

37. Papen, H., R. von Berg, I. Hinkel, B. Thoene, and H. Rennenberg. 1989. Heterotrophic nitrification by *Alcaligenes faecalis:* NO_2^-, NO_3^-, N_2O, and NO production in exponentially growing cultures. Appl. Environ. Microbiol. **55**:2068–2072.

38. Papen, H., R. von Berg. B. Hellmann, and H. Rennenberg. 1991. Einfluss von

saurer Beregnung und Kalkung auf chemolithotrophe und heterotrophe Nitrifikation in Böden des Höglwaldes. Forstw. Forsch. **39**:111–116.

39. Poth, M., and D.D. Focht. 1985. ^{15}N kinetic analysis of N_2O production by *Nitrosomonas europaea:* an examination of nitrifier denitrification. Appl. Environ. Microbiol. **49**:1134–1141.
40. Rasmussen, R.A., and M.A.K. Khalil. 1986. Atmospheric trace gases: Trends and distributions over the last decade. Science **232**:1623–1624.
41. Remde, A., and R. Conrad. 1991. Role of nitrification and denitrification for NO metabolism in soil. Biogeochemistry **12**:189–205.
42. Remde, A., F. Slemr, and R. Conrad. 1989. Microbial production and uptake of nitric oxide in soil. FEMS Microbiol. Ecol. **62**:221–230.
43. Robertson, G.P., and J.M. Tiedje. 1984. Denitrification and nitrous oxide production in successional and old-growth Michigan forests. Soil Sci. Soc. Amer. J. **48**:383–389.
44. Robertson, G.P., and J.M. Tiedje. 1987. Nitrous oxide sources in aerobic soils: nitrification, denitrification and other biological processes. Soil Biol. Biochem. **19**:187–193.
45. Rosswall, T., F. Bak, D. Baldocchi, R.J. Cicerone, R. Conrad, D.H. Ehhalt, M.K. Firestone, I.E. Galbally, V.F. Galchenko, P. Groffman, H. Papen, W.S. Reeburgh, and E. Sanhueza. 1989. What regulates production and consumption of trace gases in ecosystems: Biology or physicochemistry? In M.O. Andreae and D.S. Schimel (eds.), Dahlem Workshop Reports, Life Sciences Research Report 47, Exchange of Trace Gases Between Terrestrial Ecosystems and the Atmosphere, J. Wiley & Sons, Chichester, pp. 73–95.
46. Rudaz, A.O., D.A. Davidson, and M.K. Firestone. 1991. Sources of nitrous oxide production following wetting of dry soil. FEMS Microbiol. Ecol. **85**:117–124.
47. Schimel, J.P., M.K. Firestone, and K.S. Killham. 1984. Identification of heterotrophic nitrification in a sierran forest soil. Appl. Environ. Microbiol. **48**:802–806.
48. Schmidt, J., W. Seiler, and R. Conrad. 1988. Emission of nitrous oxide from temperate forest soils into the atmosphere. J. Atmos. Chem. **6**:95–115.
49. Seiler, W., and R. Conrad. 1981. Field measurement of natural and fertilizer-induced N_2O release rates from soils. J. Air Pollut. Control Assoc. **31**:767–772.
50. Slemr, F., and W. Seiler. 1984. Field measurements of NO and NO_2 emissions from fertilized and unfertilized soils. J. Atmos. Chem. **2**:1–24.
51. Smith, C.J., and P.M. Chalk. 1979. Factors affecting the determination of nitric oxide and nitrogen dioxide evolution from soil. Soil Sci. **128**:327–330.
52. Stams, A.J.M., E. Marinka Flameling, and E.C.L. Marnette. 1990. The importance of autotrophic versus heterotrophic oxidation of atmospheric ammonium in forest ecosystems with acid soil. FEMS Microbiol. Ecol. **74**:337–344.
53. Steudler, P.A., R.D. Bowden, J.M. Melillo, and J.D. Aber. 1989. Influence of nitrogen fertilization on methane uptake in temperate forest soils. Nature **341**: 314–316.

54. Tortoso, A.C., and G.L. Hutchinson. 1990. Contributions of autotrophic and heterotrophic nitrifiers to soil NO and N_2O emissions. Appl. Environ. Microbiol. **56**:1799–1805.

55. Williams, E.J., D.D. Parrish, M.P. Buhr, and F.C. Fehsenfeld. 1988. Measurements of soil NO_x emissions in Central Pennsylvania. J. Geophys. Res. **93**:9539–9546.

Production of Methane and Nitrous Oxide by Organic Soils within a Northern Hardwood Forest Ecosystem

Joseph B. Yavitt and Timothy J. Fahey

Department of Natural Resources, Cornell University, Ithaca, New York 14853

ABSTRACT

Methane and nitrous oxide fluxes and the processes governing these fluxes were measured in a well-drained, upland forest soil (Spodosol) and a water-logged, organic soil (Histosol) in the northern hardwood forest at the Hubbard Brook Experimental Forest in New Hampshire, USA, to determine the extent to which fluxes and processes varied between soil types. Field measurements were carried out between April and November 1990 and indicated that methane fluxes were mostly negative (net consumption of atmospheric methane) for the forest soil and positive (net emission into the atmosphere) for the organic soil. The rate of methane consumption by the forest soil was highest when the soil was relatively dry. Methane production by the organic soil seemed to be controlled by sulfate deposition in rain which may have stimulated sulfate reduction at the expense of methane production. Nitrous oxide fluxes were negligible for the forest soil and relatively high for the organic soil. Interactions of several processes seemed to control nitrous oxide production by the soils and nitrous oxide fluxes observed *in situ*. When fluxes were weighted by the areal extent of these soil types in the forest, the ecosystem became a net source of atmospheric methane, with a mean daily flux of 0.5 mg m^{-2} d^{-1}, rather than a methane sink as commonly thought and a slightly larger nitrous oxide source than expected.

Well-drained soils in temperate forests consume atmospheric methane and produce nitrous oxide.[12] The methane sink is caused by methanotrophic bacteria that can consume methane even at atmospheric concentrations.[23] Nitrous oxide production results from several diverse biological processes including both nitrification and denitrification.[8]

Extrapolating the few published rates of methane consumption and nitrous oxide production to all of the world's temperate forests suggests that the biome is a small but significant global sink for atmospheric methane[4] and a small global source for atmospheric nitrous oxide.[7] However, a potentially important error in the calculation is the assumption that all temperate forests are established on well-drained soils (e.g., Alfisols, Inceptisols, Spodosols, Ultisols), thereby constraining the soil processes that affect methane and nitrous oxide. Studies to date do not seem to recognize that organic soils (i.e., Histosols) can be an important component of nearly all temperate forests. Because organic soils remain waterlogged for much of the year, they have the capacity to produce methane through anaerobic methanogenesis[24] and large amounts of nitrous oxide through anaerobic denitrification.[17]

Organic soils are not always restricted to conspicuous wetlands. Even in steeply forested watersheds, relatively inconspicuous areas dominated by organic soils may occur in small bedrock depressions and adjacent to drainage streams. For example, we have estimated that organic soils cover approximately 5% of the forested watersheds at the Hubbard Brook Experimental Forest (HBEF) in central New Hampshire even though the watersheds are steep and slopes are generally >25%. In this study, we examined the temporal variation in methane and nitrous oxide fluxes for organic and forest soils at the HBEF and some of the processes governing those fluxes.

Materials and Methods

Site Description and Sampling Strategy

The HBEF is an extensively studied northern hardwood forest in central New Hampshire (43°57′N, 71°44′W). The 80-year-old forest is dominated by American beech (*Fagus grandifolia* Ehrh.), yellow birch (*Betula alleghaniensis* Britt.), and red maple (*Acer saccharum* Marsh.). Most of the forest soils are Typic Haplorthods and Frigiorthods of the Tunbridge, Lyman, Berkshire, Skerry, and Becket series.

The watersheds at the HBEF are steep (~25% slope) and range from 350 to 800 m elevation. Our study area straddled Bear Brook at 701 m elevation. Streambed morphology is best described as[9] a "stair-step" series of small pools connected by free-fall zones or riffles. The pool we studied was filled by an organic soil with a maximum depth of 1 m and covered by *Sphagnum* sp. mosses, although *Carex* sp. sedges are common along with alders and eastern hemlock (*Tsuga canadensis* [L.] Carr.). The organic soils at the HBEF are Typic or Terric Borohemists of the Chocorua and Greenwood series.

Sampling was conducted eight times in 1990. Sampling locations in the upland forest were determined randomly on each sampling date, and they were always within 25 m of the organic soil. Nine sampling locations were set up permanently in the organic soil in November 1989 (Figure 1).

Field Measurements of Gas Fluxes and Dissolved Gases in Soil Solutions

Gas exchange across the soil–atmosphere interface was estimated by measuring the time-series change in concentration within the head-space of open-

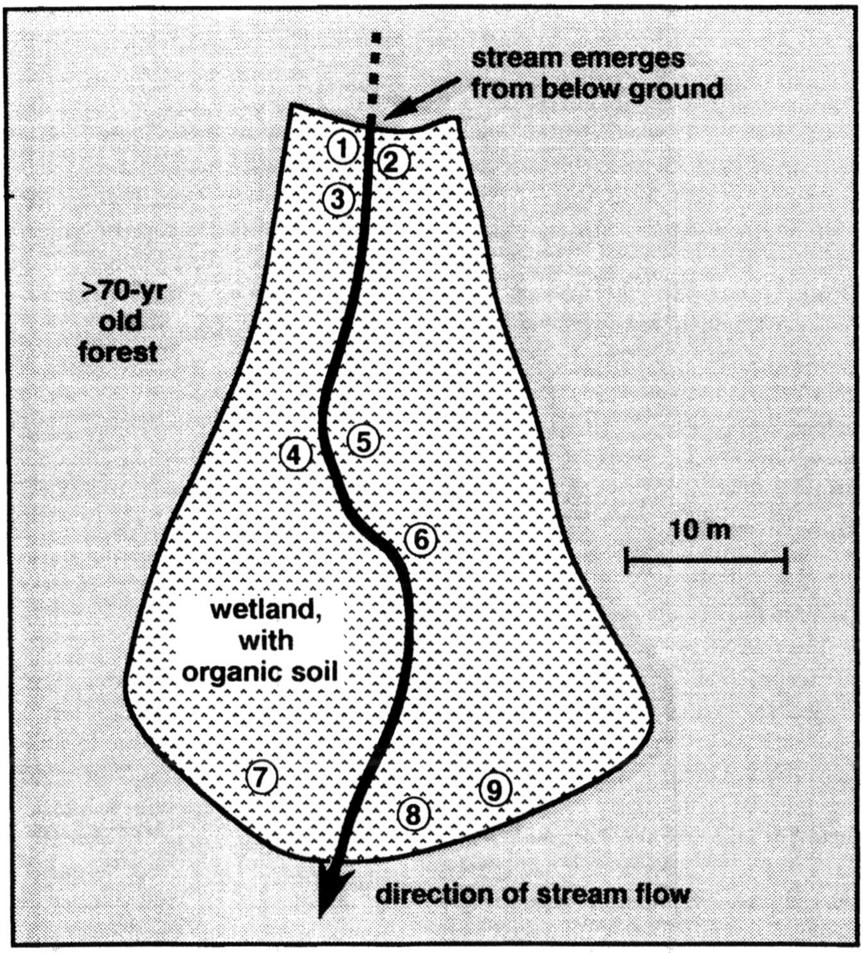

Figure 1. The study site at the Hubbard Brook Experimental Forest, New Hampshire.

bottom, welded aluminum chambers (2.5-L volume) that covered 0.082 m^2 surface area. The chambers were inserted 1 cm into the litter, and head-space air samples were collected with 30-ml polypropylene syringes, with siliconized polypropylene plungers and attached needles, 2, 7, 12, 17, and 22 min after deployment. Sampling syringes also served as storage vessels until analyses could be conducted within 24 h of collection. Leakage from the syringe was inhibited by capping the needle with a silicone stopper and by applying distilled water to the barrel to restrict diffusion of gases. Storage tests indicated that syringes filled with 1.08-, 9.8-, 99.0-, or 990-ppm methane standards remained nearly unchanged for 48 h. Syringes were disassembled after each use to allow barrels and plungers to equilibrate with ambient air.

Concentrations of dissolved gases were determined in solutions from Bear Brook at the point where the stream emerged from below ground (Figure 1). Thin soils overlying impermeable bedrock suggest that these samples represent a composite of soil water from the upland forest without any contribution from deep groundwater. Lysimeters were used to collect solutions at each of the nine sampling locations in the organic soil. The lysimeters were perforated 125-ml plastic bottles with 250-μm-mesh plankton netting fixed over the perforations. At each location, one lysimeter collected water from the 2.5-cm depth, and a second lysimeter collected water from the 17.5-cm depth. Immediately upon collection, solutions were equilibrated with an equal volume of air in a 60-mL plastic syringe by shaking vigorously for 1 min. A subsample of the equilibrated air was saved for analysis of gas concentrations by gas chromatography. In parallel samples, dissolved sulfate was determined by ion chromatography (Dionex, Inc., Sunnyvale, CA) and dissolved nitrate was determined by diazotiation after reduction to nitrite by cadmium.[1]

Laboratory Incubations of Soil Samples

The processes governing methane and nitrous oxide fluxes were investigated in laboratory incubations of forest and organic soils. Three replicate cores (5-cm diameter × 25-cm length) were randomly collected in the upland forest in September. On the day after collection, each core was divided into Oi, Oe, Oa, and mineral layers, and subsamples of each depth per core were sealed in mason jars (0.473-L volume) equipped with a septa to facilitate sampling of gases in the head space. One set of jars had laboratory air in the head space, and a parallel set had laboratory air that was amended with 7 mL of 1% CH_4 (in N_2) to test for methane limitation of aerobic methane consumption. Two additional sets of jars had an O_2-free N_2 head space that was amended with 45 mL of acetylene to facilitate nitrous oxide production via denitrification[25]; one set also received 25 mL of a 1-mM NO_3^- solution (as KNO_3) to test for nitrate limitation of denitrification. Immediately after sealing, the atmosphere in each jar was mixed by pumping five times with a 60-mL syringe before a 10-mL gas sample was withdrawn. The latter procedure was then repeated hourly for analysis of changes in gas concentrations. Make-up gas of the same concentration upon sealing was added before each sample was removed. All incubations were carried out at 19°C and they lasted 6 h.

The effect of soil moisture level on methane fluxes was investigated with the first two sets of jars. After the last gas sample was taken, the jars were opened and the soil was allowed to air dry for 36 h before the jars were sealed; one set was amended with methane, and the jars were incubated and sampled for methane concentrations as described above. After sampling the air-dried soil, the jars were opened and the soil was saturated with distilled water before the jars were again sealed; one set was amended with methane, and the jars were incubated and sampled as described above.

The organic soils were sampled in September and again in November. One core was collected adjacent to each of the nine permanent sampling locations and divided by depth (0–5-cm and 15–20-cm depths). Subsamples of each depth per core were sealed in mason jars (0.473-L volume). Two sets of jars had an O_2-free N_2 head space, one of which also received 25 mL of a 1-mM SO_4^{2-} solution (as Na_2SO_4) to test for the effect of sulfate reduction on methane production. Two additional sets of jars had an O_2-free N_2 head space that was amended with 45 mL of acetylene; one set also received 25 mL of a 1-mM NO_3^- solution (as KNO_3). Immediately after sealing, the jars were incubated and sampled for gas concentrations as described above.

After all incubations, head-space volume was determined by weight difference after filling the jars with water. Soils were filtered and dried at 105°C, and results were normalized to a dry soil basis. Head-space concentrations were corrected for gases that remained dissolved in soil pore waters using transfer equations given by Flett et al.[10] and appropriate Bunsen coefficients for methane and nitrous oxide at 19°C.

Gas Measurements

Concentrations of methane and nitrous oxide in gas samples were determined with gas chromatographs (Shimadzu Inc., model Mini-2) equipped with a flame ionization and electron capture detector, respectively. Gases were separated on Poropak-Q columns (80/100 mesh, 2 m × 3 mm) maintained at 35°C with He carrier gas (30 mL min^{-1}). Both gas chromatographs were equipped with eight-port backflush values to vent water vapor and acetylene. Gas concentrations were quantified by comparing peak areas for samples and standards. Certified standards (1.08–1054 ppm CH_4 in N_2; 0.0342–9.793 ppm N_2O in N_2; Scott Specialty Gases, Plumsteadville, PA) bracketed every 8–10 samples.

Statistical Analyses

Differences in methane and nitrous oxide fluxes between soil types were analyzed with repeated-measures analysis of variance (ANOVA). Soil type (organic versus forest soil) and sampling date ($N = 8$ for methane, $N = 4$ for nitrous oxide) were main effects, and repeated measure occurred on sampling date. Dissolved gases were not analyzed statistically because forest soil solutions were not replicated per sampling date. Differences in methane and nitrous oxide fluxes among laboratory incubations were analyzed with ANOVA or a *t*-test. Prior to statistical tests, all values were logarithmically transformed to reduce heterogeneity in variances.

Results

Field Measurements of Methane Fluxes and Dissolved Methane in Soil Solutions

Both positive (net methane emission) and negative (net methane consumption) fluxes were observed in both the upland forest soil and in the organic soil. Methane fluxes in the forest soil (Figure 2) ranged from −3.84 to 6.40 mg CH_4 m^{-2} d^{-1}, with a growing-season mean flux of −0.92 (± 0.86 SE, $N = 32$) mg CH_4 m^{-2} d^{-1}. Fluxes in the organic soil ranged from −3.36 to 328.3 mg CH_4 m^{-2} d^{-1} and averaged 27.2 (± 16.4 SE, $N = 32$) mg CH_4 m^{-2} d^{-1}. Total fluxes during the growing season were calculated by multiplying the daily mean value by 255 d which was the length of the growing season. Total fluxes were −0.235 g CH_4 m^{-2} for the forest soil and 6.936 g CH_4 m^{-2} for the organic soil. Despite the large difference in flux rates between soil types, the difference was not statistically significant ($P > 0.05$, repeated-measures analysis of variance).

Concentrations of dissolved methane in the organic soil averaged 14.6 (± 3.8 SE, $N = 144$) μM and were about two orders of magnitude higher than those in the forest soil, which averaged 0.17 (± 0.04 SE, $N = 8$) μM. It is notable that the relatively high emission rate for methane from the organic soil in July (Figure 2) was associated with higher concentrations of dissolved methane than that observed on the other dates, except in November.

Laboratory Incubations for Methane Production and Methane Consumption

Forest soils consumed (or produced) methane without any noticeable time lag after the soil was sealed in a jar (time series not shown). The unamended rates (Table 1) were highest for the organic layers (Oi, Oe, and Oa), which showed no difference ($P > 0.05$ with a one-way ANOVA) among the different soil layers. A sharply lower rate was observed for the mineral soil. When methane concentrations in the jars were well above atmospheric levels, the rate of methane consumption was more than 10-fold higher than the unamended rates for all soil moisture treatments, indicating that methane concentration in part limits the rate of methane consumption in the field. Consumption rates were similar between air-dried and field-moist soils in all cases, whereas saturated soils generally showed lower consumption rates than those for drier soils, although the differ-

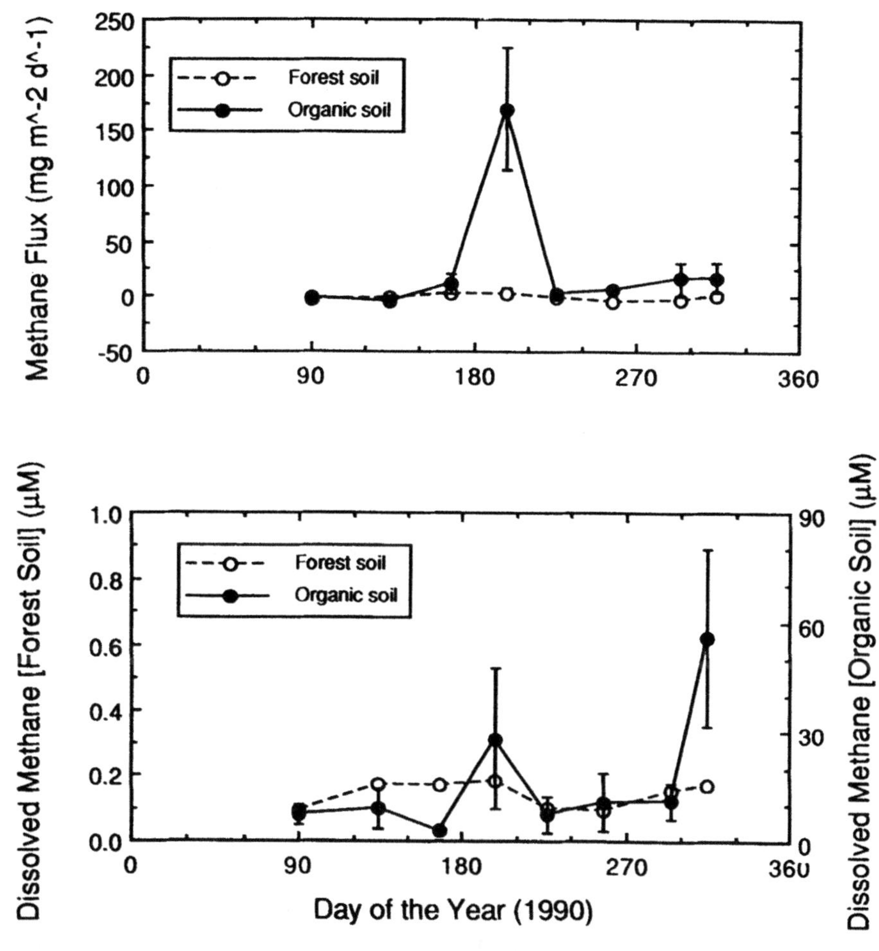

Figure 2. Methane fluxes (top) and concentrations of dissolved methane in soil solutions (bottom) in forest and organic soils at the Hubbard Brook Experimental Forest, New Hampshire. Fluxes are means of four replicate chambers. Symbols for dissolved methane are means of 18 replicates for soil water collected at two depths at each of nine sampling locations in the organic soil and a single collection in the forest soil. Bars represent one standard error.

Table 1. Consumption of methane by forest soils from the Hubbard Brook Experimental Forest, New Hampshire. Flux rates are the mean of three measurements and one standard error is given in parentheses. Final concentration for methane addition = 150 ppm.

	Consumption by Soil Layer (nmol g dry soil^{-1} h^{-1})			
Treatment	Oi	Oe	Oa	Mineral
No amendment				
Air dry	0.20 (0.10)	0.32 (0.04)	0.21 (0.17)	0.07 (0.02)
Field moist	0.29 (0.28)	0.16 (0.13)	0.30 (0.08)	0.08 (0.04)
Saturated	0.13 (0.06)	0.03 (0.01)	0.06 (0.02)	0.01 (0.01)
Plus methane				
Air dry	9.88 (7.03)	3.16 (1.51)	3.84 (1.89)	0.41 (0.09)
Field moist	7.99 (0.90)	2.90 (1.31)	3.21 (2.52)	0.75 (0.13)
Saturated	3.31 (0.66)	1.58 (0.32)	1.27 (0.18)	0.51 (0.01)

ence was not significant in all cases (i.e., Oi horizon, no amendment, $P > 0.05$ among moisture treatments with a one-way ANOVA).

Rates of methane production by the organic soil (Table 2) averaged 6.3 (± 2.2 SE, $N = 36$) nmol g dry soil^{-1} h^{-1}. We found no difference in rates between September and November collections. Sulfate-amended rates of methane production were significantly lower ($P < 0.05$ with a *t*-rest) than those for unamended peats in all cases and averaged 1.8 (± 0.7 SE, $N = 36$) nmol g dry soil^{-1} h^{-1}.

Field Measurements of Nitrous Oxide Fluxes and Dissolved Nitrous Oxide in Soil Solutions

Measurements of nitrous oxide fluxes began in August and continued thereafter (Figure 3). Both positive and negative fluxes were measured in the forest soil. When the values were averaged across the four sampling dates, the mean flux was essentially zero (< 0.01 ng N_2O–N cm^{-2} h^{-1}) for the forest soil. For the organic soil, the mean flux averaged across sampling dates was 0.70 (± 0.51 SE, $N = 16$) ng N_2O–N cm^{-2} h^{-1}. The difference in flux rates between soil types was statistically significant ($P < 0.05$) in the ANOVA. Concentrations of dissolved nitrous oxide in the organic soil averaged 0.047 (± 0.01 SE, $N = 72$) μM and were about 50% higher than those in the forest soil, which averaged 0.032 (± 0.003, $N = 8$) μM.

Table 2. Production of methane by organic soils from the Hubbard Brook Experimental Forest, New Hampshire. Flux rates are the mean of nine measurements and one standard error is given in parentheses.

	Production by Soil Depth (cm) (nmol g dry soil^{-1} h^{-1})	
Treatment	0–5	15–20
September		
No amendment	3.9 (1.8)	3.0 (1.4)
Plus sulfate	1.3 (0.5)	1.0 (0.4)
November		
No amendment	6.9 (2.8)	10.4 (8.4)
Plus sulfate	2.0 (0.7)	2.7 (1.1)

Laboratory Incubations for Nitrous Oxide Production

Aerobic incubations of forest soils consumed or produced small amounts of nitrous oxide (Table 3). Serendipitously, we found that 150 μL of CH_4 L^{-1} in the jar head space affected aerobic rates. For example, the Oi and Oe horizons consumed nitrous oxide when incubated without methane, but they produced small amounts of nitrous oxide with methane. In contrast, nitrous oxide production by the Oa and mineral soil decreased more than fivefold with methane compared to without methane. Anoxic incubations with acetylene markedly increased nitrous oxide production rates by the forest soils. Moreover, nitrate addition increased anoxic rates, especially for the organic horizons, and the increase was more than sixfold compared to unamended-anoxic rates. Overall, the highest rate of nitrous oxide production was found in the Oa horizon, but the difference among soil horizons was statistically significant only for the nitrate-amended anoxic rates ($P < 0.05$ using one-way ANOVA) and not for the unamended anoxic rates.

Anoxic unamended rates of nitrous oxide production by the organic soils were extremely low for both dates and at both depths (Table 4). Nitrate addition stimulated rates by more than two orders of magnitude in all cases. We found that nitrate-amended rates were slightly higher in November than in September, but the differences were not statistically significant ($P>0.05$ with a t-test) for both depths. Also, nitrate-amended rates were about twofold higher for the 0–5-cm depth compared to rates for the 15–20-cm depth, but the differences were not statistically significant ($P>0.05$ with a t-test).

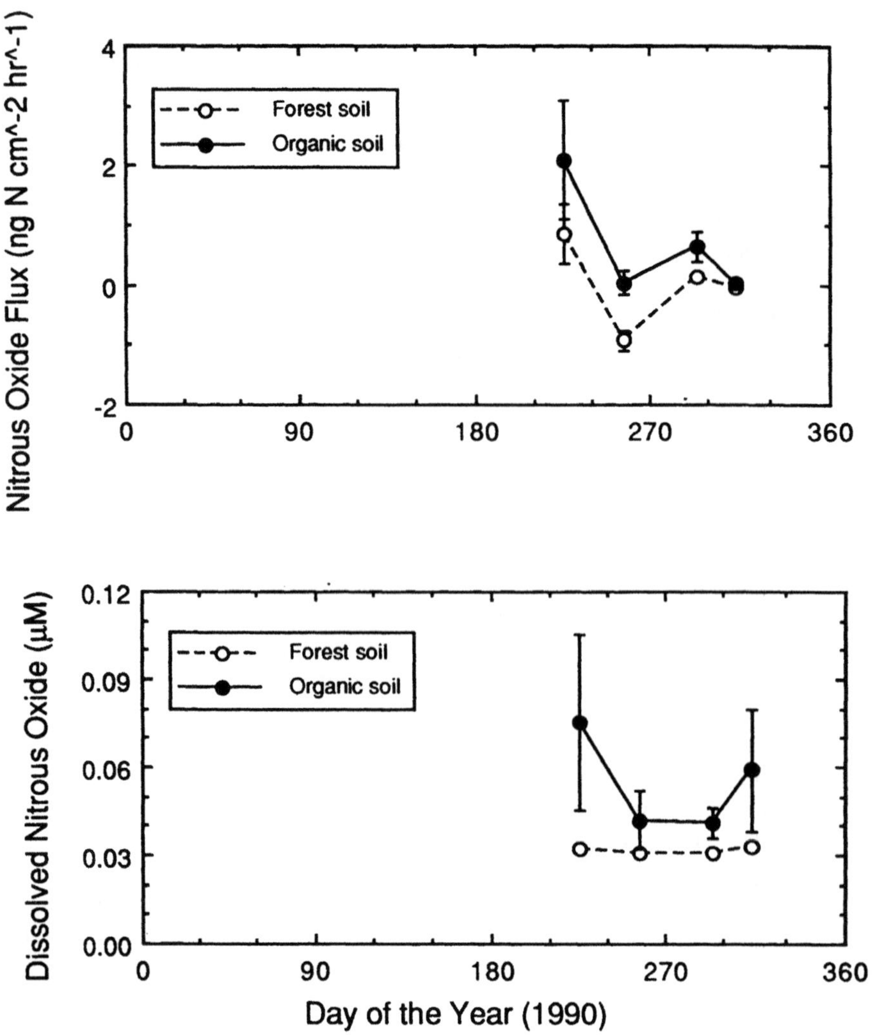

Figure 3. Nitrous oxide fluxes (top) and concentrations of dissolved nitrous oxide in soil solutions (bottom) in forest and organic soils at the Hubbard Brook Experimental Forest, New Hampshire. Fluxes are means of four replicate chambers. Symbols for dissolved nitrous oxide are means of 18 replicates for soil water collected at two depths at each of nine sampling locations in the organic soil and a single collection in the forest soil. Bars represent one standard error.

Table 3. Production (positive values) and consumption (negative values) of nitrous oxide by forest soils from the Hubbard Brook Experimental Forest, New Hampshire. Flux rates are the mean of three measurements and one standard error is given in parentheses. Final concentration for methane addition = 150 ppm.

	Production by Soil Layer ($nmol\ g\ dry\ soil^{-1}\ h^{-1}$)			
Treatment	Oi	Oe	Oa	Mineral
Aerobic incubation				
No amendment	−0.07 (0.08)	−0.02 (0.02)	0.94 (0.72)	0.06 (0.04)
Plus methane	0.33 (0.08)	0.10 (0.01)	0.04 (0.03)	0.01 (0.01)
Anoxic incubation with C_2H_2				
No amendment	5.2 (1.85)	11.1 (8.2)	32.5 (11.3)	16.5 (9.0)
Plus nitrate	31.2 (15.2)	66.7 (27.2)	316.8 (87.2)	29.2 (8.8)

Discussion

Methane Dynamics

Results of this study indicate that upland forest soils at the HBEF are a sink for atmospheric methane during the growing season (April–November). Net consumption of atmospheric methane by well-drained soil has been reported for other temperate forest ecosystems. Our mean flux of -0.92 mg CH_4 m^{-2} d^{-1} is near the low end of the values reported for other temperate forests, which generally range from low fluxes of -0.28 mg CH_4 m^{-2} d^{-1} (Ref. 12) and -1.65 mg CH_4 m^{-2} d^{-1} (Ref. 5) to the highest fluxes of -4.01 mg CH_4 m^{-2} d^{-1} (Ref. 16) and -3.46 mg CH_4 m^{-2} d^{-1} (Ref. 2).

Although our mean growing-season flux of methane was negative for the forest soil, positive fluxes occurred sporadically. Thus, it is likely that methanogenesis occurs within the soil profile. We believe that the location is probably within the mineral soil, as the organic layers showed the strongest capacity for methane oxidation (Table 1). Moreover, methanogenesis located in the mineral soil would account for our observation that the surface soil could consume atmospheric methane even though soil water (derived mostly from the mineral soil) had dissolved methane concentrations about 2.5 times higher than that expected in equilibrium with the atmosphere (0.06 μM). In contrast, Crill[5] found the highest rate of methane oxidation in the mineral layer of a woodland soil in southern New Hampshire. The contrasting results may be due to different soil types, depth of organic layers, and mineral status of the soils.

Net methane emission from the organic soil during the growing season indi-

Table 4. Production of nitrous oxide by organic soils from the Hubbard Brook Experimental Forest, New Hampshire. Flux rates are the mean of nine measurements and one standard error is given in parentheses.

	Production by Soil Depth (cm) (nmol g dry soil^{-1} h^{-1})	
Treatment	0–5	15–20
September		
No amendment	0.1 (0.1)	0.3 (0.2)
Plus nitrate	54.0 (8.4)	36.6 (7.8)
November		
No amendment	0.1 (0.1)	0.3 (0.2)
Plus nitrate	96.1 (9.9)	40.0 (8.9)

cates that the organic soils are active sites for methanogenesis and methane emission. Moreover, sporadic measurements of net consumption of atmospheric methane by the organic soil indicates that methane oxidizers are active and that their activity can occasionally outweigh that of the methanogens. These results are in agreement with previous observations that methane emission from organic soils represents the balance between methane production and methane oxidation.[23]

We found the highest emission rate for methane in the organic soil on our July sampling date. July also had the warmest soil and air temperatures, suggesting that methane emission was partially controlled by temperature. However, correlations between flux rate and either air or soil temperature were not strong and explained only 30% of the variation in the flux rate. Surprisingly, we found that the highest flux rate occurred during a relatively dry period in the summer. For example, our July sampling coincided with a period in which few rains had occurred during the preceding two weeks and flow in Bear Brook had been reduced to a trickle, compared to noticeable flow for all our other sampling times, except in November. It is possible that the soil had remained wet enough in July to support methanogenesis, so that methane was emitted into the atmosphere more easily under low- than high-flow conditions. Accordingly, high flow may have washed methane produced *in situ* into the Bear Brook where degassing or oxidation occurred. The higher dissolved methane concentrations in soil solutions in the organic soil in July than for the other samplings (Figure 1), except in November when the surface soil was frozen, further supports a relationship between water flow rate and dissolved methane concentrations.

Sulfate additions inhibited methane production by the organic soils. We be-

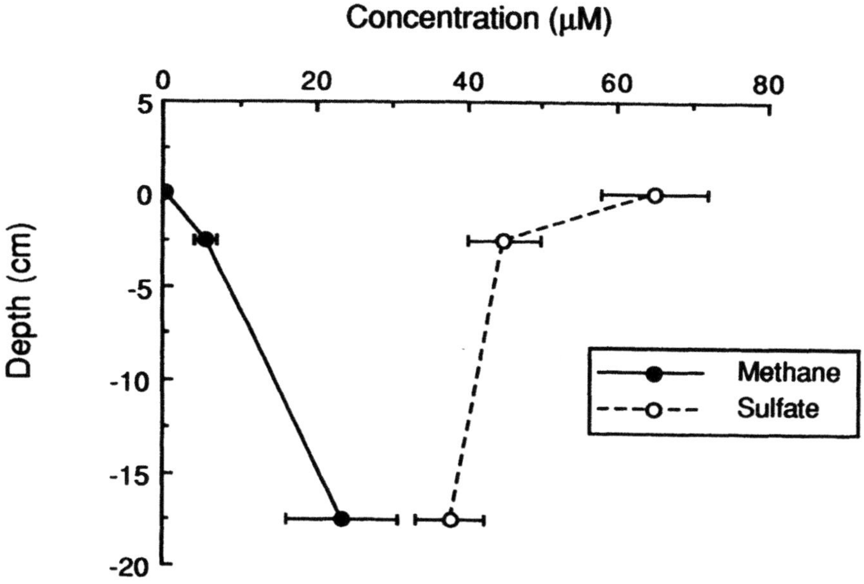

Figure 4. Concentrations of dissolved methane and sulfate versus depth in organic soils at the Hubbard Brook Experimental Forest, New Hampshire. Symbols are means of eight monthly measurements. Bars represent one standard error.

lieve the results suggest that sulfate additions may fuel microbially mediated sulfate reduction at the expense of methane production. For example, depth profiles of dissolved methane and sulfate in the lysimeter solutions (Figure 4) show that concentrations of dissolved methane are highest where sulfate is lowest, which is a typical pattern in environments where zones of maximum rates in methane production and sulfate reduction are spatially separated.[22] It is likely that sulfate reduction in our site is dependent on sulfate input via acid deposition, which, in turn, may help explain why our mean emission rate of 27.2 mg CH_4 m^{-2} d^{-1} is so much lower than the mean emission rate of 142 mg CH_4 m^{-2} d^{-1} for a physiographically similar site in Minnesota,[6] where temperatures are cooler, relative S loading via atmospheric deposition is much less, and sulfate reduction does not occur in the soil.[19]

We have evaluated the role of organic soils in methane fluxes at our site by multiplying fluxes per soil type by the proportion of type within the watershed and summing. The result is a positive annual methane flux of 0.12 g CH_4 m^{-2}. Although this value extrapolated globally would make temperate forests a negligible source of atmospheric methane, it markedly contrasts the current dogma that temperate forests are a methane sink. Moreover, our result indicates that even a small proportion of the landscape covered by an organic soil (e.g., 5%), with a relatively low emission rate for methane, can have a profound effect on total methane fluxes for the landscape.

Nitrous Oxide Dynamics

Our field measurements indicated that the upland forest soils were neither a source nor sink for atmospheric nitrous oxide. Other studies in temperate forest ecosystems have shown that well-drained soils emit nitrous oxide at mean rates as low as 0.08 ng N_2O–N cm^{-2} h^{-1} (Ref. 12) and 0.03 ng N_2O–N cm^{-2} h^{-1} (Ref. 3); however, mean emission rates as high 1.09 ng N_2O–N cm^{-2} h^{-1} (Ref. 15) and 2.77 ng N_2O–N cm^{-2} h^{-1} (Ref. 11) also have been reported. Thus, there is still uncertainty whether temperate forests are a globally important source of atmospheric nitrous oxide.

The high potential capacity for nitrous oxide production by the forest soils, at least under anoxic conditions (Table 3), suggests two hypotheses: (1) the biological processes that are responsible for producing nitrous oxide *in situ* are anaerobic and (2) nitrous oxide produced *in situ* is consumed as it diffuses through the soil, thereby limiting its emission into the atmosphere. The first hypothesis is supported by laboratory studies of anaerobic denitrification activity in forest soils and its relationship to nitrous oxide fluxes observed in the field.[13] The second hypothesis relates to field studies that have shown net consumption of atmospheric nitrous oxide by forest soils[3] and laboratory studies that have shown microbially mediated consumption of nitrous oxide by soil cores.[14] Our result that the highest capacity for nitrous oxide production was found in the Oa layer, where nitrous oxide produced would have to diffuse through the Oi and Oe layers before reaching the atmosphere, further supports the second hypothesis. Hence, a complex suit of biological processes seems to responsible for limiting nitrous oxide fluxes in the forest soil.

The mean emission rate for nitrous oxide in our organic soil is much greater than that reported by Urban et al.[18] for an organic soil in Minnesota. In fact, Urban et al.[18] found that nitrate fertilization resulted in nitrous oxide emissions of no more than 0.06 ng N_2O–N cm^{-2} h^{-1} because microbially mediated processes responsible for nitrous oxide production lost in competition with mosses for available nitrate. Our site receives a relatively high N input in bulk deposition of approximately 0.77 g m^{-2} yr^{-1} compared to the input of 0.54 g m^{-2} yr^{-1} estimated for the site studied by Urban et al.[18] which may help to explain the contrasting results. Moreover, concentrations of dissolved nitrate are usually below detection limits in pore waters of organic soils,[21] whereas we measured concentrations as high as 100 μM. It is possible that water flushed through our organic soil supplies dissolved nitrate that fuels the biological processes leading to nitrous oxide fluxes observed *in situ*.

Although we measured nitrous oxide fluxes for only 4 months, total flux for this period was 0.026 g N_2O–N m^{-2}. It is easily possible that the value is half that for the entire growing season. If true, nitrous oxide release from our organic soil represents a loss of 6.4% of N inputs.

Conclusions

A primary purpose of this study was to emphasize the role of organic soils as sources of methane and nitrous oxide in temperate forests. Organic soils can potentially occupy any small depression or basin on the landscape where water flow is retarded and organic matter decomposes under anaerobic conditions. Unfortunately, it is difficult to establish the aerial extent of these soils on most forested landscapes. Remote sensing is limited because the vegetation interferes with measurements of soil moisture or any other parameter used to differentiate saturated, organic soil from well-drained soil.[20] Typically, up to 5% of the landscape in any temperate forest, even those on relatively steep slopes, is an organic, saturated soil. Failure to account for organic soils may lead to incorrect estimates of biogenic gas fluxes across the landscape.

Acknowledgment

Research was supported by an NSF grant to G.E. Likens, F.H. Bormann, and C.T. Driscoll. We thank the aforementioned individuals for logistic support, Bob Long who advised on site selection, and the "lab crew" at Cornell for assistance with laboratory analyses.

References

1. American Public Health Association. 1981. Standard Methods for Examination of Water and Wastewater, 15th ed., American Public Health Association, Washington, DC.
2. Born, M., H. Dorr, and I. Levin. 1990. Methane consumption in aerated soils of the temperate zone. Tellus **42B**:2–8.
3. Bowden, R.D., P.A. Steudler, J.M. Melillo, and J.D. Aber. 1990. Annual nitrous oxide fluxes from temperate forest soils in the northeastern United States. J. Geophys. Res. **95**:13997–14005.

4. Cicerone, R.J., and R.S. Oremland. 1988. Biogeochemical aspects of atmospheric methane. Global Biogeochem. Cycles **2**:299–327.

5. Crill, P.M. 1991. Seasonal patterns of methane uptake and carbon dioxide release by a temperate woodland soil. Global Biogeochem. Cycles **5**:319–334.

6. Crill, P.M., K.B. Bartlett, R.C. Harriss, E. Gorham, E.S. Verry, D.I. Sebacher, L. Madzar, and W. Sanner. 1988. Methane flux from Minnesota peatlands. Global Biogeochem. Cycles **2**:371–384.

7. Davidson, E.A. 1991. Fluxes of nitrous oxide and nitric oxide from terrestrial ecosystems. In J.E. Rogers and W.B. Whitman (eds.), Microbial Production and Consumption of Greenhouse Gases: Methane, Nitrogen Oxides, and Halomethanes, American Society for Microbiology, Washington, DC, pp. 219–235.

8. Firestone, M.K., and E.A. Davidson. 1989. Microbial basis of NO and N_2O production and consumption in soil. In M.O. Andreae and D.S. Schimel (eds.), Exchange of Trace Gases between Terrestrial Ecosystems and the Atmosphere, J. Wiley and Sons, New York, pp. 7–22.

9. Fisher, S.G., and G.E. Likens. 1973. Energy flow in Bear Brook, New Hampshire: an integrative approach to stream ecosystem metabolism. Ecol. Monogr. **43**:421–439.

10. Flett, R.J., R.D. Hamilton, and N.E.R. Campbell. 1976. Aquatic acetylene reduction techniques: solutions to several problems. Can. J. Microbiol. **22**:43–51.

11. Goodroad, L.L., and D.R. Keeney. 1984. Nitrous oxide emission from forest, marsh, and prairie ecosystems. J. Environ. Qual. **13**:448–452.

12. Keller, M., T.J. Goreau, S.C. Wofsy, W.A. Kaplan, and M.B. McElroy. 1983. Production of nitrous oxide and consumption of methane by forest soils. Geophys. Res. Lett. **10**:1156–1159.

13. Robertson, G.P., and J.M. Tiedje. 1984. Denitrification and nitrous oxide production in successional and old-growth Michigan forests. Soil Sci. Soc. Amer. J. **48**:383–389.

14. Robertson, G.P., and J.M. Tiedje. 1987. Nitrous oxide sources in aerobic soils: nitrification, denitrification and other biological processes. Soil Biol. Biochem. **19**:187–193.

15. Schmidt, J., W. Seiler, and R. Conrad. 1988. Emission of nitrous oxide from temperate forest soils into atmosphere. J. Atmos. Chem. **6**:95–115.

16. Steudler, P.A., R.D. Bowden, J.M. Melillo, and J.D. Aber. 1989. Influence of nitrogen fertilization on methane uptake in temperate forest soils. Nature **341**: 314–316.

17. Tiedje, J.M., A.J. Sexstone, D.D. Myrold, and J.A. Robinson. 1982. Denitrification: ecological niches, competition, and survival. Antonie Leeuwenhoek J. Microbiol. Serol. **48**:569–583.

18. Urban, N.R., S.J. Eisenreich, and S.E. Bayley. 1988. The relative importance of denitrification and nitrate assimilation in midcontinental bogs. Limnol. Oceanogr. **33**:1611–1617.

19. Urban, N.R., S.J. Eisenreich, and D.F. Grigal. 1989. Sulfur cycling in a forested Sphagnum bog in northern Minnesota. Biogeochemistry **7**:81–109.

20. van de Griend, A.A., and E.T. Engman. 1985. Partial area hydrology and remote sensing. J. Hydrol. **81**:211–251.

21. Waughman, G.J. 1980. Chemical aspects of the ecology of some south German peatlands. J. Ecol. **68**:1025–1046.

22. Winfrey, M.R., D.G. Marty, A.J.M. Bianchi, and D.M. Ward. 1981. Vertical distribution of sulfate reduction, methane production, and bacteria in marine sediments. Geomicrobiol. J. **2**:341–362.

23. Yavitt, J.B., D.M. Downey, G.E. Lang, and A.J. Sexstone. 1990. Methane consumption in two temperate forest soils. Biogeochemistry **9**:39–52.

24. Yavitt, J.B., G.E. Lang, and R.K. Wieder. 1987. Control of carbon mineralization to CH_4 and CO_2 in anaerobic, Sphagnum-derived peat from Big Run Bog, West Virginia. Biogeochemistry **4**:141–157.

25. Yoshinari, T, and R. Knowles. 1976. Acetylene inhibition of nitrous oxide reduction by denitrifying bacteria. Biochem. Biophys. Res. Commun. **69**:705–710.

Effect of Nitrogen Fertilizer and Nitrification Inhibitors on Methane and Nitrous Oxide Fluxes in Irrigated Corn

K.F. Bronson and A.R. Mosier

USDA–ARS: Agricultural Research Service, Fort Collins, Colorado 80522

ABSTRACT

Agricultural systems contribute a large proportion of the global emissions of methane (CH_4) and nitrous oxide (N_2O) to the atmosphere. Nitrogen (N) fertilizer influences the flux rates of these gases in agroecosystems. Methane and nitrous oxide fluxes were measured periodically in an irrigated corn (*Zea mays* L.) field from time of fertilization to harvest. Fertilizer treatments included 0 and 218 kg urea–N ha^{-1}, as well as applications of the nitrification inhibitors (NIs), nitrapyrin [2-chloro-6-(trichloromethyl)-pyridine], and encapsulated calcium carbide (ECC). Brief periods of CH_4 emissions were observed after irrigations, with no effect of N or NIs. Methane consumption was not affected by urea fertilization, but was reduced 43% by the addition of ECC. The contribution, however, of changes in the consumption of CH_4 as a result of ECC addition to the net greenhouse effect was negligible (<0.1%) compared to the reduction in N_2O emissions. Nitrous oxide emissions were 16 times greater in urea-fertilized plots than in unfertilized soils. The application of NIs with urea decreased N_2O emissions 43–71%.

Methane consumption or oxidation in soils is carried out by a diverse group of methanotrophic bacteria.[40] Consumption of atmospheric CH_4 has been reported by several workers.[19,27,35,39] Changes in the sink activity of soils for CH_4, in addition to sources such as rice paddies, cattle, landfills, and biomass-burning, might explain the 1% per year increase in atmospheric CH_4. Methane is 15 times more effective than CO_2 (mass basis) in absorbing infrared radiation.[29] Nitrous oxide (N_2O) is emitted during nitrification of ammonium (NH_4^+), and as a product of denitrification of nitrate (NO_3^-).[10] Applications of N fertilizer to soil result in increases in N_2O emissions.[26,30] Nitrous oxide is 300 times (mass basis) more radiatively active than CO_2,[29] and nitrogen oxides (NO and NO_2) from N_2O oxidation may catalytically destroy ozone (O_3) in the stratosphere.[12] The proportion of applied fertilizer N that is emitted as N_2O in corn-cropped systems ranges from 1.3% (Ref. 25) to 2% (Ref. 14). Global emissions of N_2O emitted from fertilized soils is estimated to be 1.5 Tg N_2O–N/yr^{-1}.[34]

Management strategies that have been tested to reduce N_2O emissions from fertilized fields include different N sources[5,14] and the use of nitrification inhibitors (NIs).[1,20] Nitrification inhibitors retard NH_4^+ oxidation to NO_3^- and can, thereby, limit loss of N_2O during nitrification[8,23] and reduce N_2O emissions resulting from denitrification.[1] Encapsulated calcium carbide[2] (ECC) has been used as a slow-release source of acetylene (C_2H_2) (nitrification inhibitor at 0.1–10 Pa) to inhibit nitrification and reduce N_2O and N_2 emissions in flooded rice.[3,9,22] In this field study, the ability of the NIs, ECC, and nitrapyrin to reduce N_2O emissions was evaluated, and the effect of the NIs on CH_4 consumption in irrigated corn was observed.

Materials and Methods

The study was conducted with a Nunn clay loam soil (fine, montmorillonitic, mesic, Aridic, Argiustoll) at the Colorado State University experimental farm in Fort Collins. The soil (surface 30 cm) had the following chemical properties: paste pH of 7.8, 15 mg kg^{-1} organic matter, 1.5 $\mu g\ g^{-1}$ NH_4^+–N, 6 $\mu g\ g^{-1}$ NO_3^-–N, electrical conductivity of 0.7 mmho cm^{-1}, and 6.7 and 303 $\mu g\ g^{-1}$ of NH_4HCO_3–diethylenetriamine pentaacetic acid–extractable P and K, respectively. Corn had been grown at the site without N fertilization the previous two years.

Corn was planted (variety "DeKalb* 464") on 27 April 1990 in 75-cm rows. The soil was chisel-plowed prior to planting, and irrigation furrows were made with a ditcher 6 weeks after planting. Two irrigations were applied to the corn field at 3 and 29 DAF (days after fertilization).

Nitrogen fertilizer (218 kg urea–N ha^{-1}) and nitrification inhibitors were applied in the middle of the furrow at a depth of 15 cm and at 15-cm intervals on 26 June 1990. Nitrification inhibitor treatments consisted of ECC (0, 20, or 40 CaC_2 kg ha^{-1}) or nitrapyrin [2-chloro-6-(trichloromethyl)-pyridine] (0.56 L active ingredient ha^{-1}). The treatments were completely randomized within four replicate blocks. The ECC consisted of 1–2-mm CaC_2 particles that were first coated with a layer of carnauba wax, then with a layer of beeswax, and finally with a layer of shellac (40% CaC_2 and 60% coating). Unfertilized blank plots were included as controls. Additional control plots (i.e., no fertilizer N addition) were established that received 40 kg ECC ha^{-1}, making a total of 24 plots.

Two to three times a week, 1.2-L vented chambers[17] were placed in the furrows of each plot (one chamber per plot) for 1 hr beginning at 11:00 A.M.

---

*Trade names and company names are included only as a convenience to the reader and do not constitute any preferential endorsement of these products.

Forty-milliliter samples were taken from the chambers with 60-ml Monojet syringes fitted with stopcocks at 0, 30, and 60 min. Gas chromatography was used to determine concentrations of CH_4 weekly[9] and of N_2O every sample day.[24] The C_2H_2 produced by ECC in the soil was monitored by taking two 3-ml gas samples from probes placed 15 cm deep in the plots that received ECC and measured by gas chromatography.[9]

Soil temperature was monitored every sampling date by placing three thermometers 7.5 cm deep in random plots. At 2, 4, 6, and 9 weeks after fertilization, soil samples were taken to a depth of 25 cm from two furrows of each plot. Fresh soils were extracted with 2M KCl and the extracts were steam-distilled with MgO for NH_4^+ and then with MgO and Devardas alloy for NO_3^-.[6] Soil moisture was determined gravimetrically on an oven-dry basis.

Fluxes of N_2O were calculated using the nonlinear flux equation developed by Hutchinson and Mosier.[17] Calculation of CH_4 flux of was done with the linear equation of Rolston.[31]

Cumulative flux over the growing season was calculated for each plot by linearly interpolating data points and integrating the area by Simpson's Rule with the MathCad mathematics program.[21] Gas flux data were subjected to analysis of variance for each date and for the end-of-season cumulative flux by using the General Linear Models procedure in the Statistical Analysis System framework.[33]

Results and Discussion

Acetylene in Soil

Acetylene concentrations in the soil gas of plots treated with 20 and 40 kg ECC at 5 DAF reached 480 and 1475 $\mu l\ l^{-1}$, respectively (Figure 1). No C_2H_2 was detected beyond 48 DAF. High levels of C_2H_2 at the onset were undesirable and meant that the coatings on the ECC were not as effective as was required. Concentrations of C_2H_2 of 5–20 $\mu l\ l^{-1}$ in soil atmosphere would be ideal for inhibiting nitrification.[4] In laboratory incubations (data not shown), 45% and 80% inhibition of nitrification of 25 $\mu g\ g^{-1}$ NH_4^+–N was observed after 6 d in a Nunn clay loam with a C_2H_2 concentration in the head space of 5 and 20 $\mu l\ l^{-1}$, respectively.

Ammonium and Nitrate in Soil

Exchangeable NH_4^+ levels in the surface 30 cm of soil declined throughout the season in all treatments (Figure 2a) after 14 DAF. The addition of ECC to urea resulted in higher ($P < 0.05$) levels of NH_4^+ in soil than either urea alone or urea plus nitrapyrin.

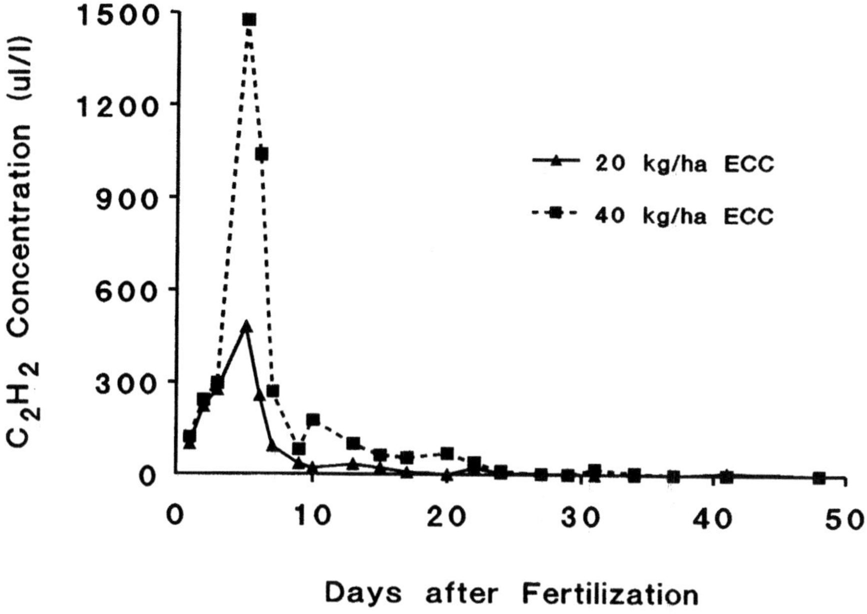

Figure 1. Acetylene concentrations at 15-cm depth of soil as affected by ECC application.

Plots that received urea alone accumulated a maximum of 37 mg kg^{-1} NO_3^-–N 14 DAF, urea plus nitrapyrin had a maximum of 28 mg kg^{-1} NO_3^-–N at 28 DAF, and the urea plus ECC plots did not accumulate more than 8 mg kg^{-1} NO_3^-–N during the first 42 DAF (Figure 2b). Nitrapyrin in this study did not inhibit nitrification as effectively as reported in other field studies.[11,38] An additional 20 mg kg^{-1} NO_3^-–N accumulated in the ECC-treated plots between 42 and 62 DAF. The corn suffered hail damage on 32 DAF and did not take up much more mineral N, as NO_3^- levels on plots with urea plus nitrapyrin and with urea alone did not decline significantly during this time.

Methane Fluxes

Brief periods of positive CH_4 fluxes were observed after irrigation on 3 and 29 DAF [Figure 3a; data for the high level of ECC plus urea are not shown, as they were not different than the low level of ECC plus urea ($P > 0.05$)], but there were no significant treatment differences ($P > 0.05$). This is in contrast to Raimbault[28] and Salvas and Taylor[32] who reported that C_2H_2 and nitrapyrin inhibited methanogenesis. Consumption of CH_4 predominated during the rest of

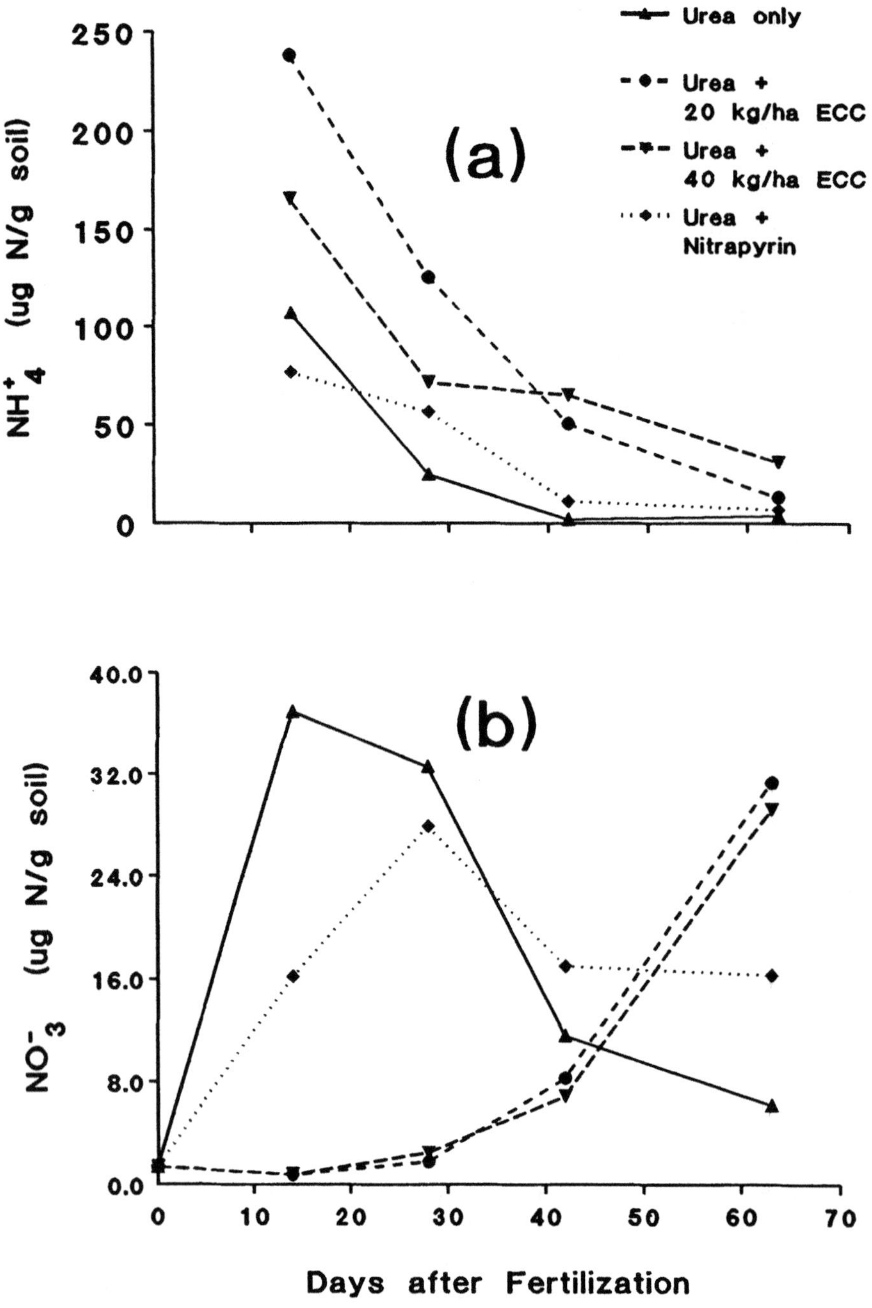

Figure 2. Exchangeable N concentration in 30-cm surface soil as affected by urea with and without ECC or nitrapyrin (a) ammonium and (b) nitrate.

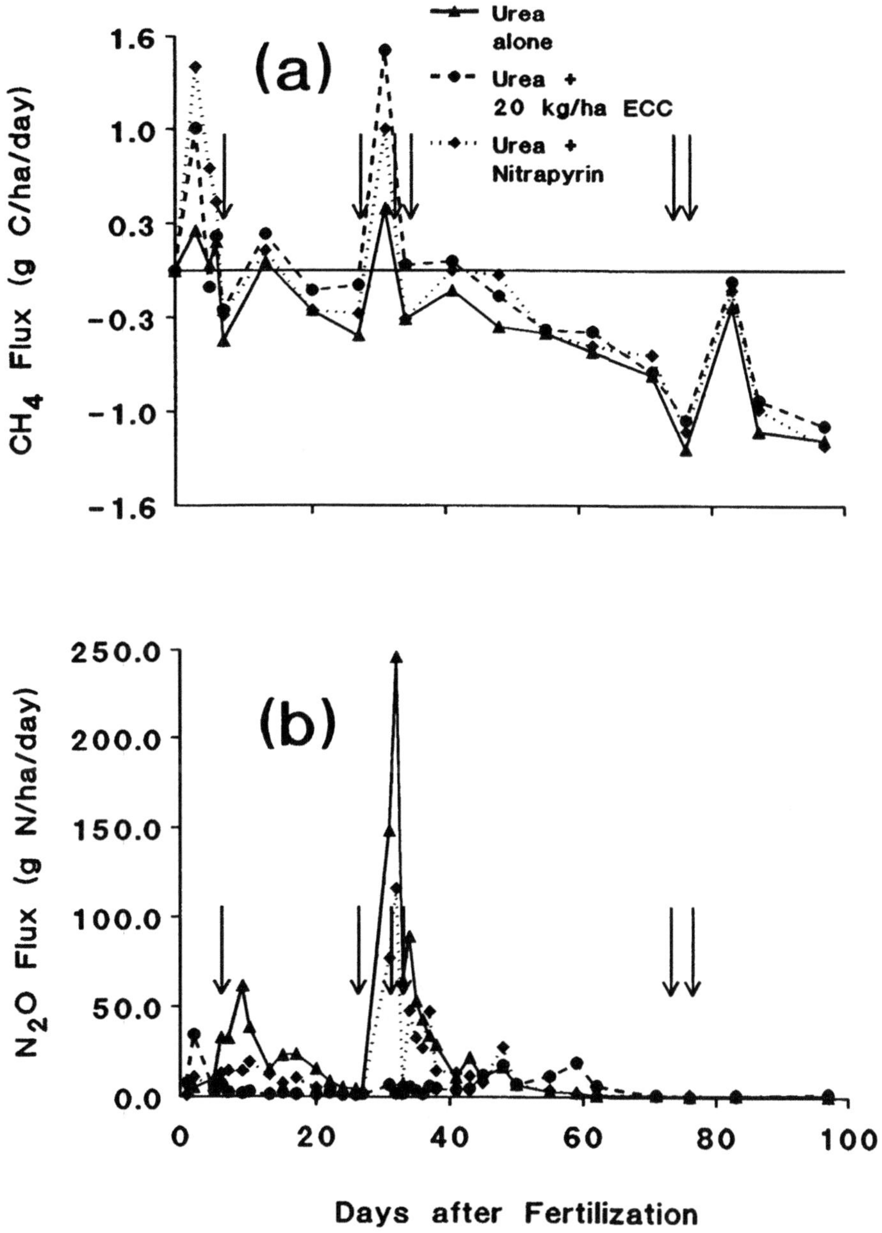

Figure 3. Trace gas fluxes from irrigated corn as affected by urea with and without ECC or nitrapyrin. (a) CH_4; (b) N_2O. Single arrows indicate an irrigation and double arrows indicate a rainfall > 25 mm.

the growing season. During the first 62 DAF, the highest CH_4 consumption rates were observed from the unfertilized control and the urea-treated plots. Methane consumption with urea alone was not different from the unfertilized control plots (data not shown) ($P > 0.05$), except at 7 DAF, when CH_4 consumption was significantly less than the control ($P < 0.05$). Other studies have indicated that NH_4^+ fertilizers inhibits CH_4 oxidation in pure culture[16,18] and in soil.[27,35]

Methane consumption was significantly inhibited by nitrapyrin and C_2H_2 ($P <$ 0.05) in this study, as has been reported in pure cultures of methanotrophic bacteria.[13,36] Acetylene produced in the soil from ECC with or without N fertilizer [data for ECC alone are not shown, as they were not significantly different from urea plus ECC ($P > 0.05$)] inhibited CH_4 consumption (compared to control and urea alone) up to 62 DAF, afterwhich no treatment differences in CH_4 consumption were observed. Methane consumption generally increased as the growing season progressed, probably due to the disappearance of C_2H_2 (Figure 1) and nitrapyrin.[37]

Nitrous Oxide Fluxes

Blockage of the N_2O to N_2 reduction step of denitrification by C_2H_2 (Ref. 41) was observed in the first 7 DAF in the ECC-treated plots (Figure 3b) when high levels of C_2H_2 were produced in the soil (Figure 1). Data for the high level of ECC plus urea are not shown, as they were not different than those for the low level of ECC plus urea ($P > 0.05$). Laboratory studies (data not shown) indicated that, in this soil, maximum blockage of the reduction of N_2O to N_2 occurs with C_2H_2 concentrations of 1000 $\mu l\ l^{-1}$. Nitrous oxide flux with the high level of ECC alone (data not shown) was not significantly different from the high level of ECC plus urea during the first 6 DAF ($P > 0.05$). This indicates that N_2O emissions from the ECC-treated plots during the first 7 DAF (when C_2H_2 production was high) came from denitrification of residual soil NO_3^- (6 $\mu g\ N\ g^{-1}$) and not from fertilizer N.

The addition of urea alone resulted in the highest N_2O emissions for most of the growing season ($P < 0.05$). Emissions of N_2O from the unfertilized control plots were low, never exceeding 2 $g\ N\ ha^{-1}\ d^{-1}$ (data not shown). Nitrous oxide fluxes from the fertilized plots peaked at 32 DAF, 2 days after the second irrigation. Flux of N_2O from urea-treated soil without NIs was 246 $g\ N\ ha^{-1}\ d^{-1}$, urea plus nitrapyrin emitted 116 $g\ N\ ha^{-1}\ d^{-1}$, and N_2O fluxes from the other plots were less than 3 $g\ N\ ha^{-1}\ d^{-1}$. (The three groups are significantly different at $P < 0.01$.) Nitrous oxide emissions declined after the hailstorm at 32 DAF, and since no further irrigations were applied, N_2O fluxes became negligible by 62 DAF.

Nitrate-N in the surface 30 cm soil was positively correlated with N_2O fluxes ($r = 0.38$, $P < 0.01$). In delaying and reducing NO_3^- formation, the NIs

Table 1. Cumulative fluxes[a] of CH_4 and N_2O–N in irrigated corn as affected by urea with and without ECC or nitrapyrin.

	Methane ($g\ CH_4$–$C\ ha^{-1}$)		Nitrous Oxide ($g\ N_2O$–$N\ ha^{-1}$)	
Treatment	avg	SD	avg	SD
Urea alone	− 43 b[b]	7	1651 a	618
Urea + nitrapyrin	− 36 ab	9	980 b	593
Urea + ECC 20 kg/ha	− 25 a	8	483 bc	128
Urea + ECC 40 kg/ha	− 26 a	8	434 bc	192
Control	− 45 b	2	108 c	29

[a]Fluxes were measured from time of fertilization (2 months after planting) to just before harvest (97 days after fertilization). The CH_4 flux and the N_2O flux consist of 19 and 35 measurement, respectively.
[b]Means in a column followed by the same letter are not significantly different at $P = 0.05$ by Duncans Mean Range Test.

indirectly reduced N_2O emissions from denitrification. Part of the N_2O emitted from urea in the absence of nitrification inhibitors was probably evolved during nitrification of NH_4^+.[5,7] However, NH_4^+–N levels in soil were not correlated with N_2O fluxes ($r = -0.06$, $P > 0.05$).

Cumulative CH_4 and N_2O Fluxes

Cumulative fluxes of CH_4 and N_2O are presented in Table 1. This clay loam soil under irrigated corn acted as a net sink for atmospheric CH_4. During the growing season, 60 g CH_4 ha^{-1} were consumed by unfertilized and urea-fertilized soil. This result that one-time N fertilizer additions to a cropped system did not affect CH_4 consumption differs from reports in forest soils[35] and in grasslands,[27] where N additions inhibited CH_4 consumption. Several years of N fertilization might have suppressed CH_4 consumption to the low levels observed in this study, such that the N additions in this experiment had no effect. Keller et al.[19] reported that cropped/cultivated tropical soils had CH_4 consumption rates that were four times less than those of tropical forest soils. On average, the CH_4 consumption rates observed in this study were about 10 times less than that measured in nearby grasslands.[27]

Acetylene produced from ECC inhibited cumulative CH_4 consumption 42–44% during the 97-day period. Nitrapyrin addition to urea reduced cumulative CH_4 consumption 22% ($P < 0.10$).

The cumulative amount of N_2O emitted during the growing season was 1480 g N_2O–N ha^{-1} from urea alone, or 0.7% of the applied fertilizer–N. Encapsulated CaC_2 controlled N_2O emissions more effectively than did nitrapyrin. The addition of urea fertilizer resulted in 13 times more N_2O–N emissions than from the unfertilized control. Several-fold increases of N_2O flux with N fertilizer application have been extensively reported (as summarized by Eichner[15]).

Net "Greenhouse" Effects

Since the NIs reduced both N_2O emissions and CH_4 consumption, any benefit of reducing the "greenhouse" effect of N_2O emissions from N fertilizer should consider the effect of NIs on CH_4 flux. Rodhe[29] calculated that, considering the 10- and 150-year decay times of CH_4 and N_2O, respectively, in the atmosphere, on a mass basis, CH_4 contributes 15 and N_2O contributes 300 times more to the "greenhouse" effect than does CO_2. Using these relative contribution factors, the mitigating effect of NIs on the net flux of N_2O and CH_4 in urea-fertilized soils was calculated on a mass basis. Although ECC addition reduced CH_4 consumption by 43%, the contribution of this change in CH_4 flux to the net greenhouse effect was negligible (<0.1%) in this system. Nitrapyrin and ECC reduced the greenhouse effect of N_2O derived from urea by 41% and 71–74%, respectively. The concept, therefore, of producing low levels of C_2H_2 in soil with ECC to inhibit nitrification and reduce N_2O emissions has potential, but the effect of C_2H_2 on CH_4 consumption needs to be monitored carefully as well.

Acknowledgments

The authors thank Sue Crookall and John Stromberger for their capable technical assistance. We are grateful also for the use of the Colorado State University Agronomy Farm and the help of Reginald Koll.

References

1. Aulakh, M.S., D.A. Rennie, and E.A. Paul. 1984. Acetylene and N-Serve effects upon N_2O emissions from NH_4^+ and NO_3^- treated soils under aerobic and anaerobic conditions. Soil Biol. Biochem. **16**:351–356.
2. Banerjee, N.K., and A.R. Mosier. 1989. Coated calcium carbide as nitrification inhibitor in upland and flooded soils. J. Indian Soc. Soil Sci. **37**:306–313.
3. Banerjee, N.K., A.R. Mosier, K.S. Uppal, and N.N. Goswami. 1990. Use of encapsulated calcium carbide to reduce denitrification losses from urea-fertilized flooded rice. Mitteilungen der Deutschen Bodenkundlichen Gesellschaft. **60**:245–248.

4. Berg P, L. Klemedtsson, and T. Rosswall. 1982. Inhibitory effect of low partial pressures of acetylene on nitrification. Soil Biol. Biochem. **14**:301–303.
5. Breitenbeck, G.A., A.M. Blackmer, and J.M. Bremner. 1980. Effects of different nitrogen fertilizers on emission of nitrous oxide from soil. Geophys. Res. Lett. **7**:85–88.
6. Bremner, J.M. 1965. Inorganic forms of nitrogen. In C.A. Black et al. (ed.), Methods of Soil Analysis. Part 2, American Society of Agronomy, Madison, WI, pp. 1179–1237.
7. Bremner, J.M., and A.M. Blackmer. 1978. Nitrous oxide: Emission from soils during nitrification of fertilizer nitrogen. Science **199**:295–296.
8. Bremner, J.M., and A.M. Blackmer. 1979. Effects of acetylene and soil water content on emission of nitrous oxide from soils. Nature **280**:380–381.
9. Bronson, K.F., and A.R. Mosier. 1991. Effect of encapsulated calcium carbide on dinitrogen, nitrous oxide, methane, and carbon dioxide emissions from flooded rice. Biol. Fertil. Soils. **11**:116–120.
10. Bouwman A.F. 1990. Soils and the Greenhouse Effect, J. Wiley and Sons, Chichester.
11. Chancy, H.F., and E.J. Kamprath. 1982. Effect of nitrapyrin on N response of corn on sandy soils. Agron. J. **74**:565–569.
12. Crutzen, P.J. 1976. Upper limits on atmospheric ozone reductions following increased application of fixed nitrogen to the soil. Geophys. Res. Lett. **3**:169–172.
13. Dalton H. 1977. Ammonia oxidation by the methane oxidising bacterium Methylococcus capsulatus Strain Bath. Arch. Microbiol. **114**:273–279.
14. Duxbury, J.M., and P.K. McConnaughey. 1986. Effect of fertilizer source and nitrous oxide emissions in a maize-field. Soil Sci. Soc. Amer. J. **50**:644–648.
15. Eichner, M.J. 1990. Nitrous oxide emissions from fertilized soils: Summary of available data. J. Environ. Qual. **19**:272–280.
16. Ferenci, T., T. Strom, and J.R. Quale. 1975. Oxidation of carbon monoxide and methane by Pseudomonas methanica. J. Gen. Microbiol. **91**:79–91.
17. Hutchinson, G.L., and A.R. Mosier. 1981. Improved soil cover method for field measurement of nitrous oxide fluxes. Soil Sci. Soc. Amer. J. **45**:311–316.
18. Jones R.D., and R.Y. Morita. 1983. Methane oxidation by Nitrosococcus oceanus and Nitrosomonas europaea. Appl. Environ. Microbiol. **45**:401–410.
19. Keller M., M.E. Mitre, and R.F. Stallard. 1990. Consumption of atmospheric methane in soils of central Panama: effects of agricultural development. Global Biogeochem. Cycles **4**:21–27.
20. Magalhaes, A., P.M. Chalk, and W.M. Strong. 1984. Effect of nitrapyrin and nitrous oxide emission from fallow soils fertilized with anhydrous ammonia. Fert. Res. **5**:411–421.
21. MathSoft, Inc. 1987. MathCad 2.0, MicroSoft Inc. Cambridge, MA.
22. Mohanty, S.K., and A.R. Mosier. 1990. Nitrification–Denitrification in flooded

rice soils. In Transactions of the 14th International Congress of Soil Science, Vol. IV, Kyoto, pp. 326–331.

23. Mosier, A.R. 1980. Acetylene inhibition of ammonium oxidation in soil. Soil Biol. Biochem. **12**:443–444.

24. Mosier, A.R., and L. Mack. 1980. Gas chromatographic system for precise, rapid analysis of N_2O. Soil Sci. Soc. Amer. J. **44**:1121–1123.

25. Mosier, A.R., and G.L. Hutchinson. 1981. Nitrous oxide emissions from cropped fields. J. Environ. Qual. **10**:169–173.

26. Mosier, A.R., W.D. Guenzi, and E.E. Schweizer. 1986. Soil losses of dinitrogen and nitrous oxide from irrigated crops in Northeastern Colorado. Soil Sci. Soc. Amer. J. **50**:344–348.

27. Mosier A., D. Schimel, D. Valentine, K. Bronson, and W. Parton. 1991. Methane and nitrous oxide fluxes in native, fertilized and cultivated grasslands. Nature **350**:330–332.

28. Raimbault, M. 1975. Etude de l'influence inhibitrice de l'acetylene sur la formation biologique du methane dans un sol de riziere. Ann. Microbiol. (Inst. Pasteur) **126A**:247–258.

29. Rodhe, H. 1990. A comparison of the contribution of various gases to the greenhouse effect. Science **248**:1217–1219.

30. Rolston, D.E., M. Fried, and D.A. Goldhamer. 1976. Denitrification measured directly from nitrogen and nitrous oxide gas fluxes. Soil Sci. Soc. Amer. J. **40**: 259–266.

31. Rolston D.E. 1986. Gas flux. In A. Klute (ed.), Methods of Soil Analysis. Part 1, Physical and mineralogical methods, American Society of Agronomy, Madison, WI, pp. 1103–1119.

32. Salvas, P.L., and B.F. Taylor. 1980. Blockage of methanogenesis in marine sediments by the nitrification inhibitor 2-chloro-6-(trichloromethyl) pyridine (nitrapyrin or N-Serve). Curr. Microbiol. **4**:305–308.

33. SAS Institute Inc. 1985. SAS users guide: Statistics. Version 5, Statistical Analysis System Institute Inc., Cary, NC.

34. Seiler, W., and R. Conrad. 1987. Contribution of tropical ecosystems to the global budgets of trace gases, especially CH_4, H_2, CO and N_2O. In R.E. Dickinson (ed.), Geophysiology of Amazonia. Vegetation and Climate Interactions, J. Wiley and Sons, New York, pp. 113–160.

35. Steudler P.A., R.D. Bowden, J.M. Melillo, and J.D. Aber. 1989. Influence of nitrogen fertilization on methane uptake in temperate forest soils. Nature **341**: 314–316.

36. Topp E., and R. Knowles. 1984. Effects of nitrapyrin [2-chloro-6-(trichloromethyl)pyridine] on the obligate methanotroph Methylosinus trichosporium OB3b. Appl. Environ. Microbiol. **47**:258–262.

37. Touchton, J.T., R.G. Hoeft, L.F. Welch, and W.L. Argyilan. 1979. Loss of nitrapyrin from soils as affected by pH and temperature. Agron. J. **71**:865–869.

38. Walters, D.T., and G.L. Malzer. 1990. Nitrogen management and nitrification inhibitor effects on nitrogen-15 urea: I. Yield and fertilizer use efficiency. Soil Sci. Soc. Amer. J. **54**:115–122.

39. Whalen S.C., and W.S. Reeburgh. 1990. Consumption of atmospheric methane by tundra soils. Nature **346**:160–162.

40. Whittenbury R., K.C. Phillips, and J.F. Wilkinson. 1970. Enrichment, isolation and some properties of methane-utilizing bacteria. J. Gen. Microbiol. **61**:205–218.

41. Yoshinari, T., R. Hynes, and R. Knowles. 1977. Acetylene inhibition of nitrous oxide reduction and measurement of denitrification and nitrogen fixation in soil. Soil Biol. Biochem. **9**:177–183.

Nitric and Nitrous Oxide Evolution from Managed Subtropical Grassland

G.L. Hutchinson

USDA, Agricultural Research Service, Fort Collins, Colorado 80522

G.P. Livingston

NASA Ames Research Center, Ecosystem Science and Technology Branch, Moffett Field, California 94035

E.A. Brams

Prairie View A&M University, Agricultural Research Complex, Prairie View, Texas 77446

ABSTRACT

Humid subtropical grasslands represent an important fraction of the Earth's savannas that are believed to be the largest contributor to the global soil source of NO_x. They may also serve as a model for predicting N oxide emissions from the rapidly increasing area of pastures formed by clearing tropical forests that represent more than half the global soil source of N_2O. We used a chamber technique to monitor NO and N_2O emissions from ungrazed Bermuda grass (*Cynodon dactylon*) pastures subject to minimum or intensive cultural management on sandy loam in humid, subtropical southern Texas. Emission of both gases was larger from intensively managed pasture and apparently resulted from the activity of nitrifying bacteria rather than denitrifiers. A large fraction of the emitted NO and N_2O was lost in brief intense bursts that followed fertilization or precipitation and especially a combination of these two. Annual N_2O evolution totaled 0.6 and 1.2 kg N ha^{-1} from plots under the two management schemes, somewhat smaller than the high fluxes characteristic of tropical regions. During the 9-week early summer period over which it was measured, NO emission averaged 7.2 times greater than N_2O emission and accounted for 3.2% of added fertilizer N. If the NO evolved during this period represents the same fraction of annual emissions as was measured for N_2O, its evolution from plots under the two management schemes would have totaled about 3 and 8 kg N ha^{-1} yr^{-1}, placing humid, subtropical grasslands among the largest biogenic NO sources and supporting the hypothesis that savanna regions dominate the global soil source of NO_x.

Introduction

Gaseous N oxides, N_2O and NO_x ($NO + NO_2$), are radiatively, chemically, and ecologically important trace atmospheric constituents.[42] For example, N_2O absorbs outgoing planetary IR radiation with wavelengths not removed by atmospheric CO_2 or H_2O and has been estimated to contribute 5% of the overall anthropogenic greenhouse effect.[34] NO_x has no direct effects on the Earth's radiation balance, but is chemically very active. It plays a critical role in establishing the abundance of tropospheric oxidants, e.g., O_3, which is radiatively active, and OH, which is involved in the photochemical removal of other greenhouse gases from the atmosphere, including CH_4.[42] Eventually, NO_x is itself photochemically oxidized to HNO_3, the fastest-growing component of acidic deposition.[28] Although N_2O is chemically inert in the troposphere, it is readily transported to the stratosphere where it is involved in destruction of the O_3 that protects life forms on Earth from incoming solar UV radiation. Other ecologically important functions of gaseous N oxides include that the emission, transport, and subsequent redeposition of NO_x may result in substantial N redistribution both within and among natural and disturbed ecosystems, thereby contributing to the unaccounted losses (or gains) typically observed in soil N balance studies.[42] Because nitrification and denitrification in soil are prominent sources of atmospheric N oxides, there is need to determine the integrated magnitudes of these processes across space and time and to identify principal contributors among the world's major biomes.

Although they occupy only about a tenth of the Earth's land area,[8] tropical forests are believed to contribute over half the global soil source of N_2O.[10] Compared to temperate and boreal systems, the same forests also have high NO emission rates,[5,26] but the meager available data suggest that their contribution to the global soil source of NO_x may be much smaller than that of savanna regions,[24,25] which have only slightly greater geographical extent. Humid subtropical grasslands represent a sizable fraction of the latter category and may have particular importance to the global atmospheric NO_x budget because they have typically been subject to intense anthropogenic disturbance. Moreover, because tropical forests are being cleared at an increasing rate and converted to pastures that may have larger[29] or smaller[14] N_2O emission rates, our ability to predict these soils N oxide emissions following conversion might be enhanced by a better understanding of N cycling processes in existing humid subtropical pastures. Toward that end, we monitored N oxide emissions from a managed Bermuda grass (*Cynodon dactylon*) pasture on sandy loam in humid, subtropical southern Texas and compared the responses of NO and N_2O evolution to changes in soil N pool sizes and transformation rates, to selected environmental parameters, and to various cultural practices such as harvest and fertilization. Experimental results have been published elsewhere[9,17,18,20]; the purpose of this chapter is to summarize the entire dataset and comment on its implications to the theme of this symposium.

Methods and Materials

The experiment was conducted in an established 30-ha Bermuda grass pasture located on well-drained, very uniform Kenney sandy loam (a member of the loamy, siliceous, thermic, Grossarenic Paleudalfs). Selected soil and climatic parameters are listed in Table 1. The two treatments imposed on 10-m × 30-m rectangular plots in the pasture were minimum cultural management (spring harvest followed by a single annual maintenance application of N and P fertilizers) and intensive cultural management (harvest and fertilization repeated on a 9-week cycle throughout the growing season). Each treatment was replicated four times in a completely randomized block design. The source of fertilizer N was $(NH_4)_2SO_4$, which was always surface-applied 5 d after grass harvest. The amount applied to intensively managed plots on each fertilization date was that recommended for maximum protein production based on standard soil tests

Table 1. Selected soil and climatic data for the experimental site.

Parameter	Value
Soil (0–15 cm)	
pH (varied with management)	6.1–7.5
CEC	4.8 cmol kg^{-1}
Bulk density	1.6 g cm^{-3}
Infiltration rate	200 mm h^{-1}
Organic matter	17 g kg^{-1}
Sand	770 g kg^{-1}
Silt	130 g kg^{-1}
Clay (predominantly kaolinite)	90 g kg^{-1}
Water retention @ 0 kPa suction	25.0%
@ 10 kPa suction	9.9%
@ 33 kPa suction	6.1%
@ 100 kPa suction	4.3%
@ 1000 kPa suction	4.0%
Climate	
Rainfall (mean annual)	1020 mm
Air temperature	26.7°C
(mean annual max.)	
(mean annual min.)	13.0°C
Growing season	304 d

Table 2. Seasonal N fertilizer application rate and NO and N_2O evolution from Bermuda grass pasture under minimum (min) or intensive (int) cultural management.

		Fertilizer ($kg\ N\ ha^{-1}$)		N_2O Flux ($kg\ N\ ha^{-1}$)		NO Flux ($kg\ N\ ha^{-1}$)	
Season	Interval (d)	min	int	min	int	min	int
Late summer	63	0	117	0.22	0.48	—	—
Fall	63	0	82	0.02	0.09	—	—
Winter	105	0	0	0.05	0.10	—	—
Spring	63	112	112	0.13	0.15	—	—
Early summer	63	0	52	0.15	0.35	0.69	2.37
Totals	357	112	363	0.6	1.2	3.[a]	8.[a]

[a]Projected amount assuming that the ratio of early summer to annual emissions was the same for NO as was measured for N_2O.

(Table 2); an identical amount was applied to plots under minimum cultural management on the early spring fertilization date only. Meteorological data were recorded at a station near the experimental plots.

Field Measurements

Soil emission of N_2O was measured periodically for 1 year at four locations within each plot. To begin each measurement, a vented, cylindrical chamber was mounted atop a permanently installed ring that had been driven into the soil, and the two were sealed together by overlapping the seam between them with an external large rubber band. The enclosures (about 25 L total volume) were constructed from rigid polyvinyl chloride pipe using design criteria suggested by Hutchinson and Mosier,[21] then insulated with polyurethane foam and covered with reflective aluminized polyester film to minimize internal heating by solar radiation. Accumulation of N_2O beneath each enclosure after 10 and 20 min was determined by drawing 30-ml air samples through the chambers' sampling ports using 60-ml polypropylene syringes fitted with nylon stopcocks. The N_2O concentration at the time of each chamber's installation was assumed the same as that of an ambient air sample collected at the field site. All samples were collected between 1100 and 1200 local time and transported to the laboratory in an insulated container for analysis within 12 h by gas chromatography using electron capture detection.[31] The nonlinear equation proposed by Hutchinson and Mosier[21] was adopted to calculate the flux when the data met criteria established by Anthony and Hutchinson.[4]

Nitric oxide flux from the soil area defined by one of the permanent rings in each plot was measured using a similar chamber modified to recirculate air from beneath the enclosure through a luminol-based NO_x detector. Because the instrument is sensitive only to NO_2, the 1.4-L min^{-1} sample airstream passed first through a CrO_3 converter to oxidize NO to NO_2, then through the analyzer, and finally through a scrubber to remove remaining NO, NO_2, and H_2O before returning to the enclosure. After correcting for the dilution caused by returning NO_x-free air, the flux of NO was estimated by linear regression using concentration data recorded once per minute during the period 2–6 min following chamber installation. Because NO emitted by soil is rapidly oxidized by ambient O_3 to NO_2, which is strongly sorbed by both soil and plant surfaces, no readings were taken during the first 2 min after chamber installation to allow time for ambient O_3 and NO_2 captured in the enclosure to be destroyed or sorbed. Periodic checks confirmed that the concentrations of both gases declined to near zero within this period, which we also interpreted as evidence that soil emission of NO_2 was negligible. Unlike the case for N_2O, soil NO emission measurements were limited to the 9-week period following early summer harvest and fertilization of intensively managed plots; the measurements were completed between 1400 and 1600 local time.

The scheduled frequency of NO and N_2O flux measurements was highest immediately following harvest and fertilization because these management inputs were expected to cause highest soil N transformation rates and, therefore, highest N oxide emission rates. Each day that gas fluxes were determined, we recorded the outputs of soil water/temperature sensors buried at 2- and 10-cm depths in each plot, collected duplicate soil samples from 0- to 2-cm and 2- to 10-cm depths in each plot, and subsampled duplicate polyethylene bags of soil that had been buried at 2- and 10-cm depths in each plot on the day of fertilization. Soil in the buried bags was a subsample of the 0- to 2-cm and 2- to 10-cm samples taken the day prior to fertilization and was thoroughly mixed with sufficient N to simulate each plot's fertilization rate; changes in its inorganic N concentrations were used to estimate the fertilizer's net nitrification rate without confoundment by plant uptake or leaching. All soil samples were immediately frozen until they could be extracted for analysis. Soil NH_4^+, NO_2^-, and NO_3^- were extracted by shaking with 1M KCl (1 : 5 soil to solution ratio) on a wrist action shaker for 1 h. The extracts were filtered through glass fiber filters[38] and analyzed using a continuous flow analytical system.

Laboratory Studies

To facilitate understanding the processes involved in the production and emission of gaseous N oxides in the field, we conducted a 10-d aerobic laboratory incubation experiment using soil collected in mid-winter from plots under minimum cultural management. Several samples taken from 0- to 10-cm depths were combined, partially air-dried to permit passage through a 2-mm sieve, and then stored at 2.8% water content for about 1 month in a refrigerator at 4°C. All identifiable pieces of fresh plant residue that passed through the sieve were removed by handpicking.

The incubation combined three N treatments with three soil water levels in a repeated measures factorial design with two replicates per treatment. Initial soil water content was 4, 7, or 10% by weight, equivalent to water potentials of approximately −1000, −30, and −10 kPa, respectively (Table 1). The N treatments included a control and two with a N amendment, one of which also received nitrapyrin [2-chloro-6-(trichloromethyl)-pyridine]. The source of N was NH_4NO_3 (50 mg N kg^{-1}), chosen because it provided substrate to both nitrifiers and denitrifiers. Nitrapyrin was used to inhibit chemoautotrophic oxidation of NH_4^+ to NO_2^- without influencing the activity of nontarget microbial groups, thereby permitting evaluation of the relative contributions of nitrifying versus denitrifying microorganisms to NO and N_2O emissions from the experimental soil under each water treatment. Minimum effective concentration of the inhibitor (30 mg kg^{-1}) was selected in a preliminary experiment where nitrapyrin concentration was the only variable (data not shown).

We used the automated flow-through soil incubation system described by Hutchinson and Andre,[16] except the air flow rate through each incubation jar was reduced to 60 ml min^{-1}, and we eliminated the water reservoir they used for final humidification of incoming air. As a result, air entering the jars was only about 75% saturated and extracted approximately 0.5 g H_2O d^{-1} from each jar. To compensate for this loss and to measure the soil's response to rewetting, we added distilled water (dropwise with a pipette) after the seventh day of incubation to return each sample to its initial water content. Incoming air was blended from NO_x-free bottled gas mixtures to contain ambient atmospheric concentrations of O_2, CO_2, and N_2O. Incubation temperature was 25°C.

To begin the experiment, 156 g (dry wt.) of soil in each incubation jar was wet by misting with distilled water from a hand-operated pump spray bottle while mixing on a large platform balance covered with waxed brown paper. Finely ground dry NH_4NO_3 and nitrapyrin were then added using 0.5 g talcum as a carrier as described by Anderson and Domsch.[3] After the talcum/chemical mixtures were thoroughly incorporated by vigorous stirring and shaking, 10 g (dry wt.) of treated soil was weighed into each of five small sample vials, and the remaining soil was spread evenly across the bottom of each jar. Soil in the jar bottom served as a continuous, undisturbed source of the measured gases, whereas that in the small vials provided convenient, preweighed subsamples for periodic chemical analyses; soil depth was about 1 cm in both cases. Four sample vials were positioned in the incubation jars as described by Hutchinson and Andre,[16] and soil in the fifth vial was immediately extracted to determine initial inorganic soil N concentrations. The other vials were removed and extracted after 1, 2, 4, and 7 d of incubation; after 10 d, subsamples were removed from the bottom of each incubation jar for final chemical analyses and pH measurement.[30] Extraction and analytical techniques were the same as described for samples from the field plots.

Soil NO and CO_2 emission rates were determined by monitoring the concentrations of these gases in the air that flushed each jar. Exhaust air was also analyzed for N_2O after 1.5, 3, 6, 9, 12, 24, 48, 72, 168, 171, and 240 h of incubation. Periodic analysis of the exhaust air for NO_2 never revealed a detectable concentration of this gas. Minimum detectable NO and NO_2 fluxes were about 0.03 to 0.04 μg N kg^{-1} h^{-1} and for N_2O, about 0.2 to 0.5 μg N kg^{-1} h^{-1}. For both field and laboratory data, statistical significance was assumed when $P < 0.05$.

Results

Soil N_2O emission rates, soil temperature at 2-cm depth, and precipitation for the entire year are shown in Figure 1. Highest N_2O fluxes generally occurred during the warmer seasons immediately after the first rain following N fertilization. The most striking example occurred following application of 117 kg N ha^{-1} to intensively managed plots on 9 August 1988. Soil temperature at 2 cm was about 34°C, and 6.4 mm rain fell both on the day of fertilization and again 3 d later. The flux of N_2O rose rapidly to 44 g N ha^{-1} d^{-1} on the third day following fertilization, then fell just as sharply to about 10 g N ha^{-1} d^{-1} where it remained relatively constant for about 3 weeks. Emissions from plots under minimum cultural management during this period were only about half those measured from intensively managed plots. A similar but smaller response was observed when 25 mm rain fell 9 d after surface application of 52 kg N ha^{-1} to intensively managed plots on 30 May 1989. In this case, the peak emission rate may not have been measured because it likely occurred between scheduled sampling dates 15 d and 19 d after harvest (1 d and 5 d after the rain).

Inadequate sampling frequency probably also prevented observing the response of N_2O emissions to the first rain following surface application of 82 kg N ha^{-1} to intensively managed plots on 11 October 1988. No rain fell for 20 d following that fertilization, and the next scheduled sampling occurred 8 d later. Nevertheless, there was still a significant difference between N_2O fluxes from plots under the two management schemes. Lower soil temperature probably contributed to the smaller rates of N_2O emission measured during this period. Finally, there was little response of soil N_2O emissions to the 112 kg N ha^{-1} applied to all plots on 29 March 1989 until soil temperature warmed significantly about a month later. Total N_2O emissions from the two treatments during the four 9-week periods following each harvest of the grass on intensively managed plots,

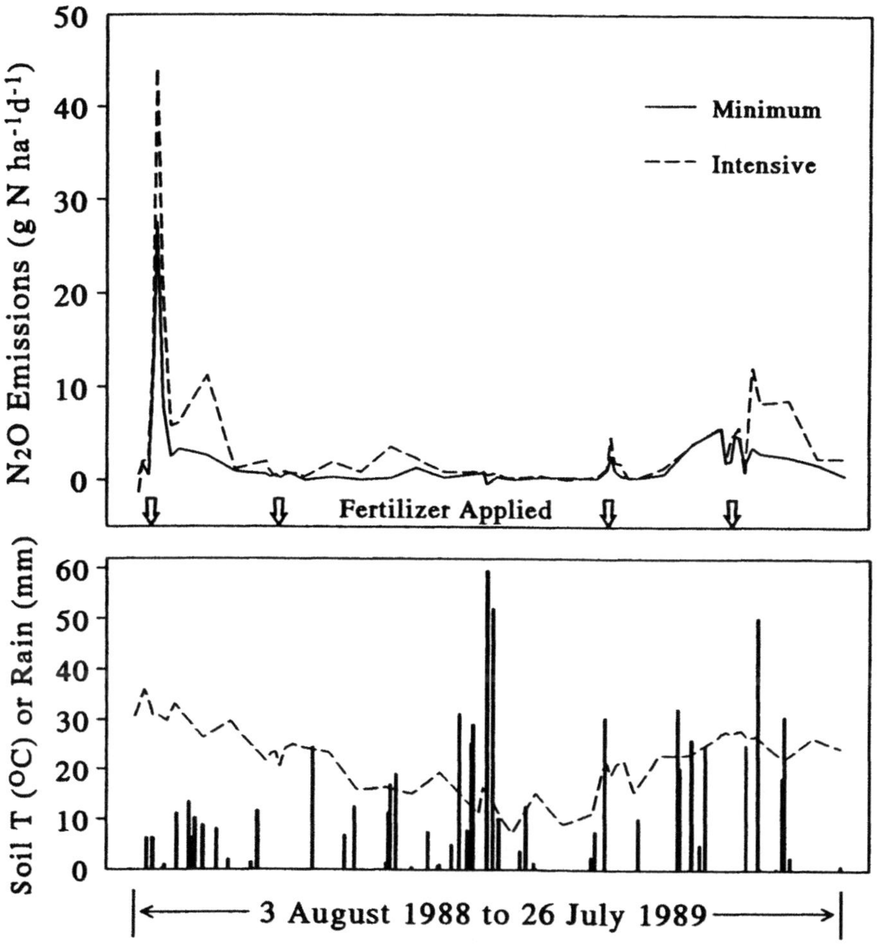

Figure 1. Nitrous oxide flux from Bermuda grass pasture subject to minimum or intensive cultural management as influenced by N fertilization, precipitation, and seasonal changes in soil temperature measured at 2-cm depth. Fluxes of N_2O represent the means of quadruplicate measurements in each of four replicate plots (n = 16). Soil temperatures (broken line) are means of measurements made in the four replicate plots of both treatments (n = 8). Dates of N fertilization are indicated by pointers on the abscissa. [Adapted from Brams et al. (Ref. 9) with the permission of J. Wiley and Sons.]

as well as the 103-d winter period when there was little grass growth, are listed in Table 2; the totals were estimated by integrating the areas under appropriate segments of the N_2O emissions curves in Figure 1.

In addition to the seasonal effects and fertilization/rainfall responses described above, soil N_2O emissions from plots under both minimum and intensive cultural management exhibited other shorter-term features related to changes in their inorganic N pool sizes and transformation rates. Because these relations were similar in all seasons, soil N data for only one of the 9-week periods is presented here. The early summer period was chosen because it is the only one for which NO emission measurements are also available. For this period, data collection commenced 24 May 1989, the day before harvest, and continued through 26 July.

Early Summer Field Data

Early summer rates of NO and N_2O evolution from the field plots under both minimum and intensive cultural management are presented in Figure 2a along with changes in soil NH_4^+ and NO_3^- concentrations at both sampling depths (Figures 2b and 2c). The NO_2^- concentrations of both bulk soil samples and buried bag subsamples never exceeded 0.05 mg N kg^{-1} and are not presented. Soil temperature measurements (Figure 1) showed no significant long-term trends that might account for the observed changes in inorganic soil N concentrations or soil N oxide emission rates during this period.

Emission of NO, but not N_2O, was apparently stimulated by clipping and removing the grass from plots under intensive cultural management (Figure 2a). The nearly twofold increase in NO evolution rate from the day preceding to the day following harvest was statistically significant and was reinforced by an additional 77% increase over the next 3 d. Nitric oxide emissions then remained nearly constant at about 17 g N ha^{-1} d^{-1} (two to three times the emission rate of plots under minimum cultural management) until at least the eleventh day after harvest. Before the next scheduled set of measurements on day 15, the effect of harvest was obscured by the much larger response to fertilization.

Soil emission of N_2O and especially NO were strongly stimulated by application of N fertilizer (Figure 2a). Although the fertilizer was applied on day 5 following harvest, its effect on N oxide evolution was delayed 9 d until the next rain (Figure 1) dissolved and moved the surface-applied fertilizer into the soil. Presence of the $(NH_4)_2SO_4$ is also not reflected in the data in Figure 2b or 2c until after rainfall because most of it was apparently included in the surface litter that was purposely brushed aside prior to soil sampling. The 25-mm rain on day 14 following harvest moved a substantial fraction of the fertilizer into the 2–10-cm soil layer, where the NH_4^+ concentration increased from 5.7 to 15 mg N kg^{-1}, compared to an increase from 10 to 26 mg N kg^{-1} in the surface layer. Ammonium concentrations then decreased to prefertilization levels over a 1–

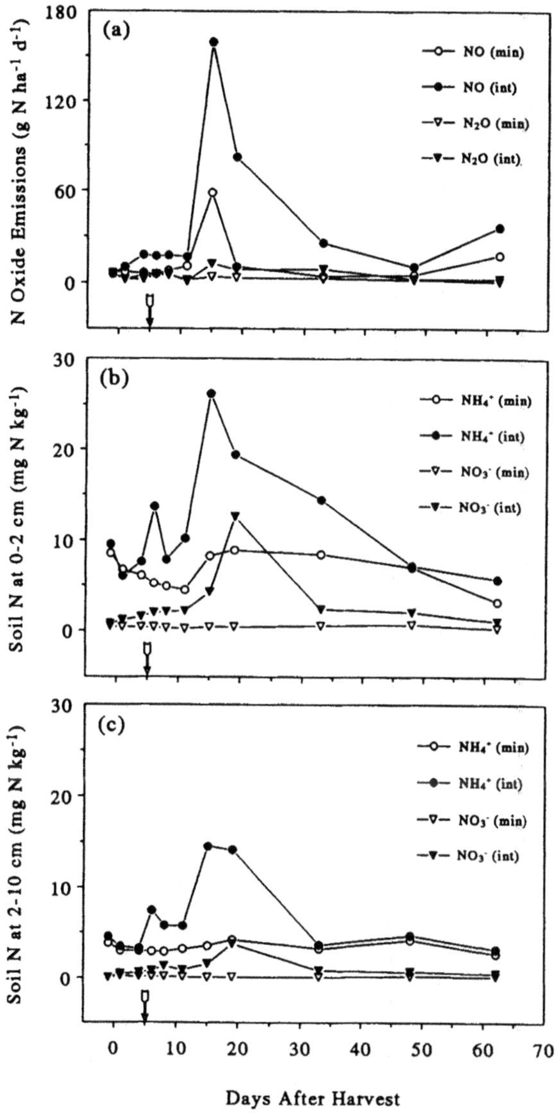

Figure 2. Early summer NO and N_2O emission rates (a) and soil NH_4^+ and NO_3^- concentrations at 0–2-cm depth (b) and 2–10-cm depth (c) in Bermuda grass pasture subject to minimum (min) or intensive (int) cultural management. Fluxes of N_2O represent the means of quadruplicate measurements, and for NO single measurements, in each of four replicate plots (n = 16 or 4); data points in (b) and (c) represent the means of duplicate samples from each of the four replicate plots (n = 8). The date of N fertilization is indicated by a pointer on the abscissa. [Adapted from Hutchinson and Brams (Ref. 17) with the permission of the *Journal of Geophysical Research.*]

2-week period in the lower soil layer and a 3–4-week period in the surface layer. Rapid accumulation of NO_3^- in both soil layers during the early part of this period confirms that microbially mediated nitrification was at least partially responsible for the observed reduction in NH_4^+ concentration. The sharp reduction in NO_3^- concentration in both layers between scheduled sampling dates 19 and 33 d after harvest was probably due both to root uptake and to leaching by the 50-mm rain on day 20. Soil NO_3^- concentration in plots under minimum cultural management remained consistently very low at both sampling depths.

The net nitrification rate of fertilizer NH_4^+ is shown in Figure 3. These data were computed from changes in the $NO_2^- + NO_3^-$ concentrations of soil in the buried bags and represent mean rates of net nitrification from one sampling date to the next, so they were plotted midway between the two dates. Because $(NH_4)_2SO_4$ was thoroughly mixed with soil placed in the bags buried in intensively managed plots on the day of fertilization, its nitrification was not subject to the same 9-d delay as fertilizer applied to the surface of the same plots. Nitrification rates at 2- and 10-cm depths rose to peaks of 9.0 and 4.8 mg N kg^{-1} d^{-1}, respectively, after 6 d, after which they decreased rapidly at first, then more slowly as the process neared completion after 3–4 weeks. Continual increases in the total inorganic N concentrations of the buried bags reflect the soil's nontrivial capacity for N mineralization; mean net mineralization rates computed from the data for bags buried 2 cm beneath the surface of plots under minimum and intensive cultural management were 0.5 and 0.9 mg N kg^{-1} d^{-1}, respectively, and for bags buried at the 10-cm depth, 0.3 and 0.6 mg N kg^{-1} d^{-1}, respectively.

The response of soil NO emissions to fertilization followed by rainfall was both large and rapid. After only 1 d the emission rate from plots under intensive cultural management had increased to 160 g N ha^{-1} d^{-1}, nearly an order of magnitude greater than the rate measured prior to rainfall. Thereafter, NO evolution from these plots declined slowly over a 3–4-week period to the levels observed before harvest and fertilization. The coincident and unexpectedly large increase in NO evolution from plots under minimum cultural management on day 15 had much shorter duration. Peak NO emissions of 58 g N ha^{-1} d^{-1} had decreased by day 19 to the level measured prior to the rain-induced increase in soil NH_4^+ concentration apparently responsible for enhancing NO emissions. The statistically significant increase in NO evolution from intensively managed plots to 36 g N ha^{-1} d^{-1} on the final sampling day of the early summer period is believed to have resulted from a similar burst of mineralization/nitrification activity induced by the 0.8-mm rain shower that fell on very dry soil just prior to sampling. Emission of NO from plots under minimum cultural management responded to the same rain shower, but with the smaller magnitude expected of a soil/plant system under greater N stress.

Emission of N_2O from intensively managed plots followed a pattern similar to that of NO, but on a much smaller scale. For example, the peak emission rate on

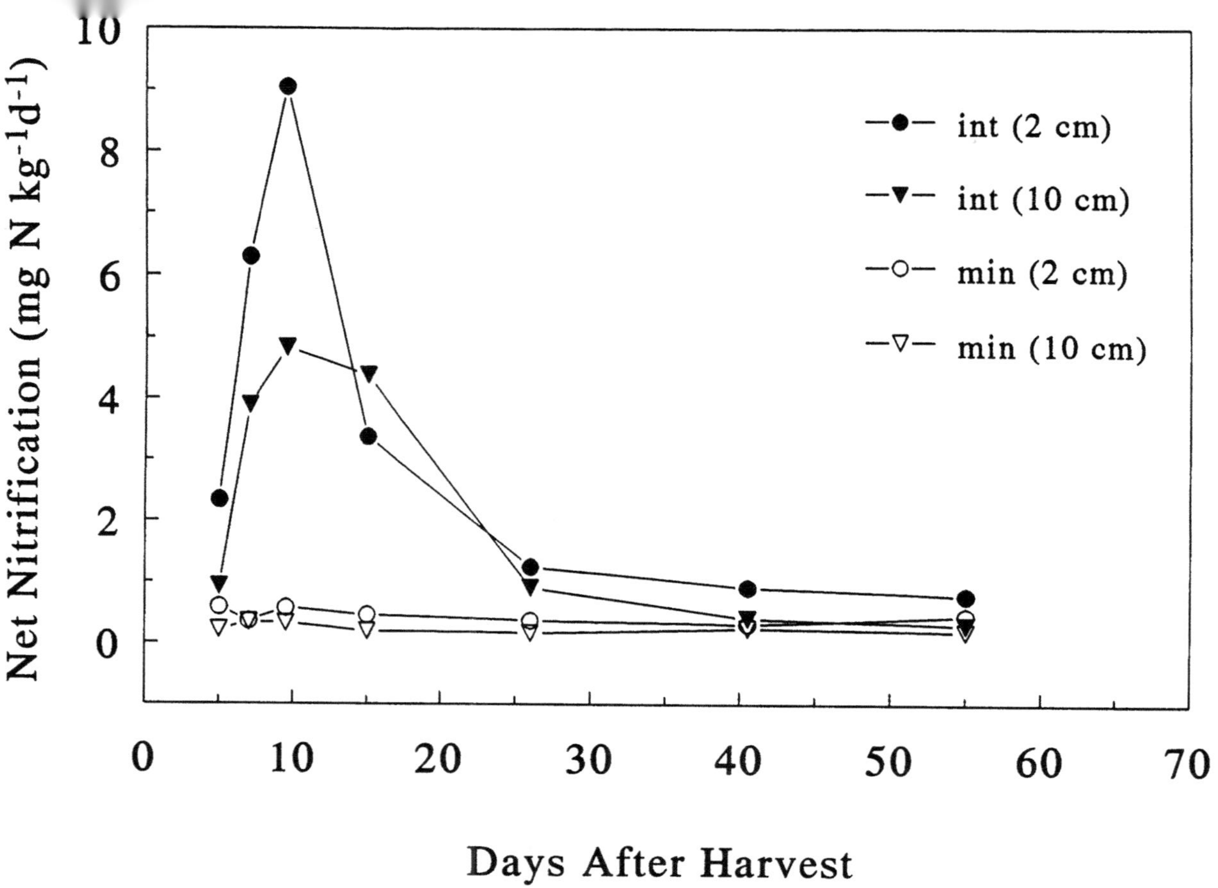

Figure 3. Nitrification rates measured in soil samples taken from 0- to 2-cm and 2- to 10-cm depths and then buried in polyethylene bags at 2- and 10-cm depths, respectively, in a Bermuda grass pasture subject to minimum (min) or intensive (int) cultural management. Data points represent the mean change in soil NO_2^- + NO_3^- concentration in duplicate bags buried at each depth in each of the four replicate plots and are plotted at the midpoint of the period over which the change was determined (n = 8). [Adapted from Hutchinson and Brams (Ref. 17) with the permission of the *Journal of Geophysical Research.*]

day 15 following harvest was only 12 g N ha^{-1} d^{-1}, compared to 160 g N ha^{-1} d^{-1} for NO. The evolution of N_2O from plots under minimum cultural management was frequently not significantly different from zero. The ratio of NO to N_2O emission rates over the 11 sampling days averaged 6.7 ± 2.9 (SEM) for plots under minimum cultural management and 7.7 ± 2.1 (SEM) for plots under intensive cultural management. Highest ratios were recorded on scheduled sampling dates 15 d and 62 d after early summer harvest, both of which immediately followed precipitation.

Laboratory Studies

Because the air that flushed each incubation jar extracted about 0.5 g H_2O d^{-1}, soil water content in the medium and high water treatments after 7 d incubation fell below the initial water content of the next drier treatment, whereas that of the low water treatment declined to 1.4% by weight (data not shown). As a result, CO_2 and N oxide emission rates were measured over the entire range of water contents from 10% to 1.4% at some time during the experiment (equivalent to water potentials of about −10 kPa to much less than −1000 kPa); (Table 1). The flush of CO_2 observed following addition of water at the beginning of the experiment (data not shown) was similar to that observed by other authors.[6] Measured rates of CO_2 evolution varied among the water treatments in the same order as soil water content, but did not differ among N treatments. The evolution rates declined rapidly during the first day of incubation, then much more slowly through the seventh day; the rate from the low water treatment became not significantly different from zero during the fifth day. An additional flush of CO_2 was observed from all treatments following rewetting after 7 d incubation.

The response of soil N transformations, and therefore gaseous N oxide evolution, to the addition of water was more complex than that of soil respiration. Figure 4 compares the NO emission rates and NH_4^+ and NO_3^- concentrations of soil under all three water treatments without added NH_4NO_3 or nitrapyrin, whereas Figure 5 shows the effect of these two soil amendments on the same parameters for the medium water treatment only. The emission of NO from unfertilized soil rose rapidly for the first 3–4 h of incubation, then more slowly to peaks after 30 h of 1.43, 1.30, and 1.21 μg N kg^{-1} h^{-1} for the low, medium, and high water treatments, respectively (Figure 4a). Differences among the three water treatments during this period were consistent, statistically significant, and probably reflect the effect of water content on soil gas diffusion rates; that is, the NO yield of nitrification increases with its ease of escape to the atmosphere.[11,16,32] After 30 h, NO evolution declined rapidly until the end of the fourth day. Thereafter, NO emissions from the high water treatment remained relatively constant, averaging 0.18 μg N kg^{-1} h^{-1}. Emissions from the medium water treatment averaged 0.16 μg N kg^{-1} h^{-1} except for the 12-h period after rewetting,

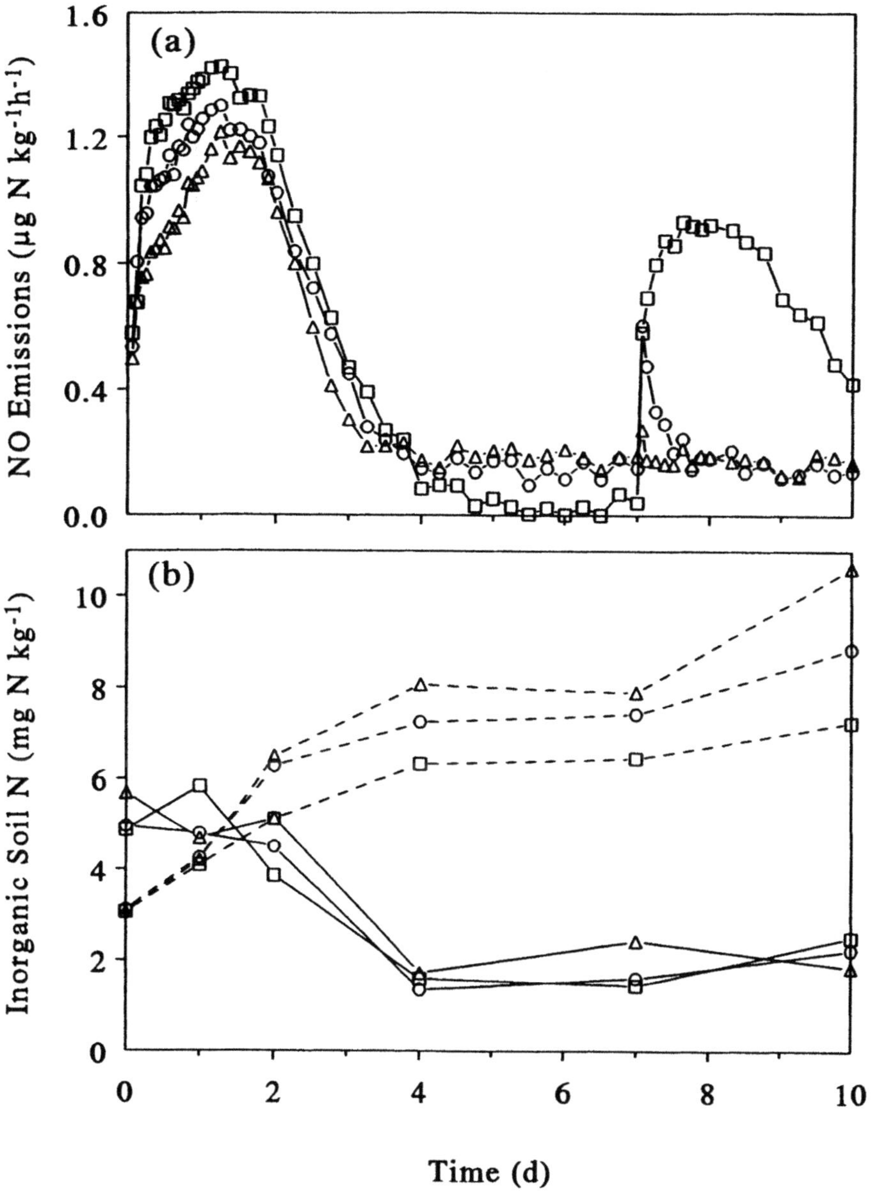

Figure 4. Nitric oxide emission rates (a) and NH_4^+ and NO_3^- concentrations (b) of soil under low (squares), medium (circles), and high (triangles) water treatments during 10 d aerobic incubation with no N amendment. In (b), symbols joined by solid lines represent NH_4^+ concentrations, whereas symbols joined by dashed lines represent NO_3^- concentrations. Data points represent the means of two replicates. [Adapted from Hutchinson et al. (Ref. 20) with the permission of *Soil Biology and Biochemistry*.]

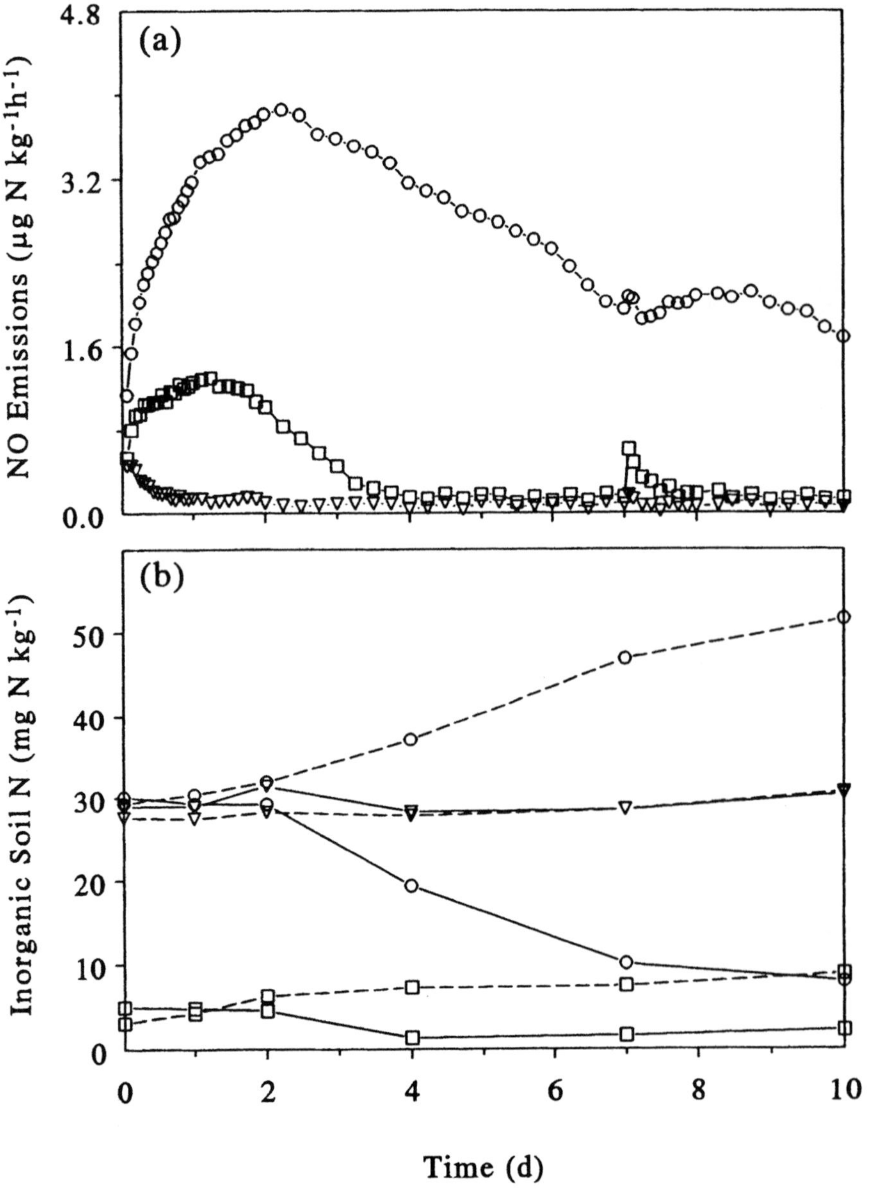

Figure 5. Nitric oxide emission rates (a) and NH_4^+ and NO_3^- concentrations (b) of soil treated with no N amendment (squares), NH_4NO_3 (circles), or NH_4NO_3 + nitrapyrin (triangles) during 10 d aerobic incubation under the medium water treatment (7% initial water content). In (b), symbols joined by solid lines represent NH_4^+ concentrations, whereas symbols joined by dashed lines represent NO_3^- concentrations. Data points represent the means of two replicates. [Adapted from Hutchinson et al. (Ref. 20) with the permission of *Soil Biology and Biochemistry*.]

during which there was a second burst of NO evolution that was much smaller and shorter than the first. In contrast, NO emissions from the low water treatment continued declining after the fourth day, becoming not significantly different from zero at about the same time that CO_2 evolution from this treatment also approached zero. When this dry soil was rewet to its initial water content (4%), a second large burst of NO emissions occurred; peak emissions were 35% lower, but durations of the two bursts were nearly identical.

When NH_4NO_3 was added to soil with 7% initial water content (Figure 5a), the peak rate of NO emission was delayed an additional 30 h and elevated about threefold; thereafter, emissions generally followed declining NH_4^+ availability. In contrast, the temporal dependence of NO emissions from NH_4NO_3-treated soil with 4% initial water content (data not shown) was nearly identical to that of the corresponding unfertilized soil (Figure 4a), but amplitudes of the two emission peaks were about doubled. Soil under this water treatment apparently became dry enough to affect the survival or metabolism of the organisms responsible for NO production. When nitrapyrin was added with the NH_4NO_3 (Figure 5a), NO evolution was immediately reduced and continued to decline for about 16 h before stabilizing at about 0.09 μg N kg^{-1} h^{-1} for the remainder of the incubation. The delay in achieving maximum inhibition probably represents the time required for this sparingly soluble chemical to dissolve and diffuse to every colony of NH_4^+ oxidizers.

Inorganic soil N concentrations shown in Figures 4b and 5b reflect the progress of microbially mediated nitrification. The NH_4^+ concentrations of samples not treated with nitrapyrin decreased while their NO_3^- concentrations increased by corresponding amounts over the same time periods. There were no significant changes in the NH_4^+ or NO_3^- concentrations of nitrapyrin-treated soil during 10 d of aerobic incubation, and CO_2 evolution was not significantly affected by the inhibitor, thus confirming that the chemical behaved as both a potent and specific inhibitor of chemoautotrophic NH_4^+ oxidation. Soil NO_2^- concentrations never exceeded 0.09 mg N kg^{-1}, and although they were higher after 1, 2, 4, 7, and 10 d incubation than in samples extracted immediately after initial wetting, there were no other significant differences among sampling times or N and water treatments. Rates of N_2O evolution from all treatments (data not shown) were generally about an order of magnitude smaller than NO emission rates and often did not exceed our minimum detectable N_2O flux.

Discussion

Biochemical Source of Gaseous N Oxides

Both field and laboratory results suggest that chemoautotrophic NH_4^+-oxidizing bacteria were the predominant source of gaseous N oxides over the entire range of soil water potentials included in our investigations, thus confirming and extending recent work of other authors.[1,23,27,32,39] Strongest support for this conclusion included (i) emission of NO from NH_4NO_3-treated soil was nearly eliminated by nitrapyrin at all three soil water levels in the incubation study and (ii) temporal variation in the N oxide emission rates shown in Figure 2a was strikingly similar to that of the net nitrification rates plotted in Figure 3, after allowing for the offset in time between nitrification of the fertilizer in buried bags versus that applied to the plot surface. Because there was no response of N oxide emissions to NH_4NO_3 in the presence of nitrapyrin during soil incubation and no correspondence of enhanced N oxide emission rates with periods when both the soil's water content and its NO_3^- concentration were elevated in the field, we concluded that heterotrophic denitrification made no significant contribution to measured emissions of NO or N_2O. Apparently, the supply of O_2 in this well-drained sandy loam was never sufficiently limited to favor the use of NO_3^-, rather than O_2 (which is preferred), as terminal electron acceptor during respiratory electron transport phosphorylation by the facultative anaerobes primarily responsible for denitrification. It should be emphasized, however, that this observation may not be repeated at similar water potentials in soils with finer texture, higher organic C availability, or more compact structure. The low rate of NO evolution from incubated nitrapyrin-treated soil may have resulted from heterotrophic nitrifier activity.[33,35]

Except for the brief burst of N oxide emissions that usually followed the addition of water to very dry soil (discussed later), NO and N_2O evolution generally followed changes in soil NH_4^+ concentration, thereby providing additional evidence for the importance of nitrification as the source of gaseous N oxides. For example, postburst rates of NO emission from soil at the two higher water contents included in the incubation experiment were strongly correlated with soil NH_4^+ concentration (R^2 = 0.71 and 0.70 for control and NH_4NO_3-treated soil, respectively). Emission of NO from the field plots was also strongly correlated with soil NH_4^+ concentration (R^2 = 0.69) and much more weakly related to soil NO_3^- concentration (R^2 = 0.34). Conversely, Williams et al.[43] reported that rates of NO emission from several diverse sites in Pennsylvania were more strongly correlated with soil NO_3^- concentration than NH_4^+ concentration. These conflicting results may be partially explained by site differences in the biochemical source of gaseous N oxides; that is, denitrification did not contribute to production of either the NO or N_2O measured in our experiments, but may have contributed to soil N oxide emissions at the Pennsylvania sites. Alternatively, because the correlation reported by Williams et al.[43] included data from diverse ecosystem types, soil NO_3^- concentration may have served simply as a surrogate for the relative abundance of N supply. In this case, the correlation of N oxide emission rates with soil NO_3^- concentrations may have resulted from high nitrification rates rather than high denitrification potential.[19]

Response to Management

Gaseous N oxide emissions from this humid subtropical grassland varied with the type and intensity of management activity. For example, early summer soil emission of NO, but not N_2O, was apparently enhanced by clipping and removing the grass from plots under intensive cultural management (Figure 2a). The lack of response of N_2O emissions to harvest was repeated in other seasons; that is, there were no significant differences between N_2O fluxes measured the day preceding and day following harvests that began any of the 9-week periods. Although the observed enhancement of NO emissions could have resulted from chemical or microbial transformation of N contained in exudates from the cut grass stems or in detritus that fell to the soil surface during harvest,[7] it is unlikely that this source would have persisted more than a day or two in the hot, dry weather that prevailed following harvest. Instead, we believe that the elevated NO emission rates were related to the increase in soil temperature that accompanied eliminating the shading provided by the dense grass canopy on intensively managed plots. Harvesting changed these plots from cooler to warmer than the unharvested plots under minimum cultural management, which would be expected to stimulate activity of the chemoautotrophic nitrifiers responsible for NO production. Other authors[37,43] have reported that NO emission from aerobic soil was extremely sensitive to changes in temperature, apparently much more sensitive than emission of N_2O.

The evolution of both NO and N_2O was stimulated by application of an ammoniacal N amendment. Except during the spring period when both treatments were fertilized similarly, the N_2O flux from field plots under intensive cultural management was significantly higher than from plots under minimum cultural management (Table 2). Other data in Table 2 suggest that the flux of NO was even more strongly influenced. During the 9-week early summer period over which it was measured, 0.69 and 2.37 kg NO–N ha^{-1} were lost from plots under minimum and intensive cultural management, respectively. The difference amounted to 3.2% of the 52 kg N ha^{-1} applied as $(NH_4)_2SO_4$, nearly an order of magnitude greater than the fraction of applied fertilizer lost as N_2O. Reports of such large fractional conversion of fertilizer N to NO_x are atypical, but not absent from existing literature; other examples include 5.4% conversion of urea applied to bare loamy sand in Spain,[37] 11% conversion of NH_4NO_3 applied to bare fine sandy loam in Canada,[36] and 10% conversion of $(NH_4)_2SO_4$ added to sandy loam in an aerobic laboratory soil incubation study.[40] The greater importance of NO, compared to N_2O, emissions as a vehicle for fertilizer N loss from this soil was confirmed during laboratory incubation. For example, the 0.63 mg N kg^{-1} lost as NO from NH_4NO_3-treated soil under the medium water treatment (Figure 5a) comprised about 96% of total gaseous N oxide emissions during 10 d incubation and accounted for 2.0% of the applied NH_4^+–N.

Response to Seasonal Change

Although many soil microorganisms involved in N transformations are believed to remain metabolically active throughout the range of temperatures plotted in Figure 1,[13] soil N_2O emission rates shown in the figure were consistently low during the cooler seasons. Apparently, warmer soil temperature was necessary to allow the existing population to respond to perturbations such as fertilization or rainfall, and especially to a combination of these two. For example, rain immediately followed early spring fertilization, and soil water content remained elevated for several weeks thereafter (data not shown), but NH_4^+ oxidation and the resulting emission of gaseous N oxides proceeded very slowly until soil temperature began to rise after a lengthy period of cool, cloudy weather.

Despite its importance, elevated soil temperature had no effect on gaseous N oxide emission rates unless all other regulators of microbial growth were also favorable. For example, soil N oxide evolution was not substantially influenced by fertilization during the warm early summer period until the next rain dissolved and moved the surface-applied $(NH_4)_2SO_4$ into soil wet enough to support microbial growth (Figure 2a). In addition to its universal requirement by all life processes, water strongly influences both the solution phase and gas phase transport rates of reactants, intermediates, and products of N transformations in soil,[19] and its addition to very dry soil often triggers an unusually large burst of gaseous N oxide evolution. Because of this complexity, seasonal changes in soil water content have great importance, but are difficult to interpret.

Response to Wetting Dry Soil

A large, but short-lived, burst of N oxide emissions not directly related to management, soil N pool sizes, or seasonal temperature change immediately followed wetting very dry soil both in the field and during laboratory incubation. Two examples in Figure 2 were described earlier. The first was the burst of NO evolution from plots under minimum cultural management that coincided with the response of intensively managed plots to the first rain following fertilization. Although no N had been applied to the former plots for over 9 weeks, their surface layer NH_4^+ concentration increased from 4.5 to 8.3 mg N kg^{-1} following the rain on day 14, and NO evolution rose briefly from 11 to 58 g N ha^{-1} d^{-1}. The second example occurred on the last day of the early summer period when NO evolution from plots under both management schemes was enhanced by a similar burst of mineralization/nitrification activity stimulated by 0.8-mm rain that fell on very dry soil just prior to sampling. Although the rain shower was small and did not cause a measurable increase in soil NH_4^+ or NO_3^- concentration, even smaller rains have been shown to have this consequence. For example, Williams and Fehsenfeld[41] measured an immediate tenfold increase in the NO emission rate of very dry native shortgrass prairie in Colorado following precipitation less than 0.3 mm.

During soil incubation in the laboratory a large burst of NO emissions occurred in response to the second addition of water only under the lowest water treatment (Figure 4a), indicating that the occurrence of such a response depended on the degree of dryness attained, rather than the amount of time elapsed, since the previous burst. To elicit a wetting response, it appeared that the soil had to become sufficiently dry to influence the survival or metabolic activity of NO-producing organisms, such as occurred in the low water treatment where both CO_2 and NO emissions approached zero during the last 60 h prior to rewetting. In the medium water treatment where mean soil water content decreased to 3.1% after 7 d of incubation, there was no similar approach to zero by CO_2 or NO emissions, yet there was a tenuous response of NO emissions to rewetting (Figures 4a, 5a). Apparently, desiccation effects on the NO-producing organisms were just beginning to materialize. Soil water content in this case was similar to its preexperiment storage level of 2.8% at which there was a large response of NO emissions to the initial addition of water, thereby suggesting a potential interaction between exposure time and exposure concentration in establishing the conditions prerequisite for supporting a burst of gaseous N oxide emissions following the next addition of water. Alternatively, other environmental limitations on microbial growth (e.g., long-term exposure to low temperature) may have substituted for limited water availability in predisposing the organisms to a burst of gaseous N oxide evolution when favorable growth conditions returned at the start of the experiment.

In contrast to results described earlier, water-induced bursts of N oxide emissions bore no relation to soil NH_4^+ concentration. For example, there were no measurable differences in NH_4^+ concentration among unfertilized soils under the incubation study's three water treatments that might explain differences in their responses to either addition of water (Figure 4a). Nevertheless, the bursts were eliminated by nitrapyrin, indicating involvement by autotrophic NH_4^+ oxidizers. Davidson[11] proposed that the burst of NO emissions induced by wetting dry soil may result from chemodenitrification of NO_2^- in transit from NH_4^+ oxidizers to NO_2^- consumers (primarily *Nitrobacter*). Because soil NO_2^- concentrations measured in our study showed no significant differences among N or water treatments, they add no support for Davidson's[11] hypothesis, but neither do they eliminate this possibility because (i) soil sampling times may not have been appropriate to reveal the dynamics of rapidly changing soil NO_2^- pool sizes and transformation rates, (ii) bulk soil analyses may have provided a poor measure of microsite NO_2^- concentrations near NH_4^+-oxidizer colonies, and (iii) although soil pH decreased in proportion to the amount of nitrification that occurred, even the minimum observed bulk soil pH of 6.9 was high compared to that where chemodenitrification has been shown to be an important process.

Microbial NO : N_2O and NO : NO_3^- Production Ratios

The ratio of NO to N_2O emission rates over the 11 early summer sampling days averaged 7.2 ± 1.7 (SEM); ratios for the individual treatments were not significantly different. Although inadequate analytical sensitivity to N_2O precluded precise determination of the NO : N_2O emissions ratio during soil incubation, its value was even larger. Such high ratios are typical where nitrification is the predominant source of both gases, whether measured in the field,[37] during laboratory soil incubation,[11,22,39] or in pure culture studies.[1,27] Somewhat lower ratios are usually observed where denitrification is also a significant source. For example, soil NO evolution was only about three times greater than N_2O evolution from a tropical rainforest where it was demonstrated that denitrification contributed to N oxide production,[26] and Anderson and Levine[2] found only a twofold difference between annual NO and N_2O emissions from a Virginia corn (*Zea mays*) field where N_2O evolution was detectable only when soil water content approached or exceeded field capacity.

In situations where the availability of O_2 is even more restricted, denitrification would be expected to become the predominant source of gaseous N oxides, resulting in an NO : N_2O emissions ratio usually less than one.[12] Reported exceptions include that NO was the principal denitrification product in soil columns under anaerobic flow conditions,[23] in low-density cell suspensions of *Pseudomonas perfectomarina* that were highly sparged to remove gaseous products,[44] and in a shallow layer of laboratory-incubated soil at −100 kPa water potential where O_2 was removed from the airstream sweeping the incubation jar head space.[40] The condition common to these three datasets that distinguishes them from the body of denitrification literature is that the process was initiated without restricting rapid equilibration of gaseous products with the ambient atmosphere. In the natural soil environment, denitrification generally occurs only when O_2 availability is restricted by high soil water content, which also restricts the diffusion rates of other gases in soil. The resulting increase in time required for NO transport to the soil surface, combined with its instability toward further reduction, allows very little of this gas to escape.[19]

The fraction of NH_4^+ oxidized by chemoautotrophic nitrifiers that is lost to the atmosphere as NO has also been reported to vary with soil gas transport rates.[11,16,32] In our incubation study, NO : NO_3^- production ratios were 0.035, 0.017, and 0.012 for unfertilized soil with initial water contents of 4, 7, and 10%, respectively. When NH_4NO_3 was added, the ratios were higher, but again spanned about a threefold range: 0.053, 0.029, and 0.019, respectively. Because we found no evidence that O_2 was limiting under any of our experimental treatments, we assumed that gas diffusion rates were not severely restricted even under the highest water treatment, so these data stress just how critical is the transport dependence of the NO yield of nitrification in soil. They further suggest that any factor with influence on a soil's gas diffusion rates will also affect both the magnitude and composition of its gaseous N oxide emissions. The transport dependence of NO yield probably reflects the propensity of this gas for rapid oxidation to NO_2^- in aerobic environments, reduction to N_2O in anaerobic environments, and assimilation in N-limited environments.

Summary and Perspective

Soil emission of NO, but not N_2O, was enhanced by early summer harvest of the grass on plots under intensive cultural management, probably because eliminating the shading provided by the dense grass canopy changed these plots from cooler to warmer than the unharvested plots under minimum cultural management, thereby stimulating the activity of organisms responsible for NO production. The evolution of both gases, but especially NO, was strongly stimulated by NH_4^+ fertilization. Except during the spring period when both treatments were fertilized similarly, the N_2O flux from intensively managed plots was significantly higher than from plots under minimum cultural management. Although it was not demonstrated, grazing subtropical grass pastures probably simulates the effect of harvest, and the ensuing deposition of animal waste likely yields an N oxide emissions response analogous to N fertilization.

Because soil N oxide evolution paralleled the nitrification of NH_4^+ applied to the field plots and was virtually eliminated by nitrapyrin in the laboratory incubation experiment, we concluded that the gases were produced by chemoautotrophic nitrifying bacteria. The same organisms were also directly or indirectly responsible for the large, but short-lived, burst of N oxide emissions that immediately followed wetting of very dry soil in both field and laboratory studies. There was no evidence that heterotrophic denitrification contributed to the NO or N_2O emissions we measured, even following large precipitation events in the field or during incubation of the soil at about -10 kPa water potential. Although this observation may not be repeated in soils with finer texture, higher organic C availability, or more compact structure, most subtropical grassland soils are not markedly dissimilar from the well-drained sandy loam at our research site.[15]

Emission of N_2O totaled 0.6 and 1.2 kg N ha^{-1} from plots under minimum and intensive cultural management, respectively (Table 2), somewhat smaller than the high fluxes characteristic of tropical forests.[10] Therefore, if the N cycling processes at our research site could be considered a model for those that might eventually develop on similar well-drained, coarse-textured soils where tropical forests were converted to pasture, high N_2O evolution from this biome might be reduced by the conversion. Elevated N_2O emissions from Amazonian forests converted to pasture[29] appear not to support this possibility; that is, annual fluxes were estimated at 1.9 and 5.7 kg N ha^{-1} for paired tropical forest and pasture sites, respectively. However, these fluxes were measured on wet, infertile oxisols with comparatively fine texture.

During the early summer period over which it was measured, emission of NO at our research site averaged 7.2 times greater than that of N_2O and accounted for 3.2% of the 52 kg N ha^{-1} applied as $(NH_4)_2SO_4$. If the ratio of early summer to annual emissions were the same for NO as was measured for N_2O, annual NO evolution from plots under minimum and intensive cultural management would have totaled about 3 and 8 kg N ha^{-1}, respectively. Emissions of this magnitude would place humid, subtropical grasslands among the largest biogenic NO sources, thereby supporting the hypothesis that savanna regions are the largest contributor to the global soil source of NO_x.[10,24,25]

Highest N oxide fluxes during the unusually dry year at our research site occurred during the warmer seasons immediately after the first rain that followed NH_4^+ fertilization. After a brief peak, the emission rates declined rapidly, but if soil temperature and water content favored active microbial growth, N oxide evolution remained higher than from plots under minimum cultural management until the soil's NH_4^+ concentration returned to prefertilization levels. Like most savanna regions, our site was also subject to multiple intense wetting and drying cycles that enhanced its potential for N oxide evolution via the burst of microbial activity that typically follows wetting very dry soil. Gaseous N oxide emission rates during one of these bursts may be up to 1000-fold higher than those preceding or following the burst, so the quantity of soil N lost during the brief duration of such an event may approach or exceed the total amount emitted during the much longer periods between times that the soil dries enough to support another emissions burst in response to the next addition of water.[42] Data reported here suggest that the bursts make a larger contribution to total NO than N_2O emissions; that is, NO : N_2O emissions ratios measured immediately following precipitation events not recently preceded by fertilization were nearly double the mean ratio over all sampling dates. This observation may also explain the unusually high fractional conversion of NH_4^+ fertilizer to NO_x that we observed.

The episodic nature of soil NO and N_2O emissions, particularly for subtropical grassland and other savanna regions, confounds the estimation of seasonal or annual fluxes by extrapolating measurements made over short time periods (as done for NO in Table 2) or by integrating periodic measurements made over longer time periods (as done for N_2O in Table 2). Better estimates must ultimately depend on simulation modeling, so the long-range goal of soil NO and N_2O exchange measurements must be to capture the exchange rates in terms of their basic physical, chemical, and biological controllers, as we attempted to do in our experiments. However, the parameters that best predict NO and N_2O emissions are likely to depend on the scale of interest. To understand the variation of N oxide emissions at any particular measurement site, for example, the distributions of roots, plant residues, bands of fertilizer, and small-scale topographic features must all be considered because they contribute to determining the local C, N, and O_2 supplies. To understand variation within a landscape, soil texture and drainage class are often useful, presumably as surrogates of O_2 availability. Finally, to improve global budgets of NO and N_2O sources and sinks, global databases of soil orders, climate, and vegetation type are likely to prove useful. Of all the factors known to be related to soil N oxide emissions, N and O_2 availabilities are the most widely applicable.[19] Recognition of the relationships between these important cellular controllers and their field, landscape, and global-scale manifestations reflects the progress made in this area of research. The challenge facing future research is to integrate knowledge of process controls at all scales to better understand contributions of all the Earth's major biomes to the global soil sources of N_2O and NO_x.

Acknowledgments

This work was supported by NASA Grants NAG 2-400 and NAG 2-655 and by USDA Specific Cooperative Agreement No. 58-82HW-7-26 with Prairie View A&M University.

References

1. Anderson, I.C., and J.S. Levine. 1986. Relative rates of nitric oxide and nitrous oxide production by nitrifiers, denitrifiers, and nitrate respirers. Appl. Environ. Microbiol. **51**:938–945.
2. Anderson, I.C., and J.S. Levine. 1987. Simultaneous field measurements of biogenic emissions of nitric oxide and nitrous oxide. J. Geophys. Res. **92**:965–976.
3. Anderson, J.P.E., and K.H. Domsch. 1973. Quantification of bacterial and fungal contributions to soil respiration. Arch. Mikrobiol. **93**:113–127.
4. Anthony, W.H., and G.L. Hutchinson. 1990. Soil cover measurement of N_2O flux: linear vs. nonlinear equations. Agron. Abstr., American Society of Agronomy, Madison, WI, p. 243.
5. Bakwin, P.S., S.C. Wofsy, S. Fan, M. Keller, S.E. Trumbore, and J.M. da Costa. 1990. Emission of nitric oxide (NO) from tropical forest soils and exchange of NO between the forest canopy and atmospheric boundary layers. J. Geophys. Res. **95**:16755–16764.
6. Birch, H.F. 1960. Nitrification in soils after different periods of dryness. Plant Soil **12**:81–96.

7. Bleakley, B.H., and J.M. Tiedje. 1982. Nitrous oxide production by organisms other than nitrifiers or denitrifiers. Appl. Environ. Microbiol. **44**:1342–1348.

8. Bolin, B., E.T. Degens, P. Duvigneaud, and S. Kempe. 1979. The global biogeochemical carbon cycle. In B. Bolin, E.T. Degens, S. Kempe, and P. Ketner (ed.), The Global Carbon Cycle, J. Wiley and Sons, New York, pp. 1–53.

9. Brams, E.A., G.L. Hutchinson, W.H. Anthony, and G.P. Livingston. 1990. Seasonal nitrous oxide emissions from an intensively-managed, humid, subtropical grass pasture. In A.F. Bouwman (ed.), Soils and the Greenhouse Effect, J. Wiley and Sons, Chichester, pp. 481–487.

10. Davidson, E.A. 1991. Fluxes of nitrous oxide and nitric oxide from terrestrial ecosystems. In J.E. Rogers and W.B. Whitman (ed.), Microbial Production and Consumption of Greenhouse Gases: Methane, Nitrogen Oxides, and Halomethanes, American Society for Microbiology, Washington, DC, pp. 219–235.

11. Davidson, E.A. 1992. Sources of nitric oxide and nitrous oxide following wetting of dry soil. Soil Sci. Soc. Amer. J. **56**:95–102.

12. Davidson, E.A. 1993. Soil water content and the ratio of nitrous oxide to nitric oxide emitted from soil, this volume.

13. Focht, D.D., and W. Verstraete. 1977. Biochemical ecology of nitrification and denitrification. Adv. Microb. Ecol. **1**:135–214.

14. Goreau, T.J., and W.Z. de Mello. 1987. Effects of deforestation on sources and sinks of atmospheric carbon dioxide, nitrous oxide, and methane from Central Amazonian soils and biota during the dry season: a preliminary study. In D. Athie, T.E. Lovejoy, and P.M. Oyens (ed.), Proceedings of the Workshop on Biogeochemistry of Tropical Rain Forests: Problems for Research, Centro de Energia Nuclear na Agricultura, Universidade de São Paulo, São Paulo, Brazil.

15. Harris, D.R. 1980. Tropical savanna environments: definition, distribution, diversity, and development. In D.R. Harris (ed.), Human Ecology in Savanna Environments, Academic Press, London, pp. 3–27.

16. Hutchinson, G.L., and C.E. Andre. 1989. Flow-through incubation system for monitoring aerobic soil nitric and nitrous oxide emissions. Soil Sci. Soc. Amer. J. **53**:1068–1074.

17. Hutchinson, G.L., and E.A. Brams. 1992. NO vs. N_2O emissions from an NH_4^+-amended Bermuda grass pasture. J. Geophys. Res. **97**:9889–9896.

18. Hutchinson, G.L., E.A. Brams, G.P. Livingston, W.H. Anthony, Y.P. Chang, and J. Trevino. Microbial, environmental, and management controls on nitrous oxide emission from a Bermuda grass pasture. Submitted for publication.

19. Hutchinson, G.L., and E.A. Davidson. 1993. Processes for production and consumption of gaseous nitrogen oxides in soil. In L.A. Harper, A.R. Mosier, J.M. Duxbury, and D.E. Rolston (ed.), Agricultural Ecosystem Effects on Trace Gases and Global Climate Change, ASA Spec. Publ. no. 55, American Society of Agronomy, Madison, WI (in press).

20. Hutchinson, G.L., W.D. Guenzi, and G.P. Livingston. 1993. Soil water controls on aerobic soil emission of gaseous nitrogen oxides. Soil Biol. Biochem. **25**:1–9.

21. Hutchinson, G.L., and A.R. Mosier. 1981. Improved soil cover method for field measurement of nitrous oxide fluxes. Soil Sci. Soc. Amer. J. **45**:311–316.

22. Hutchinson, G.L., A.C. Tortoso, and W.D. Guenzi. 1988. Comparison of NO vs. N_2O production during aerobic and anaerobic soil incubations. Agron. Abstr., American Society of Agronomy, Madison, WI, p. 218.

23. Johansson, C., and I.E. Galbally. 1984. Production of nitric oxide in loam under aerobic and anaerobic conditions. Appl. Environ. Microbiol. **47**:1284–1289.

24. Johansson, C., H. Rodhe, and E. Sanhueza. 1988. Emission of NO in a tropical savanna and a cloud forest during the dry season. J. Geophys. Res. **93**:7180–7192.

25. Johansson, C., and E. Sanhueza. 1988. Emission of NO from savanna soils during rainy season. J. Geophys. Res. **93**:14193–14198.

26. Kaplan, W.A., S.C. Wofsy, M. Keller, and J.M. da Costa. 1988. Emission of NO and deposition of O_3 in a tropical forest system. J. Geophys. Res. **93**:1389–1395.

27. Lipschultz, F., O.C. Zafiriou, S.C. Wofsy, M.B. McElroy, F.W. Valois, and S.W. Watson. 1981. Production of NO and N_2O by soil nitrifying bacteria. Nature **294**: 641–643.

28. Logan, J.A. 1983. Nitrogen oxides in the troposphere: global and regional budgets. J. Geophys. Res. **88**:10785–10807.

29. Luizão, F., P. Matson, G. Livingston, R. Luizão, and P. Vitousek. 1989. Nitrous oxide flux following tropical land clearing. Global Biogeochem. Cycles **3**:281–285.

30. McLean E.O. 1982. Soil pH and lime requirement. In A.L. Page, R.H. Miller, and D.R. Keeney, (ed.), Methods of Soil Analysis, Part 2, 2nd ed., Agron. Monogr. 9, American Society of Agronomy and Soil Science Society of America, Madison, WI, pp. 199–224.

31. Mosier, A.R., and L. Mack. 1980. Gas chromatographic system for precise, rapid analysis of nitrous oxide. Soil Sci. Soc. Amer. J. **44**:1121–1123.

32. Remde, A., F. Slemr, and R. Conrad. 1989. Microbial production and uptake of nitric oxide in soil. FEMS Microbiol. Ecol. **62**:221–230.

33. Robertson G.P., and J.M. Tiedje. 1987. Nitrous oxide sources in aerobic soils: nitrification, denitrification and other biological processes. Soil Biol. Biochem. **19**:187–193.

34. Rodhe, H. 1990. A comparison of the contribution of various gases to the greenhouse effect. Science **248**:1217–1219.

35. Schimel J.P., M.K. Firestone, and K.S. Killham. 1984. Identification of heterotrophic nitrification in a Sierran forest soil. Appl. Environ. Microbiol. **48**:802–806.

36. Shepherd, M.F., S. Barzetti, and D.R. Hastie. 1991. The production of atmospheric NO_x and N_2O from a fertilized agricultural soil. Atmos. Environ. **25A**: 1961–1969.

37. Slemr, F., and W. Seiler. 1984. Field measurements of NO and NO_2 emissions from fertilized and unfertilized soils. J. Atmos. Chem. **2**:1–24.

38. Sparrow, S.D., and D.T. Masiak. 1987. Errors in analyses for ammonium and nitrate caused by contamination from filter papers. Soil Sci. Soc. Amer. J. **51**:107–110.

39. Tortoso, A.C., and G.L. Hutchinson. 1990. Contributions of autotrophic and heterotrophic nitrifiers to soil NO and N_2O emissions. Appl. Environ. Microbiol. **56**:1799–1805.

40. Tortoso, A.C., G.L. Hutchinson, and W.D. Guenzi. 1986. Nitric and nitrous oxide emissions during nitrification and denitrification in soil. Agron. Abstr., American Society of Agronomy, Madison, WI, pp. 190–191.

41. Williams, E.J., and F.C. Fehsenfeld. 1991. Measurement of soil nitrogen oxide emissions at three North American ecosystems. J. Geophys. Res. **96**:1033–1042.

42. Williams, E.J., G.L. Hutchinson, and F.C. Fehsenfeld. 1992. NO_x and N_2O emissions from soil. Global Biogeochem. Cycles **6**:351–388.

43. Williams, E.J., D.D. Parrish, M.P. Buhr, F.C. Fehsenfeld, and R. Fall. 1988. Measurement of soil NO_x emissions in central Pennsylvania. J. Geophys. Res. **93**:9539–9546.

44. Zafiriou, O.C., Q.S. Hanley, and G. Snyder. 1989. Nitric oxide and nitrous oxide production and cycling during dissimilatory nitrite reduction by *Pseudomonas perfectomarina*. J. Biol. Chem. **264**:5694–5699.

Mechanisms Controlling Methane Emission from Wetland Rice Fields

Ralf Conrad

Max-Planck-Institut für Terrestrische Mikrobiologie, D-3550 Marburg/Lahn, Germany

ABSTRACT

Wetland rice fields are an important source in the global budget of atmospheric CH_4 and, thus, have a significant impact on climate and on atmospheric photochemistry. Methane emission rates from rice fields vary greatly with field site, management, time of day, and season. Field and laboratory studies of CH_4 turnover in paddy soil are reviewed with respect to the mechanisms that control the emission of CH_4 into the atmosphere, i.e., CH_4 production, CH_4 diffusion, CH_4 oxidation, and interaction of CH_4 turnover with nutrients such as nitrogen, iron, and sulfur compounds. Methane production involves a complex anaerobic microbial community that degrades organic matter via various intermediates to CO_2 and CH_4. The rice aerenchyma allows the diffusion of O_2 into the rhizosphere and, thus, provides oxic microsites within the anoxic submerged soil. This allows the oxidation of CH_4 and makes the involvement of other aerobic bacteria in the turnover of CH_4 possible. The rice aerenchyma also provides the predominant route for escape of CH_4 from the soil into the atmosphere and may tap bubbles that constitute CH_4 reservoirs in the submerged soil.

Introduction

Atmospheric methane plays an important role as "greenhouse gas" and as reactant in atmospheric photochemistry[17] and is presently increasing with about 1% per year.[32,63] The reason for the CH_4 increase is probably due to both decreasing sink (oxidation by atmospheric OH) and increasing source strengths.[7] Methane emission from wet rice fields has been identified as one of the most important sources in the budget of atmospheric CH_4. This fact has been acknowledged for years, although there are substantial differences in the actual estimates of the source strength.[3,7,18,20,33,60,61] There is an urgent need to understand the functioning of CH_4 emission from rice fields in greater detail because it may be possible to manage the fields to reduce their impact on the atmosphere.

Field Measurements

Field Sites

Field studies have shown that CH_4 emission rates from rice fields are not uniform but vary greatly with the field site, the management, the time of day, and the season. Intensive field studies have been conducted in California,[8,9] Spain,[62] Italy,[26,57] Japan,[78] China,[34,60] Texas,[53–55] and Louisiana.[38] It should be noted that—with a few exceptions (Japan and China)—most of the studies were not done in the major rice-growing countries. Rice fields in South East Asia and in South Asia which are the most important rice-growing areas have so far not been studied.[39,60] Moreover, there are substantial differences in the basic conditions of wet rice cultivation, for example, rain-fed rice and irrigated rice, which so far have not been considered in studying the CH_4 flux from rice fields.

Management

Methane emission is strongly influenced by management of the flooding water. The field studies generally have shown that CH_4 emission does not start before the fields have been submerged, and stops when the fields are drained for harvest. Sufficient water apparently is necessary to ensure the anoxic conditions necessary for the methanogenic microbial community.

Application of organic material (rice straw, compost, or manure) generally results in enhanced CH_4 emission rates from the treated fields.[2,54,55,57,60,78] However, the fertilization of the fields with mineral nitrogen gives results of effects on CH_4 emission that are somewhat ambiguous. Observations range from stimulatory over neutral to inhibitory effects.[8,37,38,57] Inhibition of CH_4 emission by use of potassium nitrate[37] or ammonium sulfate[57] as fertilizer may be explained by the inhibitory effects caused by enhanced nitrate or sulfate reduction in favor of methanogenesis. However, definite proof that this inhibitory mechanism is really taking place under field conditions is lacking. Moreover, inhibition by ammonium sulfate has only been observed in Italian fields,[57] whereas fertilization of Chinese fields with potassium sulfate showed no effect.[60] The inhibitory effect of urea that was occasionally observed[37,57] has no mechanistic explanation so far. In some cases, fertilized versus unfertilized fields showed reduced CH_4 emission during the early season and enhanced CH_4 emission during the late season, resulting in an overall effect that was zero.[57] The reasons for all these diverse observations are presently unknown and require more detailed knowledge of the mechanisms involved in CH_4 emission.

Seasonal Rhythms

Methane emission rates from flooded rice fields generally change with the season. The seasonal patterns appear to follow some kind of scheme, although there are quite substantial differences in the details. In general, however, the CH_4 emission rates show up to three distinct maxima accompanying the development of the rice plants. Although the mechanisms producing the seasonal maxima of CH_4 emission have not been proven in detail, we may assume that the first maximum is mainly due to decomposition of soil organic matter left from the past season, the second maximum is related to the reproductive growth of the plants, and the last maximum is related to the aging of the plants. One or two of the maxima are sometimes lacking. Furthermore, the relative sizes of the three maxima can be completely different in the various field sites. The reasons for these differences are unknown. More discussion may be found in the literature.[25,53,55,57]

Diurnal Rhythms

Methane emission rates often, but not always, change with time of day. Diurnal changes were substantial in Italian,[26] Chinese,[60] and Texan paddy fields,[55] but were minor or absent in Californian,[8] Spanish,[62] or Japanese paddy fields.[78] In Italy, diurnal changes of CH_4 emission peaked during the afternoon and were more or less related to temperature changes in the soil, although the relative change in temperature was not sufficient to explain the relative change in CH_4 emission rate.[59] In China, however, CH_4 emission rates exhibited their maxima during nighttime for still unknown reasons.[60]

Mechanisms Involved in Methane Emission

There are four different types of mechanisms that may be important for the control of the CH_4 flux from the rice fields to the atmosphere: CH_4 production, CH_4 diffusion, CH_4 oxidation, and interaction of CH_4 turnover with nutrients.

Production of CH_4

The pathways and microbial processes leading to CH_4 production and their controlling mechanisms have recently been reviewed.[10] Production of CH_4 starts some time after submerged conditions have been established. Organic matter in soil is decomposed by microorganisms using in consecutive steps as oxidants the available O_2, NO_3^-, Mn^{4+}, Fe^{3+}, and SO_4^{2-} until CO_2 reduction to CH_4 can begin.[28,45,46,79–81] The time period between flooding and start of CH_4 production thus depends on soil chemical conditions and temperature and varies between a few days[40] and several weeks.[28]

High concentrations of soil organic matter and high temperatures accelerate the production of CH_4, whereas high concentrations of FeOOH, MnO_2, or nitrate delay it. The consecutive reduction processes are paralleled by a decrease of the redox potential (E_h). Under most conditions, the E_h of paddy soils is a function of the Fe^{2+}/Fe^{3+} reaction and depends on the concentration of free Fe^{2+} and the pH. Methane production apparently does not start before most Fe^{3+} is reduced to Fe^{2+}.[28] The redox potential (measured with a platinum electrode) has then usually dropped to values of 0 to −250 mV (−250 mV only in the presence of traces of sulfide) and the pH has reached a steady state around neutrality when CH_4 production starts. This does not exclude, however, that intensive CH_4 production is taking place at a measured E_h of zero.

It should be noted that E_h measurements with a platinum electrode are problematic since the obtained results are not necessarily consistent with the theoretical values calculated from the concentrations of oxidized and reduced species and do not necessarily reflect the redox conditions that are experienced by the microorganisms (for a discussion, see Refs. 4 and 65). Moreover, the molecular basis of a requirement for low redox potentials by methanogens is hardly understood. It is possible that some of the oxidized substrates and ions in soil are able to poise the intracellular electron potential of methanogens in a manner similar to that shown recently for *Propionibacterium freudenreichii*[21,22] or that the redox potential regulates the expression of key enzymes similar as shown for the fumarate reductase of *Escherichia coli*.[71]

Interestingly, the onset of CH_4 production apparently does not depend on the growth of methanogenic bacteria. The numbers of viable methanogenic bacteria that were present in the dry and oxic paddy soil before submergence were as high as those determined during active methanogenesis.[40] Obviously, the methanogens are able to survive the dry and oxic soil conditions, although formation of spores or resting stages has not been observed among these bacteria. The numbers of methanogens were also found to be relatively constant during the entire growing season.[58] Members of the genera *Methanobacterium* and *Methanosarcina* seem to be the most important methanogens in paddy soil.[13,49]

As soon as CH_4 production begins and CO_2 reduction becomes the major electron sink in the anoxic paddy soil, CH_4 production is highly dependent on the availability of degradable organic matter. Addition of organic matter generally has a stimulatory effect on CH_4 production (see above). However, fertilization with uncomposted organic matter (e.g., rice straw) enhanced CH_4 production much more than fertilization with organic matter that had already been composted (e.g., compost).[78] The decomposition pattern of ^{14}C-labeled rice straw shows that 80–90% of ^{14}C is lost during the first year. The decomposition of the residues then follows a logarithmic decrease with a half-life of about 2 years.[43] Hence, it is likely that most of the rice straw or other plant residues have already been decomposed when CH_4 emission peaks during the high and late season (see above).

It is likely that CH_4 production during the high and late season depends on fresh organic matter, e.g., algae in the flood water or rice plant biomass. In fact, a positive correlation was found between CH_4 production and underground biomass.[53] The correlation was better with live than with dead rice plant biomass, indicating that the exudations by roots may be more important than root decay. Sass et al.[53] suggested that CH_4 production depends on the distribution of rice plant roots and spreads downward and outward from the base of the plant as root growth progresses during the season. They also observed a correlation between solar radiation, grain yield, and CH_4 emission and suggested that CH_4 production is somehow coupled to photosynthesis, probably via root exudation.[55]

Unfortunately, we know almost nothing about the molecular composition of root exudates or decay products and about pathways and bacteria involved in their degradation. Field studies in Italian rice fields showed that acetate is the quantitatively most important intermediate in methanogenesis[58] (also Rothfuss and Conrad, in preparation). Acetate concentrations in paddy soil reached millimolar levels. Decomposition of acetate in slurries of paddy soil exhibited turnover times between 8 and 24 h and was exclusively due to methanogenesis.[35,66]

Acetate is predominantly produced by fermentation of organic matter, e.g., carbohydrates. In slurries of Italian paddy soil, ^{14}C-labeled glucose was degraded with a turnover time of 4–16 min.[35] Degradation was caused by fermenting bacteria using the glycolytic pathway. Acetate and CO_2 were the quantitatively most important products of glucose fermentation. Hydrogen, propionate, butyrate, ethanol, succinate, and lactate accounted for less than 1–8% each of the glucose utilized. Acetate and CO_2 were produced in a molar ratio of 2.5 : 1 indicating a substantial contribution of the homoacetogenic type of glucose fermentation. A psychrotrophic homoacetogen, *Acetobacterium carbinolicum*, was isolated from paddy soil and was shown to contribute to the turnover of H_2.[13] Production of ^{14}C–acetate from $^{14}CO_2$ was demonstrated.[66] Hence, chemolithotrophic production of acetate seems to play a small but significant role in addition to fermentative acetogenesis. The degradation of glucose and the flow of carbon as observed in slurries of Italian rice paddy soil is shown in Figure 1.

Next to acetate, H_2/CO_2 is the most important substrate for CH_4 production. Studies with soil sampled in the rice field showed a higher percentage (20–50%) of CH_4 produced from H_2 than laboratory studies with slurries prepared from dry soil lumps[14,58] (5–25%) (Rothfuss and Conrad, in preparation). The reasons are unknown, but the effect may be due to the higher heterogeneity and the larger variability of degradable substrates in fresh field samples than in laboratory slurries. Most of the methanogenically used H_2 is not a free intermediate, but appears to be transferred directly between syntrophic H_2-producing and methanogenic bacteria within microaggregates (for review, see Refs. 11 and 12). As a consequence, CH_4 production would be limited by the supply of fatty acids or alcohols to the H_2-syntrophic methanogenic bacterial associations rather than by the supply of H_2 itself.

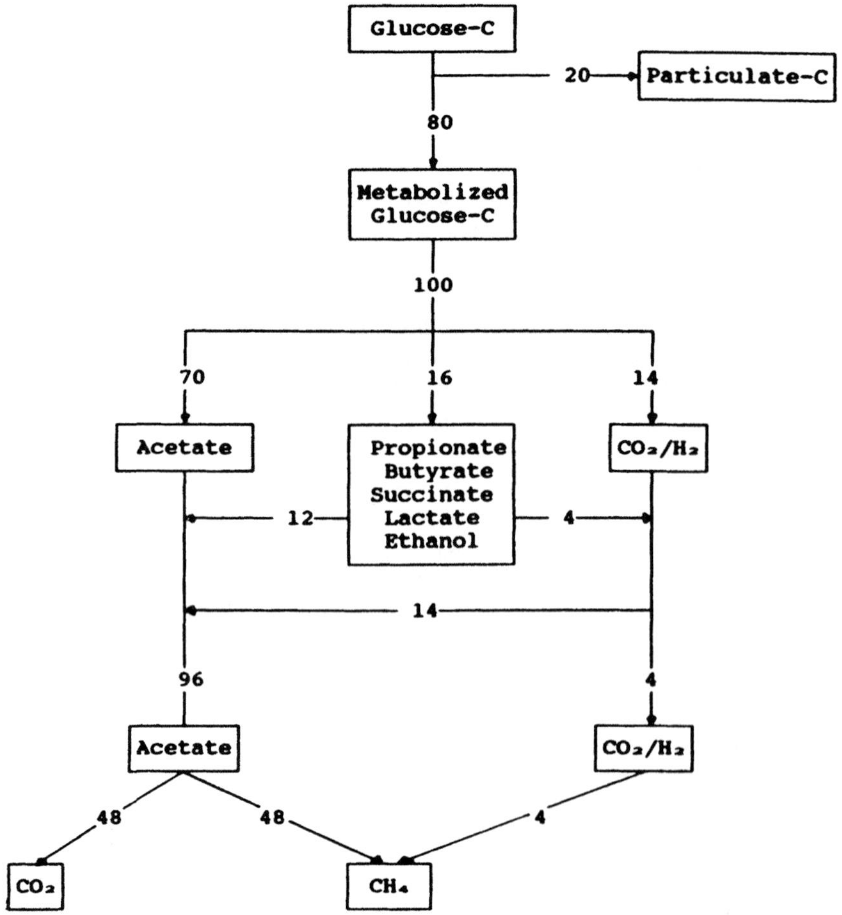

Figure 1. Flow scheme of glucose carbon to CH_4 and CO_2 via different intermediates in slurries of methanogenic paddy soil from Italian rice fields. The percentage distribution of carbon into the different routes was compiled from Krumböck and Conrad.[35]

The pathways leading to the production of CH_4 apparently can be rather complex and, besides the methanogens themselves, involve a large number of different metabolic groups of bacteria. It is the individual microorganism that catalyzes a particular reaction and that is under control of abiotic and biotic variables. We certainly need to learn more about all these microorganisms and their reactions in paddy soil to understand how CH_4 production is controlled in the rice fields. It is likely that large differences exist among the various field sites or within the season.

Diffusion of CH_4

There is no doubt that CH_4 is produced in the anoxic submerged paddy soil and eventually reaches the atmosphere. Three different routes are possible: (i) diffusion across the soil–water interface into the flooding water, and from there across the water–air interface into the atmosphere; (ii) ebullition into the atmosphere after formation of gas bubbles with sufficient buoyancy; (iii) diffusion into the roots followed by diffusion through the aerenchyma of the rice plants and, finally, from the shoots and leaves into the atmosphere.

Field and laboratory studies have consistently shown that CH_4 reaches the atmosphere predominantly via the rice plants.[8,19,24,25,58,62] Diffusion through the flooding water usually is less than 1% of the total flux. Ebullition is usually less than 10%, except when plants are too small to accommodate transport of the produced CH_4.[58] Apparently, diffusion through the plants is the general path of most gases that are released from rice fields into the atmosphere. In addition to CH_4, plant vascular transport was shown to be important for the emission of H_2,[56] CO,[16] and N_2O plus N_2[41] from rice fields. The effect of different varieties of rice cultivars on the efficiency of gas transport is largely unknown.

Without continuous CH_4 production, CH_4 emission will gradually decrease until the produced CH_4 has been depleted. CH_4 can evade as long as a diffusion gradient to the atmosphere exists, i.e., a reservoir of CH_4 in the soil. Gas reservoirs indeed exist in anoxic paddy soil. Entrapment of CH_4 in the paddy soil can easily be observed in the form of gas bubbles. Gas bubbles even occur in densely vegetated plots where enhanced diffusion via the rice plants should be possible. The total amount of CH_4 in the soil, either in a dissolved state or as gas bubbles, increases with season.[38,53] Production rates of CH_4 in the upper 7.5 cm of a rice field in Texas were so high that the CH_4 pore water concentration was turned over within just 2 h.[53] It is quite likely that production of CH_4 may be greater than emission of CH_4 so that CH_4 accumulates in the soil. Over short periods, the CH_4 emission rate would then be a function of the properties of the CH_4 reservoir rather than of CH_4 production itself. Removal of gas bubbles from soil planted with rice indeed resulted in decreased CH_4 emission rates for about 24 h until the originally observed rates were reestablished.[25]

The composition of gas bubbles entrapped in paddy soil is not constant. Mixing ratios of CH_4 in gas bubbles from Italian rice fields varied between <1 and 85%, on the average $28 \pm 20\%$.[56] Inside the plants, CH_4 mixing ratios are about 0.1–1.0%,[24,53] indicating that the concentration gradient between soil and plant's aerenchyma extends over about one order of magnitude. Methane mixing ratios of bubbles recovered from vegetated fields were about half of those recovered from unvegetated fields,[25] indicating that the plants exchange CH_4 against other gases (e.g., O_2 or N_2) or promote oxidation of CH_4 (see below). The gas composition of the bubbles also changes during or after the ebullition event; that is, the gas that is trapped in funnels generally has a lower CH_4 content than the gas that still sits inside the soil.[24]

Soil gas reservoirs in the form of CH_4 bubbles may be important to attenuate the changes in CH_4 pore water concentrations that are caused by changes in CH_4 turnover. This might explain the lack of diurnal changes in CH_4 flux observed at several occasions. In addition, rice roots are able to take up gaseous CH_4,[44] so that gas bubbles which are in direct contact to the roots may facilitate the diffusion of CH_4 from the soil into the plant's aerenchyma (Figure 2). Hence, gas reservoirs may be important for the control of CH_4 flux on a short-term basis. Detailed studies, however, are lacking. On a long-term basis, on the other hand, the fate of the produced CH_4 can only be its release into the atmosphere or its oxidation to CO_2.

Oxidation of CH_4

Oxidation of CH_4 is catalyzed by methanotrophic bacteria which are able to use the reaction energy for growth.[1] Methane can also be co-oxidized by ammonium-oxidizing nitrifying bacteria, although at a lower rate without supporting growth.[27,29,74] Bacteria of both groups are present in paddy soil (Bender and Conrad, in preparation), but their relative importance for CH_4 oxidation in soil is unknown. However, both bacterial groups need O_2 for CH_4 oxidation and, thus, are restricted to the oxic zones within the submerged paddy soil. The importance of anaerobic CH_4 oxidation by still unknown microorganisms has conclusively been demonstrated in anoxic marine sediments and hypersaline water,[7] but not in anoxic freshwater environments, where oxidation of $^{14}CH_4$ to $^{14}CO_2$ seems to be a minor reaction of the methanogenic bacteria.[82,83]

Studies with oxygen microelectrodes in rice field microcosms show that O_2 concentrations decrease to zero within the top 1–2 mm of the paddy soil, but show a second maximum within the root mat at about 20–40 mm depth. Highly resolved profiles of dissolved CH_4 show minimum concentrations in the oxic zones indicating oxidation of CH_4 (Ref. 15; also Frenzel, Rothfuss, and Conrad, submitted for publication). Oxidation of CH_4 was demonstrated in the oxic soil surface layer by measuring the CH_4 flux from soil cores under an oxic (air) versus an anoxic (N_2) gas head space. The flux under anaerobic conditions was equivalent to the rate of CH_4 production, whereas the flux under aerobic conditions was equivalent to the rate of CH_4 production minus that of CH_4 oxidation in the surface layer; about 80% of the CH_4 was oxidized during its passage through this layer.[15]

However, the diffusive flux of CH_4 through the soil surface is of minor importance compared to the flux through the plant vascular system. Here, CH_4 has to pass through the oxic microzones around the roots. At the time being, there is no direct experimental approach to quantify the CH_4 flux to and through

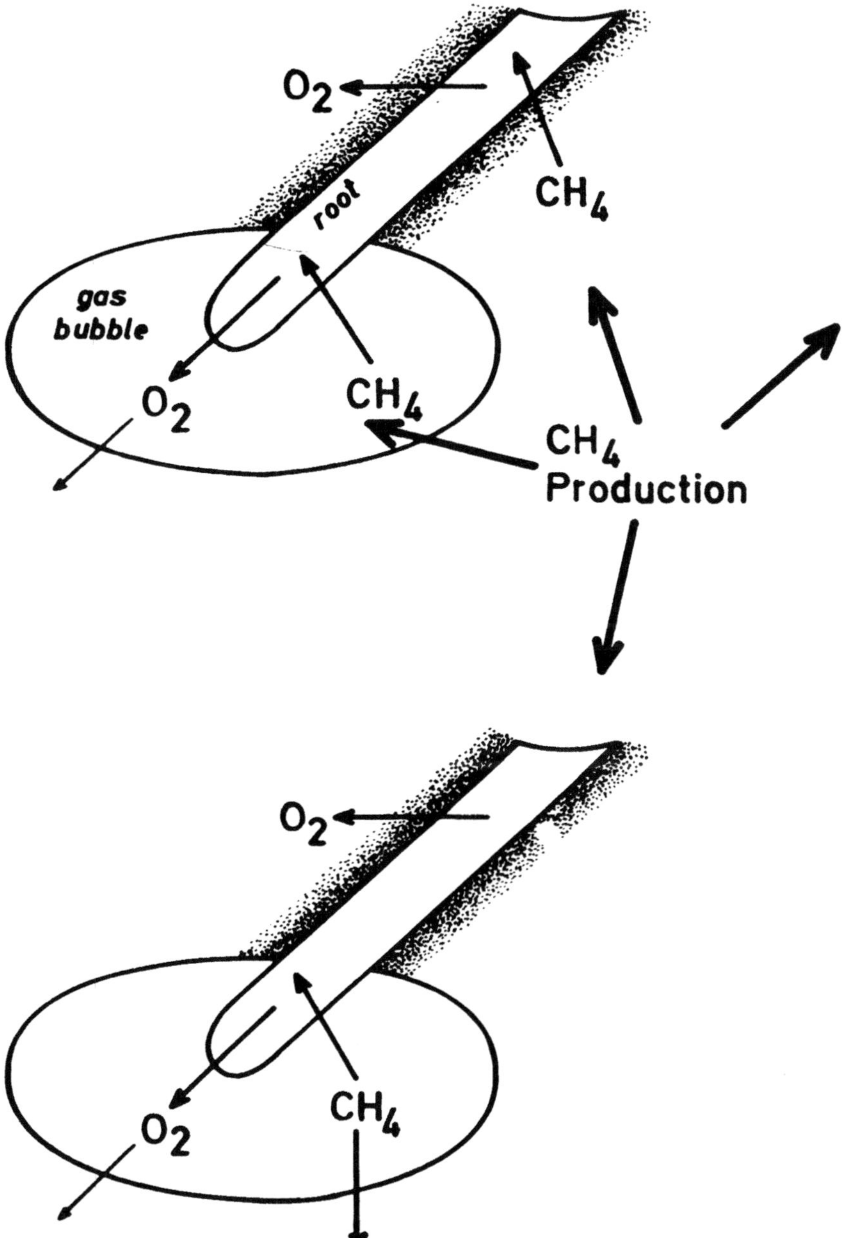

Figure 2. Conceptual model of the function of gas bubbles as CH_4 reservoirs for the emission of CH_4 via plant vascular transport. Gas bubbles arise when large amounts of CH_4 are produced. Some of the bubbles are in contact with the rice roots and allow the diffusion of CH_4 from the bubble into the gas phase of the root aerenchyma (upper scheme). This diffusion still operates and CH_4 emission continues from the bubble reservoir even when CH_4 production in the soil has stopped (lower scheme).

these microzones to assess the actual rates of CH_4 oxidation. Several indirect approaches indicate, however, that substantial amounts of CH_4 are oxidized. Comparison of CH_4 production rates in the soil profile with CH_4 emission rates indicates that 75–90% of the produced CH_4 may be oxidized.[24,58] However, this approach may be biased by overestimating CH_4 production due to sampling artifacts and by imbalances between CH_4 production and emission due to attenuation by possible gas-bubble reservoirs in the soil (see above). Balancing CH_4 accumulation rates (dissolved plus bubbles) within the rice field with CH_4 emission rates indeed resulted in somewhat lower CH_4 oxidation rates accounting for 9–40% of the produced CH_4.[24] Studies on rice microcosms which were incubated under anoxic and oxic atmospheres and under atmospheres containing acetylene as inhibitor of CH_4 oxidation confirmed active CH_4 oxidation indicating that 40–70% of the produced CH_4 were oxidized.[24,25] Measurements of O_2 microprofiles in rice microcosms incubated under light and dark conditions indicated that the O_2 in the rhizosphere originated from vascular transport of atmospheric O_2 rather than from photosynthetic O_2 production (Frenzel, Rothfuss, and Conrad, submitted for publication).

Another approach to study CH_4 oxidation under field conditions is the measurement of $\delta^{13}C$ in CH_4 before and after release from the paddy soil. Although the CH_4 that is released from methanogenic bacteria should be highly depleted in ^{13}C, the CH_4 that is left over by the methanotrophic bacteria should again be enriched in ^{13}C. It was found that $\delta^{13}C$ of organic matter increased by about 7‰ (from −23 to −16‰) with increasing depth in a Philippine rice soil indicating loss of light carbon through methanogenesis.[42] Methane emitted from rice fields is depleted in $\delta^{13}C$, showing mean values of −50 to −67‰.[68,73] These values did not show a significant trend during the season and showed the same depletion of ^{13}C for CH_4 emitted from vegetated and unvegetated soil.[68] The mean $\delta^{13}C$ in the methane emitted from paddy fields was also similar to the values observed from other wetlands, ruminants, termites, or landfills wherever bacterial methanogenesis takes place.[48] The $\delta^{13}C$ of bubble CH_4, on the other hand, tended to slightly increase from about −63‰ to −57‰ during the season.[72] The presently available isotopic data, thus, do not provide clear evidence for CH_4 oxidation. However, isotopic fractionation depends on many factors. In addition to CH_4 oxidation, fractionation could, for example, take place during the transport of CH_4 into the plant's aerenchyma system as indicated by recent observations (Bender, Whiticar, and Conrad, unpublished). Altogether, a systematic study of the isotopic composition of CH_4 along its diffusive path through the soil surface or through the plant's gas vascular system is lacking. The isotopic composition of CH_4 in the soil profile is also unknown. Measurement of δD in addition to $\delta^{13}C$ of CH_4 may significantly increase the information concerning the origin of CH_4 and the transformation processes involved in CH_4 emission.[76,77]

The oxidation of CH_4 in paddy soil is mainly controlled by the availability of

O_2. The influence of plant metabolism on the availability of O_2 in the rhizosphere is not well-known. Stomata opening seems not to be a controlling factor for O_2 transport in aquatic rice plants.[44] Mass flow of air in the submerged organs of rice plants may be driven by differential dissolution of O_2 versus CO_2.[50] The availability of O_2 in the rhizosphere depends on the root aerenchyma volume, the plant age, and plant variety.[69] It also depends on plant photosynthesis, plant respiration and respiration by the microorganisms in the rhizosphere. Particular microorganisms apparently also exert effects by excreting phytohormone-like substances that alter the diffusive resistance of the roots.[70]

Interaction of Nutrients with CH_4 Turnover

The presence of nutrients (e.g., nitrogen, sulfur, or iron) in the paddy soil may influence the turnover and the emission of CH_4 by affecting (i) the rice plants, (ii) the methanogenic microbial community, and/or (iii) the CH_4-oxidizing bacteria.

Nutrients (N, P, K, Fe, etc.) are essential for plant growth. Enhanced plant growth will also result in increased root biomass. Root biomass is positively correlated with CH_4 production (see above). However, increased root biomass may also enhance CH_4 transport through the rice plants' gas vascular system. Together with enhanced photosynthetic O_2 production it may also enhance O_2 transport to the rhizosphere. The nutritional status (e.g., K deficiency) of the rice plants seems to affect the integrity of the epidermal and cortical cells of the roots.[47] However, nothing is known about how CH_4 turnover is affected by the metabolic status of roots.

Methane production in anoxic soil may be affected by nutrients, especially by oxidized ones; for example, CH_4 production is inhibited by nitrate.[5,36] This inhibition may either be due to redox effects on the methanogenic bacteria (see above) or/and be the result of a competition between denitrifying bacteria and members of the methanogenic microbial community for common electron donors (see Ref. 10). Nitrate itself is rapidly depleted in anoxic paddy soil mainly by denitrification and to a smaller extent by nitrate reduction to ammonia.[6] However, nitrate is apparently formed by nitrification in the oxic zones of the soil surface and of the rhizosphere, thus supporting continuous recycling of oxidized nitrogen compounds.[31,51,52] The co-existence of nitrification and denitrification in an anoxic soil with oxic microniches probably results in large temporal and spatial fluctuations in soil nitrate and nitrite concentrations, especially when the rice fields are fertilized with urea or ammonium. It is unknown how CH_4 production is affected by these fluctuations.

Similar to nitrification–denitrification, cycles of Fe^{2+} oxidation–Fe^{3+} reduction, H_2S oxidation–SO_4^{2-} reduction and of other oxidation–reduction cycles probably exist in rice field soil; for example, H_2S-oxidizing *Beggiatoa*[30] and

H_2-oxidizing Knallgas bacteria[23] seem to be associated with rice roots indicating that all kinds of reduced compounds may be oxidized within the oxic zones of the rhizosphere. Consequently, oxidized compounds may be available in the vicinity of roots long after the establishment of reduced conditions in the bulk of the soil. The effects of these compounds, especially of nitrate, nitrite, N_2O, Fe^{3+}, sulfate, thiosulfate, elemental sulfur, etc., on the CH_4-producing microbial community of rice fields are unknown. The role of organic electron acceptors and of oxidation–reduction cycles of organic compounds is also unknown.

Methanotrophic bacteria may also be affected by nutrients. Oxidizable nutrients (NH_4^+, Fe^{2+}, H_2S, S^0) as well as organic substrates will consume O_2 that is required by the CH_4-oxidizing bacteria. Ammonium and a large number of other compounds may inhibit CH_4 oxidation by serving as alternate substrates of the methane monooxygenase[64]; for example, CH_4 oxidation in the oxic surface layer of rice field soil was inhibited after application of ammonium.[15] The interactions between nitrification and CH_4 oxidation in the rhizosphere are so far unknown. The same is true for other oxidizable substrates. A conceptual model of the interaction of nutrients with the turnover of CH_4 in paddy soil is shown in Figure 3.

It should also be remembered that the availability of ammonium is not only a matter of fertilizer treatment or mobilization of organic nitrogen, but is also due to biological nitrogen fixation. It wetland rice fields, N_2 fixation is mainly brought about by two groups of microorganisms: (i) phototrophic free-living or symbiotic cyanobacteria and (ii) anaerobic or microaerobic, heterotrophic bacteria.[75] Free-living N_2-fixing cyanobacteria, e.g., *Anabaena, Nostoc, Plectonema,* etc., and the water fern *Azolla* which lives in symbiosis with *Anabaena* populate the flood water. Management of these N_2 fixers is a well-established practice in rice-growing countries. However, N_2 fixation in association with the rice roots is also taking place.[67] This type of N_2 fixation may be especially important for the nitrification–denitrification in the rhizosphere and, thus, may also affect the turnover of CH_4.

Conclusion

The rice field, especially during the high season when rice plants have grown, with all its possible organic and inorganic substrates, redox cycles, degradation pathways, complex microbial communities, and potential microstructures, provides a challenging subject for research. This research is necessary to understand the CH_4 fluxes on a theoretical background and to provide the basis to manage them in a way that takes care of the atmosphere and possible climatic change. The key problems for mitigation of CH_4 emission from rice fields are (i) how to reduce the production of CH_4 and/or (ii) how to enhance the oxidation of CH_4.

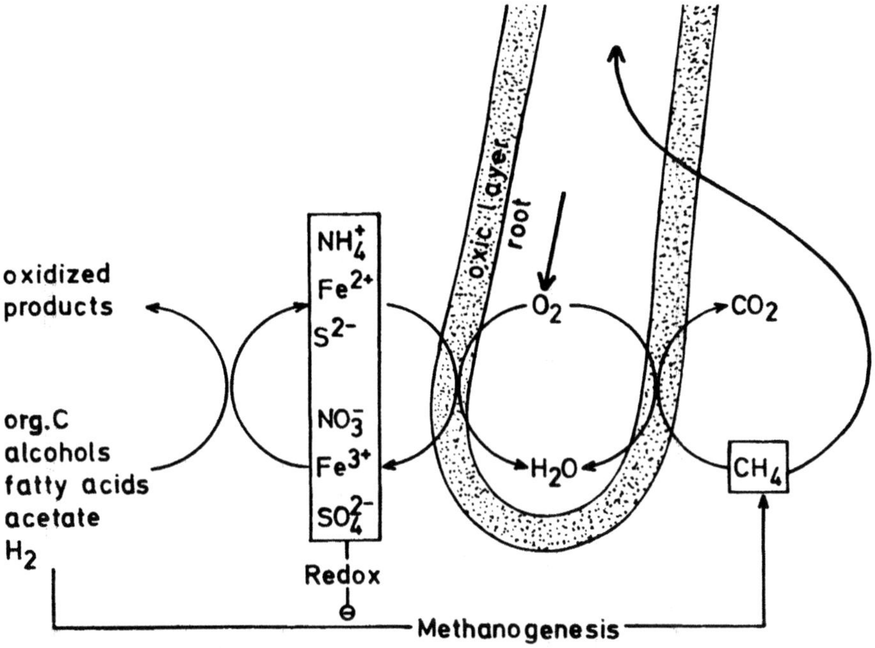

Figure 3. Conceptual model of the interaction of nutrient cycling with the production and oxidation of CH_4 in the rice rhizosphere.

For solving these problems, we especially need detailed knowledge about the interaction of the rice plants with their soil environment, e.g., root exudation, vascular gas transport, and biogeochemical cycles in the rhizosphere.

Acknowledgments

I thank Dr. F. Bak and Dr. P. Frenzel for discussion and critically reading the manuscript. The experimental work of the author was financially supported by the Deutsche Forschungsgemeinschaft and the Fonds der Chemischen Industrie.

References

1. Anthony, C. 1986. Bacterial oxidation of methane and methanol. Adv. Microb. Physiol. **27**:113–210.

2. Araragi, M., and B. Tangcham. 1979. Effect of rice straw on the composition of volatile soil gas and microflora in the tropical paddy soil. Soil Sci. Plant Nutr. **25**:283–295.

3. Aselmann, I., and P.J. Crutzen. 1989. Global distribution of natural freshwater wetlands and rice paddies, their net primary productivity, seasonality and possible methane emissions. J. Atmos. Chem. **8**:307–358.

4. Bartlett, R.J. 1986. Soil redox behaviour. In D.L. Sparks (ed.), Soil Physical Chemistry, CRC Press, Boca Raton, FL, pp. 179–207.

5. Bollag, J.M., and S.T. Czlonkowski. 1973. Inhibition of methane formation in soil by various nitrogen-containing compounds. Soil Biol. Biochem. **5**:673–678.

6. Buresh, R.J., and W.H. Patrick, Jr. 1978. Nitrate reduction to ammonium in anaerobic soil. Soil Sci. Soc. Amer. J. **42**:913–918.

7. Cicerone, R.J., and R.S. Oremland. 1988. Biogeochemical aspects of atmospheric methane. Global Biogeochem. Cycles **2**:299–327.

8. Cicerone, R.J., and J.D. Shetter. 1981. Sources of atmospheric methane: measurements in rice paddies and a discussion. J. Geophys. Res. **86**:7203–7209.

9. Cicerone, R.J., J.D. Shetter, and C.C. Delwiche. 1983. Seasonal variation of methane flux from a California rice paddy. J. Geophys. Res. **88**:11022–11024.

10. Conrad, R. 1989. Control of methane production in terrestrial ecosystems. In M.O. Andreae and D.S. Schimel (eds.), Exchange of Trace Gases between Terrestrial Ecosystems and the Atmosphere. Dahlem Konferenzen, J. Wiley and Sons, Chichester, pp. 39–58.

11. Conrad, R. 1989. Activity of methanogenic bacteria in anoxic sediments: Role of H_2-syntrophic methanogenic bacterial associations. In T. Hattori, Y. Ishida, Y. Maruyama, R.Y. Morita, and A. Uchida (eds.), Recent Advances in Microbial Ecology, Japan Scientific Societies Press, Tokyo, pp. 118–122.

12. Conrad, R. 1993. Anaerobic hydrogen metabolism in aquatic sediments. In D.D. Adams, S.P. Seitzinger, and P.M. Crill (eds.), Cycling of Reduced Gases in the Hydrosphere, Schweizerbart'sche Verlagsbuchhandlung, Stuttgart (in press).

13. Conrad, R., F. Bak, H.J. Seitz, B. Thebrath, H.P. Mayer, and H. Schütz. 1989. Hydrogen turnover by psychrotrophic homoacetogenic and mesophilic methanogenic bacteria in anoxic paddy soil and lake sediment. FEMS Microbiol. Ecol. **62**:285–294.

14. Conrad, R., H.P. Mayer, and M. Wüst. 1989. Temporal change of gas metabolism by hydrogen-syntrophic methanogenic bacterial associations in anoxic paddy soil. FEMS Microbiol. Ecol. **62**:265–274.

15. Conrad, R., and F. Rothfuss. 1991. Methane oxidation in the soil surface layer of a flooded rice field and the effect of ammonium. Biol. Fertil. Soils **12**:28–32.

16. Conrad, R., H. Schütz, and W. Seiler. 1988. Emission of carbon monoxide from submerged rice fields into the atmosphere. Atmos. Environ. **22**:821–823.

17. Crutzen, P.J. 1983. Atmospheric interactions. Homogeneous gas reactions of C, N, and S containing compounds. In B. Bolin and R.B. Cook (eds.), The Major Bio-

geochemical Cycles and Their Interactions, SCOPE 21, Wiley, New York, pp. 67–114.

18. Crutzen, P.J. 1987. Role of the tropics in atmospheric chemistry. In R.E. Dickinson (ed.), The Geophysiology of Amazonia, J. Wiley and Sons, New York, pp. 107–130.
19. DeBont, J.A.M., K.K. Lee, and D.F. Bouldin. 1978. Bacterial oxidation of methane in a rice paddy. Ecol. Bull. (Stockholm) **26**:91–96.
20. Ehhalt, D.H., and U. Schmidt. 1978. Sources and sinks of atmospheric methane. Pageoph **116**:452–464.
21. Emde, R., and B. Schink. 1990. Enhanced propionate formation by *Propionibacterium freudenreichii* subsp. *freudenreichii* in a three-electrode amperometric culture system. Appl. Environ. Microbiol. **56**:2771–2776.
22. Emde, R., and B. Schink. 1990. Oxidation of glycerol, lactate, and propionate by *Propionibacterium freudenreichii* in a poised-potential amperometric culture system. Arch. Microbiol. **153**:506–512.
23. Gowda, T.K.S., and I. Watanabe. 1985. Variation in the incidence of H_2-oxidizing chemolithotrophic bacteria in rice grown under different cultivation conditions. Plant and Soil **85**:97–105.
24. Holzapfel-Pschorn, A., R. Conrad, and W. Seiler. 1985. Production, oxidation and emission of methane in rice paddies. FEMS Microbiol. Ecol. **31**:343–351.
25. Holzapfel-Pschorn, A., R. Conrad, and W. Seiler. 1986. Effects of vegetation on the emission of methane from submerged paddy soil. Plant and Soil **92**:223–233.
26. Holzapfel-Pschorn, A., and W. Seiler. 1986. Methane emission during a cultivation period from an Italian rice paddy. J. Geophys. Res. **91**:11803–11814.
27. Hyman, M.R., and P.M. Wood. 1983. Methane oxidation by *Nitrosomonas europaea*. Biochem. J. **212**:31–37.
28. Inubushi, K., H. Wada, and Y. Takai. 1984. Easily decomposable organic matter in paddy soil. 4. Relationship between reduction process and organic matter decomposition. Soil Sci. Plant Nutr. **30**:189–198.
29. Jones, R.D., and R.Y. Morita. 1983. Methane oxidation by *Nitrosococcus oceanus* and *Nitrosomonas europaea*. Appl. Environ. Microbiol. **45**:401–410.
30. Joshi, M.M., and J.P. Hollis. 1977. Interaction of *Beggiatoa* and rice plant: Detoxification of hydrogen sulfide in the rice rhizosphere. Science **195**:179–180.
31. Katyal, J.C., M.F. Carter, and P.L.G. Vlek. 1988. Nitrification activity in submerged soils and its relation to denitrification loss. Biol. Fertil. Soils **7**:16–22.
32. Khalil, M.A.K., and R.A. Rasmussen. 1983. Sources, sinks, and seasonal cycles of atmospheric methane. J. Geophys. Res. **88**:5131–5144.
33. Khalil, M.A.K., and R.A. Rasmussen. 1990. Constraints on the global sources of methane and an analysis of recent budgets. Tellus **42B**:229–236.
34. Khalil, M.A.K., R.A. Rasmussen, M.X. Wang, and L. Ren. 1991. Methane emissions from rice fields in China. Environ. Sci. Technol. **25**:979–981.

35. Krumböck, M., and R. Conrad. 1991. Metabolism of position-labelled glucose in anoxic methanogenic paddy soil and lake sediment. FEMS Microbiol. Ecol. **85**: 247–256.

36. Laskowski, D., and J.T. Morgan. 1967. The effect of nitrate and nitrous oxide on hydrogen and methane accumulation in anaerobically-incubated soils. Plant and Soil **27**:357–368.

37. Lindau, C.W., R.D. Delaune, W.H. Patrick, and P.K. Bollich. 1990. Fertilizer effects on dinitrogen, nitrous oxide, and methane emissions from lowland rice. Soil Sci. Soc. Amer. J. **54**:1789–1794.

38. Lindau, C.W., W.H. Patrick, R.D. DeLaune, and K.R. Reddy. 1990. Rate of accumulation and emission of N_2, N_2O and CH_4 from a flooded rice soil. Plant and Soil **129**:269–276.

39. Matthews, E., I. Fung, and J. Lerner. 1991. Methane emission from rice cultivation: geographic and seasonal distribution of cultivated areas and emissions. Global Biogeochem. Cycles **5**:3–24.

40. Mayer, H.P., and R. Conrad. 1990. Factors influencing the population of methanogenic bacteria and the initiation of methane production upon flooding of paddy soil. FEMS Microbiol. Ecol. **73**:103–112.

41. Mosier, A.R., S.K. Mohanty, A. Bhadrachalam, and S.P. Chakravorti. 1990. Evolution of dinitrogen and nitrous oxide from soil to the atmosphere through rice plants. Biol. Fertil. Soils **9**:61–67.

42. Neue, H.U., P. Becker-Heidmann, and H.W. Scharpenseel. 1990. Organic matter dynamics, soil properties, and cultural practices in rice lands and their relationship to methane production. In A.F. Bouwman (ed.), Soils and the Greenhouse Effect, J. Wiley and Sons, Chichester, pp. 457–466.

43. Neue, H.U., and H.W. Scharpenseel. 1987. Decomposition pattern of ^{14}C-labeled rice straw in aerobic and submerged rice soils of the Philippines. Sci. Total Environ. **62**:431–434.

44. Nouchi, I., S. Mariko, and K. Aoki. 1990. Mechanisms of methane transport from the rhizosphere to the atmosphere through rice plants. Plant Physiol. **94**:59–66.

45. Ponnamperuma, F.N. 1981. Some aspects of the physical chemistry of paddy soils. In Academia Sinica (ed.), Proceedings of Symposium on Paddy Soil, Science Press-Springer, Beijing, pp. 59–94.

46. Ponnamperuma, F.N. 1985. Chemical kinetics of wetland rice soils relative to soil fertility. In Wetland Soils: Characterization, Classification and Utilization, Proceedings of Workshop IRRI 1984, IRRI, Los Banos, Philippines, pp. 71–89.

47. Prade, K., and G. Trolldenier. 1990. Denitrification in the rhizosphere of rice and wheat seedlings as influenced by plant K status, air-filled porosity and substrate organic matter. Soil Biol. Biochem. **22**:769–773.

48. Quay, P.D., S.L. King, J.M. Landsdown, and D.O. Wilbur. 1988. Isotopic composition of methane released from wetlands: Implications for the increase in atmospheric methane. Global Biogeochem. Cycles **2**:385–397.

49. Rajagopal, B.S., N. Belay, and L. Daniels. 1988. Isolation and characterization of methanogenic bacteria from rice paddies. FEMS Microbiol. Ecol. **53**:153–158.

50. Raskin, I., and H. Kende. 1985. Mechanism of aeration in rice. Science **228**: 327–329.

51. Reddy, K.R., and W.H. Patrick, Jr. 1986. Fate of fertilizer nitrogen in the rice root zone. Soil Sci. Soc. Amer. J. **50**:649–651.

52. Reddy, K.R., W.H. Patrick Jr., and C.W. Lindau. 1989. Nitrification–denitrification at the plant root–sediment interface in wetlands. Limnol. Oceanogr. **34**:1004–1013.

53. Sass, R.L., F.M. Fisher, P.A. Harcombe, and F.T. Turner. 1990. Methane production and emission in a Texas rice field. Global Biogeochem. Cycles **4**:47–68.

54. Sass, R.L., F.M. Fisher, P.A. Harcombe, and F.T. Turner. 1991. Mitigation of methane emissions from rice fields: Possible adverse effects of incorporated rice straw. Global Biogeochem. Cycles **5**:275–287.

55. Sass, R.L., F.M. Fisher, F.T. Turner, and M.F. Jund. 1991. Methane emission from rice fields as influenced by solar radiation, temperature, and straw incorporation. Global Biogeochem. Cycles **5**:335–350.

56. Schütz, H., R. Conrad, S. Goodwin, and W. Seiler. 1988. Emission of hydrogen from deep and shallow freshwater environments. Biogeochem. **5**:295–311.

57. Schütz, H., A. Holzapfel-Pschorn, R. Conrad, H. Rennenberg, and W. Seiler. 1989. A 3-year continuous records on the influence of daytime, season, and fertilizer treatment on methane emission rates from an Italian rice paddy. J. Geophys. Res. **94**:16405–16416.

58. Schütz, H., W. Seiler, and R. Conrad. 1989. Processes involved in formation and emission of methane in rice paddies. Biogeochem. **7**:33–53.

59. Schütz, H., W. Seiler, and R. Conrad. 1990. Influence of soil temperature on methane emission from rice paddy fields. Biogeochem. **11**:77–95.

60. Schütz, H., W. Seiler, and H. Rennenberg. 1990. Soil and land use related sources and sinks of methane (CH_4) in the context of the global methane budget. In A.F. Bouwman (ed.), Soils and the Greenhouse Effect, J. Wiley and Sons, Chichester, pp. 269–285.

61. Seiler, W. 1984. Contribution of biological processes to the global budget of CH_4 in the atmosphere. In M.J. Klug and C.A. Reddy (eds.), Current Perspectives in Microbial Ecology, American Society for Microbiology, Washington DC, pp. 468–477.

62. Seiler, W., A. Holzapfel-Pschorn, R. Conrad, and D. Scharffe. 1984. Methane emission from rice paddies. J. Atmos. Chem. **1**:241–268.

63. Steele, L.P., P.J. Fraser, R.A. Rasmussen, M.A.K. Khalil, T.J. Conway, A.J. Crawford, R.H. Gammon, K.A. Masarie, and K.W. Thoning. 1987. The global distribution of methane in the troposphere. J. Atmos. Chem. **5**:125–171.

64. Stirling, D.I., and H. Dalton. 1979. The fortuitous oxidation and cometabolism of various carbon compounds by whole-cell suspensions of *Methylococcus capsulatus* (Bath). FEMS Microbiol. Lett. **5**:315–318.

65. Stumm, W. 1984. Interpretation and measurement of redox intensity in natural waters. Schweiz. Z. Hydrol. **46**:291–296.

66. Thebrath, B., H.P. Mayer, and R. Conrad. 1992. Bicarbonate-dependent production and methanogenic consumption of acetate in anoxic paddy soil. FEMS Microbiol. Ecol. **86**:295–302.

67. Trolldenier, G. 1987. Estimation of associative nitrogen fixation in relation to water regime and plant nutrition in a long-term pot experiment with rice (*Oryza sativa* L.). Biol. Fertil. Soils **5**:133–140.

68. Tyler, S.C., P.R. Zimmerman, C. Cumberbatch, J.P. Greenberg, C. Westberg, and J.P.E.C. Darlington. 1988. Measurements and interpretation of $\delta^{13}C$ of methane from termites, rice paddies and wetlands in Kenya. Global Biogeochem. Cycles **2**:341–355.

69. Ueckert, J., T. Hurek, I. Fendrik, and E.G. Niemann. 1990. Radial gas diffusion from roots of rice (*Oryza sativa* L.) and Kallar grass (*Leptochloa fusca* L. Knuth), and effects of inoculation with *Azospirillum brasilense* Cd. Plant and Soil **122**: 59–65.

70. Ueckert, J., E.G. Niemann, and I. Fendrik. 1990. Leakage of gases from inoculated and sterile grass roots. Symbiosis **9**:125–128.

71. Unden, G., M. Trageser, and A. Duchene. 1990. Effects of positive redox potentials ($>+400$ mV) on the expression of anaerobic respiratory enzymes in *Escherichia coli*. Molec. Microbiol. **4**:315–319.

72. Uzaki, M., H. Mizutani, and E. Wada. 1991. Carbon isotope composition of CH_4 from rice paddies in Japan. Biogeochemistry **13**:159–175.

73. Wahlen, M., N. Tanaka, R. Henry, B. Deck, J. Zeglen, J.S. Vogel, J. Southon, A. Shemesh, R. Fairbanks, and W. Broecker. 1989. Carbon-14 in methane sources and in atmospheric methane: The contribution from fossil carbon. Science **245**:286–290.

74. Ward, B.B. 1987. Kinetic studies on ammonia and methane oxidation by *Nitrosococcus oceanus*. Arch. Microbiol. **147**:126–133.

75. Watanabe, I. 1986. Nitrogen fixation by non-legumes in tropical agriculture with special reference to wetland rice. Plant and Soil **90**:343–357.

76. Whiticar, M.J., E. Faber, and M. Schoell. 1986. Biogenic methane formation in marine and freshwater environments: CO_2 reduction vs. acetate fermentation—isotopic evidence. Geochim. Cosmochim. Acta **50**:693–709.

77. Whiticar, M.J., E. Faber, and M. Schoell. 1990. A geochemical perspective of natural gas and atmospheric methane. Org. Geochem. **16**:531–547.

78. Yagi, K., and K. Minami. 1990. Effect of organic matter application on methane emission from some Japanese paddy fields. Soil Sci. Plant Nutr. **36**:599–610.

79. Yamane, I., and K. Sato. 1963. Decomposition of organic acids and gas formation in flooded soil. Soil Sci. Plant Nutr. **9**:32–36.

80. Yamane, I., and K. Sato. 1964. Decomposition of glucose and gas formation in flooded soil. Soil Sci. Plant Nutr. **10**:35–41.

81. Yamane, I., and K. Sato. 1967. Effect of temperature on the decomposition of organic substances in flooded soil. Soil Sci. Plant Nutr. **13**:94–100.

82. Zehnder, A.J.B., and T.D. Brock. 1979. Methane formation and methane oxidation by methanogenic bacteria. J. Bacteriol. **137**:420–432.

83. Zehnder, A.J.B., and T.D. Brock. 1980. Anaerobic methane oxidation: occurrence and ecology. Appl. Environ. Microbiol. **39**:194–204.

Mechanism of Methane Transport by Rice Plants

*Isamu Nouchi and Shigeru Mariko**

National Institute of Agro-Environmental Sciences, Kannondai, Tsukuba, Ibaraki 305, Japan

ABSTRACT

To clarify the mechanism of CH_4 transport through rice plants, CH_4 formation in flooded soil and emission through rice plants were examined by pot culture (soil and hydroponic culture) experiments. Methane concentration in surface water above soil was extremely low as compared to that in soil water. Methane concentration in vegetated soil was about one-third of that in unvegetated pots. The emission rate was 20 times higher in vegetated than in unvegetated pots. In addition, there was a linear relationship between CH_4 emission rate from rice plants and CH_4 concentration in the culture solution with a high CH_4 concentration. Methane transport capacity of rice plants depended mainly on plant size. These results indicate that rice plants have a large capacity for CH_4 transport and that rice plants play a primary role for CH_4 flux from paddy fields.

Methane was not emitted from stomata. Air bubbles or droplets of cupric sulfate were observed on leaf sheaths at the lower leaf position by injection of air or cupric sulfate solution into the medullary cavity of rice plants. Micropores which are different from stomata were observed at the abaxial epidermis of leaf sheath by scanning electron microscopy. These results indicate that the micropores are the main site of CH_4 release from rice plants. The present results indicate that the seasonal peak of CH_4 flux from paddy fields results from a combination of CH_4 concentration in flooded soil and rice plant growth since CH_4 formed in the rhizosphere was transported to the atmosphere through the rice body by a physical process.

Introduction

Methane (CH_4) is one of the "greenhouse effect" gases and may be a major participant in future warming of the Earth's surface and atmosphere. The concentration of global atmospheric CH_4 is now 1.7 ppmv.[10] During the last decade,

*Present address: National Institute for Environmental Studies, Onogawa, Tsukuba, Ibaraki 305, Japan.

atmospheric CH_4 has increased by about 1% per year.[2] If the current rate of increase is maintained, future concentrations will reach 2.8 ppmv in the next 50 years.[10] Rice paddy fields are known to be a major source of CH_4 and the area of rice paddy fields in the world averaged over the last 35 years has increased 1.6% per year.[6] Although a full explanation of the increasing concentration of atmospheric CH_4 remains uncertain, an increasing area of rice paddy fields in the world can be an important factor in recent shifts of the atmospheric CH_4 balance. Estimated total annual CH_4 emission from rice paddy fields ranged from 25×10^{12} to 170×10^{12} g/yr.[15] The InterGovernmental Panel on Climate Change (IPCC) reported that paddy fields could contribute one-fifth of the globally generated CH_4.[15] Some researchers proposed that CH_4 emitted from paddy fields is transported mostly through plant bodies rather than by diffusion across the water–air interfaces in bubbles and by molecular diffusion.[3,6,7]

Our present understanding of the mechanism of CH_4 emission from paddy fields is still meager. One of the current objectives in CH_4 emission studies is to establish a strategy for depressing the CH_4 emission from paddy fields. It is, thus, important to elucidate the mechanism of CH_4 transport in those ecosystems. The mechanism for the emission has been reported to consist of CH_4 formation in flooded soil and then CH_4 transport through rice plants from soil to the atmosphere.

Methane emission can be explored by two fundamental approaches: (a) to determine a quantitative relationship between CH_4 formation in flooded soil by methanogenic bacteria under anaerobic conditions and the role of rice plants in CH_4 transport from flooded soil to the atmosphere and (b) to elucidate CH_4 transport processes in rice plants. We have already elucidated several of the processes and factors controlling CH_4 emission.[9,11] First, the CH_4 emission rate from rice plants was directly proportional to CH_4 concentration dissolved in soil water. Second, the main sites of CH_4 release from rice plants were micropores in the leaf sheath rather than stomata in the leaf blade. Third, the CH_4 dissolved in soil water was transported to the atmosphere through the rice plant body by molecular diffusion. The present report is based on our two previous papers concerning CH_4 transport from the rhizosphere to the atmosphere through rice plants[9,11] with an addition of newly obtained results.

Materials and Methods

Plant Culture

Rice seeds (*Oryza sativa* L. cv. Koshihikari) were sown on a floating net in a container of Kimura's B culture solution.[1] When seedlings developed four leaves (about 1 month after sowing), they were transplanted to plastic Wagner pots (200 cm^2 each in soil surface area) containing soil or hydroponic cultures. Each planting regime was three seedlings (a hill) per pot in soil culture and one seedling per pot in hydroponic culture.

Relationship Between CH_4 Emission Rate of Rice Plant and CH4 Concentration in Soil Water in Soil Culture

The pots were filled with air-dried fine gray lowland soil supplemented with 0.5 g N, 0.5 g P_2O_5, 0.5 g K_2O, and 12 g of rice straw per pot on 12 November 1988. They were flooded and then left in the greenhouse under natural light environment for 5 days. Seedlings were then transplanted in four pots out of eight. All the pots were kept in the greenhouse for 8 days. Thereafter, the pots were transported into a naturally lit, environmentally controlled glass chamber, with a temperature of 31°C at 75 ± 5% relative humidity. The experimental regime consisted of four pots with plants and four with no plants. During a growing period of 86 days (until 12 February 1989), several measurements were made for CH_4 concentration in soil water and CH_4 emission rate from each pot. Since the rice plants grew in winter under short-day conditions, they attained the heading at the end of the period.

Relationship Between CH_4 Emission Rate of Rice Plants and CH_4 Concentration in Culture Solution

The apparatus for the measurement of CH_4 emission rate from rice plants and CH_4 concentration in culture solution consists of a 1-L glass vessel for the root zone, with two sampling ports, and a magnetic stirrer.[11] Culture solution with a high CH_4 concentration was prepared by bubbling air with a particular concentration of CH_4 into Kimura's B culture solution for at least 3 h with an air stone placed at the bottom of the vessel. The solution in the vessel was slowly stirred during the measurements. Water samples (10 μl each) were taken with a microsyringe through a silicon septum of the sampling port. A rice plant was placed in the vessel and held in place by sealing at the base of culm with modeling clay and a rubber stopper divided into two pieces.

Measurements for CH_4 Emission Rate

Methane emission rates from rice plants grown in soil and hydroponic cultures were determined with a large acrylic cylindrical chamber (14.5 cm in diameter; 50 cm in height) for soil culture and a small acrylic cylindrical chamber (3 cm in diameter; 50 cm in height) or an acrylic rectangular chamber (22 cm length; 22 cm width; 38 cm height) for culture solution. Ambient air stored in 5-m^3 or 500-L Tedlar bag was continuously introduced into the large cylindrical chamber at a flow rate of 8.0 L min^{-1}, the rectangular chamber at 1.5 L min^{-1}, and the small cylindrical chamber at 1.0 L min^{-1} by a pump. Outlet air from the chambers was collected in a 1-L Tedlar bag for 1 min. At the same time, inlet air was also taken as a control air sample. Methane emission rate was calculated as the difference in CH_4 concentration between outlet and inlet air samples.

All the measurements for CH_4 emission rates were conducted in an artificially lit, environmentally controlled chamber with 15 metal halide lamps (400 W). Environmental temperatures were 25 ± 0.5°C and light levels about 500 µmole m^{-2} s^{-1} at the center of the chamber at plant height level.

Profile of CH_4 Concentration in Surface Water Above Soil and Soil Water in Beakers

The soil in a rice paddy field was an Andosol fertilized with 9.0 g of N, 9.0 g of P_2O_5, 9.0 g of K_2O, and 0.6 kg of rice straw per square meter, and flooded on 3 April 1991. About 1 kg of the flooded soil was sampled from the paddy field, put into three 2-L glass beakers (soil depth: 8 cm) and submerged. Three such beakers containing soil were left 3 months at room temperatures. Water samples (30 µl each) from surface water above soil were collected with a microsyringe at every 1-cm layer from a 6-cm water column above soil surface in the beaker. Soil water (5 µl) was sampled through a Teflon tube connected to the highly permeable filter cups buried horizontally at 1 and 4 cm in the beaker soil.

Collection of Soil Water

Soil water extractors were constructed using 8-mm-diam by 100-mm-long hollow cylinders of highly permeable sintered polyethylene filter cup.[12] The polyethylene filter cups were connected to Teflon tubes (1 mm in internal diameter) and were buried horizontally in the soil. Soil water was extracted based on the siphon theory. The exposed ends of Teflon tubes were capped with stoppers except at the time of sampling.

Application of Abscisic Acid to Rice Plants

Both leaf blades and leaf sheaths of rice plants were treated completely by 1 mM abscisic acid (ABA) solution with a brush and a hand spray. Stomatal resistance and transpiration rates were measured with a super porometer (LI-COR, LI-1600) in the artificially lit, environmentally controlled chamber.

Injection of Air or Cupric Sulfate Solution into the Medullary Cavity of Rice

The measurements of air release from shoots by injecting air with a 50-ml syringe were carried out according to the method of Higuchi et al.[4] and Nouchi et al.[11] A main stem of rice was cut off near the base of roots and submerged. Air was injected into the medullary cavity and release of the air through the shoot was monitored by bubbling of air. Similarly, cupric sulfate solution (20%) was also injected the medullary cavity with no submersion. Cupric sulfate ($CuSO_4$) was used as the injection solution for this experiment because the copper atom is one of the lesser elements in rice tissues and particles of cupric sulfate can be easily detected with an X-ray microanalyzer.

Observation of Surface Structure of Leaf Sheath with a Scanning Electron Microscope

For scanning electron microscopy, some sections of the leaf sheath releasing air bubbles or droplets of cupric sulfate were cut off with a razor blade. The sample pieces were freeze-dried, coated with gold, and examined with a scanning electron microscope (Hitachi, X-650) attached with or without an energy-dispersive X-ray microanalyzer (Horiba, EMAX-2200). Observations and photographs with the scanning electron microscope were made at 15 keV. Line analyses for the element marker of copper by the energy-dispersive X-ray microanalyzer were carried out at an accelerating voltage of 20 keV and with an analysis of 200 s.

Quantification of CH_4 Concentration

Methane concentration was determined by a gas chromatograph equipped with hydrogen flame ionization detectors (Shimadzu, GC-9A), a gas sampler (Shimadzu, MGS-4), and an integrator (Shimadzu, Chromatopac C-R3B). Separations of CH_4 gas were accomplished at 90°C using a glass column (3 mm in diameter, 2 m in length) packed with a molecular sieve 5A (60–80 mesh). The amount of CH_4 dissolved in water was determined by using a similar glass column packed with activated carbon (80–100 mesh). In each measurement, the gas chromatograph was calibrated by using 10 $\mu l\ L^{-1}$ standard gas from a pressurized cylinder.

Statistical Procedures

All statistical comparisons among mean values were made using ANOVA (Scheffe F-test) with a $P < 0.05$ level of significance.

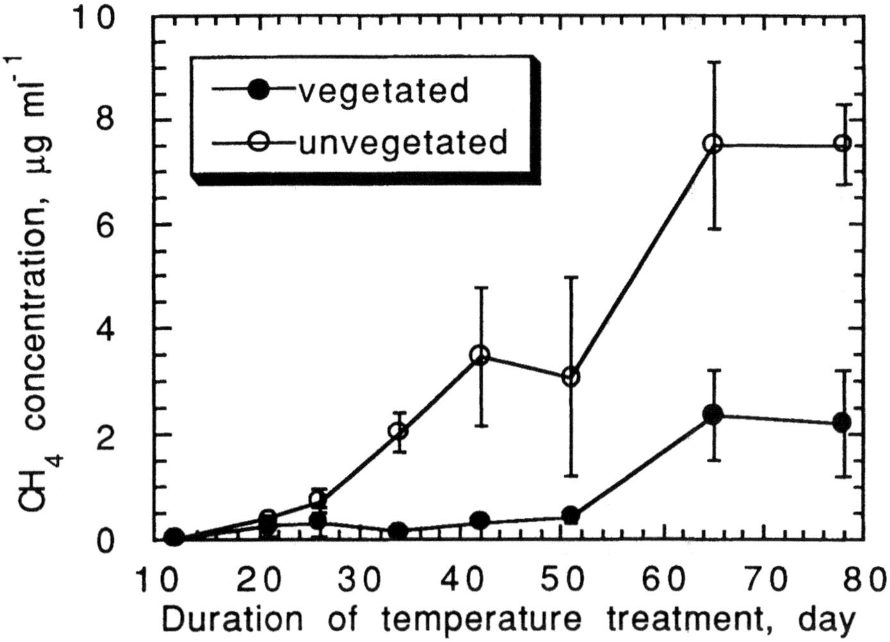

Figure 1. Time course of methane concentration in the water of pot-vegetated rice plants and unvegetated pots for 78 days at 31°C in a naturally lit, environmentally controlled glass chamber. Values are mean and SD of four replicates. Vertical bars smaller than symbols are not shown.

Results

Relationship Between CH_4 Concentration in Soil Water and Rice Plant CH_4 Emission Rate in the Soil Culture

Methane concentrations in the soil water were significantly lower in vegetated pots than in unvegetated pots (Figure 1). The CH_4 concentrations in unvegetated soil were three times higher than those in vegetated soil. Unvegetated pots exhibited considerably reduced CH_4 emission (Figure 2). There was a parallel relationship between CH_4 emission rate and CH_4 concentration in the soil water.

Relationship Between Rice Plant CH_4 Emission Rate and CH_4 Concentration in the Culture Solution

To further clarify the relationship between rice plant CH_4 emission rates and CH_4 concentrations in substrata, CH_4 emission rates were measured from a rice

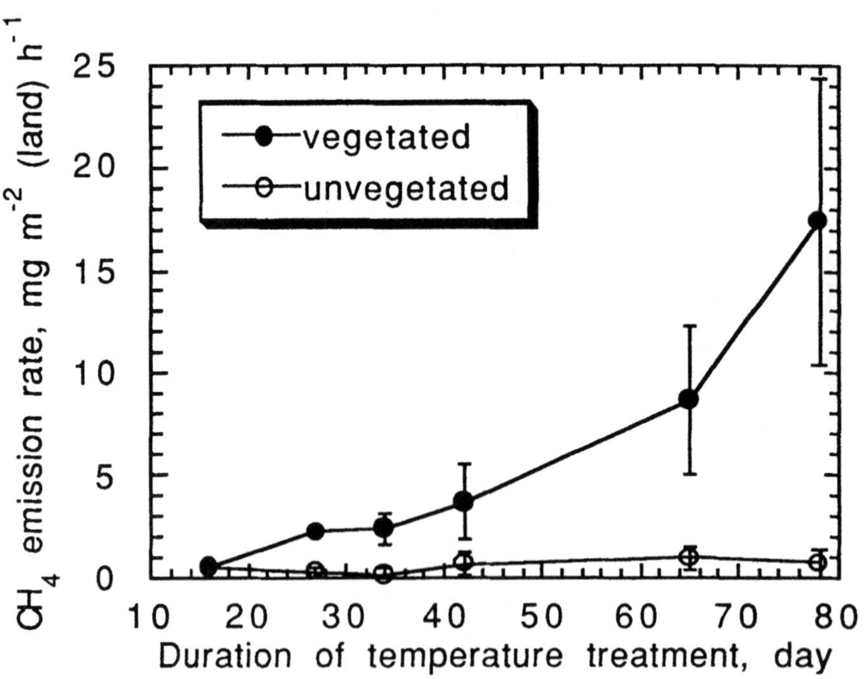

Figure 2. Time course of methane emission rate from pot-vegetated rice plants and unvegetate pots for 78 days at 31°C in a naturally lit, environmentally controlled glass chamber. Values are mean and SD of four replicates. Vertical bars smaller than symbols are not shown.

shoot having three leaves whose root had been exposed to a culture solution with a high CH_4 concentration. Figure 3 shows the maximum values of CH_4 emission rate from a shoot whose roots were exposed to four concentrations of CH_4 in the culture solution (prepared by bubbling 10, 40, 70, and 100% CH_4 gas through the culture solution). There was a linear relationship between CH_4 emission rates and CH_4 concentrations in the culture solution.

Differences in CH_4 Emission Rates Among Rice Plant Sizes

A rice plant with three or nine tillers was placed in the 1-L vessel and covered by the rectangular chamber. Consecutive measurements of CH_4 emission rate from the rice plants and of CH_4 concentration in the culture solutions were determined over about 60 h. Methane emission rate from rice plants decreased almost in parallel with CH_4 concentration in the solution. As a result, there was

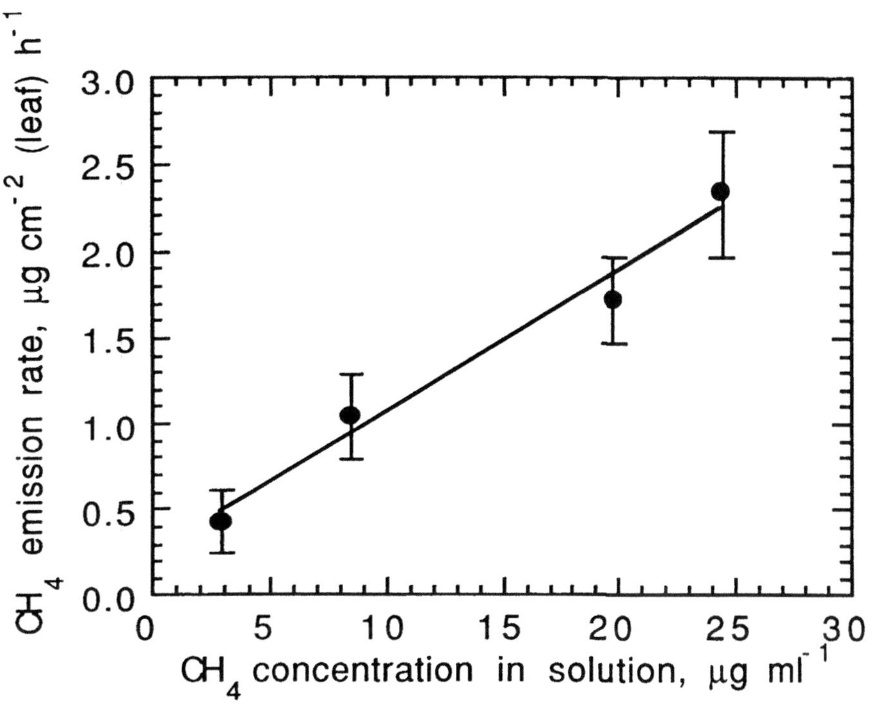

Figure 3. Relationship between methane concentration in water and methane emission rate from rice plants.

a linear relationship between both decreasing patterns, with a high positive correlation coefficient (Figure 4). The CH_4 emission rate from rice plants with nine tillers was larger than those with three tillers.

Profile of CH_4 Concentration in Surface Water Above Flooded Soil and Soil Water

The profiles of CH_4 concentrations in surface water above soil and in soil water were measured on 4 June, 24 June, and 25 July, 1991. The profile of CH_4 concentrations in 2-L beakers had almost the same patterns during each of those three measurements. Figure 5 shows a typical profile of CH_4 concentration in the surface water layer and in the soil water layer on 24 June. The concentration of CH_4 dissolved in soil water was higher at the 4-cm depth than that at the 1-cm depth. On the other hand, CH_4 was almost nonexistent in the surface water layer

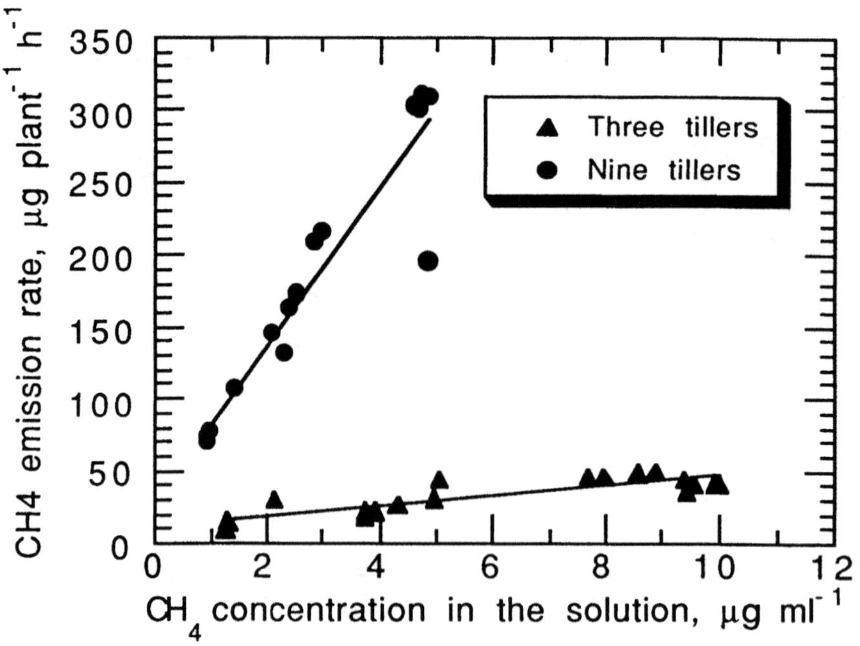

Figure 4. Difference in methane emission rate between rice plant size. A rice plant with three tillers or nine tillers placed in the 1-L vessel and consecutive measures of methane emission rate from rice plants and of methane concentration in the culture solution were determined over 60 h.

above soil. The concentration of CH_4 dissolved at the 4-cm depth in the rice straw-treated soil in the beaker from 4 June to 25 July ranged between 1.3 and 3.4 $\mu g\ ml^{-1}$.

Relationship Between CH_4 Emission Rate and Stomatal Diffusion Resistance by Application of ABA

To clarify the relationship between stomatal opening and CH_4 emission through rice plants, CH_4 emission rates were measured prior to and after applications of ABA. As shown in Table 1, stomatal diffusion resistance increased to 3.1 times and transpiration rate decreased to one-third of the values prior to the treatment by an application of ABA. On the other hand, CH_4 emission rates were almost the same as those prior to the application of ABA.

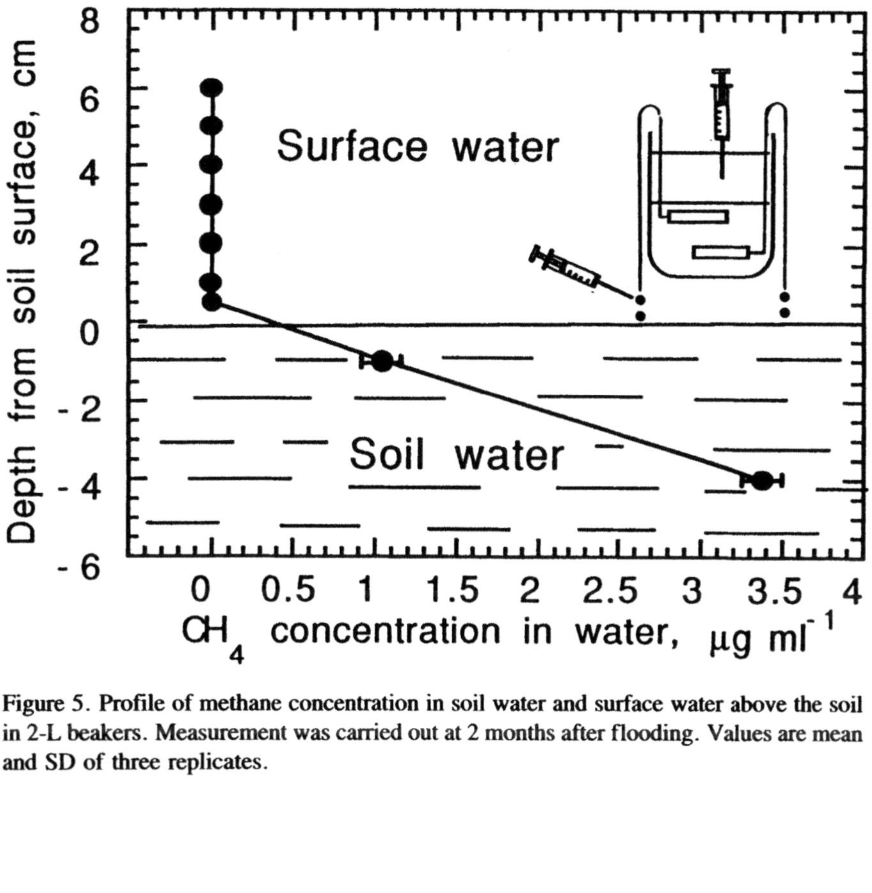

Figure 5. Profile of methane concentration in soil water and surface water above the soil in 2-L beakers. Measurement was carried out at 2 months after flooding. Values are mean and SD of three replicates.

Table 1. Effects of ABA application on CH_4 emission rate, stomatal resistance, and transpiration rate.

Treatment	CH_4 Emission Rate [mg m^{-2} (land) h^{-1}]	Stomatal Resistance (s cm^{-1})	Transpiration Rate [g m^{-2} (leaf) h^{-1}]
Prior to ABA	0.287 ± 0.118^a	3.01 ± 0.09^a	13.9 ± 0.5^a
After ABA	0.264 ± 0.053^a	9.34 ± 1.68^b	5.3 ± 1.2^b

Note: CH_4 emission rate were measured three pots in soil culture. Stomatal resistance and transpiration rate were measured 10 leaf blades at each pot. Values are mean ± SD. The values followed by different letters are significantly different between treatments ($P < 0.05$).

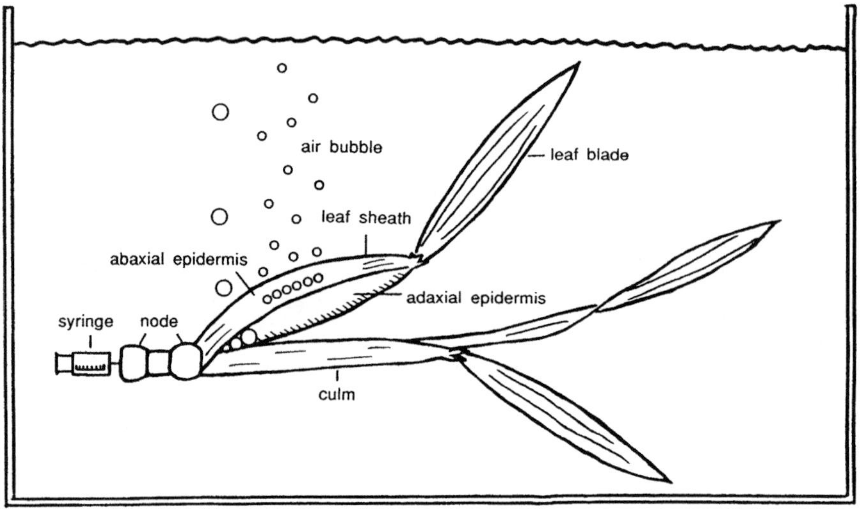

Figure 6. Release of air bubbles from shoots by injection of air into medullary cavity of rice plants.

Site of Appearance of Air Bubbles by Injection of Air into the Medullary Cavity of Rice

As shown in Figure 6, many air bubbles were released from both the abaxial epidermis of the lower portion of leaf sheath and the top gap between the epidermis of culm and leaf sheath. Small air bubbles from the abaxial epidermis of the lower portion of leaf sheath were released in a line. On the other hand, relatively large air bubbles from the top gap were released near the junction of the nodal plate and leaf sheath when the leaf sheath was stripped off carefully from the culm.

Examination of Surface Structure of Abaxial Epidermis of Leaf Sheath with Scanning Electron Microscope

A scanning electron microscope was used to examine the surface structure of the portions of shoot where air bubbles appeared during air injection. Micropores shown in Figure 7 were observed in these portions. Hook-shaped micropores, 4 μm in diameter, were arranged regularly on a vein with a mean interval of about 80 μm from each other. Furthermore, many particles of cupric sulfate were detected with the scanning electron microscope attached to the X-ray microanalyzer only in portions of the vein where the micropores were found (Figure 8).

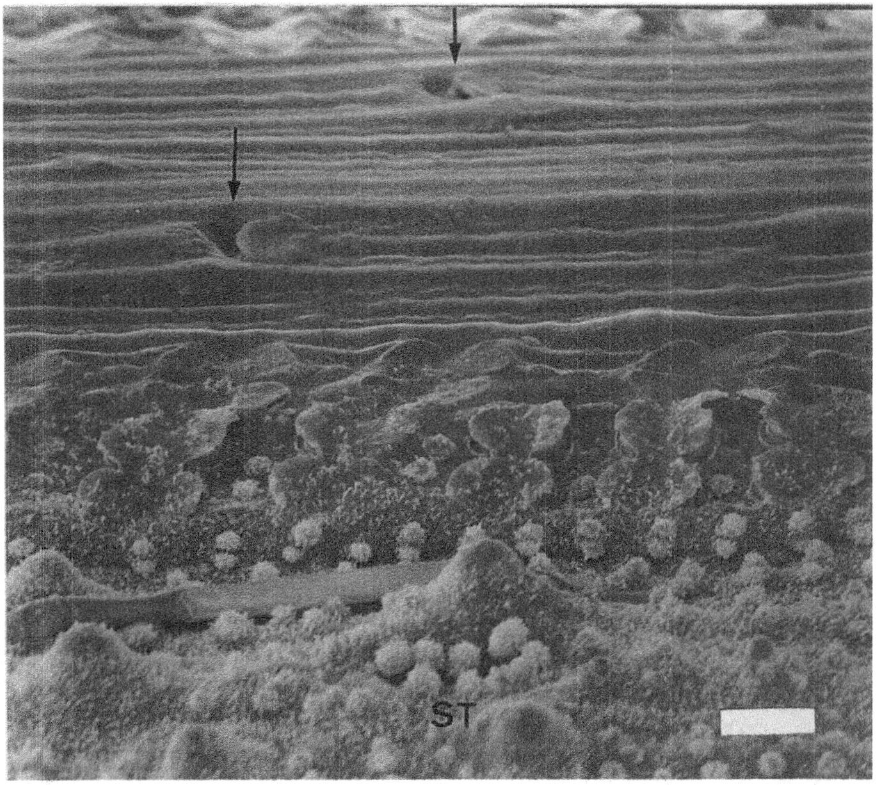

Figure 7. Scanning electron micrograph of micropores which are thought to be the site of methane release in the abaxial epidermis of the lower portion of the leaf sheath. The arrow indicates micropores which exist in a vein. ST, stomata, Scale bar = 10 μm, ×1100.

Discussion

The present results demonstrate that water in flooded soil similar to those found in rice paddies can contain a large amount of CH_4. The maximum concentration of CH_4 dissolved in soil water at 5 cm depth in the pots was 8 $\mu g\ ml^{-1}$ (Figure 1). Holzapfel-Pschorn and Seiler[6] reported that the concentrations of CH_4 in soil water were about 30–100 times higher than those in paddy water above soil surface. We also found that the CH_4 concentration in surface water above the soil was almost negligible, whereas soil water had a large amount of CH_4 (Figure 4). Holzapfel-Pschorn et al.[5] reported that up to 80% of the CH_4 produced in paddy soil does not reach the atmosphere and that most of CH_4 is

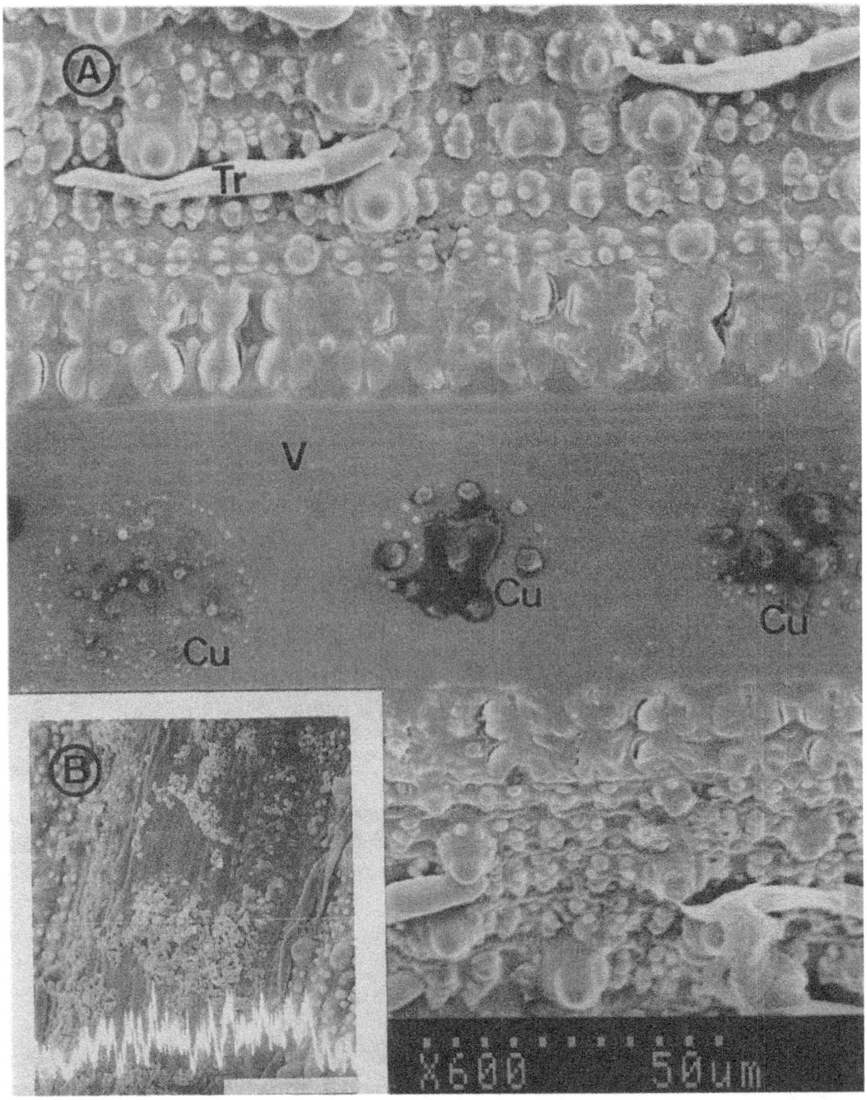

Figure 8. Scanning electron micrograph of particles of cupric sulfate in a vein on the abaxial epidermis of the lower position of the leaf sheath. Many particles of cupric sulfate ($CuSO_4$) were detected with the scanning electron microscope attached to the X-ray microanalizer (B) only on the area of the vein where the micropores existed (A). Line analysis for copper with the energy-dispersive X-ray microanalizer were carried out an accelerating voltage of 20 keV and with an analysis time of 200 s. The spectrum of copper in a fine horizontal line indicates the relative amount of copper (B). Cu, cupric sulfate; V, vein; Tr, trichome. (A) Scale bar = 50 μm, ×600; (B) scale bar = 50 μm, ×700.

oxidized by CH_4-oxidizing (methanotrophic) bacteria whenever the gas passes through an oxic zone, i.e., the oxic paddy surface layer or the oxic zone surrounding the rice roots. We also conclude that CH_4 from paddy fields is mostly emitted to the atmosphere through rice plants since little CH_4 emission from unvegetated pots (Figure 2) and negligible CH_4 concentrations in surface water above soil were found (Figure 5).

Absorption and release sites of gases are generally considered to be the stomata. However, CH_4 emission rates through rice plants were not affected by applications of ABA, which closes stomata, to both leaf blades and leaf sheaths (Table 1). This result indicates that the stomata do not play an essential part in CH_4 release from rice plants and that different pathway and release sites for CH_4 must be explored. Indeed, Nouchi et al.[11] demonstrated, by the chamber method, that CH_4 is mostly released from culm which is an aggregation of leaf sheaths. In addition, many air bubbles were released from both the abaxial epidermis of the lower portion of leaf sheath and near the junction of the nodal plate and leaf sheath at the lower position by air injection into medullary cavity of rice plants (Figure 6). Although the release of air and droplets ($CuSO_4$) from the mentioned area of leaf sheath after pressurizing the system by injection of air or cupric sulfate solution is not physiologically relevant, it is a useful method to detect the leakiest sites for gaseous substances. Furthermore, micropores, which are distinct from stomata, were observed in that portion of leaf sheath with the scanning electron microscope (Figure 7). Many particles of cupric sulfate were also observed only on the area of the veins where the micropores existed (Figure 8). These results indicate the possibility that the micropores are the primary sites of CH_4 release from rice plants. We do not know whether other hydrophytes have such micropores for gas release.

Nouchi et al.[11] have already proposed a mechanism of CH_4 transport through rice plants. First, the dissolved CH_4 in soil water surrounding the roots diffuses into the surface water of the roots, and on to the cell-wall water of the root cortex, driven by the concentration gradient between the soil water surrounding the roots and the lysigenous intercellular spaces in the roots. Methane is then gasified in the root cortex and transported to the shoots via the lysigenous intercellular spaces and aerenchyma. Eventually, CH_4 is released mainly through micropores in leaf sheath at the lower leaf position. The hypothetical pathway of CH_4 transport through rice plants is shown in Figure 9.

Seasonal changes in CH_4 flux from rice-vegetated pots appear to result from changes in rates of soil CH_4 formation (Figure 1). Since CH_4 dissolved in soil cannot easily escape to the atmosphere through overlying surface water, an unvegetated pot accumulates a larger amount of soil CH_4 than a vegetated pot (Figure 1). Methane transport capacity appears dependent on plant size (Figure 4), particularly root size, and is unsaturated over the range of CH_4 concentration observed in the actual paddy soil (Figure 3). These results indicate that rice

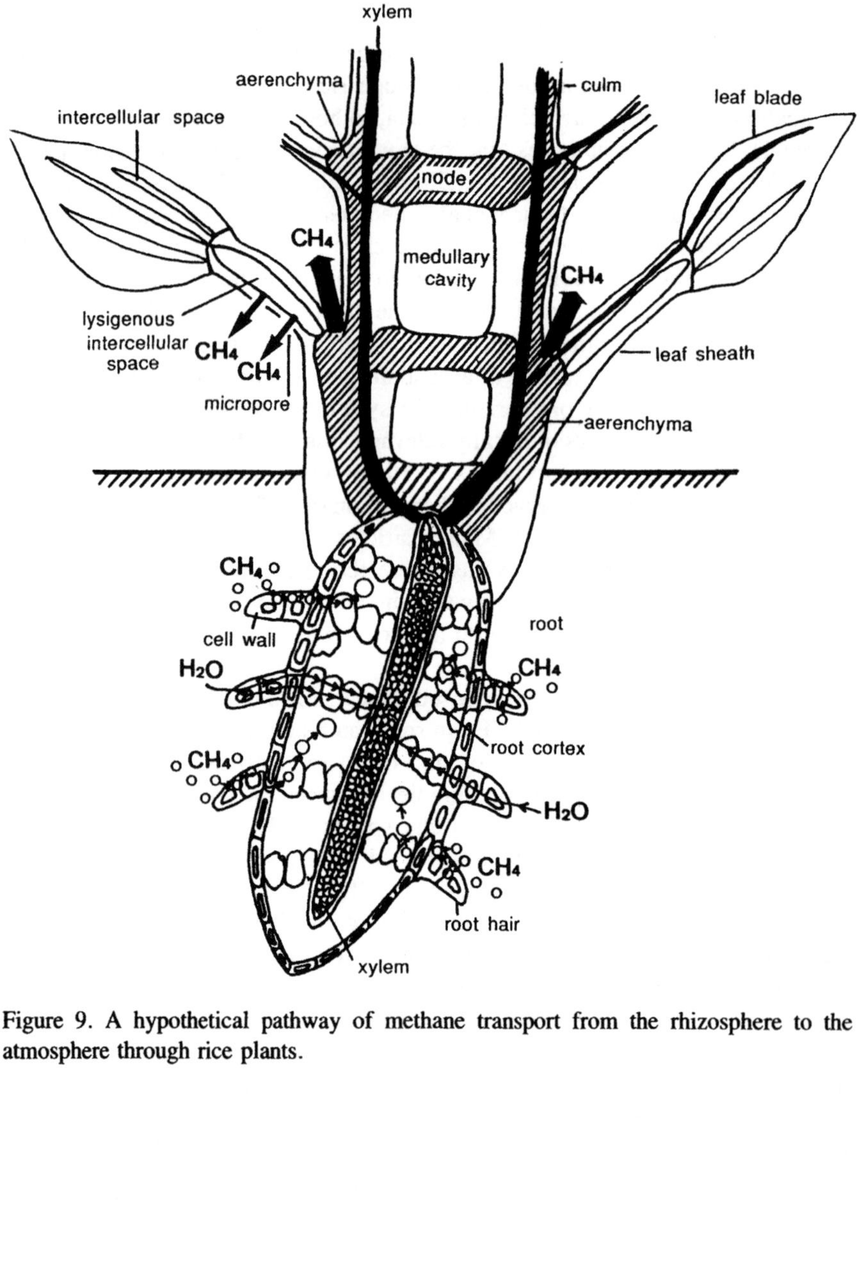

Figure 9. A hypothetical pathway of methane transport from the rhizosphere to the atmosphere through rice plants.

plants play a primary role in controlling the CH_4 flux from paddy fields.[3,6,7] The organic-rich flooded soils support high rates of CH_4 formation. Actual transport rates appear to be controlled by plant size.

Since CH_4 formed in flooded soil is transported to the atmosphere through rice plants by physical processes (molecular diffusion) depending on the concentration gradient between the soil water surrounding the root and the lysigenous intercellular spaces in the roots, no physiological and biological functions in rice plants seem to be involved. Therefore, we must establish techniques to depress CH_4 formation in paddy soil to lower fluxes to the atmosphere. For example, a possible means for depressing CH_4 output from paddy fields is (a) conversion from rice straw to compost application,[9,14] (b) appropriate application of sulfate fertilizer,[8,13] and (c) a better irrigation system.[14]

Acknowledgments

We wish to thank Dr. N. Owa and Dr. Y. Harazono, National Institute of Agro-Environmental Sciences, for their co-operation; Dr. K. Aoki, Tokyo Metropolitan Research Institute for Environmental Protection, for his valuable suggestions; and Dr. Seishi Isobe, National Institute of Agro-Environmental Sciences, for the critical reading of the manuscript.

References

1. Baba, I., and Y. Takahashi. 1956. Water and Sand Culture Methods. In Y. Togari et al. (eds.), Experimental Methods in Crop Science, Association of Agricultural Techniques, Tokyo, pp. 157–185 (in Japanese).
2. Blake, D.R., and F.S. Rowland. 1988. Continuing worldwide increase in tropospheric methane, 1978 to 1987. Science **239**:1129–1131.
3. Cicerone, R.J., and J.D. Shetter. 1981. Sources of atmospheric methane: Measurements in rice paddies and a discussion. J. Geophys. Res. **86C**:7203–7209.
4. Higuchi, T., K. Yoda, and K. Tensho. 1984. Further evidence for gaseous CO_2 transport in relation to root uptake of CO_2 in rice plant. Soil Sci. Plant Nutr. **30**:125–136.
5. Holzapfel-Pschorn, A., R. Conrad, and W. Seiler. 1985. Production, oxidation and emission of methane in rice paddies. FEMS Microbiol. Ecol. **31**:345–351.
6. Holzapfel-Pschorn, A., and W. Seiler. 1986. Methane emission during a cultivation period from an Italian rice paddy. J. Geophys. Res. **91D**:11803–11814.
7. Holzapfel-Pschorn, A., R. Conrad, and W. Seiler. 1986. Effects of vegetation on the emission of methane from submerged paddy soil. Plant and Soil **92**:223–233.

8. Hori, K., K. Inubushi, S. Matsumoto, and H. Wada. 1990. Competition for acetic acid between methane formation and sulfate reduction in the paddy soil. Nippon Dojo-Hiryo gaku Zasshi (Japan. J. Soil Sci. Plant Nutr.) **61**:572–578 (in Japanese with English summary).

9. Mariko, S., Y. Harazono, N. Owa, and I. Nouchi. 1991. Methane in flooded soil water and the emission through rice plants to the atmosphere. Environ. Exp. Bot. **31**:343–350.

10. Mitchell, J.F.B. 1989. The "greenhouse" effect and climate change. Rev. Geophys. **27**:115–139.

11. Nouchi, I., S. Mariko, and K. Aoki. 1990. Mechanism of methane transport from the rhizosphere to the atmosphere through rice plants. Plant Physiol. **94**:59–66.

12. Suzuki, D. 1982. Improvement of highly permeable sintered polyethylene filter cup for soil water and soil air. Nippon Dojo-Hiryo gaku Zasshi (Japan. J. Soil Sci. Plant Nutr.) **54**:253–254 (in Japanese).

13. Tauer, R., K. Jungerman, and K. Decker. 1977. Energy conservation in chemotrophic anaerobic bacteria. Bacteriol. Rev. **41**:100–180.

14. Yagi, K., and K. Minami. 1990. Effects of organic matter application on methane emission from some Japanese paddy fields. Soil Sci. Plant Nutr. **36**:599–610.

15. Watoson, R.T., H. Rodhe, H. Oeschger, and U. Siegenthaler. 1990. Greenhouse gases and aerosols. In J.T. Houghton, G.J. Jenkins, and J.J. Ephraums (eds.), Climate Change: The IPCC Scientific Assessment, Inter-Governmental Panel on Climate Change, Cambridge University Press, New York, pp. 1–40.

Spatial and Temporal Variations of Methane Flux from a Rice Paddy Field

Kazuyuki Yagi and Katsuyuki Minami

Department of Environmental Management, National Institute of Agro-Environmental Sciences, Tsukuba 305, Japan

ABSTRACT

Methane (CH_4) flux from paddy fields to the atmosphere were studied in a Japanese paddy field for three cultivation periods by using a closed-chamber method. Spatial variations of CH_4 flux, having CV range from 19 to 47%, were observed within the individual plots when a chamber having a cross-sectional area of 0.09 m^2 was used. CH_4 fluxes showed a diurnal variation which closely correlated with soil temperature in the surface layer. The flux was doubled with the increase in the soil temperature from 25 to 35°C at the depth of 2 cm. Pronounced seasonal variations of CH_4 flux were observed during cultivation periods. The variations of CH_4 flux during a cultivation period were related to the change of soil E_h at a depth of 2 cm. Annual emission rates of CH_4 from the mineral (control) and the rice straw plot were in the ranges 8.2–40.6 and 27.0–43.1 g m^{-2}, respectively, in the three cultivation periods. The soil E_h had a significant influence on the annual variation of the CH_4 flux, resulting in large interannual variations of CH_4 emission. These results indicate that variation of the redox potential in surface soil layer, which results from water management, has a strong influence on CH_4 emission from paddy fields.

Introduction

The concentration of atmospheric methane (CH_4) has been increasing at about 1% per year in recent years.[2,19] CH_4 has a strong influence on the atmospheric chemistry.[21] Because of its infrared absorption spectrum, CH_4 is also an important greenhouse gas in the climate system. The rapid increase, therefore, could have significant environmental consequences.[14,22]

Atmospheric CH_4 is produced both by natural and by anthropogenic processes.[5] The increasing input of CH_4 from the anthropogenic sources into the atmosphere, which combined with a possible decrease in the concentration of tropospheric OH radical, is considered to yield the observed increase in the atmospheric CH_4. However, the quantitative importance of each of the factors contributing to the observed increase is not well-known at present.[23]

Of the wide variety of sources for the atmospheric CH_4, rice paddy fields are considered as an important source because of the recent increase in their harvest area in the world.[7] Cicerone and Oremland[5] estimated the global emission rate from paddy fields to range from 60 to 170 Tg yr^{-1}, which accounts for about 10–30% of the total emission from all sources. This figure is mainly based on the field measurements of CH_4 flux from paddy fields in California,[3,4] Spain,[18] and Italy.[8] Additional studies in Italy[16] and in other locations[9,13,15,24] have been conducted since these estimates were made by Cicerone and Oremland. The measurements in various location of the world show that there are large temporal variations of CH_4 flux and that the flux differed markedly with soil types or application of organic matter and mineral fertilizer. The wide variations in CH_4 flux indicate that the flux is critically dependent on several factors, including climate, characteristics of soil and paddy, and agricultural practices. Therefore, to revise the estimation of the global CH_4 emission from paddy fields, we need to determine these parameters as well as to gather the *in situ* data over the world.

Here, we report the results of field measurements of CH_4 flux from a Japanese paddy field over 2 years. First, we show the spatial variabilities of the flux in uniform plots to evaluate size of chamber and replication of the measurements. Then, we examine factors influencing temporal variations of CH_4 flux. In a previous article,[24] we have examined the effect of organic matter application on the flux in this paddy field. Here, we focus on effect of changes of temperature and redox potential in soil on temporal variation of CH_4 flux from paddy field.

Study Site and Methods

Study Site

Field study was carried out in an experimental paddy field of the Ryugasaki Branch of the Ibaraki Agricultural Experiment Station (at latitude 35°61′N, longitude 140°13′E), located in part of the intensive rice cultivation area in the alluvial lowland of Tone River. The soil of the paddy field is classified as Gley soil (Eutric Gleysols). Two plots, the mineral and the rice straw plot, with an individual size of 100 m^2 were used for the flux measurements. At these plots, a long-term field experiment on fertilization including mineral fertilizer with/without rice straw has been performed since 1978. Percolation rates of these plots were about 10 mm d^{-1}. Soil samples were collected from each plot in April 1989. Some physical and chemical properties of the soils are listed in Table 1.

Rice (*Oryza sativa*, type *japonica*, cultivar Koshihikari) was cultivated according to conventional methods of the area. To the rice straw plot, rice straw from the previous growing season was cut and plowed back at a rate of 600 g m^{-2}. Prior to flooding, the compound mineral fertilizer (ammonium–

Table 1. Properties of the soil collected before flooding in 1989.

	Mineral Plot (Control)	Straw Plot
Texture	SCL	SCL
pH (H_2O) (10 : 25)	5.9	6.1
Total C (%)	1.36	1.60
Total N (%)	0.13	0.15
C/N	10	11
Readily mineral. C (mg/g)	0.118	0.180
Free Fe (mg/g)	8.67	8.54

potassium–chloride–phosphate) was incorporated into both the mineral (control) and the rice straw plots at rates of 6.0 g N m^{-2}, 6.0 g P_2O_5 m^{-2}, and 6.0 g K_2O m^{-2}. The fields were flooded on 5 May 1989 and 8 May 1990. Rice seedlings at the three to four leaf stage were transplanted on 10 May, spaced at 15 × 30 cm. Paddy water level was maintained between 0 and 10 cm after flooding until the middle of June. At the end of June, the paddy was drained for 3–5 days. This practice, known as midsummer-drainage, is commonly performed over Japan for supplying molecular oxygen to the root of the rice plants. After that, the irrigation water was supplied every 3–5 days until the maturing stage of rice plants. The mineral fertilizer at rates of 3.0 g N m^{-2} and 3.0 g K_2O m^{-2} was topdressed to both of the plots about 20 days prior to heading of the rice plants. The field was usually drained about 20 days before harvest, but the mineral plot was drained just before harvest in 1990. The rice was harvested on 20 September 1989 and 10 September 1990, yielding about 550 g m^{-2}. There were no significant differences in the growth of rice plants and yield between the mineral and the rice straw plot.

Flux Measurements

The closed-chamber method was used for the measurements of the CH_4 flux from paddy fields into the atmosphere.[24] The chamber and the sampling system is schematically illustrated in Figure 1. We used different sizes of open-bottom chambers for the measurements. One has cross-sectional area of 0.09 m^2 and height of 1.0 m. This chamber covers two rice plants. Another has 0.36 m^2 and 0.5 or 1.0 m and covers eight rice plants. The chambers were constructed from polycarbonate plates. Two polyvinylchloride angle bars (1.0 m long) are attached to the bottom of the chambers. Below the ceiling of the chamber, a fan

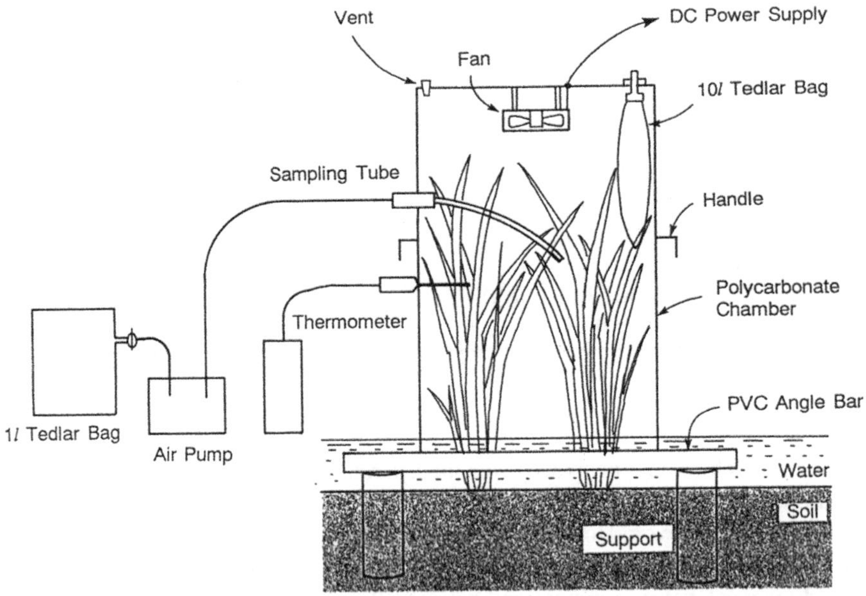

Figure 1. Schematic illustration of chamber and sampling system for the measurements of CH_4 flux from paddy fields.

and a 10-L Tedlar Bag (polyvinyl fluoride bag) were attached for mixing the air and for regulating pressure inside the chamber, respectively.

The CH_4 flux was determined by measuring the temporal increase of the CH_4 concentration of the air inside the chamber during 30 min. The measurements were usually performed once a week throughout the flooding periods. Two to five sampling sites were selected in each plot at random locations just after transplanting. At each site, four permanent supports for the chamber were inserted in soil to avoid perturbation of the surface paddy soil. At the start of the measurement, the chamber was installed on the sampling site with the PVC bars placing on the supports. The air inside the chamber was drawn through sampling tube by an air pump (Shibata, MP-2N) at a flow rate of 2 l/min, and 1 L of the air was collected in a 1-L Tedlar Bag. Air sampling was usually performed twice (15 and 30 min) after the first sampling to detect possible bubbling transportation of CH_4 from paddy soil. All the measurements were performed between 10 A.M. and 1 P.M. except for the measurements of diurnal variation of the flux.

The CH_4 concentration of the samples collected in the Tedlar Bags was determined by using a gas chromatograph (Shimadzu GC-12A) equipped with a gas sampler (Shimadzu MGS-5) and a flame ionization detector. Five milliliters of air sample was injected and separated on a molecular sieve 5A column (2.6 mm inside diameter × 2.0 m) with 40 l/min N_2 of carrier flow at 70°C. Calibration was performed by using CH_4 standard gases at concentrations of 10 and 100 ppmv (Takachiho Kogyou Co., Ltd.).

Temperature and E_h Measurements

Temperature of air and paddy soil at depth of 0, 2, 5, and 10 cm was continuously recorded during the cultivation periods by using Pt-resistance thermometers. The redox potential (E_h) of paddy soil was measured by using Pt-tipped electrodes (Hirose Rika Co., Ltd.) and an ORP meter (Toa RM-1K). For the measurements of soil E_h, the electrodes were inserted into the soil at depth of 2 and 5 cm, and kept in place throughout the cultivation periods. The potential measured against AgCl reference electrode was transformed into that against a hydrogen electrode. All the E_h measurements were made in triplicate in each depth.

Results

Spatial Variation of CH_4 Flux

Spatial variability of CH_4 flux was measured by using a chamber having cross-sectional area of 0.09 m^2. Figure 2 shows the variability in the mineral and the rice straw plots. The fluxes from three to five different sites in a plot were measured on 27 June, 11 July, and 9 August 1989. There were large spatial variations of the CH_4 flux. The coefficient of variation (CV) ranged from 19 to 47%. The temporal increase of CH_4 concentration inside the chamber was almost linear in these measurements, suggesting little contribution of bubbling transportation of CH_4. When a chamber with 0.36 m^2 was used, CV did not exceed 28% in 63 flux measurements during the whole cultivation periods over 2 years, except when bubbling transport was observed during the early stage of rice growing.

Diurnal Variation of CH_4 Flux

Several diurnal measurements of CH_4 flux were performed in the cultivation periods. Figure 3 shows the data obtained during 17–18 July 1990, along with the variation of soil temperature. Although the CH_4 flux from the rice straw plot was two times higher than that from the mineral plot at all times during that

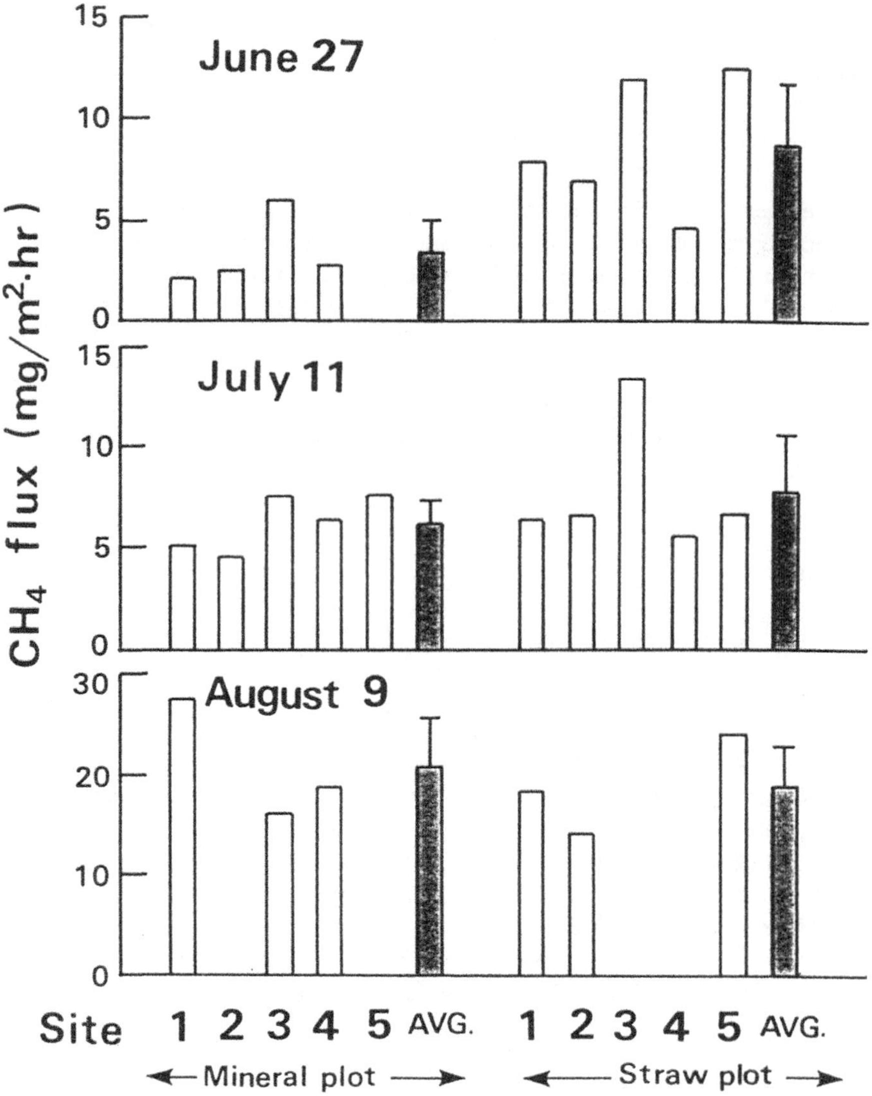

Figure 2. Spatial variations of CH_4 flux from the mineral and the rice straw plots of Ryugasaki paddy field measured by using a chamber having a cross-sectional area of 0.09 m^2.

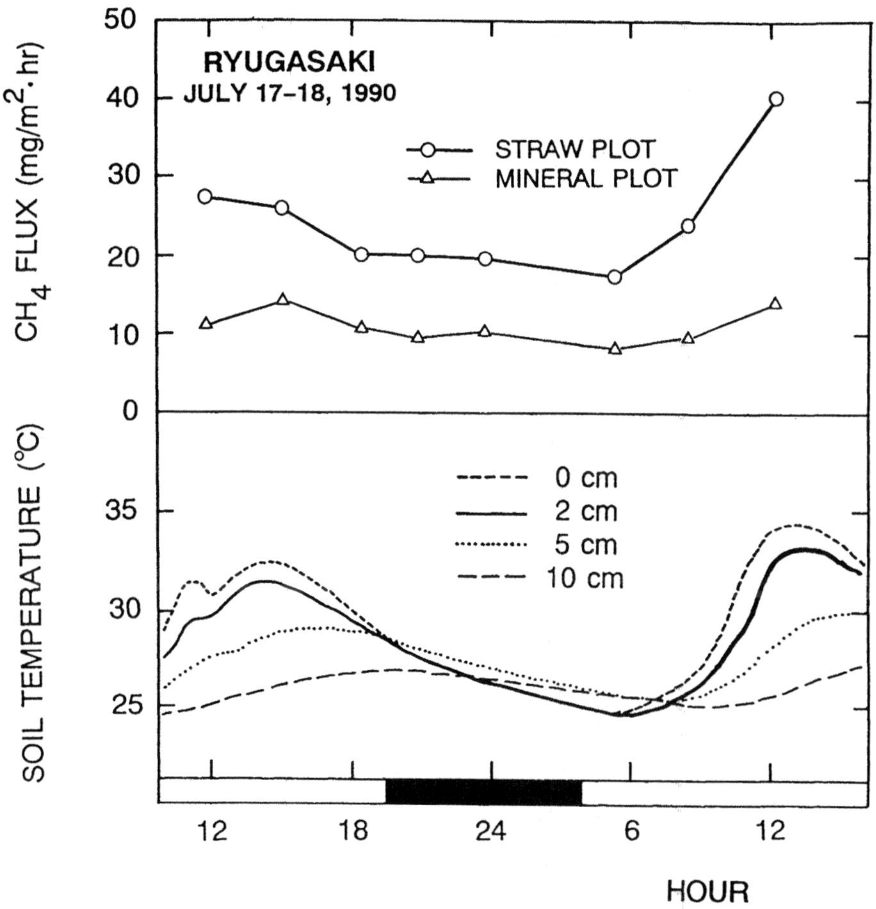

Figure 3. Diurnal variations of CH_4 flux and soil temperature in Ryugasaki paddy field during 17–18 July 1990.

period, similar diurnal variability was observed in the two plots. The minimum occurred in early morning and the maximum occurred in late afternoon. The diurnal pattern of the other measurements were almost identical with Figure 3. The variation of the flux corresponded to the variation of soil temperature in the surface layer.

The correlation between soil temperature at depths of 2, 5, and 10 cm and the CH_4 flux is shown in Figure 4. The flux was most highly correlated with the temperature at 2 cm both in the mineral and in the rice straw plot. The flux was doubled with the increase in the temperature from 25 to 35°C. The correlation with temperature at 10 cm was very weak.

Seasonal Variation of CH_4 Flux

Seasonal variations of CH_4 flux, soil temperature, and soil E_h during the cultivation periods in 1989 and 1990 are shown in Figures 5a–d. In all cases, pronounced variation of the CH_4 flux was observed. The flux began to increase 3–4 weeks after flooding, a period marked by decrease in soil E_h. From the end of July to the beginning of August, corresponding to the heading period of rice plants, the flux showed a rapid rate of increase. The maximum flux was observed in the beginning of August. During these periods, soil temperature was higher. After irrigation was interrupted and the paddy field was intermittently drained at the end of August, the flux rapidly dropped. The average soil temperature at a depth of 2 cm during the flooding period was 22.7°C and 23.5°C in 1989 and 1990, respectively.

Soil E_h at a depth of 5 cm gradually decreased in first 2 months. After that, it was maintained between −200 and −300 mV until drainage. Soil pH was about 6.0 prior to flooding and ranged from 6.1 to 6.8 during the flooding periods. The period of CH_4 emission from the paddy field to the atmosphere corresponded closely to the period of low redox potential in the paddy soil at a depth of 5 cm. The E_h at a depth of 2 cm, however, showed very strong variations by the end of July, except in the case of the mineral plot in 1990. At the period that the E_h increased, CH_4 did not increase or often decreased. During midsummer drainage in the end of June, the E_h at 2 cm apparently increased.

Table 2 summarizes the total amount of CH_4 emission from mineral and the rice straw plots during each cultivation period. The data obtained in 1988[24] are also listed together. The result shows significant interannual variations of annual emission rate of CH_4 from the paddy field. Annual emission rates of CH_4 from the mineral and the rice straw plots in 1990 were more than two times as much as those from each plot in 1989. The soil E_h, especially the data at 2 cm depth, shows that the development of anaerobic conditions in the paddy soil of the experimental plots was much advanced in 1990 rather than in 1989.

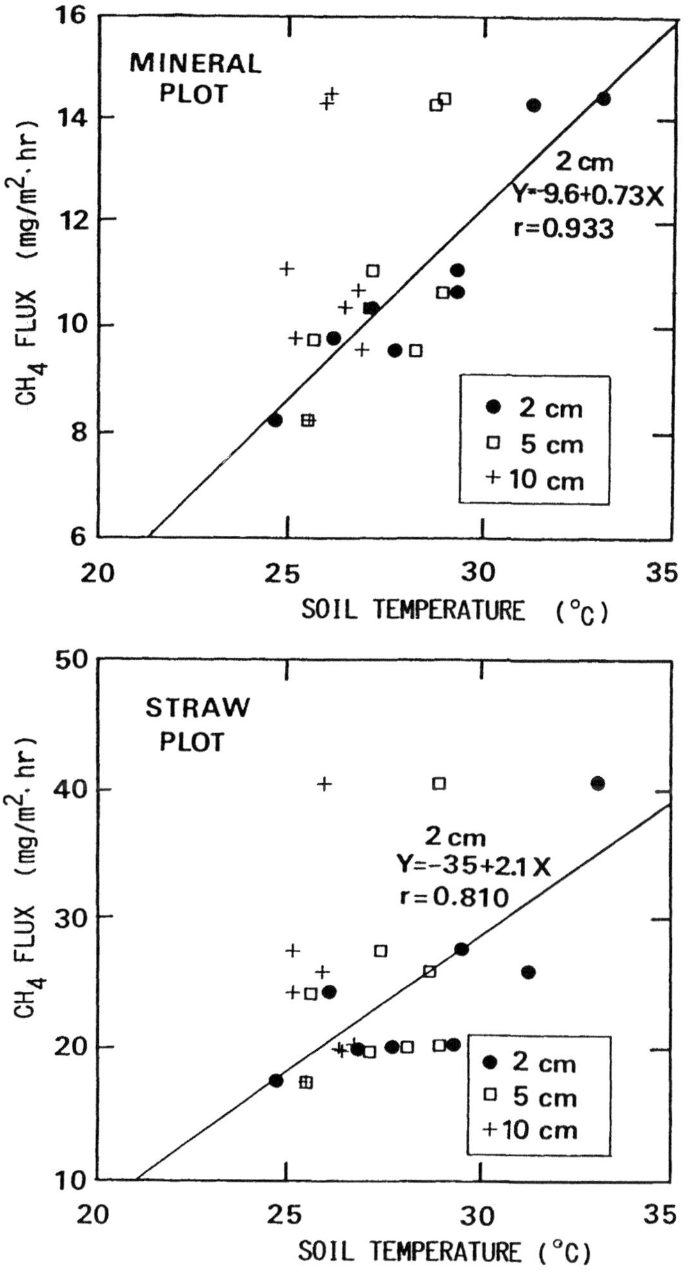

Figure 4. Correlation between soil temperature at depths of 2, 5, and 10 cm and CH_4 flux in the Ryugasaki paddy field. Data were obtained during 17–18 July 1990.

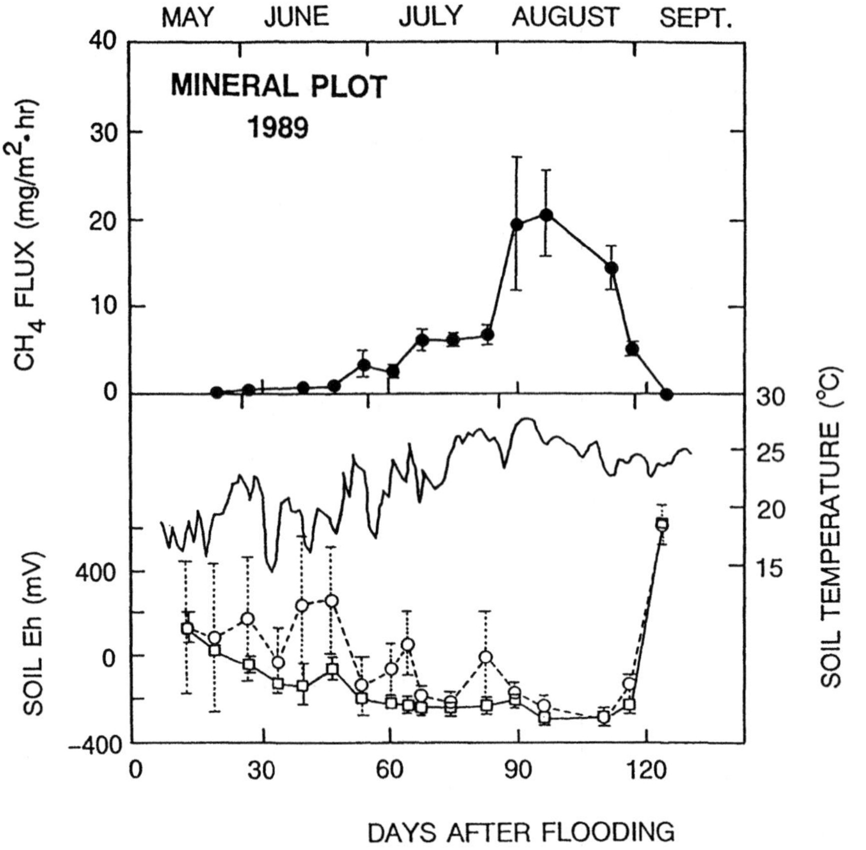

Figure 5. Seasonal variations of CH_4 flux, daily mean temperature of soil at a depth of 2 cm, and soil E_h at depth of 2 (broken line) and 5 cm (solid line): (a) the mineral plot in 1989, (b) the mineral plot in 1990, (c) the rice straw plot in 1989, (d) the rice straw plot in 1990.

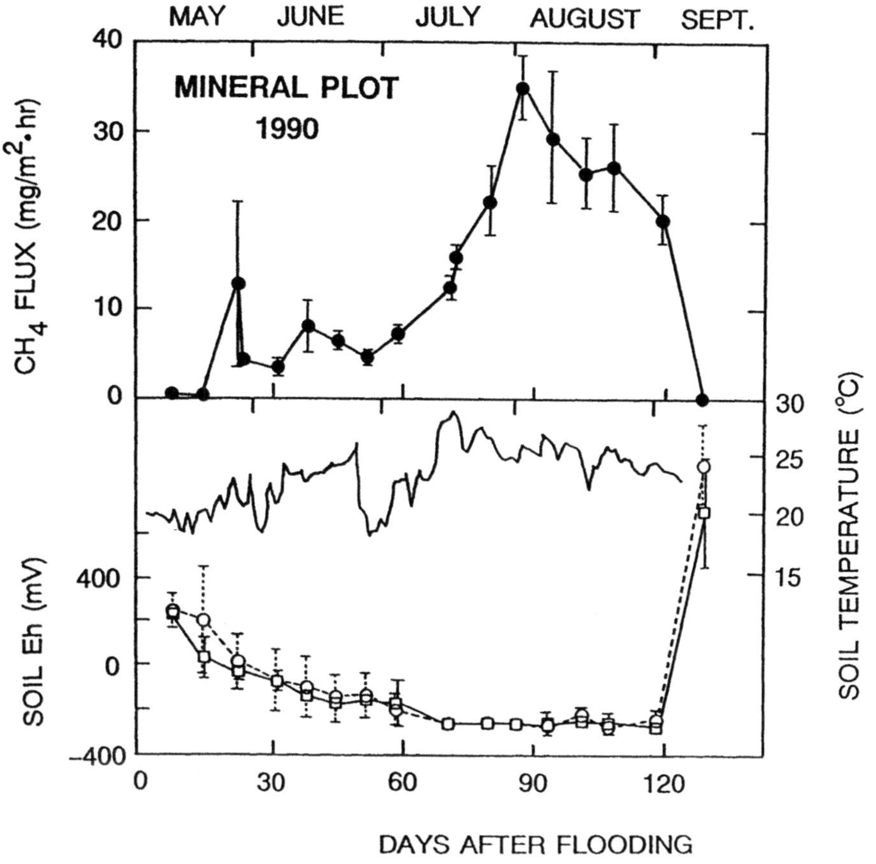

Figure 5b.

Discussion

Site spatial variability is one of the greatest problems in using chamber methods to measure trace gas flux from soil.[12] It is indicative that such spatial variability was caused by the complex nature of the soil environment. In our experimental plots, there were no significant differences in soil texture and structure, growth of rice plants, and paddy water regime within a plot. Therefore, heterogeneous distribution of organic matter, such as stubbles from the previous growing season, may contribute to the spatial variabilities. Very large spatial variations were reported for measurement of CH_4 flux from wetland[11] and denitrification gas flux from upland field.[6] Although replication was a relatively

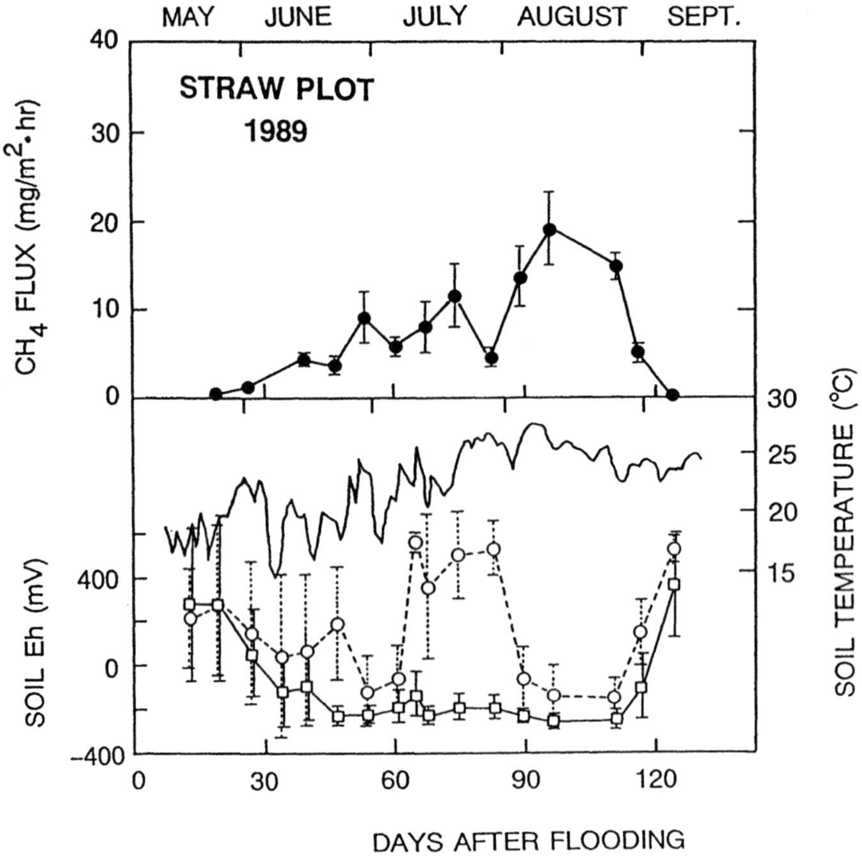

Figure 5c.

small number, the spatial variability of CH_4 flux from the experimental paddy field at Ryugasaki was smaller than that observed in the previous studies, even in the case of using the small chamber having a cross-sectional area of 0.09 m^2. In the lowland rice system, paddy soils are well-mixed by a series of puddling operations before planting the rice plant. It is likely that this soil management practice causes relatively homogeneous soil environment in paddy fields. Our results suggest that we can detect CH_4 flux from an uniform plot of paddy field with a relatively small number of replicative measurements if an appropriate size of chamber was used.

The effect of soil temperature on CH_4 emission from rice paddy fields was

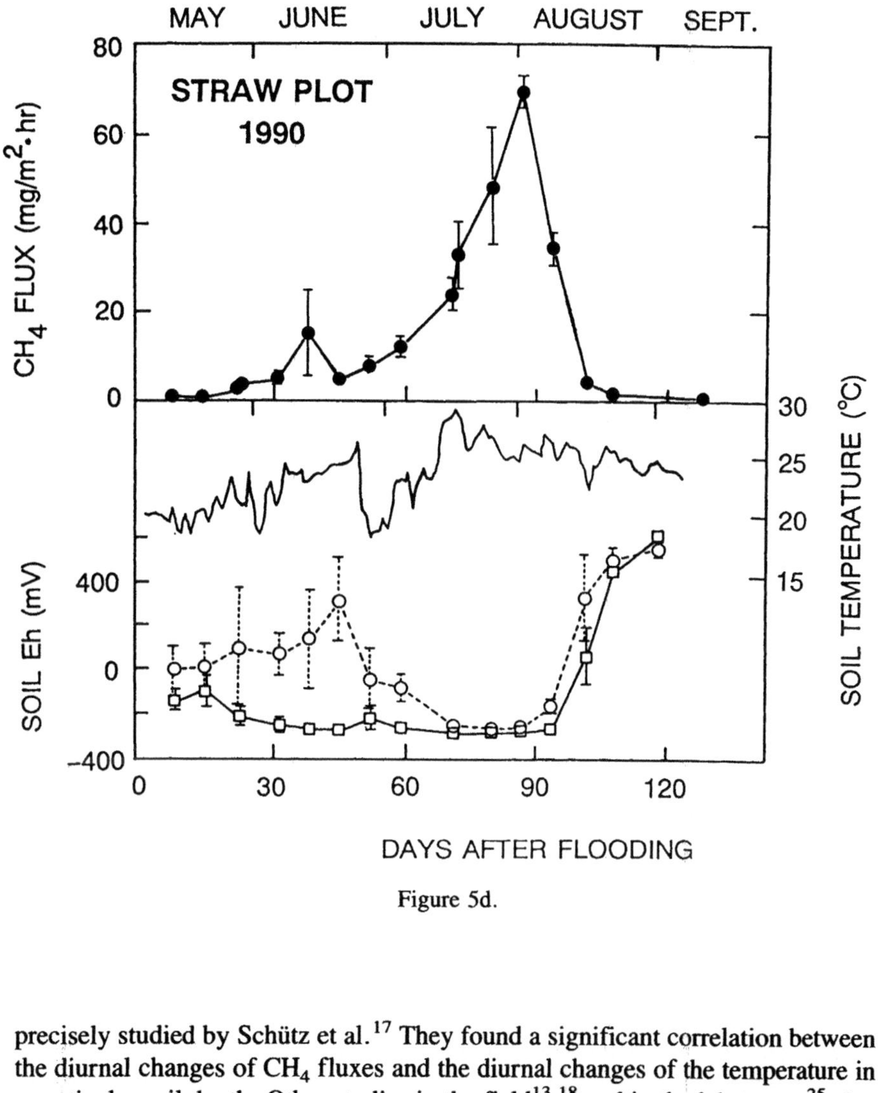

Figure 5d.

precisely studied by Schütz et al.[17] They found a significant correlation between the diurnal changes of CH_4 fluxes and the diurnal changes of the temperature in a particular soil depth. Other studies in the field[13,18] and in the laboratory[25] also showed considerable dependence of CH_4 emission and production on soil temperature. We obtained almost the same result in the Japanese paddy field as these previous studies.

On the other hand, it was reported that there was not a close correlation between seasonal variation of CH_4 flux from paddy fields and soil temperature.[17] In Ryugasaki, the maximum value of daily mean temperature of soil at 2 cm depth was observed at the end of July 1990. However, CH_4 fluxes at that period

Table 2. CH_4 flux during the cultivation periods and annual emission rates from the mineral and the rice straw plot in Ryugasaki paddy field. The data in 1988[24] are listed together.

		Flux (mg m^{-2} h^{-1})		
Plot	Year	Avg.	Max.	Annual emission (g/m^{-2})
Mineral	1988	2.9	13.3	8.2
(control)	1989	6.3	20.8	18.0
	1990	14.0	35.0	40.6
Rice straw	1988	9.6	32.6	27.0
	1989	6.2	19.0	19.9
	1990	15.4	70.1	43.1

were lower than those at the beginning of August, as shown in Figure 5. These results indicate that seasonal variation of CH_4 flux from paddy fields is controlled not only by soil temperature but also by other factors concerning production and emission of CH_4 in paddy fields.

Japanese paddy fields are characterized by a fairly low water table. Detailed water managements are performed there to yield a rich harvest. Soil E_h at a depth of 2 cm in the Ryugasaki paddy field increased very often during a cultivation period. The maximum CH_4 flux was observed at the period when soil E_h became low, and the annual emission rates were closely related to the E_h level. Consumption of molecular oxygen and decrease in soil E_h by waterlogging are essential for the activities of methanogenic bacteria in paddy soil.[20] Therefore, partially oxidation of soil by intermittent irrigation causes a strong inhibition of methanogenesis. In addition, bacterial oxidation of CH_4 produced in soil may be activated by water management. It seems that these effects influence the CH_4 flux from the Ryugasaki paddy field.

In most of the present estimates of global emission of CH_4 from paddy fields,[1,10,16] CH_4 emission rates were calculated not for the flooding period of paddy fields but for whole cultivation periods of rice plants, and the effects of water management and redox potential in soil are not well-considered. There are many types of paddy fields and agricultural practices all over the world, and the water regime would vary from place to place. As shown in our study, the seasonal emission rates of CH_4 from paddy fields greatly depend on the water regime of the paddy fields. To revise the global estimates, therefore, factors concerning water management should be most carefully considered.

Acknowledgments

We are very grateful to the staff of the Ryugasaki Blanch of the Ibaraki Agricultural Experiment Station for allowing us to use the paddy field.

References

1. Aselman, I., and P.J. Crutzen. 1989. Global distribution of natural freshwater wetlands and rice paddies, their net primary productivity, seasonality and possible methane emission. Atmos. Chem. **8**:307–358.
2. Blake, D.R., and F.S. Rowland. 1988. Continuing worldwide increase in tropospheric methane, 1978 to 1987. Science **239**:1129–1131.
3. Cicerone, R.J., and J.D. Shetter. 1981. Sources of atmospheric methane: measurements in rice paddies and a discuss. J. Geophys. Res. **86**:7203–7209.
4. Cicerone, R.J., J.D. Shetter, and C.C. Delwiche. 1983. Seasonal variation of methane flux from a California rice paddy. J. Geophys. Res. **88**:11022–11024.
5. Cicerone, R.J., and R.S. Oremland. 1988. Biogeochemical aspects of atmospheric methane. Global Biogeochem. Cycles **2**:299–327.
6. Folorunso, O.A., and D.E. Rolston. 1984. Spatial variability of field-measured denitrification gas fluxes. Soil Sci. Soc. Amer. J. **48**:1214–1219.
7. Food and Agriculture Organization. 1954–1986. Production yearbook.
8. Holzapfel-Pschorn, A., and W. Seiler. 1986. Methane emission during a cultivation period from an Italian rice paddy. J. Geophys. Res. **91**:11803–11814.
9. Khalil, M.A.K., and R.A. Rasmussen, M.-X. Wang, and L. Ren. 1991. Methane emission from rice fields in China. Environ. Sci. Tech. **25**:979–981.
10. Matthews, E., I. Fung, and J. Lerner. 1991. Methane emission from rice cultivation: geographic and seasonal distribution of cultivated areas and emissions. Global Biogeochem. Cycles **5**:3–24.
11. Moore, T., N. Roulet, and R. Knowles. 1990. Spatial and temporal variations of methane flux from subarctic/northern boreal fens. Global Biogeochem. Cycles **4**: 29–46.
12. Mosier, A.R. 1990. Gas flux measurement techniques with special reference to techniques suitable for measurements over large ecologically uniform areas. In A.F. Bouwman (ed.), Soils and the Greenhouse Effect, J. Wiley and Sons, Chichester, pp. 289–301.
13. Parashar, D.C., J. Rai, P.K. Gupta, and N. Singh. 1991. Parameters affecting methane emission from paddy fields. Indian J. Radio Space Phys. **20**:12–17.
14. Ramanathan, V., R.J. Cicerone, H.B. Singh, and J.T. Kiehl. 1985. Trace gas trends and their potential role in climate change. J. Geophys. Res. **90**:5547–5566.
15. Sass, R.L., F.M. Fisher, P.A. Harcombe, and F.T. Turner. 1990. Methane production and emission in a Texas rice field. Global Biogeochem. Cycles **4**:47–68.
16. Schütz, H., A. Holzapfel-Pschorn, R. Conrad, H. Rennenberg, and W. Seiler. 1989. A 3-year continuous record on the influence of daytime, season, and fertilizer treatment on methane emission rates from an Italian rice paddy. J. Geophys. Res. **94**:16405–16416.
17. Schütz, H., W. Seiler, and R. Conrad. 1990. Influence of soil temperature on methane emission from rice paddy fields. Biogeochem. **11**:77–95.

18. Seiler, W., A. Holzapfel-Pschorn, R. Conrad, and D. Scharffe. 1984. Methane emission from rice paddies. J. Atmos. Chem. **1**:241–268.

19. Steel, L.P., P.J. Fraser, R.A. Rasmussen, M.A.K. Khalil, T.J. Conway, A.J. Crawford, R.H. Gammon, K.A. Masarie, and K.W. Thoning. 1987. The global distribution of methane in the troposphere. J. Atmos. Chem. **5**:125–171.

20. Takai, Y., and T. Kamura. 1966. The mechanism of reduction in waterlogged paddy soil. Folia Microbiol. **11**:304–313.

21. Thompson, A.M., and R.J. Cicerone. 1986. Possible perturbations to atmospheric CO, CH_4, and OH. J. Geophys. Res. **91**:10853–10864.

22. Wang, W.C., Y.L. Yung, A.A. Lacis, J.E. Mo, and J.E. Hansen. 1976. Greenhouse effects due to man-made perturbations of trace gases. Science **194**:685–690.

23. Watson, R.T., H. Rodhe, H. Oeschger, and U. Siegenthaler. 1990. Greenhouse gases and aerosols. In J.T. Houghton, G.J. Jenkins, and J.J. Ephraums (eds.), Climate Change, the IPCC Scientific Assessment, Cambridge University Press, Cambridge, pp. 1–40.

24. Yagi, K., and K. Minami. 1990. Effect of organic matter application on methane emission from some Japanese paddy fields. Soil Sci. Plant Nutr. **36**:599–610.

25. Yamane, I., and K. Sato. 1961. Effect of temperature on the formation of gases and ammonium nitrogen in the water-logged soils. Sci. Rep. Res. Inst. Tohoku Univ. D **12**:1–10.

Soil Water Content and the Ratio of Nitrous Oxide to Nitric Oxide Emitted from Soil

Eric A. Davidson

Woods Hole Research Center, Woods Hole, Massachusetts 02543

ABSTRACT

Nitrification and denitrification are oxidative and reductive microbial processes, respectively, that occur simultaneously at different microsites within the soil, and each process produces both nitric oxide (NO) and nitrous oxide (N_2O). Although much is known about regulation of these processes in the laboratory, the salient factors appropriate for use in regional and global models of N trace gas emissions have not been identified. However, trends exist in the growing global database of N_2O and NO emissions that are consistent with soil water content functioning as an important controller of the ratio of N_2O to NO emitted from soils. Below about 60% water-filled pore space (WFPS), soil microbial processes are often limited by diffusion of organic-C and inorganic-N substrates in soil water films, whereas diffusion of O_2 in the soil atmosphere limits aerobic processes when WFPS > 60% (60% WFPS is roughly equivalent to field capacity in many soils). In laboratory studies with nitrification inhibitors, the aerobic process, chemoautotrophic nitrification, was the dominant source of both N_2O and NO when soil water content was < 60% WFPS, and NO emissions exceeded N_2O emissions. When WFPS > 60%, the anaerobic process, denitrification, was the dominant source, and N_2O emissions were 10 or more times greater than NO emissions. Emission ratios of N_2O to NO were < 1 when WFPS < 60% and were >> 1 when WFPS > 60%. Few field studies report WFPS, but even gravimetric soil water content, which does not account for textural differences among soils, crudely predicted emission ratios of N_2O to NO. Prediction of ratios of emissions of N_2O and NO based on WFPS (or a related parameter of soil water content) would be most useful when combined with ecosystem models of N cycling dynamics that predict total N gas emissions from indices of N availability or site fertility.

The global budget of atmospheric N_2O is badly out of balance, and the global budget of NO is poorly characterized.[8] Soils are important sources of NO and N_2O, and increased soil emissions resulting from changing land uses may account for some of the increasing atmospheric burden of N_2O.[31] Although data on

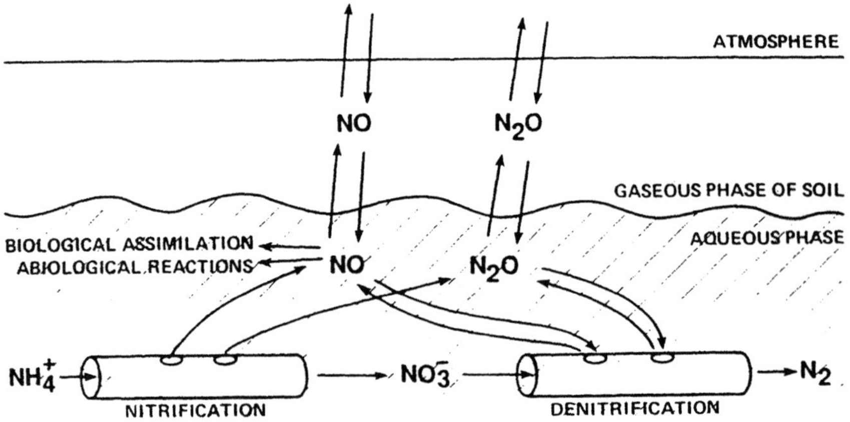

Figure 1. Conceptual model of controls on N gas emissions from soil via nitrification and denitrification, using the metaphor of a leaky pipe. The rate of N flowing through a pipe is analogous to the rate of N cycling within the soil. The sizes of the holes in the pipe through which N_2O and NO "leak" are analogous to controls on the relative proportions of these two gases emitted from the soil. Reproduced from Davidson[8] with permission from The American Society for Microbiology.

soil emissions of NO and N_2O in ecosystems throughout the world are gradually accumulating,[8,24,31] many gaps and sampling biases exist among the sites studied, and it is unlikely that a statistically valid sampling of the global variation in NO and N_2O emissions will be forthcoming. A more promising approach to estimate present and future soil emissions of NO and N_2O on regional and global scales may be to develop our understanding of processes that affect emission rates, so that spatial and temporal variation in emissions can be modeled using global databases of soil types, vegetation types, land uses, greenness indices, and climate as driving variables. Numerous environmental factors that regulate N gas production via nitrification and denitrification in soils have been noted,[15] but the appropriate factors needed in robust predictive models at regional and global scales have yet to be identified.

Firestone and Davidson[15] used a metaphor of fluid flowing through a leaky pipe to describe a conceptual framework for approaching process regulation of NO and N_2O emissions from soils (Figure 1). Rates of N cycling within ecosystems are analogous to N flowing through the pipes. At a coarse level, differences among ecosystems in rates of N cycling may account for differences in N gas emissions. High rates of emissions of N gases are expected in ecosystems where N is a readily available nutrient relative to other factors limiting plant and microbial growth. Indeed, N fertilization in temperate agricultural soils generally results in increased emissions of NO [24] and N_2O.[14] Nitrogen also cycles rapidly in many tropical forests,[47] and tropical soils are a globally important source of both NO [3,26] and N_2O.[27,31,39]

Although rates of N cycling may help predict the magnitude of total emissions of N gases, some soils produce more NO than N_2O, whereas the opposite is true for other soils. A second level of regulation is needed to distinguish between NO and N_2O as end products of nitrification and denitrification. Using the leaky pipe metaphor, this level of regulation affects the sizes of the holes in the pipes through which NO and N_2O "leak" (Figure 1). Numerous factors that affect the ratio of N_2O to NO have been identified, including acidity, temperature, sulfide content, organic-C availability, NO_3^- availability, and O_2 availability,[15] but the factors most useful for global and regional modeling are not yet clear.

More work is needed to identify appropriate and feasible variables for modeling these two levels of regulation of N gas emissions from soils. However, much is already known about these processes, and preliminary attempts to identify candidate driving variables for regional and global models can be made. In this chapter, I focus on the second level of regulation—that which determines the ratio of N_2O to NO emitted from soil. I propose that soil water content is a useful predictor of this ratio at regional and global scales.

Expressions of Soil Water

An excellent discussion of expressions of soil water is given by Papendick and Campbell,[36] which will be reviewed only briefly here. Gravimetric soil water content (θ_g) and volumetric soil water content (θ_v) have units of g H_2O/g oven-dry soil and cm^3 H_2O/cm^3 soil, respectively, and are measures of the water that volatilizes when soil is dried at 105°C. Matric potential (Ψ_m) is a measure of the energy of water in contact with the soil matrix relative to the energy of free water. Expressed in units of pressure (MPa), Ψ_m is affected by both total soil water content and soil texture. A soil is said to be saturated when all pore spaces of the soil are filled with water. Field capacity is defined as the soil water content after saturated soil has drained freely. At field capacity, the so-called "micropores" are generally water-filled, whereas the so-called "macropores" are mostly air-filled. Field capacity is often operationally defined as −0.033 MPa or −0.010 MPa ψ_m, although ψ_m at field capacity varies somewhat with soil texture. "Water-holding capacity" is another term sometimes used interchangeably with field capacity and sometimes used interchangeably with water content at saturation. Finally, water-filled pore space (WFPS), usually expressed as a percentage, is the ratio of volumetric soil water content to total porosity of the soil

($100 \times \theta_v/\epsilon$, where ϵ = cm^3 pore space/cm^3 soil). Soil at 100% WFPS is saturated.

Background Theory

Characteristics of NO versus N_2O

Two differences between NO and N_2O are pertinent to this discussion. First, NO is a more oxidized form of N than is N_2O. Therefore, if a simple relation holds between soil redox potential and gaseous-N production, NO may be a more important product under oxidizing conditions, whereas N_2O would be expected under somewhat more reducing conditions. Second, N_2O contains a double bond between two N atoms, rendering it a far more stable molecule than NO. The only major sink of N_2O is further reduction to N_2 by denitrifying bacteria, which occurs at lower redox potentials than does N_2O production ($E_h < +250$ mV at pH 7; ref. 45). In contrast, NO can be reduced by denitrifiers under more moderate reducing conditions; it may also be assimilated as an N source by bacteria or even by plants, and it can react with O_2 or O_3 to produce NO_2. It is also possible that NO or NO^+ adsorption onto clay surfaces could be important.[33] Therefore, several processes of NO consumption may be important in controlling net NO emissions from soil, whereas processes of N_2O consumption are less diverse and may be less important in many soils.

Effects of Soil Water

Two distinct mechanisms relate soil water content to the regulation of nitrification and denitrification. First, the supply of NH_4^+ and NO_2^- substrates to nitrifying bacteria and NO_3^- and NO_2^- substrates to denitrifying bacteria occurs via diffusion of these ions in soil water films. The thickness of water films affects the rate of diffusion to nitrifying and denitrifying bacteria.[36]

Second, the presence of water in soil pore spaces is the dominant controller of gaseous diffusion within the soil atmosphere. The rate of O_2 diffusion through water is about 10,000 times slower than its rate of diffusion in air. Oxygen must ultimately enter the liquid phase before it is consumed by microorganisms living in soil water films, and this phase change may be rate limiting in some cases.[43] However, the bulk of O_2 movement within soil occurs through air-filled pores, so the amount of water filling the pore spaces affects the flux of O_2 into the soil. Similarly, soil water content also affects the flux of NO, N_2O, and CO_2 out of the soil.

The supply of O_2 is an important controller of both nitrification and denitrification. Nitrification requires O_2 for the oxidation of NH_4^+ to NO_2^- and NO_2^-

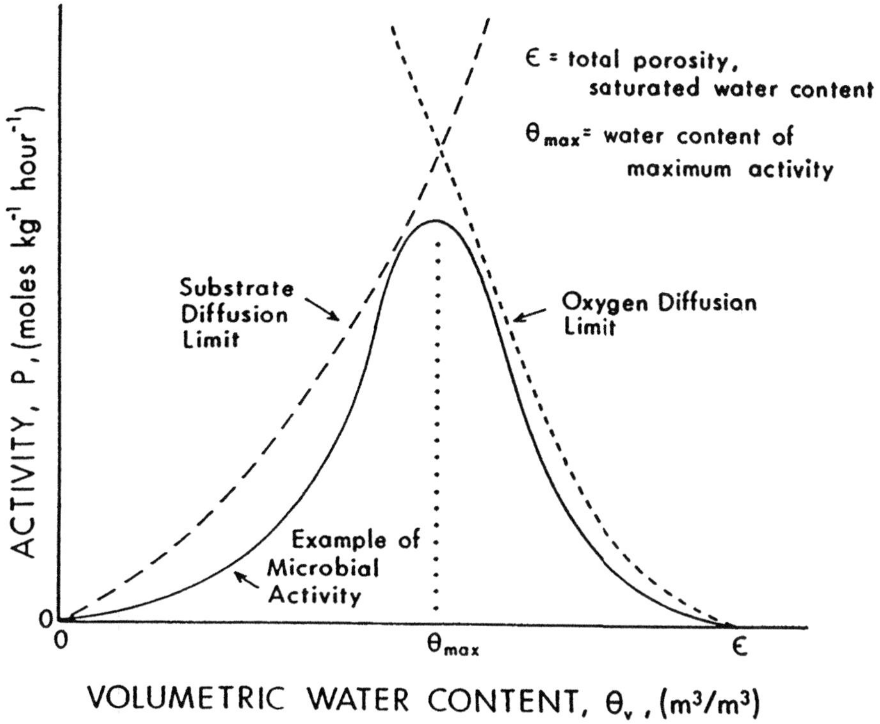

Figure 2. Conceptual plot of microbial activity as a function of soil water content. "P" stands for the rate of "production" for each microbial process. Reproduced from Skopp et al.[44] with permission from the authors and from Soil Science Society of America Journal.

to NO_3^-. Biological denitrification involves reduction of nitrogen oxides coupled to electron transport phosphorylation and generally occurs in the absence of O_2 or at low pO_2. The supply of O_2 also affects the relative proportions of end products of both nitrification and denitrification.

Linn and Doran[29] noted that diffusion of organic-C substrates in water films is probably rate limiting for microbial respiration in dry soils, whereas diffusion of O_2 is rate limiting for microbial respiration in wet soils. Similarly, chemoautotrophic nitrification could also be limited by diffusion of NH_4^+ substrate in water films of dry soil and by diffusion of O_2 in wet soil. Skopp et al.[44] used volumetric soil water content (θ_v) to predict diffusion of substrates in soil solution, and they used θ_v/ϵ (which equals WFPS) to predict diffusion of gases (Figure 2). The optimal soil water content was generally about 60% WFPS for

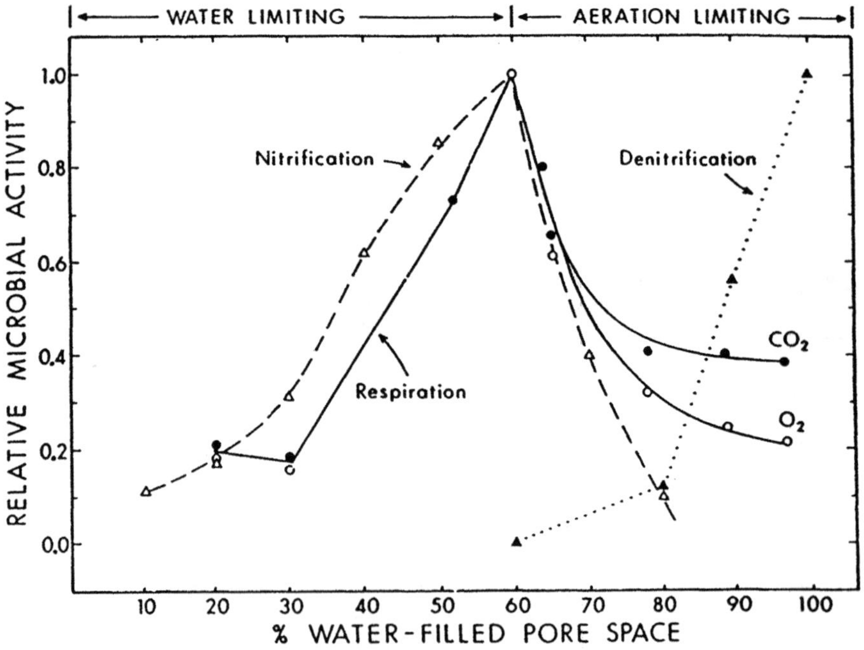

Figure 3. The relation between water-filled pore space (%WFPS = 100 × θ_v/ϵ) and relative amounts of nitrification, denitrification, O_2 uptake by microbial respiration, and CO_2 production by microbial respiration. Reproduced from Linn and Doran[29] with permission from the authors and from Soil Science Society of America Journal.

aerobic processes, such as microbial respiration and nitrification, and was > 80% WFPS for the anaerobic process of denitrification (Figure 3). The optimal value of 60% WFPS for aerobic processes represents the intersection of increasing availability of organic-C and inorganic-N and decreasing availability of O_2 with increasing soil water content (cf. Figures 2 and 3). In model simulations, the optimal value of WFPS varies depending on empirical constants used to characterize diffusion in a given soil, but optimal values of about 60% WFPS were remarkably consistent for heterotrophic respiration in a variety of actual soils examined, indicating the robustness of this parameter.[13,29,44]

Comparing Soils of Different Texture

The WFPS parameter has the additional advantage of being largely comparable among soils of different texture. The parameters θ_g and θ_v cannot be compared easily among soils because a value, such as 0.2, may be relatively dry for a fine-textured soil and relatively wet for a coarse-textured soil. Soil matric potential accounts for the effects of soil texture directly, but ψ_m is a more useful indicator of water availability to living organisms than it is a predictor of diffusion of substrates through gas and liquid phases.[44] In contrast, WFPS is directly related to diffusivity. WFPS also accounts for variation in total porosity (ϵ) among soils as a result of compaction, although it does not indicate pore size distribution.[13] Where gaseous diffusion is known to be an important controlling factor, as is the case for N gas emissions via nitrification and denitrification, WFPS appears to be the most appropriate parameter to express soil water content.

Sources of N_2O and NO in Soil

The microbial and chemical processes in soil that produce N_2O and NO have been reviewed in detail.[8,15,21] The major biological sources of N_2O and NO production in soil are nitrification and denitrification. Abiological reactions of NO_2^- with soil organic matter and self-decomposition reactions of HNO_2 may be an important source of NO in some soils and are known collectively as "chemodenitrification".[35] Nitrification and chemodenitrification could be linked if nitrifying bacteria produce NO_2^- and H^+, and these products then react to produce NO via chemodenitrification.[7]

The most important pathway of N_2O production by nitrifying bacteria (specifically NH_4^+ oxidizing bacteria) is the reduction of NO_2^- to N_2O, apparently as a result of limited O_2 availability.[17,18,38] Because NO_2^- is used as an electron acceptor by nitrifying bacteria in a reductive process similar to denitrification, this process has become known as nitrifier denitrification. This seemingly self-contradictory term implies that some O_2 is needed for NH_4^+ oxidation, and that O_2 must be at least partially limiting to promote reduction of NO_2^-. Hence, N_2O production via nitrification should peak at intermediate values of WFPS, where sufficient O_2 exists to permit NH_4^+ oxidation to NO_2^-, but where O_2 limitation at some microsites results in reduction of NO_2^- to N_2O.

The rate of denitrification increases with increasing WFPS and decreasing O_2 availability. Although the rate of denitrification peaks at 100% WFPS (Figure 3), N_2O production by denitrifiers probably peaks somewhere between 80% and 100% WFPS because some N_2O is reduced to N_2 when WFPS $>$ 80%. In flooded soils, where WFPS = 100%, denitrification largely proceeds all the way to N_2.[34]

The pathway by which nitrifiers produce NO is less clear than for production of N_2O. As with N_2O, NO could be produced from reduction of NO_2^- in "nitrifier denitrification." Nitric oxide could also be a by-product of an intermediate in the oxidative pathway of NH_4^+ to NO_2^-.[20] For nitrifier denitrification, NO production should be inversely related with pO_2, whereas the purely oxidative pathway of NO production should be positively related or unaffected by pO_2. However, conflicting data exist in the literature regarding the effects of pO_2 on NO production in pure cultures of nitrifying bacteria.[1,30] Perhaps more than one pathway may exist for NO production via nitrification.

Denitrification can be a source of NO, but denitrification is also an important sink of NO at redox conditions commonly found in soils. Consumption of NO by denitrification has been shown to be an important controller of net NO emissions from soils[16,41] and aqueous solutions.[6,48] In these studies, net emissions of NO were low when slow rates of NO transport permitted consumption of most of the NO before it escaped to the atmosphere. Soil water contents that impede O_2 diffusion, thus enhancing denitrification, would also impede NO diffusion, thus enhancing NO consumption. However, estimates of the compensation point for production and consumption of NO via denitrification indicate that a positive flux from the soil to the atmosphere is possible.[16,41] A significant net source of NO from denitrification has been deduced from observations of enhanced emissions following surface application of NO_3^-.[3,23,25,42] The denitrification source of NO in these soils was probably very near the surface, where some anaerobic microsites permitted denitrification, but where the diffusive path length for NO escape to the atmosphere was relatively short.

Nitrification and denitrification have been distinguished as sources of NO by use of nitrapyrin[46] and acetylene[7,9] as inhibitors of nitrification in laboratory incubations of soils. In all cases, nitrification was the dominant source, and the denitrification source was either trivial or below detection limits. However, this technique of selective inhibition of nitrification has not been applied yet to soils from sites where a denitrification source of NO has been deduced from responses to NO_3^- addition.

The Use of WFPS as a Predictor of Emissions of N Trace Gases

Nitrification Versus Denitrification as a Source of N_2O

Simple linear regression analysis is not appropriate for proper interpretation of the effect of water-filled pore space on N gas emissions. Nonlinear responses of nitrification and denitrification to variation in WFPS are expected (Figure 3), and optima for production of NO and N_2O should occur at intermediate values of WFPS (Figure 2). In a study of emissions and sources of N_2O during a 12-month period in a disturbed forested watershed of western North Carolina, USA, a linear fit between WFPS and N_2O produced via nitrification was not apparent (Figure 4a). Other factors, such as seasonal variation in soil temperature (6–22°C), seasonal variation of N mineralization, and variation among study sites in slope position, also contributed to variation in N_2O production,[10,11] including

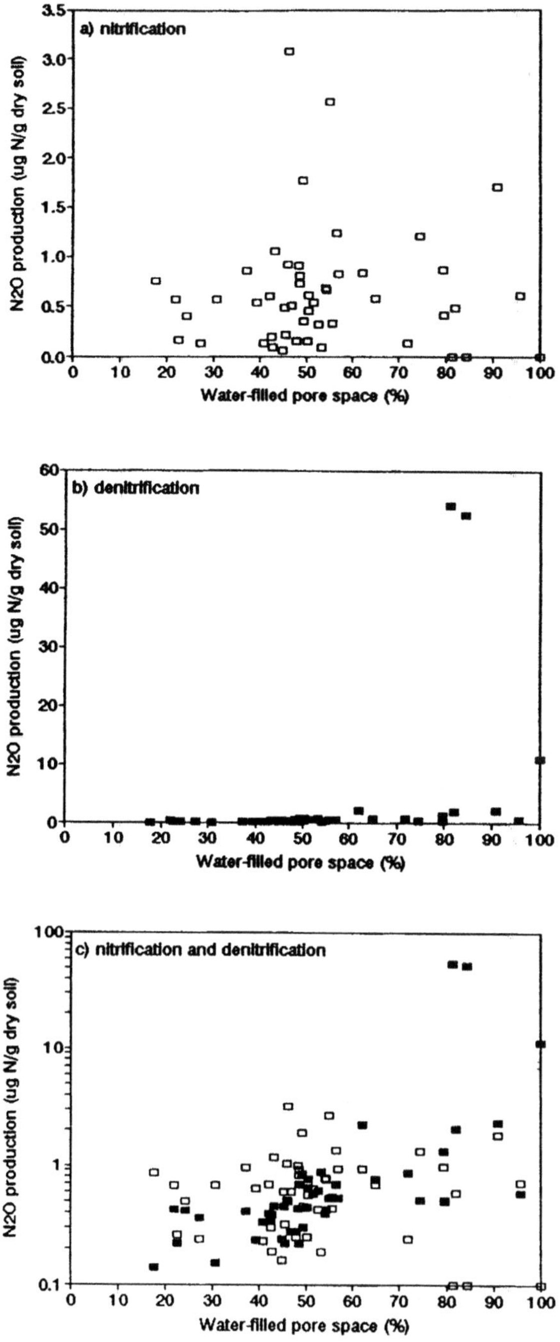

Figure 4. The relation between WFPS and N_2O production via nitrification (a) and denitrification (b) in laboratory incubations of soils sampled throughout a year in a young mixed hardwood forest of western North Carolina, USA. Both sources are also shown on a logarithmic scale (c). Open squares = nitrification; closed squares = denitrification. Adapted from Davidson et al.[11] with permission from Applied and Environmental Microbiology.

numerous instances of negligible production of N_2O. When production of N_2O was substantial, however, the highest production rates occurred at values of WFPS that correspond to expected optima. The three highest rates of N_2O production via nitrification occurred at WFPS near 50% (Figure 4a), whereas the highest rates of N_2O production via denitrification occurred when WFPS > 80% (Figure 4b). Moreover, when the nitrification and denitrification sources are compared directly, the nitrification source is generally greater when WFPS < 60% and the denitrification source is generally greater when WFPS > 60% (Figure 4c). These data agree with identification of 60% WFPS as a critical transition point between the importance of aerobic versus anaerobic processes (Figures 2 and 3).

Emissions of NO versus N_2O as a Function of Soil Water Content

Intact soil cores sampled during the dry season from an annual grassland of central California, USA[7] and from a drought-deciduous forest near the Pacific Coast of Mexico[9] were incubated in the laboratory at three or four water contents. For both soils, the highest rate of NO production was observed at intermediate soil water contents that were below field capacity, whereas N_2O production increased with increasing soil water content and increased to very high rates when field capacity was exceeded (Table 1). The ratio of N_2O to NO production increased with increasing soil water content, remaining below 1 when soil water content was below field capacity and increasing to about 10 or more when the soil water content was above field capacity. The silt loam soil of the grassland maintained its structure, so that the cores were easily kept intact, and WFPS could be calculated from measurements of bulk density and θ_g. Unfortunately, some of the coarser-textured soil from the drought-deciduous forest was lost from the tops and bottoms of the cores, preventing accurate measurement of bulk density and WFPS. For the grassland soil, NO production was highest at 56% WFPS, which is near the predicted optimum for an aerobic process such as nitrification (Figures 2 and 3). In contrast, N_2O production was highest at 74% WFPS, which is consistent with an anaerobic process such as denitrification. Acetylene inhibition also indicated that nitrification was the dominant source of both NO and N_2O at the intermediate soil water contents of both soils, whereas denitrification was the dominant source of N_2O at the highest soil water content.[7,9]

Field capacity corresponds roughly to the transition point between aerobic and anaerobic processes at about 60% WFPS (Figures 2 and 3). A textbook description of a soil with good structure suggests that the pore size distribution should be about 50% macropores and 50% micropores.[4] At field capacity, all of the micropores and a few of the macropores would be water-filled, resulting in a WFPS of just over 50%.

Table 1. Production of N_2O and NO in laboratory incubations of intact soil cores sampled during the dry season and wetted to soil water contents above and below field capacity.

Soil Water Content			Net N gas Production		
Gravimetric (θ_g) (g H_2O/g dry soil)	Water-filled pore space (%)	Field capacity	N_2O	NO	N_2O/NO
			(ng-g^{-1}-h^{-1})		
Annual grassland of California[a]					
0.03[b]	6	Below	0.00	0.04	0.00
0.30	56	Below	0.97	1.61	0.60
0.37	75	Above	4.32	0.44	9.82
Drought-deciduous forest of Mexico[c]					
0.02[b]	ND[d]	Below	0.00	0.01	0.00
0.14	ND	Below	0.39	0.72	0.54
0.27	ND	Below	0.53	0.57	0.93
0.45	ND	Above	16.9	0.32	53.8

[a]For site description, see Davidson.[7]
[b]Ambient soil water content at the time of sampling in the dry season.
[c]For site description, see Davidson et al.[12]
[d]Not Determined. See text for explanation.

Unfortunately, most field studies report only θ_g and do not report WFPS or bulk density (from which WFPS could be calculated). Anderson and Levine[2] reported only θ_g for their study of soil emissions of NO and N_2O made throughout a year at an agricultural site in Virginia, USA, although they also provided the information that $\theta_g = 0.21$ at field capacity for this soil. Hence, it can be seen that the ratio N_2O : NO was over 100 when θ_g exceeded field capacity (Figure 5). Unlike the data from the California grassland and the Mexican forest (Table 1), however, the ratio of N_2O to NO emissions was often above 1 when θ_g was below field capacity (Figure 5). However, this dataset may be biased toward plots that had high N_2O emissions because the analytical sensitivity for N_2O was low (Iris Anderson, personal communication) and ratios N_2O : NO can be calculated only when N_2O emissions were above the N_2O sensitivity threshold.

In most studies, field capacity is not reported, so only θ_g can be used for comparisons across sites, and θ_g does not account for differences in texture among soils. However, fine-textured soils generally retain more water than do coarse-textured soils, and fine-textured soils are often found in regions with humid climates. Therefore, θ_g can be viewed as a very crude index of ecosystem wetness. Although soils in humid climates usually drain to water contents below field capacity between precipitation events, they are probably closer to and above

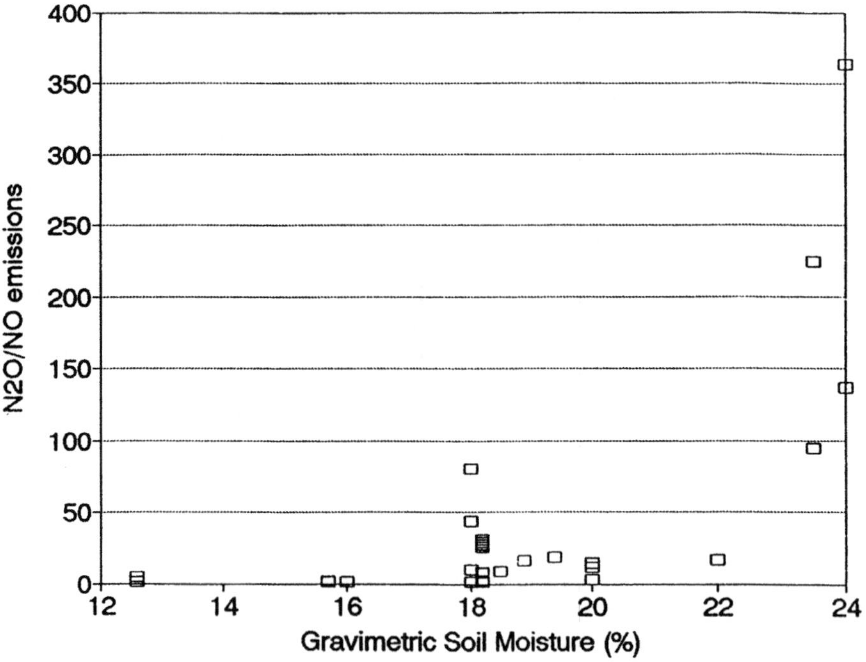

Figure 5. The relation between gravimetric soil water content (θ_g) and the ratio of emissions of N_2O and NO measured throughout a year at an agricultural site in Virginia, USA. Field capacity for this soil was reported as $\theta_g = 0.21$. Ratios of emissions of N_2O and NO are calculated from the original data published by Anderson and Levine.[2]

field capacity more often than are soils from arid climates. As crude as this index is, it nevertheless reveals that the ratio of N_2O to NO emissions is less than 1 for sites where $\theta_g < 0.12$ and that the ratio exceeds 1 at wetter sites (Figure 6). Points 6 and 7 were excluded from a least-squares linear regression because, as noted above, the calculated ratios do not include several study plots where N_2O emissions were below a relatively high sensitivity threshold. The results of the regression analysis on the remaining points should be interpreted with caution because the fit is strongly influenced by a single point representing a very wet, clayey soil of a rainforest in Costa Rica (Figure 6). Nevertheless, the fit is good (ratio of $N_2O/NO = 10.1\theta_g - 0.25$; $R^2 = 0.93$) and the distinction between sites dominated by NO emissions and those dominated by N_2O emissions is clear. This relation is especially remarkable considering that θ_g is not expected to be a good parameter for comparing soil water contents of soils with widely varying texture. Use of a more appropriate parameter, such as WFPS, would presumably

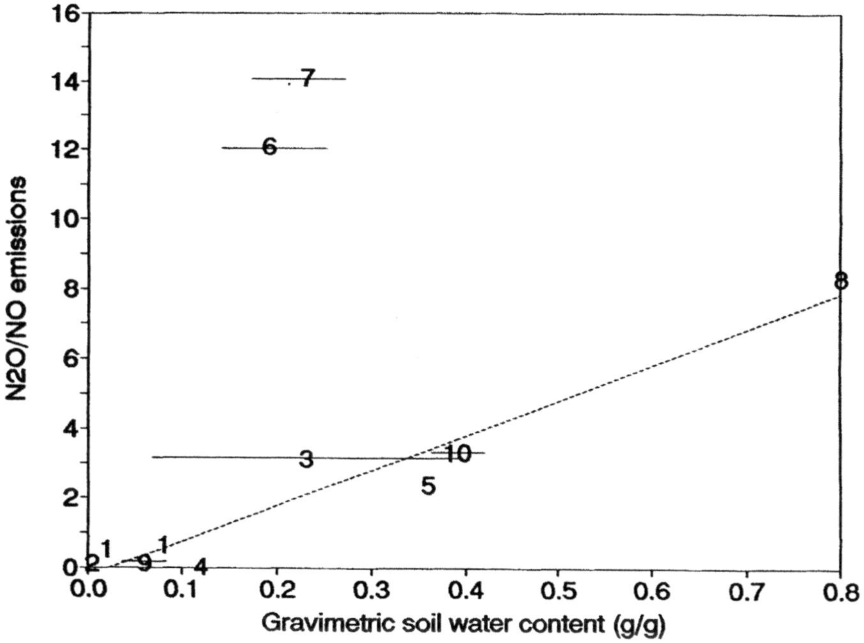

Figure 6. The relation between gravimetric soil water content (θ_g) and the ratio of emissions of N_2O and NO in a variety of ecosystems. Where a range of soil water contents was reported, a horizontal bar indicates the range and the symbol marks either the midpoint or the arithmetic average of the reported soil water contents. The ratio of emissions was calculated from either the reported mean or median flux of NO and N_2O from each site. Symbols identify data from the following studies: 1 = Mexican drought-deciduous forest during dry and wet seasons[9]; 2 = Venezuelan savanna on sandy soil[19,25]; 3 = Venezuelan savanna on sandy clay loam soil[42]; 4 = Brazilian humid tropical forest on sandy soil[3,32]; 5 = Brazilian humid tropical forest on clayey soil[3,32]; 6 = agricultural field in Virginia, USA[2]; 7 = agricultural field in Colorado, USA[2]; 8 = Costa Rican rainforest[28]; 9 = managed pasture in Texas, USA[22]; and 10 = perennial grassland of Kansas, USA (Coyne, Dunkin, and Firestone, University of California, Berkeley, unpublished data). The dotted line indicates the least-squares simple linear regression that omits points 6 and 7 (see text). The regression equation is ratio of $N_2O/NO = 10.1\theta_g - 0.25$; $R^2 = 0.93$.

provide a still better function to predict the effects of soil water content on the ratio of emissions of N trace gases.

Applications to Modeling Efforts

Simultaneous production of both NO and N_2O via either nitrification or denitrification has made identification of their sources difficult and, hence, has hindered modeling of soil emissions of these trace gases. However, a pattern has emerged that simplifies the interpretation of simultaneous oxidation and reduction reactions involving NO and N_2O. Soil emissions of NO produced via nitrification usually dominate when soil water content is below field capacity or when WFPS $< 60\%$, whereas emissions of N_2O produced via denitrification dominate in wetter soils. This is not to say that N_2O production via nitrification and NO production via denitrification should be ignored but that these processes occur at lower rates relative to the other simultaneous processes. Indeed, nitrification in soil could still be a significant global source of N_2O; the analysis presented here simply shows that NO emissions via nitrification in well-drained soils are generally greater than are N_2O emissions.

The use of a ratio of emissions of N_2O and NO is suggested as a potential modeling tool that can be used in combination with other modeling approaches to predict emission rates of NO and N_2O from soils. Several models now predict variation in N cycling dynamics at regional and global scales. For example, the CENTURY model predicts the dynamics of C and N cycling in soils[37] and has also been used to estimate N_2O fluxes at a regional scale.[5] The Terrestrial Ecosystem Model[40] predicts rates of N cycling at regional and global scales, although it does not yet include N trace gases. Both N_2O and NO emissions should be positively related to rates of N cycling, but they should respond differently to variation in soil water content. Hence, these models should be applicable as predictors of total N gas emissions based on N cycling dynamics, and a second function relating WFPS (or a related index of soil wetness) to the ratio of emissions of N_2O and NO would permit more specific predictions for each gas. This two-phase approach to modeling is analogous to the leaky-pipe conceptual model (Figure 1), where the rate of N cycling through the pipes is considered as the first level of regulation, and the sizes of the holes in the pipes that control the ratios of gases act as a second level of regulation.

An equation that predicts the ratio of emissions of N_2O and NO as a function of gravimetric soil water content has been derived here from data reported in studies of eight sites. A function based on WFPS probably would be preferable, but too few studies have reported WFPS or bulk density to provide the data needed to derive this function. Adding bulk density to the list of soil parameters routinely included in site descriptions would help provide the needed data.

Whether gravimetric soil water content or water-filled pore space is used, the theoretical justification for the applicability of soil moisture as a predictor of N gas emissions is well-established, and the remarkably strong empirical relation between soil moisture and the ratio of N trace gases in the few datasets that are available offers encouragement that a function can be derived for use in regional and global models.

Acknowledgments

The author thanks John Doran, USDA-ARS, Lincoln, NE, and Chris Potter, NASA Ames Research Center, Moffett Field, CA, for helpful reviews of an earlier draft of this chapter. The author also thanks the A.W. Mellon Foundation for a grant to The Woods Hole Research Center for financial support while this manuscript was prepared.

References

1. Anderson, I.C., and J.S. Levine. 1986. Relative rates of nitric oxide and nitrous oxide production by nitrifiers, denitrifiers, and nitrate respirers. Appl. Environ. Microbiol. **51**:938–945.
2. Anderson, I.C., and J.S. Levine. 1987. Simultaneous field measurements of biogenic emissions of nitric oxide and nitrous oxide. J. Geophys. Res. **92**:965–976.
3. Bakwin, P.S., S.C. Wofsy, S.-M. Fan, M. Keller, S.E. Trumbore, and J.M. daCosta. 1990. Emission of nitric oxide (NO) from tropical forest soils and exchange of NO between the forest canopy and atmospheric boundary layers. J. Geophys. Res. **95**:16755–16764.
4. Brady, N.C. 1989. The Nature and Properties of Soils, 10th ed., MacMillan, New York.
5. Burke, I.C., D.S. Schimel, C.M. Yonker, W.J. Parton, L.A. Joyce, and W.K. Lauenroth. 1990. Regional modeling of grassland biogeochemistry using GIS. Landscape Ecol. **4**:45–54.
6. Coyne, M.S., E.A. Davidson, and M.K. Firestone. 1990. Cell density affects N-oxide evolution by denitrifiers, Agron. Abstr., American Society of Agronomy, Madison, WI., p. 246.
7. Davidson, E.A. 1992. Sources of nitric oxide and nitrous oxide following wetting of dry soil. Soil Sci. Soc. Amer. J. **56**:95–102.
8. Davidson, E.A. 1991. Fluxes of nitrous oxide and nitric oxide from terrestrial ecosystems. In J.E. Rogers and W.B. Whitman (eds.), Microbial Production and Consumption of Greenhouse Gases: Methane, Nitrogen Oxides, and halomethanes, American Society for Microbiology, Washington, DC, pp. 219–235.
9. Davidson, E.A., P.A. Matson, P.M. Vitousek, R. Riley, K. Dunkin, G. García-

Méndez, and J.M. Maass. 1993. Processes regulating soil emissions of NO and N_2O in a seasonally dry tropical forest. Ecology, in press.

10. Davidson, E.A., and W.T. Swank. 1986. Environmental parameters regulating gaseous nitrogen losses from two forested ecosystems via nitrification and denitrification. Appl. Environ. Microbiol. **52**:1287–1292.

11. Davidson, E.A., W.T. Swank, and T.O. Perry. 1986. Distinguishing between nitrification and denitrification as sources of N_2O from soil. Appl. Environ. Microbiol. **52**:1280–1286.

12. Davidson, E.A., P.M. Vitousek, P.A. Matson, R. Riley, G. García-Méndez, and J.M. Maass. 1991. Soil Emissions of nitric oxide in a seasonally dry tropical forest of México. J. Geophys. Res. **96**:15439–15445.

13. Doran, J.W., L.N. Mielke, and J.F. Power. 1990. Microbial activity as regulated by soil water-filled pore space. Transactions of the 14th International Congress of Soil Science. Symposium III-3; Ecology of Soil Microorganism in the Microhabital Environments, 1990, Kyoto, Japan, August 12–18, pp. 94–99.

14. Eichner, M.J. 1990. Nitrous oxide emissions from fertilized soils: Summary of available data. J. Environ. Qual. **19**:272–280.

15. Firestone, M.K., and E.A. Davidson. 1989. Microbiological basis of NO and N_2O production and consumption in soil. In M.O. Andreae and D.S. Schimel (eds.), Exchange of Trace Gases Between Terrestrial Ecosystems and the Atmosphere, J. Wiley and Sons, New York, pp. 7–21.

16. Galbally, I.E., and C. Johansson. 1989. A model relating laboratory measurements of rates of nitric oxide production and field measurements of nitric oxide emission from soils. J. Geophys. Res. **94**:6473–6480.

17. Goodroad, L.L., and D.R. Keeney. 1984. Nitrous oxide production in aerobic soils under varying pH, temperature, and water content. Soil Biol. Biochem. **16**:39–43.

18. Goreau, T.J., W.A. Kaplan, S.C. Wofsy, M.B. McElroy, F.W. Valios, and S.W. Watson. 1980. Production of NO_2^- and N_2O by nitrifying bacteria at reduced concentrations of oxygen. Appl. Environ. Microbiol. **40**:526–532.

19. Hao, W.M., D. Scharffe, P.J. Crutzen, and E. Sanhueza. 1988. Production of N_2O, CH_4, and CO_2 from soils in the tropical savanna during the dry season. J. Atmos. Chem. **7**:93–105.

20. Hooper, A.B. 1984. Ammonium oxidation and energy transduction in the nitrifying bacteria. In W.R. Strohl and O.H. Tuovinen (eds.), Microbial Chemoautotrophy, Ohio State University Press, Columbus, pp. 133–167.

21. Hutchinson, G.L., and E.A. Davidson. 1992. Processes for production and consumption of gaseous nitrogen oxides in soil. In L.A. Lowry, J.M. Duxbury, A.R. Mosier, and D.E. Rolston (eds.), Agricultural Ecosystem Effects on Trace Gases and Global Climate Change, Agronomy Society of America, Special Publication Number 55, Madison, WI, in press.

22. Hutchinson, G.L., G.P. Livingston, and E.A. Brams. 1992. Nitric and nitrous oxide evolution from managed subtropical grassland, this volume.

23. Johansson, C. 1984. Field measurements of emission of nitric oxide from fertilized and unfertilized forest soils in Sweden. J. Atmos. Chem. **1**:429–442.

24. Johansson, C. 1989. Fluxes of NO_x above soil and vegetation. In M.O. Andreae and D.S. Schimel (eds.), Exchange of Trace Gases Between Terrestrial Ecosystems and the Atmosphere, J. Wiley and Sons, New York, pp. 229–246.

25. Johansson, C., H. Rodhe, and E. Sanhueza. 1988. Emission of NO in a tropical savanna and a cloud forest during the dry season. J. Geophys. Res. **93**:7180–7192.

26. Johansson, C., and E. Sanhueza. 1988. Emission of NO from Savanna soils during rainy season. J. Geophys. Res. **93**:14193–14198.

27. Keller, M., W.A. Kaplan, and S.C. Wofsy. 1986. Emissions of N_2O, CH_4, and CO_2 from tropical soils. J. Geophys. Res.**91**:1179–11802.

28. Keller, M. and W.A. Reiners. 1991. A seasonal study of the emissions of nitrous oxide and nitric oxide from soil in a wet forest area in Costa Rica: examining the effects of land use change. EOS **72**:110.

29. Linn, D.M., and J.W. Doran. 1984. Effect of water-filled pore space on carbon dioxide and nitrous oxide production in tilled and nontilled soils. Soil Sci. Soc. Amer. J. **48**:1267–1272.

30. Lipschultz, F., O.C. Zafiriou, S.C. Wofsy, M.B. McElroy, F.W. Valios, and S.W. Watson. 1981. Production of NO and N_2O by soil nitrifying bacteria. Nature **294**: 641–643.

31. Matson, P.A., and P.M. Vitousek. 1990. Ecosystem approach to a global nitrous oxide budget. Bioscience **40**:667–672.

32. Matson, P.A., P.M. Vitousek, G.P. Livingston, and N.A. Swanberg. 1990. Sources of variation in nitrous oxide flux from amazonian ecosystems. J. Geophys. Res. **95**:16789–16798.

33. Mortland, M.M. 1965. Nitric oxide adsorption by clay minerals. Soil Sci. Soc. Amer. Proc. **29**:514–519.

34. Mosier, A.R., S.K. Mohanty, A. Bhadrachalam, and S.P. Chakravorti. 1990. Evolution of dinitrogen and nitrous oxide from the soil to the atmosphere through rice plants. Biol. Fert. Soils **9**:61–67.

35. Nelson, D.W. 1982. Gaseous losses of nitrogen other than through denitrification. In F.J. Stevenson (ed.), Nitrogen in Agricultural Soils, American Society of Agronomy, Madison, WI, pp. 327–364.

36. Papendick, R.I., and G.S. Campbell. 1981. Theory and measurement of water potential. In J.F. Parr, W.R. Gardner, and L.F. Elliot (eds.), Water Potential Relations in Soil Microbiology, Soil Science Society of America Special Publications Number 9, Madison, WI, pp. 1–22.

37. Parton, W.J., D.S. Schimel, C.V. Cole, and D.S. Ojima. 1987. Analysis of factors controlling soil organic matter levels in Great Plains Grasslands. Soil Sci. Soc. Amer. J. **51**:1173–1179.

38. Poth, M., and D.D. Focht. 1985. ^{15}N kinetic analysis of N_2O production by *Nitrosomonas europaea*: an examination of nitrifier denitrification. Appl. Environ. Microbiol. **49**:1134–1141.

39. Prinn, R., D. Cunnold, R. Rasmussen, P. Simmonds, F. Alyea, A. Crawford, P. Fraser, and R. Rosen. 1990. Atmospheric emissions and trends of nitrous oxide deduced from 10 years of ALE-GAGE data. J. Geophys. Res. **95**:18369–18385.

40. Raich, J.W., E.B. Rastetter, J.M. Melillo, D.W. Kicklighter, P.A. Steudler, B.J. Peterson, A.L. Grace, B. Moore III, and C.J. Vorosmarty. 1991. Potential net primary productivity in South America: application of a global model. Ecol. Appl. **1**:399–429.

41. Remde, A., F. Slemr, and R. Conrad. 1989. Microbial production and uptake of nitric oxide in soil. FEMS Microb. Ecol. **62**:221–230.

42. Sanhueza, E., W.M. Hao, D. Scharffe, L. Donoso, and P.J. Crutzen. 1990. N_2O and NO emissions from soils of the northern part of the Guayana Shield, Venezuela. J. Geophys. Res. **95**:22481–22488.

43. Skopp, J. 1985. Oxygen uptake and transport in soils: analysis of the air–water interfacial area. Soil Sci. Soc. Amer. J. **49**: 1327–1331.

44. Skopp, J., M.D. Jawson, and J.W. Doran. 1990. Steady-state aerobic microbial activity as a function of soil water content. Soil Sci. Soc. Amer. J. **54**:1619–1625.

45. Smith, C.J., W.F. Wright, and W.H. Patrick, Jr. 1983. The effect of soil redox potential and pH on the reduction and production of nitrous oxide. J. Environ. Qual. **12**:186–188.

46. Tortoso, A.C., and G.L. Hutchinson. 1990. Contributions of autotrophic and heterotrophic nitrifiers to soil NO and N_2O emissions. Appl. Environ. Microbiol. **56**:1799–1805.

47. Vitousek, P.M., and R.L. Sanford. 1986. Nutrient cycling in moist tropical forests. Ann. Rev. Ecol. Syst. **17**:137–167.

48. Zafiriou, O.C., Q.S. Hanley, and G. Snyder. 1989. Nitric oxide and nitrous oxide production and cycling during dissimilatory nitrite reduction by *Pseudomonas perfectomarina*. J. Biol. Chem. **10**:5694–5699.

Respiration of Cultivated Histosols in Field and Laboratory Measurements and the Relationships between Respiration and Soil Properties

T.J. Hokkanen and J. Silvola

Department of Biology, University of Joensuu, SF-80101 Joensuu, Finland

ABSTRACT

Rapid decomposition of cultivated histosols, especially peat soils, may release a substantial load of CO_2 into the atmosphere. We compared three methods to measure soil respiration in cultivated histosols and determined the relationships between select bulk soil chemical characteristics and respiration. Soil respiration was measured in the field and from large intact cores and sieved soil in the laboratory. The ratio of average rates between intact cores and field respiration was 1.04. The ratio was rate-dependent and increased from ~0.5 to 1.5 with respiration rate decreasing from 1300 to 300 mg CO_2 m^{-2} h^{-1}. The ratio between field and sieved soil respiration varied from about 0.1 to 1 at the respiration rates mentioned above, but the variation was not regular. The ratio of average rates between sieved soil and field respiration was only 0.39. The main factors influencing respiration rate were soil moisture and temperature; soil chemical characteristics correlated with sieved soil respiration only. These data indicate that soil total CO_2 flux can most easily be estimated by field methods. Also intact cores can be used with precautions concerning core size and temperature. Sieved soil respiration underestimates total CO_2 flux due to the removal of roots and litter, but is useful in studies on bulk soil microbiological processes.

Introduction

The total area of the world's organic soils (histosols) is (3.4–4.5) × 10^{12} m^2.[10,13] In the boreal zone, the proportion of histosols to the total land area is large, and histosols have traditionally been used for forestry, agriculture, and fuel peat production. In Finland, the area of histosols (10 × 10^{10} m^2) is about one-third of the total land area and ~70% of that area is economically utilized. About 0.5 × 10^{10} m^2 of the histosols are cultivated. Increased decomposition of ditched and cultivated peat is a well-known phenomenon, which has been seen in the subsidence of the surfaces of cultivated peatlands.[3,17,19,24] Knowledge about peat mineralization is important because decomposition liberates stored nutrients, and carbon stored in peat is released into the atmosphere as CO_2 or CH_4, increasing the atmospheric load of these gases. The role of peatland utilization in connection with the atmospheric CO_2 concentration increase has been discussed by Armentano and Menges.[3]

Soil respiration (CO_2 evolution from the soil) can be measured from practically intact soil in the field using different methods[9,11,28] or from large, intact soil cores in the laboratory.[27,28] Another method is to measure respiration from sieved and mixed—very disturbed—samples in the laboratory.[2] Measurement of respiration, especially in the laboratory from homogenized samples, is an easy and rapid technique. With proper experimental planning and choice of the correct measuring method, one can usually obtain reliable information about aerobic soil metabolism. The aim of this study is to examine the comparability of three methods for measuring soil respiration and the relationships between soil respiration and soil characteristics.

Study Area

Four cultivated fields established on peatlands were sampled near Ilomantsi, Finland (62° 40′ N, 30° 50′ E) (Figure 1). The field at site 1 (soil 1) was established ~40 years ago, but it has been uncultivated for ~20 years. The peat layer is less than 30 cm and is mixed with mineral soil due to plowing. At site 2 (soil 2), the field is about 20–30 years old; cultivation ended 5 years ago. The depth of the peat layer is mainly less than 50 cm. The peat is mixed with mineral soil. From the analysis of the peat layer it was concluded, that the original mire type was a spruce swamp for soils 1 and 2.

The field at site 3 (soil 3) was established ~60 years ago, the original mire having been a wet flark fen. The depth of the peat layer was over 2 m and is still ~140 cm. This site is used for growing grass. Site 4 (soil 4) was also a very wet fen which was drained for agricultural use ~40 years ago. The peat layer is still greater than 2 m. Cultivation on site 4 ended in 1989, after which the site had been planted with birch seedlings (*Betula pendula*).

The mire type names are according to the Finnish mire classification system.[22]

Material and Methods

Respiration Measurements

CO_2 production of the soil was measured using three methods between 4 June and 16 August 1990. Soil 1 was sampled three times, soil 2 two times, and soils

Figure 1. The location of the study sites (Ilomantsi).

3 and 4 once. Sampling for the laboratory analyses occurred simultaneously with the field measurements at the same sites, resulting in a total of 55 samples.

Field Measurements

An area of 314 cm^2 on the soil surface was enclosed using a steel cylinder with a closed top (volume 3.35 dm^3). Green parts of plants were cut away before measuring. Three duplicate air samples were taken from the chamber with 1.0-ml syringes at intervals of 2 min immediately after closing the chamber. Before taking the samples, the air in the chamber was mixed by pumping with syringes. The soil temperature was measured at the depths of 1, 10, and 20 cm.

The time intervals to be used for air sampling were verified by numerous preliminary tests in the laboratory and in the field. For respiration rates of over 1000 mg CO_2 m^{-2} h^{-1}, the linear part of the CO_2 concentration rise in the chamber was at least 20–30 min. There was no consistent, extra increase in CO_2 concentration immediately after placement of the chamber on soil, and linear regression r^2 was >98.5% in all tests and field measurements.

CO_2 concentration in the samples was analyzed in the laboratory using an infrared gas analyzer (URAS 3G, Hartmann & Braun). The gas samples were led into the CO_2 analyzer via an oxygen flow (30 L h^{-1}). The analyzer was calibrated with similar gas samples prepared by a gas mixing pump. The decrease of CO_2 concentration in the syringes for different CO_2 concentrations and time intervals was tested and considered in the calculations. With a CO_2 concentration below 800 ppm and time below 6 h, leakage of CO_2 was insignificant.

Intact Cores

For laboratory measurements, samples were taken from the soil surface using a cylinder 14 cm in diameter and of 20 cm high. Green parts of plants were removed and samples were put into plastic jars of the same dimensions as stated above. In the laboratory, the samples were put into gas exchange chambers positioned in an incubator. The CO_2 exchange of samples was measured using a dynamic, continuous-flow method.[27] Standard air from outside the chamber was fed through flow regulators into the chambers and then subsequently through magnetic valves into the infrared gas analyzer. Soil respiration was measured at 5, 15, and 25°C.

Sieved Samples

Large soil cores (14-cm diameter, 10-cm height) were taken from all sites and transferred to the laboratory in plastic bags. Cores were stored in 4°C for 2–3 days until they were homogenized by hand. Litter and living roots were removed and the soil was sieved (3-mm mesh) for respiration measurements and chemical analyses.

Respiration was measured at 5.0, 15.0, and 25.0°C using 1–2 g (d.w.) samples; four replicates per each soil core. The samples were put into 56-ml glass vessels with a rubber septum. CO_2 evolution rate was determined from the difference in the CO_2 concentration in serial gas samples (1.0 ml) which were withdrawn through the septum using a syringe. The time interval between gas samples was from 12 to 24 h depending on the prevailing temperature. CO_2 concentration in the head-space air was determined with an infrared gas analyzer (URAS 3G, Hartmann & Braun) as with intact cores. Soils were preincubated in plastic bags at the selected temperature for 24 h before measurements.

Respiration results are expressed as $\mu g\ CO_2\text{–}C\ g^{-1}$ (soil dry weight) h^{-1} or $\mu g\ CO_2\text{–}C\ g^{-1}$ (organic matter) h^{-1}. To compare the methods, sieved soil results are computed to correspond with the field measurements by determining the bulk density of soils and converting respiration, via the wet weight of a 20-cm layer of soil, to mg $CO_2\ m^{-2}\ h^{-1}$.

Chemical Analyses

Soil gravimetric water content (per soil dry weight) was determined by drying soil in 105°C for 24 h. Soil organic matter (OM) was estimated in triplicate by loss of weight on ignition (550°C, 12 h).

Total organic carbon (C_{tot}) and nitrogen (N_{tot}) were determined in triplicate using a CARLO ERBA Elemental Analyzer Mod EA 1106 (Carlo-Erba Strumentazioni, Milan, Italy). Dissolved organic carbon (DOC) was determined in duplicate from water using a UNICARB Total Carbon Analyzer (Elektro-Dynamo Ltd, Helsinki, Finland) and the high-temperature combustion method[23] designed for water analysis. Each soil sample (2–4 g w.w.) was mixed in 50 ml of distilled, deionized, and filtered (0.22 μm pore size) water. The suspension was shaken in a rotary shaker (30 rpm) for 60 min. About 10 ml of the suspension was then centrifuged (10 min, 4000 rpm) and filtered through a Whatman GF/C Glass Fiber Filter (0.7 μm pore size).

Acidity (pH) was measured in duplicate from the soil : water slurry (10 g d.w. of soil : 50 ml of deionized water). Samples were shaken in a rotary shaker (30 rpm) for 60 min and decanted after a 30-min settling period. pH was measured in a different vessel with a combination glass electrode (Orion Research).

Total phosphorus and basic cations were determined in duplicate using routine analysis methods: A portion of soil was ashed and dissolved into HCl. Total phosphorus (P_{tot}) was measured spectrophotometrically and calcium (Ca_{tot}), magnesium (Mg_{tot}), and potassium (K_{tot}) were measured using an atomic absorption spectrophotometer.

Table 1. Soil chemical characteristics. Results in the same row followed by the same letter do not differ at 95% level of significance (method: 95% confidence limit).

Characteristic	Soil 1	2	3	4
Loss on ignition (%)	33 a	50 a	88 b	91 b
Total C (%)	23	34	49	51
Total C (of OM) (%)	56.2 a	58.3 b	55.9 a	55.7 a
Dissolved C (mg g^{-1})	0.49 a	0.51 a	0.93 b	0.49 a
DOC/C_{tot}(*1000)[a]	0.94 a	0.86 a	1.67 b	0.87 a
Total N (%)	1.17	1.58	2.55	1.95
Total N (of OM) (%)	3.01 b	2.76 b	2.92 b	2.14 a
C/N	18.8 a	21.2 b	19.2 ab	26.1 c
Total P (mg g^{-1})	1.54 ab	1.30 ab	2.00 b	0.88 a
Total Ca (mg g^{-1})	9.3 a	5.9 a	18.1 b	22.0 b
Total K (mg g^{-1})	0.85 a	0.31 a	1.20 a	0.30 a
Total Mg (mg g^{-1})	1.9 b	0.8 a	1.2 a	2.8 c
pH	5.32	5.08	5.27	5.91

[a]DOC = dissolved C; C_{tot} = total C; 1000 = arbitrary coefficient.

Qualitative microscopic analyses of the soils were performed applying the methods of Aaby and Berglund[1] and Tolonen.[30]

Results

Soil Characteristics

Soil 1 and soil 4 differ the most and characterize the range of variation in soil quality; soil 1 represents highly humified and less-peaty soils of drier origin and soil 4 of less-humified peat soils of wet origin. Microscopic analyses of the peat in the soils are in Table 1; Table 2 lists soil chemical characteristics. Select correlations between soil characteristics are presented in table 3.

Field Measurements Versus Intact Cores

Field soil respiration rate (mg CO_2 m^{-2} h^{-1}) varied from 190 in soil 1 to 2163 in soil 3, the averages being 660 in soil 1, 693 in soil 2, 1118 in soil 3, and 787 in soil 4. The variation within and between the soils was substantial, but could be partly explained by the varying temperature at measurement (Figure 2).

Table 2. Qualitative microscopic peat analyses from soils 1–4. Soils are compared only with each other using a three step scale (+++, ++, +). Degree of humification is > 4 in von Post scale for all soils.

	Soil			
	1	2	3	4
Degree of humification	+++	+++	++	+
Fungal hyphae	++	(+)	+++	+
Tree pollen	*Pinus* *Picea*	*Pinus* *Picea*	–	*Picea*
Plant cells and fragments	*Betula* bark, deciduous tree cells	*Betula* bark, *Eriophorum vaginatum remains*	*Betula* bark, *Menyanthes* remains, deciduous tree cells, *Carex* roots	*Eriophorum vaginatum* and *Carex* remains, wand and tree tissue
Sphagnum spores	+	+++	+++	+++
Sphagnum tissue	–	++	+++	+++
Testate amoebae	*Centropyxis* sp.	–	–	*Trigonopyxis arcula, Arcella artocrea*
Water fleas	–	–	–	*Alona affinis*

Table 3. Select correlations between some soil characteristics. Correlation coefficient/probability (in parentheses); $n = 30$.

	Total N	Total N (of OM)	Total C	Total C of OM	Soluble C
Total C	0.945	−0.501	1.000	0.133	0.622
	(0.0001)	(0.0048)	(0.0001)	(0.4827)	(0.0002)
Loss on ignition	0.940	−0.522	0.997	0.061	0.617
	(0.0001)	(0.0031)	(0.0001)	(0.7495)	(0.0003)
pH	−0.090	−0.441	−0.124	−0.418	−0.455
	(0.6379)	(0.0146)	(0.5157)	(0.0215)	(0.0115)
Total P	0.405	0.586	0.169	0.179	0.616
	(0.0265)	(0.0006)	(0.3931)	(0.3444)	(0.0003)

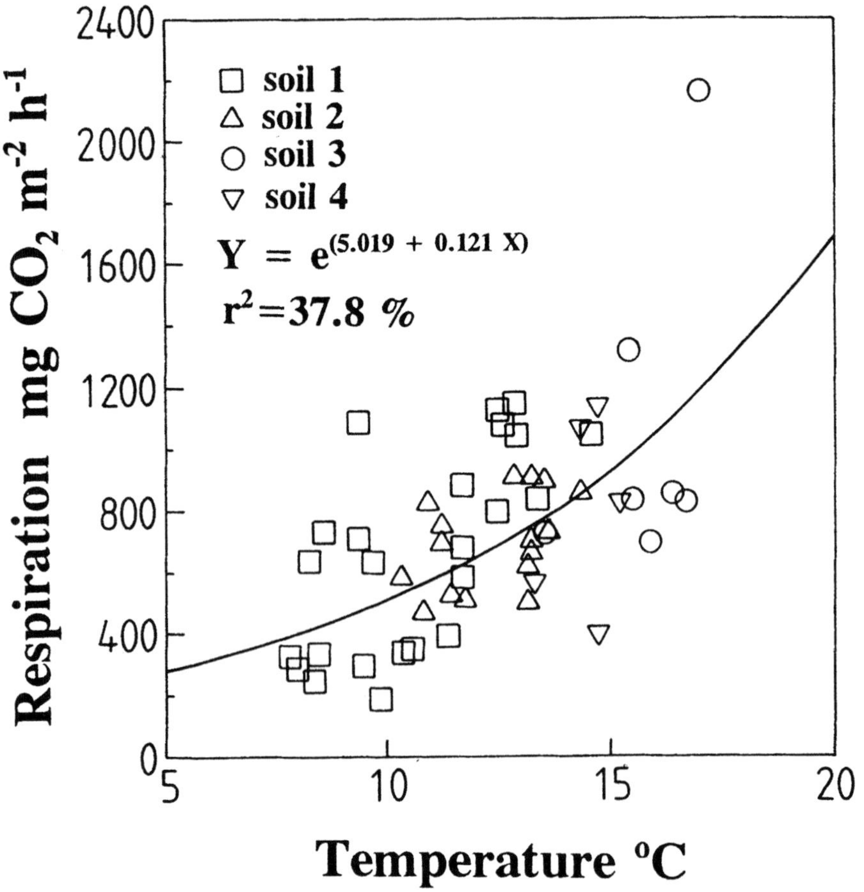

Figure 2. Soil respiration in different soils measured in the field. Temperature at the depth of 10 cm is shown on the x axis.

Intact core respiration depended more clearly on temperature (Figure 3). Respiration rate (mg CO_2 m^{-2} h^{-1}) increased from 100–300 at 5°C to 800–2000 at 25°C. At 15°C, which is a typical soil temperature during the summer at the depth of 10 cm, the following average respiration rates (mg CO_2 m^{-2} h^{-1}) were obtained from the regression curves: 630 for soil 1, 570 for soil 2, 450 for soil 3, and 380 for soil 4.

To compare the results, intact core respiration was calculated to correspond with the temperature of field measurements. The ratio of intact core to field

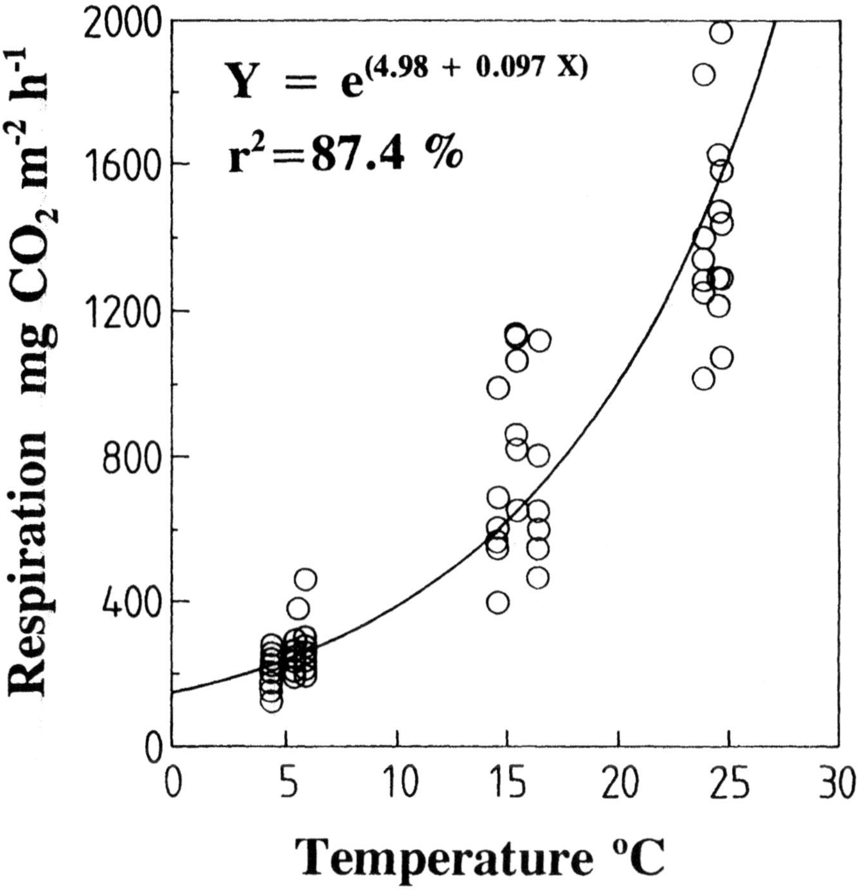

Figure 3. An example (soil 1) of respiration results obtained in laboratory measurements at different temperatures using large intact soil samples.

respiration was on the average near 1 (intact cores/field respiration = 1.04), but depended on respiration rate. When field respiration was highest, the ratio was lowest (~0.5) and vice versa (~2.75) (Figure 4).

The ratio was lower in soils 3 and 4 with a thick peat layer. However, the temperature during field measurements was more important than the thickness of the peat layer. According to the analysis of covariance, the effect of temperature was significant on the ratio (field measurement/intact cores), but that of soil was not.

Field Measurements Versus Sieved Soil

Sieved soil respiration was calculated to correspond with the temperature of field measurements and a 20-cm soil layer, and it varied from 71.6 to 837.4 mg CO_2 m^{-2} h^{-1}. The highest values were from deep peat soils 3 and 4. Respiration averages (mg CO_2 m^{-2} h^{-1}) were 190 for soil 1, 193 for soil 2, 597 for soil 3, and 419 for soil 4.

The ratio between sieved soil and field respiration varied from 0.063 to 1.12, the average being 0.39 (Figure 5). The dependence of the ratio on respiration rate was weak, but, in general, the ratio was higher in fields with a deep peat layer. Temperature during field measurements did not have a significant effect on the ratio according to the analysis of covariance.

Respiration Versus Soil Characteristics

There was a weak correlation between field respiration and soil total nitrogen ($r = 0.405$; $P < 0.026$, $n = 30$), and between field respiration and DOC ($r = 0.4341$; $P < 0.017$, $n = 30$), but no significant correlations between soil characteristics and intact core respiration were detected. Sieved soil respiration (mg CO_2 m^{-2} h^{-1} at 15°C) correlated well with soil moisture, soil organic matter, and the characteristics associated with organic matter (Table 4).

Sieved soil respiration was also calculated per soil dry weight (μg CO_2–C g^{-1} (d.w.) h^{-1} at 15°C) and the soil average was 3.1 for soil 1, 2.1 for soil 2, 8.8 for soil 3, and 8.5 for soil 4. Respiration in soils 1 and 2 was significantly lower than in soils 3 and 4. Respiration per soil dry weight had a very high correlation with soil moisture (Table 4).

Sieved soil respiration standardized for soil organic matter (μg CO_2–C g^{-1}(OM) h^{-1}) was 10.35 for soil 1, 5.1 for soil 2, 10.0 for soil 3, and 9.3 for soil 4. Soil 2 respiration was significantly less than that of soil 1. Respiration per organic matter correlated very significantly with organic matter, total K, and Mg (Table 4).

The ratio of sieved soil to field respiration correlated with soil characteristics better than that between intact cores and field measurements. The correlation between soil water content and sieved soil to field respiration ratio was 0.659 ($P < 0.0001$, $n = 30$), and also soil organic matter, pH, total carbon, total magnesium, and total calcium had a fairly weak ($r = 0.39 - 0.55$), but significant ($P < 0.007 - 0.035$) correlation with respiration ratio.

Regression Analyses

Regression analysis with stepwise variable selection was used to explain the variation in sieved soil respiration (per soil dry wt). A two-variable model

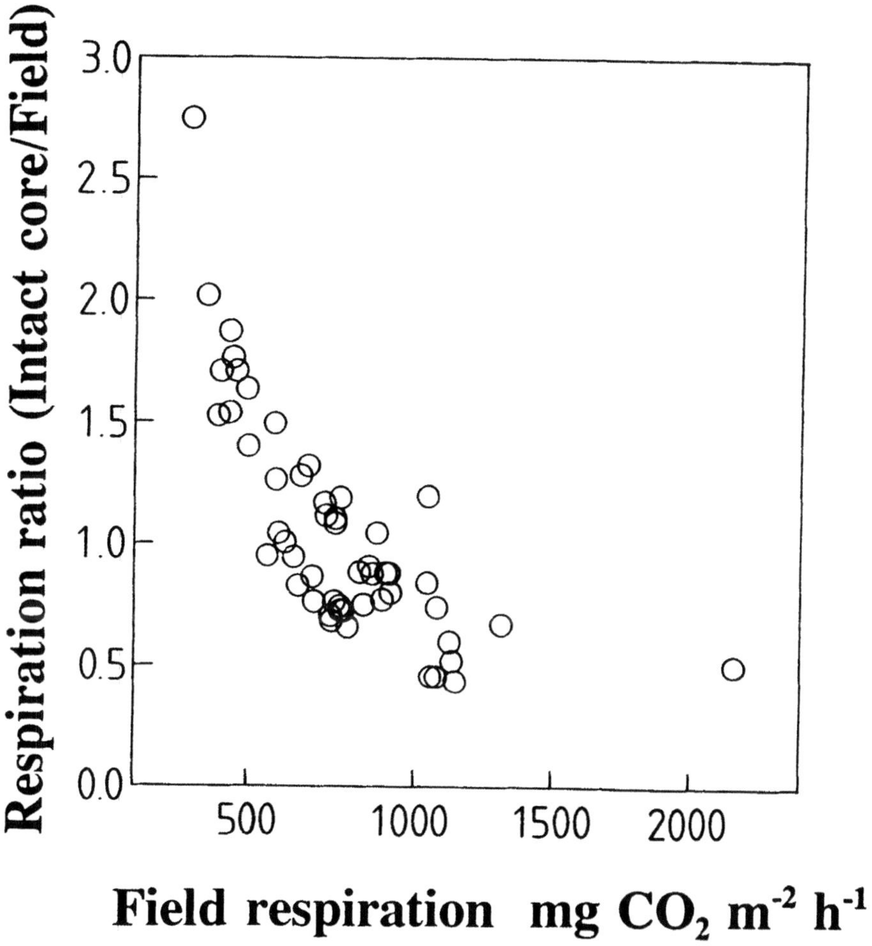

Figure 4. The ratio of respiration rates obtained from intact cores and field measurements at different respiration rates measured in the field.

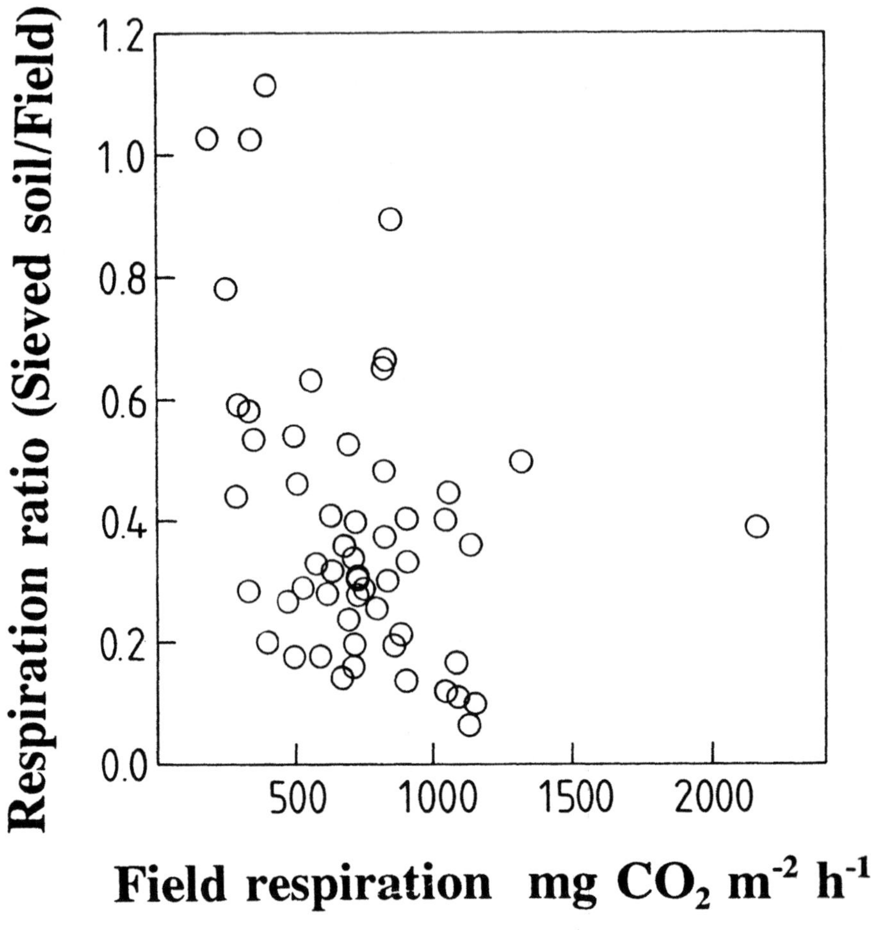

Figure 5. The ratio of respiration rates obtained from sieved soil and field measurements at different respiration rates measured in the field.

(respiration = −0.2045 + 0.0432 H_2O − 0.067 OM) explained 90.0% of the variation in respiration (Figure 6) at 15°C. Water content explained 81.6% of the variation.

Discussion

The respiration rates measured in the field, from intact cores and from sieved soil, were, in general, of the same order of magnitude as reported earlier from

Table 4. Select correlations of respiration data and some soil characteristics. Correlation coefficient/probability (in parentheses), $n = 30$.

	Respiration (per Soil Dry Weight)	Respiration (per Organic Matter)
pH	0.269	0.379
	(0.1506)	(0.0387)
Soil moisture	0.901	0.479
	(0.0001)	(0.0075)
Loss on ignition	0.575	−0.055
	(0.0009)	(0.7739)
Total C	0.548	−0.084
	(0.0017)	(0.6583)
Total N	0.524	−0.062
	(0.0029)	(0.7467)
Total K	0.563	0.683
	(0.0012)	(0.0001)
Total Mg	0.535	0.566
	(0.0023)	(0.0011)
Total Ca	0.779	0.396
	(0.0001)	(0.0302)

natural and agricultural soils using comparable methods.[8,20,21,27,28] However, the results obtained with different methods deviate from each other. Part of the intact core respiration results are substantially higher than field respiration whereas part of them is lower. Temperature and moisture[5] are likely to be significant in determining the changes in the ratio between intact cores and field respiration. The highest ratios (Figure 4) are from soil 1 in June when the mean temperature between the soil surface and the 20-cm depth differed by 6.5°C, whereas in August the difference was less than 2°C (data not shown). Core incubation, thus, changed the natural temperature gradient more in June than in August and probably overestimated the CO_2 evolution rate. Also mean soil moisture in soil 1 was higher in June (148% d.w.) than in August (113% d.w.).

On the other hand, the intact core to field respiration ratio was consistently low when field respiration rate was high. This indicates that a part of the respiration in the field originates from below the top 20 cm from the surface, which is included in core incubations. In cultivated soils as these, microbial activity is distributed more uniformly and lower in the profile than in natural, stratified soils.[33]

The reason for low sieved soil respiration as compared to field respiration is

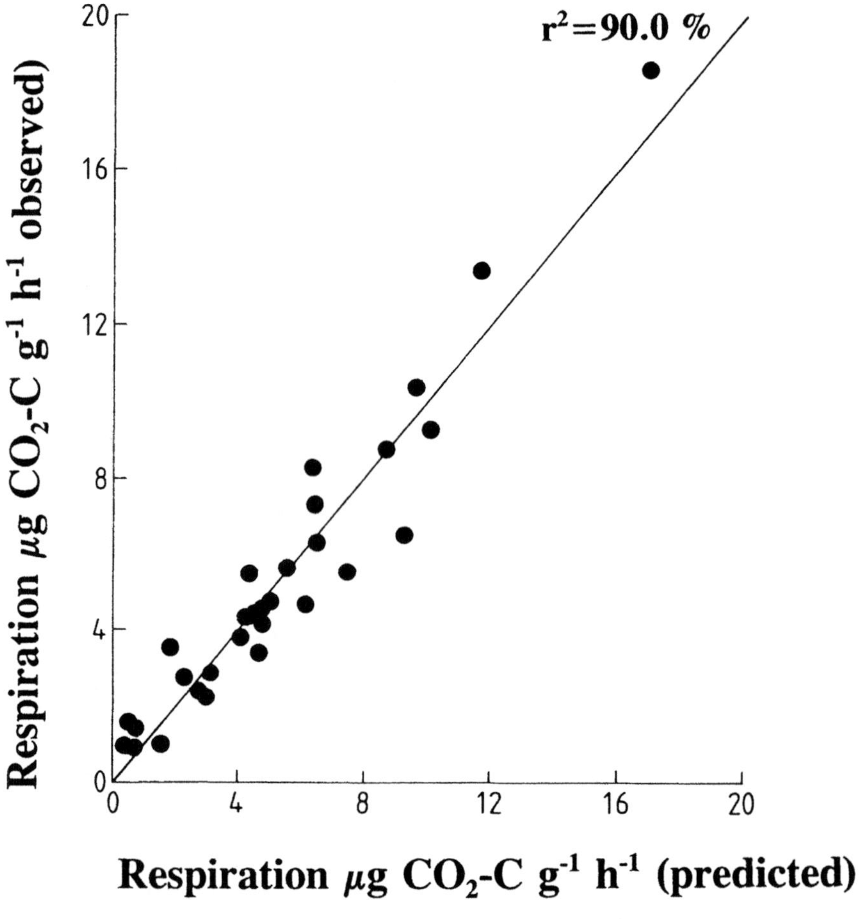

Figure 6. Predicted versus observed soil respiration at 15°C. Stepwise regression analysis model with soil water content (H_2O %, d.w. basis) and organic matter (OM %) as predicting variables was used.

the loss of the active, fresh organic matter due to sieving and removal of litter. Total CO_2 evolution from the soil originates from several sources (litter, roots, soil animals, etc.) besides bulk soil decomposition.[14] In field measurements and intact cores, these are all included, but are mostly missing from sieved soil. Probably uneven distribution of roots and litter explains the highly variable ratio between sieved soil and field respiration as compared to the ratio between intact cores and field respiration.

The lack of fresh organic matter from sieved soil also justifies why sieved soil respiration and soil properties—which were determined from sieved soil—are more tightly coupled together than soil properties and respiration by other methods. There were many significant correlations among soil properties (see also Ref. 32) and between soil properties and respiration, but soil water content explained over 80% of the variation in respiration: the drier the soil, the lower the respiration. The component effect of organic matter was, however, opposite: increase in OM content decreased respiration. Temperature and moisture[5,12,16] are generally the most important factors governing respiration in boreal zone soils, and it is likely that the available water limited respiration when high OM samples were measured in July. This can be concluded from the regression analysis (see also Ref. 18). Moreover, highly humified peat retains less water than poorly humified peat,[6] and drying of peat has been shown to be very rapid when water content is 100–200% (d.w.).[7] Thus, even rapid changes are possible in the water content and respiration[18] if the samples desiccate during sieving and incubation.

The differences in respiration among bulk soils seem to be very little. When standardized for organic matter, only the respiration of soil 2 is different from the others. The apparent reasons are the differences in the age and quality[26] of organic matter due to, e.g., cultivation,[4,15,25,29,31] and mixing of peat with underlying mineral soil which may retard or accelerate decomposition.[8] In soil 2, the high OM carbon concentration is indicative of low OM activity. This result is very similar to what was obtained from a series of soils with differing OM carbon concentrations (Hokkanen and Silvola, unpublished data).

Conclusions

Total CO_2 flux from soil can be most easily measured by using the field method. Measuring soil respiration from large intact soil cores in the laboratory gives fairly good estimates of field respiration rate if the soil is sampled at a low enough depth and the temperature gradient in the field at sampling is not high. The intact core method is, however, laborious. The sieved soil method underestimates CO_2 flux if the fresh organic matter has been removed, but obviously the results can be improved, if the respiration of removed organic matter is measured separately. Sieved soil respiration should preferably not be used for total soil CO_2 flux measurements, but, for example, in studying the relationships between soil characteristics and the decomposition process in conditions where moisture and temperature have been standardized.

Acknowledgments

The authors wish to thank M.Sc. Urpo Ahlholm and Miss Eine Ihanus for their reliable field work and Lic. For. Helen Jozefek-Ruhkala for the English language revision. The study was supported by a University of Joensuu research assistantship for Timo J. Hokkanen.

References

1. Aaby, B., and B.E. Berglund. 1986. Characterization of peat and lake deposits. In B. Berglund (ed.), Handbook of Holocene Palaeoecology and Palaeohydrology, J. Wiley and Sons, Chichester, pp. 645–666.
2. Anderson, J.P.E. 1982. Soil Respiration. In A.L. Page, R.H. Miller, and D.R. Keeney (eds.), Methods of Soil Analysis, Part 2. Chemical and Microbiological Properties, Agronomy Monograph No. 9 (2nd ed.), American Society of Agronomy, Madison, WI, pp. 831–871.
3. Armentano, T.V., and E.S. Menges. 1986. Patterns of change in the carbon balance of organic soil–wetlands of the temperate zone. J. Ecol. **74**:755–774.
4. Arshad, M.A., M. Schnitzer, D.A. Angers, and J.A. Ripmeester. 1990. Effects of till vs no-till on the quality of soil organic matter. Soil Biol. Biochem. **22**:595–599.
5. Bunnell, F.L., D.E.N. Tait, P.W. Flanagan, and K. Van Cleve. 1977. Microbial respiration and substrate weight loss—I. A general model of the influences of abiotic variables. Soil Biol. Biochem. **9**:33–40.
6. Dunberg, A., H. Krantz, and B. Lindström. 1986. Water retention characteristics of peat under non-compressed equilibrium conditions. In Basic Studies of Waterbinding in Peat, Report 1983–1986 (STEV 1461 191), Institute of Surface Chemistry, National Laboratory of Agricultural Chemistry, University of Agricultural Science, University of Umeå, pp. 197–204.
7. Dunberg, A., H. Krantz, and B. Lindström. 1986. Studies of peat drying under controlled conditions. In Basic Studies of Waterbinding in Peat. Report 1983–1986 (STEV 1461 191), Institute of Surface Chemistry, National Laboratory of Agricultural Chemistry, University of Agricultural Science, University of Umeå, pp. 187–196.
8. Farrell, E.P., and J.G. McDonnell. 1986. Decomposition in man modified peat soils. Int. Peat J. **1**:99–111.

9. Haber, W. 1959. Ökologische Untersuchung der Bodenatmung. Flora **146**:109–157.

10. Kivinen, E., and P. Pakarinen. 1981. Geographical distribution of peat resources and major peatland complex types in the world. Ann. Acad. Scient. Fenn. Ser. A III Geol.-Geogr. **132**:1–28.

11. Kosonen, M. 1969. CO_2 production in relation to temperature and plant mass. Oikos **20**:335–343.

12. Kowalenko, C.G., K.C. Ivarson, and D.R. Cameron. 1978. Effect of moisture content, temperature and nitrogen fertilization on carbon dioxide evolution from field soils. Soil Biol. Biochem. **10**:417–423.

13. Lucas, R.E. 1982. Organic soils (Histosols). Formation, Distribution, Physical and Chemical Properties and Management for Crop Production, Research Report 435, Michigan State Univ. Agric. Exp. Station and Cooperative Ext. Service, East Lansing, MI.

14. Lynch, J.M., and J.M. Whipps. 1991. Substrate flow in the rhizosphere. In D.L. Keister and P.B. Cregan (eds.), The Rhizosphere and Plant Growth, Kluwer Academic Publishers, The Netherlands, pp. 15–24.

15. Malmer, N., and E. Holm. 1984. Variation in the C/N-quotient of peat in relation to decomposition rate and age determination with ^{210}Pb. Oikos **43**:171–182.

16. Mathes, K., and Th. Schriefer. 1985. Soil respiration during secondary succession: influence of temperature and moisture. Soil Biol. Biochem. **17**:205–211.

17. McAfee, M. 1989. Aeration studies on peat soil 1. The effects of water-table depth and crop cover. Int. Peat J. **3**:157–174.

18. Mikola, J., and J. Komppula 1981. Respiratory activity of milled peat: a rapid method of measurement and observations on the effects of moisture, temperature, pH and peat quality. Soil Sci. **131**:156–162.

19. Nesterenko, I.M. 1976. Subsidence and wearing out of peat soils as a result of reclamation and agricultural utilization of marshlands. Proc. Fifth Int. Peat Congress I, Poznan, Poland, 232.

20. Rangeley, A., and R. Knowles. 1988. Nitrogen transformations in a Scottish peat soil under laboratory conditions. Soil Biol. Biochem. **20**:385–391.

21. Rochette, P., R.L. Desjardins, and E. Pattey 1991. Spatial and temporal variability of soil respiration in agricultural fields. Can. J. Soil Sci. **71**:186–196.

22. Ruuhijärvi, R. 1983. The Finnish mire types and their regional distribution. In A.J.P. Gore (ed.), Regional Studies. Ecosystems of the World 4B Mires: Swamp, Bog, Fen and Moor, Elsevier, Amsterdam, pp. 47–67.

23. Salonen, K. 1979. A versatile method for the rapid and accurate determination of carbon by high temperature combustion. Limnol. Oceanogr. **24**:177–183.

24. Schothorst, C.J. 1976. Subsidence of low moor peat soils in the western Netherland. Proc. Fifth Int. Peat Congress I, Poznan, Poland, pp. 206–217.

25. Schulten, H.-R., R. Hempfling, K. Haider, F.F. Gröblinghoff, H.-D. Ludemann,

and R. Frund. 1990. Characterization of cultivation effects on soil organic matter. Z. Pflanzenernähr. Bodenk. **153**:97–105.

26. Silvola, J., and U. Ahlholm. 1989. Effects of moisture and temperature on the decomposition of milled and sod peat. In S.A. Spigarelli (ed.), Int. Symposium on Peat/Peatland Characteristics and Uses, Bemidji State Univ., May 16–20, 1989, Center for Environmental Studies, Bemidji, MN, pp. 192–203.
27. Silvola, J., and S. Heikkinen. 1979. CO_2 exchange in the Empetrum nigrum-Sphagnum fuscum community. Oecologia (Berlin) **37**:273–283.
28. Silvola, J., J. Välijoki, and H. Aaltonen. 1985. Effect of draining and fertilization on soil respiration at three ameliorated peatland sites. Acta For. Fenn. **191**:1–32.
29. Srivastava, S.C., and J.S. Singh. 1989. Effect of cultivation on microbial carbon and nitrogen in dry tropical forest soil. Biol. Fertil. Soils **8**:343–348.
30. Tolonen, K. 1986. Rhizopod analysis. In B. Berglund (ed.), Handbook of Holocene Palaeoecology and Palaeohydrology, J. Wiley and Sons, Chichester, pp. 645–666.
31. Tyson, K.C., D.H. Roberts, C.R. Clement, and E.A. Garwood. 1990. Comparison of crop yields and soil conditions during 30 years under annual tillage or grazed pasture. J. Agric. Sci. **115**:29–40.
32. Vaughan, D., and B.G. Ord. 1985. Soil organic matter—A perspective on its nature, extraction, turnover and role in soil fertility. In D. Vaughan and R.E. Malcolm (eds.), Soil Organic Matter and Biological Activity, Developments in Plant and Soil Sciences Vol. 16, Martinus Nijhoff/Dr W. Junk Publishers, Dordrecht, pp. 1–35.
33. Woods, L.E. 1989. Active organic matter distribution in the surface 15 cm of undisturbed and cultivated soil. Biol. Fertil. Soils **8**:271–278.

Emission of Biogenic Sulfur Gases from Rice Paddies in Japan

Katsuyuki Minami, Ken-Ichi Kanda, and Haruo Tsuruta

Department of Environmental Management, National Institute of Agro-Environmental Sciences, Tsukuba 305, Japan

ABSTRACT

Emissions of biogenic sulfur gases from waterlogged paddy soils were studied in the laboratory, in paddy lysimeters, and in a paddy field. The laboratory studies involved measuring the gases produced when each of 5 different paddy soils were treated with 17 sulfur-containing compounds. Emissions of hydrogen sulfide (H_2S), carbonyl sulfide (COS), methylmercaptan (CH_3SH), dimethyl sulfide (CH_2SCH_3 or DMS), carbon disulfide (CS_2), and dimethyl disulfide (CH_3SSCH_3 or DMDS) were detected. Emission rates of DMS from the paddy lysimeter ranged from 4.1 to 7.3 mg S m^{-2} yr^{-1}. Emissions from lysimeters having a Gray Lowland soil were similar to those from lysimeters having a Light-colored Andosol, but emissions from lysimeters without percolation of water were greater than from lysimeters with water percolating at a rate of 20 mm per day. Annual emissions of DMS ranged from 2.4 to 3.7 mg S m^{-2} on plots within a paddy field. Diurnal and seasonal patterns of DMS evolution indicate that DMS emissions were greatly influenced by the activity of rice plants. Using our experimental data and the global harvested area of rice fields, we estimate that about 0.004 to 0.01 Tg S are emitted per year from paddy fields in the world.

Introduction

Interest in the atmospheric sulfur (S) budget has developed rapidly because anthropogenic activities are increasing the rates of emission of S gases into the atmosphere. These increases are important because S gases are oxidized in the atmosphere with the formation of sulfate, the resulting contributes to environmental problems associated with acid rain. There also is evidence that S gases may alter the optical properties of clouds which could affect global climate.[9]

Much attention has focused on anthropogenic emission of sulfur dioxide (SO_2) formed by combustion of coal and oil. These emissions have increased at a rate of approximately 2.4% per year during recent years. The annual anthropogenic

Table 1. Analyses of soils used.

Soil	Soil Group	Org C (%)	Total N (%)	Sand (%)	Clay (%)	pH	CEC[a] (me/100 g)
Fujimi	Gray Lowland soil	0.77	0.02	95	2	4.7	5
Nyuzen	Gray Lowland soil	0.95	0.07	78	7	6.8	6
Toyama	Gray Lowland soil	1.44	0.13	70	13	5.3	8
Isehara	Andosol	5.22	0.53	61	12	5.9	26
Iwanuma	Peat soil	7.63	0.58	37	35	5.6	25

[a]CEC: cation exchange capacity.

emission rate for this gas is currently estimated to be about 80 TgS.[5,30,32] Because natural emissions of S gases are estimated to be of similar total magnitude, we need to know more about natural sources of S gases to determine the extent to which human activity disturbs the atmospheric S cycle.

Large uncertainties remain concerning the chemical speciation and the magnitude of natural emission of S gases to the atmosphere. It has been well-documented that dimethylsulfide (DMS) is most important in oceans, which have been estimated to contribute 40 [4] or 15 TgS [12] to the atmosphere as S gases each year. The annual amounts of S evolved from soil have been estimated to be from 7 to 77 TgS.[6] Ecosystems studied include soil[1,2,28] vegetation,[13] soil and vegetation,[14,19,20] and fresh water.[8,27] Very little is known about emissions from paddy fields, which are abundant in many parts of the world.

The purpose of this chapter is to summarize the results of recent measurements of emissions of biogenic S gases from rice paddies in Japan. Emissions were measured in laboratory studies to learn more about the kinds and sources of S gases produced. Emissions were measured in paddy lysimeters and in paddy fields to gain better information about the amounts of DMS, carbonyl sulfide (COS), and carbon disulfide (CS_2) emitted, the seasonal and diurnal patterns of emission, and the factors affecting emissions of these gases under field conditions.

Experimental Procedures

Laboratory Studies

Samples of surface soils from paddy fields were collected to give a wide range of properties (Table 1). Before use, each sample was air-dried, ground to pass a 2-mm sieve, and amended with organic fertilizers or reagent-grade compounds. The organic fertilizers were rice straw, rice straw compost, bark compost, city

Table 2. Analyses of the organic materials used (oven-dry basis).

Material	Organic Carbon (%)	Total Nitrogen (%)	Total Sulfur (%)
Rice straw	40.7	0.58	0.035
Rice straw compost	30.1	1.63	0.305
Bark compost	55.0	0.28	0.036
City refuse compost 1	39.1	1.78	0.254
City refuse compost 2	32.8	2.41	0.178
Cattle manure	40.6	2.50	0.276
Sewage sludge 1	16.2	2.23	0.286
Sewage sludge 2	19.3	2.08	0.365

refuse compost, cattle manure, and sewage sludge (Table 2). Each was applied at a rate of 10 mg air-dried material per gram of soil. The regent-grade compounds were methionine, cystine, cysteine, methionine sulfoxide, methionine sulfone, taurine, ethionine, lanthionine, djenkolic acid, ammonium sulfate, potassium sulfate, potassium thiocyanate, sodium thiosulfate, thiamine nitrate, biotin, thioctic acid, and lipoic acid. Each was applied at a rate of 500 μg of S per gram of soil.

Emissions of volatile S gases were studied by placing 10-g samples of the soils in conical flasks (~67 ml), adding 10 ml of water to provide waterlogged conditions, sealing the flask with a silicone rubber septum, incubating the soils at 30°C for 12 days, and using previously described[21–23] gas-chromatographic techniques to determine volatile S gases in the flasks at various times during the incubation. It should be noted that the gas-sampling procedure involved shaking the flask to release any gases trapped within the soil.

DMS Emission from Paddy Lysimeters

Rice paddy lysimeter studies were performed using four lysimeters (2 m × 2.5 m, depth 1 m) at the National Institute of Agro-Environmental Sciences, Tsukuba, Japan. Two lysimeters contained a Gray Lowland soil and two contained a Light-colored Andosol. One lysimeter for each soil had a water percolation rate of 0 mm per day and the other had a water percolation rate of 20 mm per day. Rice plants were transplanted on 28 April and harvested on 26 September. Soil, air, and water temperature and solar radiation strength were monitored during the experimental period.

The chamber method described by Yagi and Minami[33] was used to measure

Table 3. Rates of mineral fertilizer and rice straw application and soil properties of the individual plots.

Plot	Mineral Fertilizer N	P_2O_5 (kg/ha)	K_2O	Rice straw (t/ha)	Soil Properties Total C	Total N (%)	Total S
No-N	0	60	90	0	1.4	0.11	0.040
Mineral	90	60	90	0	1.3	0.10	0.035
Straw	90	60	90	6	1.6	0.11	0.042

DMS fluxes from the lysimeters. DMS fluxes were determined by measuring temporal increases in the DMS mixing ratio of the air in the chamber. Details of the sampling and the gas-chromatographic techniques used in the study have been reported.[16]

DMS, COS, and CS_2 Emissions from a Paddy Field

Field studies were carried out from May 1990 to May 1991 in rice paddies at the Ryugasaki Branch of Ibaraki Agricultural Experiment Station, Japan. The soil in the paddy field is classified as Gray soil (Eutric Gleysols). The measurements were performed at three plots, each having a different long-term history of fertilization. These long-term fertilization practices are mineral fertilizer without nitrogen (No-N), mineral fertilizer (Mineral), and mineral fertilizer with rice straw application (Straw). Table 3 shows the rates of mineral fertilizer and rice straw application as well as soil properties of the individual plots.

The closed-chamber method described by Kanda and Minami[18] was used for the measurements of DMS, COS, and CS_2 fluxes from rice paddy field to the atmosphere. The accuracy of this procedure was confirmed by the linear temporal increase of the S gases concentration in the air within the chamber for 60 min. Analytical methods and calculation of annual emission rate of sulfur fluxes were described by Kanda et al.[17]

Results and Discussion

Sources of Gaseous Sulfur Compounds Produced in Paddy Soils

Table 4 shows that the following six volatile sulfur compounds were identified as products of microbial decomposition in waterlogged paddy soils incubated

Table 4. Volatile sulfur compounds evolved from paddy soils incubated under waterlogged condition for 12 days (ng/10 g soil).

Soil	H_2S	COS	CH_3SH	$(CH_3)_2S$	CS_2	$(CH_3)_2S_2$
Fujimi	0	8(2)	54(8)	110(9)	0	163(10)
Nyuzen	0	4(2)	8(2)	80(4)	2(2)	50(2)
Toyama	0	4(2)	0	7(4)	3(2)	0
Isehara	0	3(2)	0	200(4)	0	50(4)
Iwanuma	680(4)	30(4)	900(8)	1050(4)	0	51(4)

Note: Soil (10 g) was incubated (30°C) under waterlogged (10 ml water) conditions. Figures in parentheses indicate the day the maximum amount of sulfur compound evolved.

under laboratory conditions: hydrogen sulfide (H_2S), carbonyl sulfide (COS), methyl mercaptan (CH_3SH), dimethyl sulfide (CH_3SCH_3), carbon disulfide (CS_2), and dimethyl disulfide (CH_3SSCH_3)[24,25] Of these volatile sulfur compounds, H_2S and CH_3SH were not detected under paddy field conditions. They also were not detected in the laboratory studies when the incubation flasks were not shaken before sampling the gas phase. It appears likely that the volatile sulfur compounds contained in bubbles within soil were released to the vessel atmosphere during the shaking and that H_2S and CH_3SH in the gas phase would have been oxidized and diminished rapidly if the samples were incubated after shaking because their lifetime is very short in aerobic soil atmosphere.[24]

There are few reports in the literature of research concerning volatilization of sulfur from paddy soils treated with composts, sludges, manures, and plant materials. The volatile sulfur compounds evolved when Nyuzen soil was treated with such materials and incubated under waterlogged conditions are shown in Table 5.[24,25] The amounts of all sulfur compounds evolved from the treated soils were consistently larger than the amount evolved from the untreated soils. There was a tendency that, except for the rice straw, the amounts of COS evolved from soil treated with organic materials were correlated with the total sulfur content of the materials ($r = 0.76^*$).[22]

Studies to identify sources of six volatile sulfur compounds released from the soils under waterlogged conditions are presented in Table 6. Data for lanthionine and djenkolic acid are not given since little is known concerning the existence of these amino acids in soils.[21–25]

Numerous studies have shown that microorganisms can produce H_2S by reduction of sulfate. Many heterotrophic microorganisms can convert sulfur in organic substrates to H_2S under aerobic and anaerobic conditions.[7,15]. Although we also detected evolution of H_2S from paddy soils under waterlogged conditions after treatment with sulfate, cystine, and cysteine, it took 30 days for sulfate to

Table 5. Evolution of sulfur compounds from Nyuzen soil under waterlogged conditions treated with organic materials (ng/10 g soil).

Material	H_2S	COS	CH_3SH	$(CH_3)_2S$	CS_2	$(CH_3)_2S_2$
None	0	4(2)	8(2)	80(4)	2(2)	50(2)
City refuse compost	340(4)	15(4)	390(2)	287(2)	4(2)	120(2)
Bark compost	16(6)	6(4)	24(2)	62(4)	2(2)	43(2)
Sewage sludge	80(10)	18(4)	134(2)	823(2)	5(2)	120(2)
Rice straw	580(4)	47(4)	27(2)	300(2)	3(2)	51(2)
Rice straw compost	0	14(2)	0	80(3)	6(2)	0
Cattle manure	64(4)	17(4)	10(2)	100(2)	tr[a](2)	0

[a]Less than 2 ng/10 g soil.
Note: Soil (10 g) treated with material (100 mg) was incubated (30°C) under waterlogged (10 ml water) conditions for 12 days. Figures in parentheses indicate the day the maximum amount of compounds evolved. The data indicate the maximum amount of volatile sulfur compounds evolved on incubation.

yield detectable evolution of H_2S. Banwart and Bremner[7] have already reported that CH_3SH, CH_3SCH_3, and CH_3SSCH_3 are evolved from soils treated with methionine, methionine sulfoxide, methionine sulfone, or S-methyl cysteine. We had the same outcome (Table 6).

The sources of COS and CS_2 are not well-known. Banwart and Bremner reported that COS is derived from soils treated with lanthionine or djenkolic acid[7]. We detected evolution of COS from the soils treated with 5 of 17 S-containing compounds tested: cystine, cysteine, thiocyanate, lanthionine, or djenkolic acid.[21,22]

We detected evolution of CS_2 from the soils treated with 5 of 17 S-containing compounds tested: cystine, cysteine, thiosulfate, lanthionine, or djenkolic acid.[23] Although it has been reported that CS_2 is produced by decomposition of cystine, cysteine, lanthionine, or djenkolic acid,[7] our data indicated that sulfur in thiosulfate must also be considered as a potential source of CS_2 released from paddy soils under waterlogged conditions. Thiosulfates are found in soils.[3,31]

Dimethyl Sulfide Emission from Paddy Lysimeters

Figure 1 shows the diurnal variation of dimethyl sulfide (DMS) fluxes from the paddy lysimeters without water percolation (0 mm per day) on the Gray Lowland soil and the Light-colored Andosol. The fluxes of both soils were higher in the daytime than in the night. The measurements showed that there are relationships between DMS fluxes and solar radiation strength, DMS fluxes and

Table 6. Volatile sulfur compounds evolved from paddy soils incubated under waterlogged conditions after treatment with different of sulfur compounds.

	Volatiled S Compounds Evolved					
Form of S Added	H_2S	COS	CH_3SH	$(CH_3)_2S$	CS_2	$(CH_3)_2S_2$
Sulfate	+[a]	−	−	−	−	−
Thiosulfate	−	−	−	−	+	−
Methionine	−	−	+	+	−	+
Cystine	+	+	−	−	+	−
Cysteine	+	+	−	−	+	−
Thiocyanate	−	+	−	−	−	−

[a]It took 30 days to detect the evolution of H_2S after the addition of sulfate.
Note: Soil (10 g) treated with S-containing compounds (500 ppm S) was incubated (30°C) under waterlogged (10 ml water) conditions for 12 days.

air temperature, and DMS fluxes and water temperature. Soil type had little influence on DMS fluxes. The finding that DMS fluxes are best correlated with solar radiation intensity implies that rice plants have a large influence on DMS fluxes. The DMS fluxes varied from 0.2 to 2.4 $\mu g\ S\ m^{-2}\ h^{-1}$ during 38 h.[16] Staubes et al.[28] observed diurnal variation of DMS fluxes from upland soil in Germany and found a relationship between the diurnal change of the DMS fluxes and soil surface temperature. Cooper et al.[10,11] also reported that fluxes of DMS in a diurnal cycle were positively correlated to solar radiation and soil temperature in a coastal wetland.

Figure 2 shows the seasonal variation in DMS fluxes from the paddy lysimeters without water percolation on the Gray Lowland soil and the Light-colored Andosol. The DMS fluxes increased with the growing of rice plants, and the maximum rate was observed just before the heading period. After the heading period, the fluxes gradually decreased. The flux variation is probably due to the change of solar radiation strength, temperature, the activity of rice plants, and perhaps other factors during the growing season. The fluxes increased just after harvesting. This increase was probably due to diffusion of DMS occluded in soil atmosphere and plant debris-derived DMS through the cut aerenchyma system of the rice plants. There was no significant difference in the DMS flux between the Gray Lowland soil and the Light-colored Andosol, in spite of the difference observed in soil E_h.[16]

Our data of DMS flux from paddy fields with or without rice suggest that DMS emission is strongly linked to rice plant activity.[16] Fall et al.[13] found that emission rates of DMS from corn, wheat, and alfalfa increased with increasing air temperature and solar radiation under laboratory conditions. We also found

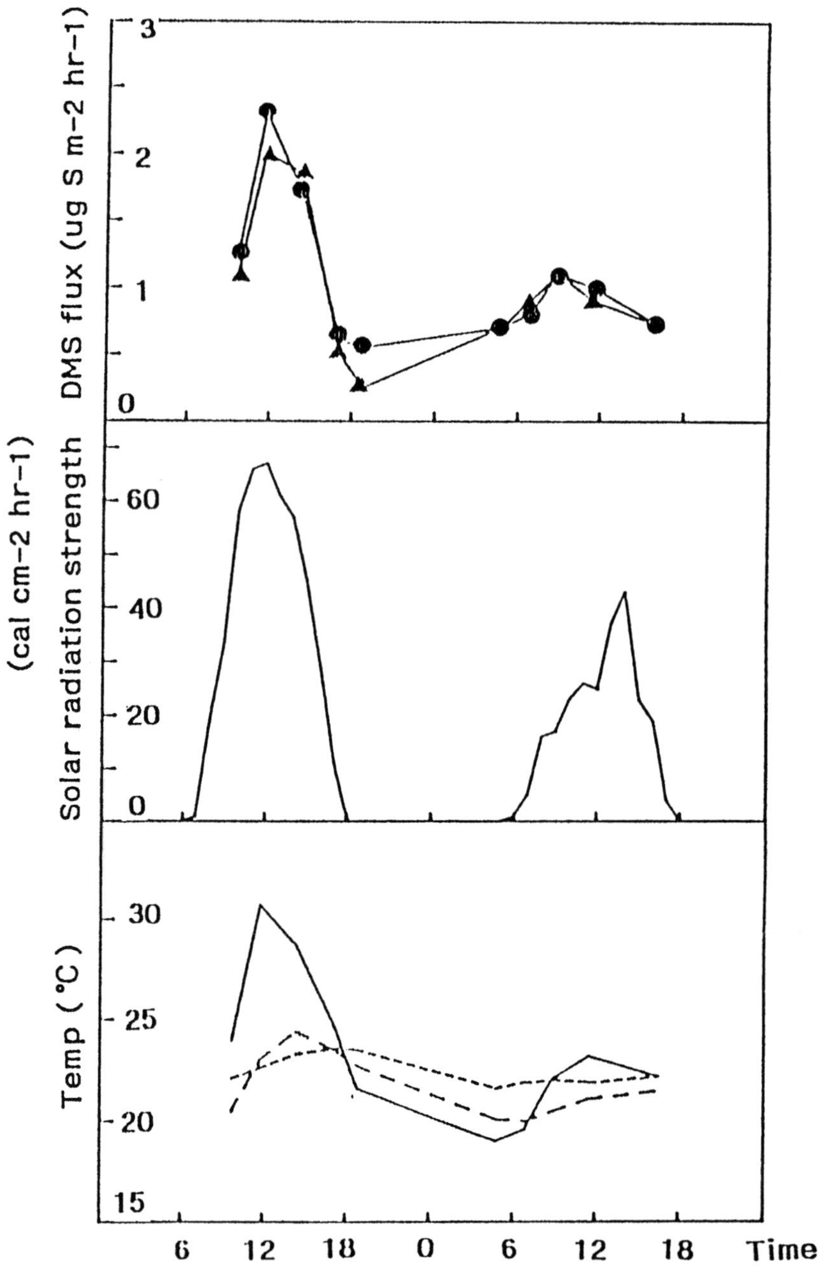

Figure 1. Diurnal variation of DMS fluxes, solar radiation flux, and temperature of air, water, and soil in paddy lysimeters on 20–21 September 1989. ● : Gray Lowland soil; ▲: Light-colored Andosol. Temperature (Air: —; soil of 5 cm depth:- - -; water:-- --).

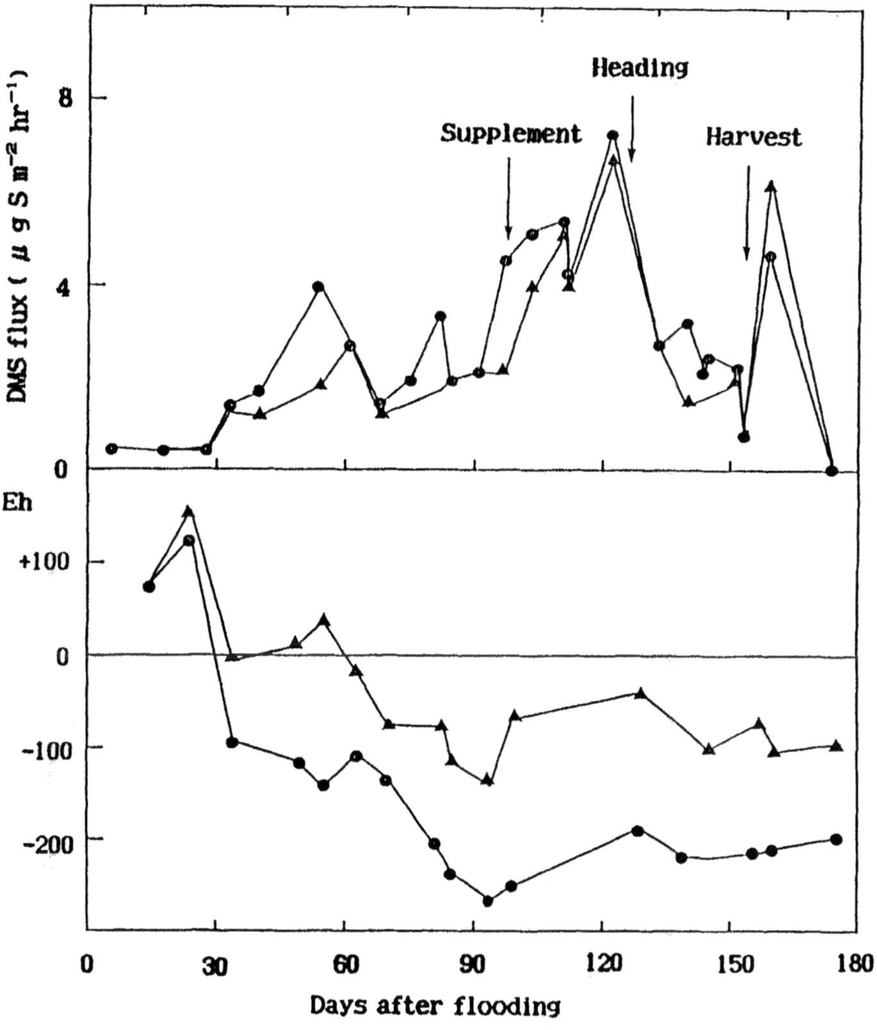

Figure 2. Seasonal variation of DMS fluxes and soil E_h in two paddy lysimeter fields after flooding. The soil E_h was measured at 5 cm depth. ● : Gray Lowland soil; ▲: Light-colored Andosol.

Table 7. Annual emission rates of DMS from the paddy lysimeters.

Soil	Water percolation[a]	Annual Emission Rate (mg S m^{-2})
Gray Lowland	0	7.3
	20	4.3
Light-colored	0	6.1
Andosol	20	4.1

[a]Water percolation rates of the lysimeters were adjusted to 0 and 20 mm per day during the observation period by periodically opening water stopping valves installed at the bottom of the lysimeters.

that the emission rate of DMS from rice paddies increased with increasing air temperature under Gray Lowland soil and Light-colored Andosol.[16]

By integration of the observed DMS emission rates, we calculated total DMS emissions from individual plots during the experimental period as shown in Table 7.[16] Emission rates of DMS from the plots ranged from 4.1 to 7.3 mg S m^{-2} yr^{-1}. Although there was no significant difference in the DMS emission rates between the Gray Lowland soil and the Light-colored Andosol, the emission rates of the 0-mm percolated plot was higher than the emission rates from the plot with 20 mm per day percolation.

Staubes et al.[28] reported that the mean annual emission rate of DMS was about 3 mg S m^{-2} for upland soils in Germany. The annual maximum emission rate of DMS from coastal wetland in Florida and Massachusetts, USA was 2870 mg S m^{-2}.[10,29] Our data on DMS emission rates from a rice paddy field were almost same as that for upland soils in Germany, but were lower than the emission rate from coastal wetlands, which contain plants that have high concentrations of DMS precursors DMSP.

DMS, COS, and CS_2 Emissions from a Paddy Field

Figure 3 shows the seasonal variation of total sulfur fluxes (DMS, COS, and CS_2) from No-N, Mineral, and Straw plots during 1990 to 1991.[18] We found clear seasonal variation of DMS fluxes from the three plots. Among the three sulfur gases, the flux of DMS was most important. Only small fluxes of CS_2 were observed, and COS was only observed in rice paddies during cropping periods. DMS flux increased with the size of rice plants and the maximum rate was observed at just after heading day. After then, the emission rates rapidly decreased. It seems likely, therefore, that DMS emission rates from rice paddy fields are influenced by activities of rice plants. The sulfur fluxes increased for

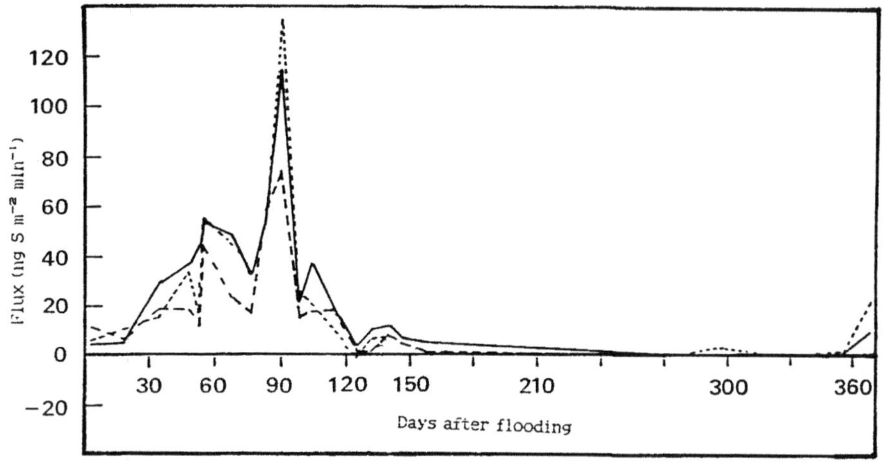

Figure 3. Seasonal variation of emission rates of sulfur gases (DMS+COS+CS_2) in paddy fields treated with mineral fertilizer without nitrogen (no nitrogen:---), mineral fertilizer (mineral; – –), and mineral fertilizer with rice straw (straw;----).

a short period just after harvesting. These increases probably are due to rapid decomposition of easily decomposable organic matter of rice residue by soil microorganisms and diffusion of gases through cut aerenchyma. However, the exact mechanism needs to be determined.

Table 8 shows that the annual emission rates of sulfur ranged from 2.8 to 4.9 mg S m^{-2} for the individual plots in the paddy field.[18] Application of nitrogen and rice straw accelerated the activity of soil microorganisms and rice plant and probably caused the increase in DMS, COS, and CS_2 emissions compared with no-nitrogen plots. The sulfur fluxes observed in this rice paddy field was slightly less than the average of 8 mg S m^{-2} yr^{-1} for upland soils in Germany.[15]

The harvested area of paddy rice is about 145 × 10^6 ha of which 95% is located in the Far East (*FAO Quarterly Bulletin of Statistics*. 1989). This area corresponds to approximately 9.5% of total global cultivated area. Using our experimental data and the global harvested area of rice fields, we estimate that about 0.004–0.01 Tg S are emitted annually from paddy fields in the world. These results suggest that sulfur emissions from paddy fields are unimportant in a global basis.

Acknowledgments

We thank Prof. A.M. Blackmer for his helpful suggestions in the preparation of this chapter, and also thank Mr. K. Yagi for his helpful suggestions.

Table 8. Estimation of annual sulfur emission rates from plots in a paddy field (mg S m^{-2}).

	During Cropping (1990, June–Sept.)			After Harvest (1990, Sept.–1991, June)			Annual (1990, June–1991, June)		
Gas	No-N	Mineral	Straw	No-N	Mineral	Straw	No-N	Mineral	Straw
DMS	2.2	3.2	3.0	0.3	0.6	0.3	2.4	3.7	3.3
COS	−0.1	−0.3	−0.1	0.2	0.1	0.9	0.0	−0.2	0.7
CS_2	0.2	0.2	0.2	0.2	0.2	0.8	0.4	0.4	0.9
Total	2.3	3.1	3.0	0.5	0.8	1.9	2.8	2.8	4.9

References

1. Adams, D.F., S.O. Farwell, M.R. Pack, and W. Bawesberger. 1979. Preliminary measurements of biogenic sulfur-containing gas emissions from soils. J. Air Pollut. Contr. Assoc. **29**:380–383.
2. Adams, D.F., S.O. Farwell, M.R. Pack, and E. Robinson. 1981. Biogenic sulfur gas emissions from soils in Eastern and Southeastern United States. J. Air Pollut. Contr. Assoc. **31**:1083–1089.
3. Alexander, M. 1977. Introduction to Soil Microbiology, 2nd ed, J. Wiley and Sons, New York.
4. Andreae, M.O., and H. Raemdonck. 1983. Dimethyl sulfide in the surface ocean and the marine atmosphere. A global view. Science **221**:744–747.
5. Andreae, M.O. 1989. The global biogeochemical sulphur cycle. A review. In B. Moore (ed.), Trace Gases and the Biosphere, University of Arizona Press, Tempe.
6. Aneja, V.P. 1990. Natural sulfur emissions into the atmosphere. J. Air Waste Manag. Assoc. **40**:469–476.
7. Banwart, W.L., and J.M. Bremner. 1975. Formation of volatile sulfur compounds by microbial decomposition of sulfur-containing amino acids in soils. Soil Biol. Biochem. **7**:359–364.
8. Bechard, M.J., and W.R. Rayburn. 1979. Volatile organic sulfides from freshwater algae. J. Phycol. **15**:379–383.
9. Charlson, R.J., J.E. Lovelock, M.O. Andreae, and S.G. Warren. 1987. Oceanic phytoplankton, atmospheric sulfur, cloud albedo and climate. Nature **326**:655–661.
10. Cooper, D.J., W.Z. de Mello, W.J. Cooper, R.G. Zika, E.S. Salzman, J.K. Prospero, and D.L. Savoie. 1987. Short term variability in biogenic sulfur emissions from a Florida Spartina alterniflora marsh. Atmos. Environ. **21**:7–12.
11. Cooper, W.J., D.J. Cooper, E.S. Saltzman, W.Z. de Mello, D.L. Savoie, R.G.

Zika, and J.M. Prospero. 1987. Emission of biogenic sulphur compounds from several wetland soils in Florida. Atmos. Environ. **21**:1491–1495.

12. Erickson, D.J., S.J. Ghan, and J.E. Penner. 1990. Global ocean-to-atmosphere dimethyl sulfide flux. J. Geophys. Res. **95**:7543–7552.
13. Fall, R., D.L. Alberitten, F.C. Fehsenfeld, W.C. Kuster, and P.O. Goldan. 1988. Laboratory studies of some environmental variables controlling sulfur emissions from plants. J. Atmos. Chem. **6**:341–362.
14. Goldan, P.D., W.C. Kuster, D.L. Albritton, and F.C. Fehsenfeld. 1987. The measurement of natural sulfur emissions from soils and vegetation: Three sites in the eastern United States revisited. J. Atmos. Chem. **5**:439–467.
15. Kadota, H., and Y. Ishida. 1972. Production of volatile sulfur compounds by microorganisms. Ann. Rev. Microbiol. **26**:127–138.
16. Kanda, K., and Minami, K. 1992. Measurement of dimethyl sulfide emission from lysimeter paddy fields. Ecol. Bull. 42.
17. Kanda, K., H. Tsuruta, and K. Minami. 1992. Emission of dimethyl sulfide, carbonyl sulfide and carbon disulfide from paddy fields. Soil Sci. Plant Nutr. **38**:709–716.
18. Kanda, K., and K. Minami. 1991. Measurement of dimethyl sulfide (DMS) emission from lysimeter paddy fields. Japan. J. Soil Sci. Plant Nutr. **62**:35–40 (in Japanese with English summary).
19. Lamb, B., H. Westberg, G. Allwine, L. Bamesberger, and A. Guenther. 1987. Measurement of biogenic sulfur emissions from soils and vegetation: Application of dynamic enclosure methods with Natusch filter and GC/FPD analysis. J. Atmos. Chem. **5**:469–491.
20. Mactaggart, D.L., D.F. Adams, and S.O. Farwell. 1987. Measurement of biogenic sulfur emissions from soils and vegetation using dynamic enclosure methods: Total sulfur gas emissions via MFC/FD/FPD determinations. J. Atmos. Chem. **5**:417–437.
21. Minami, K., and S. Fukushi. 1981. Detection of carbonyl sulfide among gases produced by the decomposition of cystine in paddy soils. Soil Sci. Plant Nutr. **27**:105–109.
22. Minami, K., and S. Fukushi. 1981. Volatilization of carbonyl sulfide from paddy soils treated with sulfur-containing substances. Soil Sci. Plant Nutr. **27**:339–345.
23. Minami, K., and S. Fukushi. 1981. Detection of carbon disulfide among the gases produced by thiosulfate and tetrathionate addition to soils. Soil Sci. Plant Nutr. **27**:541–543.
24. Minami, K., K. Okayama, and S. Fukushi. 1981. Volatile sulfur compounds evolved from paddy soils treated with organic materials. Japan. J Soil Sci. Plant Nutr. **52**:375–380 (in Japanese with English summary).
25. Minami, K. 1982. Volatilization of sulfur from paddy soils. Japan Agricul. Res. Qual. **15**:167–171.
26. Morra, M.J., and W.A. Dick, 1989. Hydrogen sulfide production from cysteine (cystine) in soil. Soil Sci. Soc. Amer. J. **53**:440–444.

27. Nriagu, J.O., and D.A. Holdway. 1989. Production and release of dimethyl sulfide from the Great Lakes. Tellus **41**:161–169.

28. Staubes, R., H.W., Georgii, and G. Ockelmann. 1989. Flux of COS, DMS and CS2 from various soils in Germany. Tellus **41**:305–313.

29. Steudler, P.A., and B.J. Peterson. 1984. Contribution of gaseous sulphur from salt marshes to the global sulphur cycle. Nature **311**:455–457.

30. Varhelyi, G. 1985. Continental and global sulfur budgets—1. Anthropogenic SO2 emissions. Atmos. Environ. **19**:1029–1040.

31. Wainwright, M. 1978. Distribution of sulfur oxidation products in soils and *Acer Pseudoplanuts* L. growing close to sources of atmosphere. Environ. Pollut. **17**: 153–160.

32. Watson, R.T., H. Rodhe, H. Oeschger, and U. Siegenthaler. 1990. Greenhouse gases and aerosols. In J.H. Houghton et al., (eds.), Climate Change, the IPCC Scientific Assessment, Cambridge University Press, Cambridge.

33. Yagi, K., and K. Minami. 1990. Effects of mineral fertilizer and organic matter applications on the emission of methane from some Japanese paddy fields. Soil Sci. Plant Nutr. **36**:599–610.

Emissions of Ammonia, Nitrous Oxide, and Methane from Cattle Slurry

O. Oenema, G.L. Velthof, and D.W. Bussink

Nutrient Management Institute, 9750 RA Haren, The Netherlands

ABSTRACT

Concomitant production of N_2O and CH_4 was measured during storage of unamended and HNO_3-amended cattle slurries at 15°C for 6 months. Emissions of NH_3, N_2O, and CH_4 were measured after surface application of these slurries to grassland in the field and under greenhouse conditions. Nitrous oxide was presumably the predominant end product of denitrification in the anoxic slurries. During storage, production of N_2O was relatively low in unamended and HNO_3-amended slurries of pH $\leq$ 5.0, but very high in amended slurries of pH 6.0. Production was most rapid at the slurry–atmosphere interface. Accordingly, more N_2O was released when the HNO_3-amended slurries were incubated for 2 days under ambient air than under O_2-free N_2. Frequent agitation and addition of H_2O_2 as an oxidant decreased N_2O production. Addition of HNO_3 inhibited methanogenesis almost completely. About 3 L CH_4 kg^{-1} was produced in unamended slurry and less than 3 ml kg^{-1} in HNO_3-amended slurry during storage for 6 months. After surface application, mean emission of NH_3 was more than 10 times lower from HNO_3-amended slurry of pH 4.5 than from unamended slurry. Negligible amounts of N_2O were emitted from unamended slurry, 50–450 μg N m^{-2} h^{-1} from slurry of pH 4.5, and 0.5–6 mg N m^{-2} h^{-1} from slurry of pH 6.0 for 5 to 7 days after surface application. Irrigation after surface application decreased emission of NH_3 but increased emission of N_2O.

Introduction

Ammonia (NH_3) plays a key role in aerosol chemistry[24] and may significantly contribute to acidification and eutrophication of land and water after its deposition.[20] A major source of NH_3 is urea from animal excretions. It has been estimated that the NH_3 emission from animal slurries in The Netherlands was 233 Gg or 12 g m^{-2} agricultural land in 1986.[1] This is about 90% of the total Dutch NH_3 emission into the atmosphere. As from 1987, a number of measures have been introduced to reduce the total NH_3 flux into the atmosphere by 70% or more by the year 2000. Such measures include increased storage of slurries in covered basins and application of these slurries to agricultural plots in a proper way and at an appropriate time with minimal losses.[6] However, increased storage time will increase methanogenesis and methane (CH_4) release from slurries.[25] The formation of CH_4 may contribute to the increasing atmospheric concentration of this trace gas and could contribute to global warming.[14] Amendment of animal slurries with nitric acid (HNO_3) to about pH 4.5 is another measure that has been introduced to reduce NH_3 emissions. Addition of HNO_3 may effectively decrease NH_3 emissions from slurries during storage and after surface application due to the shift in the $NH_3 : NH_4^+$ equilibrium. It also increases the fertilizer value of the slurry. However, animal slurries are highly anoxic, and under such conditions, the added NO_3^- may be denitrified to N_2O and N_2. The trace gas nitrous oxide (N_2O) may also have implications by its possible contribution to global warming[22] and the catalytic destruction of stratospheric ozone.[4] We have studied the effects of storage time and amendment with HNO_3 on the volatilization of NH_3, N_2O, and CH_4 from cattle slurries in laboratory, greenhouse, and field experiments.

Materials and Methods

Materials

Fresh cattle slurry, less than 1 week old, was used in the incubation study. Various 1–6-month-old cattle slurries were used in the field and greenhouse experiments. The slurries were obtained from local dairy farms.

Incubation Study

Homogenized samples of 12 kg of fresh cattle slurry in 42 PVC containers (26 L) were acidified to pH 6.0, 5.0, 4.5, 4.0, 3.5, and 3.0 by slow addition of 12.4M of HNO_3 under moderate stirring. To prevent excessive foam formation, the concentrated HNO_3 was added intermittently during several days at a mean temperature of 2°C. The pH of the slurry was 8.4, and the dry matter content was 114 g kg^{-1} at the start of the experiment. Containers with slurry were incubated in a climate-conditioned room at 15 ± 0.5°C for 6 months. At each pH level, there were three stirring treatments in duplicate. Slurries of treatment A were gently stirred with a rudder blade at 90 revolutions min^{-1}, for 1 min twice a week. This allowed sampling from homogeneous slurries. Slurries of treatment B were agitated at 240 revolutions min^{-1}, for 5 min twice a week, to stimulate aerobic biological oxidation. Slurries of treatment C received 7.5 mmol H_2O_2 kg^{-1} as oxidant, twice a week, and were agitated as for treatment B.

Changes in pH and E_h of the slurries were recorded twice a week. Additional

HNO_3 was added to the slurries if the target pH was exceeded by more than 0.15 units. Concentrations of dissolved and exchangeable NH_4^+ and NO_3^- in the slurries were measured in 20-ml subsamples. The subsamples were transferred to 50-ml polypropylene tubes containing 25 ml of 1.0M KCl. After shaking for 15 min, the slurries were centrifuged at 4000 revolutions min^{-1} for 15 min. The supernatant was diluted 100 to 250 times and analyzed by means of a Technicon auto-analyzer. Concentrations of N_2O in the slurries were determined in 5-ml subsamples transferred into 70-ml glass vials with rubber caps. Vials were agitated for 5 min and the concentration increase in the head space was analyzed via gas chromatography.

Fluxes of N_2O and CH_4 during storage were calculated from the changes in their concentrations in the head space after the containers were sealed off, usually for 0.5 to 2 h, but sometimes for 2 to 4 days. Air samples were taken with 12-ml polypropylene syringes through rubber septa in the caps at regular time intervals. Samples were analyzed by gas chromatography. During the last month, we also used a photoacoustic infrared Multi-gas Monitor 1302 from Bruel & Kjaer, directly attached to the containers.

Flux Experiments

A micrometeorological mass balance method[7] was used to measure NH_3 losses from cattle slurries surface-applied to grassland. This method assumes that the NH_3 loss is proportional to the difference in the vertically integrated product of NH_3 concentration and wind speed between the center and the upwind boundary of a measuring plot. Slurries were applied at rates of 1–4 kg m^{-2} with a computer-controlled applicator specifically designed for uniform application of slurries to small experimental fields. The measuring procedure is described in more detail elsewhere (D.W. Bussink, unpublished results).

Enclosures were used for the determination of N_2O and CH_4 emissions from surface-applied slurries to grassland (cf. Hutchinson and Mosier[10]). In the field, circular closed chambers (ID 0.6 m, ht, 0.2 m) were placed on the plots. In an attempt to circumvent the large spatial variability of natural emissions of N_2O in the field, emissions were also measured in greenhouse experiments. Slurries were applied by pipette to 6-months-old grass sod on homogenized sandy soil in 20-L pots. Closed chambers (ID 0.25 m; ht. 0.15 m) were placed over the pots for 1 or 3 h each day, for about seven consecutive days. Samples of the gas inside the chambers were taken with 12-ml polypropylene syringes, usually 0, 30, 60, and 120 min after the chambers were installed, and analyzed by gas chromatography.

Analytical Procedures

Redox potential (E_h) and pH were measured electrochemically. NO_3^- and NH_4^+ were analyzed by standard auto-analyzer methods (Technicon; US No.

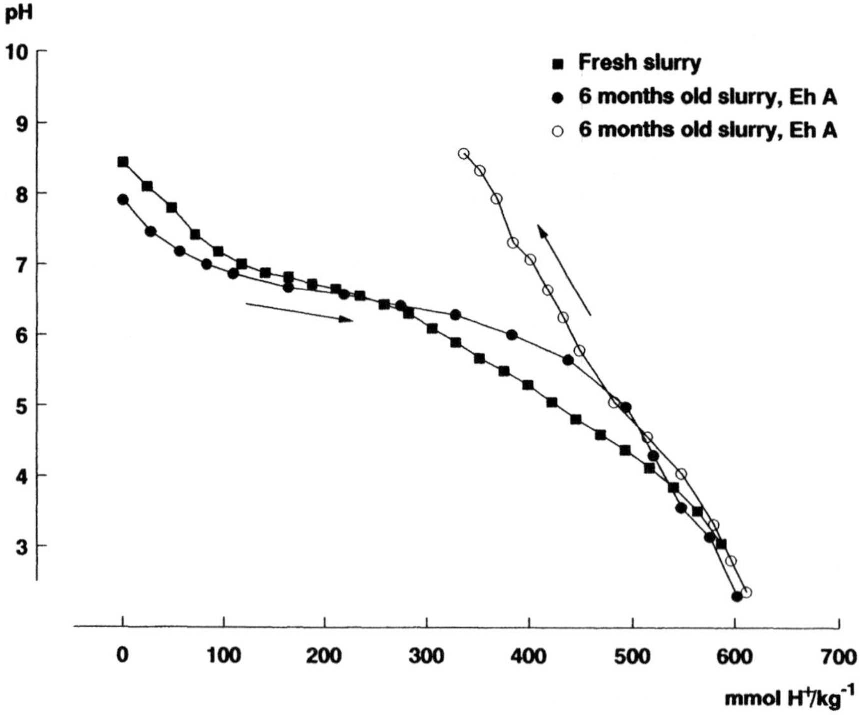

Figure 1. Titration curves of cattle slurries, fresh and after 6 months storage. The reverse titration curve using 4M NaOH of the cattle slurry after 6 months storage is also given.

824-87T and US No. 795-86T, respectively). Gas samples were analyzed via a PU 4400 gas chromatograph using a 27.5-m Porapak Q wide bore column and an electron capture detector (ECD) for the analysis of N_2O and CO_2, and a flame ionization detector (FID) for CH_4.

Results

The mean H^+ buffering capacity of the fresh cattle slurry was about 0.1 M kg^{-1} per unit of pH. Buffering was highest within the range of pH 7.0 to 6.5 (Figure 1). Ageing of the slurry for 6 months did not alter the buffering capacity significantly. The reverse titration indicated that OH^- buffering was less than H^+ buffering, suggesting that relatively small amounts of H^+ needed to be consumed to let the pH rise to values conducive to NH_3 volatilization. Because

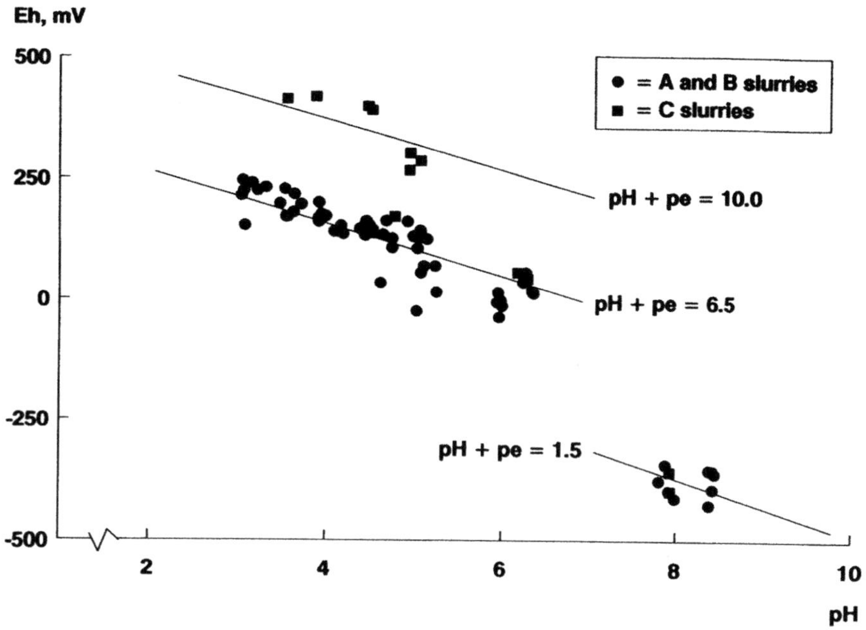

Figure 2. Relationship between pH and E_h in slurries. Three clusters with different pe+pH values have been distinguished (pe = 0.0169 E_h). Only results of the measurements in fresh cattle slurry and in slurry stored for 6 months have been plotted.

of the high H^+ buffering capacity of the slurry, the enrichment with NO_3^-–N was high. Concentrations of NO_3^-–N ranged from about 4.5 g kg^{-1} at pH 6.0 to 8.0 g kg^{-1} at pH 3.0.

The relationship between pH and E_h is shown in Figure 2. Three clusters of pe + pH were distinguished, with pe ≈0.016g E_h[19]:

a. Unamended slurries with a pe + pH of about 1.5.

b. Amended slurries with a target pH ≤ 5.0 and stirring treatments A and B, and slurries of pH 6.0 and treatment C, all with a pe + pH of about 6.5. A and B slurries of target pH 6.0 were generally somewhat more reduced and had a mean pe + pH of about 5.0.

c. Amended slurries of target pH ≤ 5.0 and treatment C, with a pe + pH of about 10.0.

The clustering clearly indicates that additions of HNO_3 and H_2O_2 significantly

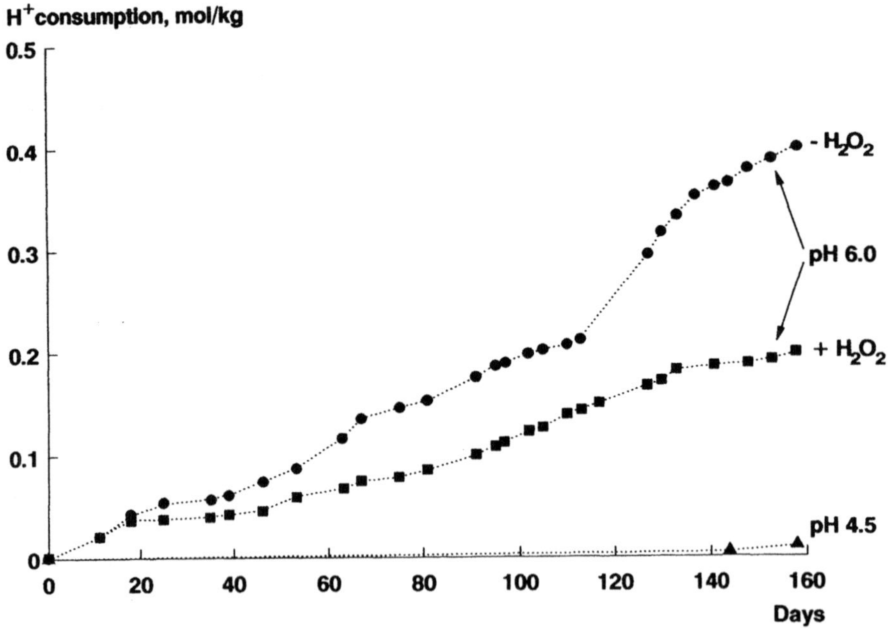

Figure 3. Cumulative H^+ consumption by slurries of pH 4.5 and stirring treatment B and by slurries of pH 6.0 and stirring treatment A (—H_2O_2) and C (+H_2O_2) for 160 days.

increased E_h. By contrast, strong agitation for 5 min twice a week did not significantly increase E_h.

During storage, rates of H^+ consumption were relatively low in unamended slurries and in amended slurries of target pH $\leq$ 5.0, but high in slurries of target pH 6.0. Agitation slightly but significantly decreased H^+ consumption. Regular addition of H_2O_2 to slurries of pH 6.0 decreased the H^+ consumption rate from a mean of 2 mmol kg^{-1} d^{-1} to about 1.2 mmol kg^{-1} d^{-1} (Figure 3). Note that between day 112 and day 126, the pH of the treatment without H_2O_2 was allowed to rise to pH 8.0 and that, as a consequence, the H^+ consumption rate was tripled between day 112 and day 138 (Figure 3). Consumption of H^+ was negligible in slurries of pH $\leq$ 5.0 during the first 140 days. Thereafter, H^+ consumption was increased to about 0.5 mmol kg^{-1} d^{-1}, but not in slurries of stirring treatment C.

Typically, the rate of H^+ consumption was fastest in surface layers during incubation. This was most pronounced in slurries of pH 6.0 and stirring treatment A (Figure 4) because denitrification was highest in this treatment. Within a few days, the pH of the surface layer increased to about 8, due to denitrification and the associated production of alkalinity. Meanwhile, E_h and NO_3^-–N concentra-

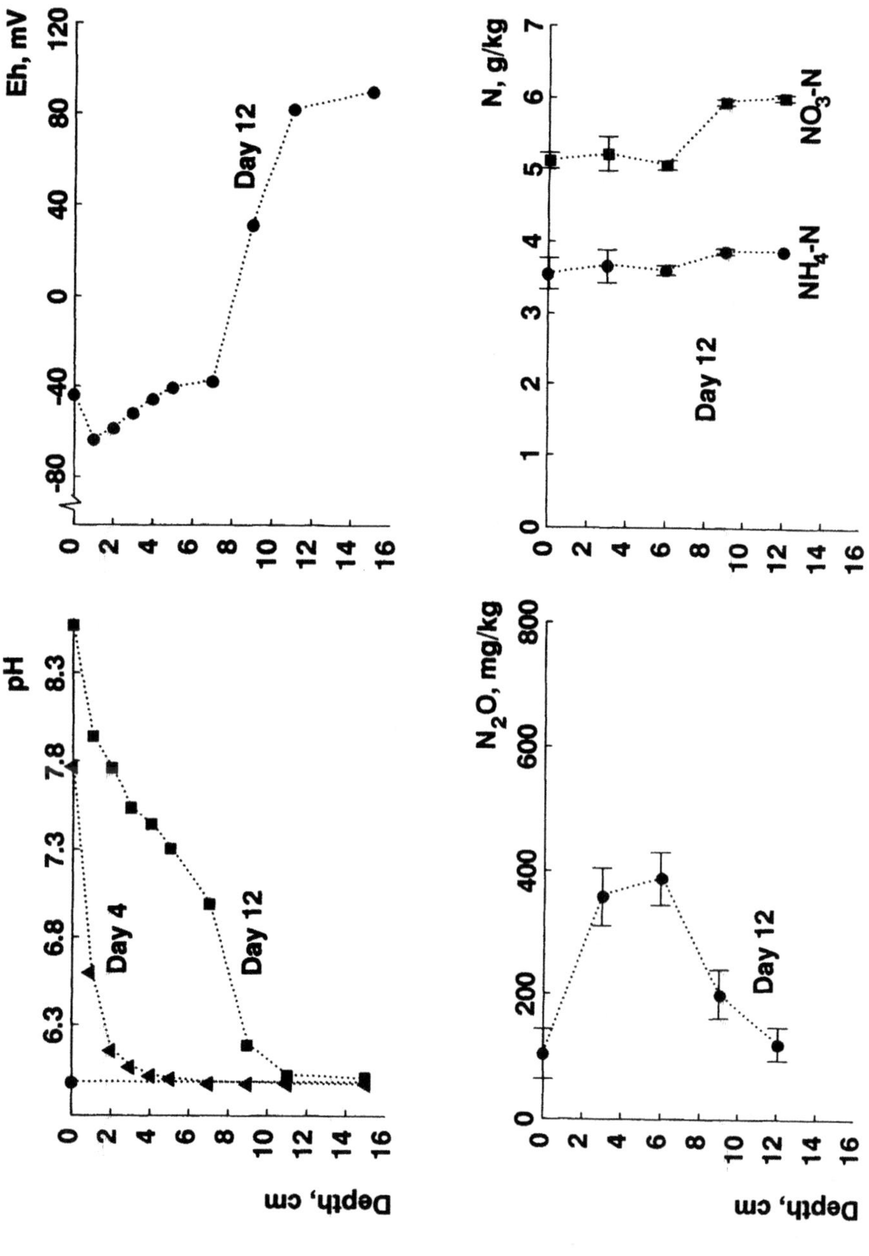

Figure 4. Depth profiles of pH, E_h, N_2O, NH_4, and NO_3^- in slurries of pH 6.0 and stirring treatment A 12 days after the slurry was homogenized.

Table 1. Diffusive N_2O and CH_4 fluxes from unamended and HNO_3-amended slurries during storage for up to 6 months. Means and standard deviations. Number of measurements in brackets.

Slurry	N_2O–N Flux (μg kg^{-1} h^{-1})				CH_4Flux (μg kg^{-1} h^{-1})			
Unamended	<0.01			(12)	572	±	120	(4)
pH 6.0	702	±	677	(24)	0.34	±	0.46	(4)
pH 5.0	1.03	±	2.64	(14)	0.20	±	0.26	(4)
pH 4.5	0.30	±	0.34	(16)	0.58	±	0.38	(4)
pH 4.0	0.25	±	0.28	(22)	0.18	±	0.22	(4)
pH 3.5	0.25	±	0.17	(7)	0.24	±	0.30	(4)
pH 3.0	0.21	±	0.12	(33)	0.07	±	0.09	(4)

tion were decreased in the surface layer and extremely high N_2O concentrations were built up at depths of 3 to 6 cm in the slurry after 12 days. NH_4^+ concentrations showed essentially no variation with depth, suggesting that the nitrification rate was relatively slow, yet, some of the N_2O in the slurry may have been released from the nitrification of NH_4^+. To determine whether the increased denitrification activity and N_2O production in the surface layer was due to the presence of O_2 in the head space, a 0.5-cm-thick layer of homogenized slurry of pH 4.5 was incubated in glass vials under ambient air or under an O_2-free N_2 atmosphere. Mean N_2O production was 14.7 μg N kg^{-1} d^{-1} in the O_2-free vials and 23.5 μg N-kg^{-1} d^{-1} in vials with ambient air, indeed suggesting that N_2O production in amended slurry was, directly or indirectly, stimulated by molecular O_2. During the sixth and last month of incubation we observed that the development of pH, E_h, N_2O and NO_3^- profiles versus depth became less pronounced because the denitrification rate was increased in the subsurface layers.

Regular agitation homogenized the slurries and also precluded the buildup of high N_2O concentration. Generally, 3 days after agitation, N_2O concentrations were up to 100 mg N kg^{-1} in slurries of pH 6.0 and only 0–to 20 μg N kg^{-1} in slurries of pH $\leq$ 5.0. Differences between pH levels 5 and 3 and among stirring treatments A, B, and C were much smaller than between slurries of pH 6.0 and pH $\leq$ 5.0. Hence, the interval between pH 5.0 and pH 6.0 was critical in the control of N_2O production and denitrification.

Mean diffusive N_2O fluxes during incubation were about three orders of magnitude higher from slurries of pH 6.0 than from those of pH $\leq$ 5.0 (Table 1). Diffusive fluxes of CH_4 were high from unamended slurries but negligible from amended slurries. Emissions of CO_2 also strongly decreased with a decrease in pH of the slurry, indicating that microbial activity was much less in amended slurries. Mass flow of N_2O was estimated from the rapid concentration increase

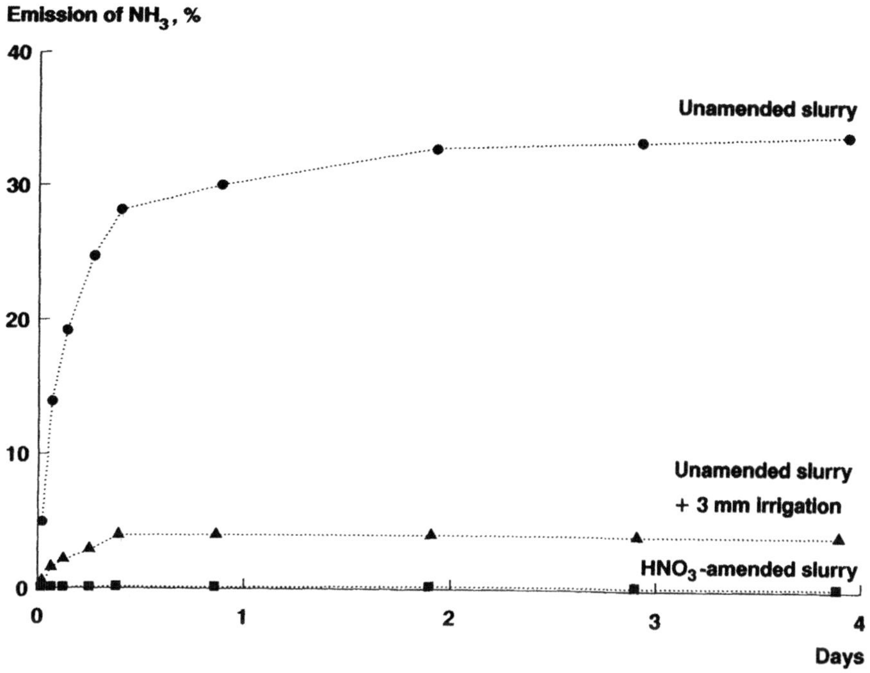

Figure 5. Emission of NH_3 from unamended slurry without and with immediate irrigation of 3 mm after surface application and from HNO_3-amended slurry of pH 4.5 to grassland. The application rate was 0.8 kg m^{-2}. Data of June 1990.

in the head space during agitation. The contribution of mass flow to the total flux of N_2O during storage was only about 10% in slurries of pH 6.0 but up to 400% in slurries of pH ≤ 5.0.

Lowering the pH to about 4.5 was effective in reducing volatile NH_3 losses from slurry in the field (Figure 5). Cumulative NH_3 loss from low-pH slurry was up to 167 times lower than that from untreated slurry after surface application to grassland. The volatilization of NH_3 from surface-applied slurries depends very much on weather conditions, i.e., temperature and rainfall. For example, irrigation immediately after surface application of untreated slurry reduced NH_3 volatilization by a factor of 2 to 6 (Figure 5).

N_2O emissions from amended slurries were increased by irrigation. Without irrigation, N_2O fluxes from slurry of pH 4.5 were below the detection limit except on day 4 (Figure 6a). High N_2O fluxes were measured from surface-applied slurry of pH 6.0, especially on day 1 (Figure 6b). The initial high N_2O

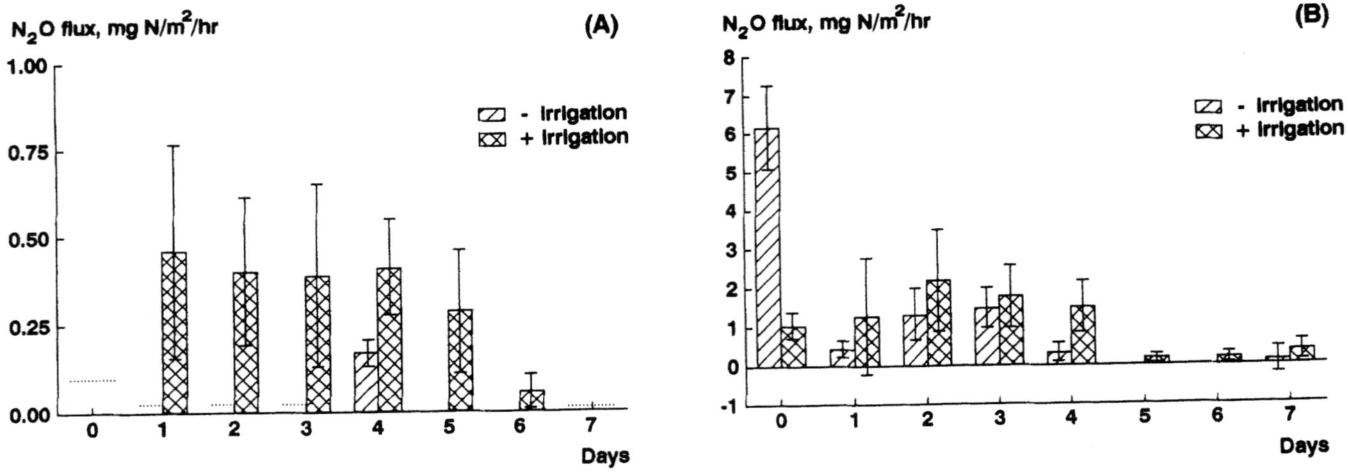

Figure 6. Emission of N_2O from slurry of pH 4.5 (a) and pH 6.0 (b) after surface application at a rate of 1.27 kg m^{-2} to pots (ID 25 cm) with *Lolium perenne* L. Half of the pots were irrigated with 5 mm water each day. Emission from pots without slurry was subtracted from pots with slurry to obtain the slurry-derived emission. Dotted horizontal lines indicate the detection limit. Means and standard deviations of four replicates. Data of June 1991.

flux from pH 6.0 slurry was related to its high N_2O concentration; the pH 6.0 slurry contained 6.3 mg N_2O kg^{-1} and the pH 4.5 slurry only 6.6 μg kg^{-1} at the time of application. Apparently, much of the N_2O stored in the slurry was emitted during the first hour after slurry application.

Discussion

Previous studies have indicated that lowering the pH of animal slurries with H_2SO_4 to pH 5.5 prior to surface application is effective in reducing NH_3 volatilization.[9,15,18] Results from this work and others (D.W. Bussink, unpublished data, 1990) indicate that a similar reduction in NH_3 volatilization may be obtained by adding HNO_3. In addition, the added NO_3^- may be efficiently utilized by the grassland, so that rates of soil acidification and amounts of unused residual NO_3^- in the soil after harvest are small if application rates are adjusted properly (Van Dijk and Oenema, unpublished data, 1990).

Addition of HNO_3 also reduced CH_4 emission from slurries during storage (Table 1). Psychrophilic digestion rates of cattle slurry at temperatures $\leq$ 15°C are generally low, but 10 L of CH_4 kg^{-1} may be produced during 6 months.[25] Assuming that slurry production is 55 kg per dairy cow per day, up to about 550 L of CH_4 may be produced per dairy cow per day. This amount is about equal to the amount that may be released per dairy cow per day during rumination.[5] In this study, the calculated mean release of CH_4 from unamended slurry was about 3 L kg^{-1} during an incubation time of 160 days. HNO_3-amended slurries produced only 1 ml kg^{-1} or less during the whole incubation period. Hence, HNO_3 addition to fresh cattle slurry is an effective inhibitor of methanogenesis in such slurries.

The major drawback of storing HNO_3-amended slurries is the high potential for denitrification and N_2O emission. In addition, if the slurry is denitrified, the resulting alkalinity rapidly increases the pH to about 8 and then the NH_3 may still volatilize. Rates of denitrification and N_2O emission were relatively low, provided that pH was kept less than 5.0 and the slurry was agitated regularly to minimize the formation of a high-pH surface layer (Table 1 and Figure 4). Clearly, lowering the pH to 6.0 was insufficient to avoid high rates of denitrification and N_2O emission during storage. In soils, the optimum pH for denitrification ranges[23] from 6.0 to 8.0, but in low-pH soils, there are microbial populations which are adapted to low pH and have their optimum at low pH.[12,16] The results of this study suggest that the optimum pH in cattle slurry is also $\geq$ 6.0 because the denitrification rate was much higher at pH 6.0 than at pH $\leq$ 5.0. Decreasing the target pH from 6.0 to 5.0 or lower reduced rates of denitrification and also N_2O emission two to three orders of magnitude, indicating that the denitrifying microbial population was not well-adapted to slightly acidic and acid environments. However, the relative increase in rate of denitrification in subsurface slurries during the sixth month of storage suggests that adaptation may have taken place. This needs further study.

Even though the nitrification process may have played a role, denitrification is thought to be the major source of N_2O because (1) rates of N_2O emissions showed a good relationship with rates of H^+ consumption and NO_3^- depletion (not shown), (2) N_2O was produced in the presence of acetylene (see below), and (3) NH_4^+ concentration decreased only slightly during incubation, suggesting that the nitrification rate was relatively slow in all treatments. The decrease in N_2O emission due to agitation and H_2O_2 addition would also suggest that denitrification and not nitrification was the dominant source of N_2O. However, dissimilatory NO_2^- reduction to NH_4^+ and mineralization of organically bound N may have supplemented the NH_4^+ pool which may lead to an underestimation of the nitrification rate. For example, Kaspar et al.[11] found that 60–70% of a small amount of $^{13}NO_3^-$ added to anoxic sewage sludge was reduced to $^{13}NH_4^+$. Such a mechanism cannot be excluded for the HNO_3-amended cattle slurries as used in our study. Hence, nitrification cannot be ruled out as a source of N_2O on the basis of a constant pool of NH_4^+ only.

The main mechanism causing the increased denitrification and N_2O production in the surface layer of the slurries is not clear. Carbon and NO_3^- were homogeneously distributed through the slurry due to the regular mixing. Consumption of bicarbonate by photosynthetic algae cannot account for the rapid pH increase at the surface either because the slurries were stored in the dark. It has been demonstrated frequently that E_h is critical in the range 200–400 mV.[13,21] However, basically all slurries in this study had an E_h below 250 mV, treatments with addition of H_2O_2 excluded (Figure 2), so that E_h cannot be a major controlling factor of denitrification in subsurface layers. Apparently, denitrifiers withstand acidic conditions more easily in the presence of molecular O_2, as is the case in the surface layer, than in the absence of O_2. After the start of denitrification and pH increase in the surface layer, the denitrification reaction may accelerate once the pH has passed some critical level. Under such conditions, all NO_3^- may be lost rapidly, as the availability of easily metabolizable carbon is not rate-limiting during the first few months of incubation.

A high NO_3^- concentration, a low pH, and moderately reducing conditions are generally considered as beneficial for the formation of N_2O relative to N_2 in the denitrification process.[8] Indeed, N_2O was presumably the predominant end product of dinitrification in the acidified slurries. Comparison of cumulative N_2O–N fluxes and cumulative NO_3^- removal indicated that essentially all NO_3^- was lost as N_2O. The predominance of N_2O over N_2 was confirmed in a batch experiment using the acetylene inhibition technique.[17] Similar amounts of N_2O were produced from slurries incubated under ambient air and under ambient air with 10% acetylene. Production of N_2O in the presence of acetylene also indicates that the nitrification process was not an important source of N_2O, as Bremner and Blackmer[2] have clearly indicated that acetylene is a potent nitrification inhibitor.

Even though microbiological denitrification appeared to be the major pathway, the chemical reduction of NO_2^- may also have contributed to NO_3^- removal. High concentrations of NO_2^- were observed frequently in slurries of pH 6.0, but in slurries of pH $\leq$ 5.0, the amounts were undetectable. In soils, the chemical reduction of NO_2^- to form various gaseous N compounds is generally favored by low pH.[3] The absence of NO_2^- in slurries of pH $\leq$ 5.0 would indeed suggest that no NO_2^- was produced or that reduction of NO_2 was faster than its formation, whereas for slurries of pH 6.0, the opposite was true. However, it is impossible to determine the contribution of chemical NO_2^- reduction to the total NO_3^- loss from these data.

The most important conclusions from this work are the following

1. Addition of HNO_3 strongly reduced rates of NH_3 volatilization and methanogenesis.
2. Emission of N_2O during storage at 15°C for up to 6 months is relatively low from HNO_3-amended slurries of pH $\leq$ 5.0, but very high from those of pH 6.0.
3. N_2O was presumably the predominant end product of denitrification in HNO_3-amended slurries.
4. A number of aspects need further study, the adaptation of denitrifiers to low pH, the rapid denitrification in the surface layers, and the effects of slurry composition.

References

1. Anonymous. 1987. Tussentijdse Evaluatie Verzuringsbeleid, Ministeries van VROM, V&W en LNV, Den Haag.
2. Bremner, J.M., and A.M. Blackmer. 1979. Effects of acetylene and soil water content on emission of nitrous oxide from soils. Nature **280**:380–381.
3. Chalk, P.M., and C.J. Smith. 1983. Chemodenitrification. In J.R. Freney and J.R. Simpson (eds.), Gaseous Loss of Nitrogen from Plant–Soil Systems, Developments in Plant and Soil Sciences Vol. 9, Martinus Nijhoff, The Hague, pp. 65–90.
4. Crutzen, P.J. 1976. Upper limits on atmospheric ozone reductions following increased application of fixed nitrogen to the soil. Geophys. Res. Lett. **88**:6708–6718.
5. Crutzen, P.J., I. Aselmann, and W. Seiler. 1986. Methane production by domestic

animals, wild ruminants, other herbivourus fauna, and humans. Tellus **38B**:271–284.

6. Den Boer, D.J., D.W. Bussink, and W.H. Prins. 1990. Measures to reduce losses of nitrogen from dairy farms with emphasis on ammonia volatilization. In N. Gaborcik, V. Krajcovic, and M. Zimkova (eds.), Soil–Plant–Animal Relationships, Proc. 13th European Grassland Federation, Banska Bystrica, pp. 347–351.
7. Denmead, O.T. 1983. Micrometeorological methods for measuring gaseous losses of nitrogen in the field. In J.R. Freney and J.R. Simpson (eds.), Gaseous Loss of Nitrogen from Plant–Soil Systems, Developments in Plant and Soil Sciences Vol. 9, Martinus Nijhoff, The Hague, pp. 133–157.
8. Firestone, M.K. 1982. Biological denitrification. In F.J. Stevenson, J.M. Bremner, R.B. Hauck, and D.R. Keeney (eds.), Nitrogen in Agricultural Soils, Agronomy No. 22, American Society of Agronomy, Madison, WI, pp. 289–326.
9. Frost, J.P., R.J. Stevens, and R.J. Laughlin. 1990. Effect of separation and acidification of cattle slurry on ammonia volatilization and the efficiency of slurry nitrogen for herbage production. J. Agric. Sci. **115**:49–56.
10. Hutchinson, G.L., and A.R. Mosier. 1981. Improved soil cover method for field measurement of nitrous oxide fluxes. Soil Sci. Soc. Amer. J. **45**:311–316.
11. Kaspar, H.F., J.M. Tiedje, and R.B. Firestone. 1981. Denitrification and dissimilatory nitrate reduction to ammonium in digested sludge. Can. J. Microbiol. **27**: 878–885.
12. Koskinen, W.C., and D.R. Keeney. 1982. Effect of pH on the rate of gaseous products of denitrification in a silt loam soil. Soil Sci. Soc. Amer. J. **46**:1165–1167.
13. Letey, J.N., N. Valoras, D.D. Focht, and J.C. Ryden. 1981. Nitrous oxide production and reduction during denitrification as affected by redox potential. Soil Sci. Soc. Amer. J. **45**:727–730.
14. Mattheus, E., and I. Fung. 1987. Global Biogeochem. Cycles **1**:61.
15. Pain, B.F., R.B. Thompson, Y.J. Rees, and J.H. Skinner. 1990. Reducing gaseous losses of nitrogen from cattle slurry applied to grassland by the use of additives. J. Sci. Food Agric. **50**:141–153.
16. Parkin, T.B., A.J. Sexstone, and J.M. Tiedje. 1985. Adaptation of denitrifying populations to low soil pH. Appl. Environ. Microbiol. **49**:1053–1056.
17. Ryden, J.C., and D.E. Rolston. 1983. The measurement of denitrification. In J.R. Freney and J.R. Simpson (eds.), Gaseous Loss of Nitrogen from Plant–Soil Systems, Developments in Plant and Soil Sciences Vol. 9, Martinus Nijhoff, The Hague, pp. 91–132.
18. Stevens, R.J., R.J. Laughlin, and J.P. Frost. 1989. Effect of acidification with sulphuric acid on the volatilization of ammonia from cow and pig slurries. J. Agric. Sci. **113**:389–395.
19. Stumm, W., and J.J. Morgan. 1981. Aquatic Chemistry, J. Wiley and Sons, New York.
20. Van Breemen, N., P.A. Burrough, E.J. Velthorst, H.F. van Dobben, Toke de Wit,

T.B. Ridder, and H.F.R. Reijnders. 1982. Soil acidification from atmospheric ammonium sulphate in forest canopy throughfall. Nature **299**:548–550.

21. Van Cleemput, O., W.J. Patrick Jr., and R.C. McIlhenny. 1976. Nitrate decomposition in flooded soil under different pH and redox conditions. Soil Sci. Soc. Amer. J. **40**:55–60.
22. Wang, W.C., Y.L. Yung, A.A. Lacis, T.M. Moe, and J.E. Hansen. 1976. Greenhouse effects due to man-made perturbations of trace gases. Science **194**:685–690.
23. Wyler, J., and C.C. Delwiche. 1954. Investigations on the denitrifying process in soil. Plant and Soil **5**:155–169.
24. Ziereis, H., and F. Arnold. 1986. Gaseous ammonia and ammonium ions in the free troposphere. Nature **321**:503–505.
25. Zeeman, G. 1991. Ph.D. Thesis, Agricultural University, Wageningen, The Netherlands.

BIOGEOCHEMISTRY OF GLOBAL CHANGE

BIOGEOCHEMISTRY OF GLOBAL CHANGE

Radiatively Active Trace Gases

Selected Papers from the Tenth International Symposium on Environmental Biogeochemistry, San Francisco, August 19-24, 1991

SPRINGER-SCIENCE+BUSINESS MEDIA, B.V.

Terrestrial Systems

Preface

Public awareness has increased about the prospect of global climatic change occurring in the coming century. Whereas broad changes in the Earth's climate have taken place in the geologic past, they have, with some significant exceptions, been of a gradual nature. Gradual change allows time for plants and animals to evolve via natural selection and to produce species capable of thriving in the new climate. Additionally, the geographic range of species already present will either expand or contract with an altered climate, depending on their ability to establish themselves in new environment.

It is now possible that anthropogenic activity may induce such a change and that this may occur at a rate too rapid for present ecosystems to adapt to or be replaced by systems suited for the new climate. Of even greater concern is whether agricultural production can be sustained to meet human needs in the face of such broad environmental change. Although there is a great disparity of views among climatologists with regard to the extent, significance, and potential disruptive effects of climate change, there is general agreement that some degree of change will occur.

The basis for this type of climatic change is directly related to the increased abundance of radiatively active trace gases in the atmosphere. "Radiatively active" refers to the capacity of certain gases, such as CO_2, CH_4, N_2O, and chlorofluorocarbons (CFCs), to absorb infrared energy reflected from sunlight incident on Earth's surface. The increased abundance of these gases derives from expanded human industrial and agricultural activities occurring over the past two centuries. Because many of these heat-trapping gases have demonstrated continuously increasing patterns of atmospheric mixing ratios over time periods ranging from decades to centuries, a "greenhouse effect" has occurred as a consequence of their buildup. In addition, some of these gases, such as CFCs and N_2O, pose a threat to the integrity of the stratospheric ozone layer. Still others, such as dimethylsulfide, although not "radiatively active," nonetheless influence climate because their oxidation products induce cloud condensation. Hence, these gases act in complex, interdependent patterns that affect atmospheric chemistry and physics. The sum of their influences on these processes, as well as on global change, is difficult to predict. In addition, their sources and fates are not well-known.

With the exception of only a few cases (e.g., CFCs), most of these gases arise from mixed sources, that is, from industry as well as agriculture, biomass-burning, and emissions from unperturbed ecosystems. Similarly, recognized sinks are diverse and include atmospheric photochemical reactions, dissolution into the oceans, and biological uptake by soil and plant communities. Understanding the relative importance of these sources and sinks to the past, present, and future global balance of these gases, as well as their influence on the biota and global habitability, poses a difficult and interdisciplinary challenge to environmental scientists and managers. Clearly, this topic encompasses significant aspects of numerous disciplines, including atmospheric chemistry, oceanography, paleoclimatology, geochemistry, soil and plant sciences, and microbiology, to name but a few. In short, this falls under the general discipline of "environmental biogeochemistry," with the scope of the problem being beyond the expertise of any individual investigator. It follows that no single conference or book can hope to cover adequately the breadth of this rapidly moving field. Nonetheless, because of the diverse and disparate perspectives of the disciplines listed above, it is essential for individual researchers to gain an appreciation of the breadth of the field. One mechanism to achieve this is by holding interdisciplinary conferences that allow a free exchange of experimental results on the topic of the biogeochemistry of radiatively active trace gases. This volume consists of invited and contributed papers presented at the Tenth International Symposium on Environmental Biogeochemistry (ISEB), held in San Francisco during the week of 19–25 August 1991. There are 45 chapters in this volume, with diverse topics ranging from geologic temporal scales of mass transfer to aspects of microbial biochemistry related to the formation of CH_4. Most of the chapters deal with experimental assessments of the fluxes of trace gases or the biological formation (or consumption) of trace gases in various ecosystems.

I conceived the idea of hosting such a meeting when I attended the Ninth ISEB in Moscow in 1989. I previously organized a meeting devoted to atmospheric methane at the 1987 American Chemical Society meeting in Denver, but I was interested in conducting a meeting having a broader theme. Inasmuch as the ISEB is an international scientific organization and because "radiatively active" trace gases are truly a problem of global concern, the ISEB would be an appropriate forum for an expanded conference. My offer was accepted, and as the ISEB meets on alternate, odd-numbered years, I returned home thinking that two years was more than sufficient to put it all together. I soon realized that time was already quite short and that I would be working on this project for the next three years, if you include editing this volume. My life was occupied with worrying about where to hold the conference, what agencies were willing to support it, mailing lists and flyers, bank accounts and bills, session topics and chairpersons, reservations and cancellations, abstracts, manuscripts, publishers, receptions, tours, banquets, and, of course, whether the San Francisco Giants would play a home game that week. (They did!) So much for my naiveté!

There are many individuals and institutions to thank for bringing this conference and book to fruition, but as the result of space restrictions, I can acknowledge only some of them. First of all, the scientific agencies of the United States Government that provided generous financial support for the event: The Geological Survey, the National Aeronautics and Space Administration, the National Oceanic and Atmospheric Administration, the Department of Energy, the Environmental Protection Agency, the Department of Agriculture, and the National Science Foundation. In particular, the National Science Foundation acted as a funding/clearinghouse for several of the above agencies, in addition to providing its own share of funds. I am indebted to the efforts of Dr. Jarvis Moyers, who successfully guided the various ISEB funds through the bureaucratic maze, and did so in a timely fashion. If for nothing else, I wish to thank him for being able to achieve all this, even in the face of my many annoying, panic-stricken phone calls! I wish to thank Jo Ann Glennon and her staff at the Cathedral Hill Hotel for making the meeting go so smoothly, and the members of the ISEB 10 National Committee for the programmatic work, and the ISEB 10 Local Committee for numerous arrangements. As all the manuscripts in this volume received two anonymous peer reviews, I thank these nameless souls for their efforts. Most important, I am deeply grateful to my associate, Ms. Francis Strohmaier, who was involved in every facet of this undertaking, and whose attention to detail and tireless secretarial efforts not only helped me to retain my sanity, but made both the conference and the publication of the book possible.

Ronald S. Oremland
Chairman, ISEB
Menlo Park, California
September 22, 1992

Sources and Migration of Methane-Rich Gas in Sedimentary Rocks on the Exmouth Plateau: Northwest Australian Continental Margin

Philip A. Meyers

Department of Geological Sciences and Center for Great Lakes and Aquatic Science, The University of Michigan, Ann Arbor, Michigan 48109-1063

Lloyd R. Snowdon

Institute of Sedimentary and Petroleum Geology, Geological Survey of Canada, Calgary, Alberta T2L 2A7 Canada

ABSTRACT

Samples of rock obtained by drilling during Ocean Drilling Program Leg 122 on the northwest Australian margin record the accumulation of sediments from Early Cretaceous to modern times. Neocomian siltstones from the Exmouth Plateau are lithologically equivalent to portions of the more deeply buried, onshore Barrow Formation, which hosts production of oil and gas. The thermally immature organic matter in the siltstones is predominantly derived from land plants. Gaseous hydrocarbons are nonetheless abundant, and their composition is dominated by methane in most of the samples. This composition mimics biogenic gas. Deeper samples (~1000 m subbottom) contain significant amounts of thermogenic C_2 to C_5 gaseous hydrocarbons but no change in organic matter character. The change in gas compositions suggests that the primary origin of methane throughout this sedimentary column is from thermal decomposition of continental organic matter and that the source is from stratigraphic units more deeply buried on the Exmouth Plateau. Preferential enrichment of methane has occurred during upward migration of gas through these fine-grained sediments.

Introduction

Ocean Drilling Program (ODP) Legs 122 and 123 investigated during 1988 the organic matter contents of sediments deposited since Early Cretaceous times on

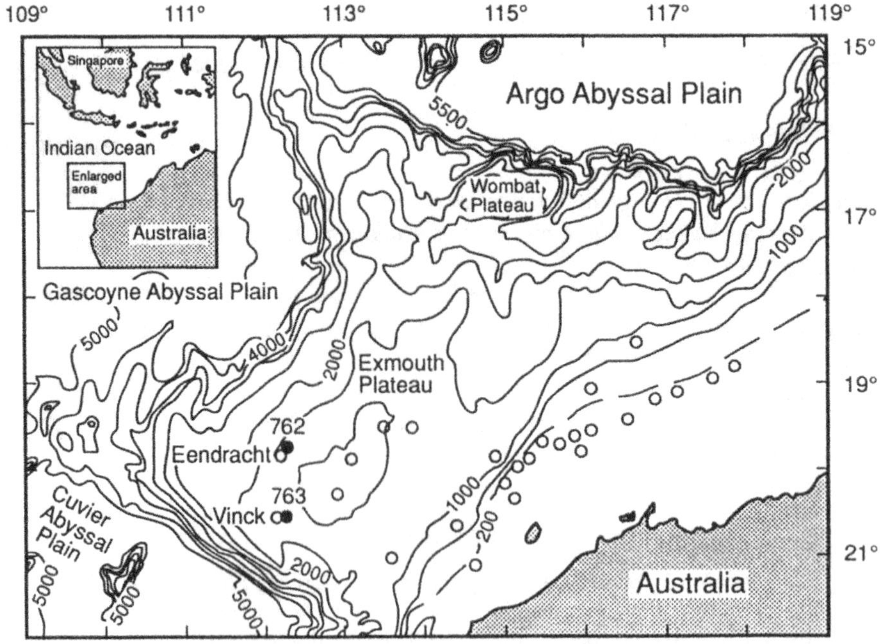

Figure 1. Locations of Sites 762 and 763 on the Exmouth Plateau, northwest Australian margin. Positions of exploratory wells are shown. Water depths given in meters.

the Exmouth Plateau on the northwestern Australian continental margin (Figure 1). The plateau is underlain by rifted and deeply subsided continental crust which is isolated from the Australian shelf by the Kangaroo Syncline.[7,15] Prior to rifting in the Jurassic, a series of shallow-water, lagoonal and reefal sedimentary sequences were deposited.[14] This series was followed by deeper-water clastic sediments which reach 10 km in thickness in some places. These sedimentary sequences consist of the equivalents of the onshore Triassic Mungaroo Formation, the Jurassic Dingo Claystone, and the fluviodeltaic Late Jurassic–Early Cretaceous Barrow Formation. Rocks from these sedimentary environments typically contain 1–2% organic carbon, mostly derived from land plants.[1,2,4,16] After tectonic isolation of the plateau from Australia in the late Early Cretaceous, this continental margin has been sediment-starved. The predominantly biogenic carbonate sediments that have been deposited since then were formed under conditions of low productivity and contain very little organic matter.

The deeper sections of the Barrow Formation have generated oil and gas. These sections are some of the sources of commercial land-based production in northwestern Australia. Noncommercial gas shows have been found by exploration drilling on the Exmouth Plateau.[1] Commercial quantities of hydrocarbons, evidently originating from the Mungaroo Formation, exist not far from the drill sites of Legs 122 and 123. Offshore production is located on the continental shelf and on Barrow Island offshore of northwestern Australia.[2]

Sediments recovered during ODP coring are routinely monitored for gaseous hydrocarbons. This monitoring is especially important when drilling in areas of known existence of petroleum reservoirs, such as on the Exmouth Plateau, inasmuch as no blow-out preventer is used during scientific ocean drilling. The results of the monitoring also provide valuable information about postdepositional decomposition of organic matter, possible microbial activity to subbottom depths of several hundred meters,[13] and migration of low-molecular-weight hydrocarbons. We present here information derived from the measurements of gaseous hydrocarbons done on board *D/V JOIDES Resolution* during ODP Leg 122.

Sampling and Analysis

Samples

Samples for head-space gas analyses were routinely taken from cores immediately after they had arrived on deck. A nominal 5-cm^3 plug of core was removed from the end of a freshly cut core section and placed in a glass vial which was then sealed with a rubber septum and a metal crimp. Totals of 113 samples over the 940 m drilled at Sites 762 and 117 samples over the 1064 m drilled at Site 763 were collected for head-space analysis.

Samples were also routinely selected for determination of total organic carbon (TOC) and for Rock-Eval analysis as drilling proceeded at Sites 762 and 763. The total number of samples collected and analyzed at Site 762 was 131; it was 268 at Site 763. Sample selection was based partly on subbottom depth and partly on stratigraphic boundaries.

Analysis

Gas analyses were performed using the shipboard head-space procedure developed for ODP samples[9,10] and refined by Emeis and Kvenvolden.[5] Septum-sealed glass vials were heated at 70°C for 1 h. Head-space gases were drawn into a syringe and injected into a Carle AGC 1000/Model 211 gas chromatograph equipped with a flame ionization detector to monitor compositions of hydrocarbon gases. Gases from samples that contained notable amounts of low-molecular-weight hydrocarbons or that were of particular lithostratigraphic interest were also analyzed using a Hewlett-Packard 5890 gas chromatograph equipped with a flame ionization detector and a thermal conductivity detector to allow measurement of hydrocarbon plus nonhydrocarbon gases. Both gas chromatographs were sufficiently sensitive to detect background concentrations of 2–3 ppm CH_4 present in laboratory air samples.

Total organic carbon (TOC) concentrations were determined onboard the *JOIDES Resolution* as part of the Rock-Eval analysis of samples using a Girdel Rock-Eval II instrument equipped with a TOC module. Samples were pyrolyzed by heating at a rate of 25°C/min between 300 and 600°C, yielding the amount of volatile hydrocarbons (S_1), the amount of thermogenic hydrocarbons (S_2), and the amount of CO_2 released during pyrolysis to 390°C (S_3). The TOC module combusts the residue of the temperature-programmed sample in air at 600°C and algorithmically determines the carbon content of this oxidation peak as well as the carbon in the pyrolysis peaks to give the TOC. These values provide the information to calculate the hydrogen index (HI = $100S_2$/TOC) and the oxygen index (OI = $100S_3$/TOC). The temperature of maximum hydrocarbon release during pyrolysis (T_{max}) is also obtained and can be interpreted as a measure of organic matter thermal maturity.[6]

Carbon and oxygen isotope measurements were done in the Stable Isotope Laboratory at The University of Michigan. Stable carbon isotope ratios of carbonate minerals were determined by acidification of samples and measurement of the $^{13}C/^{12}C$ ratios using a Finnigan MAT Model 251 mass spectrometer equipped with an automated analysis system. Organic matter present in carbonate-free samples was oxidized in a sealed preparatory system, and the $^{13}C/^{12}C$ ratios of the CO_2 produced from the organic carbon were measured with a Finnigan Delta S mass spectrometer. National Bureau of Standards carbon isotope standards were routinely and frequently used to calibrate the instruments. Results are reported relative to the PDB standard.

Results and Discussion

Organic Matter and Stratigraphic Units

Coring at Sites 762 and 763 on the Exmouth Plateau passed through an upper section of Quaternary, Tertiary, and Upper Cretaceous oozes and chalks with very low-TOC contents (Figure 2), through an early Aptian dark-colored calcareous claystone equivalent to the onshore Muderong Shale and into Neocomian black-colored siltstones equivalent to the onshore Barrow Formation. Organic carbon content is generally inversely related to the carbonate carbon content in the strata of the Exmouth Plateau.[16] Almost all of the samples of carbonate ooze and chalk contain less than 0.1% TOC, whereas nearly all of the clastic samples

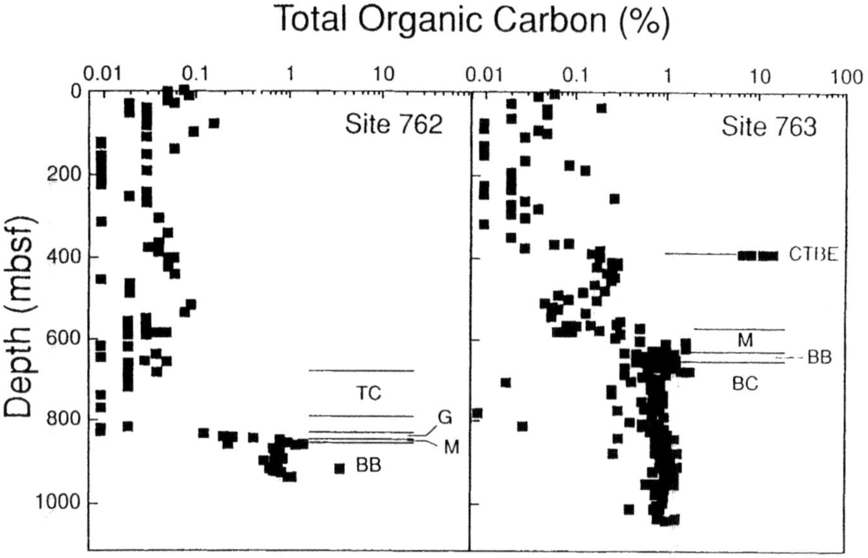

Figure 2. Total organic carbon versus depth for Sites 762 and 763 on the Exmouth Plateau. Rock units equivalent to strata identified on northwestern Australia are indicated: Toolonga calcilutite (TC); Gearle siltstone (G); Muderong Formation (M); Barrow "B" Formation (BB); Barrow "C" Formation (BC). An unnamed unit exists between 780 and 815 mbsf in the Site 762 sedimentary column. The Cenomanian–Turonian Boundary Event (CTBE) appears with its characteristic TOC enhancement at 380 mbsf in the Site 763 column.

have more than 0.5% TOC. The siltstones equivalent to the Barrow Formation typically contain about 1% TOC (Figure 2), with a few samples having up to 1.5% at Site 762 and up to 2% TOC at Site 763.

Rock-Eval data from Sites 762 and 763 are generally similar (Figure 3), yet a significant difference exists in samples from the two locations. Samples from Site 762 with >0.5% TOC show a constant HI of between 50 and 100, typical of Type III, higher land-plant organic matter. Samples from Site 763, on the other hand, show a small but definite increase in HI with increasing TOC content. Two factors contribute to this difference. First, the sediment column at Site 763 has an expanded and more complete Lower Cretaceous section than present at Site 762. TOC concentrations are somewhat higher in sediments from the equivalents of the onshore Muderong Formation and Barrow Group B sediments of Site 763 (Figure 2). The organic matter in these strata contains a higher proportion of marine organic matter, with its typically higher HI values, than in other lithostratigraphic units.[16] Second, drilling at Site 763 encountered older

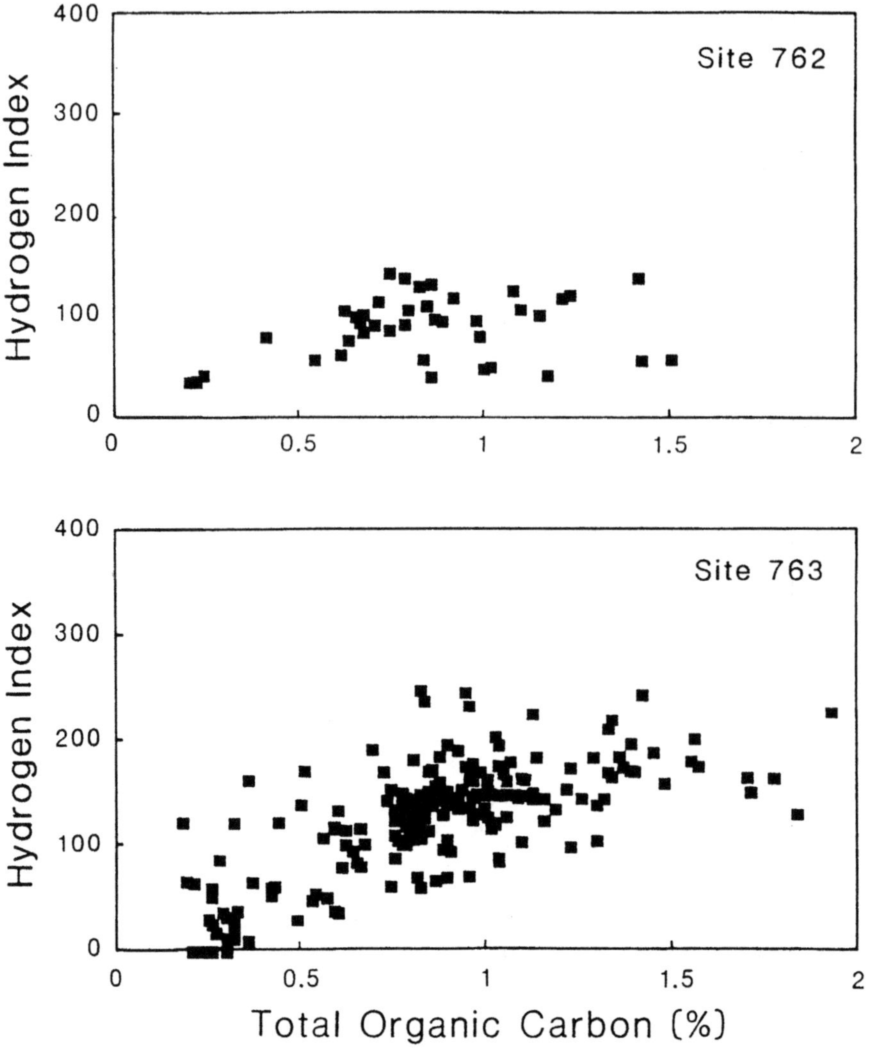

Figure 3. Rock-Eval hydrogen index versus total organic carbon for Sites 762 and 763 on Exmouth Plateau.

sediments than at Site 762, and these were deposited more rapidly and in shallower water than sediments laid down later. Less degradation of organic matter occurred, and HI values were better preserved.

The level of thermal maturity indicated by the Rock-Eval T_{max} parameter for the deltaic Barrow Formation siltstones is immature to marginally mature. T_{max} values are about 410–425°C (equivalent to a vitrinite reflectance of 0.4 to 0.5% R_0) for Site 762 with no discernable trend over the relatively short depth interval of samples from the Barrow Formation.[8] In contrast, T_{max} values increase at Site 763 over a 400-m interval from about 422°C to about 430°C (equivalent to about 0.45–0.60% R_0). This trend may represent a true increase in thermal maturation with depth, or it alternatively reflects a shift in the type of organic matter. Thermal maturation modeling done at the Geological Survey of Canada using the Institut Francais du Petrole software package MATOIL indicates that the heat flow at this site would need to have exceeded 120 mW/m^2 since Neocomian times to attain a vitrinite reflectance in excess of 0.5% R_0. Physical properties measurements[8] indicate a contemporary heat flow of about 75 mW/m^2 for the uppermost 150 m of Site 763, which is far too low to have achieved the thermal maturity indicated by the Rock-Eval T_{max} data. Instead of in situ thermal maturation, the increasing T_{max} values may indicate the presence of an increasingly larger contribution of recycled, geologically old, detrital organic matter in the progressively deeper deltaic siltstones. A larger proportion of land-derived organic matter is consistent with sedimentological evidence that the older siltstones were deposited in shallower waters closer to ancient shorelines.[8]

Gas Concentrations and Compositions

Head-space concentrations of methane were as high as 100,000 ppm in samples from Lower Cretaceous sections of Sites 762 and 763 (Figure 4). These concentrations represent the compositions of methane dissolved in sediment pore waters in these two sites and sorbed on the mineral phases of these samples; the methane is neither free nor overpressured.[8] Calculations of percent pore water saturation indicate that concentrations of 100,000 ppm correspond to approximately 50% saturation at the in situ pressures, temperatures, and pore water salinities in these drill holes.[8] This percentage saturation assumes that all the gas is dissolved in the pore waters, which were near seawater salinity. The presence of any sorbed gases would lessen this percentage and diminish further the probability of overpressuring.

The head-space gas concentrations are low in the upper 350 m of Site 762 and 200 m of Site 763 before reaching their maxima in deeper strata (Figure 4). The rapid rise in concentrations correspond to lithologic changes from sediments containing 85–90% carbonate to 75–80% carbonate.[7] Concentrations drop at the next major lithologic transition in these sediments—the upper boundary of the

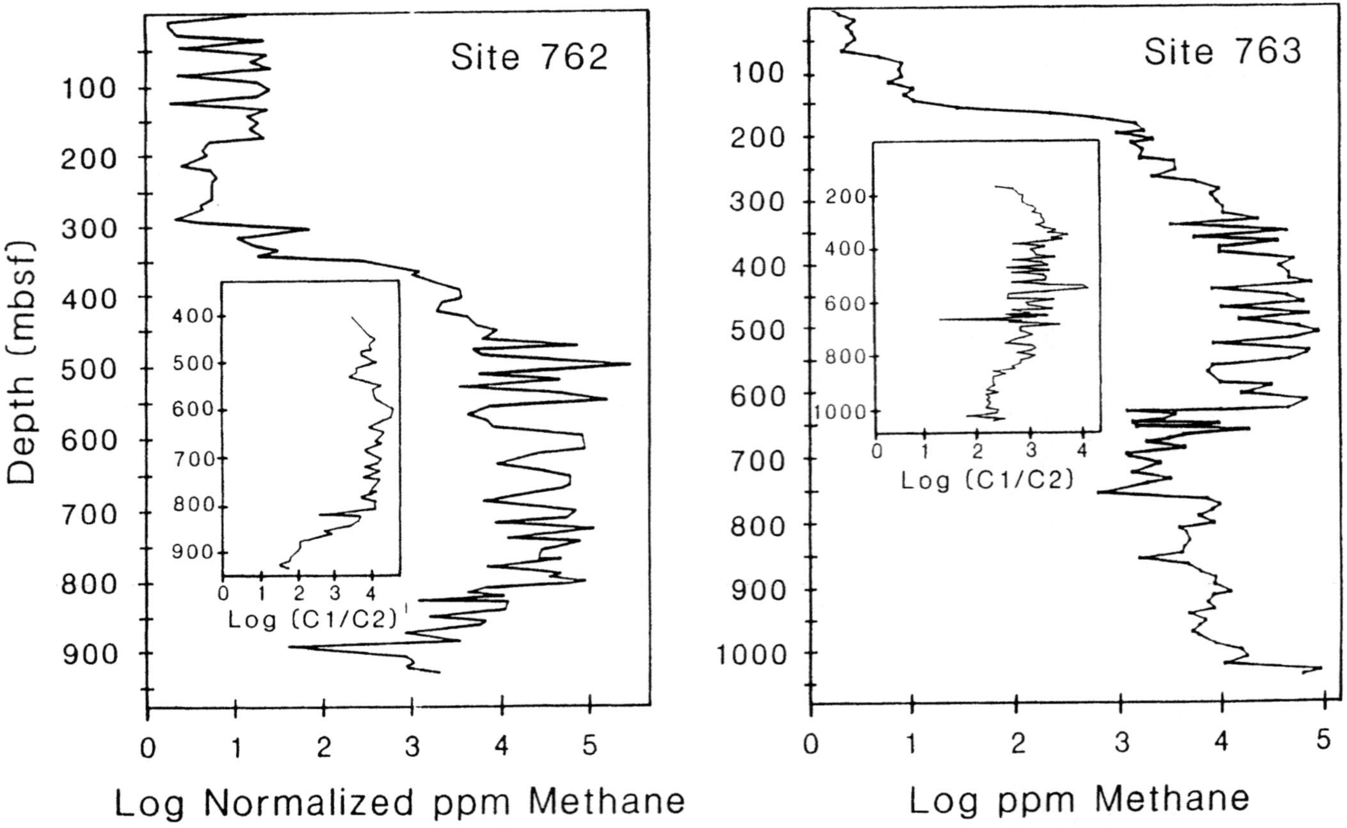

Figure 4. Methane concentrations in head-space gases generated by heating sediment and rock samples from Sites 762 and 763 on the Exmouth Plateau. A log concentration scale is necessary for the wide range in concentrations. Data from Site 762 have been normalized to a standard 5 cm^3 sample size; those from Site 763 have not been adjusted for this small (± 10%) correction. Methane/ethane ratios are expressed as C_1/C_2 ratios.

relatively porous siltstones equivalent to the Barrow Formation at ~820 m at Site 762 and at ~650 m at Site 763.

Methane/ethane ratios of the head-space gases were high, averaging 10,000 through the zone of maximum methane concentration (Figure 4). High methane/ethane ratios usually indicate gas which has originated from microbial methanogenesis,[3] although high ratios can also arise from thermal decomposition of gas-prone, continental organic matter.[17] When methane concentrations decrease in the deeper sections at both sites, methane/ethane ratios also diminish (Figure 4). Head-space gas from some of these deeper samples was analyzed by flame ionization detector gas chromatography. The results in Table 1 show that higher-molecular-weight gaseous hydrocarbons are important components of the gases in these samples. These C_3-C_5 components were not detectable in shallower samples. Their appearance in the siltstones deep in these sites suggests that the source of the gaseous hydrocarbons was being approached.

We interpret the high concentrations of head-space gas at Sites 762 and 763 to have resulted from migrated thermogenic gas derived from a deep-seated source on the Exmouth Plateau. Possible sources are the Triassic Mungaroo Formation

Table 1. Composition of head-space gases from Early Cretaceous black silty claystones from the Exmouth Plateau. Equivalent landward formations are given. Analyses done onboard *D/V JOIDES Resolution* using a Hewlett-Packard 5890 FID gas chromatograph.

Sample	Depth (mbsf)	TOC (%)	C_1 (ppm)	C_2 (ppm)	C_3 (ppm)	n-C_4 (ppm)	n-C_5 (ppm)	C_1/C_2
Muderong Fm, Aptian								
122-763B-45X-5,0-2 cm	604.5	1.71	9818	24	12	4	8	409
Barrow Group B, Valanginian								
122-762C-86X-4,0-2 cm	887.0	0.71	4618	29	12	6	0	159
Barrow Group C, Berriasian								
122-762C-89X-2,0-2 cm	912.5	0.67	710	17	16	4	0	41
122-762C-89X-5,0-2 cm	917.0	0.68	426	14	18	6	9	26
122-762C-90X-2,0-2 cm	922.0	0.87	35	12	11	4	0	3
122-762C-90X-4,0-2 cm	925.0	1.10	589	26	20	8	9	23
122-763C-30R-6,0-2 cm	882.6	1.19	6642	33	6	4	7	189
122-763C-44R-7,0-2 cm	1017.1	0.95	6711	95	12	6	7	71
122-763C-45R-7,0-2 cm	1026.6	1.30	28658	168	12	6	3	171
122-763C-46R-5,0-2 cm	1033.1	1.39	10543	148	8	4	5	71

or coal seams deposited in Jurassic coastal swamps. Vertical migration is inferred from (a) gas chimneys evident on seismic profiles,[8] (b) coincidence of high methane concentrations and very low organic carbon concentrations, and (c) the change in methane concentrations by three orders of magnitude over depth intervals of ~100 m. We further infer that most of the gas migration is by pore-space diffusion and is driven by concentration gradients. Gas evidently is trapped in the low-porosity clay-containing carbonates between 400 and 800 mbsf at Site 762 and 200–600 mbsf at Site 763, but escapes easily from the shallower, high-carbonate sediments at both locations. Methane head-space concentrations rose abruptly near the bottom of Site 763 (Figure 4), and gaseous hydrocarbon compositions began to contain substantial contributions of higher-molecular-weight components (Table 1). Drilling at this depth was believed to be approaching the Dingo Claystone, the postulated caprock above the Mungaroo Formation. The change in gas concentrations and compositions signaled dangerous proximity to the postulated source of the migrated hydrocarbons, and drilling was terminated.

The high methane/ethane ratios reflect both the origin of the pore–water gases and the effects of diffusional migration. The gas source is believed to be from thermal decomposition of Type III (woody or coaly) organic matter more deeply buried on the Exmouth Plateau. Preferential migration of methane relative to higher-molecular-weight hydrocarbons probably enhanced the ratios. These heavier hydrocarbon components become more abundant deeper in the section (Figure 4, Table 1), consistent with the effects of migration-induced gaseous hydrocarbon fractionation.[11,12]

Isotopic Evidence of Gas Migration

Direct measurements of the stable carbon isotope ratios of pore–space gas were precluded by the small absolute amounts of gas available and by the lack of suitable instrumentation on the ship. Carbon isotope ratios in organic matter and in carbonate contents of deep rock samples from Sites 762 and 763 nonetheless support the contention that gas migration took place in these strata. Organic carbon $\delta^{13}C$ data in Table 2 are consistent with a continental origin for most of the organic matter contained in these Lower Cretaceous clastic rocks. The organic carbon stable isotope ratios are uniformly typical of those of C3 land plants, ~−27‰. Inorganic carbon isotope values, in contrast, have a surprisingly broad range, from essentially 0‰ to −22‰. The light values are unexpected, inasmuch as most carbonates have ratios close to marine carbonate values, about 0 to +2‰. The isotopically light carbonate values are believed to result from oxidation of organic carbon and incorporation of the CO_2 so produced into carbonates. The three samples which have isotopically light carbonate also have higher carbonate concentrations. Augmentation of the original carbonate contents of these largely

Table 2. Carbon isotopic contents of Neocomian rock samples from Ocean Drilling Program Sites 762 and 763 on the Exmouth Plateau, northwest Australia. Unusually light carbonate values suggest that some of the carbonates in the Valanginian samples is derived from oxidation of organic matter.

Sample and Description	Depth (mbsf)	$CaCO_3$ (%)	TOC (%)	$\delta^{13}C$ (‰) Inorg.	$\delta^{13}C$ (‰) Organic
122-763B-45X-3, 140-150cm Aptian black silty claystone (Muderong Fm Equiv.)	602.9	12.7	1.70	-0.9	-28.4
122-763C-6R-1, 140-150cm Valanginian dark gray silty claystone (Barrow Group B equiv.)	662.0	12.4	0.89	-8.0	-28.3
122-762C-82X-3, 140-150cm Valanginian dark gray silty claystone (Barrow Group B equiv.)	857.9	11.7	1.14	-9.4	-26.7
122-762C-89X-4, 140-150cm Berriasian dark gray silty claystone (Barrow Group C equiv.)	916.5	14.2	0.77* (0.15)	-19.1	-27.5
122-763C-41R-4, 140-150cm Berriasian dark gray clayey siltstone (Barrow Group C equiv.)	985.5	3.3*. (0.9)	0.98* (0.09)	-0.4	n.d.

*Not determined on this sample; mean value of samples within 10 m is substituted (SD).
n.d. = not determined.

siliciclastic sediments may have occurred by precipitation of migrated CO_2 derived from oxidation of organic carbon. This type of organic matter recycling is quite possible on the Exmouth Plateau, where large volumes of methane and associated gases evidently have migrated through the Cretaceous sediments from deeper Jurassic sources. Our analyses of deep gas samples by thermal-conductivity-detector gas chromatography showed that CO_2 was enriched 100 times over atmospheric proportions in the head-space gases.[8] The probable source of this gas is from oxidation of organic matter, which conceivably includes the migrated methane. A similar situation of organic carbon oxidation and precipitation of isotopically light carbonate has been postulated in Miocene hemipelagic sediments at ODP Sites 767 and 768 in the Celebes-Sulu Sea.[18]

Summary and Conclusions

Scientific deep-sea drilling during Leg 122 of the Ocean Drilling Program has recovered gas-rich sedimentary rocks from the Exmouth Plateau on the northwest margin of Australia. The organic matter content of these rocks is probably derived from continental plants. The gaseous hydrocarbon compositions are dominated by methane which is believed to have been generated from the thermal decomposition of deeply buried, land-derived organic matter on the Exmouth Plateau. High methane/ethane ratios reflect the origin of these pore–water gases from gas-prone Type III organic matter and the effect of preferential diffusional migration of methane relative to other thermogenic gases.

Acknowledgments

The authors are grateful for the experience (which they actually enjoyed!) of being at sea for 9 weeks during Leg 122. We thank the National Science Foundation and the International Program of Ocean Drilling for their support of the Ocean Drilling Program. Part of this research was supported by funds from NSF, administered by JOI/USSAC.

References

1. Barber, P.M. 1982. Palaeotectonic evolution and hydrocarbon genesis of the central Exmouth Plateau. Austrl. Petrol. Explor. Assoc. J. **22**:131–144.
2. Campbell, I.R., A.M. Tait, and R.F. Reiser. 1984. Barrow Island oilfield, revisited. Austrl. Petrol. Explor. Assoc. J. **24**:289–298.
3. Claypool, G.A., and K.A. Kvenvolden. 1983. Methane and other hydrocarbon gases in marine sediment. Ann. Rev. Earth Planet. Sci. **11**:299–327.

4. Cook, A.C., M. Smyth, and R.G. Vos. 1985. Source potential of Upper Triassic fluvio-deltaic systems of the Exmouth Plateau. Austrl. Petrol. Expl. Assoc. J. **25**: 204–215.

5. Emeis, K.-C., and K.A. Kvenvolden. 1986. Shipboard Organic Geochemistry on *JOIDES Resolution,* Ocean Drilling Program, College Station, TX, Tech. Note 7.

6. Espitalié, J., J.L. Laporte, M. Madec, F. Marquis, P. Leplat, J. Paulet, and A. Boutefeu. 1977. Méthode rapide de caractérisation des roches mères, de leur potential pétrolier et de leur degré d'évolution. Rev. Inst. Fr. Petrole **32**:23–42.

7. Exon, N.F., U. von Rad, and U. von Stackelberg. 1982. The geological development of the passive margins of the Exmouth Plateau off northwest Australia. Marine Geol. **47**:131–152.

8. Haq, B.U., U. von Rad, S. O'Connell, et al. 1990. Proc. ODP, Initial Reports 122, College Station, TX (Ocean Drilling Program).

9. Kvenvolden, K.A., and L.A. Barnard. 1983. Gas hydrates of the Blake Outer Ridge, Site 533, Deep Sea Drilling Project 76, In R.E. Sheridan, F.M. Gradstein, et al. (eds.), Initial Reports DSDP 76, US GPO, Washington, D.C., pp 353–365.

10. Kvenvolden, K.A., and T.J. McDonald. 1985. Gas hydrates in slope sediments of the Middle America Trench, DSDP Leg 84, In R. von Heune, J. Auvouin, et al. (eds.), Initial Reports DSDP 84, US GPO, Washington, DC, pp. 667–682.

11. Leythaeuser, D., R. Schaefer, and A. Yukler. 1982. Role of diffusion in primary migration of hydrocarbons. AAPG Bull. **66**:408–429.

12. Leythaeuser, D., R. Schaefer, and H. Pooch. 1983. Diffusion of light hydrocarbons in subsurface sedimentary rocks. AAPG Bull. **67**:889–895.

13. Meyers, P.A., and S.A. Brassell. 1985. Biogenic gases in sediments deposited since Miocene times on the Walvis Ridge, South Atlantic Ocean, In D.E. Caldwell, J.A. Brierly, and C.L. Brierly (eds.), Planetary Ecology, van Nostrand Reinhold, New York, pp. 69–80.

14. von Rad, U., J. Thurow, B.U. Haq, F. Gradstein, and J. Ludden. 1989. Triassic to Cenozoic evolution of the NW Australian continental margin and the birth of the Indian Ocean (preliminary results of ODP Legs 122 and 123). Geol. Runds. **78**: 1189–1210.

15. von Rad, U., M. Schott, N.F. Exon, J. Mutterlose, P.G. Quilty, and J.W. Thurow. 1990. Mesozoic sedimentary and volcanic rocks dredged from the northern Exmouth Plateau: petrography and microfacies. BMR J. Austrl. Geol. Geophys. **11**: 449–472.

16. Snowdon, L.R., and P.A. Meyers. 1992. Source and maturity of organic matter in sediments and rocks from Sites 759, 760, 761, and 764 (Wombat Plateau) and Sites 762 and 763 (Exmouth Plateau). In U. von Rad, B.U. Haq, et al. (eds.). Proc. ODP Scientif. Res. **122**:309–315.

17. Hunt, J.M. 1979. Pet. Geochem. 438.

18. von Breyman, M., and U. Berner. 1990. Eos **71:**1392.

Boreal Systems

Methane Emissions from Northern High-Latitude Wetlands

Robert Harriss, Karen Bartlett, Steve Frolking, and Patrick Crill

Institute for the Study of Earth, Oceans, and Space, University of New Hampshire, Durham, New Hampshire 03824

ABSTRACT

Methane emissions from northern high-latitude wetlands are an important consideration for understanding past, present, and future atmospheric concentrations of this important greenhouse gas. In this chapter we review progress on measuring methane emissions from northern wetlands and, through a model, estimate emission variability in relation to one component of climate variability. Our conclusions are as follows: (1) Methane emissions from northern wetlands are dependent on both soil moisture and temperature. The relative influence of these soil climate parameters is quite variable from one region to another, as is the magnitude of the net emission rate to the atmosphere. Some important wetland regions have not been surveyed for methane emissions (e.g., the Siberian Lowlands); further progress on defining global emissions from northern wetlands awaits field data from these areas. (2) Our preliminary modeling of the sensitivity of methane flux from northern wetlands to variability in temperature indicates that feedbacks from this source are unlikely to significantly influence rates of climate change during the initial stages of a global warming.

Introduction

In this chapter, we review progress on estimating and understanding both the magnitude of, and controls on, emissions of methane from northern high-latitude wetlands to the atmosphere. During the past decade, the number of studies of methane emission from northern wetland environments has increased substantially, especially in North America. We use our synthesized dataset on methane emission rates and associated environmental variables to address two questions related to possible effects of global climate change:

1. How are methane emissions from northern high-latitude wetlands influenced by climate variability?
2. If a long-term global warming occurs, will methane emissions from northern

wetlands be a significant feedback factor influencing future rates of climate change?

A Review of Data on Methane Emissions from Northern Wetlands

A summary of reported flux measurements from northern wetlands, covering a latitudinal range of 45° to 70°N is given in Table 1. These measurements span a range of over three orders of magnitude, from less than 1 to roughly 1940 $mgCH_4\ m^{-2}\ d^{-1}$ and include a wide variety of vegetation, moisture, and soil types. A single flux of 12,068 $mgCH_4\ m^{-2}\ d^{-1}$, roughly an order of magnitude greater than all others at this site, is reported from an Alberta, Canada beaver pond.[44] Since the degree of soil wetness strongly influences anoxia, a necessary condition for methanogenesis, soil moisture is a major factor controlling methane release in many sites studied to date (e.g., Refs. 29, 39, 42; Table 2). Wet soils, however, clearly vary greatly in emissions (Table 1). Moist-to-dry wetland soils (i.e., no standing water at the surface) can be both small sources or sinks to the atmosphere. In the Arctic, moist soils most frequently appear to be small sources, on the order of 0.6–11 $mgCH_4\ m^{-2}\ d^{-1}$, with dry soils frequently exhibiting negative fluxes (consumption of atmospheric methane) generally between -0.5 and -3 $mgCH_4\ m^{-2}\ d^{-1}$.[1,2,9,18,46,47]

To investigate how the climate regime may influence methane emissions from northern wetlands, it is useful to consider the major ecoclimatic zones separately. In the following sections, we first consider emissions from arctic and subarctic regions, then from boreal wetlands.

Arctic/Subarctic Methane Emissions

Measurements in subarctic and arctic tundra (> 60°N) are overwhelmingly from sites in Alaska—from Fairbanks in the interior where annual and multiyear measurements have been made,[32,48] from the coastal plain on the North Slope where a series of investigators have made measurements encompassing both large and small spatial scales,[18,29,39,46] and from coastal tundra on the delta of the Yukon and Kuskokwim Rivers where investigators have made measurements at a variety of scales using an array of measurement techniques during the NASA Arctic Boundary Layer Expedition (ABLE 3A).[1,9,33]

Time-series measurements from Fairbanks, from the same plots, demonstrate significant differences from year to year, at least in part controlled by temperature and moisture variations.[45,48] As shown in Table 1, annual emissions from some sites varied by nearly an order of magnitude and year-to-year changes were not consistent between sites. Differences between subsites within a habitat in any one year, presumably due primarily to spatial variability, were also considerable and ranged from roughly 2-fold to more than 35-fold. This multiyear research program demonstrates some of the natural variability in flux on scales greater than 1 year.

Measurements on the North Slope have been conducted by at least four separate groups over about a 6-year period. In general, agreement between measurements in similar habitats, in the same region, appears fairly good (e.g., Sebacher's wet coastal tundra=119 mgCH_4 m^{-2} d^{-1}, Whalen and Reeburgh's wet tundra=90 mgCH_4 m^{-2} d^{-1}, Livingston and Morrissey's meadow tundra=64.4 mgCH_4 m^{-2} d^{-1}, wet tundra=100 mgCH_4 m^{-2} d^{-1}, very wet tundra=254 mgCH_4 m^{-2} d^{-1}). Given the year-to-year variability demonstrated by measurements in Fairbanks and differences between investigators in grouping vegetation, this level of similarity suggests that most of the habitats on the North Slope are fairly well-characterized by the existing measurements.

Comparison of measurements in similar vegetation habitats in different regions of Alaska suggests that some regional differences may be present, although clearly a variety of environmental factors other than vegetation may differ between sites. For tussock tundra (Table 2), in Fairbanks: tussock tundra composite=22.4 mg m^{-2} d^{-1} [45]; in the Yukon–Kuskokwim Delta: upland tundra=2.3 and 11 mg m^{-2} d^{-1} [1,9]; and on the North Slope: moist tundra=31 mg m^{-2} d^{-1}.[46] Morrissey and Livingston[29] report tussock/intertussock fluxes for the North Slope (3.4 and 2.9 mg CH_4 m^{-2} d^{-1}, respectively), roughly an order of magnitude lower than the Whalen and Reeburgh figure of 31 mg CH_4 m^{-2} d^{-1} from moist tussock tundra. Comparable differences are found for wet gramminoid tundra; in Fairbanks: wet meadow composite=32.2 mg CH_4 m^{-2} d^{-1} [45]; in the Yukon–Kuskokwim Delta: wet meadow=144 and 29 mg CH_4 m^{-2} d^{-1} [1,9]; and on the North Slope: wet (coastal) tundra=119 and 90 mg CH_4 m^{-2} d^{-1} [39,46]; meadow tundra=64.4 mg CH_4 m^{-2} d^{-1}.[29]

The only arctic flux measurements not from Alaska were made at a series of sites in a Swedish mire (Table 1). Average fluxes from hummocks and higher areas (ombrotrophic and mixed minerotrophic/ombrotrophic) are quite similar to those from comparable Alaskan areas. The average flux from wetter minerotrophic areas (360 mg CH_4 m^{-2} d^{-1}) is higher than emissions from wet Alaskan sites, but is largely driven by high measurements on a single date.

Boreal Methane Emissions

In the boreal zone (here taken to be roughly 50°–60°N), intensively studied sites fall primarily into two areas: northern Minnesota (in and around the Marcell Forest and the Red Lake Peatlands) and southeastern/southern Canada (the Schefferville area and the Hudson Bay Lowlands). Emissions from more western sites in Canada (Alberta) have also been recently examined.[44] Significant differences appear to exist between the Minnesota and Canadian regions. Although

Table 1. A summary of methane flux measurements from northern wetlands (flux units are mgCH_4 m^{-2} d^{-1}).

Habitat	Latitude	Avg. Flux	Annual Flux ($g\ m^{-2}\ yr^{-1}$)	N	Range	Site	Meas. Period	Ref.
ARCTIC								
Subarctic mire:								
Ombrotrophic	68°	11.6	0.13	95	0.3–29	Stordalen, Sweden	June–Sept.	41, 42
Intermediate		58	1.44	25	8.6–112			
Minerotrophic		360	30.5	20	80–950			
Wet coastal tundra	63–70°	119	—	44	34–266	Alaska, North Slope & Denali	Aug.	39
Moist tundra		4.9	—	12	0.3–12			
Meadow tundra		40	—	14	9–77			
Alpine fen		289	—	6	244–344			
Boreal marsh		106	—	7	86–122			
Tussock tundra composite (1987)	65°	22.4	—	—	—	Alaska, Fairbanks	Annual	45, 48
Wet meadow composite (1987)		32.2	—	—	—			
Moss (1987)		1.5	0.5	87	0–17			
Moss (1988)		10.1	4.4	57	0–146			
Moss (1989)		29	4.8	58	0–367			
Moss (1990)		3.8	0.5	51	0–26			
Intertussock (1987)		4.9	0.6	87	0–34			
Intertussock (1988)		25.2	3.9	54	0–292			
Intertussock (1989)		12.4	4.3	57	0–145			
Intertussock (1990)		5.9	0.8	51	0–57			
Carex sedge (1987)		33.5	4.9	33	<1–105			
Carex sedge (1988)		3.2	0.8	57	0–12			

Carex sedge (1989)		23	4.3	60	0–104			
Carex sedge (1990)		451.9	60.6	51	0–2216			
Eriophorum tussock (1987)		29.5	8	86	0–167			
Tussocks (1988)		60.8	11.4	57	0–653			
Tussocks (1989)		48.8	8.1	60	0–445			
Tussocks (1990)		96.8	13.6	51	0–302			
Lakes & ponds	66–70°	21	—	6	4.6–131	Alaska, North Slope	July–Aug.	46
Alpine tundra		0.6	—	8	−0.2–6.3			
Moist tundra		31	—	42	0–159			
Wet tundra		90	—	18	0–265			
Low brush–Muskeg bog		45	—	6	12–101			
Spruce forest		4.6	—	18	−0.3–67			
Meadow & tussock tundras	68–70°	30	—	27	−1–145	Alaska, North Slope & Foothills	Aug.	18
Tussock tundra								
Tussocks	69°	3.4	—	6	—	Alaska, North Slope	Aug.	29
Intertussocks		2.9	—	8	—			
Meadow tundra		64.4	—	23	—			
Lakes & ponds		37.5	—	31	—			
High centered polygons		4.9	—	5	—			
Low cent. polygons								
Troughs		61.9	—	10	—			
Basins		46.1	—	24	—			
Rims		12.1	—	15	—			

(continued)

Table 1. *(continued)*

Habitat	Latitude	Avg. Flux	Annual Flux (g m^{-2} yr^{-1})	N	Range	Site	Meas. Period	Ref.
Dry tundra	69°	—				Alaska, North Slope	Aug.	20
Moist tundra		—						
Wet tundra		100.1						
Very wet tundra		253.9						
Water		—						
(Calc. regional mean, 1987[a])		17.3						
Wet meadow tundra	61°	144	—	73	16–426	Alaska, Yukon–Kuskokwim Delta	July–Aug.	1
Upland tundra		2.3	—	82	−2.1–18			
Large lakes[b]		3.8	—	12	—			
Small lakes[b]		77	—	32	—			
Lake vegetation		89	—	51	63–154			
(Calc. regional mean[a])		48.5						
Small lakes—bubbles (calc.)	61°	112	—	—	—	Alaska, Y–K Delta	July–Aug.	21
Dry tundra	61°	11	—	Eddy corr.	—	Alaska, Y–K Delta	July–Aug.	9
Wet meadow tundra		29	—		—			
Lake (wind at 5 m/s)		57	—		—			
(Integrated areal mean)		25	—		—			
(Integrated areal means)	61°	51	—	Eddy corr.	25–85	Alaska, Y–K Delta	July–Aug.	33

Wet moss carpet	~70°S	1.6	—	208	0.03–20.3	Antarctica, Signy Isl.	Dec.–Mar.	49
BOREAL								
Forested bogs	47°	100	—	8	19–206	Minnesota, Marcell Forest & Zerkel	Aug.	15
Nonforested bogs		306	—	24	33–1943			
Forested fens		85	—	5	3–171			
Wild rice bed		493	—	4	127–883			
Sedge meadow		664	—	1	—			
Forested bogs	47°	89	—	55	11–694	Minnesota, Marcell Forest & Red Lake	May–Aug.	4
Nonforested bogs		199	—	77	18–866			
Forested fen		142	—	12	68–263			
Nonforested fens		348	—	35	152–711			
Nonforested bog	47°	177	—	Eddy corr.	120–270	Minnesota, Marcell Forest	Aug.	43
Forested bog (hummock)	47°	21	3.5	36	2–48	Minnesota, Marcell Forest	Annual	5
Forested bog (hollow)		93	13.8	36	6–246			
Fen lagg		121	12.6	27	−1–482			
Nonforested bog		356	43.1	68	0–1056			
Nonforested fen		402	65.7	37	11–767			
Nonforested bog	54–55°	0	—	90	0	Alberta, Canada	May–Oct	44
Nonforested poor fen		1	—	90	0–22			
Nonforested rich fen		65	—	90	0–1976			
Forested rich fen		24	—	90	0–1820			
Sedge meadow		148	—	90	0–1985			
Beaver pond		518	—	90	0–12,068			

(continued)

Table 1. (*continued*)

Habitat	Latitude	Avg. Flux	Annual Flux ($g\ m^{-2}\ yr^{-1}$)	N	Range	Site	Meas. Period	Ref.
Nonforested fens	55°	30.5	—	80	0–112	Schefferville, Canada	June–Aug.	25
Nonforested fens:	55°							
Center		72	—	205	29–125	Schefferville, Canada	June–Sept.	27
Margin		30	—	195	9–65			
Pools/flooded		33	—	185	24–40			
Forest/fen margin		28	—	255	0.6–51			
Patterned nonforested fens:								
Ridges		7.5	—	110	5–9			
Pools		48	—	85	32–66			
(Calc regional mean[a])		18						
Horizontal rich fen	55°	—	3	—	−3–176	Schefferville, Canada	May–Sept. (multi-yr)	26
Horizontal poor fen		—	9.8	—	12–343			
Ribbed fen								
Ridge		—	1.3	—	1–25			
Pools		—	4.5/9.9	—	1–260			
Basin swamps	45°	—	1.2/4.2	—	−3–207	Mont St. Hilaire, Canada	May–Aug. (multi-yr)	
Domed bog								
Center		—	0.1	—	−11–10			
Margin		—	0.1	—	−10–9			

Beaver ponds	45°	90.6		56	0.9–246	Low boreal forest, S. Canada	May–Oct	35
		29.7	7.6	65	0.3–300			
		47.4		65	0.2–369			
Conifer swamps		7.1		123	−0.2–236			
		0.15	0.18	132	0.1–10			
		0.2		148	−0.2–6			
		0.2		149	−0.2–9			
Mixed hardwood swamps		1.2	0.1	141	−0.3–28			
		0.25		139	−5.8–10			
Thicket swamps		69.3	4.7	145	0–304			
		0.4		144	−0.3–37			
Marshes		1.2	0.1	72	−0.1–36			
		0.5		134	−0.3–26			
Open bog		20.6	1.7	65	−0.1–140			
Forested bog		5.8		68	−0.1–107			
Fen		3.0	0.4	62	−0.2–78			
Blanket bog								
Pool	55°	—	9.3	36	—	Moor House Nat. Res., England	Annual	3
Lawn		—	5.3	36	—			
Hummock		—	1.3	36	—			
(Integrated areal means)	55°	—	—	Eddy corr.	0–50	Hudson Bay Lowlands, Canada	July	34, 38
(Integrated areal mean)	55°	16	—	Eddy corr.	—	Kinosheo Lake, Canada	July	8
Ponds & lakes	55°	26	—	—	—	Kinosheo Lake area	July	13

(continued)

Table 1. (*continued*)

Habitat	Latitude	Avg. Flux	Annual Flux ($g\ m^{-2}\ yr^{-1}$)	N	Range	Site	Meas. Period	Ref.
Fen ponds	56°	160	—	—	—	Coast. & int. Hudson Bay		
Open water	~55°	12	1.5	—	0.2–146	Southern	June–	28,
Marshes		31	2.3	—	−2.3–274	Hudson Bay	Oct	Roulet
Shrub & treed fen		2.5	0.4	—	−2.4–32	Lowlands, Canada		pers. commun.
Open fen		7.9	0.7	—	−1.6–298			
Fen pools		133	13.8	—	21–544			
Bog pools		60	6.1	—	2.2–665			
Open bog		54	4.6	—	−1.7–1356			
Shrub-rich bog		48	4.0	—	−1.5–1627			
Treed bog		1.8	0.2	—	−1.7–66			
Conifer forest		3.3	0.2	—	−2.2–50			

[a]Calculated based on habitat-specific fluxes and regional habitat coverage data.
[b]Calculated from surface water concentrations, based on a wind speed of 5 m/s.

Table 2. Methane flux from tussock tundra (warm season only) (flux units of mg CH_4 m^{-2} d^{-1}).

Microhabitat	Water Level/Info.	Avg. Flux	Std. Dev.	Std. Error	Ref.
Ombrotrophic communities					
Hummock	471[a]	1.1	0.71	0.32	42
Interhummock	1288[a]	3	2.04	1.02	
Shallow depression	1691[a]	16.5	7.8	3.5	
Deeper depression	2540[a]	25.8	6.4	2.9	
Summer tussock tundra composite—1987	—	22.4	—	—	45
Tussocks (1987)	—	29.5	—	—	48
Tussocks (1988)	−10–9[b]	60.8	—	—	
Tussocks (1989)	−29–9[b]	48.8	—	—	
Tussocks (1990)	−12–11[b]	96.8	—	—	
Intertussocks (1987)	—	2.8	—	—	
Intertussocks (1988)	−6–15[b]	25.2	—	—	
Intertussocks (1989)	−29–10[b]	12.4	—	—	
Intertussocks (1990)	0–20[b]	5.9	—	—	
Moss (1987)	—	0.9	—	—	
Moss (1988)	−18–0[b]	10.1	—	—	
Moss (1989)	−40–4[b]	29	—	—	
Moss (1990)	−21–15[b]	3.8	—	—	
Meadow/moist tundra					
MP 298	+2 cm	32.6	—	—	39
MP 346	+2–5 cm	77.6	—	14.1	
MP 416	+3 cm	152	—	8.5	
MP 416	H_2O sat.	53.7	—	5.4	
MP 298	−30 cm	3.1	—	3.1	
MP 318	−15 cm	0.3	—	0.3	
MP 346	−13 cm	12.5	—	—	
Moist tundra	—	31	19–52[c]	—	46
Tussocks	—	3.4	—	3.1	29
Intertussocks	—	2.9	—	2.4	
Upland tundra (low tussocks)	Dry	2.3	—	1.1	1
Upland tundra (low tussocks)	Dry	11	—	3	9

[a]Avg. % water dry weight.
[b]Seasonal range in water table, all three habitat subsites.
[c]95% confidence interval.

ranges overlap, measurements reported from Minnesota are generally significantly higher than those from Canada, although much of the vegetation and general topography are similar. For example, comparing emissions from these areas from wet bogs and fens (Table 1), fluxes from Marcell and vicinity average 237 ± 48 mg CH_4 m^{-2} d^{-1} (std. error of the mean, n=13; ranging from 93 to 402 mg CH_4 m^{-2} d^{-1}); whereas fluxes from the Schefferville area average 33 ± 6 mg CH_4 m^{-2} d^{-1} (n=14; a range of 0.7–72 mg CH_4 m^{-2} d^{-1}), and that from wet bogs and fens from all sites in Canada average 29 ± 6 mg CH_4 m^{-2} d^{-1} (n=40; with a range of 0–148 mg CH_4 m^{-2} d^{-1}). Comparison of Hudson Bay Lowlands (HBL) data with data collected from the Minnesota Red Lake Peatlands reported by Crill et al.[4] (an area believed to closely resemble the HBL region but with less open water), shows that the striking regional differences are maintained. It is presently unclear why these differences occur, but these results indicate that it is difficult to extrapolate flux measurements from one region to another, even though vegetation and climate regimes may be broadly similar.

Although the current data base is somewhat tenuous, ponds and inundated areas created by beaver populations may be a significant source of additional atmospheric CH_4 (Table 1). Ponds sampled by Roulet et al.[35] suggest that release rates may be variable (average fluxes varying three-fold, from 30 to 91 mg CH_4 m^{-2} d^{-1}). It is likely that emissions are dependent on such variables as time since flooding, plant species present, and water depths. Current data suggest that fluxes are not greatly different from "natural" small ponds and surrounding wetlands, but that these areas may have shorter lifetimes as atmospheric sources since they are dependent on dam integrity for inundation. The regional importance of beaver ponds is likely to be highly variable because beaver populations exhibit large spatial and temporal changes.

Comparing measurements within one of the Minnesota sites researched at different times and by several investigators suggests that although there may be significant spatial and temporal variability, averaged measurements are in general agreement, in part because they have relatively large variance. Working in sites studied by Harriss et al.[15] in 1983 and by Crill et al.[4] in 1986, Dise[5] reported the only annual multiyear flux dataset from the boreal zone. Although there were significant differences in the magnitude and timing of emissions from year to year, there were few significant differences between the data for the same months by the various investigators. Monthly emissions varied by as much as two- to threefold at individual sites, whereas annual emissions generally varied roughly 10–30% between years.[5] Moore and Knowles[26] reported multiyear summer fluxes from a variety of Canadian sites. Average methane emissions from all sites varied from nearly twofold to somewhat more than threefold between years, similar to year-to-year differences from the Minnesota sites.

Arctic/Boreal Flux Comparisons

The presence of significant year-to-year variability and the possibility of a significant flux contribution during the winter[7] mean that the variance on "average fluxes" is quite large and, at this point, not well-known since the vast majority of flux data from the high latitudes was collected during only a single warm season. The contribution of winter emissions to annual fluxes appears likely to be highly variable both from year to year and from site to site. Dise[6] reports that emissions during the winter months of November through March under significant snowcover contributed from 4 to 21% of the annual flux. At the two Minnesota sites for which there were 2 years of winter flux data, winter emissions varied from contributing 8% to 17% and from 12% to 20% of annual fluxes. Because significant winter fluxes were not observed in Fairbanks, Alaska, in contrast to Minnesota, it is possible that they occur more frequently as winter temperatures become more moderate in the boreal region.

At several sites both in the arctic and boreal regions, small lakes and ponds appear to contribute significantly to emissions. In addition to estimated and measured diffusive fluxes from lakes in the Yukon–Kuskokwim Delta (Table 1), Martens et al.[21] calculated a significant methane release from lake edges due to bubble ebullition. Recent work in the Canadian Hudson Bay Lowlands also suggests that lakes may play a major role in regional emissions.[13,28] On the Alaskan North Slope, an area of abundant small lakes, data reported to date[29,46] indicates that fluxes are lower than those from the Yukon–Kuskokwim Delta and that emissions from lakes and ponds are not particularly high relative to the surrounding landscape of wet tundra and polygonal patterned ground (Table 1). Although no detailed seasonal measurements from lakes have been made, it is likely that the seasonal dynamics of emissions from lakes may be different than the surrounding vegetated surfaces during the fall and early spring since lake water insulates the sediments and methane production may still occur through the fall. A pulse of methane built up from fall and winter production may, therefore, occur when lake surfaces unfreeze.[36]

Arctic–boreal comparisons are difficult due to uncertainties in winter emissions, annual variability, apparent regional differences, and differences in vegetational classification and grouping. If, however, the data summarized in Table 1 are simply sorted by latitude and broad moisture class and averaged (Table 3), very general trends in the data emerge. If these measurements are reasonably representative of actual flux magnitude and variability, it would suggest, as expected, that methane emissions from moist to dry soils are low, but that they are measurable and should not be ignored. Emissions from wet arctic soils appear to be approximately equal to those in the boreal region. However, as discussed above, differences within the boreal sites measured to date complicate this conclusion, and emissions during winter months in the boreal zone may be higher.

Table 3. Average high-latitude methane fluxes (derived from Table 1). (Flux units: mg CH_4 m^{-2} d^{-1}.)

Moisture Class	Region	Avg. Flux	Std. Error of Mean	No. of Meas.	Range
Wet	Arctic	96	21	28	2.8–360
	Boreal	87	18	49	0–664
Dry	Arctic	7	2	14	0.6–29
	Boreal	10	3	5	3.3–21
Mixed[a]	Arctic	21	10	3	1.6–31
	Boreal	—	—	0	-
Lakes	Arctic	51	16	6	3.8–112
	Boreal	172	75	8	12–518

[a]Includes a mixture of wet and dry soils within the vegetation grouping.

Averaged fluxes from arctic lakes are lower than emissions from wet soils, but appear to make a significant contribution and, again, variability is high. Emissions from lakes in the boreal zone appear to be quite high, but the database is limited.

Estimating Global Methane Emissions from Northern Wetlands

Regional mean fluxes, derived from either extrapolating enclosure measurements using habitat coverage data or from larger-scale measurement techniques, such as eddy correlation, offer a way to make regional cross-comparisons. Regional average fluxes incorporate both differences in habitat-specific emissions and differences in ratios of wet and dry habitats within an area. As shown in Table 4, fluxes from the Yukon–Kuskokwim Delta and the North Slope, both classified as wet coastal tundra vegetation, cover a large range, from 10 to 51 mg CH_4 m^{-2} d^{-1}. Regional estimates from the Canadian Schefferville/Hudson Bay Lowlands area are uniformly lower than those from the Yukon–Kuskokwim Delta, although both regions have abundant areas of wetlands and many small lakes. The Schefferville estimates generally appear to be quite similar to the estimate from the North Slope. These differences reemphasize the variability in areas that emit methane and the difficulty in extrapolating even relatively large-scale measurement techniques to global scales.

Using the figures for wet soils only (Table 3) and the emission season assumptions and wetland area figures from Matthews and Fung,[23] a rough estimate of magnitude of global emissions from northern regions can be calculated. Wetland areas for arctic (above 60°N) and boreal (50°–60°N) zones are 1.468×10^{12} m^2 and 1.235×10^{12} m^2, respectively. These wetlands are predominantly forested and nonforested bogs, but also include relatively small areas of swamps.

Table 4. Mean regional methane fluxes.

Region	Site	Technique	Avg. Flux (mg CH_4 m^{-2} d^{-1})	Range (mg CH_4 m^{-2} d^{-1})
Arctic	Y–K[a] Delta—entire area	Enclosures	48.5	—
	Y–K Delta—flight paths	Aircraft eddy correlation	51	25–85
	Y–K Delta—tower area	Enclosures	27	—
	Y–K Delta—tower area	Eddy correlation	25	—
	Y–K Delta—tower area (16a)[b]	Aircraft eddy corr.	42	
	tower area (16b)	Aircraft eddy corr.	62	
	North Slope	Enclosures	17.3	—
Boreal	Schefferville area	Enclosures	18	—
	Schefferville tower area	Eddy correlation	16	—
	Schefferville flight paths	Aircraft eddy corr.	—	0–50

[a]Yukon–Kuskokwim Delta.
[b]Numbers in parentheses indicate different legs of flights.
Note: Data from Refs. 1, 8, 9, 20, 27, 33, 34, and 38.

The assumed emission periods (winter fluxes assumed to be zero) are 100 days for above 60°N and 150 days for 50°–60°N. Given this model, annual emissions from areas of wet soils in the arctic (14 Tg yr^{-1}) are somewhat lower than those from the boreal region (16 Tg yr^{-1}). Given the uncertainty in this estimate, it is likely that the two figures are essentially equal. If winter season emissions from the boreal zone are disproportionately larger than those from the arctic, differences between the two regions may be greater. Areas of wetlands calculated by Matthews and Fung are for inundated soils only, so that the calculation of emissions from moist soils in the arctic and boreal regions is difficult. In the arctic, it may be possible to estimate areas of moist to dry tundra soils by subtracting the global area of wet soils (above, 1.468×10^{12} m^2) from the global area of tundra (7.324×10^{12} m^2; ref. 22). Using the same assumptions for emission period used for wet arctic soils and an average flux of 7 mg CH_4 m^{-2} d^{-1} (Table 3), a global flux of 4 Tg yr^{-1} is calculated for relatively dry tundra. Estimation of the contributions made to atmosphere from lakes and ponds is also difficult because no global estimates of their areas have been made. Small lakes and ponds typical of the arctic may well be included in estimates of areas of wet soils.

Although clearly subject to considerable uncertainties, some of which are discussed above, our emissions estimate of roughly 34 Tg CH_4 yr^{-1} from wetland areas above 50°N (30 Tg from wet soils, 4 Tg from moist to dry tundra) is in good agreement with that of 35 Tg yr^{-1} derived from a variety of other data sources.[10] It is also in reasonable agreement with those recently calculated for tundra by Whalen and Reeburgh[48] based on their 4-year flux time series (a range of 26–99 Tg yr^{-1}, an average of either 38 or 53 Tg yr^{-1} depending on whether the high flux year is included).

Methane Flux and Environmental Controls

Although most work on methane emissions from the high latitudes finds a link between flux and such obvious environmental factors as water table level and soil temperature, in most cases these are qualitative linkages rather than quantitative. In large measure, this is due to the large number of environmental factors that affect the production, consumption, and release of methane from soils. Dise[7] reports a statistical approach to determining quantitative relationships among water level, seasonal soil temperature, and flux for a series of bogs and fens that covered a gradient in wetness in northern Minnesota. The water table was found to explain 62% of the variability in flux. The addition of annual soil temperature change to the regression equation resulted in accounting for 90% of the annual variability. An additional variable that explained some of the residual uncertainty in this model was the Von Post peat humification index of the sites, a measure of the degree of decomposition. When specific sites were examined separately, soil temperature rather than water level explained most of the variability. Moore et al.[27] found that although flux was only poorly correlated with water table and temperature at any one of their sites, examination of the entire data set indicated a relationship between these variables and flux.

Aside from Svensson,[41] who reports a quantitative relationship between flux and soil moisture (as % dry weight), statistically significant correlations between flux and environmental variables are generally between flux and soil temperature. Figure 1 illustrates data from Minnesota, Sweden, and Alaska for which a significant correlation between flux and temperature was reported. Soil temperatures are from 2 to 3 cm for the Swedish data, 10 cm for the Minnesota data, and 15 cm for the Alaskan data. All of the data clearly fall along a similar line, although considerable variability within each of the datasets may be present. Even though soil temperatures from Alaska are from a slightly greater depth and, therefore, would be expected to be somewhat less variable, the Alaskan data covers a much smaller range in both temperature and flux than the other two datasets. Data from the drier Alaskan upland tundra is offset from the other,

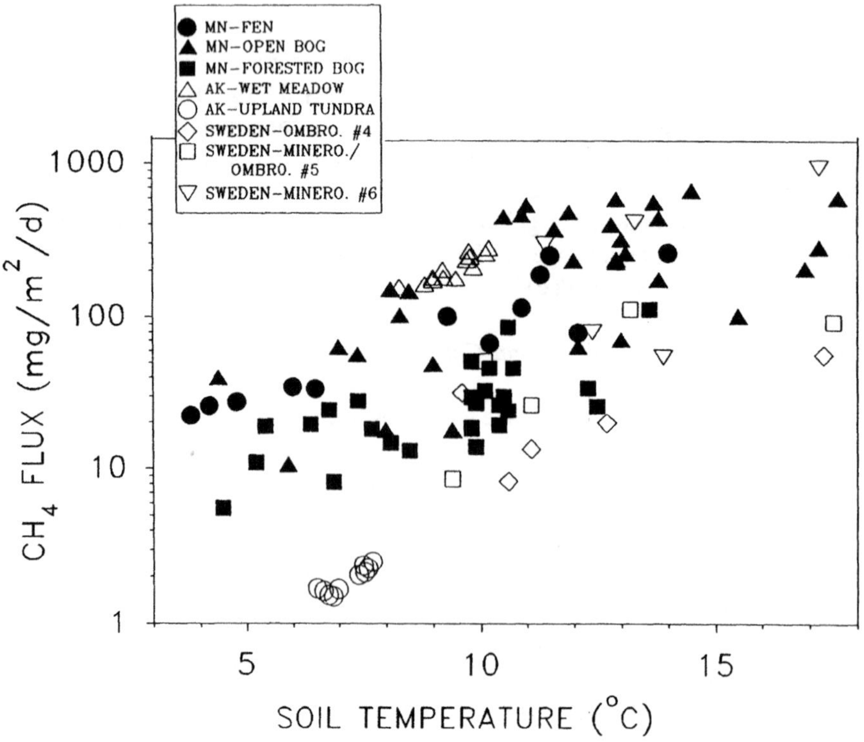

Figure 1. Methane flux versus soil temperature for eight different datasets.

wetter environments but still appears to parallel the trend of logarithmically increasing flux with increasing temperature. Figure 2 illustrates all of the reported quantitative regressions of flux with temperature. Although a sense of the variability in the original data is not conveyed, this figure more clearly demonstrates the striking similarity in slopes for these relationships. This similarity is true both for habitats within a single region (in Minnesota, for fen, open bog, and forested fen; for the Yukon–Kuskokwim Delta, for upland and wet meadow tundra) and across arctic and boreal zones. Although this similarity may be a function of the limited database to date, it may also imply a more functional relationship between temperature and methane release across a variety of high-latitude environments. If such is the case and additional data is found to fall within the range illustrated in Figure 2, it may be possible to model flux to temperature on relatively large scales. Note should be made, however, that these flux/temperature data are seasonal datasets in which additional variables are also changing. An exercise in modeling emission response to temperature variability is more fully explored below.

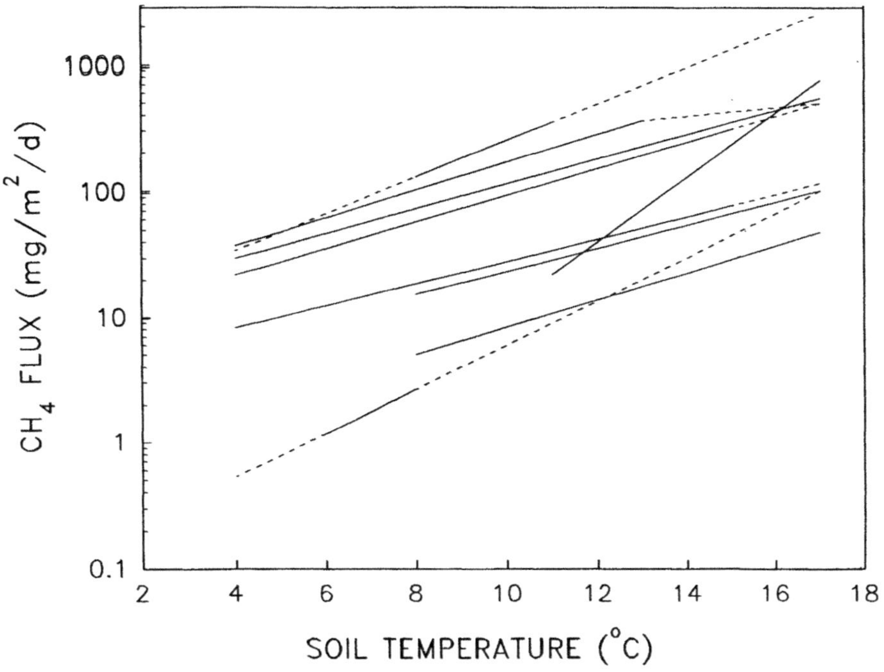

Figure 2. Regression lines calculated from the individual data sets illustrated in Figure 1. Solid lines indicate the temperature range over which the regression equation was derived. Data from Refs. 1, 4, 20, and 42.

In summary, we have shown that methane emissions from northern wetlands are influenced by both soil temperature and moisture. Although other variables like soil organic content and quality are undoubtedly important, it seems certain that a warmer, wetter climate would enhance methane emissions from most sites and a cooling would reduce emissions. A transition from wet to dry soils could change a site from a net methane source to a sink.

It is clear that additional survey measurements of methane flux are needed in areas like the vast Siberian Lowlands since we have shown that extrapolations from one region to another are unreliable. Time-series measurements to obtain seasonal variations in methane emissions at individual sites have also proven very important for understanding links to climate variability and the extent of natural variation in flux.

Modeling the Response of Methane Emissions to Climate Variability

The possibility that increasing concentrations of atmospheric greenhouse gases related to human activities will result in a global warming has been widely discussed (e.g., see Ref. 31 for a recent review). An unknown factor in the global warming scenario is the degree to which feedbacks from changes in biogeochemical, photochemical, ecological, or physical variables in response to initial warming will modify rates of climate change.[19] Because there are a large number of potential feedbacks, it seems prudent to use simple models to explore the sign and order of magnitude of specific cases to set priorities for further research (e.g., Refs. 11 and 12). In this spirit, we present our preliminary results of modeling studies of the sensitivity of methane emissions from northern wetlands to regional warming in the following subsections.

A Simple Model of Methane Flux Dependence on Air Temperature at Two Northern Sites

In this section, we describe the results of a modeling study of the methane flux response to changing air temperature at the Marcell Experimental Forest, Minnesota (47°32′N, 93°28′W) and at the center of the Yukon–Kuskokwim River Delta, Alaska (60°45′N, 161°50′W). These sites were chosen because we have relatively long-term climate records for both sites and some data on the temperature dependence of methane emissions (Figures 1 and 2). Our model uses air-temperature data and heat-diffusion principles to calculate soil temperatures. The methane flux–soil temperature relationships shown in Figure 2 are assumed to extend over a slightly larger temperature range than covered by field measurements.

To explore the sensitivity of methane emissions to variations in temperature, we elected to use historical climate records to characterize climate variability. For example, it is clear from the studies of climate records that there have been short-term "warm" periods which were 1–2°C higher than normal, providing a model for the initial stages of a global warming.

Our climate data were derived by combining the 1951–1980 monthly mean air temperatures for Bethel, Alaska and Grand Rapids, Minnesota[37] with monthly air-temperature anomalies for the same areas for 1900–1987 compiled by Hansen and Lebedeff.[14] Because methane emissions from northern wetlands are predominantly a summer season phenomenon (e.g., Ref. 45) and the flux–temperature relationships used in the model are summer data, we will focus on the summer temperatures. Figure 3 shows the constructed summer (mean of June, July, and August) temperatures for our Alaska and Minnesota sites for 1900 through 1987.

In applying this model to these two sites, we consider five scenarios for each site:

1. the "mean year" (not an actual year but a year made up of the 1951–1980 monthly means);

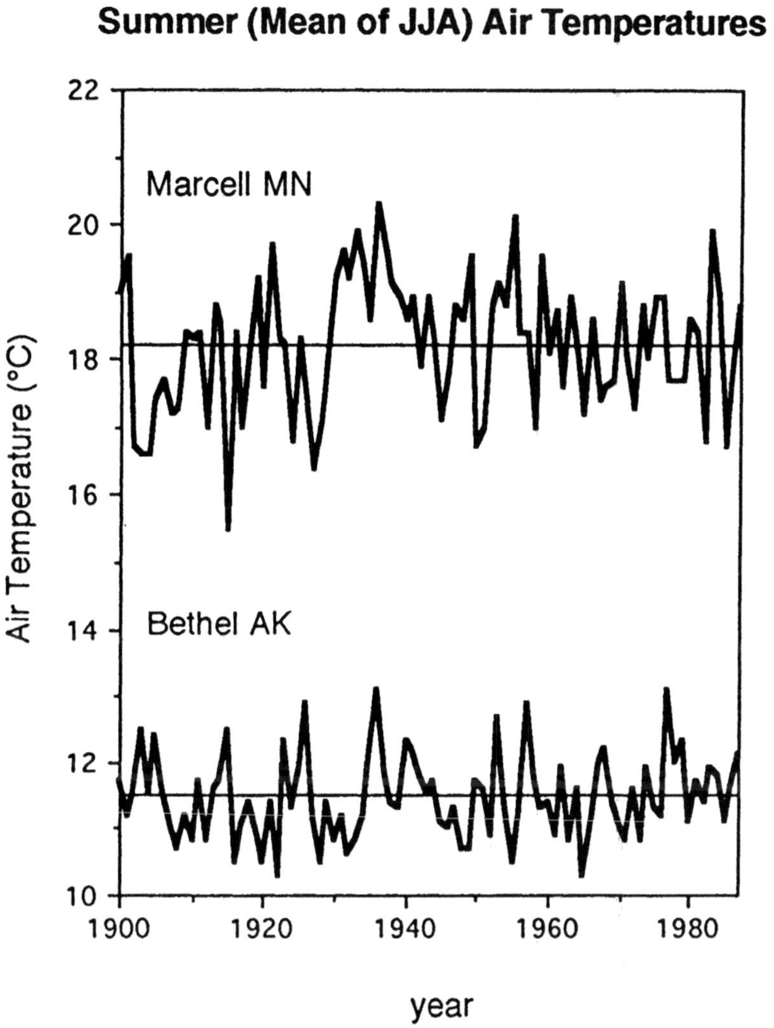

Figure 3. Mean summer (JJA) air temperatures for Bethel, Alaska and Marcell, Minnesota constructed from the 1951–1980 monthly means[37] and the monthly anomalies compiled by Hansen and Lebedeff.[14] The 1951–1980 summer means are shown as horizontal lines.

2. the year with the warmest summer (June, July, August);
3. the year with the coolest summer;
4. the warmest half-decade (five consecutive years with the warmest average of the five summers);
5. the coolest half-decade.

Soil temperature over a season is driven primarily by the soil surface energy balance. The surface air temperature is often a good proxy for the soil surface temperature (which is rarely measured), and thus for the surface energy balance, especially when averaging over days for the seasonal signal.[40] We use a simple model of soil temperature driven by surface air temperature, to be coupled with the soil temperature–methane flux relationships to give us a model of methane flux based on air temperature.

Vertical heat transfer in soils can be modeled by the heat diffusion equation

(1) $$\frac{\partial T}{\partial t} = \alpha \frac{\partial^2 T}{\partial z^2},$$

where T is the soil temperature (°C), t is time (s), z is depth in the soil (cm), and α is the soil thermal diffusivity ($cm^2\ s^{-1}$). This equation assumes that the soil thermal properties are uniform with depth (up to several meters).

If a sinusoidal soil surface temperature

(2) $$T_{surf}(t) = T_{mean} + T^* \sin(\Omega t + \phi)$$

is applied to a uniform soil profile, then the soil temperature at depth is given by

(3) $$T(z,t) = T_{mean} + T^* \exp\left(-\frac{z}{D}\right) \sin\left(\Omega t + \phi - \frac{z}{D}\right),$$

where T_{mean} is the mean temperature of the period of interest (e.g., diurnal or annual), T^* is the amplitude of the temperature oscillation, Ω is the frequency of the oscillation, ϕ is a phase shift (related to the time of maximum temperature), and D is the damping depth (e-folding depth) of the oscillation, where

(4) $$D = \left(\frac{2\alpha}{\Omega}\right)^{1/2}.$$

Equation (4) shows that the damping depth of the surface signal is shorter for higher-frequency oscillations (larger Ω). Choosing $\alpha=0.001\ cm^2\ s^{-1}$ as a typical thermal diffusivity for wet peat,[16] the damping depth for diurnal oscillations is about 5 cm, and for annual oscillations it is about 100 cm. Since the diurnal and annual temperature oscillations have roughly the same amplitude (on the order of 10°C), below about 10 cm in wet peat most temperature variation is due to the annual signal.

To make use of the available temperature data, we fit a sinusoidal function to the mean monthly air-temperature data and consider that function to represent the mean daily air temperature. To more closely fit the air-temperature data during the summer months, we use an annual sine wave plus the first two harmonics (periods = 183 days and 91.5 days). These harmonics are also sine functions and so will propagate into the soil in the same manner, though with damping depths that decrease with the square root of the frequency. So the modeled air temperature is given by

$$T_{air} = T_{mean} + \sum_{n=1,3} T_n \sin(\Omega_n + \phi_n), \tag{5}$$

where T_n is the amplitude of the annual (n=1) or harmonic (n=2, 3) waves, Ω_n is the frequency, and ϕ_n is the phase shift for each wave.

Combining eqs. 1 and 5, the soil temperature profile is given by

$$T(z,t) = T_{mean} + \sum_{n=1,3} T_n \exp\left(-\frac{z}{D_n}\right) \sin\left(\Omega_n + \phi_n - \frac{z}{D_n}\right). \tag{6}$$

The phase shifts, ϕ_n, are given by the sinusoidal fit to the monthly mean air temperatures. Figure 4 shows an example of the monthly air-temperature data and the sinusoidal fit to the data. Note that we concentrate on fitting to the warm months, when methane fluxes are high, at the expense of potential poor fits to the winter months, when methane fluxes are low or nil.

Methane flux measurements near Bethel, Alaska[1] and in the Marcell Forest[4] correlate with soil temperature (Figure 2). We use fits to these correlations to empirically predict methane fluxes from modeled soil temperatures. The flux/temperature equations used are as follows:

BETHEL (wet meadow)

$$\log(F) = 0.9594 + 0.1442T_{15}, \tag{7}$$

MARCELL

$$\ln(F) = 67.9-[17893(T_{10} + 273.2)], \tag{8}$$

where F is the methane flux (mg CH_4 m^{-2} d^{-1}), T_{15} is the daily mean soil temperature (°C) at 15 cm depth, and T_{10} is the daily mean soil temperature (°C) at 10 cm depth. We assume that when the soil temperature at depth is equal to or below 0°C, there is no methane flux. Combining historical temperature records discussed above with Eqs. (6)–(8), we can estimate methane fluxes for various years.

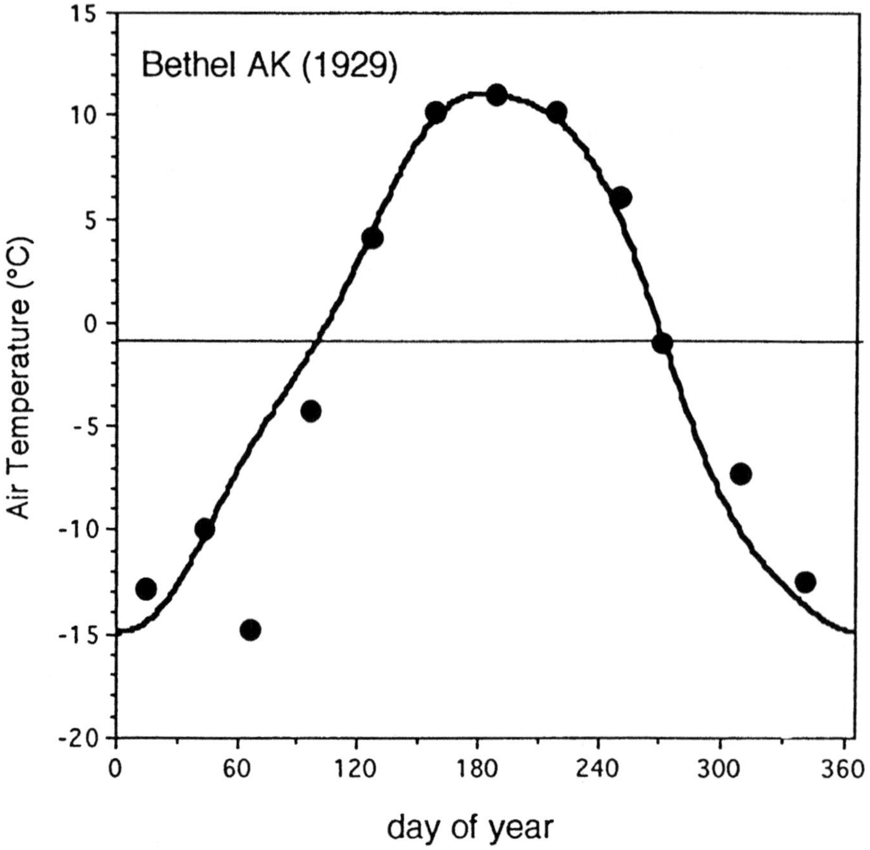

Figure 4. The modeled daily air temperature for Bethel, Alaska in 1929 (solid line) and the monthly means for that year (solid circles).[14,37] The daily air temperature is composed of an annual sine wave and two harmonics (see text). The fit was weighted to the summer months.

Discussion of Model Results

Extreme Years

For both Alaska and Minnesota, the annual modeled methane flux for the year with the warmest summer is more than twice as large as the year with the coolest summer (Figure 5a). The warmest summer annual flux is more than 50% greater than the mean year flux for both sites, and the coolest summer annual flux is about 70% of the mean year flux for both sites. The temperature changes from

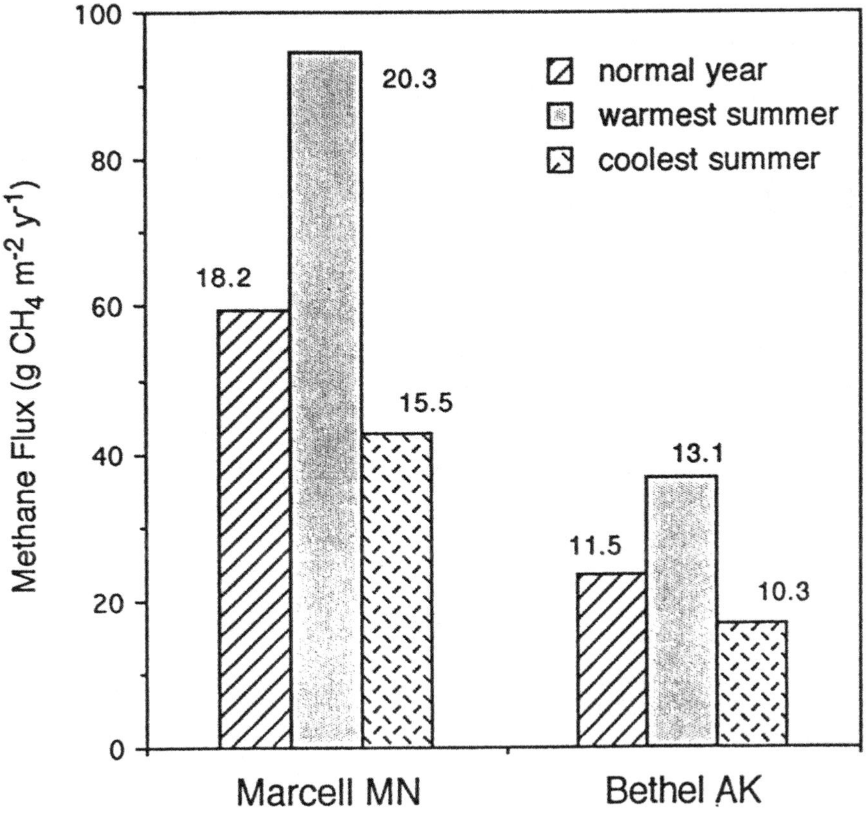

Figure 5. (a). Modeled methane fluxes (in g CH_4 m^{-2} y^{-1}) for normal and extreme years for sites in Alaska and Minnesota. Numbers by each bar are summer mean temperatures (°C). Warm summers are: Minnesota—1936; Alaska—1977. Cool summers are: Minnesota—1915; Alaska—1922. (b) Modeled methane fluxes for warmest and coolest half-decades for sites in Alaska and Minnesota. Numbers by each bar are mean summer temperatures for 5 years (°C). Warm half-decades are: Minnesota—1933–1937; Alaska—1977–1981. Cool half-decades are: Minnesota—1924–1928; Alaska—1929–1933.

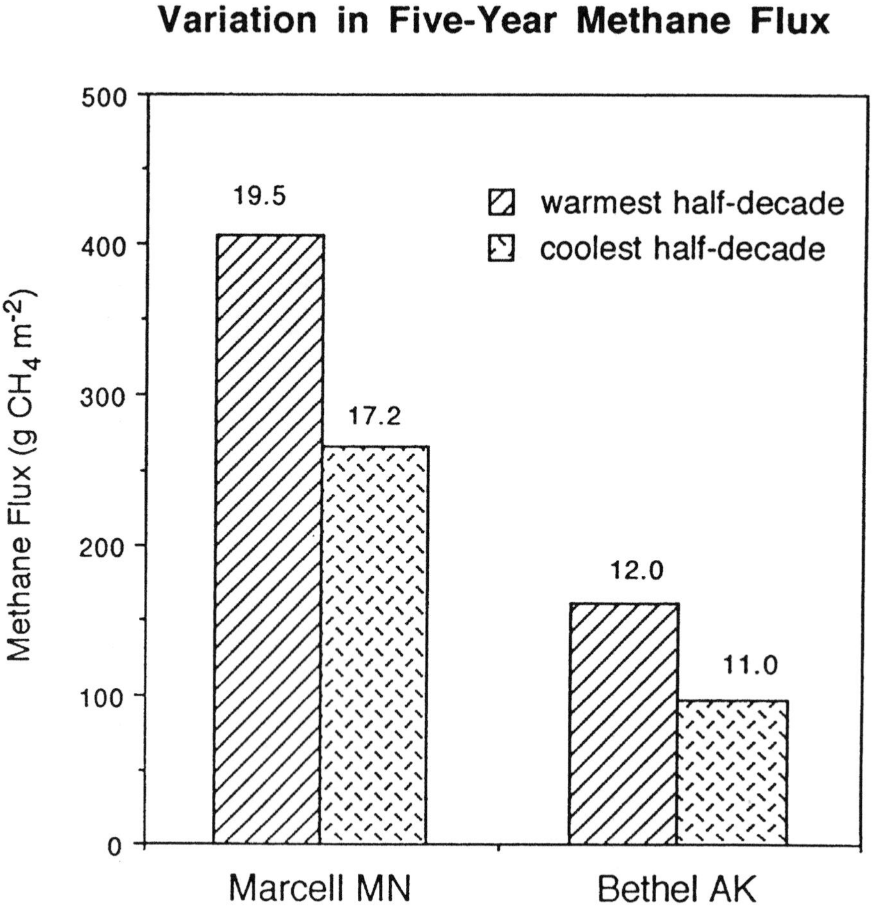

Figure 5b.

the normal summer to the warmest summer are 1.6°C for Alaska and 1.9°C for Minnesota; both changes are well within predicted greenhouse warming over the next century.[31] These two wetland systems appear to be quite sensitive to temperature change (both predicted for the future and realized in the past)—assuming the soil moisture regime does not change.

FIVE CONSECUTIVE YEARS

When the model is run for five consecutive years (each year is modeled individually and the 5 years are summed), there is still quite a large difference in predicted methane flux between warm and cool periods (Figure 5b). For Bethel, the modeled warm period (1977–1981) produced 67% more methane flux than the modeled cool period (1929–1933). For Marcell, the modeled warm period (1933–1937) produced 54% more methane flux than the modeled cool period (1924–1928). As the time period that the fluxes are integrated over increases, the difference between warm and cool periods decreases, but it is still quite pronounced over a 5-year period.

Scaling Up

To create a strong signal in the global atmospheric methane budget, large regions of the boreal and arctic wetlands will have to simultaneously experience similar climate deviations. For example, a summer that is very warm over eastern Canada and very cool over the Siberian Lowlands may have no net effect on the global methane budget. In this subsection, we use our model to examine the large-scale variability of methane flux from northern wetlands due to temperature variations over this past century.

As an initial study of the large-scale climate-induced variability of methane fluxes from northern wetlands, we consider the effects of temperature variability only, although clearly variability in precipitation and, therefore, soil moisture in these wetlands is a critical control on emissions. We group northern wetlands into eight regions: the Siberian Lowlands, the Hudson Bay Lowlands, the Fenno-Soviet Lowlands, the Alaskan North Slope, the Yukon–Kuskokwim Delta, the boundary waters area of Minnesota and Manitoba, Central Alaska, and Nouveau Quebec (see Figure 6). These regions represent most (but not all) of the fresh-water wetlands in the boreal and arctic regions. They are also regions for which at least some methane flux studies have been reported (except for the Siberian Lowlands, which because of their vast size, cannot be ignored).

For each of these regions, a total inundated wetland area is determined from the fractional inundation $1° \times 1°$ dataset compiled by Matthews and Fung,[23] taking into account that the area of a $1° \times 1°$ grid cell goes as the cosine of its latitude. For each of the regions, we assign a representative wet site methane flux and an estimated flux season length. For the Siberian Lowlands, for which we know of no flux data, we assign a flux equal to the mean of the other seven regions. Table 5 lists the regions, their delineations, inundation areas, representative fluxes, season lengths, and net regional methane flux for 1 year (the product of the previous three numbers). Note that regional emissions for these eight areas in the model vary significantly from the wet soil averages listed in Table 3 and that the total emission of 17 Tg is roughly 57% of the global northern wet soil estimate. Differences in total emissions arise both because the model does not include all northern wetland areas and because values listed in Table 3

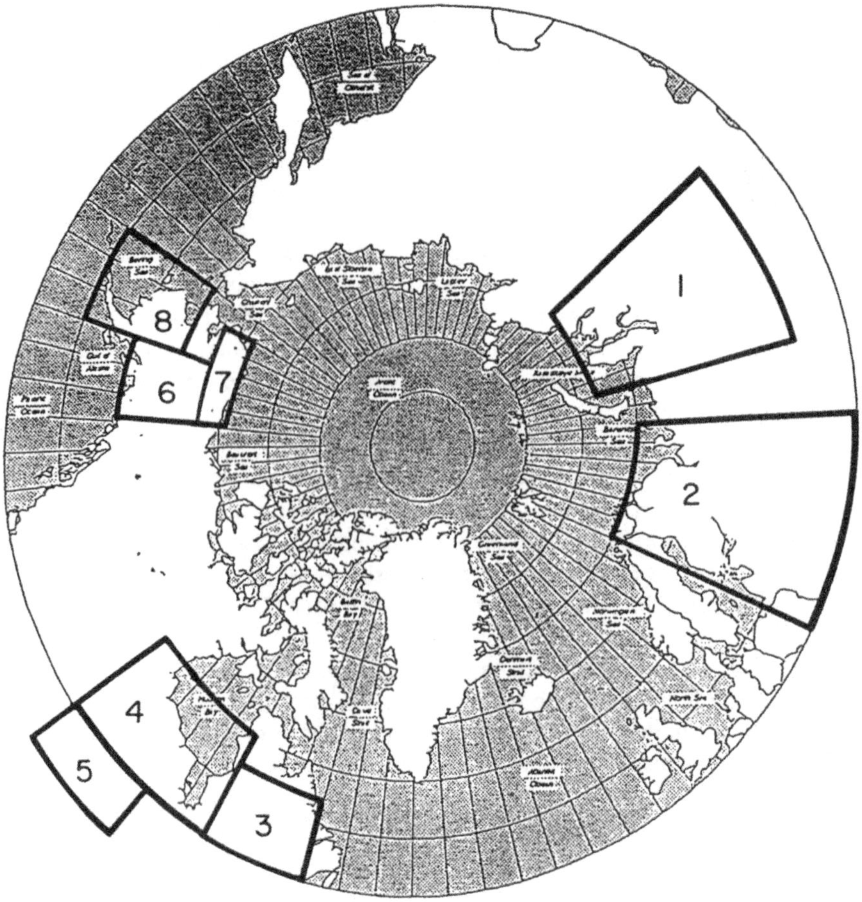

1) Siberian Lowlands
2) Fenno-Soviet Lowlands
3) Nouveau Quebec
4) Hudson Bay Lowlands
5) Boundary Waters Area
6) Central Alaska
7) North Slope Alaska
8) Yukon-Kuskokwim Delta

Figure 6. Delineation of the eight regional wetland areas modeled.

Table 5. Northern wetland regions data.

Region	Latitude Longitude	Area (10^9 m^2 (Ref. 23)	Rep. CH_4 flux (mg CH_4 m^{-2} d^{-1})	Season Length (days)	Regional CH_4 flux (Tg yr^{-1})
Siberian Lowlands	55–73N 60–90E	723	80.1[a]	120	7.0
Fenno-Sov. Lowlands	50–70N 20–50E	634	57.8[b]	120	4.4
Hudson Bay Lowlands	50–60N 75–100W	342	24.0 (Ref. 28)	120	1.0
Boundary Waters Area	45–50N 85–100W	98.0	188.0 (Refs. 4, 5)	150	2.8
North Slope Alaska	68–71N 140–165W	71.2	103.0 (Refs. 20, 39, 46)	100	0.73
Central Alaska	60–68N 140–155W	46.7	32.2 (Ref. 45)	120	0.18
Yukon–Kusk. Delta	55–65N 155–170W	56.5	86.5 (Ref. 1)	120	0.59
Nouveau Quebec	50–57N 60–75W	30.5	70.0[b] (Ref. 27)	120	0.26

[a]No data available, value is mean of values for other seven regions.
[b]P. Crill, unpublished data.

are simple averages, rather than regionally weighted averages, of reported emissions.

From the temperature anomaly data of Hansen and Lebedeff,[14] we construct mean summer (JJA) air-temperature curves for a site in each of the eight regions for 1900–1987 (see Figure 7). Hansen and Lebedeff,[14] in their analysis of air-temperature anomalies, find that for northern latitudes (>45°N), annual mean air-temperature anomalies are well-correlated for stations separated by less than about 1200 km and that the correlation improved as the station separation decreased (see their Figure 3). This result would indicate that for these regions, any climate deviation that occurred would prevail over much of any one of these extensive wetland regions, and one site's data should be fairly representative of the entire region. Most summer temperature anomalies are within about 1°C of the 1951–1980 summer normals, with occasional 2°C anomalies. No region has summer temperature behaviors greatly different from any of the others. Figure 8 shows the net summer temperature anomaly for the eight sites by taking the average of them; since the eight sites are not all in phase, the net effect is reduced to variations usually within 0.5°C of the norm. There was an extended warm period from about 1935 to 1945 for these eight sites, with all but the Siberian Lowlands' site showing warm temperatures for most of this period.

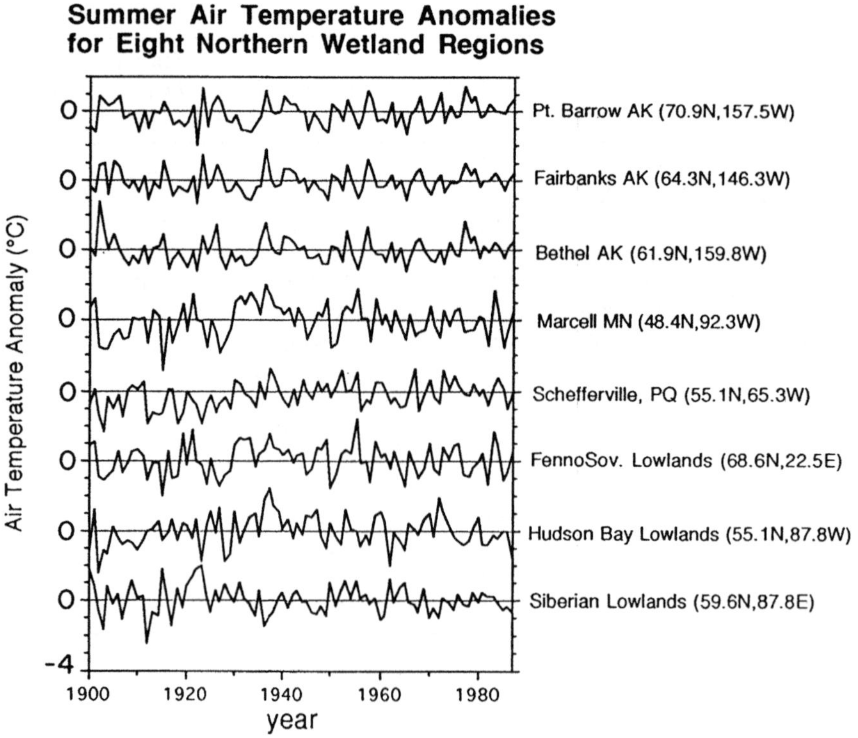

Figure 7. Summer air-temperature anomalies[14] for a site in each of the eight regions of Figure 6. Latitude and longitude refer to the center of the grid cell (200 km × 200 km) which the anomaly represents.

By taking the derivative of methane flux, F, with respect to soil temperature, T, in Eq. (8) above, we can calculate the flux variability with temperature change as

$$\frac{d(F)}{dT} = F\left(\frac{17893}{(T_{10} + 273.2)^2}\right). \tag{9}$$

Using 10°C as a representative soil temperature for most northern wetlands, this gives a relative flux change, for a one degree change in temperature, of

$$\frac{dF/dT}{F} = 0.22. \tag{10}$$

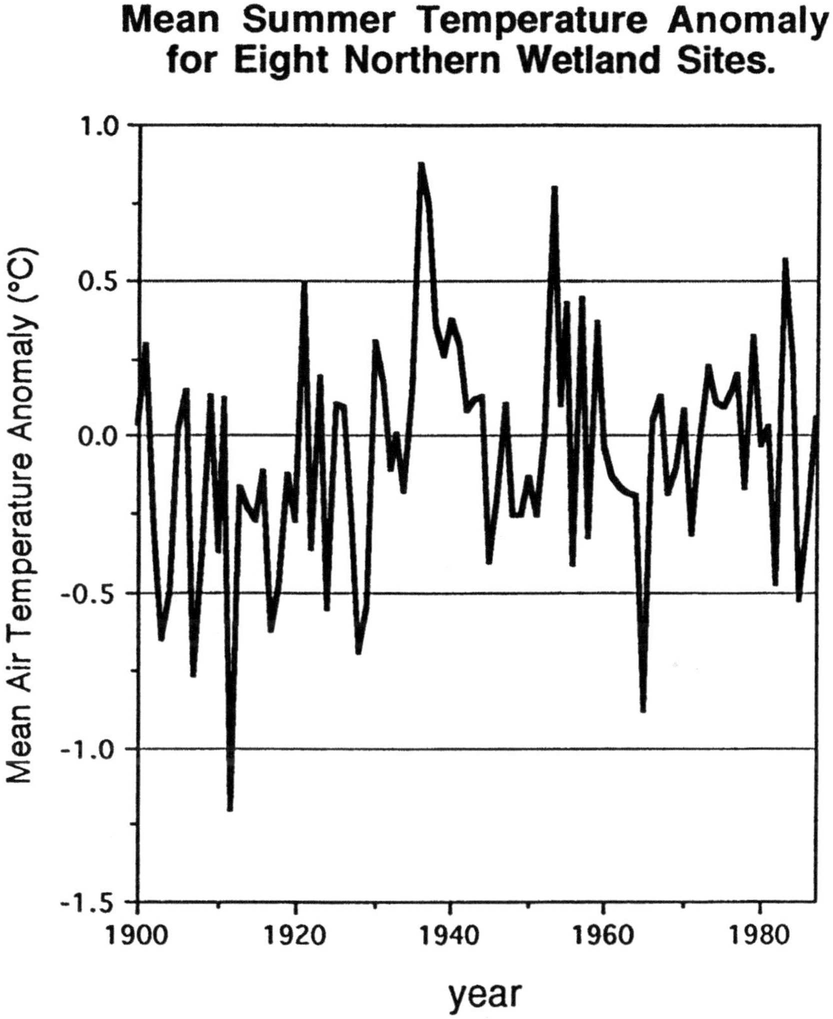

Figure 8. Net air-temperature anomaly for the eight sites of Figure 7.

A net methane flux change for each region is then given by

$$\delta F = \left(\frac{dF/dT}{F}\right) AF_{rep} SL \delta T, \tag{11}$$

where A is the region's area (10^3 km^2), F_{rep} is the region's representative methane flux (mg CH_4 m^{-2} d^{-1}), SL is the region's season length (d), δT is the

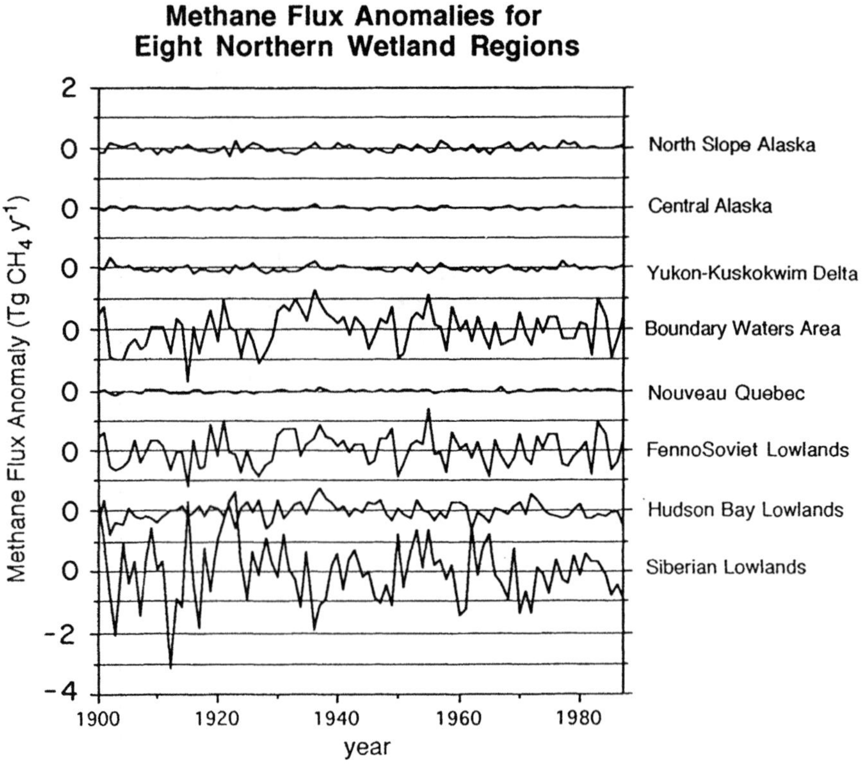

Figure 9. Annual methane flux anomaly (in Tg CH_4) for the eight northern wetland regions. See text for details.

region's summer temperature anomaly (°C), and δF is the region's variation from normal methane flux (Mg CH_4), or methane flux anomaly, for the year. Figure 9 shows this methane flux anomaly for the eight northern wetland regions modeled. The greatest flux anomaly occurs in the Siberian Lowlands due to their large area and assumed moderate representative flux strength. The large Hudson Bay Lowlands have a much smaller anomaly because of the low methane fluxes reported by Moore et al.[28] for this region (Table 1). The Boundary Waters Area of Minnesota and Manitoba have large flux anomalies due to their high reported fluxes[4,5] and their longer season length. In addition, Dise[6] reports significant winter fluxes from this region, although these are not accounted for in this model. The Fenno-Soviet Lowlands have a large anomaly due to their large areas and moderate flux strengths. As discussed above, all regions have roughly the same temperature variations (magnitude but not timing), so no one region's methane flux anomaly is enhanced or reduced with respect to the others due to climate variability.

Figure 10 shows the sum of these eight regional anomalies, or the net northern wetland modeled methane flux anomaly. Superimposed on this curve is a 9-year running mean of the net flux anomaly to account for the integrating effects of methane's lifetime in the atmosphere. In an extreme year, the anomaly (positive or negative) can be as much as 5 Tg CH_4, or 29% of the total normal flux of the model. Most variations are about 0–2 Tg, or 0–12% of the total model flux. The

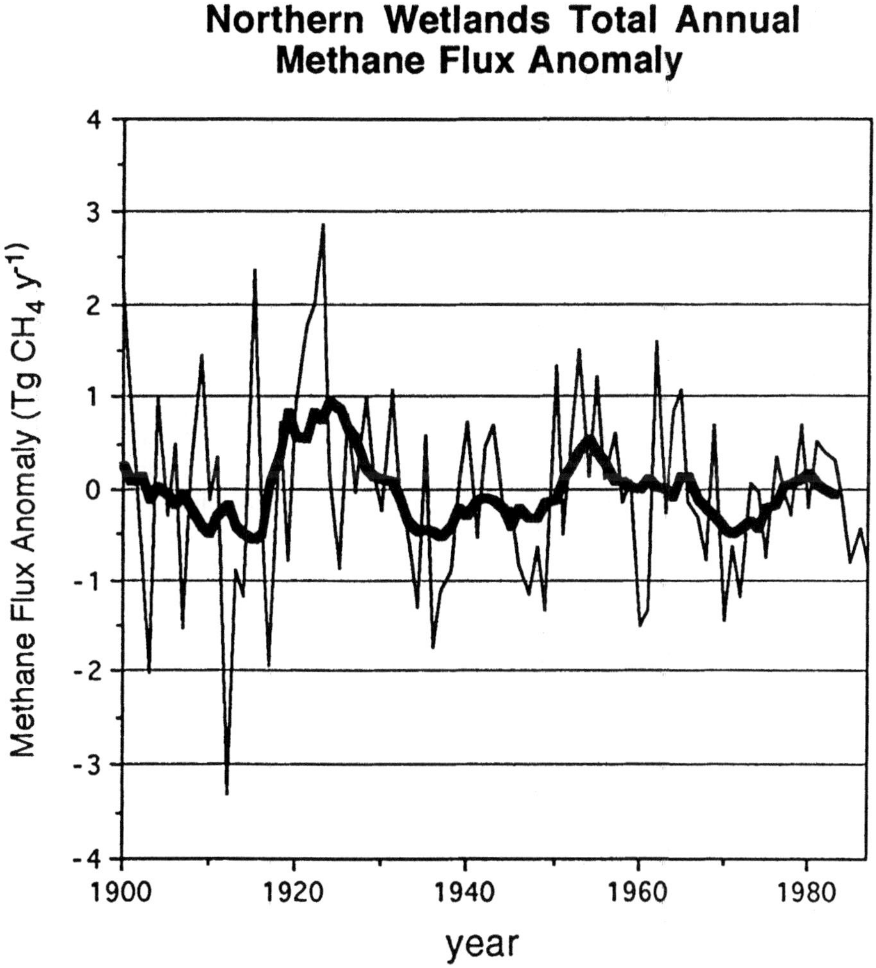

Figure 10. Total annual methane flux anomaly (in Tg CH_4). This is the sum of the eight curves in Figure 9. The heavy line is a simple 9-year running mean of this total flux anomaly.

Table 6. Boreal temperature change predictions. Average temperature change (°C).

Model	Summer (JJA)	Winter (DJF)	Annual
GISS	+2 to +4	+5 to +12	+4.2
GFDL	+4 to +8	+6 to +15	+4.0
NCAR	+0 to +4	+6 to +10	+4.0
UKMO	+5 to +6	+8 to +10	+5.2

Source: Reference 24.

9-year smoothed curve is usually within 1.0 Tg, or about 6% of the modeled normal flux. Scaling this up to our current estimate of about 30 Tg CH_4 yr^{-1} from northern wetlands, the annual temperature-driven methane flux anomaly would be about 0–4 Tg. The 9-year smoothed anomaly would vary by about 0–2 Tg. Note that there does not appear to be any trend in the pattern of emissions over the period of our model (1900–1987).

Future Climate Change and Northern Wetlands

There is a general agreement among the current generation of Global Climate Models (GCMs) that increasing greenhouse gases in the atmosphere will cause a warmer climate and that high northern latitudes will experience the greatest warming.[24] Table 6 summarizes the boreal and subarctic zone (global 50°–70°N) temperature change predictions of four GCMs for a doubling of CO_2 in the atmosphere. As the table indicates, summer temperature changes are generally expected to be less than the annual average (as winter temperature changes are greater). An average GCM predicted summer temperature change for the boreal regions is about 3–5°C. This is about two to three times the observed change for the warm and cool half-decades for Bethel and Marcell, and as large as any year's summer anomaly for the eight northern sites modeled above. In the half-decade models of Alaska and Minnesota, our modeled methane flux increased by about 50%, so we would expect a potential doubling of methane flux for the GCM predicted climate changes, assuming constant soil moisture.

There is less agreement among the models as to how precipitation will change for a doubling of CO_2, especially for the summer months. Some models predict wetter summers and some drier summers.[24] Wetland soil moisture and water table level are important controls on both methane production and oxidation (see Table 2 for a comparison of wet and dry site fluxes), and a more complete model of methane fluxes will have to incorporate soil moisture and its changes. Because high soil moisture is required for significant methane production, changes in precipitation may have a greater effect on emissions than changes in temperature.

Summary

Our review and assessment of current literature (to late 1991) on methane emissions from northern high latitude wetlands leads to the following conclusions:

1. Emissions from above 50°N total 34 Tg CH_4 yr^{-1} (wet soils, 30 Tg CH_4 yr^{-1}; moist to dry tundra, 4 Tg yr^{-1}). Although this estimate is lower than those made initially for the region based on smaller databases,[23,39] it is in agreement with recent estimates.[1,10,48] If the total emission of methane to the atmosphere from all sources is on the order of 500 Tg yr^{-1},[10] then high-latitude wetlands contribute roughly 7%. Significant uncertainties remain in the calculation of this global estimate. Large wetland areas exist from which there are no data, and our modeling of flux and soil temperatures suggest that of these areas, the Siberian Lowlands may be particularly important in making global flux estimates. Although northern wetlands make an important contribution to atmospheric methane, anthropogenic sources and a decrease in atmospheric oxidizing capacity due to increased air pollution are now thought to be the dominant factors determining the concentration and variability of this important atmospheric gas.[17] Our preliminary modeling indicates that emissions from northern wetlands have not exhibited any significant trends since the turn of the century.
2. Northern high-latitude wetlands are a potential source of atmospheric methane under warmer, wetter climate conditions. However, although our modeling studies indicate that the early response of northern wetlands to an increased frequency of warm summers would increase the flux of methane to the atmosphere, the magnitude of the increase would be too small to have a quantifiable effect on the global atmospheric concentration of methane. Moisture levels in the soil column will be an important factor determining how much of the methane produced under warmer climates is destroyed by microbial oxidation processes in the soil column before reaching the atmosphere. Because climate models are presently unable to predict regional changes in soil moisture which might accompany a global warming, speculation on mid- to long-term potential feedbacks associated with methane release from soil reservoirs (current methanogenesis as well as possible releases from thawing permafrost or methane hydrate layers) should not be used in policy assessments. Improved regional-scale longer-term predictive capabilities for soil moisture and a better understanding of processes determining methane oxidation rates will be required for progress in evaluating methane feedbacks to a global warming in northern high-latitude landscapes.

Acknowledgments

We are indebted to numerous members of the methane flux community who have generously shared their data with us prior to publication. This work was supported by the NASA Biospherics Research Program and the NASA Interdisciplinary Research Program. Steve Frolking was supported under the NASA Graduate Student Researcher Program.

References

1. Bartlett, K.B., P.M. Crill, R.L. Sass, R.C. Harriss, and N.B. Dise. 1992. Methane emissions from tundra environments in the Yukon–Kuskokwim Delta, Alaska. J. Geophys. Res. **97**:16,645–16,660.
2. Born, M., H. Dorr, and I. Levin. 1990. Methane consumption in aerated soils of the temperate zone. Tellus **42B**:2–8.
3. Clymo, R.S., and E.J.F. Reddaway. 1971. Productivity of *Sphagnum* (Bog-moss) and peat accumulation. Hidrobiologia **12**:181–192.
4. Crill, P.M., K.B. Bartlett, R.C. Harriss, E. Gorham, E.S. Verry, D.I. Sebacher, L. Madzar, and W. Sanner. 1988. Methane flux from Minnesota peatlands. Global Biogeochem. Cycles **2**:371–384.
5. Dise, N.B. Methane emission from Minnesota peatlands: Spatial and seasonal variability. Global Biogeochem. Cycles. (in press)
6. Dise, N.B. Winter fluxes of methane from Minnesota peatlands. Biogeochemistry (in press).
7. Dise, N.B. 1991. Methane emission from peatlands in northern Minnesota, Ph.D. dissertation, University of Minnesota.
8. Edwards, G., H.H. Neumann, G. den Hartog, O. Thurtell, and G.E. Kidd. 1991. Eddy correlation measurements of methane fluxes using a tunable diode laser at the Kinosheo Lake tower site during the Northern Wetlands Experiment. EOS **72**:85.
9. Fan, S-M., S.C. Wofsy, P.S. Bakwin, D.J. Jacob, S.M. Anderson, P.L. Kebabian, J.B. McManus, C.E. Kolb, and D.R. Fitzjarrald. 1992. Micrometeorological measurements of CH_4 and CO_2 exchange between the atmosphere and the arctic tundra, J. Geophys. Res. **97**:16,627–16,643.
10. Fung, I, J. John, J. Lerner, E. Matthews, M. Prather, L.P. Steele, and P.J. Fraser. 1991. Three-dimensional model synthesis of the global methane cycle. J. Geophys. Res. **96**:13033–13065.
11. Guthrie, P. 1986. Biological methanogenesis and the CO_2 greenhouse effect. J. Geophys. Res. **91**:10847–10851.
12. Hameed, S., and R.D. Cess. 1983. Impact of a global warming on biospheric sources of methane and its climatic consequences, Tellus **35B**:1–7.
13. Hamilton, J.D., C.A. Kelly, and J.W.M. Rudd. 1991. Methane and carbon dioxide flux from ponds and lakes of the Hudson Bay Lowlands. EOS **72**:84.

14. Hansen, J. and S. Lebedeff. 1987. Global trends of measured surface air temperature. J. Geophys. Res. **92**:13345–13372.

15. Harriss, R.C., E. Gorham, D.I. Sebacher, K.B. Bartlett, and P.A. Flebbe. 1985. Methane flux from northern peatlands. Nature **315**:652–654.

16. Hillel, D. 1980. Fundamentals of Soil Physics, Academic Press, New York.

17. Khalil, M.A.K., and R.A. Rasmussen. 1990. Atmospheric methane: Recent global trends. Environ. Sci. Technol. **24**:549–553.

18. King, S.L., P.D. Quay, and J.M. Lansdown. 1989. $^{13}C/^{12}C$ kinetic isotope effect for soil oxidation of methane at ambient atmospheric concentrations. J. Geophys. Res. **94**:18273–18277.

19. Lashof, D. 1989. The dynamic greenhouse: Feedback processes that may influence future concentrations of atmospheric trace gases and climatic change. Climatic Change **14**:213–242.

20. Livingston, G.P., and L.A. Morrissey. Methane emissions from Alaska arctic tundra in response to climatic change. In Proc. Int. Conf. on the Role of the Polar Regions in Global Change.

21. Martens, C.S., C.A. Kelley, J.P. Chanton, and W. Showers. 1992. Carbon and hydrogen isotopic characterization of methane from wetlands and lakes of the Yukon-Kuskokwim Delta. J. Geophys. Res. **97**:16,689–16,701.

22. Matthews, E. 1983. Global vegetation and land use: New high-resolution data bases for climate studies. J. Climate Appl. Meteorol. **22**:474–487.

23. Matthews, E., and I. Fung. 1987. Methane emission from natural wetlands: Global distribution, area, and environmental characteristics of sources. Global Biogeochem. Cycles **1**:61–86.

24. Mitchell, J.F.B. 1989. The greenhouse effect and climate change. Rev. Geophysics **24**:115–139.

25. Moore, T.R., and R. Knowles. 1987. Methane and carbon dioxide evolution from subarctic fens. Can. J. Soil Sci. **67**:77–81.

26. Moore, T.R., and R. Knowles. 1990. Methane emissions from fen, bog, and swamp peatlands in Quebec. Biogeochemistry **11**:45–61.

27. Moore, T.R., N. Roulet, and R. Knowles. 1990. Spatial and temporal variations of methane flux from subarctic/northern boreal fens. Global Biogeochem. Cycles **4**:29–46.

28. Moore, T.R., A. Heyes, S. Holland, W.R. Rouse, N.T. Roulet, and L. Klinger. 1991. Spatial and temporal variations of methane emissions in the Hudson Bay Lowlands. EOS **72**:84.

29. Morrissey, L.A., and G.P. Livingston. 1992. Methane emissions from Alaska arctic tundra: An assessment of local spatial variability. J. Geophys. Res. **97**:16,661–16,670.

30. Mosier, A.R. 1989. Chamber and isotope techniques. In M.O. Andreae and D.S. Schimel (eds.), Exchange of Trace Gases Between Terrestrial Ecosystems and the Atmosphere. Dahlem Workshop Repts., Life Sci. Res. Rept. #47, J. Wiley and Sons, New York, pp. 175–188.

31. National Academy of Sciences. 1991. Policy Implications of Greenhouse Warming. National Academy Press, Washington, D.C.

32. Reeburgh, W.S., and S.C. Whalen. High latitude ecosystems as CH_4 sources. In Trace Gas Exchange in a Global Perspective. SCOPE/IGBP (in press).

33. Ritter, J.A., J.D.W. Barrick, G.W. Sachse, G.L. Gregory, M.A. Woerner, C.E. Watson, G.F. Hill, and J.E. Collins. 1992. Airborne flux measurements of trace species in an arctic boundary layer. J. Geophys. Res. **97**:16,601–16,625.

34. Ritter, J.A., C. Watson, J. Barrick, G. Sachse, J. Collins, G. Gregory, B. Anderson, and M. Woerner. 1991. Airborne boundary-layer measurements of heat, moisture, CH_4, CO, and O_3 fluxes over Canadian boreal forest and northern wetland regions. EOS **72**:84.

35. Roulet, N.T., R. Ash, and T.R. Moore. 1992. Low boreal wetlands as a source of atmospheric methane. J. Geophys. Res. **97**:3739–3749.

36. Rudd, J.W.M., and R.D. Hamilton. 1978. Methane cycling in a eutrophic shield lake and its effects on whole lake metabolism. Limnol. Oceanogr. **23**:337–348.

37. Ruffner, J.A. 1985. Climates of the States, 3rd ed., Gales Research Co., Detroit, MI.

38. Schiff, H.I., D.R. Karecki, F.J. Lubkin, R. Eng, and G.I. Mackay. 1991. A tunable diode laser system for CH_4 flux measurements from a small aircraft. EOS **72**:78.

39. Sebacher, D.I., R.C. Harriss, K.B. Bartlett, S.M. Sebacher, and S.S. Grice. 1986. Atmospheric methane sources: Alaskan tundra bogs, an alpine fen, and a subarctic boreal marsh. Tellus **38B**:1–10.

40. Sellers, W.D. 1965. Physical Climatology, University of Chicago Press, Chicago.

41. Svensson, B.H. 1976. Methane production in tundra peat. In H.G. Schegel, G. Gottschalk, and N. Pfennig (eds.), Microbial Production and Utilization of Gases (H_2, CH_4, CO), E. Goltze, Gottingen, pp. 135–139.

42. Svensson, B.H., and T. Rosswall. 1984. In situ methane production from acid peat in plant communities with different moisture regimes in a subarctic mire. Oikos **43**:341–350.

43. Verma, S.B., F.G. Ullman, D. Billesbach, R.J. Clement, J. Kim, and E.S. Verry. 1992. Eddy correlation measurements of methane flux in a northern peatland ecosystem. Bound. Layer Meteorol. **58**:289–304.

44. Vitt, D., S. Bayley, T. Jin, L. Halsey, B. Parker, and R. Craik. 1990. Methane and carbon dioxide production from wetlands in boreal Alberta. Rept. on Contract #90-0270 to Alberta Environment.

45. Whalen, S.C., and W.S. Reeburgh. 1988. A methane flux time series for tundra environments. Global Biogeochem. Cycles **2**:399–409.

46. Whalen, S.C., and W.S. Reeburgh. 1990. A methane flux transect along the trans-Alaska pipeline haul road. Tellus **42B**:237–249.

47. Whalen, S.C., and W.S. Reeburgh. 1990. Consumption of atmospheric methane to sub-ambient concentrations by tundra soils. Nature **346**:160–162.

48. Whalen, S.C. and W.S. Reeburgh. 1992. Interannual variations in tundra methane emissions: A four-year time-series at fixed sites. Global Biogeochem. Cycles **6**:139–159.

49. Yarrington, M.R., and D.D. Wynn-Williams. 1985. Methanogenesis and the anaerobic microbiology of a wet moss community at Signy Island. In W.R. Siegfried, P.R. Condy, and R.M. Laws (eds.), Antarctic Nutrient Cycles and Food Webs, Springer-Verlag, Berlin, pp. 134–139.

Studies of Permafrost and Gas-Hydrates as Possible Sources of Atmospheric Methane at High Latitudes

Keith A. Kvenvolden, Timothy S. Collett, and Thomas D. Lorenson

U.S. Geological Survey, Menlo Park, California 94025

ABSTRACT

The United Nations Intergovernmental Panel on Climate Change has identified permafrost and gas-hydrates as important high-latitude methane sources. Two preliminary studies that address onshore and nearshore permafrost and gas-hydrates as possible methane sources are summarized here: (1) In a search for evidence for possible decomposing gas-hydrates, the contents of methane and other hydrocarbon gases were measured in 17 surface-sediment samples from Harrison Bay, Alaska. This bay is situated near an onshore area of known gas-hydrate occurrences. Geophysical surveys indicate offshore permafrost and gas anomalies in the sediment. However, our sediment samples did not contain unusually high methane contents (4.4–20.2 μL per liter of interstitial water), and so these samples provided no compelling evidence for gas-hydrate decomposition. (2) The gas composition of ice in permafrost was measured at two sites near Fairbanks, Alaska. In the air of a tunnel where ice in permafrost is subliming, methane contents of about 13 parts per million by volume (ppmv) were measured, whereas outside the entrance to the tunnel, methane contents were only 1.7 ppmv. One ice sample from the tunnel contained about 40 μL of methane per liter of melted ice (water). At the second site, samples from a core of permafrost contained as much as 6300 μL of microbially derived methane ($\delta^{13}C = -74‰$) per liter of melted ice (water). These preliminary results indicate that the methane content of shallow permafrost can vary. Under conditions of global warming, thawing permafrost may, indeed, be a high-latitude source of methane.

Methane is the most abundant organic compound in the Earth's atmosphere. Its concentration is currently increasing at a rate of about 0.9% per year (summarized by Cicerone and Oremland.[2] The increase began about 200 years ago, as demonstrated by analyses of air trapped in ice cores from Greenland and Antarctica,[20] and the contemporary atmospheric concentration of methane of about 1.7 ppmv exceeds any values observed during the past 160,000 years of geologic time.[21,23]

The infrared-absorption properties of methane make it an important greenhouse gas. Atmospheric methane influences the Earth's climate system, and the predicted future increase in atmospheric methane concentration is likely to contribute significantly to future climatic change.[2] It has been estimated that about 12% of the added greenhouse warming during the past 10 years has come from atmospheric methane.[7]

Major sources and budgets of atmospheric methane were summarized by Khalil and Rasmussen.[10] Two budgets that generally agree in their broad features are those of Khalil and Rasmussen[9] and Cicerone and Oremland.[2] The total emission of methane is estimated in these two budgets at about 550 Tg yr^{-1} (Tg = teragram = 10^{12} g), with 60–70% coming from anthropogenic emissions and the rest from natural sources.

The United Nations Intergovernmental Panel on Climate Change[7] has identified two areas, the tropics and high latitudes, where sources of methane may become important in the future. At high latitudes, the panel suggested that climatic warming could increase the flux of methane from three sources: (1) methane trapped in permafrost, (2) decaying organic matter in permafrost, and (3) decomposition of methane hydrates. Because few data exist relating to these possible methane sources, we have undertaken two studies to begin to assess the methane content of permafrost and gas-hydrates. First, to test for possible leakage of methane from decomposing gas-hydrates, methane contents were measured in bottom sediment of Harrison Bay, an area offshore from known continental gas-hydrate deposits. Second, methane contents were measured in permafrost ice near Fairbanks, Alaska.

Arctic Gas-Hydrates as a Source of Methane

Natural-gas-hydrates are a potential source of enormous volumes of methane. These solid substances are composed of rigid cages of water molecules that contain gas molecules, mainly methane. Gas-hydrates occur only within certain ranges of temperature and pressure; outside these ranges, they decompose, releasing methane. The Arctic is particularly well-endowed with gas-hydrates because conditions for their occurrence are met in three distinct regions: (1) offshore in sediment of the continental slopes, (2) onshore in areas of continuous permafrost, and (3) on the nearshore Continental Shelf, where relict permafrost has persisted since times of lower sea level. Evidence for gas-hydrates in these regions of the North American Arctic was reviewed by Kvenvolden and Grantz.[14] Offshore gas-hydrates are inferred mainly from marine seismic surveys in the Beaufort Sea. Onshore gas-hydrates in Canada and Alaska are identified mainly on the basis of well logs and a single pressure–core–barrel test. The presence of gas-hydrates on the nearshore Continental Shelf was noted by

Weaver and Stewart[24] from well logs in the Mackenzie Delta. The nearshore Continental Shelf elsewhere in the Arctic is also assumed to have permafrost-associated gas-hydrates.

Of these three regions where gas-hydrates are known to occur, the nearshore Continental Shelf may be of most interest at present because gas-hydrates there may already be destabilizing because of changing temperature conditions.[11] The basic cause of thermal destabilization of the gas-hydrates in this region is the marine transgression that has taken place during at least the past 10 ka. Rising sea level has brought near-freezing seawater to an even-colder submerging surface. As the shelf is inundated by the advancing sea, the surface temperature of the shelf increases by about 10°C.[16] As a result, permafrost and any associated gas-hydrates slowly decompose over time scales measured in thousands of years, owing to heat conducted downward from the sea floor and upward from the Earth's interior. Decomposition of gas-hydrates releases methane which may eventually reach the atmosphere. Gas-hydrates offshore in continental-slope sediment and gas-hydrates onshore likely have not yet been affected by such a significant temperature change and so are probably not a current source of methane.

Of interest is the question of how much methane could come from decomposing gas-hydrates. Estimates have varied; for example, Revelle[22] calculated that 640 Tg yr^{-1} of methane would be released from destabilized gas-hydrates, on the assumption that all gas-hydrates worldwide would be unstable. This estimate is unrealistically high in comparison with the total methane emission of about 550 Tg yr^{-1} from all sources, as indicated by Khalil and Rasmussen[9] and Cicerone and Oremland.[2] Bell[1] also calculated a rate of methane release of 8000 Tg yr^{-1} from a postulated gas-hydrate layer in the Arctic; this estimate included an erroneous assumption of areal extent that was corrected by Revelle.[22] Bell's corrected estimate is about 160 Tg yr^{-1} of methane release from gas-hydrate. Kvenvolden[11] concluded, on the basis of observed gas-hydrate distributions, that the current rate of methane release must be significantly less than the minimum of previous estimates, that is, less than 160 Tg yr^{-1}. Cicerone and Oremland[2] suggested that the rate of methane release from gas-hydrates probably does not exceed 100 Tg yr^{-1} and is more likely in the range of 5 Tg yr^{-1}, a rate they indicate as highly questionable. A review of Arctic gas-hydrates as a source of methane[13] agrees that a rate of methane release from gas-hydrates of 3–5 Tg yr^{-1} is reasonable, given current information. If these poorly constrained estimates are valid, then gas-hydrates may contribute only about 1% to the current total emission of methane to the atmosphere of about 550 Tg yr^{-1}.

Methane in Sediment of Harrison Bay

If, indeed, methane from decomposing gas-hydrates is released from sediment of Arctic continental shelves, then unusually high concentrations of methane might be expected in the shelf sediment. To test this idea, we measured the methane contents of surface sediment of Harrison Bay, a major indentation on the north coast of Alaska (Figure 2). Southeast of Harrison Bay are onshore deposits of gas hydrates (Figure 1) that were studied by Collett et al.[3] These deposits extend northward at least to the coast at Milne Point and probably continue offshore in shelf sediment where relict permafrost is present. North of Harrison Bay, beyond the shelf break (Figure 1), lie seismically inferred gas-hydrates in sediment of the continental slope.[5] Thus, Harrison Bay is situated in between two major gas-hydrate occurrences in northern Alaska.

Seismic surveys in Harrison Bay indicate the presence of shallow gas. For example, gas associated with faults in central and outer Harrison Bay was mapped by Craig and Thrasher,[4] whereas Neave and Sellmann[18] showed large nearshore areas inferred to contain gas in inner Harrison Bay (Figure 2) at sediment depths of 20–400 m. They suggested that this subsurface gas may have come from the decomposition of gas-hydrates. The presence of shallow gas anomalies and the nearness of mapped gas-hydrate deposits onshore made Harrison Bay a reasonable target for measurement of the methane contents of bottom sediment.

Procedure

A total of 17 surface-sediment samples were obtained in Harrison Bay. Sample sites were located along a main transect southeast from Cape Halkett at intervals of about 10 km and on two short transects that crossed the main transect near the endpoints (Figure 3). Two samples came from Weller Bank, a topographic high about 35 km north of the main transect. Sampling from the spring ice cover was carried out by augering through an average of 2.1 m of ice and driving or augering a 45.8-cm-long core barrel into the sea floor. This core barrel, 8.3 cm in diameter, was connected to a crossbar or hammer by sections of 1.5-m-long hollow drill string to a maximum length of 13.8 m. This length set the limit of the water depth from which samples could be collected.

The core-barrel assembly was lowered to the sea floor and rotated until penetration ceased. Auguring worked well for the stiff clay and silt most commonly penetrated, but the sandy sediment at Weller Bank was best sampled by using a straight-nose bit and hammer. Total core lengths ranged from 10 to 30 cm. After the sediment core was extruded, a section, generally about 10 cm long, was cut from the middle of the core, and the rest of the core was discarded. This section was then placed in a 1-quart gas-extraction can. Seawater was added, and a 100-cm^3 head space was left in the can. About 1 g of sodium azide was used to inhibit any methane production during storage. The can was sealed, and the head space was purged with helium through septa-covered ports in the can. Then, the can was frozen until the sample could be analyzed about a week later. Sample

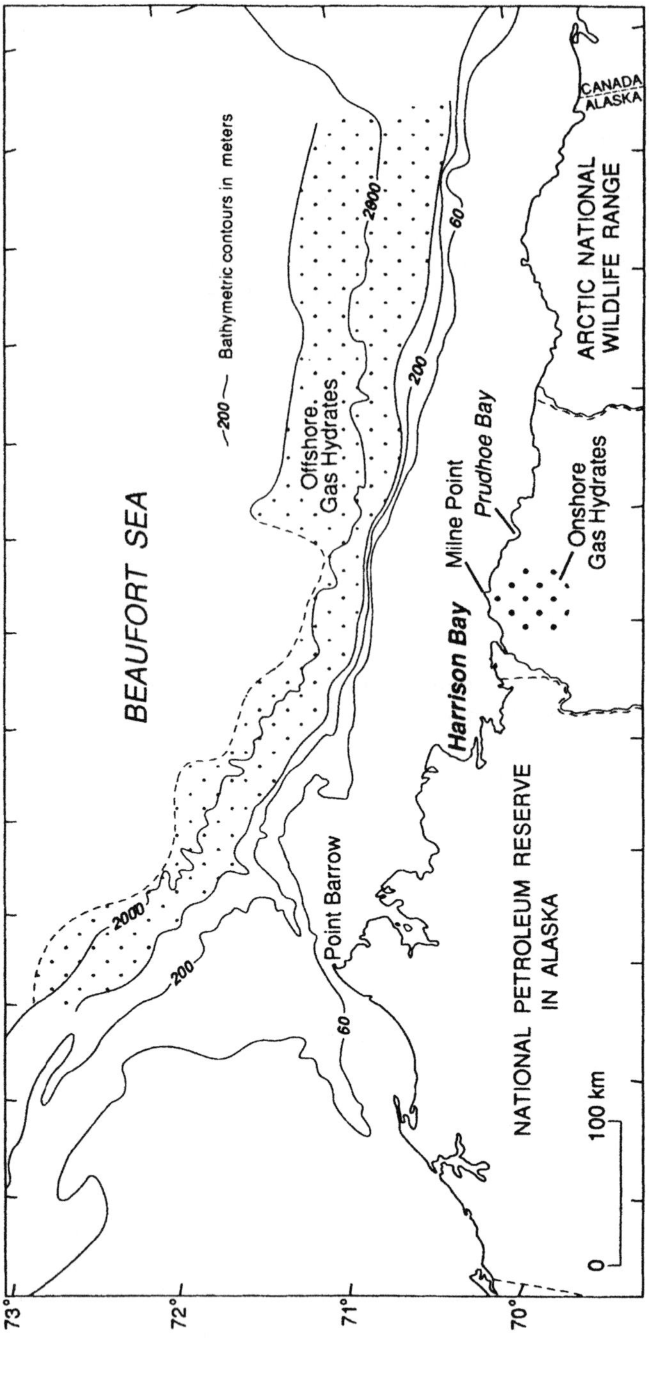

Figure 1. Harrison Bay, Alaska, showing locations of onshore and offshore gas-hydrate deposits.

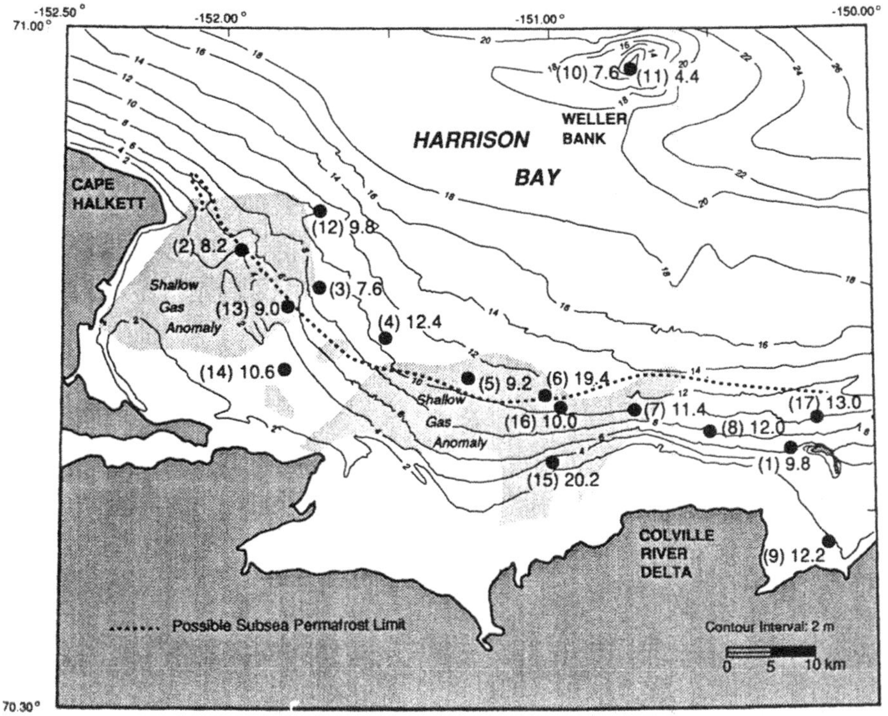

Figure 2. Harrison Bay, Alaska, showing locations of sampling sites (in parentheses) and methane contents (in microliters of methane per liter of water) in bottom sediment at these sites.

cans were thawed and shaken, and a portion of the sealed head space was analyzed for hydrocarbon gases by gas chromatography, according to the procedures of Kvenvolden and Redden.[15]

Harrison Bay Results

Methane contents were originally calculated in microliters per liter of wet sediment and then converted to microliters per liter (μL/L) of sample water (Table 1). For this conversion, a water content of 40% was assumed, on the basis of engineering studies in Harrison Bay by Lee et al.[17]; this water content means that about 50% of the wet volume of the sample is water. Hydrocarbon gases from methane through the butanes are present in these samples, and methane is about 10 to 100 times more abundant than any of the other hydrocarbons. Only meth-

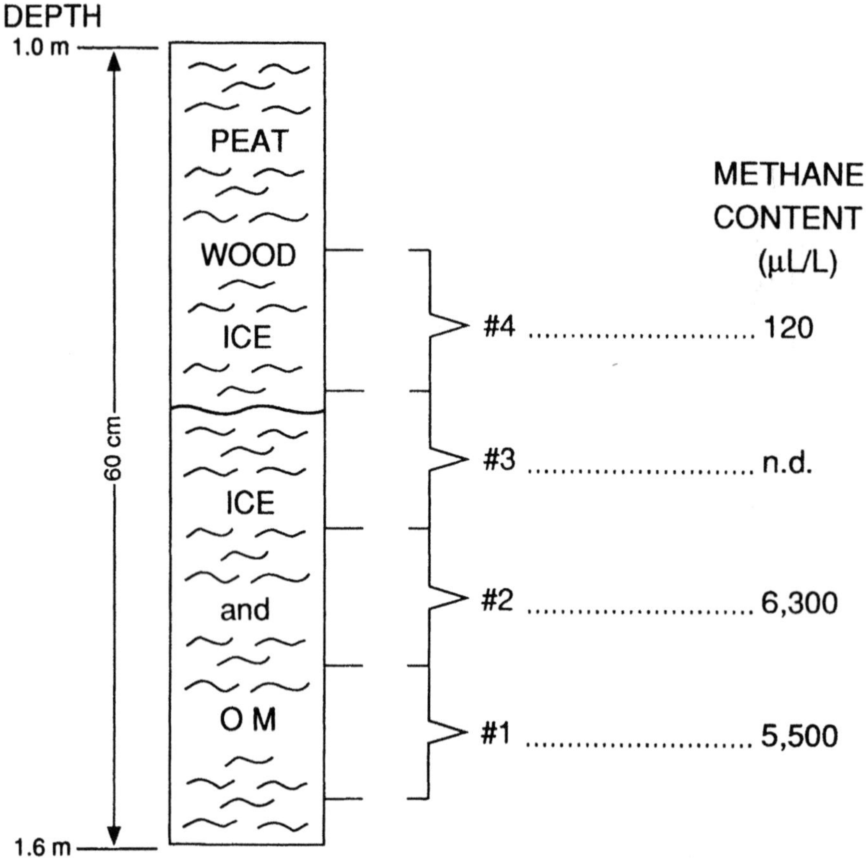

Figure 3. Diagram showing samples taken for gas analysis in a permafrost core drilled near the University of Alaska campus at Fairbanks, Alaska. Methane contents are in microliters of methane per liter of melted ice (water). OM, organic matter; n.d., not determined.

ane is considered here. The methane content ranged from 4.4 to 20.2 μL/L of interstitial water (Figure 2). These contents are similar to those observed in other bottom sediment surveys such as in the Bering Sea[15] and Southern Pacific Ocean.[12] Therefore, the methane contents of bottom sediment in Harrison Bay do not appear to be anomalously high. However, our sampling program was very limited (17 samples) and reached a maximum depth of only 30 cm. If methane from decomposing gas-hydrates accounts for the shallow gas anomalies (Figure 2) observed by Neave and Sellmann,[18] our results suggest that this methane is not seeping to the surface, at least not at our sampling sites.

Table 1. Methane in sediment samples from Harrison Bay, Alaska.

Sample No.	Location	Lithology	Water Depth (m)	Subbottom Sample Interval (cm)	Methane Concentration (µL/L sed.)	Methane Concentration (µL/L water)
1	70°33.9'N 150°15.0'W	Med. fn. sand	6.7	5-15	4.9	9.8
2	70°46.3'N 151°57.2'W	Med. fn. sand	4.9	5-15	4.1	8.2
3	70°44.0'N 151°47.7'W	Fn. sand, sandy clay	9.5	5-15	3.8	7.6
4	70°40.9'N 150°29.8'W	Stiff blk. clay	11.9	8-18	6.2	12.4
5	70°38.4'N 150°15.2'W	Fn. sand	13.1	3-10	4.6	9.2
6	70°37.2'N 150°59.9'W	Med. fn. sand w/ shells	13.7	8-15	9.7	19.4
7	70°36.3'N 150°44.4'W	Sandy blk. silt	13.7	13-23	5.7	11.4
8	70°35.0'N 150°30.1'W	Stiff blk. silty clay	11.9	5-15	6.0	12.0
9	70°28.1'N 150°08.3'W	Stiff silty clay	3.4	5-15	6.6	12.2
10	70°57.5'N 150°44.5'W	Coarse med. sand	11.0	25-30	3.8	7.6
11	70°57.5'N 150°44.5'W	Coarse med. sand	11.0	8-18	2.2	4.4
12	70°48.7'N 150°42.6'W	Stiff blk. clay	10.7	0-10	4.9	9.8
13	70°42.8'N 150°48.5'W	Fn. brn. sand w/ shells	5.5	5-13	4.5	9.0
14	70°38.9'N 150°49.0'W	Fn. sand and sandy silt	4.0	3-15	5.3	10.6
15	70°33.2'N 150°59.7'W	V. fn. sand and shells	7.3	5-15	10.1	20.2
16	70°36.6'N 150°58.0'W	Stiff blk. silty clay	12.2	5-15	5.0	10.0
17	70°35.8'N 150°10.1'W	V. fn. sand and silt	10.7	5-15	6.5	13.0

Methane in Permafrost

Although the United Nations Intergovernmental Panel on Climate Change indicated that methane trapped in permafrost could contribute to climatic warming, we have been unable to find published documentation on the methane content of permafrost. Therefore, we analyzed for methane in some permafrost samples of opportunity obtained near Fairbanks, Alaska.

Permafrost Core

A permafrost core was obtained from T. Osterkamp of the Geophysical Institute, University of Alaska, Fairbanks, from a drill site directly east of the University of Alaska campus. We received a segment of the core taken from beneath the active layer at a depth of about 1.0–1.6 m. The core segment was composed of ice and organic matter in the lower part and of peaty organic matter and wood fibers in the upper part. Four samples were taken of this core, as shown in Figure 3. Samples 1 through 3 were place directly in gas-extraction cans, with sodium azide added as a microbial growth inhibitor; sample 4 was processed in a manner similar to the samples from Harrison Bay, in that water was added to establish a 100-cm^3 head space. Sample 5 was a blank used to test the methane background in the analytical procedure. The samples were frozen for transit and thawed for analysis later; the analytical procedures followed those described previously.

The analytical results are listed in Table 2 and illustrated in Figure 3. Samples 1 and 2 from the lower part of the core have high methane contents of 5500 and 6300 μL/L of water, respectively. The carbon-isotopic composition of this methane was measured to be −74.0 and −74.2‰, relative to the Peedee Belemnite (PDB) standard (Global Geochemistry Corp.), suggesting a microbial origin for this gas. These contents are 100 to 1000 times higher than those measured in Harrison Bay sediment. The magnitude of these preliminary results is sufficiently encouraging to warrant a more detailed study of permafrost as a possible source of methane in global change.

Permafrost Tunnel

We also measured methane contents in permafrost in a tunnel located about 16 km north of Fairbanks near Fox, Alaska. This tunnel was excavated between 1963 and 1968, primarily by the U.S. Army Corps of Engineers, Cold Regions Research and Engineering Laboratory (CRREL), and currently is operated jointly by CRREL and the University of Alaska. The tunnel has been used for research in excavating through permafrost and in evaluating permafrost deformation.

Table 2. Methane in permafrost.

Sample No.	Location	Description	Methane (ppmv)	Methane (μL/L water)
1	E. of Univ. Campus, Fairbanks, Alaska	Ground ice w/ minor organic matter	6400	5500
2	"	Ground ice w/ minor organic matter	3900	6300
3	"	Ground ice and organic matter	1800	nd
4	"	Frozen peat	600	120
5	"	Water blank	1.3	0.3
6	Permafrost Tunnel, Fox, Alaska	Wedge ice	14	12
7	"	Wedge ice	50	59
8	"	Pond ice	13	17
9	"	Massive ice	14	nd

nd = not determined

Figure 4 shows an isometric view of the tunnel, which consists of a 109-m-long adit (horizontal passage from surface) cut into perennially frozen silt and gravel, and a 62-m-long winze (inclined shaft) sloping 12° downward to bedrock.[19] Silt is the main depositional material in the tunnel; the silt contains abundant ground ice as pore fillings, lenses, wedges, and buried pond ice.[6]

Samples of air were collected in gas-tight syringes within and outside the tunnel. These samples were analyzed by S. Whalen and W. Reeburgh of the Institute of Marine Science, University of Alaska, Fairbanks. The results are shown in Figure 4. Triplicate analyses were done at two locations in the winze, and duplicate analyses at the front end of the adit near the tunnel entrance. The methane content in three samples of tunnel air averaged 12.92 ± 0.14 parts per million by volume (ppmv). In the airlock at the entrance, the methane content measured 4.24 ppmv, and outside the tunnel 1.70 ppmv, about the same as the contemporary global average atmospheric methane concentration.[2] Permafrost ice in the tunnel is subliming as air exchange takes place during tunnel operations.[8] As a result of this sublimation, methane apparently is being released from the permafrost. This preliminary study suggests that permafrost is a source of atmospheric methane.

We also collected 3.8-cm-diameter cores of wedge or pond ice, using a portable electric drill. Ice was cored from the tunnel walls at four places (X's, Figure 4). At each place, cores of ice were placed in gas-extraction cans, and the cans were sealed within the tunnel; however, there was no opportunity to purge the head space of the cans with helium. The cans were frozen for shipment and later thawed for analysis. A portion of the gas in each can was analyzed by gas chromatography; the results are listed in Table 2. The methane contents of samples 6, 8, and 9 are approximately the same as those in the air of the tunnel. Thus, these ice samples did not release significant methane upon extraction, although sample 7 did have a higher methane content. The contribution of methane from ice in this sample is calculated to be about 40 μL/L of water. This single result and those obtained from the permafrost core from Fairbanks indicate that the methane content of permafrost varies widely. More samples are needed to show the range of methane contents that can be expected.

Summary

Two studies were carried out concerning methane in gas-hydrates and permafrost. In the first study, the Arctic Continental Shelf was postulated to be a critical region for investigating gas-hydrate decomposition because gas-hydrates in this region have been undergoing significant thermal change during transgression of the Arctic Ocean over the Continental Shelf. To test this idea, we attempted to determine whether methane anomalies could be found in bottom

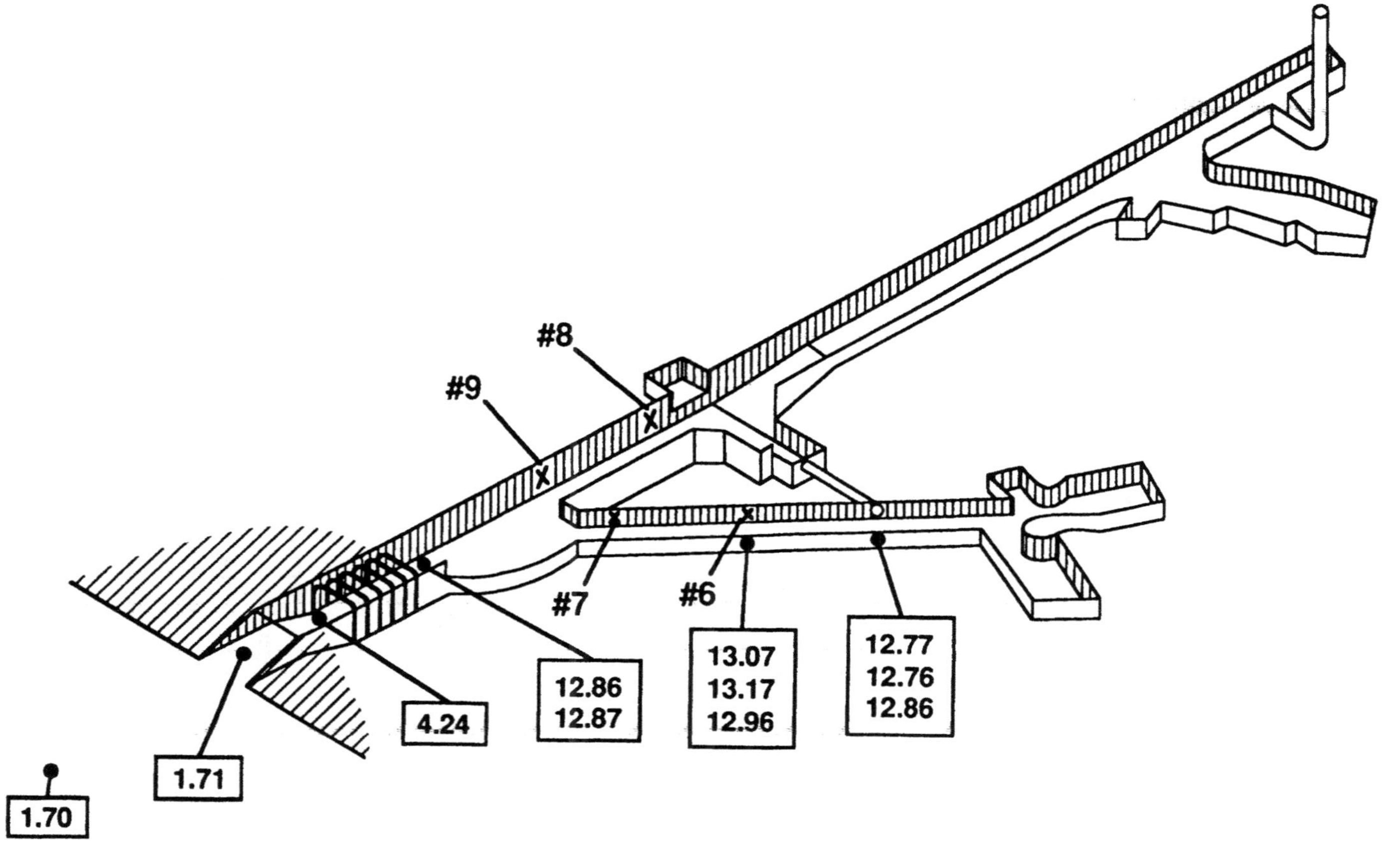

Figure 4. Isometric view of CRREL/University of Alaska permafrost tunnel located at Fox, Alaska, showing places where ice samples were recovered (X's) and methane contents (in ppmv) in air inside and outside the tunnel (dots) were determined. Adit and winze are 109 and 62 m long, respectively.

sediment of Harrison Bay. This study area was chosen because of its nearness to mapped gas-hydrate deposits onshore and the presence of seismic anomalies indicating shallow gas in the sediment. The observed methane contents of this sediment are similar to those measured elsewhere in other surveys. No anomalous amounts of methane were found; our results provide no direct evidence for gas-hydrate decomposition in this area. In the second study, methane contents were measured in samples of permafrost and in the air of a tunnel into permafrost. Results from our limited sample set suggest that the methane contents of permafrost range from essentially zero to as high as 6 mL/L of water. Furthermore, methane can apparently be released from permafrost ice by sublimation. Our observations suggest that permafrost can be a source of atmospheric methane under conditions of global warming.

Acknowledgments

We thank E. Reimnitz of the U.S. Geological Survey for his guidance in sampling sediments from the ice cover of Harrison Bay, Alaska; T.E. Osterkamp of the University of Alaska, Fairbanks, for providing a permafrost core for methane analysis; and R.S. Williams, Jr. of the U.S. Geological Survey for his help in sampling the permafrost tunnel. We greatly appreciate the methane analyses of air inside and outside the permafrost tunnel by W.S. Reeburgh and S.C. Whalen of the University of Alaska, Fairbanks. This research was supported in part by the U.S. Geological Survey's Global Change and Climate History Program.

References

1. Bell, P.R. 1983. Methane hydrate and the carbon dioxide question. In W.C. Clark (ed.), Carbon Dioxide Review, 1982, Oxford University Press, New York, pp. 401–406.
2. Cicerone, R.J., and R.S. Oremland. 1988. Biogeochemical aspects of atmospheric methane. Global Biogeochem. Cycles **2**:299–327.
3. Collett, T.S., K.A. Kvenvolden, and L.B. Magoon. 1990. Characterization of hydrocarbon gas within the stratigraphic interval of gas-hydrate stability on the North Slope of Alaska, U.S.A. Appl. Geochem. **5**:279–287.
4. Craig, J.D., and G.P. Thrasher. 1982. Environmental geology of Harrison Bay, northern Alaska. U.S. Geological Survey Open-File Report 82–35.
5. Grantz, A., D.A. Dinter, E.R. Hill, S.D. May, R.H. McMullin, R.L. Phillips, and E. Reimnitz. 1982. Geologic framework, hydrocarbon potential, and environmental conditions for exploration and development of proposed oil and gas lease sale 87 in

the Beaufort and northeast Chukchi Seas. U.S. Geological Survey Open-File Report 82–482.

6. Hamilton, T.D., J.L. Craig, and P.V. Sellmann. 1988. The Fox permafrost tunnel: a late Quaternary geologic record in central Alaska. Bull. Geol. Soc. Amer. **100**: 948–969.
7. Houghton, J.T., G.T. Jenkins, and J.J. Ephraums (eds.). 1990. Climate Change, Cambridge University Press, Cambridge.
8. Johansen, N.I., P.C. Chalich, and E.W. Wellen. 1981. Sublimation and its control in the CRREL permafrost tunnel. CRREL Special Report 81–8.
9. Khalil, M.A.K., and R.A. Rasmussen. 1983. Sources, sinks, and seasonal cycles of atmospheric methane. J. Geophy. Res. **88**:C9:5131–5144.
10. Khalil, M.A.K., and R.A. Rasmussen. 1990. Constraints on the global sources of methane and an analysis of recent budgets. Tellus **42B**:229–236.
11. Kvenvolden, K.A. 1988. Methane hydrates and global climate. Global Biogeochem. Cycles **2**:221–229.
12. Kvenvolden, K.A. 1988. Hydrocarbon gas in sediment of the southern Pacific Ocean. Geo-Mar. Lett. **8**:179–187.
13. Kvenvolden, K.A. 1991. A review of Arctic gas hydrates as a source of methane in global change. In G. Weller, C.L. Wilson, and B.A.B. Severin (eds.), International Conference on the Role of the Polar Regions in Global Change: Proceedings of a Conference Held June 11–15, 1990 at the University of Alaska Fairbanks, Vol. 2, Geophysical Institute and Center for Global Change and Arctic System Research, University of Alaska, Fairbanks, pp. 696–701.
14. Kvenvolden, K.A., and A. Grantz. 1990. Gas hydrates of the Arctic Ocean region. In A. Grantz, L. Johnson, and J.F. Sweeney (eds.), The Arctic Ocean region. The Geology of North America, Vol. 50, Geological Society of America, Boulder, CO, pp. 539–549.
15. Kvenvolden, K.A., and G.D. Redden. 1980. Hydrocarbon gas in sediment from the shelf, slope, and basin of the Bering Sea. Geochim. Cosmochim. Acta **44**:1145–1150.
16. Lachenbruch, A.H., J.H. Sass, B.V. Marshall, and T.H. Moses, Jr. 1982. Permafrost, heat flow, and the geothermal regime at Prudhoe Bay, Alaska. J. Geophy. Res. **87**:B11:9301–9316.
17. Lee, H.J., W.J. Winters, and E.J. Chamberlain. 1985. Geotechnical properties and freeze/thaw consolidation behavior of sediment from the Beaufort Sea, Alaska. U.S. Geological Survey Open-File Report 85–612.
18. Neave, G.K., and P.V. Sellmann. 1982. Subsea permafrost in Harrison Bay, Alaska. CRREL Report 82–24.
19. Pewe, T.L., and R.D. Reger (eds.). 1989. Quaternary geology and permafrost along the Richardson and Glen highways between Fairbanks and Anchorage, Alaska. 28th International Geological Congress, Field Trip Guidebook T102, Washington DC.

20. Rasmussen, R.A., and M.A.K. Khalil. 1984. Atmospheric methane in the recent and ancient atmospheres: Concentrations, trends, and interhemispheric gradient. J. Geophy. Res. **89**:D7:11599–11605.

21. Raynaud, D., J. Chappellaz, J.M. Barnola, Y.S. Korotkevich, and C. Lorius. 1988. Climatic and CH_4 cycle implications of glacial-interglacial CH_4 change in the Vostok ice core. Nature **333**:655–657.

22. Revelle, R.R. 1983. Methane hydrates in continental slope sediments and increasing atmospheric carbon dioxide. Changing Climates, National Academy Press, Washington, DC, pp. 252–261.

23. Stauffer, B., E. Lochbronner, H. Oeschger, and J. Schwander. 1988. Methane concentration in the glacial atmosphere was only half that of the preindustrial Holocene. Nature **332**:812–814.

24. Weaver, J.S., and J.M. Stewart. 1982. *In situ* hydrates under the Beaufort Sea Shelf. In M.H. French (ed.), Proceedings, Fourth Canadian Permafrost Conference, 1981 (The Roger J.E. Brown Memorial Volume), National Research Council of Canada, Ottawa, pp. 312–319.

Aquatic Systems

The Oceanic Global Methane Cycle

M.V. Ivanov

Institute for Microbiology, Moscow, Russia

A. Yu. Lein

Institute of Geochemistry and Analytical Chemistry, Moscow, Russia

V.F. Galchenko

Institute for Microbiology, Moscow, Russia

ABSTRACT

In the global oceanic carbon cycle, there are three main methane sources. Biogenically formed CH_4 in the reduced sediments of shelf and continental slopes ($\delta^{13}C$ from −58.2 to −87.0‰). "Fossil" CH_4 coming from fluid and gas seeps ($\delta^{13}C$ from −43.0 to −71.0‰). Abiogenically produced CH_4 coming from hydrothermal vents of Mid Oceanic Ridges and Back Arc spreading rift zones ($\delta^{13}C$ from −15.0 to −18.2‰). The main part of oceanic methane is oxidized by methanotrophic bacteria with formation of CO_2 and carbonate minerals. Biomass of methanotrophic bacteria serves as an additional source of organic matter. But some part of the CH_4 from shallow sediments and from gas seeps can enter to the atmosphere.

Introduction

Three main sources of methane play important roles in the global methane cycle of the ocean.

1. Biogenic methane formed by methanogenic bacteria in anoxic sediments of the shelf and continental slope. Its $\delta^{13}C$ value varies from −58.2 to −87.0‰.[1,11,26,36]
2. Methane from "cold" gas seeps has $\delta^{13}C$ values from −43.0 to −71.0‰.[8,22,25,30]

Table 1. Annual oceanic methane flux to the atmosphere.

Flux of CH_4 (Tg)	Percent of Total Global Emission	Refs.
10.0	1.8	10, 18
4.0–6.7	—	13
1.3–16.6	—	14
65.0[a]	5.0	39
7.0–71.0[a]	3.1–18.0	38
18.0–34.0[a]	3.0–3.6	6

[a]Including CH_4 emission from lakes.

3. Abiogenic methane coming from hydrothermal vents of rift zones on oceanic ridges has $\delta^{13}C$ values from −15.0 to −18.2‰.[12,28,31]

Most of the methane formed in sediments is oxidized by methanotrophic bacteria in the upper layers of sediments, whereas a considerable part of methane from hydrothermal vents and cold seeps is oxidized in the water column.[18] In both vents and cold seeps, the biomass of methanotrophic bacteria serves as an additional source of organic matter and contributes to nutrition of marine invertebrates.[16,17,31] The fractionation of stable carbon isotopes takes place during processes of microbial formation and oxidation of methane. The general result is enrichment of biogenic methane and products of its oxidation (organic matter of microbial biomass, CO_2, and carbonate minerals) with isotope ^{12}C.[16,17,21,28–30] Some portion of biogenic methane from shallow water sediments and coming from seeps can enter the atmosphere and thus affect the global atmospheric radiation, oxidation, and water budgets. The quantitative evaluation of this methane flux differ at least by one order of magnitude (Table 1).

Oceanic methane can be divided into three main groups according to its origin (Table 2):

a. the biogenic methane formed in the Holocene sediments of the world's oceans;
b. the fossil methane entering the ocean from the sedimentary deposits of different age;
c. the abiogenic methane released from the rift zones with the hydrothermal fluids.

The aim of this chapter is to review our results and data from literature on the production and oxidation of these different types of oceanic methane.

Table 2. Main types of oceanic methane.

Type of Methane	Depth of Ocean (m)	$\delta^{13}C$ of Methane (‰)	$\delta^{13}C$ of Methane Derived $CaCO^3$ (‰)	Refs.
Biogenic methane produced in holocene sediments	0–4000	−58.2 to −87.0	−27.0 to −42	1,11,20,26,35
Fossil methane from sedimentary rocks: Biogenic mixed biogenic and abiogenic	100–3800	−43.0 to −71.0	−26.0 to −66.7	8,21,24,29,37
Methane of hydrothermal vents of rift zones: Abiogenic, mixed abiogenic, and biogenic	1560–5000	−15.0 to −18.2 −39.6 to −50.8	−3.6 to −4.7 −9.7 to −11.0	12,28,31,41,42

Materials and Methods

These investigations were begun in the Laboratory of Microbial Biogeochemistry of the USSR Academy of Sciences in 1973 and have been carried out by a large group of microbiologists, geochemists, and oceanologists using research ships of the academy and other organizations. A variety of methods have been used in the investigations of the distribution and geochemical activity of methanogens and methanotrophs.[3–5,15] Bacterial numbers were studied by incubation on liquid nutrient media and, more recently, with indirect immunofluorescence techniques.[3,23,34]

Radioisotopic methods were used for studies of rate processes. We used $^{14}CO_2$ and $^{14}CH_4COOH$ as the main substrates to study methanogenesis. In some cases, we also used ^{14}C methanol and ^{14}C methylamine. Concurrently, we investigated the stable carbon isotopic composition of methane, CO_2, and carbonates in both the pore water and solid phase.[20,26,29] The sediment samples were taken by a bottom dredge and geological cores. Specimens were analyzed from the sediments surface to the depth of several meters.

Results

Biogenic Methane in Holocene Sediments

Microbial methanogenesis in the anoxic sediments of the continental ocean border has been studied in most detail.[3–5,19,20,25,26,30] The immunofluorescence technique has allowed us to show that methanogens are a significant part of the microbial population in reduced oceanic sediments. They reach up to 1–10 million cells per gram of wet sediment, whereas incubation methods reveal only thousands or tens of thousands.[34]

Interesting regularities were found during studies of the vertical distribution of methanogenesis in the sediment (Figure 1). First, we did not confirm the widely held idea of methanogenesis inhibition in the active sulfate reduction zone in the upper horizons of the sea sediments. In many sediments, sulfate reduction as methanogenesis peaks practically coincided (Figure 1a). In other cases (Figure 1c), the methanogenesis rate is practically the same throughout the total thickness of the sediments studied. And, finally, in some sediment, the methanogenesis rate increased toward the bottom of the sediment column (Figure 1b).

It is important to stress that in all sediments studied (as thick as several meters) we were not able to reach the "bottom" of the active methanogenesis zone. Hence, our estimates of the rate of microbial methanogenesis should be considered minimal values. According to some indirect data, we think that the zone of active methanogenesis in faintly lithified anoxic sediments may extend to sedimentary depths of hundred of meters.

As mentioned above, in all the experiments we used two labeled substrates: $^{14}CO_2$ and $^{14}CH_3COOH$. In most cases, the bulk of methane in the sediments resulted from CO_2 reduction. It is known from our laboratory experiments with pure cultures of methanogens that methane formed from CO_2 reduction is more enriched in ^{12}C than methane produced from acetate. These fractionation data are in good agreement with the observations of the methane $\delta^{13}C$ values in the bottom sediments (Table 3). Usually it is characterized by consistently more negative $\delta^{13}C$ values, indicating that CO_2 reduction is the basic mechanism of its genesis.

The mean values of the methanogenesis rate in different parts of the shelf and continental slope of the world's oceans are quite similar (Table 4). This allows us to make a preliminary estimation of the global methane production in the anoxic sediments, keeping in mind the shelf and continental slope areas. Such a calculation shows that the yearly methane production in the upper 1-m-thick interval of shelf and continental slope sediments is about one million tons.

The bulk of methane formed in the holocene sediments is oxidized by methanotrophic organisms.[16,20,35] It is indicated by drastic changes of methane concentrations at the interface of the bottom sediments and near-bottom waters. As a result, methane formed in the shelf sediments and in those of the near-continental ocean are not a significant source of atmospheric methane.

However, during the past few years, two groups of new facts have appeared concerning methane transport from the bottom sediments to the sea surface as streams of methane bubbles. Investigations by Namsaraev[34] in the shallow bays of Cuba and the Sea of Japan have shown that this kind of methane transport

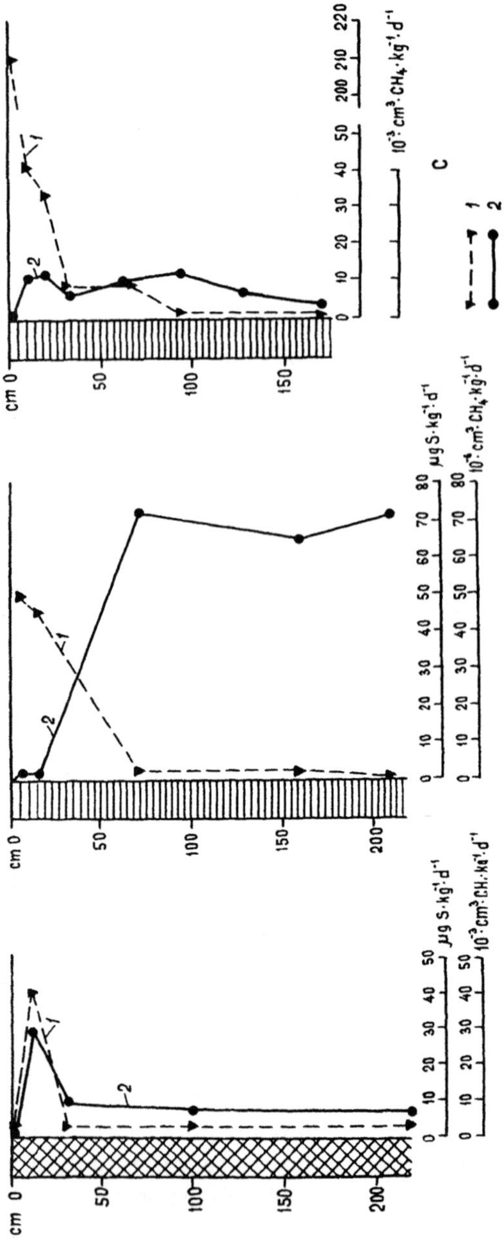

Figure 1. Distribution of the rate of sulfate reduction (SR) and methane formation (MF) in shelf (a,c) and continental slope (b) sediments. (1) SR; (2) MF; (a) the Black Sea, depth 86 m (Ref. 19); (b) Pacific Ocean near Peru, depth 600 m (Ref. 25); (c) the Black Sea, depth 93 m (Ref. 19).

Table 3. Isotopic composition of methane in holocene sediments.

Region	Horizon (cm)	$\delta^{13}C$ CH_4(‰)
Black Sea	0–200	−60 to −82
Baltic Sea	0–75	−64 to −85
Bering Sea	0–35	−70 to −80
California Bay	0–300	−76

Table 4. Daily CH_4 production in the upper horizon (0–100 cm) of holocene sediments.

Zone	Region	CH_4 (mg $C/m^2/d$)
Shelf	Black Sea	10.7
	Bering Sea	14.1
	South Chinese Sea	11.4
Continental Slope	Bering Sea	22.5
	South Chinese Sea	5.0
	Aden Bay	16.4
	Arabian Sea	19.4

exists. Methanogenesis in shallow sediments (Table 5) exceeds the values obtained for the shelf sediments, and the shallow depth of the reservoirs studied favors methane bubbles springing directly into the atmosphere.

Special attention should be paid to the fact that in the case of organically polluted shallow sediments, the rate of methanogenesis increases by severalfold.[33,34] Thus, shallow sediments can serve as a source of the biogenic methane for the atmosphere, and this methane stream will increase if pollution of littoral ocean parts continues.

Fossil Methane Flux from Cold Seeps

Another essential source of the atmospheric methane may be so-called cold methane seeps which have been studied intensively during the past years.[2,7,9,21,29,32] the methane seeps have been found at many points of the continental margin of the ocean at depths from 100 up to 3800 m (Table 1). In many cases, methane bubble streams can be seen from submarines. These observations have been confirmed by numerous echograms on which the methane bubble stream is seen as vertical continuous or discontinuous signals (Figure 2). The cross-sectional diameter of these signals is sometimes equal to tens of meters.[32]

In Figure 2, it can be seen that the gaseous methane stream can be observed

Table 5. Daily CH_4 production in the upper horizon (0–100 cm) of shallow sediments (Ref. 34).

Region	Water Depth (m)	T (°C)	CH_4 mg C m^{-2} d^{-1}
Sea of Japan			
Vostok Bay	6–26	14–16	333
Troika Bay	1–22	15–17	598
Carribian Sea			
Batabano Bay	3.5–9	30–32	348
Salado Lagoon	1.5–8	28–29	395
Santa-Fe mangrove	0.2	30–32	10,201

from its point of release at a depth of several hundred meters at the bottom of the Black Sea. In many cases, the methane bubbles reach the sea surface and methane directly enters the atmosphere.[32] The observation of methane seeps reaching the sea surface has been reported visually from the research ships in the Black Sea, the Northern Sea, and the Gulf of Mexico.

Unfortunately, quantification of methane entering the ocean water from the cold seeps and, moreover, of that reaching the atmosphere is not possible because of the scarcity of observations. But even the minimal value of the methane stream from the active seep in the subduction zone near the Oregon coast (Table 6) given by Suess[40] exceeds by three orders of magnitude the total biogenic methane productivity values in the shelf and continental slope sediments (see Table 4). In Table 6, our data are given on the evaluation of two gas emission points in the northwest part of the Black Sea. The values are close to those of Carson et al.,[9] although that seep was located in a subduction zone and dealt with the gas emission in a lower shelf region of the Black Sea.

A rather intensive carbon stream through the methane seeps is also indicated by widely spread diagenetic carbonate precipitation at the release points.[19,32,36] In some cases diagenetic calcite, aragonite, or dolomite form concretions or cement in the upper horizons of the sediments.[30] In other locations, concretions form on the surface of sediment carbonate plates, tubes, coral-, and tree-like structures.[21] Inside these structures (Figure 3), there are cavities and channels from which gas bubbles emanate. The bubbles have a significant methane content (up to 80%).

We have studied the genesis of these methane-derived carbonates at the release points of two seeps. One was found at the bottom of the Sea of Okhotsk at a depth of 800 m. The diagenetic carbonates were represented by nodules and slabs on the surface of bottom sediments.[29] Another seep was found at the depth of 239 m at the border between the shelf and continental slope in the northwest part of the Black Sea.[21] On the surface of the methane-derived carbonate sediments a

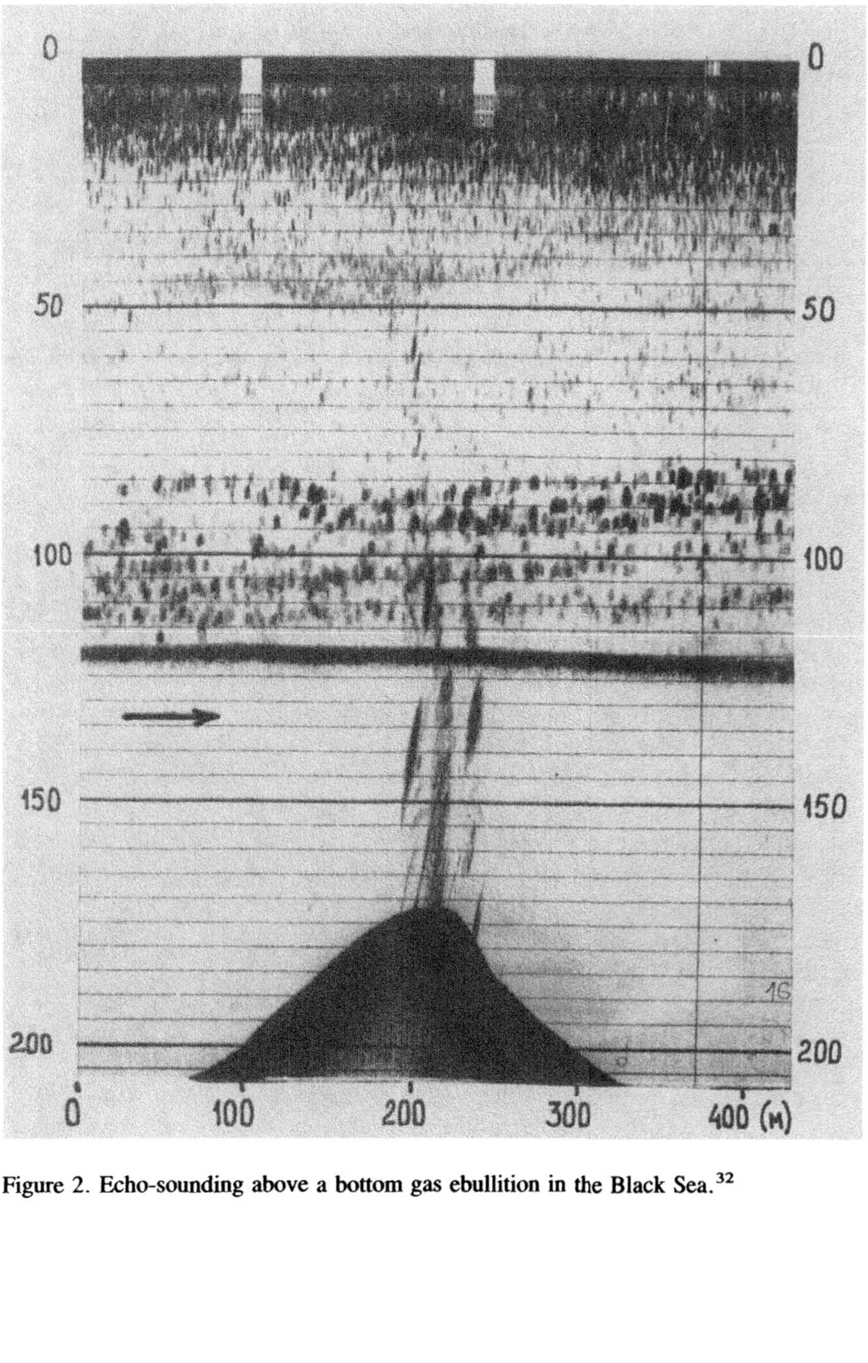

Figure 2. Echo-sounding above a bottom gas ebullition in the Black Sea.[32]

Table 6. Methane flux from cold seeps (mg CH_4 m^{-2} d^{-1}).

Region	Depth (m)	CH_4 Flux	Refs.
Pacific Ocean near cost of Oregon, Alvin Point 1428	2050	146	40
Pacific Ocean near cost of Oregon, Alvin Point 1900	2100	1336 235 235	40
Cascadia accretionary complex	—	13	9
Black Sea, st. 1	239	7	This work
Black Sea, st. 2	176	7	This work

Figure 3. Aragonite chimney (h = 30 cm) from a methane seep from the Black Sea bottom, depth 239 m.

coral-like construction about 40 cm high was formed (Figure 3). The isotopic data indicate that in both cases the carbonate minerals were formed by the oxidation of methane (Table 7).

Microbiological investigations and experiments with $^{14}CH_4$ support this conclusion. In the bottom sediments, including the carbonate nodules of the sea of Okhotsk, and on the surface of the Black Sea aragonite structures, huge masses of methanotrophic bacteria have been found. Experiments with $^{14}CH_4$ have shown rather high intensities of methane oxidation as well as of organic substance synthesis in the methanotrophic biomass. During this last process, an additional fractionation of the carbon isotopes takes place which results in organic matter with a $\delta^{13}C$ value up to −83.8‰. This organic matter is derived from methane which has a $\delta^{13}C$ signature of −58.2‰ (Table 7).

The dissolved methane content in the near-bottom water close to the seeps in the Black Sea exceeds the background values by two orders of magnitude and the seep effect is seen at least 100 m upward. Therefore, in the total water column here intensive microbial methane oxidation and methanotrophic bacterial biomass production occurs (Table 8).

The stable carbon isotopic composition of methane in different seeps varies within the broad limits of the $\delta^{13}C$ values from −43.0 to −71.0‰ (Table 2). Evidently, methane from different seeps can be of different origin. In the Sea of Okhotsk, the methane stream was formed during partial destruction of the gas-hydrate field found in the holocene sediments. In the Black Sea seep, the methane carbon was from 3 to 5.5 thousand years old as was shown by the ^{14}C content.[21] In some methane seeps near the Oregon coast, a high content of nonmethane hydrocarbon gases was found which indicated a thermocatalytic methane origin.[36]

Further investigation is necessary to solve the problem of methane origin in different seeps as well as for quantitative evaluation of this methane contribution to both the oceanic and global carbon cycles.

Table 7. Isotopic composition ($\delta^{13}C$‰) of methane, methane-derived carbonate minerals, and organic carbon from cold seeps.

	$\delta^{13}C$(‰)		
Region	CH_4	$CaCO_3$	C org.
Sea of Okhotsk, depth—800 m	−54.6	−47.5 to −49.2 ($n = 7$)[a]	—
Black Sea, depth—239 m	−58.2	−32.5 to −40.4 ($n = 13$)[a]	−75.6 to −83.8 ($n = 6$)[a]

[a]Number of samples.

Table 8. Methane oxidation and biomass production by methanotrophs in the Black Sea water above a cold seep (depth 230 m).

Depth (m)	CH_4 (ml L^{-1})	CH_4 oxidation (mg C L^{-1} d^{-1})	Biomass Production (mg C L^{-1} d^{-1})
130	14	14	6
140	25	36	16
160	44	91	70
170	67	98	45
180	83	70	43
200	100	85	46
221	111	103	50
225	145	368	62

Methane Cycle in the Ocean Rift Zones

Abiogenic methane is a regular component of high-temperature fluids released at the hydrothermal fields in the rift zones of the Atlantic and Pacific Oceans. Methane concentrations in fluids of the middle ocean ridges vary within very narrow limits (from 80 to 120 mM per kg of solution), and its isotopic composition ($\delta^{13}C$ from -15.0 to -18.2‰) is an indication of a high-temperature genesis of this methane (see Table 2).

A much higher methane concentration was found in fluids of the near-continental rift in the Guayas basin (California Bay), where the hydrothermal fluids pass through the sediments and become enriched in biogenic methane. This biogenic CH_4 component is also seen from the anomalously light isotopic composition of methane $\delta^{13}C$ which varies from -39.6 to -50.8‰ (Table 2).

Dilution of the hydrothermal fluid with the sea water results in the oxidation of reduced compounds, such as H_2S, NH_4, and methane,[15,23] carried out with participation of chemoautotrophic and methanotrophic bacteria.[16,17,31] We have studied participation of methanotrophic bacteria in this process in the hydrothermal fields of the deep water rift zones (Table 9).

The water column up to 300–400 m over the hydrothermal fields is enriched significantly in methane. Maximal rates of microbial methane oxidation (up to tens of micrograms of CH_4 L^{-1} d^{-1}) were found in the water samples taken near the hydrothermal vents (Table 9). In the water column peaks of the total number of microbes, methanotrophs and methane oxidation rate coincide with the peak of maximum methane concentration (Figure 4).

Analysis of data on the methane distribution over the hydrothermal vents and oxidation activity of methanotrophic bacteria allows us to draw three important conclusions characterizing the methane cycle in the ocean rift zones.

Table 9. Methane content and rate of methane oxidation in water column above active hydrothermal fields.

Samples	CH_4 Content ($ml\ L^{-1}$)	Rate of CH_4 Oxidation ($mg\ C\ L^{-1}\ d^{-1}$)
Oceanic water	4 to 8 $\times 10^{-6}$	0
TAG, 12–14 m above active field	2.4×10^{-3}	—
TAG, 200–400 m above active field	10.5×10^{-5}	1.1×10^{-6}
Guaymas Basin, near hydrothermal vent	8.5	68
Guaymas Basin, 100–300 m above active field	7.1×10^{-3}	36×10^{-3}
Manus Basin, near hydrothermal vent	1.5	1.4
Manus Basin, 50 m above active field	72.5×10^{-3}	53×10^{-3}

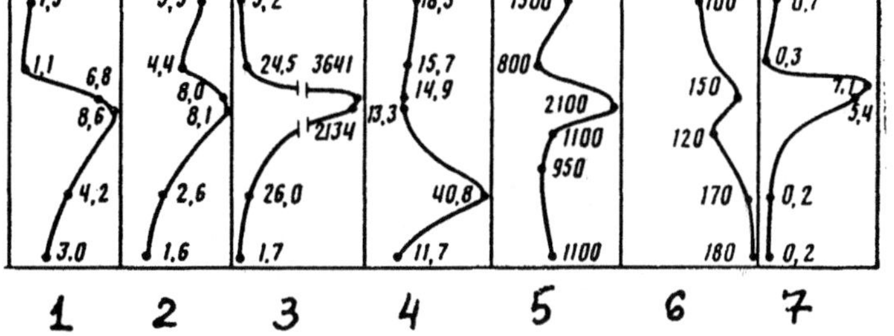

Figure 4. Some biogeochemical and geochemical measurements from the water column above hydrothermal vent in the Guaymas Basin. Number (cell/ml) of (1) methanotrophic bacteria (10^3), and (2) common bacteria number (10^4); the rate (10^{-3} mg C L^{-1} d^{-1}) of (3) CH_4 oxidizing and (4) CO_2 assimilation; (5) optical density; contents of (6) Mn, ng/L and (7) CH_4, ml/L.

1. Hot hydrothermal solutions released in different parts of the ocean are characterized by rather low concentrations of abiogenic methane, enriched in ^{13}C.
2. The region of anomalously high methane concentrations in the near-bottom waters extends 200–400 m up into the water column. Taking into consideration the fact that even near-continental rift valleys are over 1000 m deep, we have no reason to think that abiogenic methane of the rift zones can contribute to the atmospheric methane pool.
3. Methane oxidation in the hydrothermal vents is carried out by methanotrophic bacteria. These bacteria are widely distributed in the sediments and near-bottom waters of the release zones of hydrothermal fluids. They are also found in specialized tissues of many animals inhabiting the hydrothermal fields. Utilizing methane as their only energy source, the methanotrophs form biomass which is consumed by heterotrophic organisms and, thus, enter into the oceanic food chains.

According to our opinion, from three different sources of oceanic methane discussed, only the seep methane and the biogenic methane from the littoral ocean zones can reach the atmospheric reservoir and become involved in the global methane cycle.

References

1. Alexeev, F.A., and V.S. Lebedev. 1975. Carbon isotopic composition of CO_2 and CH_4 of the Black Sea bottom sediments. In Dissipated Gases, VNIIYaGG, Moscow, pp. 49–53.
2. Anderson, A.L., and W.R. Bryant. 1990. Gassy sediment occurrence and properties: Northern Gulf of Mexico. Geo. Marine Lett. **10**:209–220.
3. Belyaev, S.S., and Z.I. Finkelstein. 1976. Anaerobic gas forming bacteria in bottom sediments of Californian Gulf and Pacific Ocean. In Biogeochemistry of Diagenesis of Pacific Ocean sediments, Nauka, Moscow, pp. 75–82.
4. Belyaev, S.S., and M.V. Ivanov. 1975. Radioisotopic method for determination of bacterial methanogenesis rate. Microbiology **44**(1):166–178.
5. Belyaev, S.S., A. Yu. Lein, M.V. Ivanov. 1981. Role of methane-producing and sulfate-reducing bacteria in the destruction of organic matter. In Biogeochemistry of Ancient and Modern Environments, Australian Academy of Science, Canberra, pp. 235–242.
6. Blake, D.R., 1984. Increasing concentrations of atmospheric methane, 1979–1983. Ph.D Thesis. University of California, Irvine, p. 213.

7. Brooks, J.M., M.C. Kennicutt, R.R. Fay, and T.J. McDonald. 1984. Thermogenic gas hydrates in the Gulf of Mexico. Science **225**:1135.

8. Brooks, J.M., M.C. Kennicutt, C.R. Fisher, et al. 1987. Deep sea hydrocarbon seep communities: evidence for energy and nutritional carbon sources. Science **238**:1138.

9. Carson, B., E. Suess, and J.C. Strasser. 1990. Fluid flow and mass flux determinations at vent sites on the Cascadia Margin Accretionary Prism. J. Geophys. Res. **95**(B6):8891–8897.

10. Cicerone, R.J., and R.S. Oremland. 1988. Biogeochemical aspects of atmospheric methane. Global Biogeochem. Cycles 2:299–325.

11. Claypool, G.E., and I.R. Kaplan. 1974. The origin and distribution of methane in marine sediments. in I.R. Kaplan (ed.), Natural Gases in Marine Sediments, Plenum, New York, pp. 99–139.

12. Craig, H., J.A. Welhan, K. Kim, et al. 1980. Geochemical studies of the 21°N EPR hydrothermal fluids (Abstr.). EOS **61**:992.

13. Ehhalt, D.H. 1974. The atmospheric cycle of methane. Tellus **26**:58–70.

14. Ehhalt, D.H., and U. Schmidt. 1978. Sources and sinks of atmospheric methane. Paleogeochemistry **116**:452–464.

15. Evans, W.C., L.D. White, and J.B. Rapp. 1988. Geochemistry of some gases in hydrothermal fluids from the southern Juan de Fuca Ridge. J. Geophys. Res. **93**(B12):15305–15313.

16. Galcheno, V.F., A. Yu. Lein, E.M. Galimov, and M.V. Ivanov. 1988. Methanotrophic bacteria—symbionts as an original link of food chain in the ocean. DAN USSR **300**(3):717–720.

17. Galchenko, V.F., S.V. Galkin, A. Yu. Lein, et al. 1988. The role of bacteria-symbionts in the nutrition of invertebrates from the regions of active underwater hydrotherms. Oceanologya **28**(6):1020–1031.

18. Galchenko, V.F.A. Yu. Lein, and M.V. Ivanov. 1989. Biological sinks of methane. M.O. Andreae, and D.S. Shimel (eds.), in Exchange of Trace Gases between Terrestrial Ecosystems and the Atmosphere, Dahlem Conference, J. Wiley and Sons, New York, pp. 59–71.

19. Ivanov, M.V. M.B. Vainstein, V.F. Galchenko, et al. 1983. Distribution and geochemical activity of bacteria in sediments of the western part of the Black Sea. In Geochemical Processes in the Western Part of the Black Sea, Bulgarian Academy of Sciences, Sofia, pp. 150–181.

20. Ivanov, M.V., and A. Yu. Lein. 1989. Quantitative evaluation of biogenic methane generation and oxidation in oceanic sediments. In P. Brimblecombe and A. Lein (eds.), Evolution of Global Biogeochemical Sulfur Cycle, J. Wiley and Sons, New York, pp. 159–169.

21. Ivanov, M.V., G.G. Polycarpov, A. Yu. Lein, et al. 1991. Biogeochemistry of carbon cycle on the Black Sea region of CH_4 gas seeps. DAN USSR 320, N.5. 1235–1240 in Russian.

22. Kim, K.R., J.A. Welhan, and H. Craig. 1984. The hydrothermal vent fields at 13°N and 11°N on the East Pacific Rise: Alvin, EOS **65**:973.

23. Kondratenko, G.P., Yu. I. Nikolenko, L.V. Bezrukova, et al. 1981. Identification of methanotrophic bacteria by immunofluorescence method. Microbiologya **50**(2): 320–325.

24. Kulm, L.D., E. Suess, J.C. Moore, et al. 1986. Oregon subduction zone: venting, fauna and carbonates. Science **231**:561–566.

25. Laurinavichus, K.S., V.G. Tokarev, and Yu. N. Mchensky. 1976. The rate of microbiological CH_4 formation in the sediments of the Pacific Ocean and the South-China Sea. In Geochemical Activity of Microorganisms in the Pacific Ocean Sediments, Biological Centre, Puschino, pp. 179–186.

26. Lein, A. Yu., B.B. Namsaraev, V. Ya Trotsyuk, and M.V. Ivanov. 1981. Bacterial methanogenesis in Holocene sediments of the Baltic Sea. Geomicrobiolog. J. **2**(4): 299–315.

27. Lein, A. Yu. 1983. Organic carbon consumption during the processes of organic matter mineralization in modern oceanic sediments. Geochimya **11**:1631–1639.

28. Lein, A. Yu., V.F. Galchenko, V.A. Grinenko, et al. 1988. Mineral composition and geochemistry of rocks with bacterial overgrowths from the submarine hydrothermal constructs. Geochemya **9**:1235–1248.

29. Lein, A. Yu., V.F. Galchenko, B.G. Pokrovsky, et al. 1989. Marine carbonate nodules as a result of processes of microbial oxidizing of gas-hydrate methane of the Sea of Okhotsk. Geochemya **10**:1396–1406.

30. Lein, A. Yu. 1989. Interaction of the sulfur and carbon cycles in recent marine sediments. In P. Brimblecombe and A. Yu. Lein (eds.), Evolution of the Global Biogeochemical Sulfur Cycle, SCOPE 39, J. Wiley and Sons, New York, pp. 143–149.

31. Lein, A. Yu., and V.F. Galchenko. 1990. The primary production of chemosynthesis at m. Axia of Juan de Fuca ridge. In Geology and hydrothermal deposits of Juan de Fuca Ridge, Nauka, Moscow, pp. 146–157.

32. Polycarpov, G.G., V.N. Egorov, A.I. Nezdanov, et al. 1989. The active gas seeping from the Black Sea escape. DAN USSR. Ser. B. Geol., Chem. and Biol. Sci. **12**:13–16.

33. Namsaraev, B.B, M. Lugioyo, M. Bellota, et al. 1990. The destruction of organic matter by microorganisms in bottom sediments of the Batabano Bay in Cuba. Microbiologya **59**(5):903–911.

34. Namsaraev, B.B. 1993. The distribution of methanogens in sea sediments. Microbiologya (in press).

35. Reeburg, W.S. 1989. Coupling of the carbon and sulfur cycles through anaerobic methane oxidation. In P. Brimlecombe and A. Lein (eds.), Evolution of the Global Biogeochemical Surfur Cycle, SCOPE 39. J. Wiley and Sons, Chichester, pp. 149–158.

36. Ritger, S., and E. Carson. 1987. Methane-derived authigenic carbonates formed by

subduction-induced pore-water expulsion along the Oregon (Washington margin). Geol. Soc. Amer. Bull. **98**:147–156.

37. Sakai, H., K. Kobayashi, X. Pichon, et al. 1990. Biogenic methane-derived carbonate precipitates recovered from 2000 to 4000 m deep slope on the Naukai Trough during the 1989 Kaiko Naukai Project. In J.P. Cadet and X. LePichou (eds.), Fluids in Subduction Zones, College de France and Ecole Normale Superien, Paris.
38. Seiler, W. 1984. Contribution of biological processes to the global budget of CH_4 in the atmosphere. In M. Klug and C. Reddy (eds.), Current Perspectives in Microbial Ecology, American Society of Microbiology, Washington, DC, pp. 468–477.
39. Sheppard, J.C., Westburg, J.F. Hopper, et al. 1982. Inventory of global methane sources and their production rates. J. Geophys. Res. **87**:1301–1312.
40. Suess, E. 1990. Role of fluid circulation in the crust and the prospects for future ocean drilling. In J.P. Cadet and X. LePichon (eds.), Fluids in Subduction Zones, College de France and Ecole Normale Superien, Paris.
41. Welhan, J.A., and H. Craig. 1983. Methane, hydrogen, and helium in hydrothermal fluids at 21°N on the East Pacific Rise. In P.A. Rona et al (eds.), Hydrothermal Processes at Sea Floor Spreading Centers, Plenum, New York, pp. 391–408.
42. Welhan, J.A., and J.E. Lupton. 1987. Light hydrocarbon gases in Guaymas basin hydrothermal fluids: thermogenic versus abiogenic origin. Amer. Assoc. Petrol. Geol. Bull. **2**:215–223.
43. Zyakun, A.M., V.A. Bondar, B.B. Namsaraev, and A.I. Nesterov. 1983. About possible use of carbon isotopic analysis for determination of microbial methane oxidation rate in natural ecosystems. Geochimiya **5**:759–765.

Global Carbon Dioxide and Methane Fluxes from Shallow-Water Marine Carbonate Frameworks

Francis J. Sansone

Oceanography Department, School of Ocean and Earth Science and Technology, University of Hawaii, Honolulu, Hawaii 96822

ABSTRACT

Suboxic and anoxic conditions are widely observed within coral reefs and other shallow-water marine carbonate frameworks. These conditions result from the importation of organic matter by the advection of seawater through the frameworks, with subsequent decomposition of these materials within the frameworks. These processes result in the buildup of end products such as CO_2 and CH_4 in the framework porewaters, which can then be transported by advection, diffusion, and turbulent mixing to the overlying and adjacent seawater. Estimates of porewater-to-seawater transport rates from several sources were multiplied by porewater concentrations to yield gross fluxes of CO_2 and CH_4 to the overlying seawater. Using published estimates of the global area of coral reefs and carbonate platforms, these flux rates were extrapolated to give global CO_2 and CH_4 fluxes of 9600 and 0.6 Tg yr^{-1}, respectively, for coral reefs, and 16 and 0.005 Tg yr^{-1} for carbonate platforms. In comparison, the global rates of net ocean–atmosphere exchange for CO_2 and CH_4 have been estimated as 350,000 and 10 Tg yr^{-1}, respectively. These results suggest that these very large sedimentary systems may be important, but not dominant, components in global carbon cycles, although more research will be needed to determine their exact role.

Introduction

The interiors of coral reefs and other marine carbonate frameworks have been generally assumed until recently to be oxic and not to play an important role in diagenetic recycling of carbon and nutrients. It has become clear from recent research, however, that the interiors of these systems are typically suboxic to anoxic.[1,22,24,31] These conditions result from the importation of organic matter by the advection of seawater into the framework, with subsequent decomposition of these materials within the framework.[23] This organic matter degradation is reflected by stoichiometrically consistent increases in end products such as total carbon dioxide ($\sum CO_2$), inorganic nutrients, alkalinity, and methane (CH_4) in the framework porewater.[30,31] Thus, it appears that organic matter decomposition within these systems is important on at least a local basis, although its significance on a global scale has not been previously examined.

A number of studies have also focused on the importance of advection, diffusion, and turbulent mixing in transporting porewater and its constituents through reefs[1,2,4,10,12,14,16,18,30,31,32] and other carbonate frame-works.[20,21,25,28,35] These processes also result in the transport of porewater with elevated diagenetic end product concentrations to the overlying seawater.[29,30,32] Thus, it is possible that porewater may be a significant source of CO_2, CH_4, nutrients, and other end products to the external environment. The very large areal extent of these systems make it important to determine their possible contribution to global carbon cycles.

The goal of this chapter is to estimate the global rates of transfer of CO_2 and CH_4 from marine carbonate frameworks to the neighboring seawater environment. Because carbonate frameworks exist in shallow waters, it is likely that large fractions of these fluxes will be transported to the atmosphere, thereby contributing to the atmospheric pools of these radiatively important gases. Although there is only limited data thus far on relevant rates and concentrations needed for these calculations, the results indicate that shallow-water marine carbonate frameworks may be significant components in global carbon cycles.

Results and Discussion

Porewater Concentrations

The compositions of reef and platform porewaters appear to be quite similar (Table 1). In this chapter reef data from only the upper 1 m will be used because vertical transport in reefs greatly exceeds horizontal transport (see below), and, thus, porewater concentrations in the uppermost area of reefs should approximate the composition of porewater reaching the framework–seawater interface. The "best guess" values shown in this chapter were selected as conservative estimates and reflect the author's perception of the relative importance of the individual field sites for the empirical data listed in the tables.

Porewater Advection Rates

To my knowledge, porewater-to-seawater transport rates have been measured only at one reef. These were measured in Checker Reef, Oahu, Hawaii, by means of hydrological modeling of data from (1) residence time estimates from rain-

Table 1. Composition of porewaters in coral reef frameworks and marine carbonate platforms.

Site	Sampling Depth (m)	ΣCO_2 (mM)	CH_4 (nM)	Refs.
Coral Reefs				
Checker Reef, Oahu, Hawaii	1	2.16–2.70	61–441	31
Davies Reef, Great Barrier Reef	≤1	2.05–2.44	<25	1,22
"Best guess" average		2.20	50	
Carbonate Platforms				
Mud Hole Submarine Springs, Florida Platform	0	2.23	101	6
Submarine frameworks, Oahu, Hawaii	19–45	2.13–2.31	109–137	24
Marathon Key, Florida Platform	305–380	2.05–2.42	1040–5000	24
"Best guess" average		2.20	200	
Tropical surface seawater	0	1.96	12	1,31

induced salinity perturbations, (2) in situ pressure sensor and manometer arrays, and (3) direct measurements of porewater transport out of the reef.[29,30,32] Transport rates can also be estimated using internal porewater velocities, data which are available for several sites. Table 2 summarizes available data on both vertical and horizontal velocities. In this chapter, however, only vertical velocities will be used to calculate fluxes because vertical advection rates appear to be generally much larger than horizontal rates.[4,30,32]

In addition to these quantitative estimates, a number of other descriptive observations have been reported on the rapid advective exchange between reefs and the adjacent seawater.[1,2,4,10,17] These studies also serve to note the highly heterogeneous character of coral reefs, in that large advective flows can be observed at fissures at the reef–seawater interface. Also, there can be significant differences in advection rates at different locations of a given reef.[1,14]

Large-scale (>100 km^2) marine carbonate platforms present a more difficult system to model because of the lack of data on the rate of exchange of porewater with the adjacent seawater. Nevertheless, porewater–seawater exchange rates can be estimated by assuming that internal vertical porewater advection is con-

Table 2. Reef porewater advection rates.

Site	Porewater Velocity ($m\ d^{-1}$) Vertical	Horizontal	Refs.
"Typical" reef	0.86–86		3
Checker Reef, Oahu, Hawaii	0.1–2.2		32
Davies Reef, Great Barrier Reef	1.4–50		14
Tague Bay Reef, St. Croix	1.7–5.2	0.086–0.259	4
Orpheus Island, Great Barrier Reef		9–120	16
"Best guess" average	4		

tinuous with exchange at the platform margins. The rates of such internal advection in platforms have been estimated by Rougerie et al. by measuring the distribution of nutrient tracers.[19,20] These data are summarized in Table 3; note that these rates are ~1000 times slower than those for coral reefs (Table 2). Other papers have described porewater advection in similar systems based on isotopic tracer data[21] and by modeling the convective movement of porewater due to thermal[8,9,28] and salinity[7,26,35] gradients; however, the lack of quantitative information in these reports prevent their use here.

Global Porewater Fluxes

Global fluxes of CO_2 and CH_4 to the adjacent seawater were calculated separately for reefs and platforms using the following formula:

$$\text{Flux} = (c_{pw} - c_{sw})v_{pw}A, \qquad (1)$$

where c_{pw} = concentration in pore water ($mole/m^3$),
c_{sw} = concentration in adjacent seawater ($mole/m^3$),
v_{pw} = porewater velocity ($m\ yr^{-1}$)
A = global area of coral reefs ($6.2 \times 10^{11}\ m^2$) or shallow-water carbonate platforms ($7.8 \times 10^{11}\ m^2$).

Only vertical porewater velocities (Table 2), along with porewater concentrations for the upper 1 m (Table 1), were used for reef calculations. The global area of shallow-water carbonate platforms was calculated as the difference between

Table 3. Advection rates of saline porewater in carbonate platforms (averaged over whole platform).

Site	Porewater Velocity ($m\ yr^{-1}$)	Refs.
"Typical" platform	0.8–4	25
Pacific atolls	1.8–88	20
Mururoa Atoll, French Polynesia	0.51–0.97	19
"Best guess" average	2.0	

Milliman's estimate of the global area of all shallow water carbonates ($1.4 \times 10^{12}\ m^2$)[13] and Smith's estimate of the global area of reefs ($6.2 \times 10^{11}\ m^2$).[27]

These data yield "best guess" global coral reef fluxes of CO_2 and CH_4 to overlying seawater of 9600 and 0.6 Tg yr^{-1}, respectively. Computed "best guess" global carbonate platform fluxes of CO_2 and CH_4 were 16 and 0.005 Tg yr^{-1}, respectively. The results are summarized in Table 4 and compared with estimates of global net ocean–atmosphere exchange rates. Table 4 also presents ranges of possible values for the computed fluxes. The lower values in the ranges were calculated using the smallest $\sum CO_2$ or CH_4 porewater concentrations in Table 1 and the lowest advection rates for each environment in Table 2 or 3; the higher values used the largest concentrations and the highest advection rates.

Unfortunately, there are few empirical data available to compare with the fluxes predicted here. However, the predicted CH_4 flux of 55 mmole $m^{-2}\ yr^{-1}$ for coral reefs (Table 4) is similar to the integrated flux of 25 mmole $m^{-2}\ yr^{-1}$ from the Kaneohe Bay (Oahu, Hawaii) barrier reef measured by Webb.[34] In contrast, the predicted flux is higher than the fluxes of 0.12–4.1 mmole $m^{-2}\ yr^{-1}$ from reef flank sediments (water depth of 14–18 m) measured by Oremland.[15] The latter data are not directly comparable, however, in that flank sediments are probably not associated with upper reef internal advection and may contain lower levels of labile organic matter due to their distance from the upper reef primary production.

To my knowledge, fluxes of CO_2 have not yet been directly measured from shallow marine carbonate frameworks. However, changes in the chemical composition of surface water as it passes over coral reefs can be used to estimate net CO_2 release from reef-associated calcification and organic matter decomposition. On a global basis, a flux of 80–260 Tg CO_2 yr^{-1} can be calculated from such data.[33] Although this estimate is largely within the range of possible values listed in Table 4 for coral reefs, it is much lower than the "best guess" value of 9600 Tg CO_2 yr^{-1}, thus suggesting that the latter may be an overestimate of

Table 4. Computed gross global gas fluxes from carbonate frameworks, and estimates of net global ocean–atmosphere gas exchange rates.[5,11] Fluxes are for "best guess" data in Table 1–3. Values in parentheses are ranges possible using all data in Tables 1–3 (see text).

	CO_2 Flux		CH_4 Flux	
System	(mole m^{-2} yr^{-1})	(Tg yr^{-1})	(mmole m^{-2} yr^{-1})	(Tg yr^{-1})
Coral reefs	350 (3.3–23,000)	9600 (90–630,000)	55 (<0.5–13,000)	0.6 (<0.005–130)
Carbonate platforms	0.5 (0.046–40)	16 (1.6–1400)	0.38 (0.045–440)	0.005 (0.0006–5.5)
Net ocean–atmosphere exchange		350,000		10

global CO_2 emissions from coral reefs. This discrepancy may be the result of autotrophic processes at the reef surface that consume a portion of the CO_2 flux from the reef interior.

As shown in Table 4, the global rates of net ocean-atmosphere exchange for CO_2 and CH_4 have been estimated as 350,000 [11] and 10 Tg yr^{-1},[5] respectively. Thus, the data presented here indicate that marine carbonate frameworks may be significant, but probably not dominant, components in global carbon cycles, although more research will be needed to determine their exact role. For example, porewater advection rates need to be measured in a wide range of reefs and platforms to provide better estimates of the range of mass fluxes possible in these rather heterogeneous systems. Also, there is clearly a need to measure actual CH_4 fluxes to the overlying seawater to determine if CH_4 oxidation is taking place at the framework–seawater interface. The latter research is currently underway at Checker Reef, Oahu.

Conclusions

The estimates presented here on global framework–seawater fluxes are severely constrained by the very limited data available on these processes. As a result, there are large ranges associated with these estimates. Nevertheless, these results suggest that shallow-water marine carbonate frameworks are important, but not dominant, components in global carbon cycles because of their large areal extent. The data also show the need for additional measurements of the rates and composition of material fluxes from reefs and other carbonate frameworks. This information will be required for an accurate determination of the global magnitude of these processes.

Acknowledgments

I thank Fred Mackenzie, Steve Smith, and Gordon Tribble for valuable discussions. This research was supported by grants OCE-8911150 and OCE-9116288, U.S. National Science Foundation, Chemical Oceanography Program. University of Hawaii School of Ocean and Earth Science and Technology Contribution No. 2758.

References

1. Buddemeier, R.W., and J.A. Oberdorfer. 1986. Internal hydrology and geochemistry of coral reefs and atoll islands: Key to diagenetic variations. In J.H. Schroeder and B.H. Purser (eds.), Reef Diagenesis, Springer-Verlag, Berlin, pp. 91–111.
2. Buddemeier, R.W., and J.A. Oberdorfer. 1988. Hydrogeology and hydrodynamics of coral reef pore waters. Proc. 6th Int. Coral Reef Symp., Australia **2**:485–490.
3. Buddemeier, R.W. (Kansas Geological Survey). 1990. Personal communication.
4. Carter, B.L., M.A. Simms, C.H. Moore, H.H. Roberts, and A. Lugo-Fernandez. 1989. Wave-forced hydrogeology and diagenetic responses in Tague Reef, St. Croix, U.S. Virgin Islands. In D. Hubbard (ed.), Marine and Terrestrial Geology of St. Croix, U.S. Virgin Islands, West Indies Laboratory, St. Croix, pp. 111–116.
5. Cicerone, R.J., and R.S. Oremland. 1988. Biogeochemical aspects of atmospheric methane. Global Biogeochem. Cycles **2**:299–327.
6. Fanning, K.A., R.H. Byrne, J.A. Breland, P.R. Betzer, W.S. Moore, R.J. Eslinger, and T.E. Pyle. 1981. Geothermal springs of West Florida continental shelf: evidence for dolomitization and radionuclide enrichment. Earth Planet. Sci. Lett. **52**:345–354.
7. Hanshaw, B.B., and W. Back. 1980. Chemical mass-wasting of the northern Yucatan Peninsula by groundwater dissolution. Geology **8**:222–224.
8. Kohout, F.A. 1965. A hypothesis concerning cyclic flow of salt water related to geothermal heating in the Floridian aquifer. Trans. New York Acad. Sci., Series 2 **28**:249–271.
9. Kohout, F.A., H.R. Henry, and J.E. Banks. 1977. Hydrogeology related to geothermal conditions of the Floridian plateau. In D.L. Smith and G.M. Griffin (eds.), The Geothermal Nature of the Floridian Plateau. Bureau of Geology, Tallahassee, FL, pp. 1–41.
10. Land, L.S., H.J. Lund, and M.L. McCullough. 1989. Dynamic circulation of interstitial seawater in a Jamaican fringing reef. Carbonates Evaporites **4**:1–7.
11. Mackenzie, F.T. (University of Hawaii) 1992. Personal communication.

12. Matthews, R.K. 1974. A process approach to diagenesis of reefs and reef associated limestones. In L.F. Laporte (ed.), Reefs in Time and Space, Society of Economic Paleontologists and Mineralogists, Special Publ. 18, Tulsa, OK, pp. 234–256.

13. Milliman, J.D. 1974. Marine Carbonates, Springer-Verlag, New York.

14. Oberdorfer, J.A., and R.W. Buddemeier. 1986. Coral-reef hydrology: Field studies of water movement within a barrier reef. Coral Reefs **5**:7–12.

15. Oremland, R.S. 1975. Methane production in shallow-water, tropical marine sediments. Appl. Microbiol. **30**:602–608.

16. Parnell, K.E. 1986. Water movement within a fringing reef flat, Orpheus Island, North Queensland, Australia. Coral Reefs **5**:1–6.

17. Pigott, J.D., and J.E. Laughlin. 1988. Carbon dioxide budget of a modern tropical lagoon. Proc. 6th Int. Coral Reef Symp., Australia **2**:569.

18. Roberts, H.H., A. Lugo, B. Carter, and M. Simms. 1988. Across-reef flux and shallow subsurface hydrology in modern coral reefs. Proc. 6th Int. Coral Reef Symp., Australia **2**:509–515.

19. Rougerie, F. 1983. Nouvelles données sur le fonctionnement interne des lagons d'atoll. C.R. Acad. Sci. Paris **297**:909–912.

20. Rougerie, F., and B. Wauthy. 1986. Le concept d'endo-upwelling dans le fonctionnement des atolls-oasis. Oceanologica Acta **9**:133–148.

21. Rougerie, F., C. Andrié, and P. Jean-Baptiste. 1991. Helium-3 inside atoll barrier reef interstitial water: A clue for geothermal endoupwelling. Geophys. Res. Lett. **18**:109–112.

22. Sansone, F.J. 1985. Methane in the reef-flat porewaters of Davies Reef, Great Barrier Reef (Australia). Proc. 5th Int. Coral Reef Symp., Tahiti **3**:415–420.

23. Sansone, F.J., G.W. Tribble, R.W. Buddemeier, and C.C. Andrews. 1988. Time and space scales of anaerobic diagenesis within a coral reef framework. Proc. 6th Int. Coral Reef Symp., Australia **3**:367–372.

24. Sansone, F.J., G.W. Tribble, C.C. Andrews, and J.P. Chanton. 1990. Anaerobic diagenesis within Recent, Pleistocene, and Eocene marine carbonate frameworks. Sedimentology **37**:997–1009.

25. Simms, M. 1984. Dolomitization by groundwater-flow systems in carbonate platforms. Trans. Gulf Coast Assoc. Geolog. Soc. **34**:411–420.

26. Smart, P.L., J.M. Dawans, and F. Whitaker. 1988. Carbonate dissolution in a modern mixing zone. Nature **335**:811–813.

27. Smith, S.V. 1978. Coral-reef area and the contributions of reefs to processes and resources of the world's oceans. Nature **273**:225–226.

28. Sproul, C.R. 1977. Spatial distribution of ground water temperature in south Florida. In D.L. Smith and G.M. Griffin (eds.), The Geothermal Nature of the Floridian Plateau, Bureau of Geology, Tallahassee, FL, pp. 65–89.

29. Tribble, G.W., F.J. Sansone, Y.-H. Li, S.V. Smith, and R.W. Buddemeier. 1988. Material fluxes from a reef framework. Proc. 6th Int. Coral Reef Symp., Australia **2**:577–582.

30. Tribble, G.W. 1990. Ph.D. thesis, University of Hawaii at Manoa.

31. Tribble, G.W., F.J. Sansone, and S.V. Smith. 1990. Stoichiometric modeling of carbon diagenesis within a coral reef framework. Geochim. Cosmochim. Acta **54**:2439–2449.

32. Tribble, G.W., F.J. Sansone, R.W. Buddemeier, and Y.-H. Li. 1992. Hydraulic exchange between a coral reef and surface seawater. Geol. Soc. Amer. Bull. **104**: 1280–1291.

33. Ware, J.R., S.V. Smith, and M.L. Reaka-Kudla. 1991. Coral reefs: Sources or sinks of atmospheric CO_2? Coral Reefs, **11**:127–130.

34. Webb, M.D. 1977. M.S. thesis, University of Hawaii at Manoa.

35. Whitaker, F.F., and P.L. Smart. 1990. Active circulation of saline ground waters in carbonate platforms: Evidence from the Great Bahama Bank. Geology **18**:200–203.

Effects of Seagrass Die-Off on Benthic Fluxes and Porewater Concentrations of ΣCO_2, ΣH_2S, and CH_4 in Florida Bay Sediments

Timothy R. Barber and Paul R. Carlson, Jr.

Florida Marine Research Institute, Department of Natural Resources, St. Petersburg, Florida 33701

ABSTRACT

Benthic fluxes and porewater concentrations of ΣCO_2, ΣH_2S, and CH_4 in living seagrass beds of Florida Bay (USA) are higher than in adjacent areas that have been recently denuded by disease. The enrichment and seasonal variation of porewater ΣCO_2, ΣH_2S, and CH_4 in seagrass sediments suggest seagrasses contribute autogenic organic matter to the sediments. The mean stoichiometric ratio of porewater $\Sigma CO_2 : \Sigma H_2S$ in both the live and dead zone is greater than 5 : 1, which is considerably higher than 2 : 1 predicted by complete oxidation of organic matter via sulfate reduction. Precipitation of sulfides is not thought to be significant, and sulfide oxidation is only a partial explanation. Methane flux across the sediment/water interface ranged from 13.6 μmol m^{-2} d^{-1} from dead seagrass areas in the winter to 341 μmol m^{-2} d^{-1} from live seagrass areas in the fall; these rates are insignificant in a global context but are sufficient to calibrate an empirical sediment–water exchange model. The flux of sulfide from live-zone sediments was generally much greater than that from denuded areas. Our results demonstrate that seagrass beds influence the biogeochemistry of the carbonate sediments in Florida Bay.

Introduction

Seagrass communities are important components of most tropical and temperate ocean margins and estuaries.[20,54] Seagrasses form highly productive systems that provide food and physical structure for a complex assemblage of organisms, serve as a nursery habitat for a number of economically important fish and shellfish species, and export detrital organic matter to other communities.[63,71,73] The complex matrix of seagrass roots and rhizomes binds sediments and reduces erosion.[4,53,58] Seagrasses also reduce bottom-current velocity, promoting sedimentation and inhibiting resuspension.[22–24,58]

Florida Bay is a triangular body of water bounded on the north by the Florida mainland, to the south and east by the Florida Keys, and to the west by the Gulf of Mexico.[57] Seagrasses previously covered more than 80% of the 1800 km^2 of surface area of Florida Bay.[72] Since 1987, the dominant seagrass species in Florida Bay, *Thalassia testudinum,* has been affected by disease. Robblee and others[55] estimate that 40 km^2 of seagrass beds have been completely denuded of *T. testudinum* and that 230 km^2 have been affected to a lesser degree by the die-off phenomenon. Most of the areas affected by *T. testudinum* die-off are located in western Florida Bay. The die-off affected the present study area, Johnson Key Basin (JKB, see Figure 1), in early 1989, denuding large areas (up to 100 m^2) of *T. testudinum.* In 1990 (9 to 20 months after the die-off episode), we studied visibly healthy *T. testudinum* beds (hereafter described as "live zones") and adjacent, denuded patches (described as "dead zones").

This study was undertaken to estimate the contribution of sediment chemical oxygen demand to community oxygen metabolism of seagrass beds affected by die-off. We measured depth profiles of porewater pH, ΣCO_2, ΣH_2S, and CH_4 from sediments with and without living seagrass. We estimated sediment–water sulfide fluxes using empirical exchange coefficients derived from CH_4 profiles and chamber flux measurements. This study also addresses the larger issue of the influence of seagrass beds on carbonate sediment chemistry.

Methods

Porewater Collection and Analyses

Porewater samples were collected and benthic-flux chambers were deployed four times in 1990 (1/19, 3/22, 7/20, and 9/21). Interstitial water was collected in opaque PVC equilibrium–dialysis samplers.[27] Each sampler had two columns of 20 rows of 10-ml cells centered at 1.5-cm intervals. The cells were filled with filtered surface water from Johnson Key Basin and were enclosed with 0.4-μm polycarbonate membrane filters. Three samplers were placed in the sediments in each zone (live and dead) of the study area. Samplers were allowed to equilibrate for 4 weeks before each benthic-flux experiment. Porewater was collected in 10-ml disposable glass syringes fitted with rubber plungers and gas-tight, serum-vial stoppers. Samples were stored in the dark at 4°C until time of analysis (usually within 48 h). No systematic loss of CO_2 or CH_4 was observed during storage.

Total carbon dioxide and methane were separated and quantified using a gas chromatograph equipped with thermal-conductivity and flame-ionization detectors in series. A modified multiple-phase equilibration technique,[21] with helium

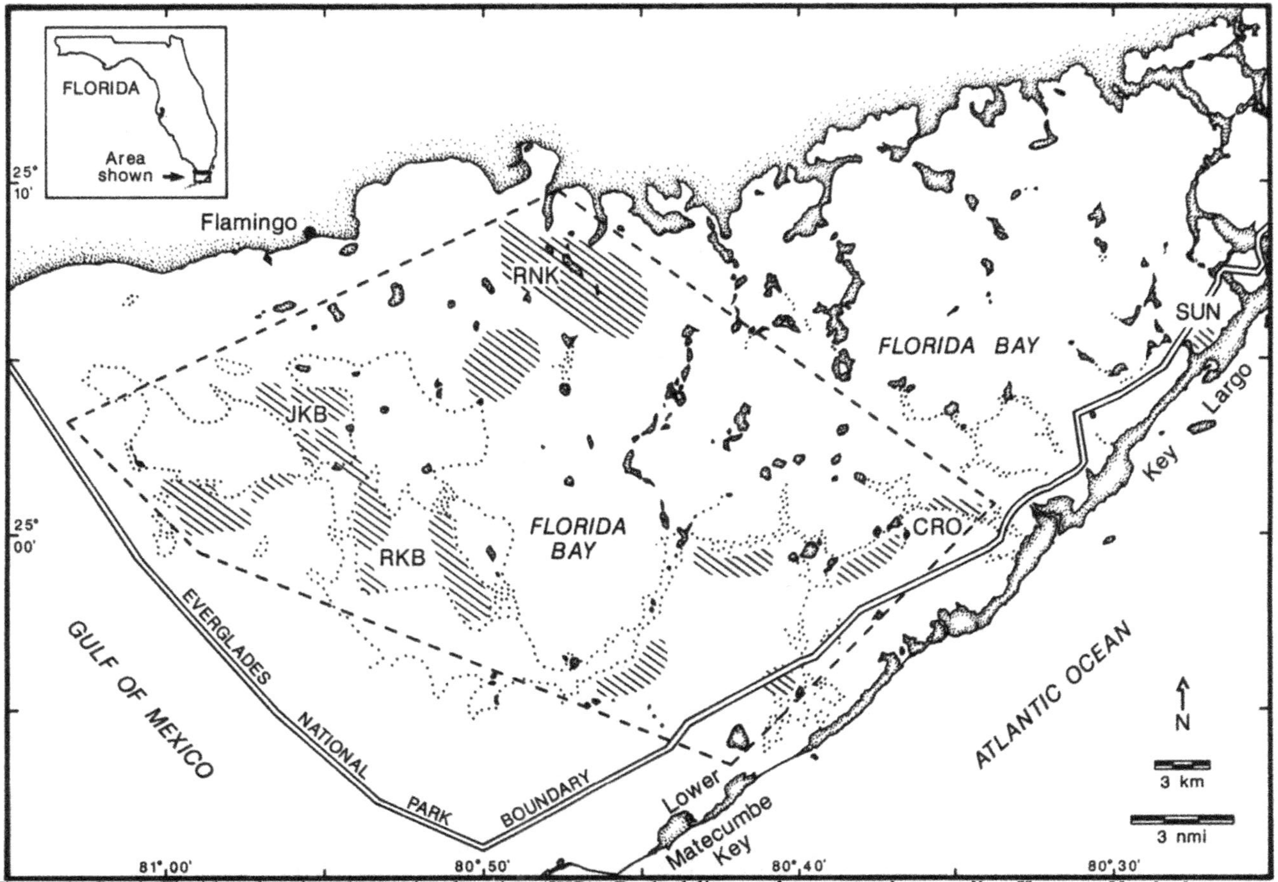

Figure 1. South Florida, showing the station location (JKB). Dashed line encloses most known die-off areas. Hatched areas are severely affected, having lost up to 95% of *T. testudinum*. Reproduced from Robblee et al.[55] with the permission of Inter-Research.

as the purge gas, was used to quantify CO_2 and CH_4 gas concentrations using the following equation:

$$M = X\left[1 + \left(\frac{\alpha A}{B}\right)\right], \quad (1)$$

where M = total volume of gas dissolved in the aqueous phase in ml, X = volume of gas transferred into the gas phase, α = Bunsen coefficient (the volume of gas that will dissolve in 1 ml of water), A = volume of aqueous phase, and B = volume of gas phase. Samples were acidified with 10% H_3PO_4 to convert bicarbonate and carbonate to CO_2. Methane and CO_2 were separated on a 6-ft × 1/8-in. Porapak Q (80/100 mesh) column, with helium as the carrier gas at 40 ml/min. The chromatograph was calibrated by injecting known volumes of standard gases (supplied by Scott Specialty Gases, Plumsteadville, Pennsylvania). Bunsen coefficients were calculated for extraction temperature and sample salinity using the equations of Weiss[67] and Wiesenburg and Guinasso.[70]

Porewater samples used to determine sulfide concentration were preserved immediately in the field with sulfide anti-oxidant buffer. Sulfide concentrations were determined potentiometrically with a silver/sulfide ion-selective electrode.[14]

A PVC corer (7.5 cm diameter) was used to collect sediment for pH profiles. Samples were analyzed immediately by inserting a Ross combination electrode (Orion Research) into the sediments through holes that had been drilled into the side of the core barrel. Values are reported on the free-hydrogen-ion scale.[46]

In Figures 2–5, each plotted value is an average of data from three replicate cores or porewater samplers taken from areas with seagrass (live) and denuded sediment zones (dead). Depths are given for the midpoint of the sampling ports. The sediment–water interface is assigned a depth value of zero, and depths are defined as positive downward. Error bars represent the standard error calculated from the range.[19]

Benthic-Flux Measurements

Four cylindrical, plexiglass chambers (60 cm diameter × 45 cm height) were used to measure sediment/water fluxes of methane. During each flux experiment, two chambers were placed over living seagrass and two chambers were placed over denuded sediments in the dead zone. Chambers were deployed in the late afternoon. Chamber edges were inserted 5–8 cm into the sediments, and the chambers were flushed for 15–30 min before closure. A submersible pump circulated the water within the chamber to maintain its homogeneity. Water samples were collected hourly at night. Dissolved-oxygen concentrations inside the chambers were monitored during the methane-flux experiments; sampling ceased and chambers were opened when dissolved-oxygen concentrations reached 1.0 mg/L. The flux per unit of bottom area was calculated for each deployment from the slope of a plot of CH_4 concentration versus time and a geometric estimate of the chamber's volume. A total of 24 chamber deployments were made during the course of this study; no flux samples were collected in the summer.

Sediment–Water Flux Model

The flux of gases and dissolved inorganic ions across the sediment–water interface can be modeled as the product of the following exchange coefficient and concentration gradient[11,26]:

$$F = K_L\,(C_W - C_0), \tag{2}$$

where F is the flux in moles $m^{-2}\,d^{-1}$, K_L is the exchange coefficient with velocity units, C_W is the porewater concentration extrapolated to the interface, and C_0 is the concentration of the bottom water. The exchange coefficient is generally controlled by the interaction between slow, diffusive transport and rapid, turbulent processes[65]; however, in some cases sediment–water exchange can be dominated by bubble transport,[44] bioturbation,[1] and/or plant-mediated flux.[15] Several predictive models have been developed, but their applications are dependent on knowledge of the local, often complex, hydrodynamic conditions. Our approach involves measuring the exchange coefficient in the field using methane as a tracer and then estimating the flux of other substances using a simple calculation based on porewater-concentration data. Exchange coefficients were corrected for the differences in diffusivities and temperatures using the relationship

$$(K_L)_Y = (K_L)_X\,(Sc_Y/Sc_X)^n, \tag{3}$$

where $(K_L)_Y$ is the exchange coefficient for a gas or dissolved inorganic ion at a specified temperature (°C); Sc is the Schmidt number (dimensionless), which is the kinematic viscosity of the medium divided by the diffusivity; X refers to methane at 25°C; and Y is the gas or dissolved inorganic ion of interest. Both experimental and theoretical evidence indicate that n is equal to $-2/3$ for rigid surfaces.[18,29,33,40]

We estimated the flux of sulfide across the sediment–water interface by multiplying an effective exchange coefficient by the sulfide concentration extrapolated to the sediment–water interface. The sulfide concentration of the oxygenated bottom water was negligible. Because the pK_1 of H_2S is 7.0,[25] a significant portion of the total dissolved sulfide at depth in the sediments should be present as hydrogen sulfide gas. At the sediment–water interface, however, bisulfide ion (HS^-) represents 85% to 95% of the total sulfide between pH 7.5 to 7.8 at 25°C.[25] The remaining sulfide is present as H_2S. The exchange coefficient was corrected for differences in diffusivities between methane and bisulfide ion using the appropriate Schmidt numbers [Eq. (3)]. The diffusion coefficient of the bisulfide ion was taken as 1.73×10^{-5} at 25°C.[41]

Results

Salinity and temperature data recorded during this study ranged from 35 to 45 psu and 22 to 30°C, respectively.

Data for porewater constituents are summarized in Figures 2–5. Although sediment pH values declined with depth in both live and dead zones on all three sampling dates, the decrease was more pronounced in live-zone sediments (Figure 2). During the study period, live-zone pH values dropped from a range of 7.5 to 7.8 at the surface to asymptotic values that ranged from 6.8 to 7.2 at 15-cm depth and decreased only slightly at greater depths. Minimum values for dead-zone sediments ranged from 7.2 to 7.4. Our live-zone pH data is consistent with the model of Ben-Yaakov;[7] pH of pore water under anoxic conditions is essentially buffered around 6.9.

Concentrations of ΣCO_2 were consistently higher in live-zone sediments than they were in dead-zone sediments (Figure 3). While ΣCO_2 concentrations in both zones were low in winter and spring and higher in summer and fall, live-zone ΣCO_2 profiles exhibited a more pronounced seasonal response than did dead-zone profiles. The depth-integrated, live-zone average of ΣCO_2 ranged from 6.95 mM in winter to 13.0 mM in summer, 8.41 mM in spring, and 12.0 mM in fall.

Sulfide profiles were similar to ΣCO_2 profiles; live-zone concentrations were consistently higher than dead-zone concentrations and exhibited a more marked seasonal response than did dead-zone values (Figure 4). Dead-zone profiles for all sampling dates were similar to one another; concentrations increased gradually from less than 0.5 mM near the surface to 1–1.5 mM at depth. In the live zone, sulfide concentrations were lowest in winter and highest in spring. Values in spring, however, had the highest variation among replicate samplers, indicating greater spatial heterogeneity. The depth-integrated average of total sulfide in the live zone ranged from 0.88 mM in winter to 3.11 mM in spring, 2.01 mM in summer, and 2.50 mM in fall.

Methane levels in the live zone were also consistently higher than in the dead zone, except during the winter when the two profiles were indistinguishable (Figure 5). The profile of CH_4 in the dead zone showed very little seasonal

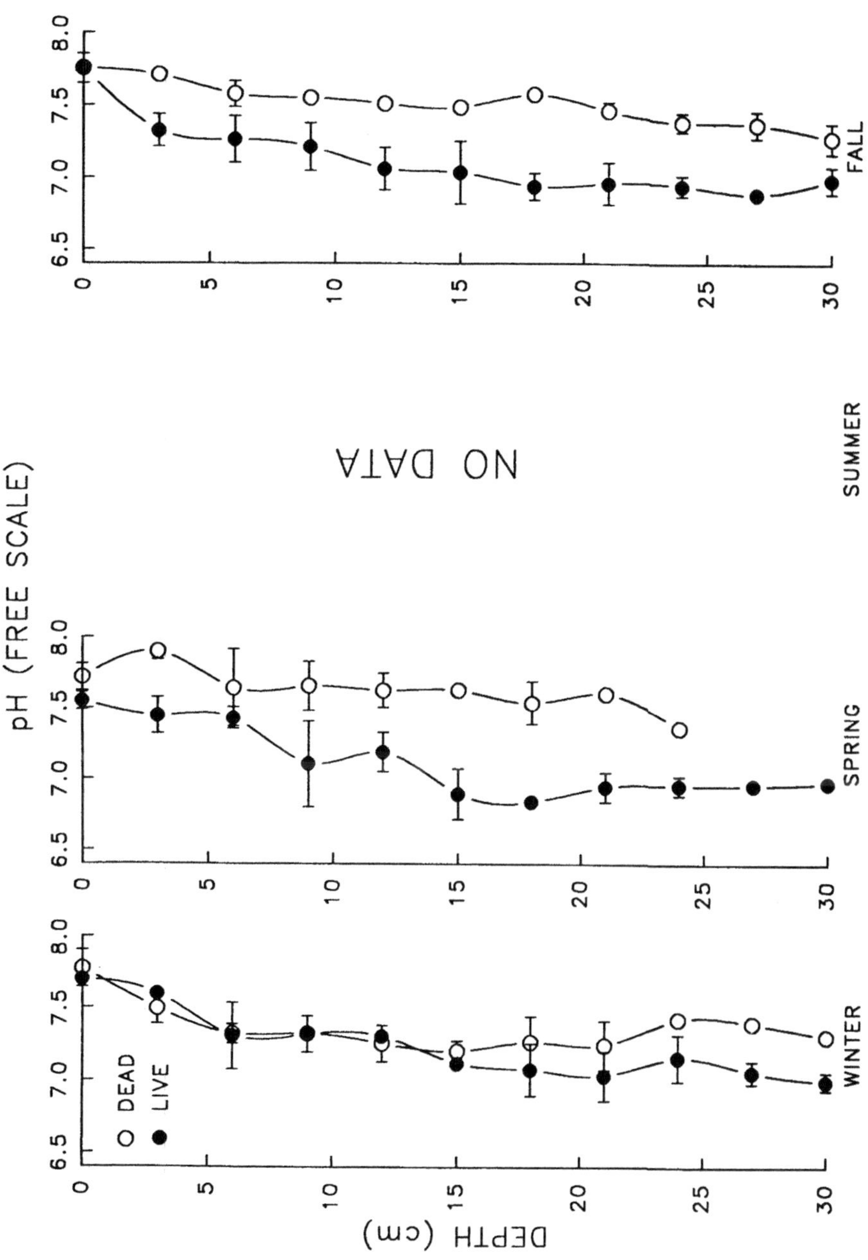

Figure 2. Porewater profiles of pH observed in JKB sediments. Measurements are reported on the free-hydrogen-ion scale. No measurements were made during the summer.

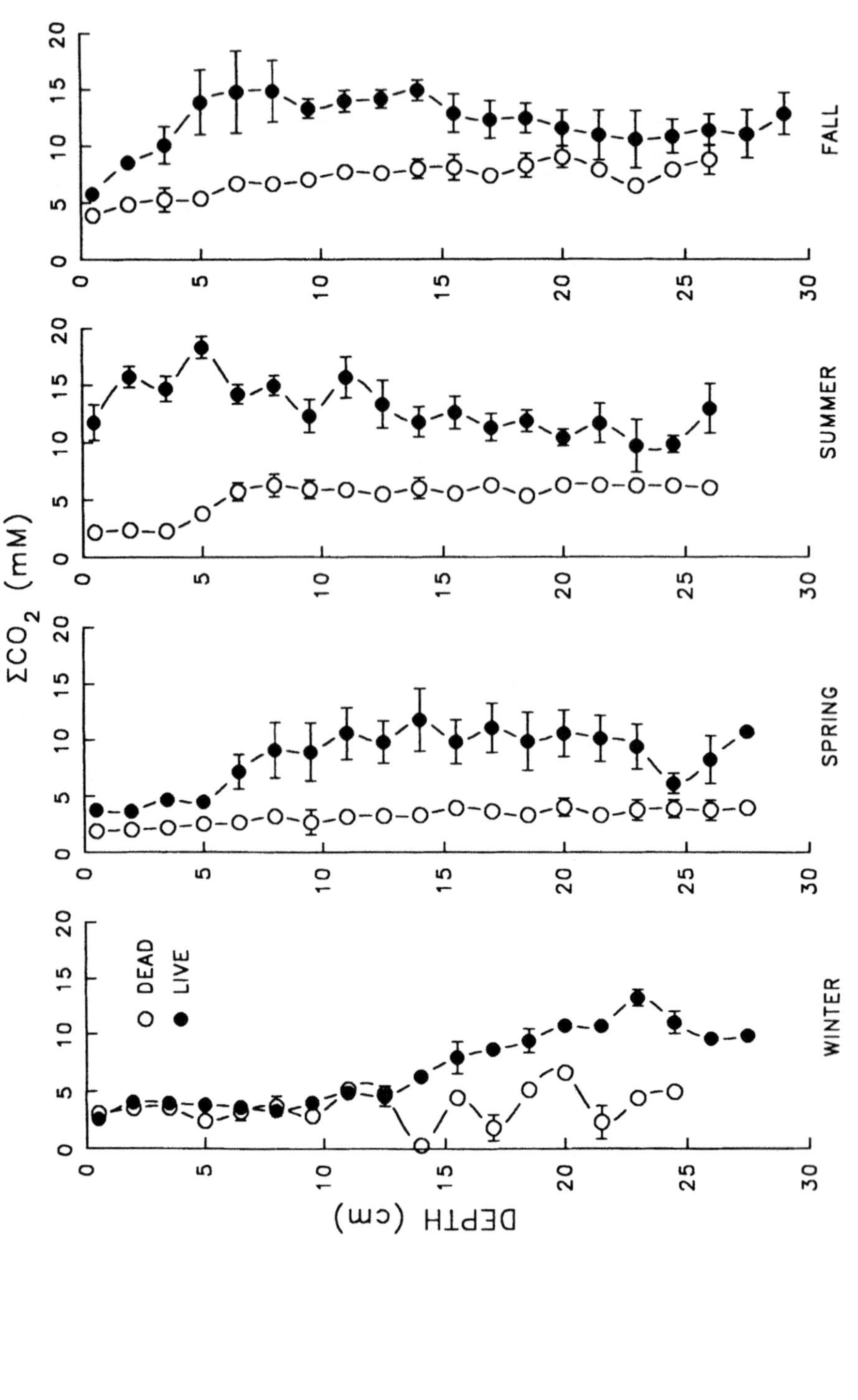

Figure 3. Porewater profiles of ΣCO_2 observed in JKB sediments.

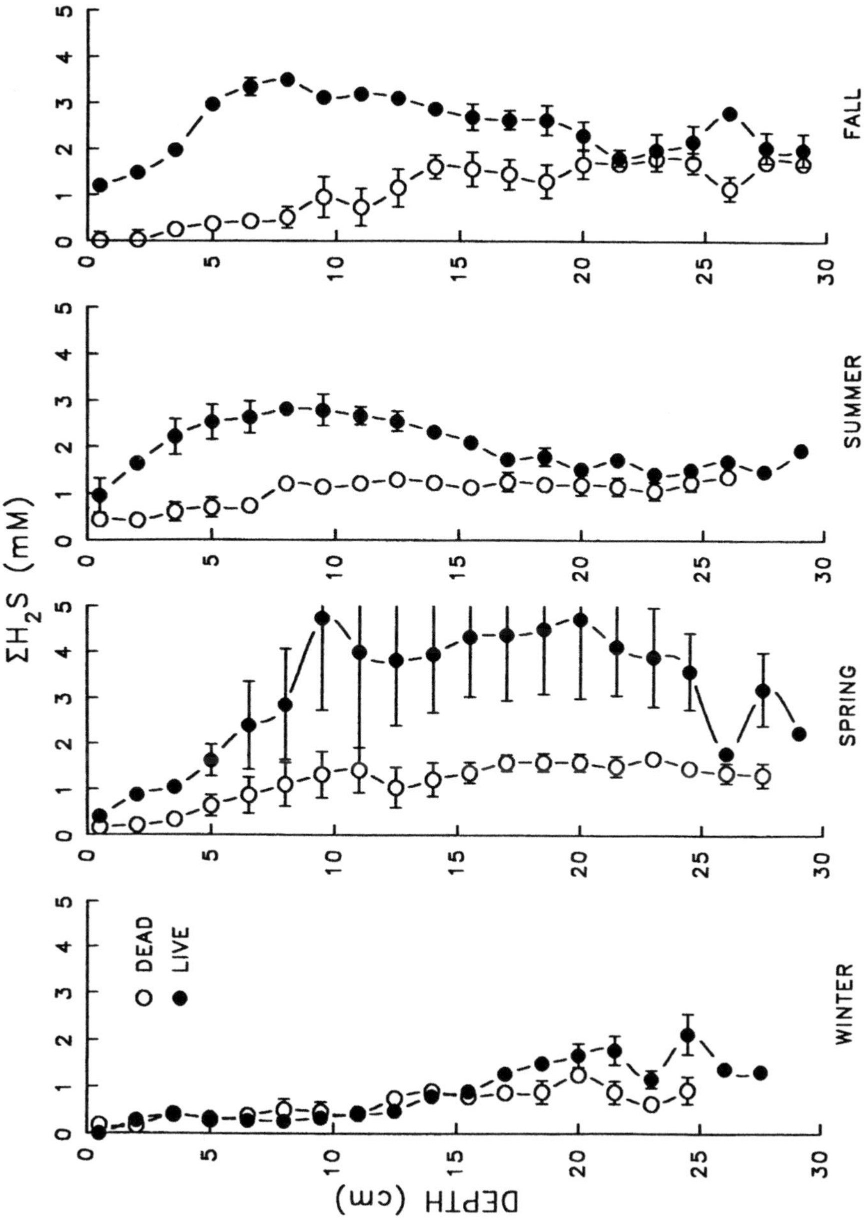

Figure 4. Porewater profiles of ΣH_2S observed in JKB sediments.

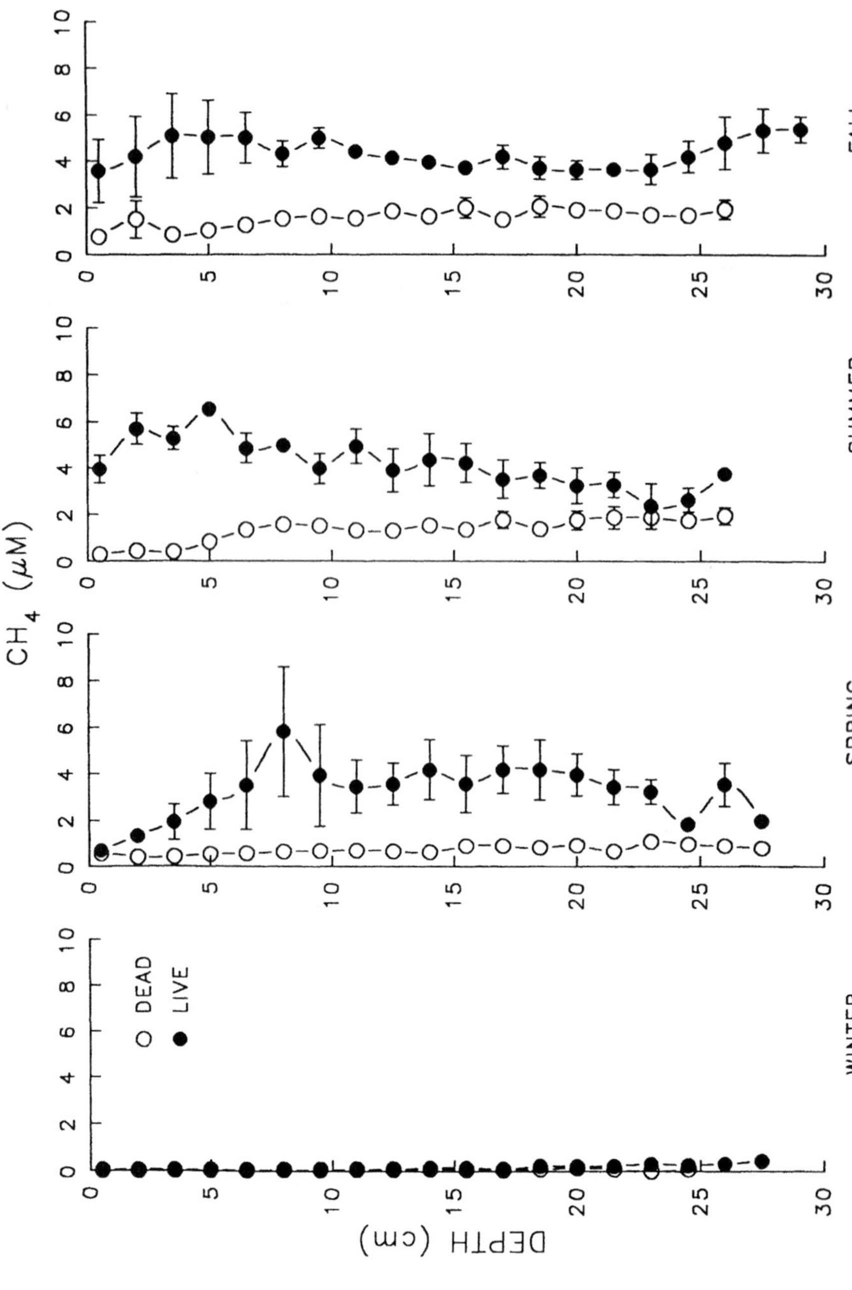

Figure 5. Porewater profiles of CH_4 observed in JKB sediments.

Table 1. Methane flux from carbonate sediments in Florida Bay.[a]

	[μmol m^{-2} d^{-1} ($\pm$SE)]	
Season	Dead Zone	Live Zone
Winter	13.6 ± 5.62	25.8 ± 5.98
Spring	22.6 ± 9.73	40.9 ± 21.6
Fall	158 ± 65.4	341 ± 121

[a]Values reported are means of four measurements (two deployments × two nights).

response; concentrations were less than 2 μM throughout the study period. Live-zone CH_4 concentrations generally varied between 3 and 5 μM throughout the sediment column in the summer and fall. Below 10 cm in the spring data, CH_4 levels were also between 3 and 5 μM; however, close to the surface the concentration fell to less than 1 μM.

Methane fluxes are listed in Table 1. The lowest mean flux was observed in the dead zone during the winter (13.6 μmol m^{-2} d^{-1}); the highest mean flux was observed in the fall from the live zone (341 μmol m^{-2} d^{-1}). Methane fluxes from the live-zone sediments were generally higher than fluxes observed from denuded areas, and the emission of methane in both zones increased during the year. Our measurements of CH_4 flux are within the range of other flux measurements reported from marine environments.[5,6,38,48] Although the emission of methane from Florida Bay sediments is insignificant with respect to the global methane cycle, it can provide a useful tracer for sediment–water exchange processes.

Discussion

Abundance and Distribution of Porewater Constituents

Enrichment and seasonal variation in porewater ΣCO_2, ΣH_2S, and CH_4 within the live zone (Figures 2–5) suggest that seagrasses are supplying the organic matter undergoing remineralization.[28] In addition to delivering particulate organic matter via sedimentation, seagrasses also contribute organic matter through mortality of roots and rhizomes and through leakage of dissolved organic carbon from living tissue.[34,68,71] The rhizoplane for *T. testudinum* in Florida Bay is between 5 and 20 cm,[66] coinciding with broad subsurface maxima in most constituents (Figures 3–5). Contrary to the conclusions of Morse et al.,[47] our data indicate *T. testudinum* is affecting sediment geochemistry in Florida Bay seagrass beds.

Organic carbon oxidation occurs primarily via sulfate reduction in anaerobic marine sediments.[13,16,30] The supply of sulfate to the microbial population is maintained via diffusion from the overlying water. Although depletion may occur at depth in the sediments, sulfate levels in this study ranged from 23.8 to 34.0 mM throughout the sediment column (maximum sampling depth in this study was 30 cm) and displayed no spatial or temporal trend (data not shown). Swart et al.[62] reported no sulfate depletion to 200 cm in similar Florida Bay sediments.

Bacterial sulfate reduction is the process by which certain low-molecular-weight organic compounds are oxidized to CO_2, as sulfate is reduced to sulfide. For example, the oxidation of the common substrate acetate may be expressed as

$$(4) \qquad CH_3COOH + SO_4^{-2} + H_2 \rightleftharpoons 2CO_2 + H_2S + 2H_2O.$$

The input of CO_2 to the interstitial water affects the carbonate system equilibria, as shown here:

$$(5) \qquad CO_2 + H_2O \rightleftharpoons HCO_3^- + H^+ \rightleftharpoons CO_3^{-2} + 2H^+.$$

Therefore, ΣCO_2 can be monitored to assess the amount of organic carbon remineralization.[3,10]

The expected molar ratio of $\Sigma CO_2 : \Sigma H_2S$ from the complete oxidation of organic matter by sulfate-reducing bacteria is 2 : 1.[25] Measurements made during the spring approximate theoretical stoichiometry; however, the ratio of $\Sigma CO_2 : \Sigma H_2S$ for most of the summer, fall, and winter samples is greater than 5 : 1 (Figure 6, bottom). The high porewater $\Sigma CO_2 : \Sigma H_2S$ ratio we observed must be due to preferential removal of dissolved sulfide or enrichment of CO_2. Sources of CO_2 other than microbial respiration are unlikely, but porewater sulfide concentrations may be influenced by oxidation, precipitation, or efflux. The lack of significant difference between live- and dead-zone $\Sigma CO_2 : \Sigma H_2S$, except for in winter (Figure 6, bottom), suggests that the mechanism responsible for altering the stoichiometric ratio is, in part, independent of the seagrass community.

Several processes may contribute to the observed variation in the $\Sigma CO_2 : \Sigma H_2S$ ratio. The sediments in Florida Bay are primarily biogenic carbonates;[9,61] significant precipitation of iron sulfides is unlikely.[8] The lack of significant change in $\Sigma CO_2 : \Sigma H_2S$ ratios with depth suggests sulfide oxidation at the sediment–water interface[69] is probably not responsible for the high $\Sigma CO_2 : \Sigma H_2S$ ratios. Another possible mechanism is plant-mediated subsurface sulfide oxidation. Seagrasses and freshwater macrophytes have well-developed systems of lacunae to support below-ground respiration requirements;[45,51,60] porewater ΣCO_2 :

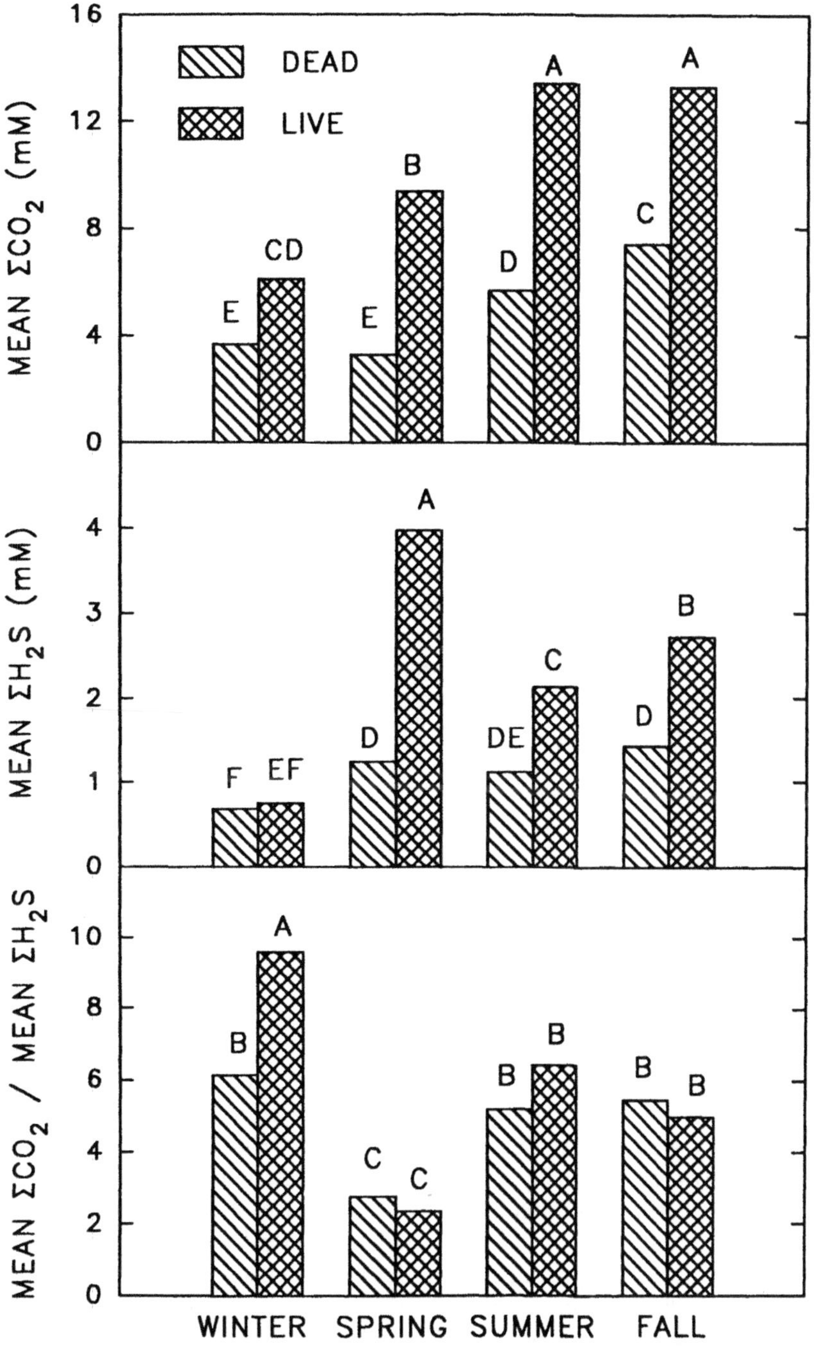

Figure 6. Histograms of mean ΣCO_2 (top), mean ΣH_2S (middle), and mean ΣCO_2: mean ΣH_2S (bottom) from porewater samples collected from 5 to 20 cm in JKB area sediments. Means with the same letter are not significantly different.

ΣH_2S ratios may be affected by O_2 leakage from the roots and rhizomes.[12,64] The significant difference between live- and dead-zone $\Sigma CO_2 : \Sigma H_2S$ in the winter (Figure 6, bottom) suggests that plant-transported O_2 may influence local sediment oxidation.[30–32] Furthermore, there is a significant decrease in live-zone porewater sulfide during the summer, despite an increase in ΣCO_2 (Figure 6). However, significant differences in $\Sigma CO_2 : \Sigma H_2S$ were observed between live and dead zones only in winter, suggesting that sediment oxidation caused by leakage from plant roots and rhizomes is overwhelmed by high sulfate-reduction rates at higher temperatures. It is apparent that plant-mediated oxidation is unable to explain the elevated $\Sigma CO_2 : \Sigma H_2S$ found in the dead-zone sediment during most of the year.

Physical transport processes may also affect the porewater $\Sigma CO_2 : \Sigma H_2S$ ratio in Florida Bay sediments. However, the diffusion coefficients of bisulfide and bicarbonate are so similar[41] that the high ratios probably do not reflect a preferential loss of HS^- via upward diffusion. The significant amount of H_2S gas at the pH values we encountered[25] might also influence the observed ratio values if there is a preferential removal mechanism for gaseous sulfide (e.g., ebullition). Our data does not permit us to adequately explain why $\Sigma CO_2 : \Sigma H_2S$ ratios greater than 5 : 1 persist throughout the sediment column in both live and dead zones.

The limited methane production observed during the study is most likely supported by noncompetitive substrates, for example, methanol, trimethylamines, and dimethylsulfide.[35–37,42,49] Methanogenesis (via acetate fermentation or CO_2 reduction) can be discounted in these sediments due to the high concentration of dissolved sulfate.[43] Energetically more efficient sulfate-reducing bacteria out-compete methanogens for acetate and H_2.[50,52] The depth distribution of porewater methane (a broad subsurface maxima coinciding with the *T. testudinum* rhizoplane) suggests that seagrasses are also the ultimate source of the organic matter supporting the CH_4 production (Figure 5).

Sediment–Water Exchange

We needed to estimate the flux of sulfide across the sediment–water interface to quantify its contribution to the total chemical oxygen demand of the system. The magnitude of the flux depends on the velocity of mass transfer (K_L) and the concentration gradient [see Eq. (2)]. For the benthic boundary layer, K_L is mostly controlled by current velocity and bottom shear. Determination of an appropriate K_L in Florida Bay was complicated by the presence of a plant canopy, which altered the current regime.[22,23] Oremland and Taylor[51] reported the presence of methane in the aerenchyma of *T. testudinum*, suggesting that sedimentary gases diffuse freely through seagrasses.[2,15,17,59]

Direct measurement of the benthic flux of CH_4 enabled us to circumvent these

Table 2. Calculation of K_L: Flux and porewater parameters.

		Flux Data			Porewater Data			
Date	Zone	n[a]	[nM/h (±SE)][b]	Chamber Height[c] (cm)	n	Porewater [nM (±SE)][d]	Bottom Water[e] [nM (±SE)]	K_L [cm/h (±SE)]
1/90	Dead	20	1.69 ± 0.70	33.4	12	15.4 ± 8.11	15.0 ± 8.9	138 ± 92.5
1/90	Live	20	3.20 ± 0.74	33.6	12	25.8 ± 11.1	15.0 ± 2.5	10.0 ± 6.60
3/90	Dead	15	2.91 ± 1.25	32.3	10	384 ± 64.6	333 ± 132	1.83 ± 0.84
3/90	Live	18	5.23 ± 2.76	32.6	8	747 ± 659	398 ± 66.2	0.49 ± 0.50
7/90	Dead	ND	ND	ND	9	384 ± 233	168 ± 36.5	ND
7/90	Live	ND	ND	ND	6	3540 ± 547	2260 ± 467	ND
9/90	Dead	16	19.6 ± 8.10	33.6	11	776 ± 224	532 ± 81.7	2.70 ± 1.35
9/90	Live	15	42.6 ± 15.1	33.4	7	3380 ± 530	2690 ± 1080	2.05 ± 0.78

[a]Total number of dissolved-CH_4 analyses used to determine the flux.
[b]Best-fit coefficient from linear model of methane buildup versus deployment time.
[c]Average chamber height (n = 4, two chambers × two nights).
[d]Porewater concentration of methane extrapolated to the sediment–water interface.
[e]Bottom-water concentration of methane.

complexities and calculate an effective exchange coefficient by dividing the CH_4 flux measured in the chambers by the concentration gradient calculated from porewater profiles (Table 2). The concentration gradient was calculated by subtracting the concentration of the bottom water from the porewater concentration at the sediment–water interface (y-intercept of a linear regression model of the top 5–10 cm of the profile in Figure 5). The porewater samplers used in this study provided profiles on the scale of centimeters. Our measurements did not allow us to resolve the thickness of the diffusive sublayer.[56] Nevertheless, we estimated sediment–water flux using a simple, empirical model. Complex hydrodynamical conditions should be considered for a more thorough investigation of the processes of sediment–water exchange.

From our model calculations, we concluded plant-mediated methane flux did not appear to be significant. Despite higher CH_4 fluxes in the live zone, the exchange coefficients calculated from the live-zone data were less than those calculated for the bare-sediment areas (Table 2). Our results suggest that gas exchange is slow between the water column and submerged macrophytes. The slow diffusivity of gases in water, the presence of a cuticle, and the lack of stomata impede mass flux across the leaf–water interface.[39]

The mean K_L for the spring and fall experiments was 1.76 cm/h (n = 4, Table 2); this value was used to estimate the sulfide flux. Exchange coefficients cal-

Table 3. Estimated HS^- flux out of carbonate sediments in Florida Bay.[a]

Season	Dead Zone [mmol m^{-2} d^{-1} (±SE)]	Live Zone [mmol m^{-2} d^{-1} (±SE)]
Winter	54.3 ± 28.3	32.6 ± 37.2
Spring	22.1 ± 34.9	203 ± 69.8
Summer	112 ± 53.1	433 ± 95.7
Fall	57.4 ± 50.4	368 ± 90.7

[a]Assuming Sc = 570 for HS^- at 25°C.

culated from the winter sampling period were rejected due to the exceedingly low levels of porewater methane. The low signal-to-noise ratio provided unreliable mass-transfer calculations.

Except during winter, the calculated sulfide flux from the vegetated zone was higher than that from the dead zone (Table 3). The flux of sulfide also exhibited seasonal variations; it was highest in the summer and lowest in the winter. These variations may reflect the dependence of the sediment–water flux of sulfide on the supply of organic matter, ambient temperature, and the physiological state of the microbial populations involved in sulfate reduction.

Conclusions

1. Benthic fluxes and porewater concentrations of ΣCO_2, ΣH_2S, and CH_4 in living seagrass beds are higher than in adjacent, recently denuded areas in Florida Bay.
2. The enrichment and seasonal variation of porewater ΣCO_2, ΣH_2S, and CH_4 in live-zone sediments suggest that seagrasses are significant sources of sedimentary organic matter and that they influence early diagenetic processes.
3. The mean stoichiometric ratio of porewater $\Sigma CO_2 : \Sigma H_2S$ in both the live and dead zones is greater than 5 : 1, which is considerably higher than the 2 : 1 predicted by complete oxidation of organic matter via sulfate reduction.
4. Methane may be used as a tracer of sediment–water exchange processes, although the sensitivity of the technique may be affected by low CH_4 concentrations and unknown methane oxidation rates.
5. The elevated flux of CH_4 from live-zone sediments was supported by a larger concentration gradient, not by a plant-mediated flux pathway.

6. The flux of sulfide into the water column is much greater from vegetated sediments; the associated chemical oxygen demand is a significant component of the total oxygen budget for seagrass beds in Florida Bay.

Acknowledgments

We thank the Florida Everglades National Park Service Research Division for use of their boats, laboratory, and housing. We particularly thank Mike Robblee for logistical support. We also thank Herman Arnold for assistance in the laboratory. Funds for this project were provided by the Florida Department of Environmental Regulation, Office of Coastal Management, using funds made available through the National Oceanic and Atmospheric Administration under the Coastal Zone Management Act of 1972, as amended.

References

1. Aller, R.C., and J.Y. Yingst. 1978. Biochemistry of tube dwellings: a study of the sedentary polychaete *Amphitrite ornata* L.J. Marine Res. **36**:210–254.
2. Armstrong, W. 1978. Root aeration in the wetland condition. In D.D. Hook and R.R.M. Crawford (eds.), Plant Life in Anaerobic Environments, Ann Arbor Science, Ann Arbor, MI, pp. 269–297.
3. Anderson, L.G., P.O.J. Hall, A. Iverfeldt, and M.M. Rutgers van der Loeff. 1986. Benthic respiration measured by total carbonate production. Limnol. Oceanogr. **31**:319–329.
4. Ball, M.M., E.A. Shinn, and K.W. Stockman. 1967. The geologic effects of Hurricane Donna in South Florida. J. Geol. **75**:583–597.
5. Bartlett, K.B., R.C. Harriss, and D.I. Sebacher. 1985. Methane flux from coastal salt marshes. J. Geophys. Res. **90**:5710–5720.
6. Bartlett, K.B., D.S. Bartlett, R.C. Harriss, and D.I. Sebacher. 1987. Methane emissions along a salt marsh salinity gradient. Biogeochem. **4**:183–202.
7. Ben-Yaakov, S. 1973. pH buffering of pore water of recent anoxic marine sediments. Limnol. Oceanogr. **18**:86–94.
8. Berner, R.A. 1984. Sedimentary pyrite formation: an update. Geochim. Cosmochim. Acta **48**:605–615.
9. Bosence, D. 1989. Surface sublittoral sediments of Florida Bay. Bull. Marine Sci. **44**:434–453.
10. Boudreau, B.P. 1987. A steady-state diagenetic model for dissolved carbonate species and pH in the porewaters of oxic and suboxic sediments. Geochim. Cosmochim. Acta **51**:1985–1996.

11. Boudreau, B.P., and N.L. Guinasso. 1982. The influence of a sublayer on the accretion, dissolution, and diagenesis at the sea floor. In K.A. Fanning and F.T. Manheim (eds.), The Dynamic Environment of the Ocean Floor, Lexington Books, Lexington, KY, pp. 115–145.

12. Caffrey, J.M., and W.M. Kemp. 1991. Seasonal and spatial patterns of oxygen production, respiration and root-rhizome release in *Potamogeton perfoliatus* L. and *Zostera marina* L. Aquat. Bot. **40**:109–128.

13. Capone, D.G., and R.P. Kiene. 1988. Comparison of microbial dynamics in marine and freshwater sediments: contrasts in an aerobic carbon catabolism. Limnol. Oceanogr. **33**:725–749.

14. Carlson, P.R. Jr., L.A. Yarbro, C.F. Zimmerman, and J.R. Montgomery. 1982. Porewater chemistry of an overwash mangrove island. Florida Sci. **46**:239–249.

15. Cicerone, R.J., and J.D. Shetter. 1981. Sources of atmospheric methane: measurements in rice paddies and a discussion. J. Geophys. Res. **86**:7203–7209.

16. Claypool, G.E., and I.R. Kaplan. 1974. The origin and distribution of methane in marine sediments. In I.R. Kaplan (ed.), Natural Gases in Marine Sediments, Plenum, New York, pp. 99–139.

17. Dacey, J.W.H., and M.J. Klug. 1979. Methane efflux from lake sediments through water lilies. Science **203**:1253–1255.

18. Deacon, E.L. 1977. Gas transfer to and across an air–water interface. Tellus **29**: 363–374.

19. Dean, R.B., and W.J. Dixon. 1951. Simplified statistics for small number of observations. Anal. Chem. **23**:636–640.

20. den Hartog, C. 1970. The Seagrasses of the World, North-Holland, Amsterdam.

21. Flett, R.J., R.D. Hamilton, and N.E.R. Campbell. 1976. Aquatic acetylene-reduction techniques: solutions to several problems. Can. J. Microbiol. **22**:43–51.

22. Fonseca, M.S., and J.S. Fisher. 1986. A comparison of canopy friction and sediment movement between four species of seagrass with reference to their ecology and restoration. Marine Ecol. Prog. Ser. **29**:15–22.

23. Fonseca, M.S., J.S. Fisher, J.C. Zieman, and G.W. Thayer. 1982. Influence of the seagrass *Zostera marina* L., on current flow. Estuarine Coastal Shelf Sci. **15**: 351–364.

24. Ginsburg, R.N., and H.A. Lowenstam. 1958. The influence of marine bottom communities on depositional environment of sediments. J. Geol. **66**:310–318.

25. Goldhaber, M.B., and I.R. Kaplan. 1974. The sedimentary sulfur cycle. In E.D. Goldberg (ed.), The Sea, Vol. 5, J. Wiley and Sons, New York, pp. 569–655.

26. Hartman, B., and D.E. Hammond. 1984. Gas exchange rates across the sediment–water and air–water interfaces in south San Francisco Bay. J. Geophys. Res. **89**: 3593–3603.

27. Hesslein, R.H. 1977. An in situ sampler for close interval pore water studies. Limnol. Oceanogr. **21**:912–914.

28. Hines, M.E., S.L. Knollmeyer, and J.B. Tugel. 1989. Sulfate reduction and other sedimentary biogeochemistry in a northern New England salt marsh. Limnol. Oceanogr. **34**:578–590.

29. Holmén, K., and P. Liss. 1984. Models for air–water gas transfer: a experimental investigation. Tellus **36**:92–100.

30. Howarth, R.W., and J.M. Teal. 1979. Sulfate reduction in a New England salt marsh. Limnol. Oceanogr. **24**:999–1013.

31. Howes, B.L., J.W.H. Dacey, and D.D. Groehringer. 1986. Factors controlling the growth form of *Spartina alterniflora*: Feedbacks between above-ground production sediment oxidation, nitrogen and salinity. J. Ecol. **74**:881–898.

32. Howes, B.L., J.W.H. Dacey, and G.M. King. 1984. Carbon flows through oxygen and sulfate reduction pathways in salt marsh sediments. Limnol. Oceanogr. **29**: 1037–1051.

33. Jähne, B., W. Huber, A. Dutzi, T. Wais, and J. Ilmberger. 1984. Wind/wave tunnel experiment on the Schmidt number and wave field dependence of air/water exchange. In W. Brutsaert and G.H. Jirka (eds.), Gas Transfer at Water Surfaces, D. Reidel, Hingham, MA, pp. 303–309.

34. Kenworthy, W.J., and G.W. Thayer. 1984. Production and decomposition of the roots and rhizomes of seagrasses, *Zostera marina* and *Thalassia testudinum*, in temperate and subtropical marine ecosystems. Bull. Marine Sci. **35**:364–379.

35. Kiene, R.P., R.S. Oremland, A. Catena, L.G. Miller, and D.G. Capone. 1986. Metabolism of reduced methylated sulfur compounds in anaerobic sediments and a pure culture of an estuarine methanogens. Appl. Environ. Microbiol. **52**:1037–1045.

36. King, G.M. 1984. Metabolism of trimethylamine, choline, and glycine betaine by sulfate-reducing and methanogenic bacteria in marine sediments. Appl. Environ. Microbiol. **48**:719–725.

37. King, G.M., M.J. Klug, and D.R. Lovley. 1983. Metabolism of acetate, methanol, and methylated amines in intertidal sediments of Lowes Cove, Maine. Appl. Environ. Microbiol. **45**:1848–1853.

38. King, G.M., and W.J. Wiebe. 1978. Methane release from soils of a Georgia salt marsh. Geochim. Cosmochim. Acta **42**:343–348.

39. Larkum, A.W.D., G. Roberts, J. Kuo, and S. Strother. 1989. Gaseous movement in seagrasses. In A.W.D. Larkum, A.J. McComb, and S.A. Sheperd (eds.), Biology of Seagrasses, Elsevier Science Publishers, Amsterdam, pp. 686–722.

40. Ledwell, J.J. 1984. The variation of the gas transfer coefficient with molecular diffusivity. In W. Brutsaert and G.H. Jirka (eds.), Gas Transfer at Water Surfaces, D. Reidel, Hingham, MA, pp. 293–302.

41. Li, V.H., and S. Gregory. 1974. Diffusion of ions in sea water and in deep-sea sediments, Geochim. Cosmochim. Acta **38**:703–714.

42. Lovley, D.R., and M.J. Klug. 1983. Methanogenesis from methanol and methylamines and acetogenesis from hydrogen and carbon dioxide in the sediments of the eutrophic lake. Appl. Environ. Microbiol. **45**:1310–1315.

43. Martens, C.S., and R.A. Berner. 1974. Methane production in the interstitial waters of sulfate depleted marine sediments. Science **85**:1167–1169.

44. Martens, C.S., and J.V. Klump. 1980. Biogeochemical cycling in an organic rich coastal marine basin. I. Methane sediment exchange processes. Geochim. Cosmochim. Acta **44**:471–490.

45. McRoy, C.P., and C. McMillan. 1977. Production ecology and physiology of seagrasses. In C.P. McRoy and C. Helfferich (eds.), Seagrass Ecosystems: A Scientific Perspective, Marcel Dekker, New York, pp. 53–87.

46. Millero, F.J. 1986. The pH of estuarine waters. Limnol. Oceanogr. **31**:839–847.

47. Morse, J.W., J.J. Zullig, R.L. Iverson, G.R. Choppin, A. Mucci, and F.J. Millero. 1987. The influence of seagrass beds on carbonate sediments in the Bahamas. Marine Chem. **22**:71–83.

48. Oremland, R.S. 1975. Methane production in shallow-water, tropical marine sediments. Appl. Microbiol. **30**:602–608.

49. Oremland, R.S., L.M. Marsh, and S. Polcin. 1982. Methane production and simultaneous sulfate reduction in anoxic salt marsh sediments. Nature (London) **296**: 143–145.

50. Oremland, R.S., and S. Polcin. 1982. Methanogenesis and sulfate reduction: competitive and noncompetitive substrates in estuarine sediments. Appl. Environ. Microbiol. **44**:1270–1276.

51. Oremland, R.S., and B.F. Taylor. 1977. Diurnal fluctuations of O_2, N_2, and CH_4 in the rhizosphere of *Thalassia testudinum*. Limnol. Oceanogr. **22**:566–570.

52. Oremland, R.S., and B.F. Taylor. 1978. Sulfate reduction and methanogenesis in marine sediments. Geochim. Cosmochim. Acta **42**:209–214.

53. Perkins, R.D., and P. Enos. 1968. Hurricane Betsy in the Florida–Bahama area—geologic effects and comparison with Hurricane Donna. J. Geol. **76**:710–717.

54. Phillips, R.C., and C.P. McRoy. 1980. Handbook of Seagrass Biology: An Ecosystem Perspective. Garland STPM Press, New York.

55. Robblee, M.B., T.R. Barber, P.R. Carlson, Jr., M.J. Durako, J.W. Fourqurean, L.K. Muehlstein, D. Porter, L.A. Yarbro, R.T. Zieman, and J.C. Zieman. 1991. Mass mortality of the tropical seagrass *Thalassia testudinum* in Florida Bay (USA). Marine Ecol. Prog. Ser. **71**:297–299.

56. Santschi, P.H., R.F. Anderson, M.Q. Fleisher, and W. Bowles. 1991. Measurements of diffusive sublayer thickness in the ocean by alabaster dissolution, and their implications for the measurements of benthic fluxes. J. Geophys. Res. **96**:10641–10657.

57. Schomer, N.S., and R.D. Drew. 1982. An ecological characterization of the lower Everglades, Florida Bay, and Florida Keys. U.S. Fish Wildlife Serv. Biol. Serv. Program FWS/OBS-82/58.1.

58. Scoffin, T.P. 1970. The trapping and binding of subtidal carbonate sediments by marine vegetation in Bimini Lagoon, Bahamas. J. Sediment. Petrol. **40**:249–273.

59. Sebacher, D.I., R.C. Harriss, and K.B. Bartlett. 1985. Methane emissions to the atmosphere through aquatic plants. J. Environ. Qual. **14**:40–46.

60. Smith, R.D., W.C. Dennison, and R.S. Alberte. 1984. Role of seagrass photosynthesis in root aerobic processes. Plant Physiol. **74**:1055–1058.

61. Stockman, K.W., R.N. Ginsburg, and E.A. Shinn. 1967. The production of lime mud by algae in South Florida. J. Sediment. Petrol. **37**:633–648.

62. Swart, P.K., D. Berler, D. McNeill, M. Guzikowski, S.A. Harrison, and E. Dedick. 1989. Interstitial water geochemistry and carbonate diagenesis in the subsurface of a holocene mud island in Florida Bay. Bull. Marine Sci. **44**:490–514.

63. Thayer, G.W., K.A. Bjorndal, J.C. Ogden, S.L. Williams, and J.C. Zieman. 1984. Role of large herbivores in seagrass communities. Estuaries **7**:351–376.

64. Thursby, G.B. 1984. Root-ended oxygen in the aquatic angiosperm *Ruppia maritima*. Marine Ecol. Prog. Ser. **16**:303–305.

65. Vanderborght, J.P., R. Wollast, and G. Billen. 1978. Kinetic models of diagenesis in disturbed sediments. 1. Mass transfer properties and silicate diagenesis. Limnol. Oceanogr. **22**:787–793.

66. Wanless, H.R., and M.G. Tagett. 1989. Origin, growth, and evolution of carbonate mudbanks in Florida Bay. Bull. Marine Sci. **44**:454–489.

67. Weiss, R.F. 1974. Carbon dioxide in water and seawater: the solubility of a non-ideal gas. Marine Chem. **2**:203–215.

68. Wetzel, R.G., and P.A. Penhale. 1979. Transport of carbon and excretion of dissolved organic carbon by leaves and roots/rhizomes in seagrasses and their epiphytes. Aquat. Bot. **6**:149–158.

69. Whitcomb, J.H., R.D. DeLaune, and W.H. Patrick, Jr. 1989. Chemical oxidation of sulfide to elemental sulfur: its possible role in marsh energy flow. Marine Chem. **26**:205–214.

70. Wiesenburg, D.A., and N.L. Guinasso, Jr. 1979. Equilibrium solubilities of methane, carbon monoxide, and hydrogen in water and seawater. J. Chem. Eng. Data **24**:356–360.

71. Zieman, J.C. 1982. The ecology of the seagrasses of south Florida: a community profile. U.S. Fish Wildlife Serv. Biol. Serv. Program FWS/OBS-82/25.

72. Zieman, J.C., J.W. Fourqurean, and R.L. Iverson. 1989. Distribution, abundance, and productivity of seagrasses and macroalgae in Florida Bay. Bull. Marine Sci. **44**:292–311.

73. Zieman, J.C., and R.T. Zieman. 1989. The ecology of seagrass meadows of the west coast of Florida: a community profile. U.S. Fish Wildlife Serv. Biol. Rep. 85(7.25).

The Oceanic Distribution of Methane and Its Flux to the Atmosphere over Southern California Waters

Frank J. Cynar and A. Aristides Yayanos

Scripps Institution of Oceanography, University of California, San Diego, La Jolla, California 92093-0202

ABSTRACT

Mapping the distribution of methane in the oceans can help to discern the influence of various processes that produce or remove it. In this chapter, we present results from surveys of the distribution of methane in the upper 300 m of the Southern California Bight and estimate its ventilation to the atmosphere. Data from 150 profiles over a period of 22 months and 6 oceanographic cruises revealed water with 0.9–530 times the atmospheric-equilibrium concentration of methane. Methane maxima in vertical profiles occurred directly below the steepest gradient of the pycnocline. The horizontal distribution of plumes of CH_4-enriched waters extended along a narrow density layer that was characterized by a maximum in hydrostatic stability. These results suggested an offshore transport of methane from the coast along isopycnal surfaces. The annual (1989–1990) estimate of the flux of CH_4 to the atmosphere was 37 mg CH_4 m^{-2}. The majority of the gas flux occurred near the coast with a maximum temporal and spatial variability located in the waters surrounding Point Conception. Extrapolating the mean flux estimates for waters of southern California to the global area of coastal oceans resulted in an annual flux of 1 Tg CH_4, or roughly one-quarter of the estimated global flux. The variability of coastal processes, seasonal or otherwise, were thought to influence the inventory of methane and its rate of exchange with the atmosphere.

Introduction

The observed supersaturation of methane in surface waters[5,9,20,28,35] is the result of an interplay of supply and removal processes. The potential sources of methane include the atmosphere, rivers, buried methane clathrates, submarine hydrocarbon seeps, anaerobic bacterial metabolism in sediments or anoxic basins, and methanogenic bacterial activity within the water column itself. The sinks of methane include microbial and chemical oxidation in the water column and sediments and ventilation to the atmosphere. The net balance of the various rates of production and loss determines the inventory of methane in the ocean, yet many of these terms cannot yet be estimated directly or without large uncertainties. In addition, the lack of sufficient methane distribution data and the dynamic nature of the marine environment often preclude an adequate analysis of the processes controlling the cycle of methane in the ocean. This chapter summarizes information collected as part of a systematic investigation into the distribution of methane across the upper 300 m of the Southern California Bight. Here, we describe the association between methane, density, and water column stability and estimate the sea–air flux of methane.

The Southern California Bight encompasses the coastal ocean from San Diego to Point Conception. The eastern boundary current of the central Pacific gyre flows southward along the California coast, forming the meandering branches and mesoscale eddies of the California Current System. A seasonal cycle of intensified northerly winds along the coast results in the upwelling of water near the coast, with a replacement flow from shoreward currents at depth.[16,23,27] Hydrographic variability in the California Current System is seasonal.[22,23] The submarine topography is partitioned by deep basins, eight islands, and several offshore banks.

Methods

Dissolved methane was determined during six oceanographic cruises of the California Cooperative Oceanic Fisheries Investigations Program (CalCOFI) aboard RV *New Horizon* and RV *David Starr Jordan* over a 22-month period. The number of stations sampled and the dates of each cruise are listed in Table 1. Methane was profiled at a subset of the hydrographic stations regularly occupied during the CalCOFI cruises; the nominal hydrographic station locations are illustrated in Figure 1. During the first three cruises (8901, 8904, 8907), only a subset of stations along the two southernmost east–west trending lines were sampled, whereas during the next three cruises (8911, 9003, 9011), a subset of the stations along all of the six lines were sampled for methane.

Water for methane analyses was obtained from bottle casts at each station. Five-liter plastic (PVC) sampling bottles, equipped with reversing thermometers, were subsampled for methane immediately after returning to the surface. Sampling depths for cruise 8901 were restricted to the upper 150 m, whereas subsequent cruises sampled the upper 300 m (Table 1). Approximately 12 depths were sampled during each cast within these depth ranges. Replicate 300-ml samples were drawn from each sampling depth, poisoned with Na-azide, and analyzed within 1 day. Methane was extracted using a vacuum extraction system and quantified using a gas chromatograph equipped with a flame-ionization

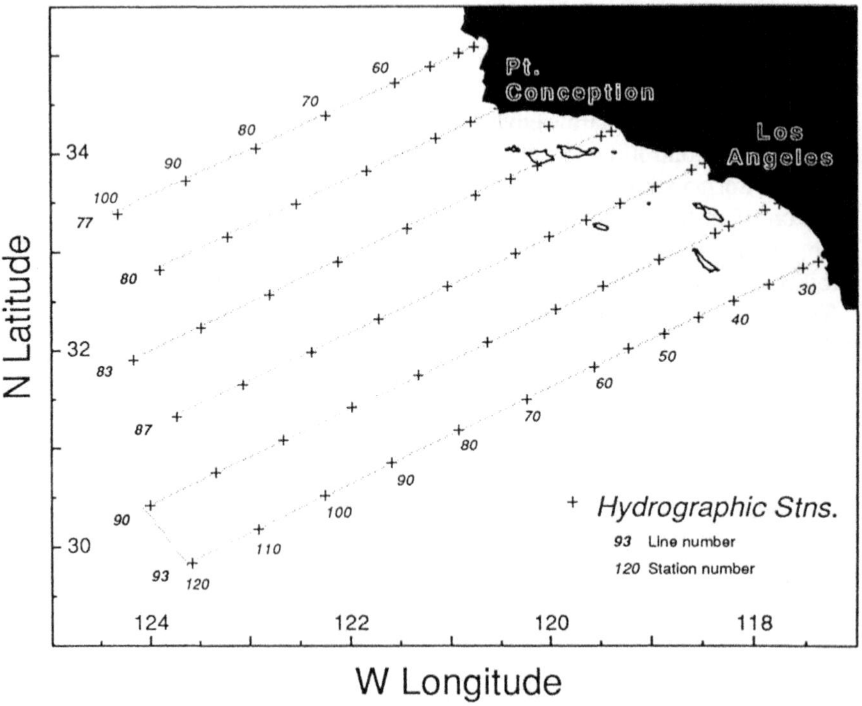

Figure 1. Nominal location of hydrographic stations regularly sampled during CalCOFI cruises. Bottle casts at a subset of the stations were analyzed for CH_4 during six cruises between Jan. 1989 and Nov. 1990.

detector; detection threshold and peak resolution were 0.1 nmol CH_4 L^{-1} during shipboard determinations. Detailed descriptions of the procedures employed in the sampling, extraction, and analysis of methane aboard ship are reported elsewhere.[11]

Salinity and temperature, in addition to dissolved O_2, nutrients, and other computed parameters, such as density and dynamic height, were determined for all depths sampled by scientists participating on the CalCOFI cruises;[30–33] the methods used to measure and compute these parameters are discussed in the references. The density of seawater was computed separately for the calculation of the buoyancy frequency using the 1980 International Equation of State algorithm.[24] The potential density (density$_{(s,t,0)}$) of a parcel of water was calculated at the in situ salinity, the potential temperature, and a relative pressure of 0 atm. For convenience, it is reported in sigma notation, defined as sigma-t = (density$_{(s,t,0)}$ − 1.0) × 10^3.[26]

Table 1. Dates, depth ranges, and the number of stations sampled for methane during six oceanographic cruises.

Cruise (yymm)	Dates	No. Stns Sampled	Nominal Depths (m)
8901	20–26 Jan. 1989	11	0–150
8904	17–22 Apr. 1989	13	0–300
8907	18–24 July 1989	17	0–300
8911	6–20 Nov. 1989	35	0–300
9003	4–19 Mar. 1990	44	0–300
9011	5–19 Nov. 1990	30	0–300

Dissolved methane is reported as an absolute concentration in Figures 2 and 3, and as the difference between the mean measured concentration and the computed concentration in equilibrium with the environmental *T* and *S*, and the concentration of methane in the atmosphere for Figures 4 and 5. The latter quantity denotes the methane saturation anomaly (MSA, nmol L^{-1}), and is defined here as

$$MSA = [CH_4]_{obs} - [CH_4]_{equ},$$

where $[CH_4]_{obs}$ = mean of replicate determinations; $[CH_4]_{equ}$ = concentration in equilibrium at in situ *T*, *S*, and an assumed atmospheric mixing ratio. Equilibrium solubilities were estimated using an atmospheric mixing ratio of 1.77 parts per million (v/v) (Ref. 2; P. Lang, NOAA, personal communication) and the appropriate solubility equations.[37] Values of MSA are reported as the mean integrated concentration over a specified depth interval.

The Brunt–Väisälä frequency (*N*) or buoyancy frequency is a measure of the static stability of the water column.[25] This parameter was computed for all depths from the temperature, salinity, and pressure data. The buoyancy frequency[26] (*N*) was computed as N^2,

$$N^2 = g\left(-\frac{1}{\rho}\frac{\delta\rho}{\delta z} - \frac{g}{C^2}\right),$$

where g = acceleration of gravity, 9.81 m s^{-2}; $\rho_{(s,t,p)}$ = in situ density (Ref. 24); z = depth; C = sound velocity calculated as a function of *T*, *S*, *P* (Ref 6).

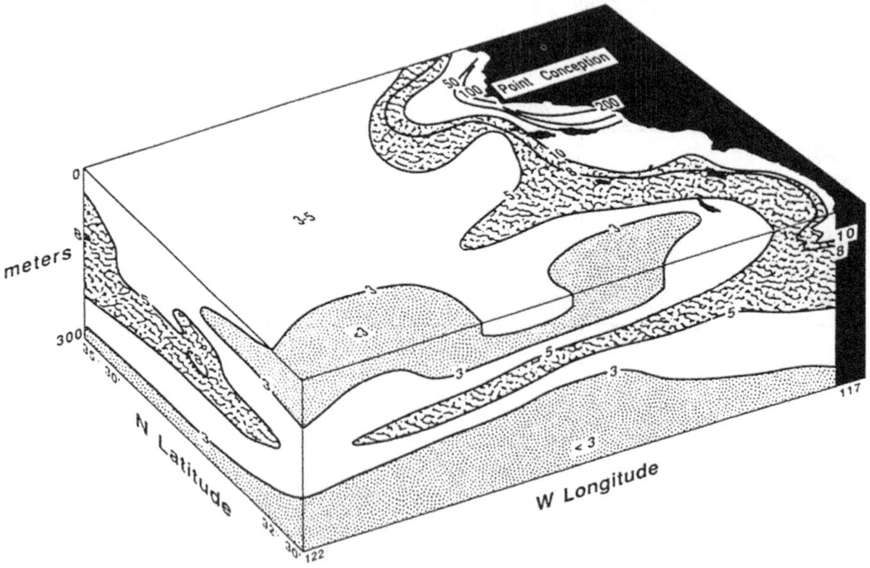

Figure 2. A three-dimensional perspective of the distribution of CH_4 across the sea surface and two perpendicular sections during March 1990. Contours along the horizontal and vertical sections denote levels of methane (nmol CH_4 L^{-1}). The contoured upper surface reflects the CH_4 concentration at the surface. The two vertical sections (0–300 m) display the concentration of CH_4 along the southern and western boundaries of the surveyed area.

Calculations of Brunt-Väisälä frequency (N) for cruises 8911, 9003, and 9011 are plotted as N^2 in Figure 3.

The location and intensity of the geostrophic currents charted in Figures 4A and 5A were computed from the prevailing temperature (T) and salinity (S) fields of each cruise, as reported.[30–33]

Sea–air fluxes of CH_4 were estimated with the boundary layer model of Liss and Merlivat.[21] This model estimates the rate of gas exchange as the product of the transfer velocity and the concentration gradient of the gas across the boundary layer. The model is expressed mathematically as

$$\text{flux} = k_w \, \Delta C,$$

where flux = sea–air flux for methane (mol cm^{-2} h^{-1}); k_w = transfer velocity (cm h^{-1}) is computed as a function of the wind speed at a height of 10 m above the sea surface and the Schmidt number for methane; ΔC = concentration gradient across the sea surface. The Schmidt number for CH_4 (Sc_{CH_4}), defined as the

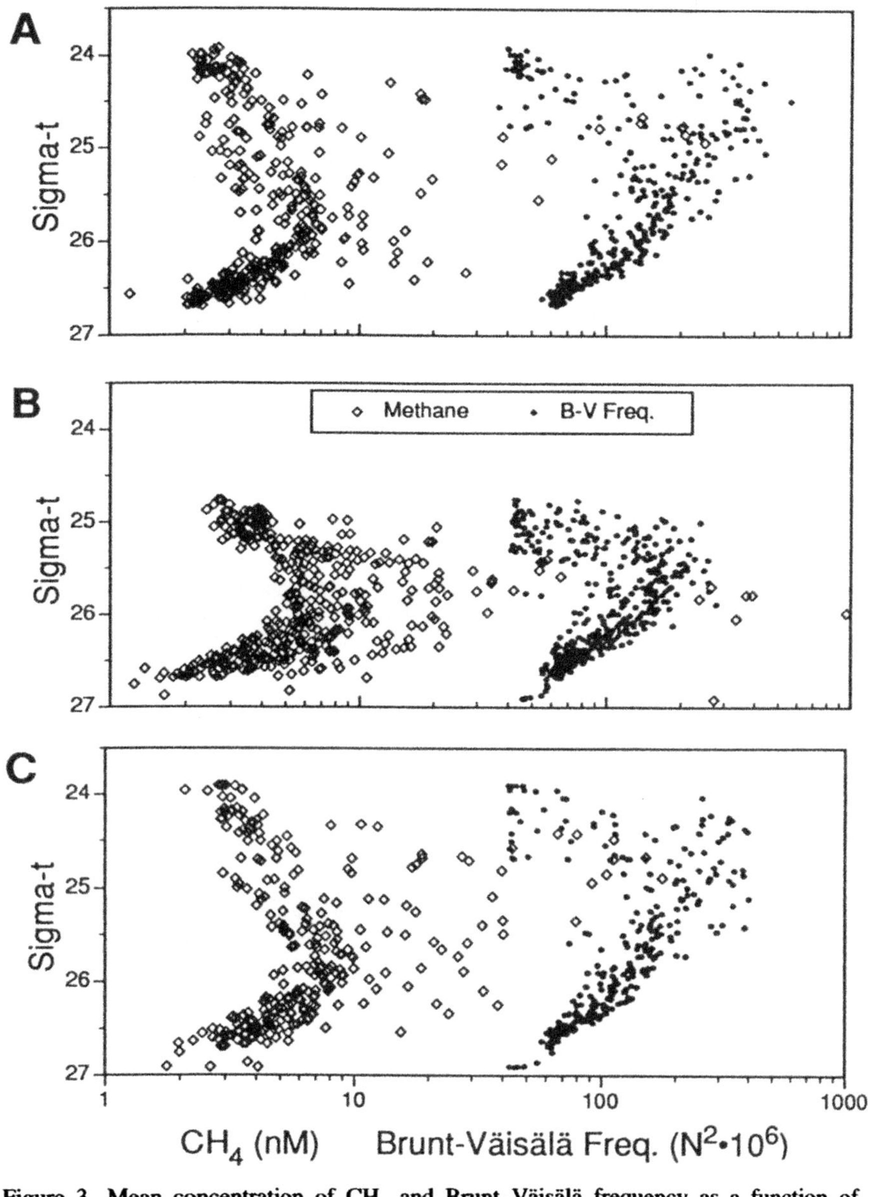

Figure 3. Mean concentration of CH_4 and Brunt–Väisälä frequency as a function of potential density (expressed as sigma-theta) for (A) cruise 8911, (B) cruise 9003, and (C) cruise 9011. The mean of replicate methane determinations (nmols L^{-1}) and Brunt–Väisälä frequency (as $N^2 \times 10^6$) are plotted on a log axis (abscissa).

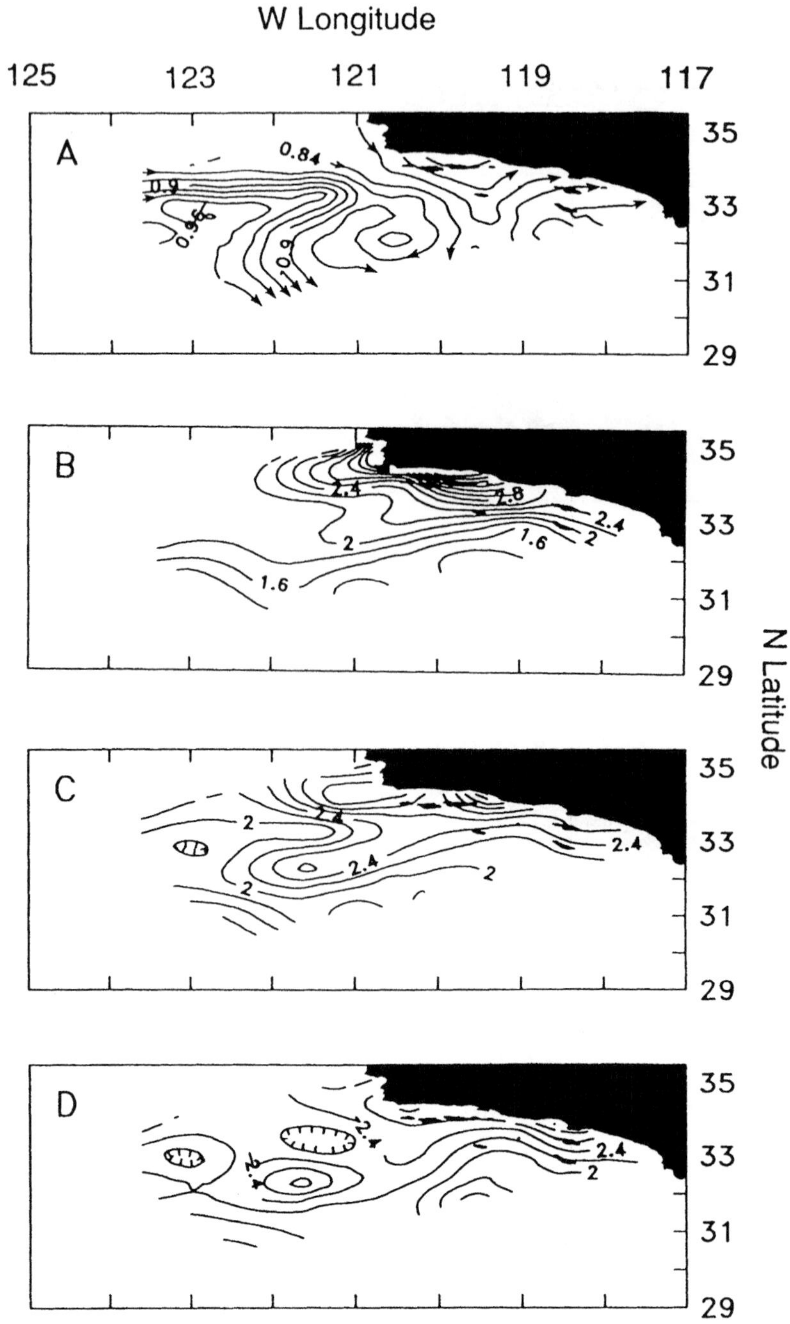

Figure 4. Dynamic topography and the horizontal distribution of the methane saturation anomaly (MSA) for three depth regions during cruise 9003. (A) Contours of equal dynamic height denote the dynamic topography of the upper 500 m (0 over 500 dB), (B) log-transform of the methane saturation anomaly (μmol CH_4 m^{-2}) for the 0–50-m depth range, (C) for the 50–100-m depth range, and (D) for the 100–150 m depth range. Current trajectories in panel A are parallel to the contours and indicated with arrows; the magnitude of current velocities increase with decreasing contour spacing.

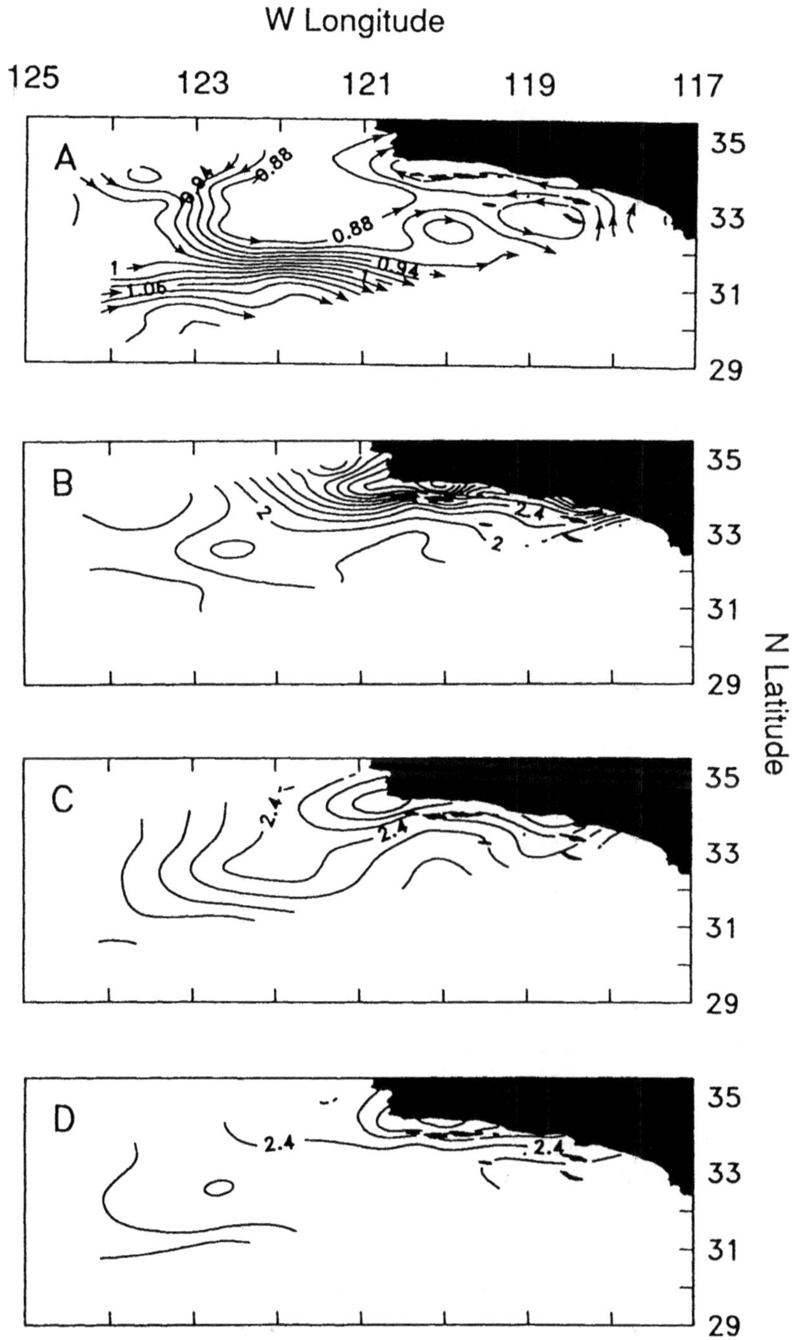

Figure 5. As in Figure 4, except for cruise 9011.

ratio of the kinematic viscosity of water to the molecular diffusivity for CH_4, was computed as a function of the surface temperature and salinity using the algorithm for viscosity[12] and the algorithm for the molecular diffusion coefficient of methane.[17] The Schmidt number was scaled as $Sc^{-2/3}$ for wind speeds < 3.6 m s^{-1} and scaled as $Sc^{-1/2}$ for higher wind speeds.[17] The variation of the transfer velocity (k_w, cm h^{-1}) for a gas with Sc = 600 to the 10-m wind speed (u_{10}, m s^{-1}) are as in Liss and Merlivat,[21]

$k_w = 0.17u_{10}$	$u_{10} \leq 3.6$ m s^{-1}	$Sc^{-0.66}$
$k_w = 2.85u_{10} - 9.65$	$3.6 \leq u_{10} \leq 13$ m s^{-1}	$Sc^{-0.5}$
$k_w = 5.9\ u_{10} - 49.3$	$u_{10} > 13$ m s^{-1}	$Sc^{-0.5}$

Seasonal variation in the concentration of methane in the atmosphere was not considered in the calculation of gas fluxes; the concentration of methane in the atmosphere was assumed a constant (1.77 ppmv). Mean wind speeds were derived as the average 10-m wind speed for each cruise and station during a 3-month period. Nearly 9000 observations from >40 years of CalCOFI surveys were used for the computation of mean wind speeds. The range of values for Sc, k_w, ΔC, and sea–air flux, computed by region and by cruise, are listed in Table 2.

Topographic plots of vertical sections (Figure 2) and horizontal surfaces (Figures 2, 4, and 5) were generated from mapping the irregularly spaced station data to a grid using a minimum curvature algorithm.[3] Maps of computer-generated contours agreed with maps drafted manually. Irregularly spaced station data (sea-surface *T*, *S*, and $[CH_4]_{obs}$) were interpolated onto uniform grids. The sea–air flux of CH_4 was calculated at each grid element and the planar areas of the grid within specified subranges were integrated over the range of fluxes. Integrated areal estimates of the sea–air flux of methane were computed for cruises 8911, 9003, and 9011.

Results

Dissolved methane varied over three orders of magnitude, from <0.2 to >1400 nmol L^{-1}. The error associated with replicate determinations of methane at sampling depths averaged <6%. One-hundred fifty stations and >3600 methane samples were extracted during the six cruises. In the subsections that follow, a subset of the vertical and horizontal distributions of methane are described as they relate to vertical and horizontal distribution patterns, water movements, and rates of gas exchange with the atmosphere.

The Vertical and Horizontal Distribution of CH_4

The distribution of methane during cruise 9003 is presented across three perpendicular faces of a rectangular box extending between 32–35°N latitude and

Table 2. Range of sea–air flux model parameters and estimates by date and region.

Cruise Region	Schmidt Number		$[CH_4]$equ. (nmol L^{-1})		Transfer Vel. (cm hr^{-1})		$[CH4]_{surf}$ (nmol L^{-1})		Sea–Air Flux (nmol cm^{-2} d^{-1})		Obs.
	min	max	min	max	min	max	min	max	min	max	*n*
90–93 Southern											
8901	782	975	2.3	2.6	0.4	8.7	3.0	4.4	0.01	0.4	11
8904	704	855	2.2	2.4	0.5	13.4	2.9	7.3	0.06	0.6	13
8907	610	754	2.1	2.3	0.5	10.0	2.7	7.0	0.01	0.2	17
8911	666	756	2.2	2.3	0.4	9.9	2.8	9.6	0.01	0.2	15
9003	820	903	2.4	2.5	4.6	19.2	2.8	18.6	0.01	0.2	18
9011	650	736	2.2	2.3	0.5	6.5	2.9	4.0	0.01	0.3	6
83–87 Middle											
8911	699	818	2.2	2.4	0.5	11.1	2.2	34	−0.03	0.4	12
9003	855	967	2.4	2.6	8.5	19.6	3.1	247	0.10	17.4	15
9011	680	787	2.2	2.4	0.5	15.3	2.7	1202	0.02	14.9	12
77–82 Northern											
8911	721	836	2.3	2.4	2.6	9.6	2.6	203	0.03	46.1	8
9003	842	1043	2.4	2.7	9.0	18.2	3.7	1121	0.20	209.	11
9011	739	818	2.3	2.4	2.6	10.0	2.9	89	0.13	19.5	12

117–122°W longitude (Figure 2). Dissolved methane at the 5-m isobath was contoured across the horizontal plane of the rectangular volume. Methane values within the 0–300-m depth range were contoured along the southern and western edges of the surveyed region. Methane ranged from <3 nmol L^{-1} in the upper mixed layer along the southern boundary to >200 nmol L^{-1} in the waters adjacent to Point Conception. The vertical sections portrayed near-surface to subsurface maxima in methane throughout the region for all cruises (data not shown). The depth and magnitude of the subsurface maxima in CH_4 increased with increasing distance from shore. Steep horizontal gradients of methane were observed along the coast with a maximum, in the range of 200–1400 nmol CH_4 L^{-1}, often located in the Santa Barbara Channel region; this area is located immediately south of Point Conception (Figure 1).

Methane, Density, and Hydrostatic Stability

The relationship between methane, density, and an estimate of water column stability for cruises 8911, 9003, and 9011 are plotted in Figures 3A–3C. These plots were drawn with less dense water (sigma-*t* = 24) at the top of the *y*-axis, which corresponds with the observation of lower density water at the surface of the ocean. Water sampled for methane during cruise 8911 (Figure 3A; *n* = 299) and 9011 (Figure 3C; *n* = 343) exhibited a range of sigma-*t* between 23.5 and 27, and for cruise 9003 (Figure 3B; *n* = 439) a sigma-*t* of 24.5–27. The relative distribution of the minimum methane values along the density axis of each panel (Figures 3A–3C) formed a crescent-shaped pattern. This pattern was clearly present for cruises 9003 and 9011, where all water with sigma-*t* = 26.0 contained >4 nmol CH_4L^{-1} (Figures 3A and 3C). During all three cruises, the highest methane concentrations, >100 nmol L^{-1}, occurred within a relatively narrow density band (Figures 3A–3C).

The Brunt–Väisälä frequency (*N*) was computed for the same water depths. The distribution of N^2 also formed crescent-shaped patterns as a function of density (Figures 3A–3C). During cruises 8911 and 9011, $N^2 > 300$ occurred in the 24–25.5 sigma-*t* range, whereas a similar pattern was not observed during cruise 9003.

Water Movements and the Distribution of Methane

Charts of dynamic topography (Figures 4A and 5A) along with maps of the methane saturation over three depth intervals (Figures 4B–4D and 5B–4D) allowed comparison of the distribution of geostrophic currents with the horizontal distribution of methane for cruises 9003 and 9011. The location and relative intensity of the geostrophic currents were charted for cruises 9003 and 9011 (Figures 4A and 5A); these charts illustrate the dominant density-driven currents in the 0–500-m layer relative to motion at 500 m. These charts, which were computed from density distributions, do not reflect the near-surface water movements generated by winds or tides. The dominant features of the dynamic topography indicated the location and relative intensity of the equatorward California Current in the offshore waters. The position of this current was variable between the cruises.

The distribution of the log-transform of the integrated methane saturation anomaly (MSA) is displayed for three depth ranges (0–50, 50–100, 100–150 m) for cruises 9003 and 9011 (Figures 4B–4D and 5B–5D). Water at saturation with respect to methane in the atmosphere was estimated to contain 2.4–2.95 (mean = 2.66 nmol CH_4 L^{-1}) for cruise 9003 and 2.15–2.95 (mean = 2.58 nmol CH_4 L^{-1}) for cruise 9011. The log MSA = 2.4 contour, portrayed in Figures 4B–4D and Figures 5B–5D divides the regions with >250 nmol of CH_4 in excess of saturation for each depth interval of the water column; the 2.4 contour is equivalent to a mean water column concentration of 5 nmol CH_4 L^{-1} in excess of saturation. The MSA integrated over the 0–50-m depth region portrayed a minimum along the southern edge (0.32 nmol CH_4 L^{-1} in excess of saturation) and a maximum along the northern coast (>900 nmol CH_4 L^{-1}) for cruise 9003 (Figure 4B). The mean excess methane in the 50–100-m depth layer ranged between 0.5 and 71 nmol CH_4 L^{-1} (Figure 4C) and between 0.6 and 19 nmol CH_4 L^{-1} in the 100–150-m region (Figure 4D). The minima and maxima MSA for cruise 9011 were 0.6 and 390, 0.8 and 43, and 1.5 and 25 nmol CH_4 L^{-1} for the 0–50, 50–100, and 100–150-m depth regions, respectively (Figures 5B–5D).

Sea–air Flux Estimates

Temperature, salinity, and methane concentration data from within 3 m of the surface, and mean wind speed at a height of 10 m were used to estimate the sea–air flux of methane at 150 stations for the six cruises. Sea–air flux estimates were grouped by latitudinal regions; the six east–west trending lines of stations shown in Figure 1 were divided into three groupings. The range of the model parameters (Schmidt number, transfer velocity, $[CH_4]_{equ}$, $[CH_4]_{surf}$) and the resulting estimates of sea–air methane flux are listed by region in tabular form for each of six cruises (Table 2); lines 90 and 93 correspond to the two southernmost transects, lines 83 and 87 correspond to the mid-region, and lines 77 and 80 correspond to the northernmost region. Estimates of the sea–air flux of CH_4 ranged from −0.05 to 435 nmol cm^{-2} d^{-1}. The spatial and temporal variation of individual flux estimates between 0.05 and 350 nmol cm^{-2} d^{-1} are portrayed on semi-log plots, by region, as a function of longitude in Figures 6A–6C.

Integrated areal estimates of the sea–air flux of methane are plotted for cruises 8911, 9003, and 9011 (Figure 7). Topographic maps of sea–air fluxes generated by interpolation of temperature, salinity, and CH_4 fields were consistent with individual flux estimates calculated from the irregularly spaced station data. Flux estimates were computed for the area of the ocean within the bounds of the CH_4 observations (187,000–213,000 km^2). Mean integrated fluxes of 6, 8, and 5 μmol CH_4 m^{-2} d^{-1} were estimated for cruises 8911, 9003, and 9011, respectively. The range of fluxes estimated for coastal and offshore areas are listed in Table 3 along with published estimates from other waters.

Discussion

Methane Inventory, Potential Sources, and Its Transport to Offshore Waters

The upper ocean off southern California contained dissolved methane in the range of 0.9–530× saturation. A subsurface layer with a maximum in methane saturation extended from the coast to >500 km offshore (Figure 2). During cruise 9003, the CH_4 content of the subsurface layer was greater in the northern

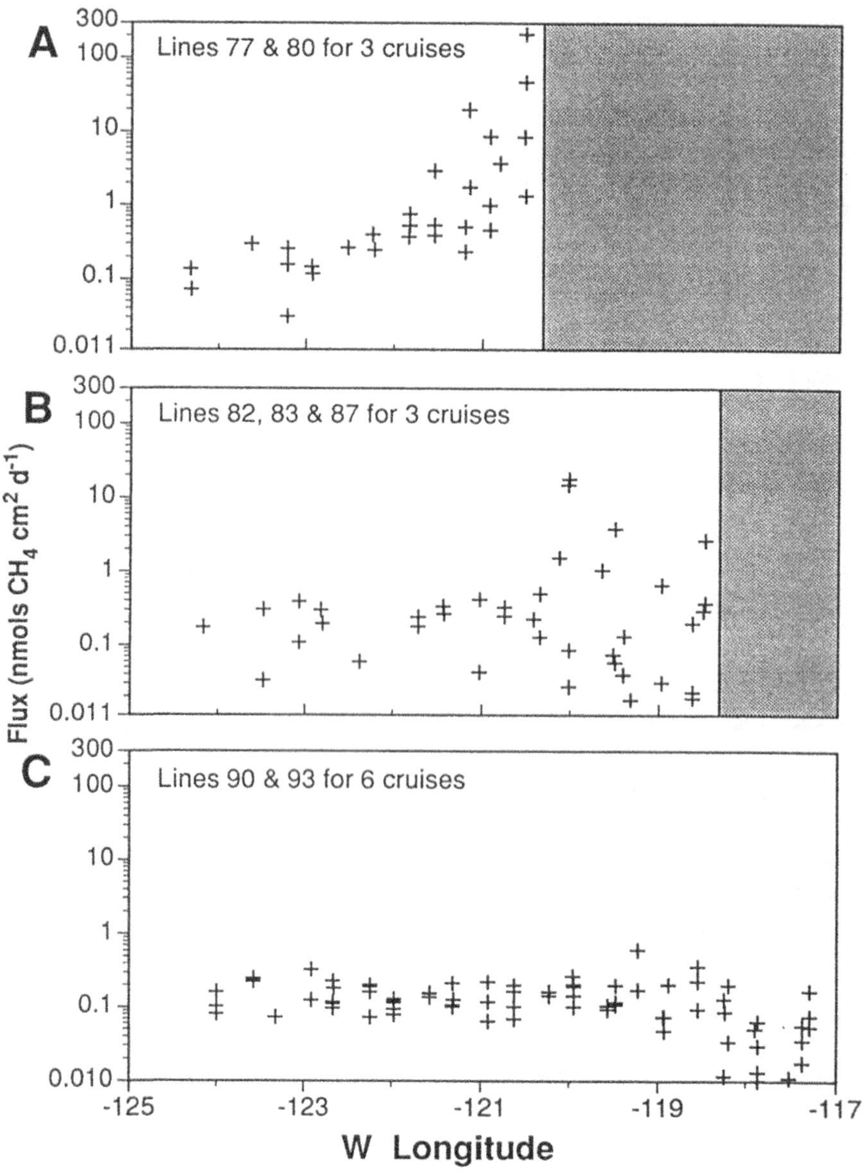

Figure 6. Sea–air flux estimates for six cruises are plotted against longitude and grouped by latitudinal region. Individual estimates of the sea–air flux (nmol CH_4 cm^{-2} d^{-1}) are plotted on a log axis (ordinate) as a function of longitudinal position (117–125°W Long.). (A) Estimates from stations in the northernmost region (lines 77, 80), (B) mid-region (lines 83, 87), and (C) southernmost region (lines 90, 93) are displayed. The shaded regions of the panels indicate land.

Table 3. Estimates of the sea–air flux of CH_4 for southern California waters and other oceans.

Region		Sea–Air Flux ($mg\ CH_4\ m^{-2}\ yr^{1}$)	Ref.
Pacific (S. California)			
Nov. 1989–Nov. 1990	mean	37	This work
Oceanic: 124°W Long.	mean	9	
Coastal: Pt. Conception	max	12,213	
S.E. Bering Sea			
May 1981	min	21	8
Oct. 1980	max	340	
	mean	140	
Caribbean Sea			
Cariaco Basin		1.3	36
Atlantic Oceans			
50°N to 35°S	mean	5	9
	max	27	
N. Atlantic: 19°N		1.3	28

region relative to the southern region. A layer of water with a minimum methane saturation was commonly present at depths >250 m and at the surface in offshore regions. This pattern was observed in the majority of profiles observed in southern California waters during cruises 8911 and 9003, as reported elsewhere.[11]

The Relationship Between Dissolved Methane and Water Column Stability

The highest levels of methane were found just below the sharpest gradient of the pycnocline. Hydrostatic stability, as defined by the Brunt–Väisälä frequency (N) or buoyancy frequency, was maximal at the pycnocline. A slight asymmetry between the location of maximal CH_4 and maximal stability (plotted as N^2) were indicated for cruises 8911, 9003, and 9011 (Figures 3A–3C). Methane maxima were characterized by a density slightly greater than the density of water exhibiting a maximum in static stability.

Above the pycnocline, the stability of the water column approached a minimum and methane saturation approached equilibrium with the atmosphere. This region corresponded to the well-mixed layer at the surface of the ocean located in offshore areas. At the surface, atmospheric and radiative processes tend to mix the upper water column, resulting in a well-mixed layer and an enhanced gas exchange at the air–sea interface.

Below the pycnocline, at depths where sigma-*t* >25.0, the N^2 and the CH_4 concentrations converged to minima; these minima lie in the 2–3 nmol CH_4 L^{-1} range for each of the three cruise dates (Figures 3A–3C). Samples with >50 nmol CH_4 L^{-1} were associated with the 24.5–25.0 density range for cruises 8911 and 9011 (Figures 3A and 3C). The crescent-like patterns generated from plots of methane as a function of density (Figures 3A–3C) suggested that physical or biological processes responsible for the maintenance of the subsurface maxima are enhanced or restricted to a discrete range of densities characterized by a maxima in hydrostatic stability.

A similar shape for the distributions of N^2 and CH_4 as a function of density was evident for the November cruises of 1989 and 1990 (Figures 3A and 3C). Although the similarity may be only fortuitous, it may suggest that the distribution of methane in the upper water column, with respect to density and hydrostatic stability, was influenced by seasonal processes.

Water Movements and the Distribution of Methane

Near-shore coastal areas exhibited the highest concentrations and the steepest gradients of CH_4 (Figures 2, 4B–4D, and 5B–5D). The steep gradients of CH_4 surrounding Point Conception suggested a coastal source of methane to the region. This interpretation is also suggested by the common solution to a set of conservation equations, in which the distributions of salt, silicate, and phosphate were treated as conserved tracers of current trajectories. This approach was used to identify the likely sources of the CH_4-enriched waters found in near-surface waters located at Point Conception. This model was not constrained by geostrophy and carried the assumption that salt, SiO_3, and PO_4 were conservative tracers of water motion over the short distances between stations (50–75 km). This approach is similar to an inverse tracer problem.[39] During cruise 8911, the common (non-negative) solutions to the set of conservation equations, indicated that approximately 90% of the water from 11 m at Point Conception was likely transported from a depth of 11 m in the Santa Barbara Channel (to the south). This result suggested a northwesterly flow, which was consistent with the prevailing current direction derived from geostrophy. During cruise 9003, the model suggested that approximately 85% of the water from a depth of 21 m at Point Conception was transported from a depth of 72 m from the north. This solution, which indicated a southeasterly flow, was also consistent with the prevailing near-shore geostrophic currents (Figure 4A). To explain the large gradients of methane in this area, a new source of methane to the water column must be assumed, and based on the distances between Point Conception and neighboring stations, the location of the required source is constrained to within a distance of 50 km to the north and 75 km to the south. Caution must be exercised in interpreting these results since more than one set of solutions can satisfy the model. The solutions derived here, however, were consistent with the pattern of independently derived geostrophic currents. The model solutions suggested that near-surface water enriched in methane was transported from relatively shallow depths. Although methane-rich water from the bottom of the Santa Barbara basin (520 m) was included as a candidate source in the models, the solutions suggested that water from this depth region did not contribute to the water properties associated with the CH_4 maximum in the near-surface waters. This result was not surprising, considering that the stably stratified water column is resistant to mixing down to such depths.

The limited spatial resolution of the station data in the region surrounding Point Conception does not permit an unambiguous description of the source. The nature of the methane source (biogenic or abiogenic) could not be directly determined from this data. Steep gradients of CH_4 in a region of petroleum seeps and offshore oil operations, however, suggested a petrogenic origin for the methane. More than 60 seeps are known in the waters between Point Conception and Santa Monica.[38] The seeps in the area off Point Conception are associated with petroliferous sedimentary rocks. Seepages in this area have been commonly found where surfical sediments were relatively thin or absent[15]; the most productive seeps are located along fault lines and near the crests of anticlines.[14] Additional interpretations of this data have suggested a seepage origin for the methane in the California waters.[11] Similar origins for the high levels of CH_4 observed in the Gulf of Mexico shelf waters have been reported.[1,4] Reports of hydrocarbons from marine sediments off Alaska,[19] and various sites in the southern Pacific Ocean[18] suggest that the occurrences of hydrocarbons in the sediments of the global oceans are not uncommon. The distribution of petroleum deposits[40] suggests that large areas of the sea bottom have the potential for containing submarine hydrocarbon seeps.

The offshore distribution of methane in the upper 150 m of the water column was consistent with a coastal source of methane. The integrated MSA for the 50–100-m depth interval of cruises 9003 and 9011 portrayed a plume of methane saturation extending several hundred kilometers off the coast (Figures 4C and 5C). The patterns observed in the 0–50- and 100–150-m depth intervals also indicated that high concentrations of methane penetrate >400 km into offshore waters. The shape of the plume outlined by the 2.4 contour during cruise 9003 (Figure 4C) appears to have been influenced by the path of the southward flowing California Current (Figure 4A). During cruise 9011, the location of the plume outlined by the 2.4 contour for the same depth region (Figure 5C) may have been influenced by a shifting of the main core of the California Current away from the coast at more northerly latitudes (Figure 5A). The apparent association of subsurface CH_4 maxima with a narrow range of densities (Figure 3) and the lateral extension of plumes of methane from the coast (Figures 4 and 5) were suggestive of transport of methane from the coast along isopycnal surfaces. The close association of methane maxima with maxima in hydrostatic stability (Figure 3) suggest lateral mixing of waters by isopycnal processes. A microbial origin of the methane, however, is also a consistent explanation for a part of the methane supersaturation often found at the pycnocline.[5,9,28,29] Recent success in the cultivation and characterization of a marine methanogenic bacterium from plankton[10] suggests that microbial activities in the water column may contribute to the supersaturation of methane observed in coastal and oceanic waters. A potential biochemical pathway mediating CH_4 production by marine bacteria in the water column is reported elsewhere (Ref. 10; Cynar, and Yayanos. 1990. p. 310, Abstr. Amer. Soc. Microbiol. 1990, Anaheim, CA).

Ventilation of Methane to the Atmosphere

Estimated methane fluxes, as a function of longitude (Figure 6), portrayed spatial and temporal variability. The northernmost region off Point Conception exhibited the greatest spatial variability (Figure 6A), whereas the southernmost region extending southwest from San Diego portrayed the smallest temporal variability (Figure 6C). The largest diffusive gas flux (209 nmol CH_4 cm^{-2} d^{-1}) was estimated for a station near the coast in the northern latitudinal region (Figure 6A). In contrast, the flux maxima in the middle and southern regions occur 150–180 km from the coast (Figures 6B and 6C). Lower estimates of the mean wind speeds in the 150–180 km off the coast are responsible for the lower estimates of diffusive fluxes. With the exception of a negative flux (transport into the ocean) for the mid-region during cruise 9011, the flux of CH_4 in offshore waters varied over a small range (10–400 pmol CH_4 cm^{-2} d^{-1}). The relatively constant range of sea–air fluxes observed in the southernmost region (Figure 6C; Table 3) suggested that the influence of seasonal processes was diminished relative to the northern area adjacent to the hypothesized source region.

The potential uncertainty of the flux estimates due to the variability of the wind speeds was estimated. Of the parameters influencing the estimation of the sea–air flux, the magnitude of uncertainty associated with the wind speed has the greatest effect. To estimate the uncertainty of the wind speed on flux estimates, the variability of >2600 measurements of wind speed were incorporated into flux estimates of methane for the March 1990 cruise. The historical, areawide variability of wind speeds during the Feb.–Apr. period was 5.5 ± 3.6 m s^{-1} (± 1 standard deviation) over a 40-yr period extending from 1950 to 1990. On average, the flux computed with the wind speed decreased by one standard deviation was approximately 0–20% of the flux computed with the mean wind speed. The flux associated with the mean wind speed increased by one standard deviation was 200–400% of the flux computed with the mean wind speed. The magnitude and the variability of fluxes computed at lower wind speeds is not mirrored by the magnitude and variability of fluxes computed at the higher range of wind speeds. The asymmetry of the variability is derived from the nonlinear relation between transfer velocity and wind speed; the Liss and Merilvat[21] model of transfer velocity relates the magnitude of the transfer velocity to wind speed by three equations (see the section "Methods"), each with a different linear response to wind speed. The uncertainty associated with estimates of the sea–air flux, based on the parameterized wind speed variability, suggests the need for cautious interpretation of these results.

The majority of the sea surface (>90%) off southern California was characterized by a sea-to-air flux of <10 μmol CH_4 m^{-2} d^{-1} for cruises 8911, 9003, and 9011 (Figure 7A). The sea surface gas fluxes >10 μmol CH_4 m^{-2} d^{-1}, however, contributed >74% of the total integrated flux for each cruise (Figure 7B). These results were consistent with the longitudinal distribution of individual flux estimates described in Figures 6A–6C. The individual area–flux curves for cruise 9003 and 9011 were similar, whereas the area–flux curve for 8911 shows a shift to slightly lower values for gas fluxes <1 μmol CH_4 m^{-2} d^{-1} (Figure 7A). The direction of the shift was not inconsistent with the seasonal pattern of upwelling along the coast. Cynar and Yayanos[11] have suggested that the upwelling of water containing elevated concentrations of methane and other nutrients near the coast may contribute to elevated CH_4 levels in offshore waters. We suggest that both physical transport of methane-rich waters and stimulated microbial production of methane, as a result of this transport, are likely to occur as a result of periods of intensified coastal upwelling.

The mean integrated flux of CH_4 ranged between a minimum of 5–6 μmol m^{-2} d^{-1} for November 1989 and 1990 and a maximum of 8 μmol m^{-2} d^{-1} for March 1990. These estimates were used to extrapolate an annual (1989–1990) estimate of 37 mg CH_4 m^{-2}. Similarly, the extrapolated annual flux for the offshore regions was 9 and the maximum near-shore flux was 12,213 mg CH_4 m^{-2} (Table 3). The magnitudes of sea–air flux estimates for southern California waters were generally greater than other estimates for the open ocean, but within the maximum range (21–340 mg CH_4 m^{-2} yr^{-1}; Table 3) observed in the Bering Sea, a shallow shelf system. The mean flux of 3.1×10^{10} g CH_4 yr^{-1} for the whole of the southern California region (area = 2×10^{11} m^2) was comparable to the 4.6×10^{10}-g CH_4 yr^{-1} flux estimated for a 3.2×10^{11}-m^2 shelf region of the Bering Sea.[8] An estimated flux of 9 mg CH_4 m^{-2} yr^{-1}, for the more oligotrophic waters off California (124°W) was nearly sevenfold greater than the sea–air fluxes estimated for the subtropical North Atlantic Ocean and relatively oligotrophic waters of the Caribbean Sea (1.3 mg CH_4 m^{-2} yr^{-1}; Ref. 36). This 9-mg CH_4 m^{-2} yr^{-1} oceanic flux estimate, however, was approximately one-third the maximum flux (27 mg CH_4 m^{-2} yr^{-1}; Table 3) reported by Conrad and Seiler[9] in the Atlantic Ocean.

Assuming a mean sea–air flux of 37 mg CH_4 m^{-2} yr^{-1} (Table 3), the 2×10^{11} m^2 of surface area of this study would generate 0.04–0.6% of the 1.3–16.6

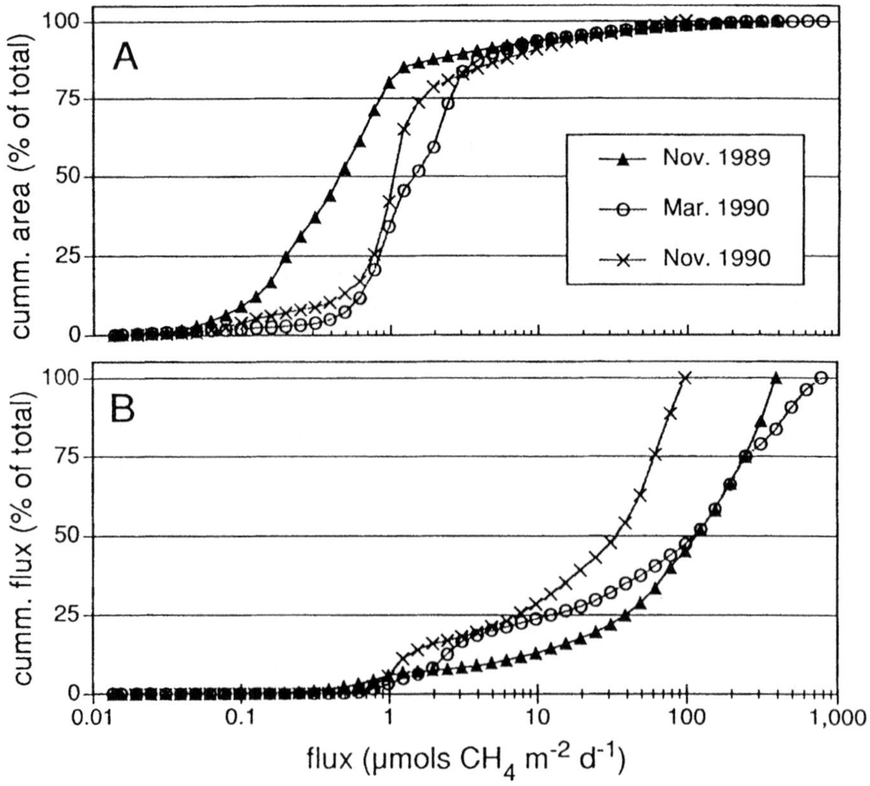

Figure 7. Areal integration of methane fluxes for cruises 8911, 9003, and 9011. (A) The cumulative area, expressed as a percent of the total of the sea surface, is plotted against the range of fluxes computed for three cruises (μmol CH_4 m^{-2} d^{-1}). (B) The cumulative methane flux, expressed as a percent of the total flux, is plotted against the range of fluxes. The shared abscissa denotes the range of methane fluxes on a log scale.

Tg CH_4 yr^{-1} flux suggested by Ehhalt.[13] Similarly, the flux estimate of 4.6×10^{10} g CH_4 yr^{-1} for the southeastern Bering Sea[8] is equivalent to an additional 0.3–3.5% of the global ocean flux. Together, southern California and the southeastern Bering Sea may be responsible for 0.3–4.1% of Ehhalt's [13] global estimate. Extrapolating the mean sea flux of 37 mg yr^{-1} to the global area of the coastal oceans (coastal ocean = 2.7×10^{13} m^2, Ref. 34), resulted in an annual flux of 1 Tg CH_4 for coastal seas. Similarly, extrapolating the 9-mg CH_4 m^{-2} yr^{-1} flux estimate for the more oligotrophic offshore waters to the global area of the open ocean (open ocean = 3.34×10^{14} m^2 Ref. 34) resulted in an annual flux of 3 Tg CH_4 for the open seas. Together, the annual global ocean source of methane was estimated at 4 Tg CH_4. Ignoring the potentially large uncertainties inherent with these extrapolations, the magnitude of the coastal flux is one-third of the open-ocean flux. If this interpretation is correct, processes that influence the production or air–sea exchange of methane in coastal environments will be important to the overall flux of methane from the oceans.

In comparison with existing estimates, the 4 Tg CH_4 yr^{-1} global ocean source computed here lies within the 1.3–16.6 Tg CH_4 yr^{-1} range estimated by Ehhalt[13] and slightly lower than the 5–20-Tg CH_4 yr^{-1} range of revised estimates.[7] In comparison to other global sources of methane, the oceans are thought to be a small source to the global atmosphere, approximately 2% of the global annual release (400–640 Tg CH_4) is attributed to the oceans.[7]

Acknowledgments

The authors express their gratitude to the captains and crews of R/V *New Horizon* and R/V *David Starr Jordan* for shipboard assistance. Several members of the CalCOFI group at Scripps Institution of Oceanography and the National Marine Fisheries Laboratory in La Jolla are gratefully acknowledged for collecting, calibrating, and processing the volumes of data. A special gratitude is extended to G. Hemingway, A. Mantyla, S. Gripp, and D. Newton for providing timely access to hydrographic and meteorological data, and Dr. M. Mullin and Dr. K. Smith for comments on the manuscript. This work was supported by a grant (to AAY) from the National Science Foundation (OCE 88-19601). Travel assistance from various funding agencies supporting the ISEB-10 is gratefully acknowledged.

References

1. Bernard, B.B., J.M. Brooks, and W.M. Sackett. 1978. Light hydrocarbons in recent Texas continental shelf and slope sediments. J. Geophys. Res. **83**:4053–4061.
2. Boden, T.A., P. Kanciruk, and M.P. Farrell. 1990. Trends '90: A compendium of data on global change. Oak Ridge National Laboratory, ORNL/CDIAC-36.
3. Briggs, I.C. 1974. Machine contouring using minimum curvature. Geophysics **39**: 39–48.
4. Brooks, J.M., B.B. Bernard, and W.M. Sackett. 1977. Input of low-molecular-weight hydrocarbons from petroleum operations into the Gulf of Mexico. In D.A. Wolfe (ed.), Fate and Effects of Petroleum Hydrocarbons in Marine Ecosystems and Organisms, Pergamon, Oxford, pp. 373–384.

5. Brooks, J.M., D.F. Reid, and B.B. Bernard. 1981. Methane in the upper water column of the Northwestern Gulf of Mexico. J. Geophys. Res. **86**:11029–11040.

6. Chen, C.-T., and F.J. Millero. 1977. Speed of sound in seawater at high pressures. J. Acoust. Soc. Amer. **62**:1129–1135.

7. Cicerone, R.J., and R.S. Oremland. 1988. Biogeochemical aspects of atmospheric methane. Global Biogeochem. Cycles **2**:299–327.

8. Cline, J.D., Charles N. Katz, and K. Kelly-Hansen. 1986. Seasonal cycles of dissolved methane in the Southeastern Bering Sea. In M.L. Sohn (ed.), Organic Marine Geochemistry, 305th ed., American Chemical Society, Washington, DC, pp. 272–296.

9. Conrad, R., and W. Seiler. 1988. Methane and hydrogen in seawater (Atlantic Ocean). Deep-Sea Res. **35**:1903–1917.

10. Cynar, F.J., and A.A. Yayanos. 1991. Enrichment and characterization of a methanogenic bacterium from the oxic upper layer of the ocean. Curr. Microbiol. **23**: 89–96.

11. Cynar, F.J., and A.A. Yayanos. 1992. The distribution of methane in the upper waters of the Southern California Bight. J. Geophys. Res. **97**:11269–11285.

12. Dietrich, G., K. Kalle, W. Krauss, and G. Siedler. 1980. General Oceanography: An Introduction. 2nd ed. J. Wiley and Sons, New York.

13. Ehhalt, D.H., and U. Schmidt. 1978. Sources and sinks of atmospheric methane. Pageoph. **116**:452–464.

14. Fischer, P.J. 1976. Late neogene-quaternary tectonics and depositional environments of the Santa Barbara Basin, California In A.E. Fritsche, H.T. Best, and W.W. Wornardt (eds.), Proceedings of the Annual Meeting of the Pacific Section of the Society of Economic Palaeontologists and Mineralogists, Society of Economic Palaentologists and Mineralogists, San Francisco, CA.

15. Fischer, P.J., and A.J. Stevenson. 1973. Natural hydrocarbon seeps along the northern shelf of the Santa Barbara Basin, California. Offshore Technology Conference, Paper 1738, Houston, TX.

16. Hickey, B.M. 1979. The California Current System—hypotheses and facts. Prog. Oceanogr. **8**:191–279.

17. Jahne, B., G. Heinz, and W. Dietrich. 1987. Measurement of the diffusion coefficients of sparingly soluble gases in water. J. Geophys. Res. **92**:10767–10776.

18. Kvenvolden, Keith A. 1988. Hydrocarbon gas in sediment of the Southern Pacific Ocean. Geo-Marine Lett. **8**:179–187.

19. Kvenvolden, K.A., and G.D. Redden. 1980. Hydrocarbon gas in sediment from the shelf, slope, and basin of the Bering Sea. Geochim. Cosmochim. Acta **44**:1145–1150.

20. Lamontagne, R.A., J.W. Swinnerton, and V.J. Linnenbom. 1974. C_1-C_4 hydrocarbons in the North and South Pacific. Tellus **26**:71–77.

21. Liss, P.S., and L. Merlivat. 1986. Air–sea gas exchange rates: Introduction and

synthesis. In P. Baut-Menard (ed.), The Role of Air–Sea Exchange in Geochemical Cycling, Reidel, Dordecht, pp. 113–127.

22. Lynn, R.J., K.A. Bliss, and L.E. Eber. 1982. Vertical and horizontal distributions of seasonal mean temperature, salinity, sigma-*t*, stability, dynamic height, oxygen and oxygen saturation in the California Current, 1950–1978. State of California Marine Research Commission, CalCOFI Atlas, 30.

23. Lynn, R.J., and J.J. Simpson. 1987. The California Current system; The seasonal variability of its physical characteristics. J. Geophys. Res. **92**:12947–12966.

24. Millero, F.J., C.-T. Chen, A. Bradshaw, and K. Schleicher. 1980. A new high pressure equation of state for seawater. Deep-Sea Res. **27**:255–264.

25. Pedlosky, J. 1987. Geophysical Fluid Dynamics, Springer-Verlag, New York, pp. 351–355.

26. Pond, S., and G.L. Pickard. 1983. Introductory Dynamical Oceanography, 2nd ed, Pergamon, Oxford, p. 329.

27. Reid, J.L., G.I. Roden, and J.G. Wyllie. 1958. Studies of the California Current System. Scripps Institution of Oceanography Progress Report 998.

28. Scranton, M.I., and P.G. Brewer. 1977. Occurrence of methane in the near-surface waters of the western subtropical North-Atlantic. Deep-Sea Res. **24**:127–138.

29. Scranton, M.I., and J.W. Farrington. 1977. Methane production in the waters off Walvis Bay. J. Geophys. Res. **82**:4947–4953.

30. Scripps Institution of Oceanography. 1989. Physical, chemical, and biological data report, CalCOFI cruise 8901, CalCOFI 8904. University of California, San Diego, Data Report SIO Ref 89–26.

31. Scripps Institution of Oceanography. 1990. Physical, chemical, and biological data report, CalCOFI cruise 8907, CalCOFI 8908, and CalCOFI cruise 8911. University of California, San Diego, Data Report SIO Ref 90–19.

32. Scripps Institution of Oceanography. 1991. Physical, chemical, and biological data report, CalCOFI cruise 9003 and CalCOFI 9004. University of California, San Diego, Data Report SIO Ref 91–4.

33. Scripps Institution of Oceanography. 1991. Physical, chemical, and biological data report, CalCOFI cruise 9007 and CalCOFI 9011. University of California, San Diego, Data Report SIO Ref 91–18.

34. Sverdrup, H.U., M.W. Johnson, and R.H. Fleming. 1942. The Oceans, Prentice-Hall, Englewood Cliffs, NJ.

35. Swinnerton, J.W., V.J. Linnenbom, and C.H. Cheek. 1969. Distribution of methane and carbon monoxide between the atmosphere and natural waters. Environ. Sci. Technol. **3**:836–838.

36. Ward, B.B., K.A. Kilpatrick, P.C. Novelli, and M.I. Scranton. 1987. Methane oxidation and methane fluxes in the ocean surface layer and deep anoxic waters. Nature **327**:226–229.

37. Wiesenburg, Denis A., and Norman L. Jr. Guinasso. 1979. Equilibrium solubilities

of methane, carbon monoxide, and hydrogen in water and sea water. J. Chem. Eng. Data **24**:356–360.

38. Wilson, R.D., P.H. Monaghan, R. Osanik, L.C. Price, and M.A. Rogers. 1972. Natural marine oil seepage. Science **184**:857–865.

39. Wunsch, C. 1989. Tracer inverse problems. In D.L.T. Anderson and J. Willebrand (ed.), Oceanic Circulation Models: Combining Data and Dynamics, Kluwer Academic, Dordecht, in cooperation with NATO.

40. Yen, T.F. 1975. Genesis and degradation of petroleum hydrocarbons in marine environments. In T.M. Church (ed.), Marine Chemistry in the Coastal Environment, American Chemical Society, Washington, DC, pp. 231–266.

The Carbon Isotope Biogeochemistry of Methane Production in Anoxic Sediments: 1. Field Observations

*Neal E. Blair, Susan E. Boehme, and W. Dale Carter, Jr.**

Department of Marine, Earth and Atmospheric Sciences, North Carolina State University, Raleigh, North Carolina 27695-8208

ABSTRACT

The natural abundance $^{13}C/^{12}C$ ratio of methane from anoxic marine and freshwater sediments in temperate climates varies seasonally. Carbon isotopic measurements of the methanogenic precursors, acetate, and dissolved inorganic carbon from the marine sediments of Cape Lookout Bight, North Carolina have been used to determine the sources of the seasonal variations at that site. Movement of the methanogenic zone over an isotopic gradient within the dissolved CO_2 pool appears to be the dominant control of the methane $^{13}C/^{12}C$ ratio from February to June. The onset of acetoclastic methane production is a second important controlling process during mid-summer. An apparent temperature dependence on the fractionation factor for CO_2 reduction may have a significant influence on the isotopic composition of methane throughout the year.

Introduction

Methane, like other carbon-containing materials, is comprised of approximately 99% ^{12}C and 1% ^{13}C.[20] Small but distinct source-related variations in the $^{13}C/^{12}C$ ratio are well-documented.[52,59] It is important to understand the processes that control the isotopic composition of methane for a variety of reasons. The $^{13}C/^{12}C$ of biogenic methane may provide information concerning the methanogenic ecosystem and the relative rates of methane-producing pathways as a function of season[7,9,15,44] or environment.[59] The increasing concentration of tropospheric methane and its potential impact on global temperature and the stratospheric mixing ratios of water vapor and chlorine radicals[8,16,25] require a better understanding of the sources and sinks of that gas. A carbon isotope

*Present address: Archer Daniels Midland Company, P.O. Box 10640, Southport, NC 28461.

budget of atmospheric methane that includes seasonal effects would be a powerful constraint on source estimates.[44,55] An understanding of the controls of the isotopic composition of biogenic methane would also aid hydrocarbon exploration as approximately 20% of the world's natural gas resources are biogenic in origin.[50]

Attempts to model the carbon isotopic composition of methane have been hampered by insufficient information concerning the rates of the methanogenic processes and the isotopic signatures of the methanogenic precursors. Early models used a Rayleigh distillation calculation to simulate the isotope effects associated with methane production in a marine sediment.[17,46] These models had three major assumptions: (1) Methane was produced only via CO_2-reduction; (2) net CO_2 consumption occurred in the methanogenic zone; and (3) the system was closed to material fluxes of CO_2 and CH_4. No information was available concerning the importance of acetate dissimilation or other pathways. Ignoring the CO_2 production that occurs during methanogenesis[57] led to a significant error in the models. Subsequent treatments, which involved freshwater systems, considered both CO_2 reduction and acetate dissimilation pathways as well as net CO_2 production in the methanogenic zone.[28,36] It was assumed that the acetate was utilized only by methanogens in the methanogenic zone, a point that is contradicted by ^{14}C-tracer studies.[12,37,60] The isotopic composition of sedimentary acetate was not measured in those studies.

This chapter reviews our attempts to model the carbon isotopic composition of biogenic methane (Ref. 5; N.E. Blair and S.E. Boehme, unpublished data). The ultimate goal of this project is to determine the source of the seasonal variations observed in the $^{13}C/^{12}C$ ratio of methane from the organic-rich marine sediments of Cape Lookout Bight, North Carolina.[44] In general, the $^{13}C/^{12}C$ ratio increased in the summer months when methane production rates were highest (Table 1). Similar seasonal variations have been observed in freshwater environments[15] and may be a common phenomenon. The general approach in this project has been to measure the natural abundance $^{13}C/^{12}C$ ratios of methanogenic precursors, CO_2, and acetate, and combine those values with estimates of the relative rates of CO_2 reduction and acetate dissimilation to simulate the seasonal variations with an open-system model. In doing so, hypotheses concerning the controls of the isotopic composition of methane are tested. In a second study reported in this volume, results from a laboratory microcosm experiment are used to test the model.[1]

Field Site

Cape Lookout Bight, North Carolina, is a 1-km^2 coastal basin located 115 km Southwest of Cape Hatteras on the Outer Banks (Figure 1; Refs. 40 and 43). Fine-grained sediment, with an organic content up to 4% dry weight,[41] accu-

Table 1. Carbon isotopic compositions of methane bubbles from Cape Lookout Bight, North Carolina.

	$\delta^{13}C$		
Month	1983[a]	1984[a]	1986[b]
February		−63.4	−60.0
April		−63.8	−61.7
May		−66.2	−60.8
June	−64.5	−64.1	
July	−62.2	−60.0	−58.5
August	−59.6	−57.6	−55.9
September	−60.3	−58.0	−58.0
October	−60.0		−58.3
November	−62.2		−59.4

[a]Data from (Ref. 44).
[b]Data from S.E. Boehme, Ph.D. dissertation, North Carolina State University, Raleigh, in progress.

mulates at a rate[11,13] of 8–12 cm/yr at the sampling station A-1. The organic matter appears to be derived from phytoplankton and seagrass debris.[32]

The rapid flux of metabolizable organic matter to the seabed results in a high rate of organic carbon remineralization.[41] Sulfate reduction, occurring in the upper 10 cm of sediment during summer months, and methanogenesis, which occurs in the underlying zone,[21] are the dominant diagenetic processes at this site and respectively account for 68 ± 20% and 32 ± 16% of the organic carbon remineralization.[41] Approximately 20–30% of the methane is produced via acetate dissimilation and the remainder of the gas is formed primarily by CO_2 reduction.[22]

Methods

All $^{13}C/^{12}C$ ratios (R) are reported in the $\delta^{13}C$ notation which is defined[20] as

$$\delta^{13}C = \left(\frac{R_{\text{sample}} - R_{\text{PDB}}}{R_{\text{PDB}}}\right) \times 10^3.$$

R_{PDB} is the carbon isotopic ratio of the international standard, Peedee Belemnite, and has the accepted value of 0.0112372.[33] The preparation of samples for isotopic analysis is discussed below in the appropriate sections.

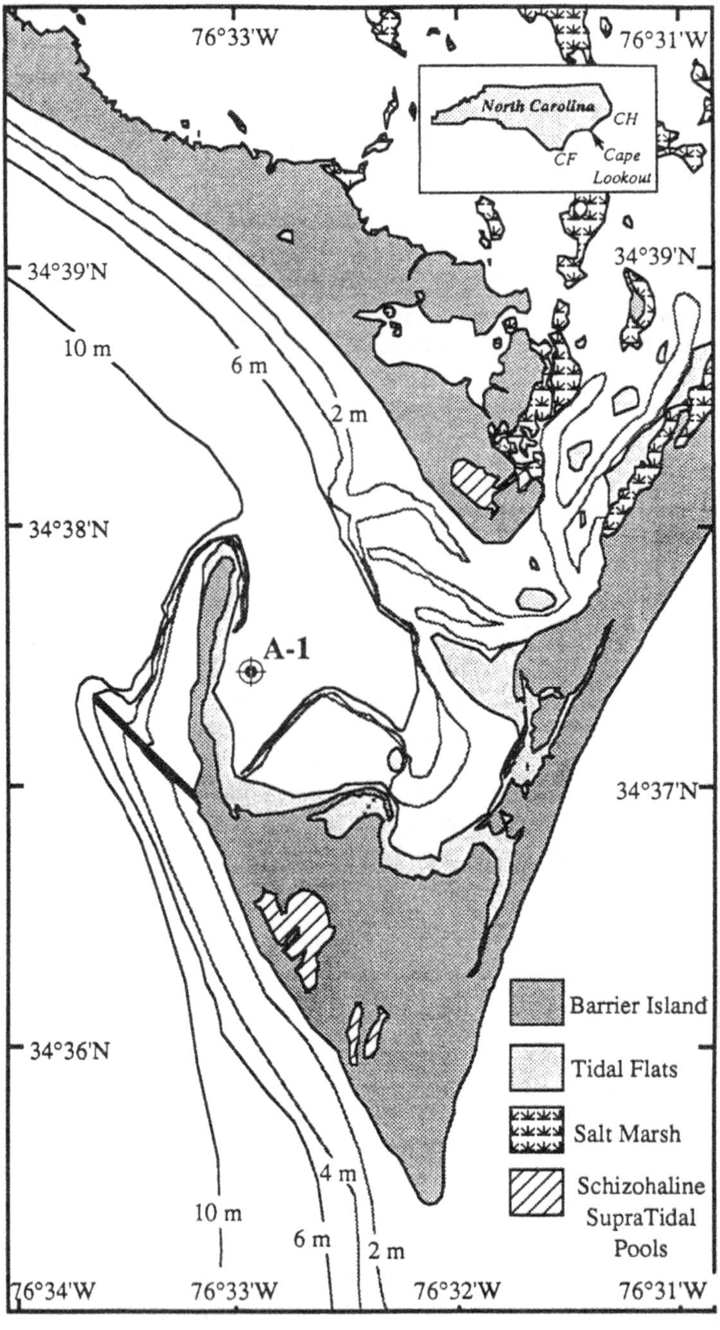

Figure 1. Cape Lookout Bight, North Carolina. The sampling station is designated A-1.

Diver-collected cores were obtained at A-1 with 9.5-cm-diam. lucite tubes. Pore water samples for sulfate and ΣCO_2 measurements were collected with a sediment press.[49] Porewater samples for acetate $\delta^{13}C$ measurements were isolated with the press or by centrifugation. The porewater samples were frozen immediately after collection and stored at $-86°C$ until analysis.

The acetate samples were treated as described previously.[5–7] The acetate fraction was isolated by a series of cryogenic distillations coupled with a preparative liquid-chromatography step. The acetate was converted to CO_2 for isotopic analysis with a gas chromatograph–combustion system.[24,39]

The isotopic analysis of the acetate methyl group was accomplished by the pyrolysis of sodium acetate.[6,45,47] A 20 : 1 mixture of NaOH and acetate (from the last distillation) was dried under N_2 at 135°C in a quartz tube (9 mm inner diameter × 20 cm long). The tube was evacuated after drying and heated to 500°C. Methane, which is derived from the methyl group, was quantitatively collected, measured, and injected into the gas chromatograph–combustion system via a Toepler pump.

One- to two-milliliter subsamples of porewater were injected into evacuated 120-ml serum bottles (Wheaton) sealed with crimped 20-mm rubber stoppers (Alltech Assoc.) and frozen until analysis for ΣCO_2 concentrations and $\delta^{13}C$ values. Immediately prior to analysis, 1 ml of 1 M phosphoric acid saturated with cupric sulfate was added to the thawed sample. The cupric sulfate was added to precipitate sulfide. The resulting CO_2 was removed from the bottle through a 23 gauge hypodermic needle connected to a vacuum line via a ¼-in. Ultratorr union (Cajon). The CO_2 was purified cryogenically, quantitated with a manometer, and sealed in a 6-mm-outer diameter pyrex tube for isotopic analysis. The analytical precision and accuracy of the ΣCO_2 extraction procedure were ±4% and ±0.5 mM respectively as determined by the measurement of CO_2 and bicarbonate standards.[62] The accuracy of the $\delta^{13}C$ measurements, as determined by the analysis of the NBS-20 carbonate standard,[19] was ±0.02 per mil.

The methane bubbles were collected from stirred sediment and stored in sealed bottles.[44] The methane was converted to CO_2 at 780°C with CuO and purified cryogenically for isotopic analysis.[24,39]

The $\delta^{13}C$ measurements of the CO_2 from the various preparations were analyzed on either a modified Nuclide 6-60 RMS (NASA—Ames Research Center) or one of two Finnigan MAT 251 mass spectrometers (NCSU Stable Isotope Laboratory and the University of Georgia Center for Applied Isotope Studies). A cross-calibration of a CO_2 standard by the three facilities produced results consistent to within 0.25 per mil. Procedural blanks were collected and used to correct the results of all analyses.

Dissolved sulfate was measured on 5 ml of porewater by the gravimetric analysis of the precipitated barium salt.[14,63] Sulfide was removed immediately after the recovery of the porewater sample by the addition of $ZnCl_2$ followed by the filtration of the zinc sulfide precipitate. The accuracy of this procedure is typically ±0.5 mM.

Results and Discussion

Ebullition is the primary mode of transport of methane from the sediments of Cape Lookout Bight, accounting for approximately 86% of the total flux of 7.4 ± 2 mol m^{-2} yr^{-1} (Refs. 41 and 42, and S.E. Boehme et al., unpublished data). The rapid bubble transport from the methanogenic zone through the overlying sediments limits the exposure of the methane to oxidizing conditions that could alter its isotopic composition.[2,18] Concordance between the $\delta^{13}C$ values of naturally released and diver-released bubbles and porewater methane supports that conclusion.[44] Thus, the $\delta^{13}C$ value of the methane is controlled primarily by its production.

In organic-rich marine sediments, methane is formed by CO_2-reduction[17,22,46] and acetate dissimilation[22,35] with the latter process accounting for 20–50% of the total production. Accordingly, the isotopic composition of methane produced will be the result of a mass balance of material from those two sources. The isotopic composition of the methane from each pathway is dependent on the isotopic composition of the precursor, CO_2 or acetate, and the fractionation factor, α, (k_{12}/k_{13}) associated with each process.

Acetate Dissimilation

Acetoclastic methanogenesis is accomplished by the *Methanosarcina* and *Methanothrix* genera.[58] Acetate is converted to acetyl-CoA, after which the carbon–carbon bond of the acetyl moiety is cleaved.[31,58] The methyl is reduced to methane and the carboxyl is oxidized to CO_2.[10,54,61] The fractionation factor associated with the formation of methane via acetate dissimilation by cultures of *Methanosarcina barkerii* is[34,64] 1.02–1.03.

The $\delta^{13}C$ value of the acetate methyl group and the in situ fractionation factor for acetate dissimilation in Cape Lookout Bight sediments have been determined via intramolecular carbon isotope measurements of acetate isolated from porewater samples.[5,7] The $\delta^{13}C$ value of the total acetate molecule ranges from −17.6 per mil in nonmethanogenic surficial sediments to −2.8 per mil in methane-producing sediments (Figure 2 and Ref. 5). Near the sediment surface, the similarity of the $\delta^{13}C$ value of the acetate to that of the average particulate organic carbon fraction (−19.1 ± 0.3,[5,7]) indicates that little net fractionation occurs during acetate cycling in the sulfate-reducing zone. However, a large fractionation occurs in the methanogenic zone which leaves the acetate enriched in ^{13}C.

Isotopic analysis of the methyl group and a mass balance calculation of the

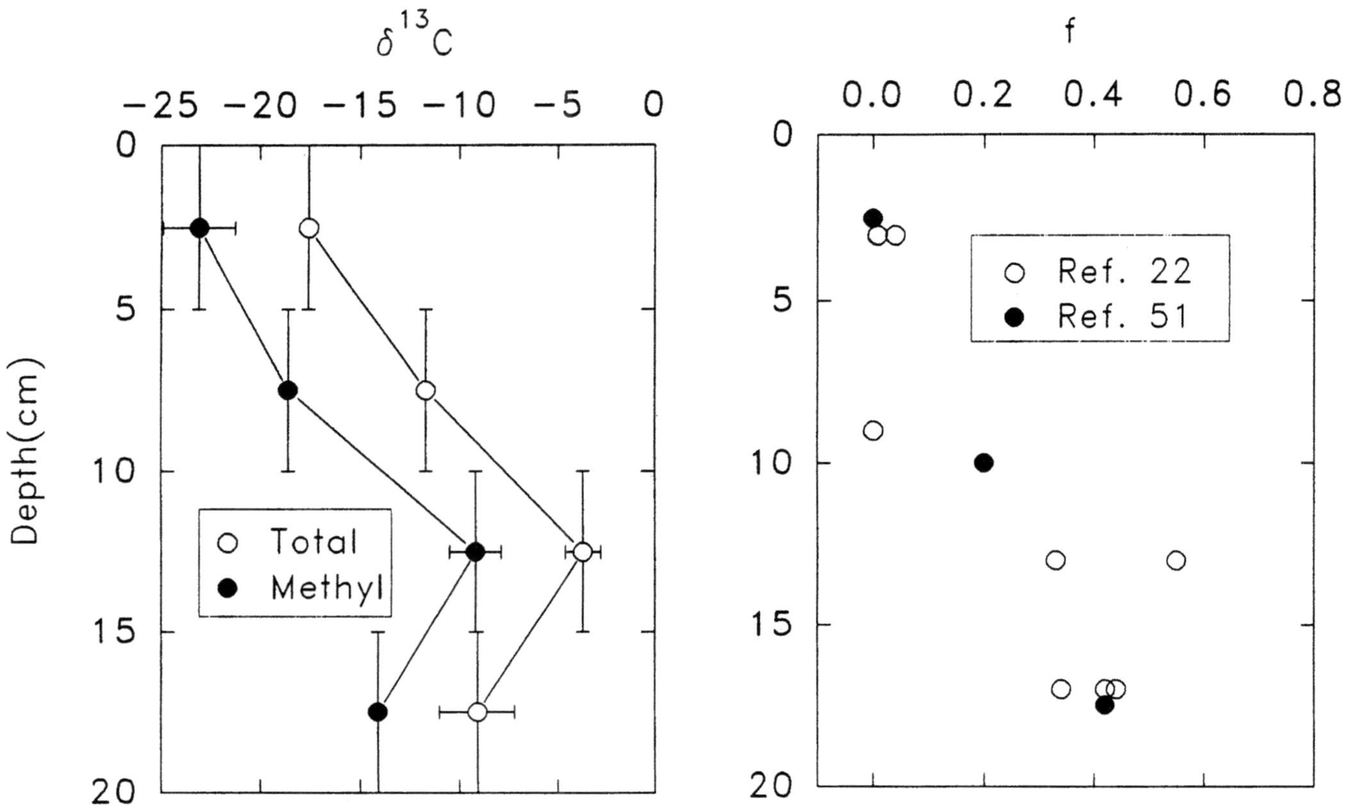

Figure 2. The $\delta^{13}C$ values of acetate and its methyl group as a function of depth. The $\delta^{13}C$ values are averages of results from 8/86 and 7/87.[5] The fraction (f) of the methyl group converted to methane as a function of depth.[22, 51] Figure adapted from Blair and Carter[5] with permission from Geochimica Cosmochimica Acta published by Pergamon Press Ltd.

$\delta^{13}C$ value of the carboxyl group indicates that the fractionation affects both carbon atoms of acetate (Figure 2). The magnitude of the ^{13}C enrichment correlates well with the parameter f, which is defined by

$$f = \frac{r_{CH_4}}{r_{CH_4} + r_{CO_2}}, \tag{1}$$

where r_{CH_4} and r_{CO_2} are the respective rates of the conversion of the acetate methyl group to CH_4 and CO_2.[5] The downcore profile of f (Figure 2) was determined from the turnover rate of U–^{14}C acetate in Cape Lookout sediments.[22,51] On average, 38 ± 11% ($f = 0.38$) of the methyl group is reduced to methane in the uppermost sediments of the sulfate-depleted zone (10–20 cm) at this site.

The in situ fractionation factor for the dissociation of the methyl group from acetate (α_{diss}) was estimated by

$$\alpha_{diss} = 1 + \frac{\delta_{obs} - \delta_{syn}}{f(\delta_{syn} + 10^3)}, \tag{2}$$

where δ_{obs} is the average $\delta^{13}C$ value of the methyl group in the 10–20-cm interval (−11.2 ± 3.0) and δ_{syn} represents the $\delta^{13}C$ value of the newly synthesized acetate.[5] The average $\delta^{13}C$ value of the methyl group in the 0–5-cm interval (−23.2 ± 2.2) was used as an estimate for δ_{syn} because, as noted earlier, the similarity of the isotopic composition of the total acetate from that interval with the particulate organic fraction suggests that little fractionation occurs during the synthesis or uptake in the surficial sediments. The assumption is made that the synthetic isotope effect is also small in the methanogenic zone. It is assumed that the fractionation factor for the conversion of acetate to CO_2 and other nonmethane products is 1.000. Using $f = 0.38 \pm 0.11$, α_{diss} was calculated to be 1.032 ± 0.014. The excellent agreement between our estimate of the in situ α_{diss} with the culture-derived values[34,64] of 1.02–1.03 indicates that our assumptions are reasonable to a first approximation.

Under steady-state conditions,[33] the $\delta^{13}C$ value of the methane produced from acetate is given by

$$\delta^{13}C(CH_4/Acet) = \delta_{syn} - (1 - f)(\alpha_{diss} - 1) \times 10^3. \tag{3}$$

Using the same values and assumptions as above for the appropriate parameters, the $\delta^{13}C$ value of the methane is calculated to be −43 ± 10 per mil. A similar ^{13}C enrichment relative to methane produced via CO_2 reduction (see below) has been observed in the laboratory microcosm experiment.[1]

Approximately, 20 and 26% of the methane is derived from acetate in the upper 30 cm of sediment at Cape Lookout in July and August, respectively, with the remainder formed via CO_2 reduction.[22] Direct measurements of the relative rates of the two methanogenic processes are not available for other months. The $\delta^{13}C$ value of the acetate methyl group in the 8–16-cm interval in June 1984 was −26 per mil,[7] which is a value more similar to that found in the sulfate-reducing sediments than in the underlying methanogenic zone. That suggests that little of the acetate was dissimilated to CH_4 and CO_2 ($f = 0$) at that time. On the other hand, one must consider the possibility that the June value is the combined result of a synthetic isotope effect, similar to that associated with acetogenesis,[30,48] and a methanogenic isotope effect. For the purposes of the model, the simpler scenario is assumed, that is, the synthetic isotope effect remains small and unchanged throughout the summer and $f = 0$ in June. Accordingly, when $f = 0$, the proportion of methane produced from acetate relative to total methane production (F) must be zero. There is no information concerning F for any other months. For the purposes of the model, $F = 0$ for all months except July and August, where it equals 0.20 and 0.26, respectively (Figure 3b).

CO_2 Reduction

$^{12}CO_2$ is selectively converted to CH_4, creating an isotopic gradient in the dissolved inorganic carbon (ΣCO_2) pool as buried sediment encounters and passes through the methanogenic zone (Ref. 46, Figure 4; Table 2). At Cape Lookout Bight, the methanogenic zone moves along the $\delta^{13}C$ gradient in response to the seasonal changes in the depth of sulfate penetration (Figure 4). In addition, the $\delta^{13}C$ profiles of ΣCO_2 respond to the seasonal changes in organic matter remineralization rates. Thus, the methanogens are exposed to different isotopic compositions of CO_2 throughout the year because of the two phenomena. The temporal CO_2 signal has been estimated by calculating the isotopic composition of CO_2 in equilibrium with HCO_3^- at the peak of the CO_2-reducing zone for each month that ΣCO_2 profiles were available (Blair and Boehme, unpublished data; S.E. Boehme, Ph.D. thesis, North Carolina State University, Raleigh, in preparation). The peak of the CO_2-reduction zone approximately coincides with the shallowmost depth where the sulfate concentration is less than 1.0 ± 0.5 mM.[21,22] The relative contributions of the CO_2, HCO_3^-, and $CO_3^=$ to the ΣCO_2 pool were estimated, assuming mutual chemical equilibrium,[53] for the average pore water pH of 6.95 (Ref. 63; N.E. Blair, unpublished results). The isotopic composition of the CO_2 was estimated by solving the following equations simultaneously:

$$\delta^{13}C(\Sigma CO_2) = x\delta^{13}C(CO_2) + y\delta^{13}C(HCO_3^-) + z\delta^{13}C(CO_3^=), \tag{4}$$

$$\alpha\left(\frac{HCO_3^-}{CO_2}\right) = \frac{10^3 + \delta^{13}C(HCO_3^-)}{10^3 + \delta^{13}C(CO_2)}, \tag{5}$$

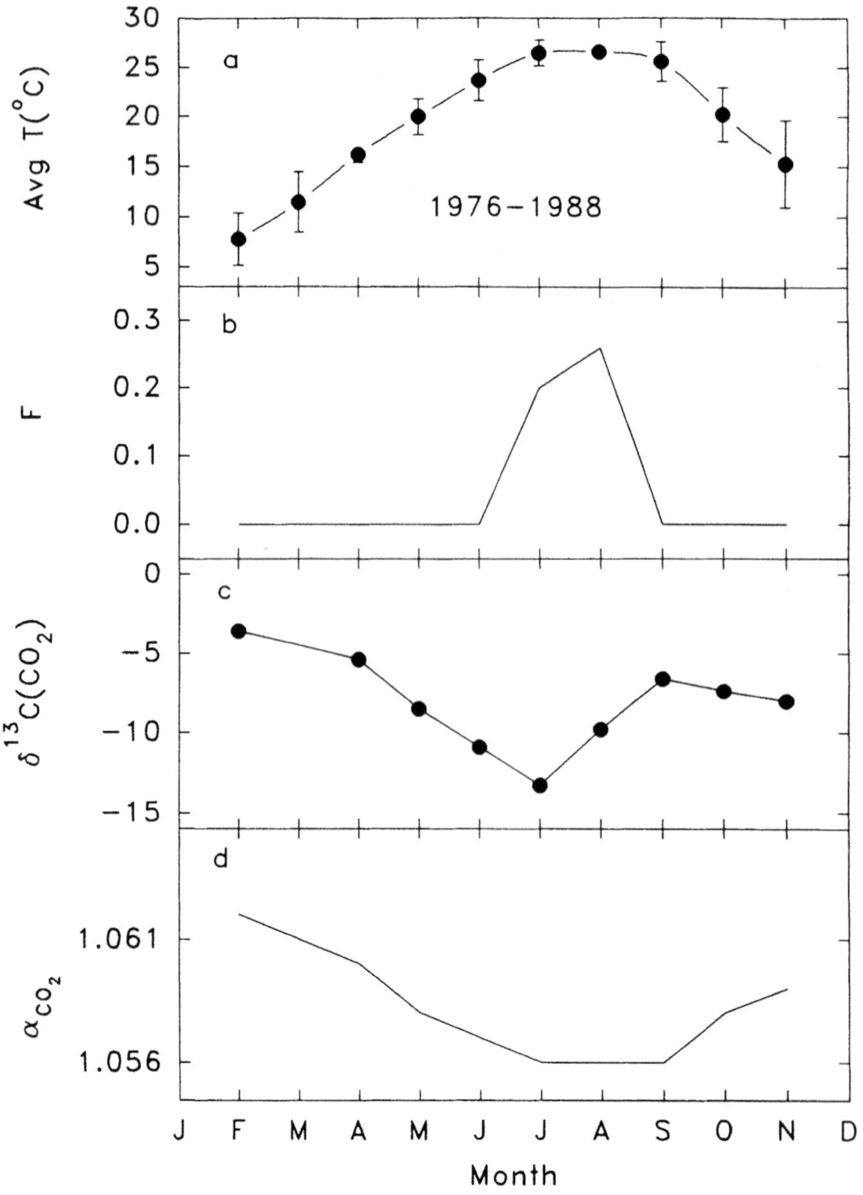

Figure 3. Model parameters as a function of time. (a) Average monthly temperature at Cape Lookout (Refs. 40, 41 and 44 and this study). (b) The fraction of methane derived from the dissimilation of acetate (*F*). (c) The $\delta^{13}C$ value of dissolved CO_2 in the methanogenic zone. (d) The fractionation factor (α) for CO_2 reduction.

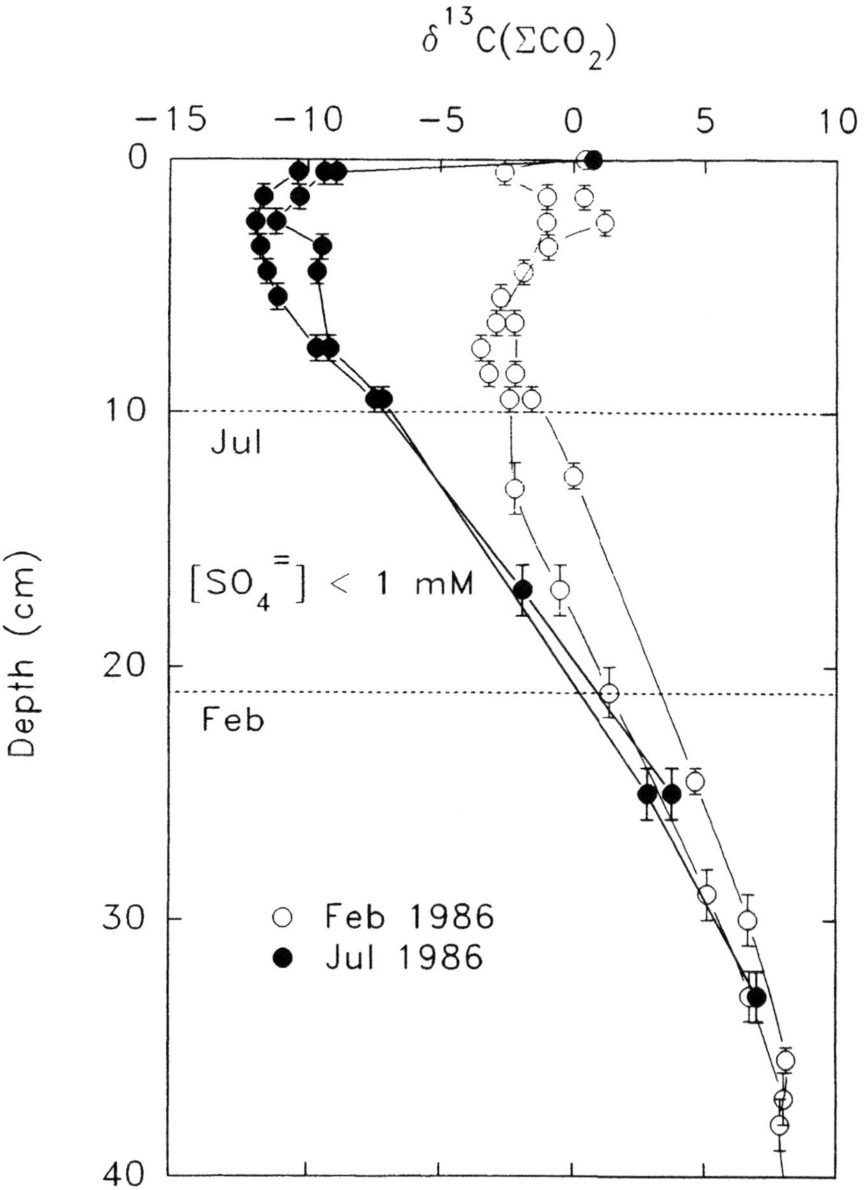

Figure 4. The $\delta^{13}C$ values of ΣCO_2 as a function of depth within the sediment from Cape Lookout. Profiles from February and July 1986 are shown. The dotted horizontal lines represent the depths at which dissolved sulfate concentrations are equal to or less than 1.0 ± 0.5 mM in February and July.

Table 2. Carbon isotopic composition ($\delta^{13}C$) of porewater ΣCO_2 from Cape Lookout Bight, North Carolina.

	2/20/86		7/1/86	
Depth (cm)	Core 1	Core 2	Core 1	Core 2
Bottom water	0.7			
0–1	−2.6		−10.4	−9.4
1–2	−1.0	0.4	−11.6	−10.3
2–3	−1.0	1.2	−12.0	−11.2
3–4		−1.0	−11.8	−9.5
4–5	−1.9		−11.5	−9.6
5–6	−2.7		−11.1	
6–7	−2.2	−2.9		
7–8		−3.5	−9.7	−9.2
8–9		−3.2		
9–10	−1.6	−2.4	−7.5	−7.2
10–12				
11–12	0.0			
12–14		−2.2		
16–18		−0.5	−1.9	
20–22		1.4		
24–25	4.6			
24–26			3.8	2.8
28–30		5.1		
29–31	6.6			
32–34		6.7		7.0
35–36	8.1			
36–38		8.0		
37–39	7.8			

(6) $$\delta^{13}C(HCO_3^-) = \delta^{13}C(CO_3^=),$$

where x, y, and z represent the fractions of each of the dissolved components. The $\delta^{13}C$ values of the HCO_3^- and $CO_3^=$ ions are assumed to be equivalent [Eq. (6)] to simplify the calculations. Theoretical studies indicate that the HCO_3^- ion may be enriched in ^{13}C by 1.4–1.7 per mil for the temperature range involved.[23] For the given pH, the $CO_3^=$ ion represents less than 1% of the ΣCO_2 pool; thus, the small isotopic difference is considered insignificant. The equilibrium fractionation factor is given by

(7) $$\ln \alpha\left(\frac{HCO_3^-}{CO_2}\right) = \left(\frac{9.552}{T}\right) - 0.0241,$$

where temperature (T) is in Kelvin.[26] The resulting CO_2 $\delta^{13}C$ values as a function of time are shown in Figure 3c. The seasonal isotopic variations are large and clearly must have a significant effect on the methane $\delta^{13}C$ values.

The fractionation factor for CO_2 reduction ranges from 1.03 to 1.06 in cultures.[3,4,27,29] The evidence for a temperature effect on the fractionation factor, while expected, is equivocal and may be dependent on culture conditions. We have attempted to estimate the in situ fractionation factor using data associated with the two temperature extremes at this site. At 7.8°C (Feb.), the $\delta^{13}C$ values of the CH_4 and CO_2 were −61.7 and −3.6, respectively. Using the equation

$$\alpha_{CO_2} = \frac{\delta^{13}C(CO_2) + 10^3}{\delta^{13}C(CH_4) + 10^3}, \tag{8}$$

α_{CO2} was calculated to be 1.062, during a period of time when it has been assumed that methane was produced predominantly via CO_2 reduction. In August (T = 26.5°C), the $\delta^{13}C$ of the methane was −57.7. Given that 26% of the methane is derived from acetate dissimilation and $\delta^{13}C(CH_4/Acet) = -43$, then the $\delta^{13}C$ value of the methane produced via CO_2 reduction $[\delta^{13}C(CH_4/CO_2)]$ should be approximately −62.9. Finally, with $\delta^{13}C(CO_2) = -9.8$, we estimate $\alpha_{CO_2} = 1.057$ at 26.5°C. Fitting the two estimates to an Arrhenius temperature dependence, one obtains

$$\ln \alpha_{CO_2} = \frac{23.0}{T} - 0.022, \tag{9}$$

where temperature is in Kelvin. The seasonal variation of α_{CO_2} is shown in Figure 3d. The isotopic composition of the methane produced from the CO_2 for the other months was calculated using

$$\delta^{13}C(CH_4/CO_2) = \left(\frac{\delta^{13}C(CO_2) + 10^3}{\alpha}\right) - 10^3 \tag{10}$$

and is shown in Figure 5a.

$\delta^{13}C(CH_4)$

The isotopic composition of the methane produced at A-1 is described simply by the mass balance relationship

$$\delta^{13}C(CH_4) = F\delta^{13}C(CH_4/Acet) + (1 - F)\,\delta^{13}C(CH_4/CO_2). \tag{11}$$

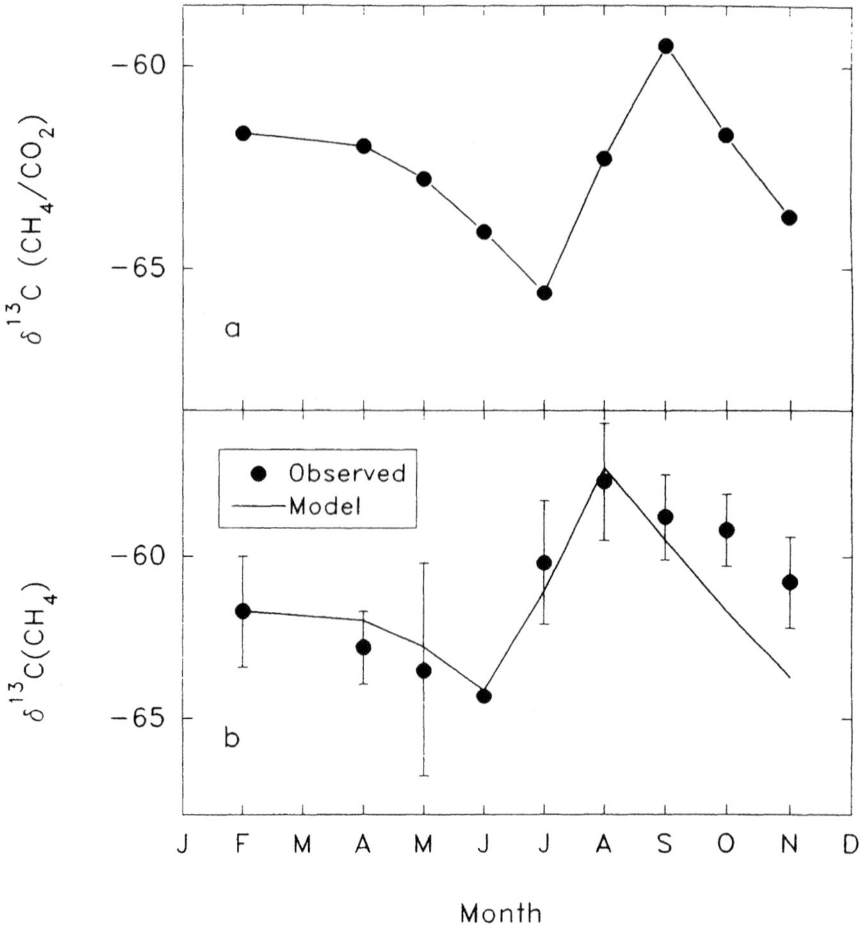

Figure 5. The measured and calculated $\delta^{13}C$ values of methane from Cape Lookout. (a) Calculated $\delta^{13}C$ values of methane produced from CO_2 reduction. (b) The monthly average measured $\delta^{13}C$ values of methane bubbles for the period 1983–1984 and 1986 (b). The calculated $\delta^{13}C$ values using the model described in the text (—).

The calculated monthly $\delta^{13}C$ values, using the parameters in Figures 3a–3d, are in excellent agreement with measured values for the period February to September (Figure 5b). The movement of the methanogenic zone over the ΣCO_2 $\delta^{13}C$ gradient and the temporal variation of α_{CO_2} are responsible for the gradual depletion of ^{13}C in the methane between February and June. The onset of acetate dissimilation in July–August coupled with the change in $\delta^{13}C$ value of the CO_2 within the methanogenic zone results in a dramatic enrichment of ^{13}C in those months. The trend is reversed when acetate dissimilation ceases in late August. The subsequent ^{13}C depletion in October and November is caused by changes in α_{CO_2} and the $\delta^{13}C$ value of the CO_2. The relatively poor fit of the model to the observed values in October and November may be because the model estimates instantaneous $\delta^{13}C$ values and the measured values represent a pooled product. The largest deviation between modeled and observed values would be expected at this time when the reservoir of methane is large and bubbling rates are low.

Implications for Freshwater Sediments

Methane produced in freshwater environments is often enriched in ^{13}C relative to biogenic gas from marine sediments.[59] It has been hypothesized that this is due to the relatively greater importance of acetate dissimilation as a methane-producing pathway in freshwater systems.[59] ^{14}C-tracer studies indicate that 50–70% of methane production is via acetoclastic processes (F = 0.5–0.7,[12,37,53,60]). Our calculations indicate that the methane derived from the acetate methyl group is enriched in ^{13}C relative to that from CO_2 reduction, thus apparently confirming the hypothesis. The ^{13}C-enrichment is a consequence of the smaller fractionation factor and large degree of conversion of the methyl group to methane [$f = 0.4$, see Eq. (3)]. In freshwater sediments, where $f = 0.7$–0.9,[12,37,53,60] the potential ^{13}C enrichment could be greater if the synthetic pathways of acetate and the associated isotope effects are comparable to those at Cape Lookout. Other investigators, using culture and modeling results, have proposed that acetate synthesis in freshwater sediments occurs by a very different process, that is, acetogenic CO_2 reduction.[30] Acetate produced in this manner would be significantly depleted in ^{13}C because the acetogenic process exhibits a large ($\alpha \cong 1.06$) isotope effect.[30,48] However, ^{14}C-tracer studies indicate that $<2\%$ of the acetate in eutrophic lake sediments is produced by CO_2 reduction.[38] The results from Ref. 1 suggest that the ecological niche of the acetogenic bacteria may be an opportunistic one. Application of the approach summarized in this chapter to freshwater systems should resolve the issue. Such work is currently underway.

Summary

The isotopic composition of methane varies seasonally at a variety of sites.[15,44] In the organic-rich marine sediments of Cape Lookout Bight, North Carolina, the carbon isotopic variations appear to be the result of three factors. During the period February–June, CO_2 reduction is the dominant methanogenic pathway and the $\delta^{13}C$ variations are driven by the movement of the methanogenic zone along an isotopic gradient within the dissolved CO_2 pool. Changes in the relative rates of CO_2 reduction and acetate dissimilation become the dominant factor from July to September. Throughout the whole time period, a temperature dependence on the fractionation factor for CO_2 reduction may play a role controlling the methane $\delta^{13}C$ value.

The Cape Lookout model is the first to combine measured isotopic compositions of both methanogenic precursors, CO_2 and acetate, with measured rates of methanogenic processes. The agreement between model predictions and observed $\delta^{13}C$ values of methane verifies estimates of in situ fractionation factors associated with acetate cycling and methanogenesis. The model also indicates that methane produced by acetate dissimilation should be enriched in ^{13}C relative to that produced via CO_2 reduction, thus verifying, in general terms, earlier hypotheses concerning the isotopic differences commonly observed between methane from freshwater and marine sediments.[59]

Acknowledgments

We thank Chris Martens and his group for their invaluable assistance throughout this project. Jeff Chanton, Dan Albert, Bob Haddad, and Carol Green provided critical field assistance. David Des Marais, Bill Showers, and Randy Culp provided mass spectrometer time. We thank Marc Alperin for his helpful discussions and two anonymous colleagues for their critical reviews. Laboratory space and boating facilities were provided by the University of North Carolina Institute of Marine Sciences, Morehead City, NC. This research was supported by NASA grant NAGW-838 and PRF grant 17527-G2.

References

1. Alperin, M.J., N.E. Blair, D.B. Albert, and T.M. Hoehler. 1993. The carbon isotope biogeochemistry of methane production in anoxic sediments. 2. A laboratory experiment. This volume, 594–605.
2. Alperin, M.J., W.S. Reeburgh, and M.J. Whiticar. 1988. Carbon and hydrogen isotope fractionation resulting from anaerobic methane oxidation. Global Biogeochem. Cycles **2**:279–288.
3. Balabane, M., E. Galimov, M. Hermann, and R. Letolle. 1987. Hydrogen and carbon isotope fractionation during experimental production of bacterial methane. Org. Geochem. **11**:115–119.
4. Belyaev, S.S., R. Wolkin, W.R. Kenealy, M.J. DeNiro, S. Epstein, and J.G.

Zeikus. 1983. Methanogenic bacteria from the Bondyuzhskoe Oil field: General characterization and analysis of stable-carbon isotopic fractionation. Appl. Environ. Microbiol. **45**:691–697.

5. Blair, N.E., and W.D. Carter, Jr. 1992. The carbon isotope biogeochemistry of acetate from a methanogenic marine sediment. Geochim. Cosmochim. Acta **56**: 1247–1258.
6. Blair, N., A. Leu, E. Muñoz, J. Olsen, E. Kwong, and D.J. Des Marais. 1985. Carbon isotopic fractionation in heterotrophic microbial metabolism. Appl. Environ. Microbiol. **50**:996–1001.
7. Blair, N.E., C.S. Martens, and D.J. Des Marais. 1987. Natural abundances of carbon isotopes in acetate from a coastal maine sediment. Science **236**:66–68.
8. Blake, D.R., and F.S. Rowland. 1988. Continuing worldwide increase in tropospheric methane, 1978 to 1987. Science **239**:1129–1131.
9. Burke, R.A., Jr., C.S. Martens, and W.M. Sackett. 1988. Seasonal variations of D/H and $^{13}C/^{12}C$ ratios of microbial methane in surface sediments. Nature **332**: 829–831.
10. Buswell, A.M., and F.W. Sollo, Jr. 1948. The mechanism of methane fermentation. J. Amer. Chem. Soc. **70**:1778–1780.
11. Canuel, E.A., C.S. Martens, and L.K. Benninger. 1990. Seasonal variations in ^{7}Be activity in the sediments of Cape Lookout Bight, North Carolina. Geochim. Cosmochim. Acta **54**:237–245.
12. Cappenberg, Th.E., and R.A. Prins. 1974. Interrelations between sulfate-reducing and methane-producing bacteria in bottom deposits of a fresh-water lake. III. Experiments with ^{14}C-labeled substrates. Antonie van Leeuwenhoek **40**:457–469.
13. Chanton, J.P., C.S. Martens, and G.W. Kipphut. 1983. Lead-210 sediment geochronology in a changing coastal environment. Geochim. Cosmochim. Acta **47**:1791–1804.
14. Chanton, J.P., C.S. Martens, and M.B. Goldhaber. 1987. Biogeochemical cycling in an organic-rich coastal marine basin. 7. Sulfur mass balance, oxygen uptake and sulfide retention. Geochim. Cosmochim. Acta **51**:1187–1199.
15. Chanton, J.P., and C.S. Martens. 1988. Seasonal variations in ebullitive flux and carbon isotopic composition of methane in a tidal freshwater estuary. Global Biogeochem. Cycles **2**:289–298.
16. Cicerone, R.J., and R.S. Oremland. 1988. Biogeochemical Aspects of Atmospheric Methane. Global Biogeochem. Cycles **2**:299–327.
17. Claypool, G.E., and I.R. Kaplan. 1974. The origin and distribution of methane in marine sediments. In I.R. Kaplan (ed.), Natural Gases in Marine Sediments, Plenum, New York, pp. 99–139.
18. Coleman, D.D., J.B. Risatti, and M. Schoell. 1981. Fractionation of carbon and hydrogen isotopes by methane-oxidizing bacteria. Geochim. Cosmochim. Acta **45**: 1033–1037.
19. Coplen, T.B., C. Kendall and J. Hopple. 1983. Comparison of stable isotope reference samples. Nature **302**:236–238.

20. Craig, H. 1953. The geochemistry of the stable carbon isotopes. Geochim. Cosmochim. Acta **3**:53–92.
21. Crill, P.M., and C.S. Martens. 1983. Spatial and temporal fluctuations of methane production in anoxic, coastal marine sediments. Limnol. Oceanogr. **28**:1117–1130.
22. Crill, P.M., and C.S. Martens. 1986. Methane production from bicarbonate and acetate in an anoxic marine sediment. Geochim. Cosmochim. Acta **50**:2089–2097.
23. Deines, P., D. Langmuir, and R.S. Harmon. 1974. Stable carbon isotope ratios and the existence of a gas phase in the evolution of carbonate ground waters. Geochim. Cosmochim. Acta **38**:1147–1164.
24. Des Marais, D.J. 1978. Variable-temperature cryogenic trap for the separation of gas mixtures. Anal. Chem. **50**:1405–1406.
25. Dickinson, R.E., and R.J. Cicerone. 1986. Future global warming from atmospheric trace gases. Nature **319**:109–115.
26. Friedman, I., and J.R. O'Neil. 1977. Compilation of stable isotope fractionation factors of geochemical interest. U.S. Geological Survey Professional Paper 440-KK.
27. Fuchs, G., R. Thauer, H. Zeigler, and W. Stichler. 1979. Carbon isotope fractionation by *Methanobacterium thermoautotrophicum*. Arch. Microbiol. **120**:135–139.
28. Games, L.M., and J.M. Hayes. 1976. On the mechanisms of CO_2 and CH_4 production in natural anaerobic environments. In J.O. Nriagu (ed.), Proc. 2nd Int. Sym. on Environ. Biogeochem., Michigan Science Press, Ann Arbor, pp. 51–73.
29. Games, L.M., J.M. Hayes, and R.P. Gunsalus. 1978. Methane-producing bacteria: natural fractionations of the stable carbon isotopes. Geochim. Cosmochim. Acta **42**:1295–1297.
30. Gelwicks, J.T., J.B. Risatti and J.M. Hayes. 1989. Carbon isotope effects associated with autotrophic acetogenesis. Org. Geochem. **14**:441–446.
31. Grahame, D.A., and T.C. Stadtman. 1987. In vitro methane and methyl coenzyme M formation from acetate: evidence that acetyl-CoA is the required intermediate activated form of acetate. Biochem. Biophys. Res. Commun. **147**:254–258.
32. Haddad, R.I., and C.S. Martens. 1987. Biogeochemical cycling in an organic-rich coastal marine basin: 9. Sources and accumulation rates of vascular plant-derived organic material. Geochim. Cosmochim. Acta **51**:2991–3001.
33. Hayes, J.M. 1983. Practice and principles of isotopic measurements in organic geochemistry. In W.G. Meinschein (ed.), Organic Geochemistry of Contemporaneous and Ancient Sediments, Society of Economic Paleontolists and Mineralogists, Great Lakes Section, Bloomington, Indiana, pp. 5-1–5-31.
34. Krzycki, J.A., W.R. Kenealy, M.J. DeNiro, and J.G. Zeikus. 1987. Stable carbon isotope fractionation by *Methanosarcina barkeri* during methanogenesis from acetate, methanol, or carbon dioxide-hydrogen. Appl. Environ. Microbiol. **53**:2597–2599.
35. Kuivila, K.M., J.W. Murray, and A.H. Devol. 1990. Methane production in the sulfate-depleted sediments of two marine basins. Geochim. Cosmochim. Acta **54**: 403–411.

36. LaZerte, B.D. 1981. The relationship between total dissolved carbon dioxide and its stable carbon isotope ratio in aquatic sediments. Geochim. Cosmochim. Acta **45**: 647–656.

37. Lovley, D.R., and M.J. Klug. 1982. Intermediary metabolism of organic matter in the sediments of a eutrophic lake. Appl. Environ. Microbiol. **43**:552–560.

38. Lovley, D.R., and M.J. Klug. 1983. Methanogenesis from methanol and methylamines and acetogenesis from hydrogen and carbon dioxide in the sediments of a eutrophic lake. Appl. Environ. Microbiol. **45**:1310–1315.

39. Matthews, D.E., and J.M. Hayes. 1978. Isotope ratio monitoring GC-MS. Anal. Chem. **50**:1465–1473.

40. Martens, C.S., and J.V. Klump. 1980. Biogeochemical cycling in an organic-rich coastal basin—I. Methane sediment–water exchange processes. Geochim. Cosmochim. Acta **44**:471–490.

41. Martens, C.S., and J.V. Klump. 1984. Biogeochemical cycling in an organic-rich coastal marine basin. 4. An organic carbon budget for sediments dominated by sulfate reduction and methanogenesis. Geochim. Cosmochim. Acta **48**:1987–2004.

42. Martens, C.S., and J.P. Chanton. 1989. Radon as a tracer of biogenic gas equilibration and transport from methane-saturated sediments. J. Geophys. Res. D3 **94**: 3451–3459.

43. Martens, C.S., G.W. Kipphut, and J.V. Klump. 1980. Sediment–water chemical exchange in the coastal zone traced by in situ radon-222 flux measurements. Science **208**:285–288.

44. Martens, C.S., N.E. Blair, C.D. Green, and D.J. Des Marais. 1986. Seasonal variations in the stable carbon isotopic signature of biogenic methane in a coastal sediment. Science **233**:1300–1303.

45. Meinschein, W.G., G.G.L. Rinaldi, J.M. Hayes, and D.A. Schoeller. 1974. Intramolecular isotopic order in biologically produced acetic acid. Biomed. Mass Spectrom. **1**:172–174.

46. Nissenbaum, A., B.J. Presley, and I.R. Kaplan. 1968. Early diagenesis in a reducing fjord, Saanich Inlet, British Columbia—I. Chemical and isotopic changes in major components of interstitial water. Geochim. Cosmochim. Acta **36**:1007–1027.

47. Oakwood, T.S., and M.R. Miller. 1950. The decarboxylation of simple fatty acids. J. Amer. Chem. Soc. **72**:1849.

48. Preuß, A., R. Schauder, G. Fuchs, and W. Stichler. 1989. Carbon isotope fractionation by autotrophic bacteria with three different CO_2 fixation pathways. Z. Naturforsch. **44c**:397–402.

49. Reeburgh, W.S. 1967. An improved interstitial water sampler. Limnol. Oceanogr. **12**:163–165.

50. Rice, D.D., and G.E. Claypool. 1981. Generation, accumulation and resource potential of biogenic gas. Bull. AAPG **65**:5–25.

51. Sansone, F.J., and C.S. Martens. 1982. Volatile fatty acid cycling in organic-rich marine sediments. Geochim. Cosmochim. Acta **46**:1575–1589.

52. Schoell, M. 1980. The hydrogen and carbon isotopic composition of methane from natural gases of various origins. Geochim. Cosmochim. Acta **44**:649–661.
53. Schutz H., W. Seiler, and R. Conrad. 1989. Processes involved in formation and emission of methane in rice paddies. Biogeochemistry **7**:33–53.
54. Stadtman, T.C., and H.A. Barker. 1949. Studies on the methane fermentation. VII. Tracer experiments on the mechanism of methane formation. J. Biochem. **21**:256–264.
55. Stevens, C.M., and F.E. Rust. 1982. The carbon isotopic composition of atmospheric methane. J. Geophys. Res. **87**:4879–4882.
56. Stumm, W., and J.J. Morgan. 1981. Aquatic Chemistry, 2nd ed., J. Wiley and Sons, New York, pp. 173–177.
57. Tarvin, D., and A.M. Buswell. 1934. The methane fermentation of organic acids and carbohydrates. J. Amer. Chem. Soc. **56**:1751–1755.
58. Thauer, R.K., D. Moller-Zinkhan, and A.M. Spormann. 1989. Biochemistry of acetate catabolism in anaerobic chemotrophic bacteria. Ann. Rev. Microbiol. **43**: 43–67.
59. Whiticar, M.J., E. Faber, and M. Schoell. 1986. Biogenic methane formation in marine and freshwater environments: CO_2 reduction vs. acetate fermentation—Isotope evidence. Geochim. Cosmochim. Acta **50**:693–709.
60. Winfrey, M.R., and J.G. Zeikus. 1979. Anaerobic metabolism of intermediate methane precursors in Lake Mendota. Appl. Environ. Microbiol. **37**:244–253.
61. Zeikus, J.G. 1983. Metabolism of one-carbon compounds by chemotrophic anaerobes. Adv. Microbiol. Physiol. **24**:215–299.
62. Boehme, S.E. 1989. M.S. Thesis, North Carolina State University, Raleigh.
63. Chanton, J.P. 1985. Ph.D. Thesis, University of North Carolina, Chapel Hill.
64. Risatti, J.B., and Hayes, J.M. 1983. Geol. Amer. Abstract Program. **15**:671.

The Carbon Isotope Biogeochemistry of Methane Production in Anoxic Sediments: 2. A Laboratory Experiment

Marc J. Alperin

Curriculum in Marine Sciences, University of North Carolina, Chapel Hill, North Carolina 27599-3300

Neal E. Blair

Department of Marine, Earth and Atmospheric Sciences, North Carolina State University, Raleigh, North Carolina 27695-8208

Daniel B. Albert and Tori M. Hoehler

Curriculum in Marine Sciences, University of North Carolina, Chapel Hill, North Carolina 27599-3300

ABSTRACT

The factors that contribute to variability in the isotopic composition of methane produced by anoxic sediment were evaluated by monitoring stable carbon isotope ratios in methane produced by a sediment microcosm. Anoxic marine sediment from Cape Lookout Bight, North Carolina, was placed in a 2-L pyrex incubation syringe for 114 d. Sediment samples were removed at weekly intervals and analyzed for sulfate, acetate, and methane concentrations as well as $^{13}C/^{12}C$ ratios in the dissolved CO_2 and methane pools. In addition, rates of sulfate reduction, methane production from acetate, and methane production from CO_2 were measured by radiotracer techniques. The isotopic composition of the methane pool varied by more than 40‰ over the course of the experiment. The changes appear to be the result of at least two factors: (1) changes in the isotopic composition of the dissolved CO_2 pool and (2) changes in the fraction of the methane production derived from acetate. A comparison of the laboratory experiment with field data suggests that similar biogeochemical processes operate to control carbon isotope ratios of methane in both undisturbed and laboratory-incubated sediment.

Introduction

Biogenic methane produced in anoxic sediments does not have a specific stable carbon isotopic composition. For example, methane produced in marine sediments is generally depleted in ^{13}C relative to methane produced in freshwater sediments.[22] In addition, there is a growing body of evidence indicating that $^{13}C/^{12}C$ ratios in methane produced in marine and freshwater sediments change on a seasonal basis.[5,6,11,14,15] The biogeochemical processes that contribute to spatial and temporal variability in the isotopic composition of biogenic methane are not well-understood.[17,20]

In Part 1,[4] Blair et al. use an open-system isotope mass balance model to reproduce seasonal variations in the $^{13}C/^{12}C$ ratio of methane produced in the organic-rich marine sediments of Cape Lookout Bight, North Carolina. Their model demonstrates that the isotopic composition of methane produced in these sediments is largely controlled by three factors: (1) seasonal changes in the isotopic composition of the dissolved CO_2 reservoir; (2) changes in the kinetic isotope effect for methane production from CO_2; and (3) changes in the relative rates of the two major methanogenic pathways, CO_2 reduction and acetate fermentation.

In this paper, we describe a laboratory experiment in which surface sediments from Cape Lookout Bight were incubated under anoxic conditions for 114 d. The purpose of the experiment was to identify the biogeochemical factors that contribute to variability in the isotopic composition of methane produced by a well-characterized sediment sample. The experiment clearly demonstrates that changes in the relative rates of methane production from CO_2 and acetate as well as changes in the isotopic composition of the dissolved CO_2 reservoir can play an important role in controlling $^{13}C/^{12}C$ ratios of methane produced in anoxic sediments.

Materials and Methods

Field Site

The sediment used in this study was collected from Cape Lookout Bight, North Carolina. The field site is described in detail in Part 1.[4] Sediment for the laboratory experiment was collected in July 1990, when temperatures (26.5°C) were near their annual maximum. Cores were collected by SCUBA diver by gently inserting plexiglass tubes into the sediment. The tubes were capped on top and bottom, gently withdrawn from the sediment, and kept at in situ temperature during transport to the laboratory.

Incubation Syringe

Sediment from the 0–3-cm depth interval was transferred to a custom-fabricated 2-L pyrex syringe.[2] The syringe design allowed for intermittent sample withdrawal without introducing a head space. Samples were taken at weekly intervals by extruding 120 ml of sediment through a 10-mm bore stopcock welded to the end of the syringe barrel. Sediment in the syringe was periodically homogenized by means of an internal mixing plate. The incubation syringe was kept in the dark at 22°C for 114 d.

Concentrations

Porewater samples for sulfate and acetate concentration measurements were obtained by centrifugation. The supernatant was filtered (0.45 μm) to remove suspended solids. Sulfate concentrations were determined by ion chromatography. Acetate concentrations were determined by high-performance liquid chromatography using precolumn derivatization with 2-nitrophenylhydrazine.[1] Whole-sediment methane concentrations were measured by head-space equilibration-gas chromatography[3] and converted to porewater concentration units by dividing by sediment porosity.

Rates

Rates of sulfate reduction, methane production from CO_2, and methane production from acetate were measured by previously described radioisotope techniques.[1,7] Whole-sediment samples were transferred to glass incubation syringes,[3] injected with the appropriate tracer ($^{35}SO_4^{2-}$, [U–^{14}C]acetate, or [^{14}C]–HCO_3^-), and incubated at 22°C for 10 min to 27 h. The actual length of the incubation was adjusted so that 1–10% of the tracer pool turned over. The incubation was terminated by sealing sediments in serum vials containing 1 N NaOH solution. The radioisotopes were quantified and turnover rates were calculated according to standard procedures.[10]

Stable Isotope Ratios

The methods used to determine $^{13}C/^{12}C$ ratios in ΣCO_2 and methane are described in Part 1.[4] Stable carbon isotope ratios are reported in standard $\delta^{13}C$ notation (see Part 1[4]).

Results and Discussion

Concentrations

Changes in sulfate, acetate, and methane concentrations during the 114-d experiment are summarized in Figure 1. The sulfate concentration decreased

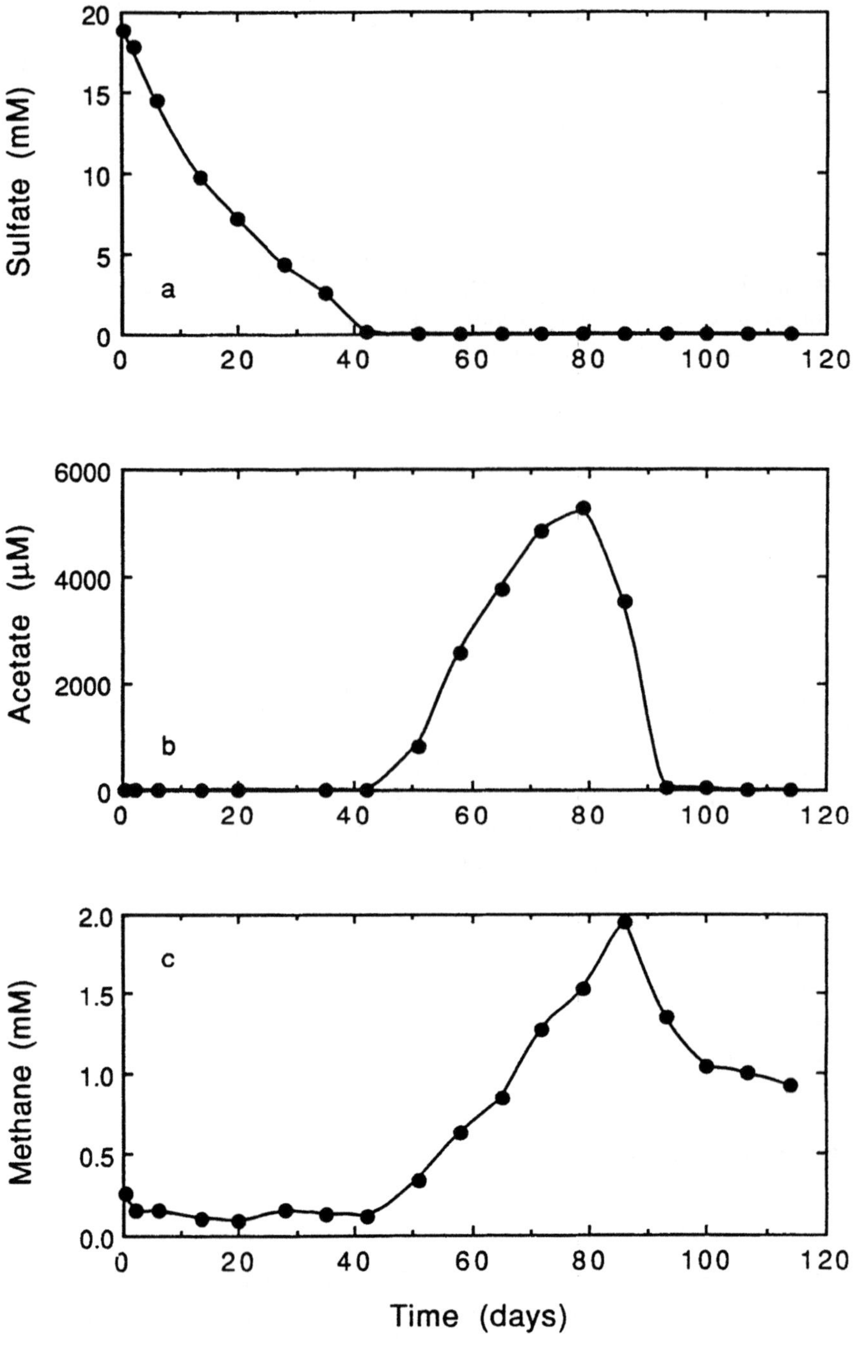

Figure 1. Changes in pore water concentrations during the 114-d experiment. (a) Sulfate; (b) acetate; (c) methane.

rapidly from an initial value of 18.6 mM to 0.02 mM after 50 d (Figure 1a). Ingvorsen et al.[9] have shown that uptake by sulfate-reducing bacteria ceases when sulfate concentrations fall below 0.005 to 0.02 mM.

Immediately following sulfate depletion, the acetate concentration increased dramatically from <10 μM to a peak concentration of 5300 μM at 80 d (Figure 1b). The accumulation of acetate was presumably caused by a reduction in the rate of acetate consumption that occurred when sulfate-reducing bacteria became sulfate-limited. After reaching a maximum, the acetate pool experienced a precipitous decline to concentrations <20 μM. The acetate depletion is consistent with the development of an active population of acetate-utilizing methanogenic bacteria to fill the ecological niche previously occupied by sulfate reducers.

The methane concentration was relatively constant (~0.2 mM) throughout the initial period when the sediment contained appreciable levels of sulfate (Figure 1c). The lack of methane accumulation in the presence of sulfate suggests that methane production from "noncompetative" substrates (i.e., compounds not metabolized by sulfate-reducing bacteria) is not a major methanogenic pathway in these sediments. Following sulfate depletion, the methane concentration began to increase (Figure 1c). Methane accumulated to concentrations in excess of saturation (1.2 mM),[23] leading to the formation of a gas phase within the incubation syringe. After 85 d, the methane concentration decreased as dissolved and gaseous methane came into equilibrium.

Rates

Changes in rates of sulfate reduction, methane production from acetate, and methane production from CO_2 are shown in Figure 2. Sulfate reduction rates were approximately 1000 μM d^{-1} at the outset of the experiment, but decreased rapidly during the initial 40-d period (Figure 2a). The falloff in sulfate reduction was presumably due to a decrease in the lability of the sediment organic matter. Sulfate reduction effectively ceased when sulfate concentrations dropped below 0.2 mM.

Rates of methane production from acetate and CO_2 were negligible during the period that sediment sulfate concentrations were greater than 0.02 mM (Figures 2b and 2c). The absence of methane production in the presence of sulfate is consistent with enzyme kinetic studies[12,19] showing that sulfate-reducing bacteria outcompete methanogens for acetate and H_2.

After the sediment became depleted in sulfate, the two methanogenic pathways followed somewhat different patterns. Methane production from acetate started out relatively slowly but accelerated sharply after 60 d, reaching a peak rate of nearly 400 μM d^{-1} (Figure 2b). The high rates of methane production from acetate occurred during the period that the sediment contained elevated acetate concentrations (Figure 1b). Methane production from acetate then de-

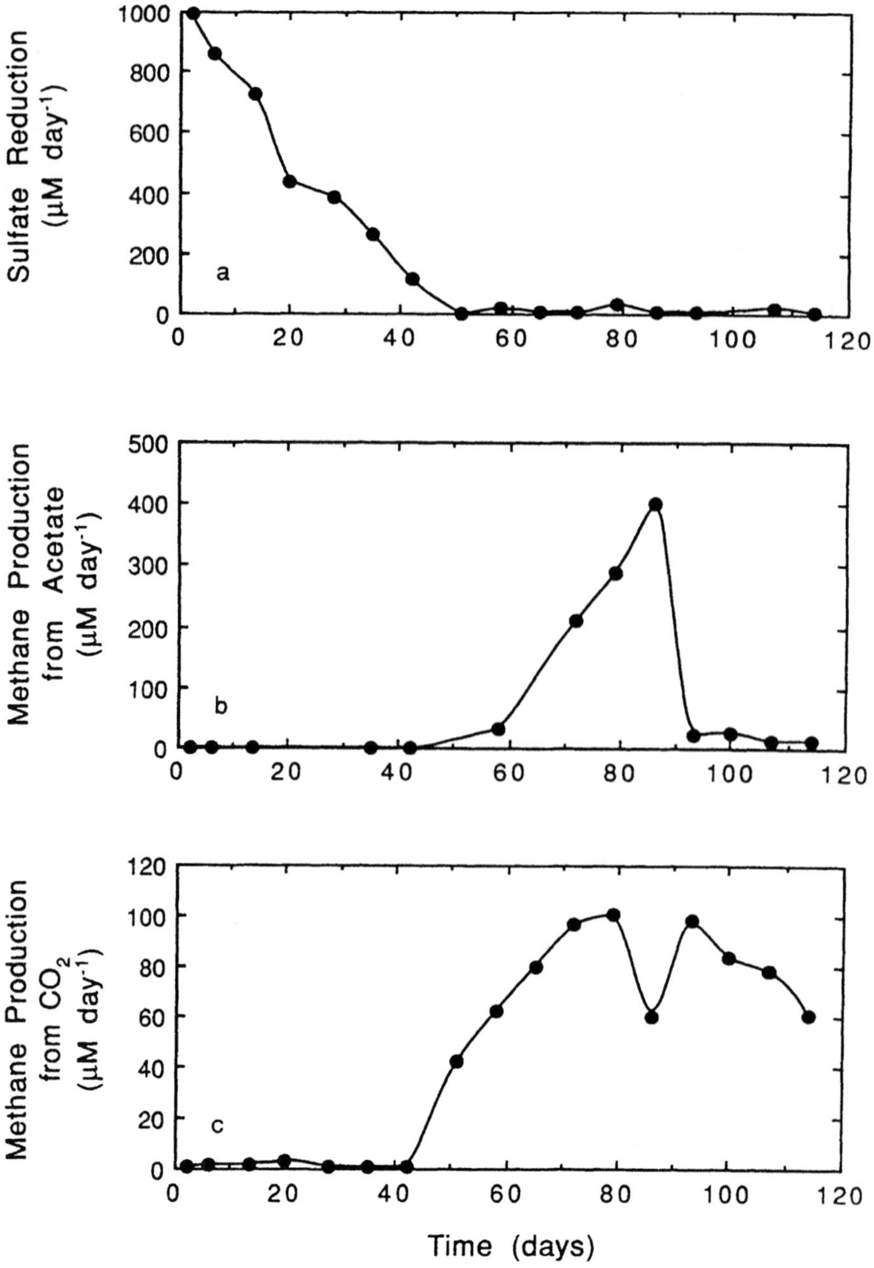

Figure 2. Changes in rates of microbial processes during the 114-d experiment. (a) Sulfate reduction; (b) methane production from acetate; (c) methane production from CO_2. The cause of the anomalously low methane production rate from CO_2 at 86 d is not known. Note that rate scales differ on each graph.

creased dramatically, the decline coinciding with the depletion of the acetate reservoir.

In contrast to methane production from acetate, CO_2 reduction began rather quickly following sulfate depletion, reached a maximum rate of about 100 $\mu M\ d^{-1}$, and then declined relatively gradually toward the end of the experiment (Figure 2c). The faster response of the CO_2-reducing methanogens following sulfate depletion is consistent with pure culture studies showing that methanogens growing on H_2–CO_2 have higher specific growth rates than methanogens growing on acetate.[21]

Stable Isotope Ratios

Changes in $\delta^{13}C(\Sigma CO_2)$ and $\delta^{13}C(CH_4)$ during the 114-d incubation period are shown in Figure 3. Throughout the period that sediment contained sulfate, $\delta^{13}C(\Sigma CO_2)$ values remained relatively constant at −13 to −14‰ (Figure 3a). Following sulfate depletion, the ΣCO_2 pool became progressively enriched in ^{13}C, possibly due to preferential uptake of $^{12}CO_2$ by CO_2-reducing methanogenic bacteria.[16]

The methane reservoir experienced large changes in isotopic composition over the course of the 114-d experiment (Figure 3b). During the initial period when sediment contained sulfate, $\delta^{13}C(CH_4)$ values remained relatively constant at −59 to −61‰. Following sulfate depletion, $\delta^{13}C(CH_4)$ values initially dropped to −88‰ before increasing to −47‰. The $\delta^{13}C(CH_4)$ values decreased slightly toward the end of the experiment.

The variations in $\delta^{13}C(CH_4)$ that occurred after sulfate depletion appear to result from at least two factors. First, the $\delta^{13}C$ of the dissolved CO_2 reservoir increased by more than 10‰ during the period of methane production (Figure 3a). This suggests that the methane produced by CO_2 reduction became enriched in ^{13}C as the experiment progressed. Second, the relative rates of methane production via acetate fermentation and CO_2 reduction changed over the course of the experiment (Figures 2b and 2c). In Part 1,[4] Blair et al. show that the methyl group of acetate and, hence, the methane derived from acetate fermentation is enriched in ^{13}C relative to methane produced by CO_2 reduction. Therefore, the $\delta^{13}C$ value of methane produced by the sediment should become more positive as the contribution from acetate fermentation increases.

The proportion of methane produced from acetate relative to total methane production (F) changed dramatically throughout the methane production period (Figure 3c). Methane production from acetate accounted for only ~30% of the total methane production immediately following sulfate depletion. During the 40–60-d period, the methane reservoir became progressively depleted in ^{13}C (Figure 3b), consistent with CO_2 reduction as the dominant methanogenic pathway.

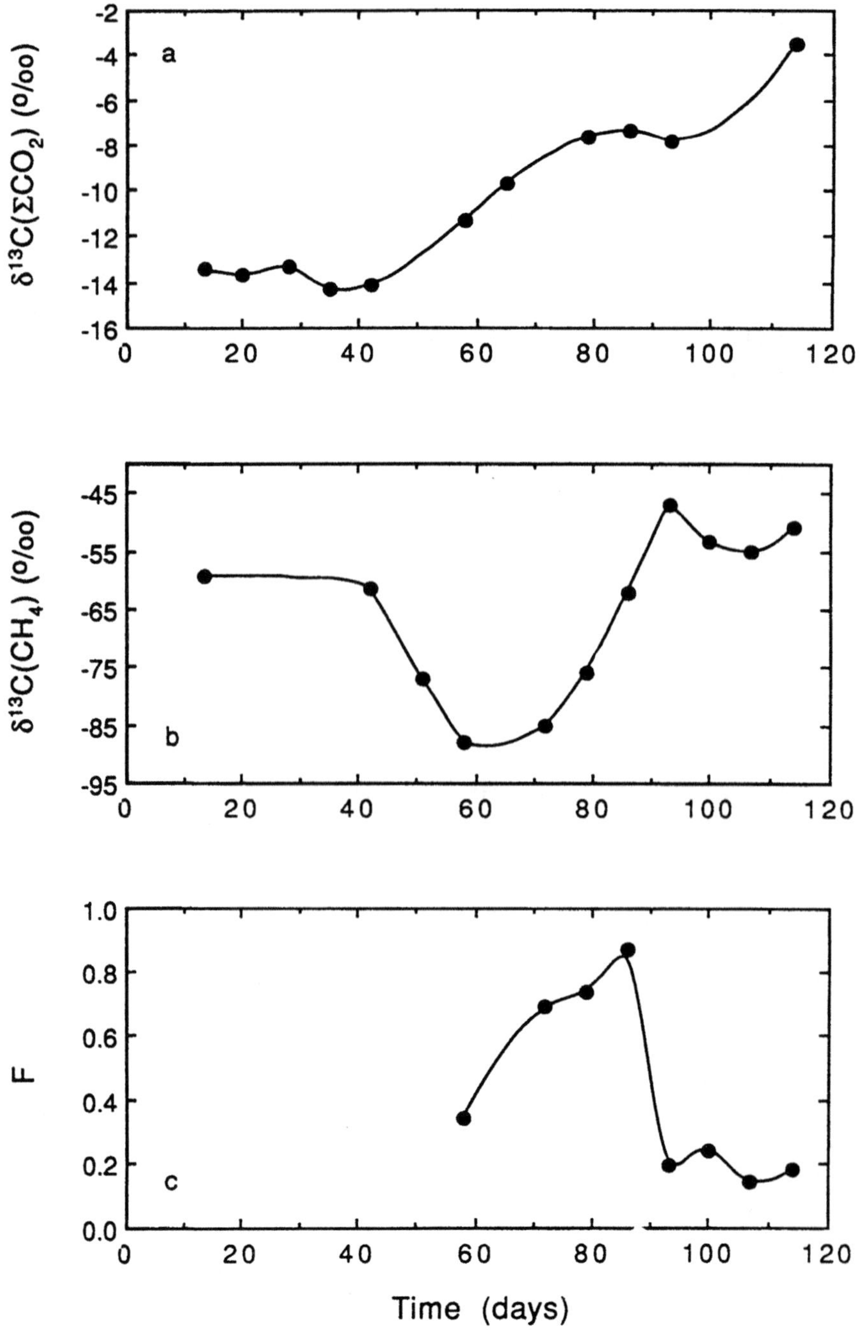

Figure 3. Changes in (a) $\delta^{13}C(\Sigma CO_2)$ and (b) $\delta^{13}C(CH_4)$ during the 114-d experiment. Samples for isotopic analysis were not taken during the first 10 d of the experiment. (c) Changes in *F* (proportion of methane produced from acetate relative to total methane production) during the 114-d experiment. *F* was calculated using the methane production rate data in Figures 2b and 2c. Note that *F* was not calculated for the initial 42 d when total methane production rates were approximately zero.

As acetate accumulated in the porewater (Figure 1b), the methanogenic pathway shifted and acetate became the major methane precursor, accounting for as much as 86% of the total methane production (Figure 3c). Acetate fermentation remained the dominant methane production pathway throughout the period that sediment contained elevated acetate concentrations. The methane pool reached a maximum $\delta^{13}C$ value of −47‰ just after the peak in acetate fermentation (Figures 3b and 3c). This level of ^{13}C enrichment is consistent with acetate as the primary source of methane. In Part 1,[4] Blair et al. calculate that methane produced by acetate dissimilation has a $\delta^{13}C$ value of -43 ± 10‰.

The fraction of the methane production derived from acetate diminished after 90 d (Figure 3c) and acetate concentrations returned to steady-state levels (Figure 1b). The increased importance of the CO_2 reduction pathway at this time is reflected in the slight decrease in $\delta^{13}C(CH_4)$ values during the final 20-d period (Figure 3b). The relatively modest decline in $\delta^{13}C(CH_4)$ is probably due to the large methane reservoir (Figure 1c) and relatively slow methane production rates (Figures 2b and 2c) during this final period.

Some of the changes in the isotopic composition of the methane pool cannot be explained in terms of changes in $\delta^{13}C(\Sigma CO_2)$ or shifts in the relative importance of acetate fermentation and CO_2 reduction. For example, the $\delta^{13}C$ of the methane pool between 50 and 80 d is less than −80‰, significantly depleted in ^{13}C relative to the calculated $\delta^{13}C$ values of methane produced from CO_2 reduction presented in Part 1[4] (−60 to −67‰). Enhanced ^{13}C depletion in methane produced by the sediment microcosm may be due to a larger isotope fractionation factor for CO_2 reduction and/or greater ^{13}C depletion in the acetate methyl group. One known mechanism for producing acetate with a methyl group highly depleted in ^{13}C is autotrophic acetogenesis.[8] However, acetogenic bacteria are thought to be incapable of competing with methanogens for H_2.[13] One might speculate that acetate synthesis via CO_2 reduction could occur during the sulfate–methane transition when competition for H_2 may be at a minimum.

Comparison of Field Study and Laboratory Experiment

The onset of acetoclastic methane production plays an important role in controlling the isotopic composition of methane in both undisturbed (see Part 1[4]) and laboratory-incubated sediment. In the laboratory experiment, the shift toward acetate fermentation can be understood as a response of the methanogen population to elevated acetate concentrations that appeared during the sulfate–methane transition. A similar increase in acetate inventory has been observed in Cape Lookout Bight sediments between May and August.[18] The mid-summer onset of acetate dissimilation described in Part 1 is also likely to be related to the presence of elevated acetate concentrations.

Summary

The laboratory experiment demonstrates that the $\delta^{13}C$ of the methane reservoir in anoxic surface sediments from Cape Lookout Bight, North Carolina, can vary from −47 to −88‰ during the course of a 114-d constant-temperature incubation. The variations in methane isotopic composition were primarily driven by changes in the $^{13}C/^{12}C$ ratio of the dissolved CO_2 reservoir and changes in the relative contribution of acetate fermentation and CO_2 reduction to the total methane production. The onset of high rates of acetoclastic methane production coincided with the appearance of large acetate inventories. A comparison of the laboratory experiment with field data suggests that seasonal variability in the isotopic composition of methane emitted from undisturbed Cape Lookout Bight sediments may be controlled by similar biogeochemical processes.

Acknowledgments

The stable carbon isotope analyses were done by Connie Crossley, Julie Gunn, and Howard Mendlovitz at North Carolina State University. Laboratory space and boating facilities were provided by the University of North Carolina Institute of Marine Sciences, Morehead City, North Carolina. This research was supported by Technical Service Contract OD4067NTEX from the Environmental Protection Agency (MJA), grant NAGW-838 from the National Aeronautics and Space Administration (NEB), and grant OCE-9017979 from the National Science Foundation (MJA, DBA).

References

1. Albert, D.B., C. Taylor, and C.S. Martens. 1993. Sulfate reduction rates and low molecular weight fatty acid concentrations in the water column and surficial sediments of the Black Sea. Deep-Sea Res. (in press).
2. Alperin, M.J., N.E. Blair, D.B. Albert, T.M. Hoehler, and C.S. Martens. 1992. Factors that control the stable carbon isotopic composition of methane produced in an anoxic marine sediment. Global Biogeochem. Cycles **6**:271–291.
3. Alperin, M.J., and W.S. Reeburgh. 1985. Inhibition experiments on anaerobic methane oxidation. Appl. Environ. Microbiol. **50**:940–945.
4. Blair, N.E., S.E. Boehme, and W.D. Carter, Jr. (1993). The carbon isotope biogeochemistry of methane production in anoxic sediments: 1. Field observations. In R.S. Oremland (ed.), Biogeochemistry of Global Change: Radiatively Active Trace Gases, Chapman and Hall, New York, pp. 574–593.
5. Burke, R.A., Jr., C.S. Martens, and W.M. Sackett. 1988. Seasonal variations of D/H and $^{13}C/^{12}C$ ratios of microbial methane in surface sediments. Nature **332**: 829–831.

6. Chanton, J.P., and C.S. Martens. 1988. Seasonal variations in ebullitive flux and carbon isotopic composition of methane in a tidal freshwater estuary. Global Biogeochem. Cycles **2**:289–298.
7. Crill, P.M., and C.S. Martens. 1986. Methane production from bicarbonate and acetate in an anoxic marine sediment. Geochim. Cosmochim. Acta **50**:2089–2097.
8. Gelwicks, J.T., J. Bruno Risatti, and J.M. Hayes. 1989. Carbon isotope effects associated with autotrophic acetogenesis. Organ. Geochem. **14**:441–446.
9. Ingvorsen, K., A.J.B. Zehnder, and B.B. Jørgensen. 1984. Kinetics of sulfate and acetate uptake by *Desulfovibrio postgatei*. Appl. Environ. Microbiol. **46**:403–408.
10. Jørgensen, B.B. 1978. A comparison of the methods for the quantification of bacterial sulfate reduction in coastal sediments. I. Measurement with radiotracer techniques. Geomicrobiology **1**:11–27.
11. Kelley, C.A., N.B. Dise, and C.S. Martens. 1992. Stable carbon isotopic characterization of seasonally varying emissions of methane from Minnesota peatlands. Global Biogeochem. Cycles **6**:263–269.
12. Kristjansson, J.K., P. Schonheit, and R.K. Thauer. 1982. Different K_s values for hydrogen of methanogen bacteria and sulfate reducing bacteria: an explanation for the apparent inhibition of methanogenesis by sulfate. Arch. Microbiol. **131**:278–282.
13. Lovely, D.R., and M.J. Klug. 1983. Methanogenesis from methanol and methylamines and acetogenesis from hydrogen and carbon dioxide in the sediments of a eutrophic lake. Appl. Environ. Microbiol. **45**:1310–1315.
14. Martens, C.S., N.E. Blair, C.D. Green, and D.J. Des Marais. 1986. Seasonal variations in the stable carbon isotopic signature of biogenic methane in a coastal sediment. Science **233**:1300–1303.
15. Martens, C.S., C.A. Kelley, J.P. Chanton, and W. Showers. 1992. Carbon and hydrogen isotopic characterization of methane from wetlands and lakes of the Yukon–Kusokwim Delta, Western Alaska. J. Geophys. Res. **97**:16,689–16,701.
16. Nissenbaum, A., B.J. Presley, and I.R. Kaplan. 1968. Early diagenesis in a reducing fjord, Saanich Inlet, British Columbia—I. Chemical and isotopic changes in major components of interstitial water. Geochim. Cosmochim. Acta **36**:1007–1027.
17. Oremland, R.S. 1988. The biogeochemistry of methanogenic bacteria. In A.J.B. Zehnder (ed.), Biology of Anaerobic Microorganisms, J. Wiley and Sons, New York, pp. 641–705.
18. Sansone, F.J., and C.S. Martens. 1982. Volatile fatty acid cycling in organic-rich marine sediments. Geochim. Cosmochim. Acta **46**:1575–1589.
19. Schonheit, P., J.K. Kristjansson, and R.K. Thauer. 1982. Kinetic mechanism for the ability of sulfate reducers to out-compete methanogens for acetate. Arch. Microbiol. **132**:285–288.
20. Tyler, S.C. 1991. The global methane budget. In J.E. Rogers and W.B. Whitman (eds.), Microbial Production and Consumption of Greenhouse Gases: Methane, Nitrogen Oxides, and Halomethanes, American Society for Microbiology, Washington, DC, pp. 7–38.

21. Vogels, G.D., K.T. Keltjens, and C. Van der Drift. 1988. Biochemistry of methane production. In A.J.B. Zehnder (ed.), Biology of Anaerobic Microorganisms. J. Wiley and Sons, New York, pp. 707–770.

22. Whiticar, M.J., E. Faber, and M. Schoell. 1986. Biogenic methane formation in marine and freshwater environments: CO_2 reduction vs. acetate fermentation—Isotope evidence. Geochim. Cosmochim. Acta **50**:693–709.

23. Yamamoto, S., J.B. Alcauskas, and T.E. Crozier. 1976. Solubility of methane in distilled water and seawater. J. Chem. Eng. Data **21**:78–80.

Oxidation of Methane by Deep-Sea Mytilids in the Gulf of Mexico

Charles R. Fisher

Department of Biology, The Pennsylvania State University, University Park, Pennsylvania 16802

ABSTRACT

At least four different species of mytilids which harbor abundant bacterial symbionts in their gills have been discovered in three general areas of the Gulf of Mexico. Two of the species harbor methanotrophic symbionts, one harbors both methanotrophic and chemoautotrophic sulfur-oxidizing symbionts, and the fourth harbors only chemoautotrophic sulfur-oxidizing symbionts. Mytilids with methanotrophic symbionts can reach very high densities and dominate the habitats in which they occur. Estimates of methane oxidation by the most shallow of these mytilid communities (500–700 m) are made based on mussel population size structures and densities, measurements of live animal methane oxidation rates and requirements for methane, in situ levels of dissolved methane, and the empirically determined relation between shell size and tissue biomass. This estimation indicates that oxidation rates between 22.4 and 42.4 g CH_4 m^{-2} d^{-1} are occurring within beds of these mytilids.

Introduction

In 1986 and 1987, three associations between methanotrophic bacteria and invertebrates were demonstrated and subsequently reported.[8,9,35] One is a very small pogonophoran tube-worm, which is found in reducing sediments in the central Skagerrak.[35] The other two documented methanotrophic symbioses are bivalves of the family mytilidae which are related to *Bathymodiolus thermophilus*, the hydrothermal vent mussel originally described from the Galapagos Rift.[18] All three of the methanotrophic symbioses were demonstrated by a combination of enzyme activity determinations, electron microscopy (which showed both the presence of symbionts and their morphologic similarity to Type 1 methanotrophs), and stable carbon isotope composition (reflecting source methane $\delta^{13}C$). In two of the cases, this evidence was supplemented by demonstration of ^{14}C–methane oxidation by the hosts,[14,34] and in one case, CH_4 consumption by intact animals and isolated gills was shown to be accompanied by elevated oxygen consumption and CO_2 production.[9,20] More recently, several other new species of mytilids have been discovered in the Gulf of Mexico. One of these new species harbors both methanotrophic and chemoautotrophic sulfur-oxidizing symbionts, as indicated by enzyme analyses, transmission electron microscopy, and elemental sulfur determination.[12,26] Not only are the two types of symbionts present in the same individuals, they are often found in the same vacuoles of the same gill cell.

Here the available data on the symbiont-containing mytilids of the Gulf of Mexico are synthesized, along with new information on very recently discovered species and subspecies. What is known of the distribution of all species is presented. Extant data on in situ densities, size distributions, and laboratory-measured methane consumption rates are combined with biomass measurements and used to estimate rates of methane consumption by communities of mussels on the sea floor.

Methods

Sample Collection

The mussels were collected either by the ALVIN or the Johnson SeaLink and placed in a temperature-insulated box for recovery to the surface. Upon recovery of the submersible, the mussels were immediately transferred to chilled seawater (7°C) and placed in a refrigerated van for dissection. The samples used for enzyme analyses were frozen in liquid nitrogen within 2 h of recovery of the submersible, and then stored at −70°C for up to 16 months prior to analysis. The relation between shell length and wet weight was determined for 132 shallow seep mytilids (species Ia). Exterior shell dimensions were measured and seawater contained inside the shells was drained by gently prying the mussels open to about 3 mm. The adductor muscles were then severed and the soft tissues removed to a preweighed bag and frozen at −20°C for subsequent determination of wet and dry weight.

Enzyme Analyses

Frozen tissues were analyzed for activity of four enzymes associated with chemoautotrophic sulfur-oxidizing or methanotrophic symbionts. A sample of gill tissue from each individual was analyzed for each enzyme. Samples of mantle tissue from three individuals were analyzed for each enzyme as control, symbiont-free tissues. The weighed tissue samples were homogenized with a ground-glass tissue grinder in 0.2 mM Tris/HCl (pH 7.5) with 1% Triton X100 (1 : 4 tissue : buffer for the methanol dehydrogenase and RuBP carboxylase assays, and 1 : 9 for the others). The homogenate was not centrifuged before use in the assays because a substantial portion of the enzyme activity has been found to be associated with the pellet in these symbioses. Crude homogenates were assayed for RuBP carboxylase (EC# 4.1.1.39) activity by the ^{14}C incorporation method[39] as modified by Felbeck[10] and methanol dehydrogenase activity by the spectrophotometric method of Anthony and Zatman.[1] ATP sulfurylase (EC# 2.7.7.4) and APS reductase (EC# 1.8.99.2) activities were assayed using the spectrophotometric methods described by Felbeck.[10]

Histology

For transmission electron microscopy (TEM), pieces of gill from freshly collected seep mytilid sp. III individuals were fixed in 3% glutaraldehyde in 0.1 M phosphate-buffered 0.35 M sucrose (pH 7.3) and stored in this fixative at 4°C for up to 2 weeks. The tissues were then washed in buffered sucrose, postfixed in 1% osmium (on ice) for 1 h, dehydrated through a graded ethanol series, and embedded in Spurrs embedding medium. Thin sections were cut using a Sorvall MT2 ultramicrotome, stained with uranyl acetate and lead citrate, and examined with a Jeol 1200 EXII transmission electron microscope.

Calculation of Community Methanotrophy Rates

Estimates of mussel in situ biomass were calculated from measurements of mussel density and population structure in both "young" and "old" mytilild beds at a site known as "Mussel Beach,"[25] measurements of percent cover by these aggregations over a 62-m^2 area,[25] and determination of the relation between shell length and tissue wet weight in over 134 mussels between 0.5 and 121.5 g wet weight (and 24 to 121 mm in shell length).

Biomass estimates were converted to methane consumption rates based on the in situ methane oxidation rate model of Kochevar et al.[20] which relates methane consumption to ambient methane concentration. To incorporate data from several individuals the authors standardized the data to "% of maximal consumption rate." As indicated by the polynomial form of the resultant equation, methane consumption was saturated at methane concentrations of about 250 μM (this equation provides the best fit for the data up to that level):

$$\%\ \text{max}\ CH_4\ \text{consumption rate} = -0.001\ [CH_4]^2 + 0.545\ [CH_4] + 0.726.$$

Using this model and published carbon content for a related mytilid,[13] the authors calculate rates of methane oxidation necessary to support net growth (from methane) by the mussels and estimate growth rates based on ambient methane concentrations.

Table 1. Species designations of *Bathymodiolus*-like mytilids from the Gulf of Mexico.

Species	Collection Sites	Depth (m)	Refs.
Seep mytilid Ia[a]	Shallow seeps	500–700	3, 6, 9, 14, 19, 20, 22–25, 28, 32
Seep mytilid Ib[a]	Alaminos Canyon	2200	4, 12, 26
Seep mytilid II	Alaminos Canyon	2200	4, 12, 26
Seep mytilid III	Shallow seeps	500–700	This work
Seep mytilid Va	Florida Escarpment	3250	5, 7, 8, 17, 29–31
Seep mytilid Vb	Florida Escarpment	3200	b

[a]Seep Mytilids Ia and Ib are very similar and may be subspecies.
[b]C. Craddock, R. Vrijenhoek, R. Lutz, and R. Turner, personal communication.

Stable Carbon Isotope Determinations

$\delta^{13}C$ values of gill tissue from seep mussel species III were determined with standard methods, using Craig-type combustion techniques,[33,36] by the Geochemical and Environmental Research Group at Texas A & M University. The tissues were freeze-dried and acidified to remove possible carbonate contamination prior to combustion. Carbon dioxide from the combusted samples was analyzed on a Finnigan Mat-251 isotope ratio mass spectrometer and reported relative to PDB.

Results and Discussion

Distribution of Symbiont-Containing Mytilids in the Gulf of Mexico

Taxonomy of the symbiont-containing deep sea mytilids is rather confusing as only one species has been assigned a specific name to date (*Bathymodiolus thermophilus* from hydrothermal vents), and there are at least four other related species in the Gulf of Mexico. However, geneticists and taxonomists working on these mytilids have divided them into several groups on the basis of both variations in the allozyme banding patterns at 24 presumptive gene loci and differences in shell morphology (C. Craddock, R. Vrijenhoek, R. Lutz, and R. Turner, personal communication). In Table 1, the taxonomy and known range of these related mytilids from the Gulf of Mexico are summarized, along with a listing of references concerning each of these mytilids. Each group represents a population which is recognizable from the others. Unique roman numerals indicate differences minimally at the species level and different letters indicate differences minimally at the subspecies level. In Table 2, the types of symbionts

Table 2. Evidence of symbiont activity in seep mytilids from the Gulf of Mexico.

Sp.	Symbiont	Enz. Act.	IM	$\delta^{13}C$ (‰)	$^{14}CH_4$	°S	Refs.
SM Ia	M	MD	+	−57.6 to −40.1	+	−	3, 9, 14
SM Ib	M	MD	nd	−48.7 to −40.3	nd	nd	12, 26
SM II	M & CA	MD, A, R	+ & −	−55.7 to −44.4	nd	+	12, 26
SM III	CA	R, A	−	−38.1 to −38.2	nd	nd	TR
SM Va	M	MD, HPS	+ & −	−76.4 to −62.3	nd	nd	5, 8, 30
SM Vb	?	nd	nd	nd	nd	nd	nd

Abbreviations: Enz. Act. = Symbiont enzyme activity detected in significant amounts; IM = stacked internal membranes visualized by transmission electron microscopy; $^{14}CH_4$ = methanotrophy confirmed by oxidation of ^{14}C-methane; °S = significant levels of elemental sulfur detected in gills; SM = seep mytilid; M = methanotroph; CA = chemoautotrophic sulfur-oxidizer; nd = not determined; MD = methanol dehydrogenase; A = ATP sulfurylase; R = RuBP carboxylase; HPS = hexulose phosphate synthase; TR = this report.

present in each group of mussels are presented along with the kinds of data used to indicate their presence.

The seep mytilids with methanotrophic symbionts are found in three main areas in the Gulf of Mexico (Figure 1). The confirmed distributions of each type is undoubtably smaller than the actual distribution of these species in the Gulf of Mexico, as confirmation is usually dependent on submersible exploration. Surface trawling is not practical at most sites both because these animals are often associated with carbonate outcrops and because of the extreme depth at some sites. The Alaminos Canyon communities were only discovered in March 1990.[4]

Symbioses in Gulf of Mexico Seep Mytilids

Seep mytilid Ib was collected at the same location as species II during ALVIN dive 2011. The levels[26] of methanol dehydrogenase present in the gills of species Ib (~7 units/g) are higher than has been reported in the gills of any other mussel, including its shallow-living closest relative, species Ia.[14] The absence of detectable RuBP carboxylase in the gills of species Ib provides further evidence that the activities reported in the co-occurring species II are not due to contaminating bacteria on the surfaces of its gills, but rather to the presence of a second symbiont in that species. As one might expect from the different autotrophic symbiont complements in their gills, the stable carbon isotope composition of the gills of these two co-occurring species are significantly different; SM Ib gill $\delta^{13}C$ = -43.9 ± 1.8‰ and SM II gill $\delta^{13}C = -48.8 \pm -1.7$‰ (mean ± SE, n =

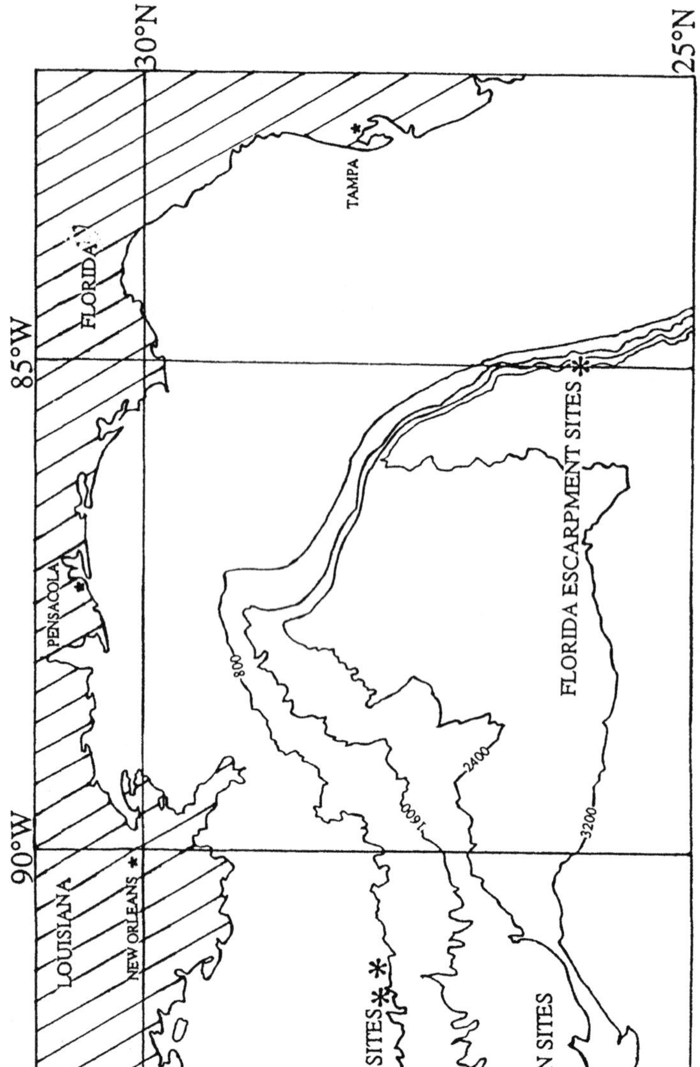

Figure 1. Locations of confirmed mytilid communities at Gulf of Mexico cold-seep sites. Depth contours in meters.

5 for each, $p = 0.02$ Student's *t*-test).[26] However, the $\delta^{13}C$ values in species II, which harbors both methanotrophic and chemoautotrophic sulfur-oxidizing symbionts are more negative than the values found in species Ib which contains only methanotrophic symbionts (Table 2). This observation is contrary to what one would first predict. However, a probable explanation is that the more negative values are a result of the chemoautotroph refixing light CO_2 released by the methanotroph which is often present in the same vacuoles of the same cells.[12]

Seep mytilid III has only recently been recognized from collections at the shallow hydrocarbon seep sites. It is rare there and constitutes less than 1% of the mussels collected from those sites. All of the 10 specimens collected to date are much smaller (up to about 5 cm total shell length) than the largest, or even the average, individuals of species Ia, with which it co-occurs. Thus, it constitutes an insignificant portion of the biomass of the shallow mussel communities.

Species III does contain symbionts which are apparently chemoautotrophic sulfur-oxidizers and not methanotrophic. Levels of RuBP carboxylase in gill tissue ranged from 0.036 to 0.335 units g^{-1} min^{-1} (mean = 0.22, SD = 0.11, $n = 5$) in species III, and methanol dehydrogenase was undetectable (gill tissue from species Ia was used as a positive control in the MDH assays). Furthermore, ATP sulfurylase was present at activities ranging from 2.2 to 10.0 units g^{-1} min^{-1} (mean = 5.6, SD = 2.5, $n = 5$) in the gill tissue of species III. These levels of both RuBP C/O and ATP sulfurylase are higher than levels reported for similarly frozen gill tissue of *Bathymodiolus thermophilus*,[14] a hydrothermal vent species with chemoautotrophic sulfur-oxidizing symbionts. Although ATP sulfurylase is not diagnostic for chemoautotrophic sulfur-oxidizing bacteria (because it is also found at low activities in some yeast and metazoans), it has proven to be an excellent indicator for the presence of chemoautotrophic sulfur-oxidizing symbionts in bivalves. ATP sulfurylase is not detectable in any bivalve tissues tested except the gills of species with chemoautotrophic sulfur-oxidizing symbionts (see Ref. 11 for review) and is not detectable in the gills of closely related mussels with only methanotrophic symbionts.[5,14] The stacked internal membranes which are characteristic of type I methanotrophs and have been observed in the methanotrophic symbionts of other mussels, are not present in the symbionts of species III. $\delta^{13}C$ values for five individuals ranged from −38.1 to −36.2‰. These values are more positive than those reported for any metazoans with methanotrophic symbionts (see Ref. 11 for review) and are in the range of $\delta^{13}C$ values reported for vestimentiferans (which also contain chemoautotrophic sulfur-oxidizing symbionts) from the same habitat.[3,15] The relatively negative $\delta^{13}C$ values of the metazoans with chemoautotrophic symbionts in this habitat are likely a reflection of ^{12}C-enriched inorganic carbon produced by free-living and symbiotic methanotrophs in the same habitat.

Therefore, seep mytilid III is not competing directly with seep mytilid Ia for an energy source. On the other hand, it may be competing for oxygen in the dense mussel beds, as species Ia has quite high respiratory demands,[9,20] and species III seems to occur on the bottom of the beds (J. Childress, personal communication and personal observation. Interestingly, species III harbors a commensal scale worm which appears similar to the commensal scale worm, *Branchipolynoe symmytilida,* found in both *B. thermophilus* and Seep mytilid Va. The other mytilids from the shallow seep sites and Alaminos Canyon do not harbor similar commensals (personal observation).

Community Methane Consumption Rates

In the three areas of the Gulf of Mexico where communities of mussels have been found, the communities are dominated by mussels with methanotrophic symbionts. Only the shallow sites have been studied sufficiently for estimates of community methane oxidation rates. At most sites, the distribution of the mussels is quite patchy, with the mussels occurring in discrete beds, apparently in the vicinity of a methane source. In fact, methane bubbles have been observed and collected from within several of these beds.[3,25] Patches around "Bush Hill" (one of the shallow sites) range in size from about 1 m^{-2} to approximately 20 m^{-2}.[41] A particularly impressive aggregation of mussels has been described surrounding a brine-filled pockmark in the northern Gulf of Mexico. At that site, live mussel coverage is continuous over an area of about 540 m^2.[24] Densities and size distributions of mussels within several discrete beds at a shallow site known as Mussel Beach were determined by Macdonald et al.[25] and are used in the calculations below.

A logarithmic regression of shell length and tissue wet weight in 120 mussels between 12 and 110 mm in length gives the allometric relation:

$$\text{Tissue wet weight (g)} = 2.88 \times 10^{-5}\ (\text{shell length (mm)})^{3.162}$$
$$(R^2 = 0.966).$$

Using this relation and the densities and size distributions for discrete "young" and "old" mussel beds given by Macdonald et al.,[25] one can calculate the biomass in mussel beds dominated by "old mussels" to be 18.25 kg m^{-2} and the biomass in the two "young mussel" beds to be 24.1 and 34.5 kg m^{-2}. A map constructed from a photomosiac of 62 m^2 of the Mussel Beach site[25] indicates that about 10.3% of the area is covered by "old mussel" beds and 10.9% by "young mussel" beds. Biomass calculations based on this coverage yield density estimates of 4.5 to 5.6 kg m^{-2} for this general area.

The shallow seep mytilid Ia has been extensively studied in the laboratory where it can be maintained for over a year, although progressive loss of symbionts over that time is indicated by the reduced methane oxidizing activity in laboratory-maintained animals.[20] Rates of methane consumption by mussels in the laboratory were highly dependent on external methane and oxygen concentrations, and maximal rates in individuals under optimal conditions ranged from 1.8 to about 14 μmol methane g^{-1} h^{-1}.[20] The authors indicate that a methane consumption rate of 4–5 μmol g^{-1} h^{-1} is "reasonable"; however, they also note rates of up to 10–13 μmol g^{-1} h^{-1} in animals tested very soon after collection. For purposes of the following calculations, I will assume a maximal possible methane oxidation rate for the mussels of 10 μmol g^{-1} h^{-1}, as it is probable that the laboratory-based measurements include animals with declining symbiont populations.

All measurements made to date of methane concentration in bottom waters among the mussels are likely to be underestimates due to the bulk water sampling methodology employed. The sampler used with the Johnson SeaLink is designed for relatively large water samples and requires large flushing volumes. Therefore, water sampled with this apparatus from among the mussels is significantly diluted with overlying ambient water and is not an accurate sample of the interstitial water between the mussels in the mussel beds. For this reason, I will use the highest published values obtained from water sampled with this system (60–70 μM CH_4 in water taken from around mussels[23]) for estimations of community methanotrophy.

According to the model of Kochever et al.,[20] mussels capable of consuming methane at a rate of 10 μmol g^{-1} h^{-1} under optimal conditions will consume methane at a rate of 3.2 μmol g^{-1} h^{-1} at an ambient methane concentration of 65 μM methane (a rate of at least 1.1 μmol g^{-1} h^{-1} is required to meet the respiratory demands of this size mussel). This suggests community oxidation rates of between 22.4 and 42.4 g CH_4 m^{-2} d^{-1} for isolated patches of mussels (and for the entire 540-m^2 patch of mussels around the brine-filled pockmark described above).* In the general area of the Mussel Beach site studied by Macdonald et al.[25] and described above, the average calculated rates of methane consumption due to the mussels is between 5.5 and 6.9 g CH_4 m^{-2} d^{-1}. Although these estimates will certainly be refined with better environmental mea-

*These calculations assume uniform weight-specific metabolic and methanotrophic rates for a size range of mussels, although allometric variation in these parameters is to be expected. The size range of mussels included in these calculations is from 2 to 120 g wet weight, and the laboratory rate measurements were conducted on animals with an average wet weight of 23 g. Although the relations between body weight and metabolic rates have not been determined for any deep sea mussels, there is an extensive literature on this subject for intertidal mussels such as *Mytilus* (see Ref. 2). Assuming an allometric relation similar to that of *Mytilus,* one can expect the total variation in weight specific metabolism over this size range of mussels to be less than onefold, and rates from 23-g animals will fall near the center of this range. Thus, errors introduced through use of this single-rate value are minimal.

surements and when ongoing in situ growth studies are completed, the estimates are adequate for "order-of-magnitude" predictions. These rates are three to four orders of magnitude higher than those reported for natural soils[27,37,40] or aquatic microbial populations.[16,21] Whalen et al.[38] report rates of 45 g m^{-2} d^{-1} for topsoil covering a retired landfill and note that these rates drop to 6.1 mg L^{-1} d^{-1} when these soils are waterlogged. They interpret this drop as being due to the difference in gas phase and aqueous molecular-diffusion rates. The high rates reported here for the aquatic communities of mytilids is most likely a reflection of the high biomass in these communities, high ambient methane concentrations, as well as advantages inherent in the symbiotic habit, such as a regulated environment in an organ bathed in circulating seawater and specialized for gas exchange (the gills).

The Alaminos Canyon sites were very recently discovered and have only been visited twice by submersible.[4] However, it appears to harbor the most concentrated biomass of any of the cold-seep communities and some of the largest mussels of this type yet reported (personal observation). Activities of methanol dehydrogenase in both species present here are as high or higher than has been reported for any other animals with methanotrophic symbionts.[26] Therefore, the above estimates of community methane consumption rates for the shallow-water sites probably represent underestimates of the consumption rates at the Alaminos Canyon sites.

Only a small percentage of the many potential sites for communities of mussels with methanotrophic symbionts have been visited by submersible. Because of the high level of industrial interest in the hydrocarbon reserves in many areas of the Gulf of Mexico and the protection of the "chemosynthetic communities" by the Mineral Management Service, continued exploration and discoveries are almost certain. The very high rates of methane oxidation these communities are capable of suggest that even with their patchy distribution, these symbioses may contribute significantly to oxidation of oceanic methane. However, more data on large-scale distribution of the communities is needed before reliable estimates can be made for the Gulf of Mexico as a whole.

Acknowledgments

This work was supported by subcontract L100094 to Mineral Management Service project RF-6899 and NSF-EAR-9158113. Ship and submersible time were provided by MMS project RF-6899 to J.M. Brooks and the NOAA National Undersea Research Program. I thank the captain and crew of the RV *Seward Johnson* and RV *Atlantis II* as well as the subcrew and pilots of the Johnson SeaLink and Alvin submersibles for their excellent support. I thank J. Vodenichar and K. Kelly for technical support, I. MacDonald for discussion of distribution of Seep Mytilid Ia, and C. Craddock, R. Vrijenhoek, R.

Lutz, R. Turner, and R. Gustafson for access to unpublished information on species differences, J.M. Brooks and the Geochemical and Environmental Research Group at Texas A & M University for conducting the $\delta^{13}C$ determinations, and J. Childress, E. Nix, J. Vodenichar, and two anonymous reviewers for critically reviewing the manuscript.

References

1. Anthony, C., and L.J. Zatman. 1965. The microbial oxidation of methanol. The alcohol dehydrogenase of *Pseudomonas* sp. M27. Biochem. J. **96**:808–812.
2. Bayne, B.L. 1976. Marine Mussels, Their Ecology and Physiology, Cambridge University Press, New York.
3. Brooks, J.M., M.C. Kennicutt, C.R. Fisher, S.A. Macko, K. Cole, J.J. Childress, R.R. Bidigare, and R.D. Vetter. 1987. Deep-sea hydrocarbon seep communities: Evidence for energy and nutritional carbon sources. Science **20**:1138–1142.
4. Brooks, J.M., D.A. Wiesenburg, H. Roberts, R.S. Carney, I.R. MacDonald, C.R. Fisher, N.L.J. Guinasso, W.W. Sager, S.J. McDonald, R.A.J. Burke, P. Aharon, and T.J. Bright. 1990. Salt, seeps and symbiosis in the Gulf of Mexico. EOS **71**:1772–1773.
5. Cary, C., B. Fry, H. Felbeck, and R.D. Vetter. 1988. Multiple trophic resources for a chemoautotrophic community at a cold water brine seep at the base of the Florida Escarpment. Marine Biol. **100**:411–418.
6. Cary, S.C., C.R. Fisher, and H. Felbeck. 1988. Mussel growth supported by methane as sole carbon and energy source. Science **240**:78–80.
7. Cavanaugh, C.M. 1985. Symbiosis of chemoautotrophic bacteria and marine invertebrates from hydrothermal vents and reducing sediments. Bull. Biol. Soc. Washington **6**:373–388.
8. Cavanaugh, C.M., P.R. Levering, J.S. Maki, M.E. Lidstrom, and R. Mitchell. 1987. Symbiosis of methylotrophic bacteria and deep-sea mussels. Nature **325**:346–348.
9. Childress, J.J., C.R. Fisher, J.M. Brooks, M.C. Kennicutt II, R. Bidigare, and A. Anderson. 1986. A methanotrophic marine molluscan symbiosis: Mussels fueled by gas. Science **233**:1306–1308.
10. Felbeck, H. 1981. Chemoautotrophic potential of the hydrothermal vent tube worm, *Riftia pachyptila* Jones (Vestimentifera). Science **213**:336–338.
11. Fisher, C.R. 1990. Chemoautotrophic and methanotrophic symbioses in marine invertebrates. Rev. Aquat. Sci. **2**:399–436.
12. Fisher, C.R., J.M. Brooks, J.S. Vodenichar, J.M. Zande, J.J. Childress, and R.A. Burke Jr. 1993. The co-occurrence of methanotrophic and chemoantrophic sulfur-oxidizing bacterial symbionts in a deep-sea mussel. Marine Ecol. (in press).
13. Fisher, C.R., J.J. Childress, A.J. Arp, J.M. Brooks, D. Distel, J.A. Favuzzi, H.

Felbeck, R. Hessler, K.S. Johnson, M.C. Kennicutt II, S.A. Macko, A. Newton, M.A. Powell, G.N. Somero, and T. Soto. 1988. Microhabitat variation in the hydrothermal vent mussel *Bathymodiolus thermophilus*, at Rose Garden vent on the Galapagos rift. Deep-Sea Res. **35**:1769–1792.

14. Fisher, C.R., J.J. Childress, R.S Oremland, and R.R. Bidigare. 1987. The importance of methane and thiosulphate in the metabolism of the symbionts of two deep-sea mussels. Marine Biol. **96**:59–71.

15. Fisher, C.R., M.C. Kennicutt II, and J.M. Brooks. 1990. Stable carbon isotopic evidence for carbon limitation in hydrothermal vent vestimentiferans. Science **247**: 1094–1096.

16. Frenzel, P., B. Thebrath, and R. Conrad. 1990. Oxidation of methane in the oxic surface layer of a deep lake sediment (Lake Constance). FEMS Microbiol. Ecol. **73**:149–158.

17. Hecker, B. 1985. Fauna from a cold sulfur seep in the Gulf of Mexico: Comparison with hydrothermal vent communities and evolutionary implications. Bull. Biol. Soc. Washington **6**:465–474.

18. Kenk, V.C., and B.R. Wilson. 1985. A new mussel (Bivalvia, Mytilidae) from hydrothermal vents in the Galapagos Rift Zone. Malacologia **26**:253–271.

19. Kennicutt, M.C., II, J.M. Brooks, R.R. Bidigare, and G.J. Denoux. 1988. Gulf of Mexico hydrocarbon seep communities—I. Regional distribution of hydrocarbon seepage and associated fauna. Deep-Sea Res. **35**:1639–1651.

20. Kochevar, R., J.J. Childress, C.R. Fisher, and E. Minnich. 1992. The methane mussel: roles of symbiont and host in the metabolic utilization of methane. Marine Biol. **112**:389–401.

21. Kuivila, K.M., J.W. Murray, A.H. Devol, M.E. Lidstrom, and C.E. Reimers. 1988. Methane cycling in the sediments of Lake Washington. Limnol. Oceanogr. **33**:571–578.

22. Lee, R.W., E.V. Thuesen, J.J. Childress, and C.R. Fisher. 1992. Ammonium and free amino acid uptake by a deep-sea mussel containing methanotrophic bacterial symbionts. Marine Biol. **113**:99–106.

23. MacDonald, I.R., G.S. Boland, J.S. Baker, J.M. Brooks, M.C. Kennicutt II, and R.R. Bidigare. 1989. Gulf of Mexico hydrocarbon seep communities. II. Spatial distribution of seep organisms and hydrocarbons at Bush Hill. Marine Biol. **101**: 235–247.

24. MacDonald, I.R., J.F.I. Reilly, N.L.J. Guinasso, J.M. Brooks, R.C. Carney, W.A. Bryant, and T.J. Bright. 1990. Chemosynthetic mussels at a brine-filled pockmark in the northern Gulf of Mexico. Science **248**:1096–1099.

25. MacDonald, I.R., W. Russell Callender, R.A. Burke, Jr., S.J. McDonald, and R.S. Carney. 1990. Fine-scale distribution of methanotrophic mussels at a Louisiana cold seep. Prog. Oceanog. **24**:15–24.

26. McDonald, S.J., D.A. Wiesenburg, C.R. Fisher, J.M. Brooks, R.A. Burke Jr., W.W. Sager, P. Aharon, and R.S. Carney. 1992. Gulf of Mexico hydrocarbon seep

communities: VIII. New deep water petroleum seep communities in Alaminos Canyon. Deep-Sea Res. (in submission).

27. Mosier, A., D. Schimel, D. Valentine, K. Bronson, and W. Parton. 1991. Methane and nitrous oxide fluxes in native, fertilized and cultivated grasslands. Nature **350**:330–332.

28. Page, H.M., C.R. Fisher, and J.J. Childress. 1990. The role of suspension-feeding in the nutritional biology of a deep-sea mussel with methanotrophic symbionts. Marine Biol. **104**:251–257.

29. Paull, C.K., B. Hecker, R. Commeau, R.P. Freeman-Lynde, C. Neumann, W.P. Corso, S. Golubic, J.E. Hook, E. Sikes, and J. Curray. 1984. Biological communities at the Florida Escarpment resemble hydrothermal vent taxa. Science **226**: 965–966.

30. Paull, C.K., A.J.T. Jull, L.J. Toolin, and T. Linick. 1985. Stable isotope evidence for chemosynthesis in an abyssal seep community. Nature **317**:709–711.

31. Paull, C.K., C.S. Martens, J.P. Chanton, A.C. Neumann, J. Coston, A.J.T. Jull, and L.J. Toolin. 1989. Old carbon in living organisms and young $CaCO_3$ cements from abyssal brine seeps. Nature **342**:166–168.

32. Powell, E.N., A.C. Morrill, and R.R. Bidigare. 1989. Catalase in sulfide- and methane-dependent macrofauna from petroleum seeps. Experientia **45**:198–200.

33. Sackett, W.M., W. Nakaparksin, and D. Dalrymple. 1970. Carbon isotope effects in methane production by thermal cracking. In G.D. Hobson and G.C. Speers (eds.), Advances in Organic Geochemistry, Pergamon Press, Oxford, pp. 37–53.

34. Schmaljohann, R. 1987. Endosymbiosen zwischen methylotrophen Bakterien und marinen Invertebraten. Forum Mikrobiologie **10**:166–171.

35. Schmaljohann, R., and H.J. Flügel. 1987. Methane-oxidizing bacteria in Pogonophora. Sarsia **72**:91–98.

36. Schoell, M., E. Farber, and M.L. Coleman. 1983. Carbon and hydrogen isotope comparisons of the NBS-22 and NBS-21 stable isotope reference materials: an inter-laboratory comparison. Organ. Geochem. **5**:3–6.

37. Whalen, S.C., and W.S. Reeburgh. 1990. Consumption of atmospheric methane by tundra soils. Nature **346**:160–162.

38. Whalen, S.C., W.S. Reeburgh, and K.A. Sandbeck. 1990. Rapid methane oxidation in a landfill cover soil. Appl. Environ. Microbiol. **56**:3405–3411.

39. Wishnick, M., and M.D. Lane. 1971. Ribulose diphosphate carboxylase from spinach leaves. Meth. Enzym. **23**:570–577.

40. Yavitt, J.B., D.M. Downey, G.E. Lang, and A.J. Sexstone. 1990. Methane consumption in two temperate forest soils. Biogeochemistry **9**:39–52.

41. Brooks, J.M., M.C. Hennicutt II, I.R. MacDonald, D.L. Wilkinson, N.L. Guinasso Jr., and R.R. Bidigare. 1989. Gulf of Mexico hydrocarbon seep communities: Part IV—Descriptions of known chemosynthetic communities. Offshore Technology Conference Proceedings 5954.

Seasonal Variations in the Isotopic Composition of Methane Associated with Aquatic Macrophytes

Jeffrey P. Chanton

Department of Oceanography, Florida State University, Tallahassee, Florida 32306

Lesley K. Smith

Center for Limnology, Department of Environmental, Organismic and Population Biology, University of Colorado Boulder, Colorado 80309-0334

ABSTRACT

Samples of methane bubbles were collected from sediments populated by emergent aquatic macrophytes during the early rising water period on the Amazon River floodplain. Methane $\delta^{13}C$ was distinctly lighter (−61.3 ± 0.3‰, n = 3) in bubbles stirred from sediments populated by the rooted C-3 plant *Oryza perinnis* than it was in bubbles from sediments containing the rooted C-4 plants *Paspalum repens* (−52.0 ± 1.1‰, n = 6) or *Echinochloa polystachya* (−48.3 ± 0.2‰, n = 4). Leaf carbon $\delta^{13}C$ was −28.2‰ in *Oryza* and −11.4‰ in *Paspalum*. In August, just after the peak flood stage of the river, the *Paspalum*, now a detached floating mat, was sampled again. Methane bubbles trapped within the floating meadow had an isotopic composition of −43.5 ± 0.1‰ (n = 3), suggesting an increase in the importance of microbial methane oxidation as the emergent aquatic macrophytes progressed through their annual life cycle from a rooted plant to a floating meadow. The life cycle of these macrophytes may be partly responsible for the observed 10‰ seasonal variation in the isotopic composition of methane emitted from the Central Amazon floodplain.

Introduction

Natural wetlands contribute substantially to the tropospheric methane inventory. Studies combining estimates of the global distribution of wetlands and published methane emission rates calculate that wetlands contribute about 100

Tg yr^{-1} or about 25% of the global biogenic source.[3,14,30] A recent update of an earlier estimate[35] of natural wetland sources of tropospheric methane has reemphasized the importance of tropical wetlands; they may contribute 50% rather than 29% of the total wetland methane emissions.[3]

The methane emission from the Amazon River floodplain is estimated to be $(2\text{–}14) \times 10^{12}$ g yr^{-1},[3,20] which is 2–10% of the total estimated biogenic source. Of the three main habitats found on the Central Amazon floodplain, open water, flooded forest, and emergent aquatic macrophytes, the macrophyte habitat has the highest emission rate on an areal basis and represents about 40% of the total methane emission from the floodplain.[3,19,20]

Seasonality on the Amazon floodplain is driven by the 10-m change in stage height of the Amazon River. As a consequence, the character of the floodplain oscillates between dry land for approximately 6 months of the year and inundated wetland over the remainder of the year. The aquatic emergent macrophytes undergo an annual cycle of growth and decay that is linked to the changing water levels.[28] During low water, their presence is minimal, whereas during high water, they may cover up to ~50% of the inundated portions of the basin.[23] Aquatic macrophytes are dominated by the C-4 grasses *Paspalum repens* and *Echinochloa polystachya*.[28,4] At the beginning of the rainy season (December), the grasses grow rooted in the soil, and as the water level rises, they become uprooted, primarily *Paspalum* and, to a lesser extent, *Echinochloa*, and form floating islands or meadows.[28]

Knowledge of the stable isotopic composition of methane is useful in constructing global atmospheric models of methane sources and sinks.[17,42,47,48] Stable isotope measurements also serve as useful tracers of processes affecting methane cycling within an individual system. Measurement of the $\delta^{13}C$ or the combination of $\delta^{13}C$ and δD (hydrogen, deuterium) of methane can reveal key information regarding methane production pathways,[6,7,51] methane oxidation,[1,12,16,31] and methane transport processes.[8,10,13]

The isotopic composition of organic carbon is dependent on the photosynthetic pathway by which it is fixed.[5,40] Generally, C-3 plants have an isotopic composition of about −28‰, whereas C-4 plants are about −14‰.[24,40,45,46] Other effects, such as the isotopic composition of the carbon dioxide pool used for photosynthetic carbon fixation, can also be important.[22] The resultant isotopic signature of the parent organic material is altered by fermentative bacterial degradation of organic matter. Kinetic isotopic fractionation results in the production of highly ^{13}C-depleted methane. The isotopic composition of biogenically produced methane ranges from −50 to −110‰ but clusters in the range of −50 to −80‰.[15,43,44] Variations in methane isotopic signature depend on the isotopic composition of the initial organic substrates and whether the methane was produced from acetate fermentation or carbon dioxide reduction,[51] which are the two main metabolic pathways for methanogenesis.

Oxidative processes can also have a profound effect on methane isotopic composition. For example, in the Central Amazon basin, methane emitted from regions covered by emergent aquatic macrophytes during rising- or high-water periods (floating meadows[28]) is enriched in ^{13}C, with a value of $-44.4 \pm 4.2‰$.[11] Methane bubbles trapped within the floating meadow matrix are also ^{13}C enriched; bubbles collected during the rising-water period ranged between -41 and $-46‰$.[11,21] These values stray from the range typically found in biogenically produced methane (-50 to $-80‰$). It has been hypothesized[11,21] that the ^{13}C-enriched methane emitted by the floating meadows was a result of a combination of the C-4 photosynthetic pathway utilized by the macrophytes and aerobic microbial oxidation of the methane trapped within the floating meadows. Methane oxidizing microbes are known to consume $^{12}CH_4$ preferentially, leaving residual methane enriched in $^{13}CH_4$.[1,16] The floating meadows offer an excellent environment for methanotrophic bacteria because methane is present within the meadow matrix and oxygen is found at concentrations ranging from 0.2 to 4.0 mg L^{-1} (Ref. 2; L. Smith, unpublished data).

Seasonal variations of the order of 10‰ in the isotopic composition of methane emitted from the Central Amazon floodplain have been reported.[21] In periods of rising water, the isotopic composition of methane emitted from the floodplain was $-52‰$, and during low water, it was $-61‰$. Devol et al.[21] suggested that this seasonal variation results from the variation in the magnitude of the changing areal extent of the macrophytes. It was our hypothesis that, in addition to the changing areal extent of the macrophytes, their physiological changes contributed to the seasonal shift in the isotopic composition of CH_4 emitted to the atmosphere from the floodplain. We hypothesized that as the macrophyte beds make the transition from being rooted in the sediments early in the rising-water season to becoming floating mats in July and August, the methane within them becomes more ^{13}C enriched due to increased methane oxidation. We do not suggest that our hypothesized mechanism was the only one controlling the observed seasonal variations from the floodplain, only that they could be an additive factor. We also wanted to evaluate the importance of the carbon fixation pathway in determining the isotopic composition of methane by simultaneously sampling methane from C-3 and C-4 plants under rooted conditions.

Methods

Sampling was conducted at Lake Calado (3°15′S, 60°34′W), a dendritic, floodplain lake with a permanent connection to the Solimões River (the portion of the Amazon River located upriver from the Negro River). The lake is located on the northern reaches of the river about 80 km west of the confluence with the Negro River (Figure 1). In response to the annual 10 m change in river stage

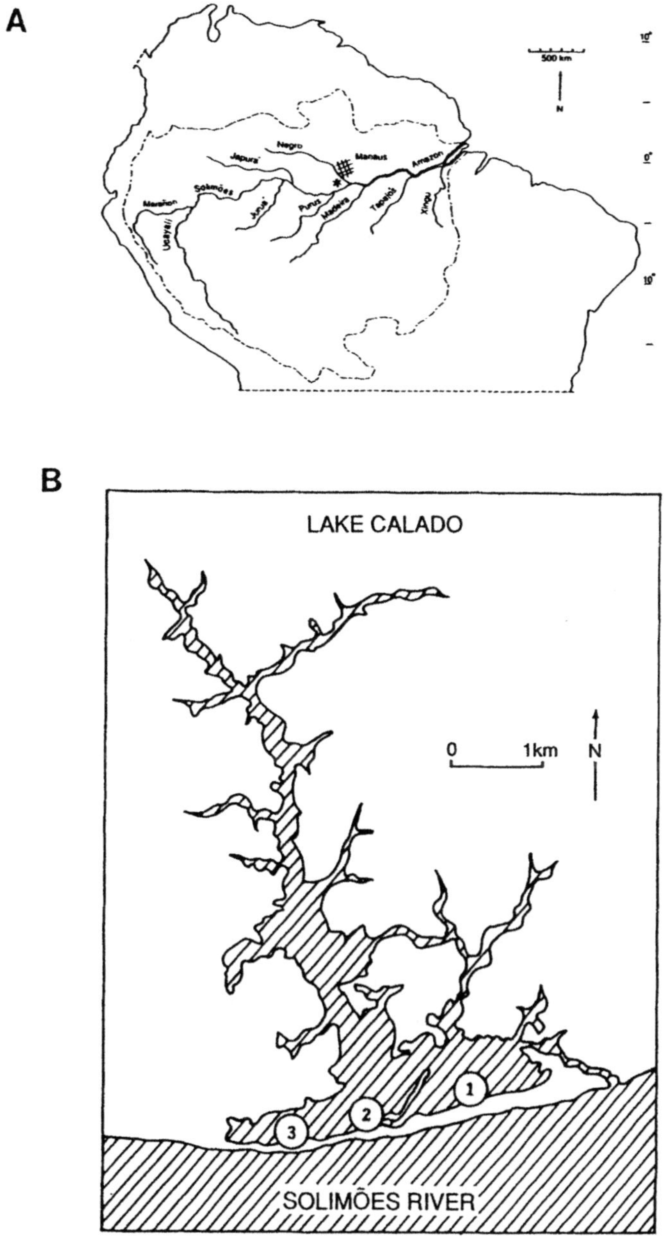

Figure 1. Lake Calado is located on the north bank of the Solimoes River, ~80 km west of the confluence of the Negro and the Amazon Rivers (panel A, marked with an asterisk). Sampling sites are indicated on the map of the lake (panel B).

height, the lake's surface area ranges from 2 to 8 km^2 and its maximum depth fluctuates between 1 and 13 m. Detailed limnological studies can be found in Refs. 18 and 36–38.

The three habitats found on the Central Amazon floodplain, open water, flooded forest, and floating emergent macrophytes, are well-represented in the lake. During low water, *Oryza* (wild rice) is the dominant macrophyte species found in the lake, and during rising water, *Paspalum repens* and *Echinochloa polystachya* dominate, accounting for 80–90% of the floating meadow biomass.[23] *Paspalum* colonizes the open-water areas, whereas *Echinochloa* remains attached to the submerged soils and develops floating stems. *Echinochloa* typically forms attached stands between the shoreline and the free-floating *Paspalum* which often occupies deeper waters.

Methane-rich bubbles were collected from three sources: (1) the sediment–root mat complex of macrophytes growing under rooted conditions (*Paspalum, Echinochloa* and *Oryza*), (2) macrophyte root mats under floating conditions (*Paspalum*), and (3) sediments underlying floating meadows of *Paspalum*. The bubbles were captured under water with a submerged inverted funnel sealed with a three-way stopcock located on the top. The bubble samples were transferred to submerged inverted vials by water displacement and sealed with butyl rubber stoppers.

Methane was prepared for isotopic analysis by combustion to CO_2 in a helium gas stream passed over 800°C copper oxide.[8] Resultant CO_2 was purified cryogenically, sealed into glass tubes, and analyzed on a Finnegan Mat Delta D Isotope Ratio Mass Spectrometer in the laboratory of Dr. Neal Blair at North Carolina State University.

The ratio of light to heavy isotopes is expressed in the "δ" notation, which indicates the depletion (negative) or the enrichment (positive) of the heavy isotope relative to the light isotope according to a known standard. The notation is defined as follows[27]:

$$\delta‰ = \left(\frac{R_{\text{sample}}}{R_{\text{standard}}} - 1\right) \times 1000,$$

where R_{sample} and R_{standard} are the $^{13}C/^{12}C$ ratios of the sample of interest and a standard respectively (Peedee belemnite, 0‰).

During the rising-water period (March 1989), all macrophytes were rooted, and gas samples were collected by stirring the sediments in which they were growing. At this time, the *Oryza* had been inundated and was beginning to decay on the sediment–water interface. Both gas and grass samples were collected from the sediments at Station 1 (see Figure 1). Station 2 was dominated by *Paspalum repens*. Gas samples were collected from the sediment–root mat. Root and leaf samples of the *Paspalum* were collected for ^{13}C analysis. Station 3 was dominated by *Echinochloa;* only gas samples were collected from the sediments–root mat at this site.

During this rising-water period, an experiment was conducted to test the hypothesis that bubbles trapped within the floating meadows would become ^{13}C enriched over time. Bubbles were stirred from the sediments of Station 2 and collected within the funnel system described above. Subsamples were collected from the funnel to determine the initial methane concentration and isotopic composition. The funnel was then lodged within the *Paspalum* mat and left for 1 week after which time the gas was resampled for methane concentration and isotopic composition.

During the high-water period, *P. repens* became unrooted and developed extensive floating meadows. Station 2 was revisited in August 1989, and bubbles were collected both from the sediments located below the floating *Paspalum* and from within the dense floating meadow.

All collected plant material was oven-dried (50°C) in Brazil and shipped to the United States for analysis. Organic C and N concentrations were determined with a Carlo Erba CHN analyzer. Organic carbon samples were analyzed for ^{13}C by Coastal Science Laboratories in Austin, Texas.

Methane concentrations were determined within 24 h by flame-ionization gas chromatography.

Results

During rising water (March), all of the macrophytes in Lake Calado were rooted, and the isotopic composition of bubbles stirred from the sediments within which they grew varied from −61 to −48‰ (Table 1). The most ^{13}C-depleted methane was found associated with the decaying *Oryza,* a C-3 pathway plant,[39] whose carbon is light compared to the other C-4 aquatic plants found in the lake (Table 2). Methane associated with the C-4 plants, *Paspalum* and *Echinochloa,*[25,41] was ^{13}C enriched at values of −52 and −48‰, respectively. Our value for the $\delta^{13}C$ of *Paspalum* leaves is similar to that obtained by others.[26] A root sample of *Paspalum* had distinctly lighter isotopic signature than did the leaf sample, although the C : N ratio was not different.

The methane that was incubated within the funnel in the rooted *Paspalum* (Station 2) underwent significant changes. Initially, the bubbles contained 72–91% methane with an isotopic composition of −53‰. At the end of the 7-d incubation period, the bubbles trapped within the funnel consisted of 8–20% methane with an isotopic composition of −46‰.

In August, the *Paspalum* had formed extensive floating meadows at Station 2, and bubbles trapped within the meadow matrix were reduced in methane con-

Table 1. Methane concentration and methane $\delta^{13}C$ (standard deviation in parentheses or 1/2 range if $n = 2$) of gas bubbles associated with rooted emergent aquatic macrophytes. Data from 1989. n = the number of replicate samples analyzed. All samples were collected from the sediments unless indicated.

Plant Sampled	Station	Date	$\%CH_4$	$\delta^{13}C$ (‰)	n
Rising water					
Oryza perennis	1	3/13	31 (19)	−61.3 (0.3)	3
Paspalum repens	2	3/10	57 (5)	−51.7 (0.9)	4
Echinochloa polystacha	3	3/23	66 (1)	−48.3 (0.2)	4
Incubation experiment					
Paspalum repens	2	3/15	81 (10)	−53.0 (0.9)	2
Paspalum repens[a]	2	3/23	13 (6)	−45.5 (0.3)	3
Falling water					
Paspalum repens[b]	2	8/19	54 (1)	−54.9 (0.2)	2
Paspalum repens[a]	2	8/19	14 (1)	−43.5 (0.1)	3

[a]Sample collected from bubbles trapped within floating meadow matrix.
[b]Sample collected from bubbles within sediments below the floating meadow.

centration (14%) when compared to bubbles collected from the sediments (54%) located below the macrophytes. In addition, the methane trapped within the meadows had also become enriched in ^{13}C (Table 1).

Discussion

Within our dataset, the $\delta^{13}C$ of CH_4 within bubbles collected from the sediments of Lake Calado reflects to a certain degree the isotopic composition of the parent material from which it was derived. For example, methane stirred from the sediments–root mat where *Oryza* was found was 9–13‰ lighter in isotopic signature than methane collected from sediments–root mat inhabited by *Paspalum* and *Echinochloa.* This indicates that the photosynthetic pathway used by the emergent aquatic macrophytes to fix carbon has a sympathetic effect on the isotopic composition of methane formed during bacterial degradation of the plant material.

Over the March to August period, the *Paspalum* went through a transition from a mat rooted in the sediment to a floating meadow. Accompanying this transition, the methane within the root mat increased from −52‰ under rooted conditions to −43‰ under floating conditions. Methane concentrations within the bubbles of the mat decreased from 51–90% under rooted conditions to 14%

Table 2. Plant organic carbon and nitrogen composition, molar C/N ratio, and $\delta^{13}C$ of organic carbon.

Plant	%C	%N	C/N	$\delta^{13}C$
Oryza leaves	34.9	1.6	25.2	−28.2
Echinochloa leaves	nd	nd	nd	−12.4[a]
Paspalum leaves	41.4	2.7	18.0	−11.4
Paspalum roots	34.5	2.2	18.1	−16.0

[a]Data from Ref. 26.

under floating conditions. We were able to mimic what occurs naturally by incubating bubbles disturbed from the sediments within the funnel lodged within the *Paspalum* mat. The initial methane composition was −53‰ and 72–91% methane, and after 1 week the incubated bubbles were similar in composition (−45.5‰, 8–20% CH_4) to the bubbles found trapped within the floating meadows sampled in August (−43.5‰, 14% CH_4).

Two hypotheses have been proposed for the origin of methane found within the floating mats[2]: in situ methanogenesis versus trapping of bubbles released by ebullition from the sediments below. Our observations of changes in methane isotopic composition both under natural and experimental conditions (described earlier) support the latter hypothesis. We suggest that methane within the floating meadows originated within the sediments below them, was released by bubble ebullition, intercepted and trapped by the floating meadows, and partially oxidized while held within the *Paspalum* mat. This scenario is illustrated in Figure 2.

There are two additional lines of evidence which may suggest that the second hypothesis is correct. First, methane emissions from regions covered by floating meadows are principally the result of ebullition,[2,11] which is unusual for vegetated habitats.[9] Second, the dissolved methane concentration which would be at equilibrium with the bubbles which were found within the floating mat (14% CH_4) is 177 μM.[49] However, maximum concentrations of dissolved methane within the floating macrophyte beds have been determined to be 16.5 ± 15 μM (L. Smith, unpublished data) and 20–25 μM,[2] which is significantly less than the predicted value. Dissolved methane and radon concentrations have been found to be at equilibrium with bubble methane concentrations in other locations.[9,12,32] The low concentrations of dissolved methane relative to equilibrium conditions indicate that the methane bubbles should be dissolving into the waters of the floating meadow root mats rather than forming from dissolved methane within the mat.

From the isotopic composition of methane in sedimentary bubbles collected below the floating mats and those trapped within the floating mat in August,

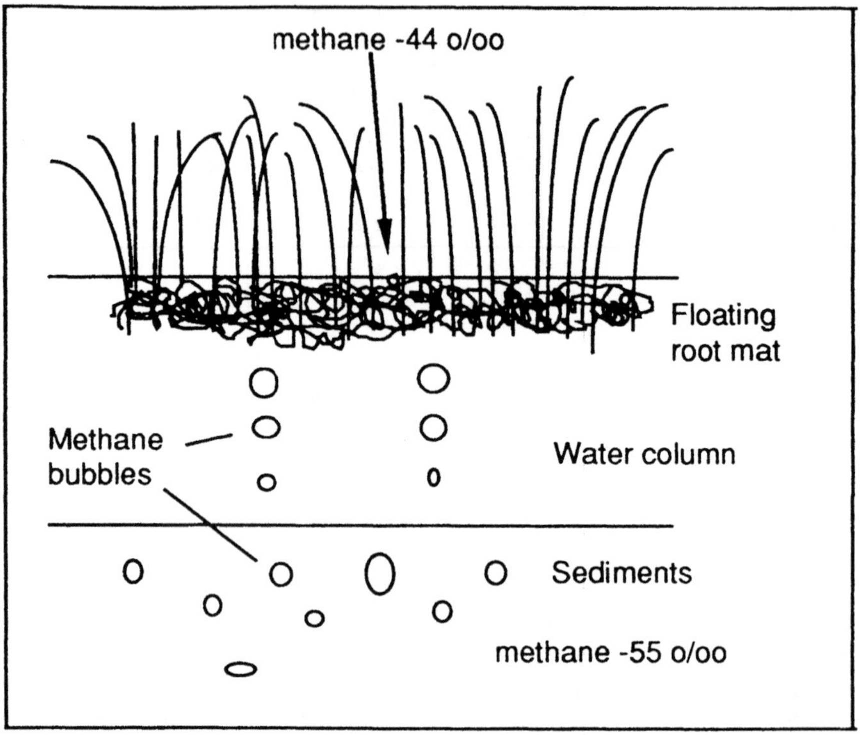

Figure 2. Our results suggest that methane is supplied to the floating meadows via gas-bubble ebullition from the sediments below. The isotopic signature of sedimentary methane in August when the mats were floating was −55‰. The isotopic composition of methane within gas bubbles trapped within the floating mat was −44‰ suggesting oxidation of 40–60% of the methane while it was trapped within the mat (see text). Depth of the water column is not drawn to scale.

−54.9 and −43.5‰, respectively (Table 1), we can offer a preliminary estimate of the fraction of methane oxidized while the bubbles were trapped within the floating mats. The isotopic composition of residual methane during oxidation can be approximated by a closed-system Rayleigh equation,[50]

$$\delta^{13}C_{CH_4} = (\delta^{13}C_0 + 1000F^{(1/\alpha - 1)}) - 1000,$$

where $\delta^{13}C_0$ is the isotopic composition of sedimentary methane, $\delta^{13}C_{CH_4}$ is the isotopic composition of methane which was sampled within the mat, α is the fractionation factor for aerobic methane oxidation, which may be between 1.013

and 1.025,[16] and F is the fraction of methane oxidized. The isotopic value of methane collected from sediments below the mat is used for the initial starting value of methane when it is first introduced to the floating mat ($\delta^{13}C_0$) because there is no isotopic fractionation during bubble release and transport.[8,9] Solving the equation for F, we suggest that between 40 and 60% (depending on α) of the methane delivered to the mat had been oxidized prior to our sample collection.

The above calculation assumes closed-system conditions within the mat, which are only approximated, because methane should be dissolving from the bubbles into the water. However, this is a reasonable approximation, as isotopic fractionation in this process is small. The equilibrium partitioning of methane between the aqueous phase and air is on the order of 0.3%.[4]

Our preliminary data set suggests that the seasonal variations in the isotopic composition of methane emitted from the floodplain may be partly due to changes in the magnitude of methane oxidation rates caused by the transition of the plants, particularly *Paspalum*, from rooted mats to floating meadows (our hypothesis) in addition to changes in the areal extent of the C-4 macrophytes as suggested by others.[21] This transition from rooted to floating conditions should result in more O_2 being supplied to the methanotrophic organisms. Some oxidation may occur within the rooted emergent aquatic macrophytes because we found that methane was ^{13}C enriched within the rooted *Echinochloa;* however, we suggest that the magnitude of methane oxidation is greater in the floating meadows than in the rooted emergent macrophytes.

Seasonal variations in the isotopic composition of methane on the order of 10‰ have been observed in temperate eutrophic marine environments,[6,33] temperate river estuaries,[8] and Alaskan lakes.[34] These studies have generally suggested that the observed seasonal variations in the isotopic composition of methane were caused by variations in production mechanisms. The Central Amazon basin appears to be unique because variations in oxidative processes appear to be an important control on the observed seasonal variation in the isotopic composition of methane emitted to the troposphere. To our knowledge, only one other study (of Minnesota Peat bogs[29]) has suggested that seasonal variations in the isotopic composition of emitted methane are due to oxidation processes. In both locations, the variations have been related to changing water levels.

Our results are consistent with previous results suggesting that floating meadows are a source of ^{13}C-enriched methane to the atmosphere.[11] They suggest that this enrichment is due both to the C-4 photosynthetic pathway that is predominantly used by the emergent aquatic macrophytes found on the Central Amazon floodplain and to methane oxidation that occurs within the dense floating-meadow assemblage. The average value for methane in freshwater environments is -60.0 ± 6.5‰.[51] The methane emitted by the floating meadows in Lake Calado appears to be outside the range of typical values.

Acknowledgments

This work was funded by NASA grant NAGW 1823 to J. Chanton and NSF grant BSR 87-06643 to T.R. Fisher and J.M. Melack. We thank Cheryl Kelley, Kelley Cox, Neal Blair, Jim Happell, and Ghislain Gerard for assistance with sample preparation and analysis, and Mark Moline and Becky Shaw for assistance with sample collection. Karen Bartlett, Patrick Crill, and Michelle Epp commented critically on the manuscript.

References

1. Barker, J.F., and P. Fritz. 1981. Carbon isotope fractionation during microbial methane oxidation. Nature **293**:289–291.
2. Bartlett, K.B., P.M. Crill, D.I. Sebacher, R.C. Harriss, J.O. Wilson, and J.M. Melack. 1988. Methane flux from the Central Amazonian Floodplain. J. Geophys. Res. **93**:1571–1582.
3. Bartlett, K.B., P.M. Crill, J.A. Bonassi, J.E. Richey, and R.C. Harriss. 1990. Methane flux from the Amazon river floodplain: Emissions during rising water. J. Geophys. Res. **95**:16773–16778.
4. Bernard, B.B., J.M. Brooks, and W.M. Sackett. 1976. Natural gas seepage in the Gulf of Mexico, Earth Planet. Sci. Lett. **31**:48–54.
5. Berry, J.A. 1989. Studies of mechanisms affecting the fractionation of carbon isotopes in photosynthesis. In J. Ehleringer, K. Nagy, and P. Rundel (eds.), Stable Isotopes in Ecological Research, Springer-Verlag, New York, pp. 82–95.
6. Burke, R.A., C.S. Martens, and W. Sackett. 1988. Seasonal variations of D/H and $^{13}C/^{12}C$ ratios of microbial methane in surface sediments. Nature **332**:829–831.
7. Burke, R.A., T.R. Barber, and W.M. Sackett. 1988. Methane flux and stable hydrogen and carbon isotope composition of sedimentary methane from the Florida Everglades. Global Biogeochem. Cycles **2**:329–340.
8. Chanton, J.P., and C.S. Martens. 1988. Seasonal variations in ebullitive flux and carbon isotopic composition of methane in a tidal freshwater estuary. Global Biogeochem. Cycles **2**:289–298.
9. Chanton, J.P., and J.W.H. Dacey. 1991. Effects of vegetation on methane flux, reservoirs, and carbon isotopic composition. In T. Sharkey, E. Holland, and H. Mooney (eds.), Trace Gas Emissions from Plants, Academic Press, San Diego, pp 65–92.
10. Chanton, J.P., G.G. Pauly, C.S. Martens, N.E. Blair, and J.W.H. Dacey. 1988. Carbon isotopic composition of methane in Florida Everglades soils and fractionation during its transport to the troposphere. Global Biogeochem. Cycles **2**:245–252.
11. Chanton, J.P., P.M. Crill, K.B. Bartlett, and C.S. Martens. 1989. Amazon Capims

(Floating Grassmats): A source of ^{13}C enriched methane to the troposphere. Geophys. Res. Lett. **16**:799–802.

12. Chanton, J.P., C.S. Martens, and C.A. Kelley. 1989b. Gas transport from methane-saturated tidal freshwater and wetland sediments. Limnol. Oceanog. **34**:807–819.

13. Chanton, J.P., C.S. Martens, C.A. Kelley, P.M. Crill, and W.J. Showers. 1992. Mechanisms of methane transport and isotope fractionation in macrophytes of Alaskan tundra lakes. J. Geophys. Res. **97**:16,681–16,688.

14. Cicerone, R.J. and R.S. Oremland. 1988. Biogeochemical aspects of atmospheric methane. Global Biogeochem. Cycles **2**:299–327.

15. Claypool, G.E., and I.R. Kaplan. 1974. The origin and distribution of methane in marine sediments. In I.R. Kaplan (ed.), Natural Gases in Marine Sediments, Plenum, New York, pp. 99–138.

16. Coleman, D.D., J.B. Risatte, and M. Schoell. 1981. Fractionation of carbon and hydrogen isotopes by methane oxidizing bacteria. Geochim. Cosmochim. Acta **45**:1033–1037.

17. Craig, H., C.C. Chou, J.A. Welhan, C.M. Stevens, and A. Engelkemeir. 1988. The isotopic composition of methane in polar ice cores. Science **242**:1535–1539.

18. Crill, P.M., K.B. Bartlett, J.O. Wilson, D.I. Sebacher, R.C. Harriss, J.M. Melack, S. MacIntyre, L. Lesack, and L. Smith-Morrill. 1988. Tropospheric methane from an Amazonian floodplain lake. J. Geophys. Res. **93**:1564–1570.

19. Devol, A.H., J.E. Richey, W.A. Clark, S.L. King, and L.A. Martinelli. 1988. Methane emissions to the troposphere from the Amazon floodplain. J. Geophys. Res. **93**:1583–1592.

20. Devol, A.H., J.E. Richey, B.R. Forsberg, and L. Martinelli. 1990. Seasonal dynamics in methane emissions from the Amazon River floodplain to the troposphere. J. Geophys. Res. **95**:16417–16426.

21. Devol, A.H., J.E. Richey, S.L. King, J. Lansdown, and L. Martinelli. 1993. Seasonal variations in the ^{13}C–CH_4 of Amazon floodplain waters. In D. Adams, S. Seitzinger, and P. Crill (eds.), Cycling of Reduced Gases in the Hydrosphere, Schweizerbart'sche Verlagsbuchhandlung, Stuttgart (in press).

22. Farquhar, G.D., M.H. O'Leary, and J.A. Berry. 1982. On the relationship between carbon isotope discrimination and the intercellular carbon dioxide concentration in leaves. Aust. J. Plant. Physiol. **9**:121–137.

23. Fisher, T.R., and M.A. Moline. 1992. Seasonal plant cover on the Amazon River floodplain with aerial videography and image analysis. In C.H. Blazquez (ed.), Proc. 13th Biennial Workshop on Color Aerial Photography in the Plant Sciences, ASPRS. Bethesda, MD, pp. 209–216.

24. Fry, B., and E. Sherr. 1984. $\delta^{13}C$ measurements as indicators of carbon flow in marine and freshwater ecosystems. Cont. Marine Sci. **27**:13–47.

25. Hatch, M.D., C.R. Slack, and H.S. Johnson. 1976. Further studies on a new pathway of photosynthetic carbon fixation in sugarcane and its occurrence in other plant species. Biochem. J. **102**:417–422.

26. Hedges, J.I., W.A. Clark, P.D. Quay, J.E. Richey, A.H. Devol, and U.M. Santos. 1986. Compositions and fluxes of particulate organic material in the Amazon River. Limnol. Oceanogr. **31**:717–738.

27. Hoefs, J. 1987. Stable Isotope Geochemistry, 3rd ed., Springer-Verlag, New York.

28. Junk, W.J. 1970. Investigations on the ecology and production-biology of the "floating meadows" (Paspalo-Echinochloetum) on the middle Amazon. I. The floating vegetation and its ecology. Amazoniana **2**:449–495.

29. Kelley, C.A., N. Dice, and C.S. Martens. 1992. Seasonal variations in methane emitted from Minnesota bogs and fens. Global Biogeochem. Cycles **6**:263–269.

30. Khalil, M.A.K., and R.A. Rasmussen. 1983. Sources sinks, and seasonal cycles of atmospheric methane, J. Geophys. Res. **88**:5131–5144.

31. King, S.L., P.D. Quay, and J.M. Lansdown. 1989. The $^{13}C/^{12}C$ kinetic isotope effect for soil oxidation of methane at ambient atmospheric concentrations. J. Geophys. Res. **94**:18273–18277.

32. Martens, C.S., and J.P. Chanton. 1989. Radon tracing and biogenic gas equilibration and transport from methane saturated sediments. J. Geophys. Res. **94**:3451–3459.

33. Martens, C.S., N.E. Blair, C.D. Green, and D.J. Des Marais. 1986. Seasonal variations in the stable carbon isotopic signature of biogenic methane in a coastal sediment. Science **233**:1300–1302.

34. Martens, C.S., C.A. Kelley, J.P. Chanton, and W.J. Showers. 1992. Carbon and hydrogen isotopic characterization of methane from wetlands and lakes of the Yukon–Kuskokwim delta, Western Alaska. J. Geophys. Res. **97**:16,689–16,702.

35. Matthews, E., and I. Fung. 1987. Methane emissions from natural wetlands: Global distribution, area and environmental characteristics of sources, Global Biogeochem. Cycles **1**:61–86.

36. MacIntyre, S., and J.M. Melack. 1984. Vertical mixing in Amazon floodplain lakes., Vern. Internat. Verein. Limnol. **22**:1283–1287.

37. MacIntyre, S., and J.M. Melack. 1988. Frequency and depth of vertical mixing in an Amazon floodplain lake (L. Calado, Brazil). Vern. Internat. Verein. Limnol. **23**:80–85.

38. Melack, J.M., and T.R. Fisher. 1983. Diel oxygen variations and their ecological implications in Amazon floodplain lakes. Arch. Hydrobiol. **98**:422–442.

39. Milthorpe, F.L., and J. Morrby. 1974. An Introduction to Crop Physiology, Cambridge University Press, Cambridge, England.

40. O'Leary, M.H. 1980. Carbon isotope fractionation in plants. Phytochemistry **20**: 533–567.

41. Piedade, M.T.F., W.J. Junk, and S.P. Long. 1991. The productivity of the C-4 grass Echinochloa Polystachya on the Amazon Floodplain. Ecology **72**:1456–1463.

42. Quay, P.D., S.L. King, J.M. Lansdown, and D.O. Wilbur. 1988. Isotopic composition of methane released from wetlands: Implications for the increase in atmospheric methane. Global Biogeochem. Cycles **2**:385–397.

43. Rice, D.D., and G.E. Claypool. 1981. Generation, accumulation and resource potential of biogenic gas. Amer. Assoc. Petrol. Geol. Bull. **67**:1199–1218.

44. Schoell, M. 1980. The hydrogen and carbon isotopic composition of methane from natural gases of various origins. Geochim. Cosmochim. Acta **44**:649–661.

45. Smith, B.N., and W.V. Brown. 1973. The Kranz syndrome in the gramineae as indicated by carbon isotope ratios. Amer. J. Bot. **60**:505–513.

46. Smith, B.N., and S. Epstein. 1971. Two categories of $^{13}C/^{12}C$ ratios for higher plants. Plant Physiol. **47**:1004–1013.

47. Stevens, C.M., and A. Engelkemeir. 1988. Stable carbon isotopic composition of methane from some natural and anthropogenic sources. J. Geophys. Res. **93**:725–733.

48. Tyler, S.C. 1989. C-13 ratios in atmospheric methane and some of its sources. In J. Ehleringer, K. Nagy, and P. Rundel (eds.), Stable Isotopes in Ecological Research, Springer-Verlag, New York, pp. 395–409.

49. Yamamoto, S., J.B. Alcauskas, and T.F. Crozier. 1976. Solubility of methane in distilled water and seawater. J. Chem. Eng. Data **21**:78–80.

50. Whiticar, G.J., and E. Faber. 1986. Methane oxidation in sediment and water column environments—Isotopic evidence. Organ. Geochem. **10**:759–768.

51. Whiticar, G.J., E. Faber, and M. Schoell. 1986. Biogenic methane formation in marine and freshwater environments: CO_2 reduction vs. acetate fermentation—Isotopic evidence. Geochim. Cosmochim. Acta **50**:693–709.

Nitrogen Fixation and Denitrification in the Intertidal and Subtidal Environments of Tomales Bay, California

Samantha B. Joye and Hans W. Paerl

Institute of Marine Sciences, University of North Carolina at Chapel Hill, Morehead City, North Carolina 28557

ABSTRACT

We have examined contemporaneous nitrogen fixation and denitrification at two intertidal and three subtidal sites in Tomales Bay, California over a seasonal cycle. At subtidal sites, denitrification dominates, whereas nitrogen fixation dominates the intertidal sites. Potential rates of denitrification in intertidal sites fluctuate and can be up to one-half the nitrogen-fixation rate; therefore, the net input of fixed nitrogen to intertidal systems varies. Intertidal and subtidal sites appear to play quantitatively distinct roles in the internal, microbially mediated nitrogen cycle in Tomales Bay. However, the limited areal extent of intertidal regions (5% of total sedimentary area) minimizes their contribution of fixed nitrogen to the bay ecosystem. This intertidal input by nitrogen fixation amounts to only 3% of the nitrogen lost during subtidal denitrification.

Introduction

Estuaries are functional transition zones between terrestrial and oceanic habitats. Biogeochemically, they function as recycling sites for nutrients and organic matter which are exchanged between the land and the sea.[23,28,35] Relatively high rates of estuarine primary production are limited by the availability of certain dissolved nutrients; frequently, fixed nitrogen serves the role of a limiting nutrient.[21–24,28,32]

Sources of fixed nitrogen to coastal environments are of allochthonous (terrestrial runoff, groundwater, tidal influx, and atmospheric inputs) or autochthonous (N_2 fixation) nature. Dissolved inorganic nitrogen (DIN) is often rapidly incorporated into biomass and quickly recycled by microbially mediated processes upon the death or senescence of an organism.[10] In addition to a recycling component, the nitrogen cycle has both microbially mediated source (N_2 fixation) and sink (denitrification) terms.[5,8,11,29] The terms ''source'' and ''sink'' are used to reflect the fact that nitrogen fixation represents a new input of fixed nitrogen to the system,[7,19] whereas denitrification represents a loss, or sink, of fixed nitrogen from the system.[32] These two processes share common environmental requirements, including source of reductant and microaerophilic conditions. Rates of these processes, when extrapolated to the ecosystem level, may act to either relieve (N_2 fixation) or enhance (denitrification) system-level nitrogen limitation.[4,8,14,34]

The nitrogenase enzyme complex, which mediates N_2 fixation, is directly inactivated by NH_4^+ but not by NO_3^- (up to 100 μM).[40,41] Because NO_3^- serves as the primary substrate for denitrifiers, the two processes may operate contiguously and contemporaneously. In systems where both processes have been assessed, denitrification rates generally exceed N_2 fixation rates.[32] Long-term concurrent assessments of denitrification and N_2 fixation on the ecosystem level have rarely been attempted.[18,36] Quantification of both processes is crucial before these biotic sources and sinks of fixed nitrogen can be fully understood.

Estuarine subtidal environments are frequently marked by high rates of denitrification.[20,32,34] In contrast, intertidal sediments and cyanobacteria-dominated microbial mats are important sites of N_2 fixation in temperate and tropical coastal environments.[4,5] Because of this apparent disparity, we sought to determine the relative importance of N_2 fixation and denitrification in the subtidal and intertidal environments of a temperate estuary, Tomales Bay, California. We also examined the influence of seasonality and diel periodicity on N_2 fixation and denitrification. The overall goal was to evaluate the roles intertidal and subtidal environments play as unique and quantitatively important components of the microbially mediated nitrogen cycle.

Methods

Study Site

Tomales Bay, California, is a shallow (mean depth = 3 m), semienclosed basin characterized by well-defined hydrological boundaries and low annual dissolved inorganic nutrient loading (Figure 1; Refs. 13, 35, and 37). The bay has distinct seasonal cycles with respect to nutrient sources, with the winter season dominated by inputs from two creeks (Walker Creek to the North and Lagunitas Creek to the South; Figure 1), and the summer season dominated by oceanic inputs via upwelling.[36,37]

Three subtidal sites were chosen (Figure 1; Refs. 6 and 38). Two of these, stations 10 and 16, lie within the inner bay region (IB), between Lagunitas Creek and station 8; the other, station 6, lies in outer bay region (OB), between station 8 and the mouth. Inner bay sediments are composed of fine-grained, organic-rich muds of low porosity, whereas OB sediments are composed of coarse-grained,

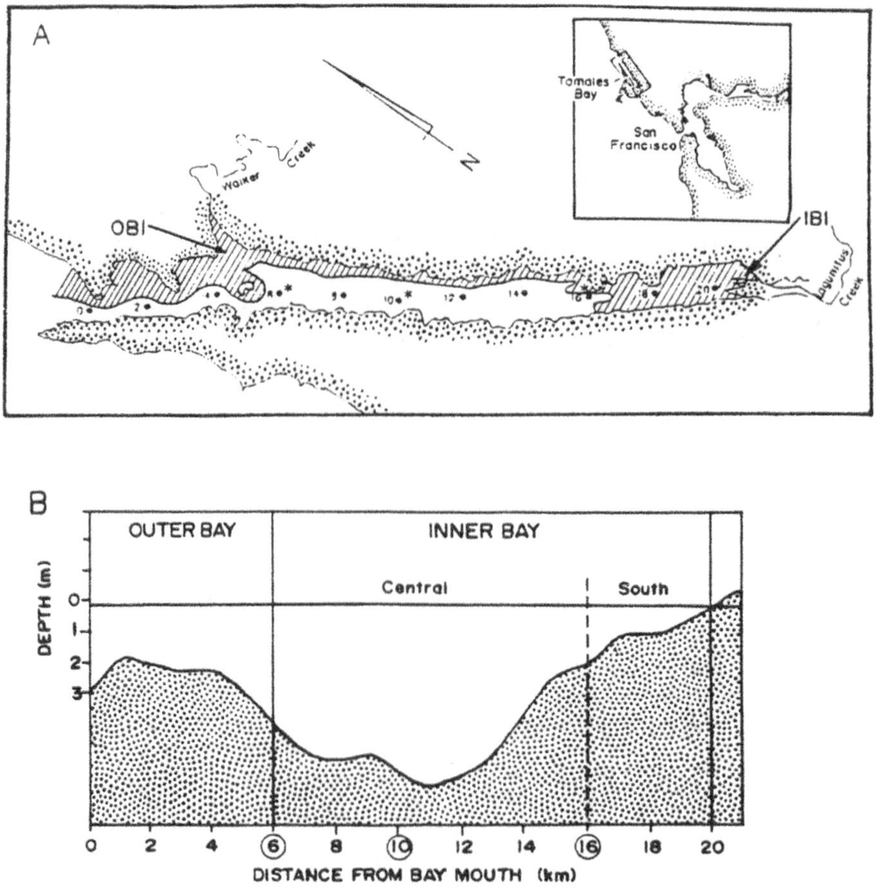

Figure 1. Map of Tomales Bay. Panel A shows a schematic of the bay highlighting the subtidal sites at stations 6, 10, and 16. Panel B shows a depth profile and demarcates the regions of the bay. OBI = outer bay intertidal site; IBI = inner bay intertidal site. Subtidal sites are circled on the bottom panel.

organic poor sands of higher porosity. The two regions are characterized by seasonally variable particulate organic matter (POM) and nutrient input sources. All sites are bioturbated to 15–40 cm with maximal faunal densities in summer and fall.

Two intertidal sites consisting of extensive cyanobacteria-dominated microbial mat communities and *Salicornia* marshes were also chosen (Figure 1). Mats were characterized by morphology, resident cyanobacteria present, and location; three

Table 1. Types of intertidal microbial mats examined.

Type	Dominant Cyanobacteria	Morphology
Mounds	*Microcoleus sp.*	• Leathery surface
	Lyngbya sp.	• Laminated
	Sprirulina sp.	• Height: 20–25 cm
Tufts	*Lyngbya sp.*	• Loosely aggregated
	Oscillatoria sp.	• Not laminated
	Spirulina sp.	• Variable height
Salicornia mats	*Microcoleus sp.*	• Well-defined
	Oscillatoria sp.	• Grow on surface in *Salicornia* marshes
	Cylindrospermum sp.	
	Spirulina sp.	• Height: 1–5 mm
	Unidentified coccoid	

distinct types were investigated (Table 1). Filamentous tufts were a dominant feature in the IB from May to September but were absent at the OB site. Exposed mounds were present year-round at both sites. Mats growing on sediment surfaces in *Salicornia* marshes were also present year-round at both sites. These three types combined to cover approximately 70% of the intertidal marsh–mudflat sediment surface area.

Intertidal N_2 Fixation and Denitrification Assessments

Bimonthly assays of intertidal N_2 fixation and denitrification were initiated in May 1990. A large mat sample (0.3 m^2) was collected by sectioning with a scalpel and sliding a flexible plastic sheet beneath the mat. The section was transported back to the laboratory and acclimated in an ambient seawater bath prior to an assay.

Contemporaneous estimates of N_2 fixation and denitrification were made on the same sample using acetylene reduction and acetylene inhibition techniques, respectively.[9,25] Small pieces of mat (1 cm^2) were placed into 38-ml serum bottles with 20 ml of ambient marsh water. Bottles were stoppered with rubber sleeve stoppers and a 7-ml volume (20%) of the head space was replaced with CaC_2-generated acetylene. Acetylene is both an alternative substrate for the nitrogenase enzyme in N_2 fixers and an inhibitor of nitrous oxide reductase in denitrifying microorganisms.[25] Triplicate light and dark (foil-covered) samples were incubated in a water bath for 2 h between 1000 and 1300 under natural irradiance and temperature conditions. After incubation, bottles were shaken for 90 s to equilibrate ethylene and nitrous oxide between aqueous and gas phases.

Gas-phase samples were collected by displacement of a 15-ml head-space sample with 15 ml of 1 M KCl; samples were stored in evacuated serum bottles for later analysis.

Potential Denitrification Assessment

Potential denitrification rates were determined by enriching with nitrate (KNO_3, 1 mM) and organic carbon (D-glucose, 1 mM) substrates in the presence of 1 g/L chloramphenicol. Chloramphenicol inhibits bacterial protein synthesis; therefore, only presynthesized enzyme activity is measured (as opposed to activity associated with de novo enzyme synthesis). Experiments were conducted in parallel with routine N_2 fixation–denitrification assays. Details of this procedure are presented elsewhere (Joye, unpublished).

Chromatographic Methods

Analysis of ethylene was performed with a Shimadzu 14 gas chromatograph equipped with a flame-ionization detector at 200°C. Subsamples were removed and direct injected onto a 2-m Poropak T column (80°C); nitrogen was used as the carrier gas. Ethylene concentrations were determined by comparison with purified ethylene standards (Scott Specialty Gases, Inc.).

Nitrous oxide was analyzed on the same gas chromatograph with an electron capture detector at 300°C. Carrier flow (5% CH_4 in 95% Ar at 30 ml/min) was directed through a 1-m Poropak Q column at 25°C, the main column (2-m Poropak Q at 90°C), and detector. Concentrations of N_2O in samples were determined by comparison with a calibration curve generated using purified N_2O standards (Scott Specialty Gases, Inc.).

Carbon-Fixation Measurements

Concurrent measurements of inorganic carbon fixation (as determined by photosynthetic ^{14}C–Na HCO_3 uptake) were performed to examine temporal relationships between CO_2 and N_2 fixation and denitrification.[2] Trace amounts of sterile ^{14}C–$NaHCO_3$ having a specific activity of 58 mCi $mmol^{-1}$ were added to 1-cm^2 squares of mat in 20 ml of seawater and incubated for 3 h. Samples were processed and prepared for liquid scintillation counting according to Paerl et al.[26] Triplicate light and dark samples, in ambient seawater as well as seawater amended with 2 μM 3-(3,4-dichlorophenyl)-1,1-dimethylurea (DCMU), were run. DCMU prevents photosystem II mediated oxygen production and, thus, provides information regarding of the effect of oxygenic photosynthesis on these processes during short-term incubations.[2]

Subtidal N_2-Fixation Assessment

Cores (50 cm length, 6.4 cm diameter) were collected using SCUBA and transported back to the laboratory within 2 h. Slurries of sediment were used to determine rates of subtidal N_2 fixation. Two 3-cc volumes of sediment were removed at 1.5-cm intervals and placed into a 38-ml bottle with 5 ml of N_2 purged (1 h at 100 ml/min) bottom water. After sealing with serum stoppers and crimping, the head space (27 ml) was flushed with ultra-high-purity N_2 for 10 min. Acetylene was then injected into the water phase, as described previously. Duplicate samples were incubated in the dark from 30 to 120 min at in situ temperature. Gas-phase samples were collected and concentrations of ethylene and N_2O were determined as described earlier.

Denitrification Assessment: Whole-Core Method

Whole-core denitrification rates were determined using the acetylene block technique.[39] A 400-μl volume of C_2H_2-saturated N_2-flushed distilled water was injected at 1-cm intervals down the core. Acetylene was added to the overlying water and gas phases to yield final concentrations of 20% (vol : vol). Control cores, to account for background N_2O production, were injected with N_2-saturated distilled water. Potential denitrification rates were determined by adding NO_3^- to the C_2H_2-saturated DDW solution to yield a final porewater concentration of 200 μM; nitrate was also added to the overlying water (200 μM). Cores were incubated in a water bath from 30 to 120 min with constant stirring of the water phase via a magnetic stirring apparatus.

After incubation, 3-cc subcores were removed as sediment was extruded. Two subcores were placed into a tared 38-ml serum bottle containing 10 ml of 1 M KCl (enzyme grade) and quickly sealed with a serum septum. Bottle weights with sediment were recorded and gas-phase samples were collected as described above.

Nutrient Extraction

Fifteen to 20 milliliters of the KCl extract was centrifuged and filtered (GF/C) into acid-washed, deionized-water-rinsed glass scintillation vials for subsequent nutrient (NH_4^+, PO_4^{3-}, and NO_3^-) analyses. Ammonium and phosphate were determined using standard methods.[27] Nitrate was determined using a cadmium-shaking technique.[15]

Denitrification Assessment: Slurry Method

Denitrification rates were routinely determined on samples collected from N_2-fixation slurries using the procedures outlined above. Data from March 1990 will be discussed to illustrate the potential for high rates of denitrification in Tomales Bay sediments.

Statistical Analyses

All statistical analyses (ANOVA) were performed using SAS Version 5.0.[31]

Results

Intertidal N_2 Fixation: Seasonality

Highest N_2-fixation rates occurred during May and September in the IB (Figure 2). *Salicornia* mats and filamentous tufts flourished during these periods. Mounds accounted for high rates in March and November (Figure 2). Although rates in mounds and tufts decreased in summer, rates in *Salicornia* mats increased (Figure 2).

Highest rates in the OB also occurred during spring (Figure 3). Rates were lowest in July, after which there was a gradual increase toward maximal rates in March. *Salicornia* mats were present year-round and showed maximal fixation rates in March; little variation was observed during the remainder of the year.

Diel Integrated Rates of Carbon and Nitrogen Fixation

Rates of inorganic carbon and nitrogen fixation were similar at the two sites; only in July was there a significant difference (IB N_2 fixation higher than OB; $p < 0.05$). At both sites, carbon-fixation rates were maximum in July and were significantly higher ($p < 0.05$) than rates in November or March. In the IB, maximal daily integrated rates of both N_2 fixation and CO_2 fixation occurred in July (Table 2). In contrast, at the OB site, maximum CO_2-fixation rates occurred in July when minimum N_2-fixation rates were observed (Table 2).

Intertidal Denitrification

Ambient denitrification rates in the intertidal environments were detectable but accounted for only a small percentage of N_2-fixation rates (0.1–6.0%). Rates ranged from 2 to 40 μmoles N_2 m^{-2} d^{-1} (Figures 4 and 5). Despite low ambient rates, potential rates were between 100 and 2000 μmoles N_2 m^{-2} d^{-1} (Figures 4 and 5). At the IB site, highest potential rates were measured in *Salicornia* mats during March (Figure 4). Both ambient and potential denitrification rates were higher in the OB (Figures 4 and 5).

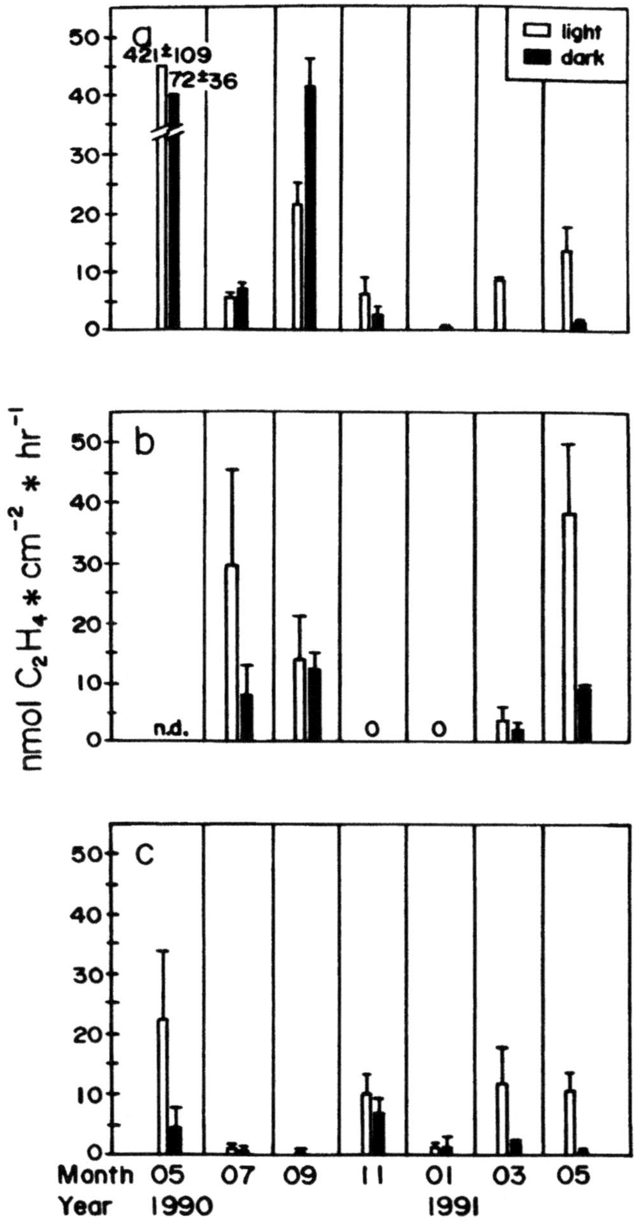

Figure 2. Inner bay intertidal nitrogen fixation rates for the three types of mats examined: (a) tufts, (b) *Salicornia* mats, (c) mounds. Dates of sampling are given at the bottom. All data are the mean of three replicates; error bars represent standard deviation of the mean. 0 = rate equals zero; n.d. = no data.

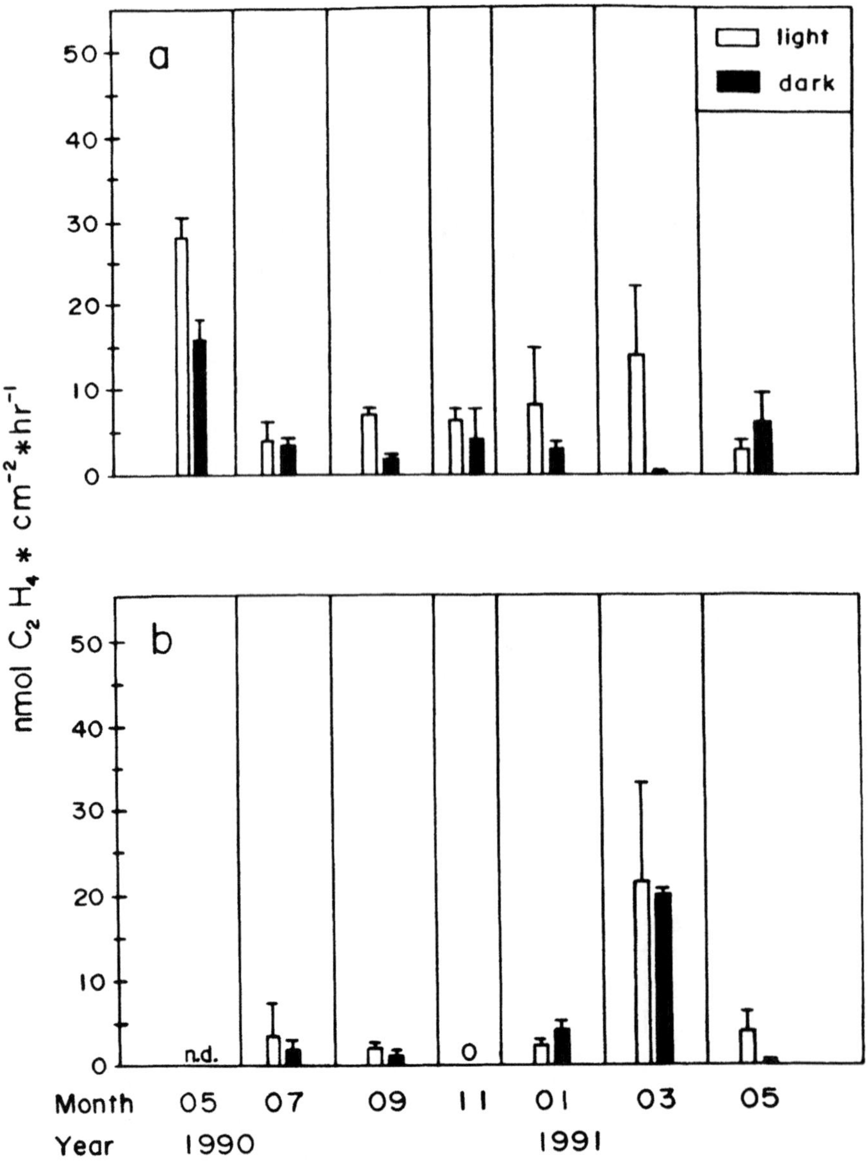

Figure 3. Outer bay intertidal nitrogen fixation for (a) mounds and (b) *Salicornia* mats. Notations same as Figure 2.

Table 2. Diel integrated rates of carbon and nitrogen fixation.

Date: Site	NFIX	CFIX	C : N[b]
Daily rates[a]			
0790: IB	5.0	52.5[c]	12.0
0790: OB	1.2[c]	72.2[c]	70.2
1190: IB	2.7[c]	12.0	5.3
1190: OB	3.3	18.6	6.7
0391: IB	4.3	26.4	6.1
0391: OB	3.9	26.5	6.9
Annual rates[d]			
IB	1.5	11.1	8.9
OB	1.0	14.3	16.5

[a]Daily rates in mg C or N m^{-2} d^{-1}.
[b]Fixed C : N ratio by atom.
[c]Statistically different ($p < 0.05$) from other rates within a group (i.e., NFIX in the IB).
[d]Average annual rates in g C or N m^{-2} d^{-1}.

Subtidal N_2 Fixation

Subtidal N_2-fixation rates were similar at the two IB sites and exhibited spring maxima (Figure 6a). Rates declined throughout the summer and fall at the mid-bay site (station 10), but the south bay site (station 16) showed an increase in November. Fixation rates at the OB site increased gradually in spring (beginning in May) and maximized in fall (September). Rates in the winter and early spring (January and March) were below detection at the OB site.

Subtidal Denitrification

Denitrification rates were highest in the IB (station 10; Figure 6b). Rates at the other two stations were high in January, but decreased dramatically thereafter (Figure 6b). Because only a limited number of in situ rate data exist, the data obtained from a stoichiometric model[40] will be discussed to gain insight regarding possible rates for the remainder of the year.

Subtidal denitrification rates were significantly stimulated by the addition of nitrate. During March, NO_3^- addition resulted in marked stimulation of denitrification—22 times unamended rates. During May, a 40-fold increase was documented. Porewater NO_3^- concentrations decreased from a high of 35 μM in January to less than 5 μM in May (Figure 7).

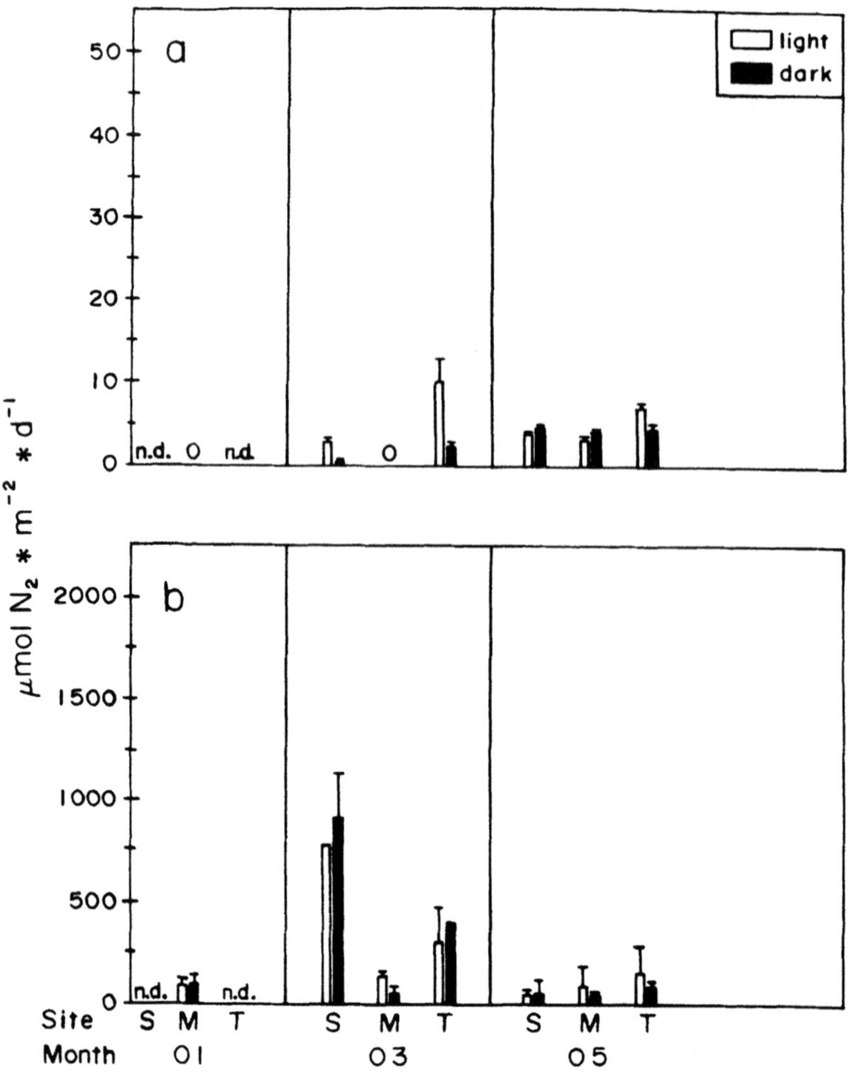

Figure 4. Actual (a) and potential (b) denitrification rates in inner bay intertidal regions. Note the difference in scale between the two panels. n.d. = no data; 0 = rate was zero. All data are mean of three replicates; error bars represent standard deviation of the mean. S = *Salicornia* mats, M = mounds, T = tufts.

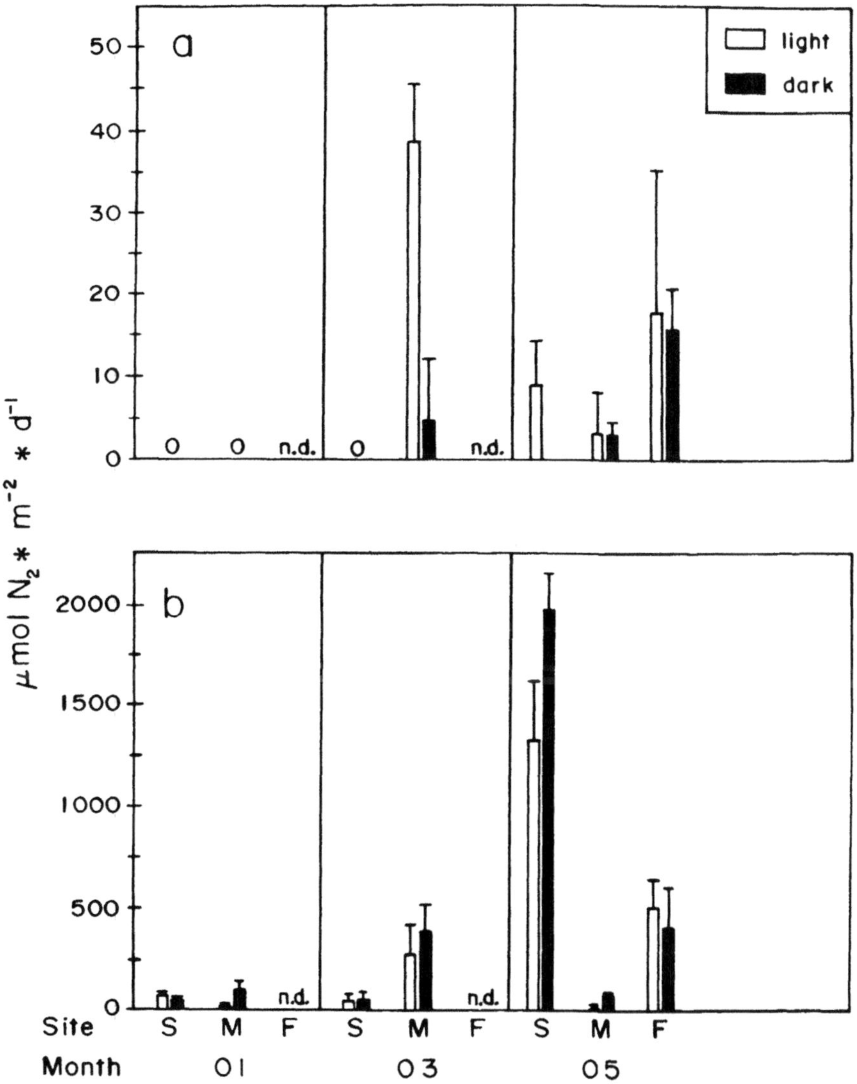

Figure 5. Actual (a) and potential (b) denitrification rates for the outer bay intertidal region. Notations same as Figure 4 except F = mounds with filamentous algal surface layer.

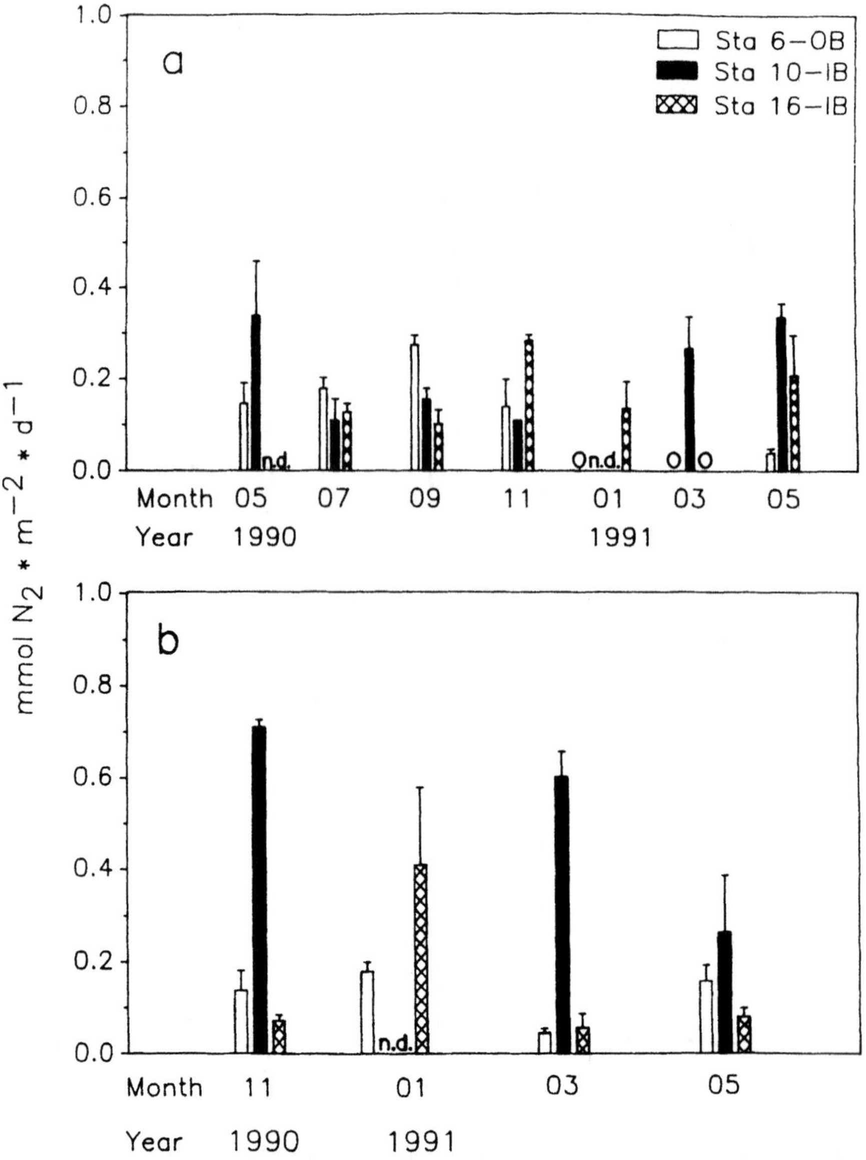

Figure 6. Subtidal nitrogen fixation and denitrification at inner and outer bay sites. In situ denitrification rates only given for November 1990–May 1991. For station locations, refer to Figure 1; 0 = rate equals zero; n.d. = no data.

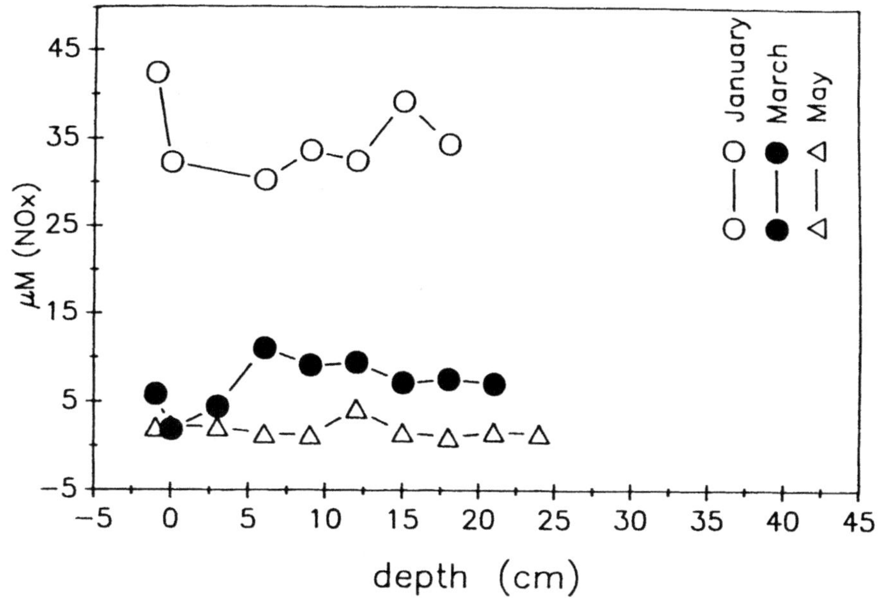

Figure 7. Pore water nitrate + nitrite (NO_x) profiles from the outer bay site in January, March, and May.

Discussion

Intertidal Environments: Diel Patterns

Microbial mats experience marked changes in O_2 gradients during daily cycles due to variable spatial–temporal patterns of oxygenic photosynthesis and respiration.[2,33] Because nitrogenase is extremely sensitive to O_2, examining the interaction of carbon fixation (oxygenic photosynthesis) and nitrogen fixation is important.[33] Integrated rates of HCO_3^- and N_2 fixation were similar at the two sites during November and March. However, during July, differences were noted. Diel integrated carbon-fixation rates were greater in the OB, whereas nitrogen-fixation rates were greater in the IB. In both the November and March samplings, no differences were noted between sites. Thus, differences occurring in July led to annual differences in the input of fixed carbon and nitrogen by microbial mats to intertidal sites: The OB site appears to account for a higher annual fixed C input, whereas the IB accounts for a higher annual fixed N input.

Comparisons of the ratios of C : N fixed and examinations of seasonal variations in that ratio, point toward potential driving forces behind the patterns of nitrogen and carbon fixation. During spring (March) and fall (Nov.), fixed C : N ratios at both sites are near Redfield values (6.6). However, during summer (July), there is an enormous increase in the C : N ratio and a significant difference between sites (IB: 12.0, OB: 70.2; Table 2). The increase at the OB site results from two factors: a decrease in N_2 fixation activity and an increase in HCO_3^- fixation activity. There is no significant decrease in IB N_2 fixation; thus, the fixed C : N ratio reflects changes in HCO_3^- fixation rate only. What factors might be responsible for altering these ratios?

During winter, storms scour the intertidal regions and much of the previous season's accretion by mat communities is eroded. During spring, productivity increases in response to rising temperature and light regimes. Production of biomass at this time appears to be fueled by "new" nitrogen. The fixed C : N ratio at this time (Table 2) suggest that mat organisms are able to keep pace with N requirements (meeting C : N demands based on carbon fixation rates) via biological N_2 fixation.

During summer, CO_2 fixation rates increase dramatically; N_2-fixation rates do not. Light and temperature conditions are well-suited to support high carbon-fixation rates at this time. High rates of O_2 production, associated with CO_2 fixation, may have contributed to relatively low N_2-fixation rates. DCMU-amended rates were three times the control rates during peak photosynthesis, suggesting that in situ O_2 production was a likely factor limiting N_2 fixation (S. Joye, unpublished data). However, this effect was short-lived and only present during periods of maximal carbon fixation. If DCMU-amended N_2-fixation rates are used to calculate integrated rates, elevated C : N ratios persist. Therefore, another source of fixed N must be available to mat organisms at this time.

Mats may obtain fixed N from external (i.e., sources other than N_2 fixation) sources during this period. Studies have recently shown that regenerated N can, in fact, be an important fixed N source to microbial mats.[3] A seasonal transition in the importance of regenerated production in mat communities might be expected. Seasonal patterns in the release of ammonium from subtidal sediments in Tomales Bay have been documented; release rates are low in spring and tend to maximize during summer.[6] A seasonal shift from dependence on new to regenerated nitrogen to support production could be responsible for the patterns seen in intertidal mats.

During fall, the C : N ratio again approximates Redfield values. At this time, HCO_3^--fixation rates have decreased. N_2-fixation rates in the OB are significantly higher, whereas rates in the IB remain unchanged. Reduced HCO_3^--fixation rates may reflect seasonal changes in temperature and light regimes. Increased N_2-fixation rates at the OB site could be due to the presence of exogenous organic matter sources (seagrass detritus) which are deposited there during fall.

Annual Inputs

Inner bay marshes appear to be a larger source of fixed nitrogen than OB marshes due to substantially higher IB N_2-fixation rates during July. Assuming that rates are similar between diel measurements, the IB marshes supply "new" nitrogen at the rate of 1.50 g N m^{-2} yr^{-1}, whereas the OB marshes supply 1.01 g N m^{-2} yr^{-1}. Inner bay marshes make up 80% of the total intertidal marsh area (3 km^2) and, therefore, dominate input of fixed nitrogen to intertidal regions. CO_2-fixation rates were higher at the OB site, amounting to 14.3 g C m^{-2} yr^{-1} as compared to 11.1 g C m^{-2} yr^{-1} at the IB site.

Denitrification

Ambient denitrification rates in the intertidal environments were usually much lower than N_2-fixation rates (DNF < 10% NF; Figures 4 and 5). Both ambient and potential rates are highest in the OB, with potential rates typically 50–200 times ambient rates. Despite low ambient rates, denitrification may be very important in intertidal systems. Given the spatial scales typical of microbial mats (μms), tight coupling between nitrification and denitrification may exist. Potential denitrification rates show that there is a large capacity for denitrification in mat communities, which suggest that denitrifying organisms are present and rely on an *in situ* nitrate supply, e.g., nitrification.[20] Acetylene also blocks nitrification[25]; therefore, in ambient treatments, nitrification and denitrification are decoupled. When nitrate supply is interrupted, denitrification rates rapidly decline. Thus, low ambient rates may be the result of an experimental artifact, the decoupling of nitrification and denitrification, rather than an accurate depiction of denitrification rates under ambient conditions. In treatments where nitrate is supplied, denitrification rates are sometimes comparable to N_2-fixation rates.

Based on data collected thus far in intertidal systems, denitrification appears to play a minor role in the net nitrogen cycle. However, the potential rate data may actually be a better representation of naturally occurring denitrification rates. Because a nitrate source is provided, the artifact associated with decoupling nitrification and denitrification in ambient treatments is removed. If potential rates are realized, denitrification could remove up to half of the nitrogen input by N_2 fixation. In either case, intertidal systems appear to function as net sources of fixed nitrogen in Tomales Bay, though the amount of that input may vary (depending on the magnitude of denitrification rates).

Subtidal N_2 Fixation

Subtidal N_2-fixation rates show seasonal patterns. These patterns may reflect variations in particulate organic matter (POM) source and delivery to the two regions. The IB receives POM primarily of autochthonous origin, whereas the OB receives an additional allochthonous POM from oceanic upwelling. Inner bay N_2-fixation rates correlate well with the spring phytoplankton bloom, whereas the OB rates correlate well with loading of upwelling-associated POM. Minima at both IB and OB sites in January most likely reflect regulation of microbial activity by temperature (6°C at sediment surface).

Denitrification

Highest rates are consistently found at the mid-bay site. During winter, denitrification activity was restricted to the upper 5 cm of the sediment column and was probably supported by bottom water nitrate. Porewater nitrate concentrations also peaked at this time (Figure 7). Decreases in bottom water and porewater NO_3^- concentration in March were accompanied by decreased denitrification rates. Whole-core additions of nitrate (OB site) resulted in increases of up to 40 times ambient rates. During March, the whole-core rate was 0.05 mmoles N_2 m^{-2} d^{-1}, whereas the amended rate was 1.04 mmole N_2 m^{-2} d^{-1}; in May, the ambient rate was 0.15 mmoles N_2 m^{-2} d^{-1}, whereas the amended rate was 6.23 mmole N_2 m^{-2} d^{-1}. It is apparent that nitrate availability during the assay determines the measured denitrification rates. Nitrification and denitrification have been shown to be coupled in estuarine sediments (Refs. 17 and 20, and references therein). Decoupling of these processes during whole-core experiments probably accounts for the low measured rates in ambient treatments. Measurements of nitrification rates in these sediments show that there is a large *in situ* nitrate supply to support substantial rates of denitrification (S. Joye, unpublished data).

Sediment slurries gave rates much higher rates than did whole-core determinations. In March, slurries yielded a rate of 1.5 mmole N_2 m^{-2} d^{-1} in the IB and 7.2 mmoles N_2 m^{-2} d^{-1} in the OB. Higher rates in slurries are not surprising, since substrates appear more homogeneously distributed and, therefore, more accessible to denitrifiers. These high rates also support the contention that Tomales Bay sediments have a high potential for denitrification.[38]

A stoichiometric nutrient flux model shows that the entire bay becomes a sink for fixed N during summer and fall.[38] This model gives an annual baywide denitrification rate of approximately 3.2 mmol N m^{-2} d^{-1} (1.56 mmol N_2). Most of the nitrogen loss occurs in the summer (estimated to be up to 10 mmol N m^{-2} d^{-1}), as a response to imported (oceanically derived) POM. N_2-fixation rates during this period (in situ rate) range between 0.15 and 0.30 mmol N m^{-2} d^{-1}. The balance is a net loss of fixed nitrogen amounting up to 9.70–9.85 mmol N m^{-2} d^{-1} during summer and fall.

Based on current results, the subtidal sediments of Tomales Bay account for a net loss of fixed nitrogen. However, the magnitude of this loss is ameliorated to some extent by subtidal N_2 fixation during certain times of the year; additional data for subsequent years are needed to confirm this. Rates of denitrification in Tomales Bay sediments are comparable to other estuaries.[32] N_2-fixation rates are higher (0.1–0.7 mmol N m^{-2} d^{-1}) than those reported by Capone[14] for subtidal environments (0.0–0.11 mmol N m^{-2} d^{-1}). The higher N_2 fixation rates reported in this study may result from extensive bioturbation. Bioturbation may stimulate N_2 fixation in a manner similar to that of plant roots by supplying solutes (organic carbon and nutrients) to microorganisms in the sediment[12] or by enhancing NH_4^+ transport away from the zones of active N_2 fixation.[1] Rates of intertidal N_2 fixation (0.28 mmol N m^{-2} d^{-1}) are typical of those reported microbial mats.[4]

On an areal basis, assuming that the IB marshes constitute 80% of the intertidal marsh area (3 km^2), we derive an annual baywide input of fixed nitrogen amounting to 2.9×10^5 mol N yr^{-1} (IB = 2.3; OB = 0.6). The average yearly subtidal N_2-fixation rate amounts to 6.8×10^5 mol N yr^{-1}. These inputs of fixed nitrogen, 9.77×10^5 mol N yr^{-1}, can be compared to the net baywide denitrification rate (using the annual average[36] of 3.2 mmol N m^{-2} d^{-1} and a subtidal area of 28 km^2) of 327×10^5 mol N yr^{-1}. The baywide N_2-fixation rate, based on areas examined and the above-mentioned conversion rates, amounts to only 3% of the baywide denitrification rate. Thus, despite a substantial input of fixed nitrogen in the intertidal and subtidal environments, the bay as a whole appears to function as a sink for fixed nitrogen.

Acknowledgments

We thank S. Vink, J. Plant, and R. Chambers for technical assistance. We also thank R. Chambers, E.A. Canuel, S.V. Smith, and two anonymous reviewers for constructive reviews of the manuscript. H. Barnes assisted with the illustrations.

This work was supported by National Science Foundation grants (OCE 88–20036, OCE 91–15706, and OCE 89–14833).

References

1. Aller, R.C. 1980. Quantifying solute distribution in the bioturbated zone of marine sediments by defining an average microenvironment. Geochim. Cosmochim. Acta **44**:1955–1965.
2. Bebout, B.M., H.W. Paerl, K.M. Crocker, and L.E. Prufert. 1987. Diel interactions of oxygenic photosynthesis and N_2-fixation (acetylene reduction) in a marine microbial mat community. Appl. Environ. Microbiol. **53**(10):2353–2362.
3. Bebout, B.M., D.E. Canfield, D.E., Des Marias, and H.W. Paerl. 1992. Nitrogen

cycling in marine microbial mats: Relative importance of microscale nitrogen remineralization and uptake to nitrogen fixation. ASLO Abstract, Winter meeting.

4. Capone, D.G. 1983. Benthic nitrogen fixation. In E.J. Carpenter and D.G. Capone (eds.), Nitrogen in the Marine Environment, Academic Press, New York, pp. 105–137.
5. Capone, D.G. 1988. Benthic nitrogen fixation. In T.H. Blackburn and J. Sorensen (eds.), Nitrogen Cycling in Coastal Marine Environments, J. Wiley and Sons, New York, pp. 85–123.
6. Dollar, S.J., S.V. Smith, S.M. Vink, S. Obreski, and J.T. Hollibaugh. 1991. Annual cycle of benthic nutrient fluxes in Tomales Bay, California and contribution of the benthos to total ecosystem metabolism. Marine Ecol. Prog. Ser. **79**:115–125.
7. Eppley, R.W., and B.J. Peterson. 1979. Particulate organic matter flux and planktonic new production in the deep ocean. Nature **282**:677–680.
8. Focht, D.D., and W. Verstrate. 1977. Biochemical ecology of nitrification and denitrification. Adv. Microb. Ecol. **1**:135–214.
9. Hardy, R.W.F., R.D. Holsten, E.K. Jackson, and R.C. Burns. 1968. The acetylene-ethylene assay for nitrogen fixation: Laboratory and field evaluation. Plant Physiol. **43**:1185–1207.
10. Harrison, W.G. 1982. Nutrient regeneration and primary productivity in the sea. In P.G. Falowski (ed.), Primary Productivity in the Sea, Plenum Press, New York, pp. 433–460.
11. Hattori, A. 1983. Denitrification and dissimilatory nitrate reduction. In E.J. Carpenter and D.G. Capone (eds.), Nitrogen in the Marine Environment, Academic Press, New York, pp. 191–232.
12. Henriksen, K., M.B. Rasmussen, and A. Jensen. 1983. Effect of bioturbation on microbial nitrogen transformations in the sediment and fluxes of ammonium and nitrate to the overlying water. Environ. Biogeochem. Ecol. Bull. **35**:193–205.
13. Hollibaugh, J.T., et al. 1988. Tomales Bay, California: a macrocosm for examining biogeochemical coupling at the land-sea interface. EOS **69**:843–845.
14. Howarth, R.W., R. Marino, J. Lane, and J. Cole. 1988. Nitrogen fixation in freshwater, estuarine, and marine ecosystems: 1. Rates and importance. Limnol. Oceanogr. **33**:669–687.
15. Jones, M. 1984. Nitrate reduction by shaking with cadmium: alternative to cadmium columns. Water Res. **18**(5):643–646.
16. Kaspar, H.F. 1983. Denitrification, nitrate reduction to ammonium and inorganic nitrogen pools in intertidal sediments. Marine Biol. **74**:133–139.
17. Kemp, W.M., P. Sampou, J. Caffrey, M. Meyer, K. Henrikson, and W.R. Boynton. 1990. Ammonium regeneration versus denitrification in Chesapeake Bay sediments. Limnol. Oceanogr. **35**:1545–1563.
18. Knowles, R. 1978. Denitrification, acetylene reduction and methane metabolism in lake sediment exposed to acetylene. Appl. Environ. Microbiol. **38**:486–493.
19. Legendre, L., and M. Gosselin. 1989. New Production and export of organic matter

to the deep ocean: Consequences of some recent discoveries. Limnol. Oceanogr. **34**(7):1374–1380.

20. Nishio, T., I. Koike, and A. Hattori. 1983. Estimates of denitrification and nitrification in coastal and estuarine sediments. Appl. Environ. Microbiol. **43**:648–653.
21. Nixon, S.W., C.A. Oviatt, and S.S. Hale. 1976. Nitrogen regeneration and the metabolism of coastal marine bottom communities. In J.M. Anderson and A. Macfayden (eds.), The Role of Terrestrial and Aquatic Organisms in Decomposition Processes, Blackwell, Boston, pp. 269–283.
22. Nixon, S.W. 1980. Between coastal marshes and coastal waters: A review of twenty years of speculation and research on the role of salt marshes in estuarine productivity and water chemistry. In P. Hamilton and K.B. MacDonald (eds.), Estuarine and Wetland Processes, Plenum Press, New York, pp. 437–525.
23. Nixon, S.W. 1981. Remineralization and nutrient cycling in coastal marine ecosystems. In B.J. Neilson and L.E. Cronin (eds.), Estuaries and Nutrients, Humana Press, Totowa, NJ, pp. 111–138.
24. Nixon, S.W., C.A. Oviatt, J. Frithsen, and B. Sullivan. 1987. Nutrients and the productivity of estuarine and coastal marine ecosystems. J. Limnol. Soc. S. Afr. **12**(1/2):43–71.
25. Oremland, R.S., and D.G. Capone. 1987. Use of specific inhibitors in microbial ecological and biogeochemical studies. In K.C. Marshall (ed.), Advances in Microbial Ecology, Vol. 10, Plenum Press, New York, pp. 285–383.
26. Paerl, H.W., B.M. Bebout, and L.E. Prufert 1989. Naturally occurring patterns of oxygenic photosynthesis and nitrogen fixation in marine microbial mats: Physiological and Ecological Ramifications. In Y. Cohen and E. Rosenberg (eds.), Microbial Mats: Physiological Ecology of Benthic Microbial Communities. American Society of Microbiology, Washington, DC.
27. Parsons, T.R., Y. Maita, and C.M. Lalli. 1984. A Manual of Chemical and Biological Methods for Seawater Analysis, Permagon Press, New York.
28. Pilson, M.E.Q. 1985. Annual cycles of nutrients and chlorophyll in Narragansett Bay, Rhode Island. J. Marine Res. **43**:849–873.
29. Postgate, J.R. 1982. Nitrogen Fixation, Studies in Biology No. 92, E. Arnold, London.
30. Ryther, J.H., and W.M. Dunstan. 1971. Nitrogen, phosphorus and eutrophication in the coastal marine environment. Science **171**:1008–1013.
31. SAS Version 5.0. 1985. SAS Institute Inc., Cary, NC.
32. Seitzinger, S.P. 1988. Denitrification in freshwater and coastal marine ecosystems: Ecological and geochemical significance. Limnol. Oceanogr. **33**:702–724.
33. Sherer, S., H. Almon, and P. Boger. 1988. Interaction of photosysthesis, respiration and nitrogen fixation in cyanobacteria. Photosyn. Res. **15**:95–114.
34. Smith, C.J., R.D. DeLaune, and W.H. Patrick, Jr. 1985. Fate of riverine nitrate entering an estuary: 1. Denitrification and nitrogen burial. Estuaries **8**:15–21.
35. Smith, S.V., et al. 1987. Stoichiometry of of C, N, P, and Si fluxes in a temperate-climate embayment. J. Marine Res. **45**:427–460.

36. Smith, S.V., J.T. Hollibaugh, S.J. Dollar, and S. Vink. 1989. Tomales Bay, California: a case for carbon controlled nitrogen cycling. Limnol. Oceanogr. **34**: 37–52.

37. Smith, S.V., and J.T. Hollibaugh. 1989. Carbon-controlled nitrogen cycling in a marine "macrocosm": an ecosystem scale model for managing cultural eutrophication. Marine Ecol. Prog. Ser. **52**:103–109.

38. Smith, S.V., J.T. Hollibaugh, S.J. Dollar, and S. Vink. 1991. Tomales Bay Metabolism: C–N–P stoichiometry and ecosystem heterotrophy at the land–sea interface. Estuarine Coast. Shelf Sci. **33**:223–257.

39. Sorensen, J. 1978. Denitrification rates in a marine sediment as measured by the acetylene inhibition technique. Appl. Environ. Microbiol. **36**:139–143.

40. Stewart, W.D.P. 1975. Nitrogen Fixation by Free-Living Microorganisms, Cambridge University Press, London.

41. Yates, M.G. 1977. Physiological aspects of nitrogen fixation. In W. Newton, J.R. Postgate, and C. Rodrigues-Barrueco (eds.), Recent Developments in Nitrogen Fixation, Academic Press, London, pp. 219–270.

The Influence of Glycine Betaine on Dimethyl Sulfide and Dimethylsulfoniopropionate Concentrations in Seawater

Ronald P. Kiene and Susan K. Service*

University of Georgia Marine Institute, Sapelo Island, Georgia 31327

ABSTRACT

We investigated whether additions of a common methylated osmolyte, glycine betaine (GBT), affected the concentrations of dimethyl sulfide (DMS) and its principal precursor, β-dimethylsulfoniopropionate (DMSP) in seawater samples. DMS concentrations were frequently elevated by additions of GBT over the concentration range 0.125–5 μM. Additions of other methylated amines including choline and trimethylamine also increased DMS concentrations but to a lesser degree than did GBT. Nonmethylated compounds such as glucose, glycine, or acrylate had no effect on DMS concentrations. GBT affected both particulate and dissolved pools of DMSP. The addition of 5 μM GBT to a natural assemblage of growing phytoplankton decreased the production of particulate DMSP but not that of chlorophyll *a*. GBT substantially inhibited the consumption of dissolved DMSP ($DMSP_{diss}$) and the production of DMS from $DMSP_{diss}$. Inhibition of $DMSP_{diss}$ consumption resulted in a buildup of this pool which was only transient in some cases. As the consumption of $DMSP_{diss}$ resumed, DMS production was stimulated. Our findings suggest DMS and DMSP concentrations may be influenced by naturally occurring compounds such as GBT.

Introduction

Dimethyl sulfide (DMS) is an important atmospheric trace gas and there is a considerable global flux of DMS from the oceans.[1,2,6,16] Interest in the flux of DMS stems from its role in atmospheric chemistry and possibly climate regulation.[3,5,8,29] The mechanisms involved and the factors controlling DMS production in seawater are of considerable interest, but are still poorly understood.

In seawater, DMS is produced primarily by the degradation of β-dimethylsulfoniopropionate (DMSP),[10,26,35] which is present in a variety of marine phy-

*Present address: Department of Marine Sciences, University of South Alabama, LSCB 25, Mobile, AL 36688.

toplankton and macroalgae.[13,22,35] DMSP is present in seawater in a particulate pool, presumably in algae and grazers, and a dissolved pool.[20,35] The origin of the dissolved pool is still a matter of speculation but it may be derived from algal exudation or by incidental leakage of cell contents when the algae are eaten or digested.[21] The dissolved pool of DMSP is labile, being degraded quickly in estuarine and oceanic waters.[24,26] Recent studies suggest that only a portion (<30%) of the DMSP degraded in estuarine and oceanic waters may be converted to DMS[24,26] with the larger fraction probably undergoing demethylation (Refs. 24 and 34; Visscher et al., unpublished data). Nonetheless, DMSP is the major precursor of DMS in seawater and, therefore, factors which affect its degradation need to be understood.

DMSP functions as an osmolyte in certain marine phytoplankton.[13–15,22,36] Marine plants also utilize a variety of quarternary ammonium compounds for osmotic purposes,[7,13–15] including the nitrogen-containing compound glycine betaine (GBT, $(CH_3)_3N^+CH_2COO^-$). A variety of marine phytoplankton produce both GBT and DMSP, with intracellular concentrations often exceeding 100 mM (Refs. 13 and 19; Kiene and Keller, unpublished data). DMSP production by organisms other than plants has not been reported, but a variety of marine invertebrates are known to contain high levels of GBT.[27] Although to our knowledge GBT has never been measured as a dissolved constituent of seawater, it is undoubtedly present for the same reasons as is dissolved DMSP: algal exudation and cellular disruption due to grazing. GBT is a C_1 substrate (containing N-methyl groups) and a functional analogue of DMSP. As such, it could influence the metabolism of C_1 substrates such as DMSP and DMS. Therefore, we investigated whether exogenous GBT had any effects on pools of DMS and DMSP in natural seawater samples.

Materials and Methods

Sample Collection

Water samples used in this study were collected at Marsh Landing Dock, located on the Duplin River, Sapelo Island, Georgia (31°23′N, 81°17′W). The Duplin River is a large tidal creek, with little direct freshwater input. Local salinity is influenced by coastal plain rivers and typically varies over the range 22–32 parts per thousand. Water for experiments was collected directly into the incubation bottles while sitting in a small boat tied to the dock. Sample bottles were held several centimeters below the surface and always faced into the tidal current while filling. The bottles were immediately placed in the dark and transported to the laboratory within 10 min.

Incubation of Samples

Experiments were carried out by incubating water samples and measuring the concentrations of DMS, $DMSP_{part}$, and $DMSP_{diss}$ over time courses which were generally less than 15 h. Not all pools were measured in each experiment. Incubation bottles were either 250-ml translucent Teflon (FEP) or 1-L polycarbonate bottles. Except where indicated, all samples were incubated in the dark. Some samples were held in polycarbonate bottles in a light incubator with fluorescent illumination at 100 microeinsteins $m^{-2}\ s^{-1}$. All samples were incubated within 2°C of the in situ temperature (25–29°C).

Experimental

Each experiment was carried out on a different date and, therefore, water samples were different for each experiment. All of the experiments tested the effects of at least one level of GBT on either DMS, DMSP, or both. The final added concentration of GBT varied depending on the experiment, but was in the range 0.125–50 μM. Other additions included N-methyl compounds such as choline, dimethylglycine, and trimethylamine each at 500 nM. Several non-methylated organic compounds (500 nM each) including glycine, glucose, and acrylate (the carbon skeleton of DMSP) were also tested for their effects relative to GBT. DMSP was added in some cases and a combined treatment of GBT plus DMSP was used to test for the effects of GBT on DMSP metabolism. Duplicate or triplicate bottles for each treatment were used. Ranges or standard deviations of the means were calculated and plotted as error bars in the figures.

Analytical

Several experiments employed the addition of approximately 500 nM DMSP or DMS to water samples. In these cases, DMS was measured in the head spaces of sealed bottles as described in Kiene.[23] Trace levels (< 50 nM) of DMS and DMSP were measured as described in Kiene and Service.[26] Briefly, small (2-ml) subsamples were filtered (Whatman GF/F or Gelman A/E) and introduced into a sparging system. The water was bubbled with helium to strip out the sulfur gases which were subsequently trapped in Teflon tubing immersed in a liquid-N_2 bath. After concentration of the sulfur gases, the cold trap was replaced with hot water and the gases swept with a helium flow into a gas chromatograph (GC). The GC was a Shimadzu GC-9A equipped with a flame-photometric detector. The column was Teflon (2 m × ⅛ in. diameter) filled with Carbopack BHT-100 or Chromosil 330 (Supelco). The oven temperature was 100°C (60°C for the Chromosil column) and the carrier was He at 60 ml-min^{-1}. Peak areas were recorded by a Shimadzu CR-6A integrator. DMS standards were prepared gravimetrically in ethylene glycol as described by Andreae and Barnard.[4] Microliter quantities of standard were injected directly into the sparging system which held 2 ml of presparged distilled water. The standard was then sparged and trapped in the same way as the samples. The square root of the peak area was a linear function of the amount of DMS injected. The minimum detectable amount of DMS ranged from 0.1 to 1 pmole per injection depending on the sensitivity of the detector, which varied during the course of the study. For typical water volumes analyzed (2 ml), this method gave detection limits for DMS of 50–500 pM. DMS standards prepared in glycol were compared to DMSP standards which were treated with base to produce DMS (as described below). This comparison between the two different standards showed agreement to within 5%.

Particulate and Dissolved DMSP

Particulate DMSP was defined as that retained on a glass fiber filter (Whatman GF/F or Gelman A/E, nominal retentions of 0.7 and 1.0 μm, respectively). Removal of particulate DMSP by these filters was quantitative and no differences between the two types of filters were observed with respect to DMSP analysis in estuarine waters. Incubation bottles were gently mixed before 10–25 ml was removed for analysis. The sample was filtered by gravity and the filters placed in 14-ml serum bottles. Two milliliters of 5 M NaOH was added and the bottles quickly capped with Teflon-faced septa. The samples were then allowed to react for at least 6 h; usually 12–24 h at 25°C. DMSP is quantitatively and stoichiometrically decomposed in strong alkali to DMS and acrylate.[9,37] The DMS evolved into the head space was measured by removing a subsample of the head space gas and injecting it directly into the GC column. DMSP standards were prepared by adding known amounts of DMSP to the vials and treating these exactly as the samples.

Dissolved DMSP was defined as that which passed through the glass fiber filters. The filtrate was poured into a small polyethylene vial and immediately frozen and stored for later analysis. Storage did not result in any loss of DMSP. For analysis, a subsample of 1 ml was injected into the sparging tube and 1 ml of 5 M NaOH added to produce DMS from the DMSP. The sample was sparged for 5 min after which time tests showed that >95% of the DMSP was reacted. The DMS was cryotrapped and quantified as described above. Precision of the dissolved DMSP determinations for natural water samples was 7% (coefficient of variation $n=5$).

Chlorophyll *a* was determined by capturing particulate material on Whatman GF/F filters, extracting these in 90% acetone and determining fluorescence as described in Strickland and Parsons.[32]

Reagents and Chemicals

Reagent DMSP · HCl was obtained from Research Plus, Inc. GBT (betaine hydrochloride, 99+%) was obtained from Aldrich. All other chemicals used were of reagent grade or better.

Results

Additions of 0.5–5 μM GBT to incubated water samples resulted in DMS concentrations which were higher than in unamended controls (Figure 1). This effect was observed at GBT concentrations as low as 125 nM (the lowest level tested) and in samples incubated in the light or dark (Figure 2). In an experiment designed to test the effects of a variety of N-methyl compounds (each at 500 nM) on DMS levels, net loss of DMS was observed in untreated samples (Figure 3). The loss of DMS was decreased by trimethylamine (TMA), choline, and GBT; with GBT, the DMS concentration increased (Figure 3). In a separate experiment, additions of nonmethylated substrates including glucose, glycine, and acrylate (each at 500 nM) did not substantially affect DMS concentrations (Figure 4), whereas equimolar amounts of GBT caused a transient increase in DMS. Although DMS levels were increased by GBT in many experiments during this study, there were occasions when no increases occurred, when increases were transient, or when increases occurred only later in the incubations (see below).

DMS consumption was unaffected by 5 μM GBT and other N-methyl compounds such as choline and trimethylamine (Figure 5). We, therefore, investigated the effects of GBT on DMS production from DMSP. Our initial experiment (Figure 6) employed additions of 500 nM DMSP, 5 μM GBT or both and revealed that when GBT and DMSP were added together DMS production was slower over the first 7 h of incubation than when DMSP alone was added. After 7 h, however, DMS production in the GBT+DMSP treatment accelerated and at 12 h, DMS levels in the presence of GBT were higher than in the DMSP only treatment. A subsequent experiment with 10-fold lower levels of DMSP and GBT showed that $DMSP_{diss}$ consumption and DMS production were partially but significantly inhibited by GBT during a 7-h incubation (Figure 7).

We sometimes observed increases in $DMSP_{diss}$ concentrations after additions of GBT (Figure 8A). The response of the $DMSP_{diss}$ pool was different for different addition levels of GBT. With 500 nM added GBT, the increase in $DMSP_{diss}$ was small (~1 nM) and transient. Both 5 and 50 μM GBT caused $DMSP_{diss}$ concentrations to increase to about 8 nM from the control level of 3–4 nM. The $DMSP_{diss}$ concentration in the 50-μM treatment did not decline over the 12-h incubation (Figure 8A), whereas the concentration in the 5-μM GBT treat-

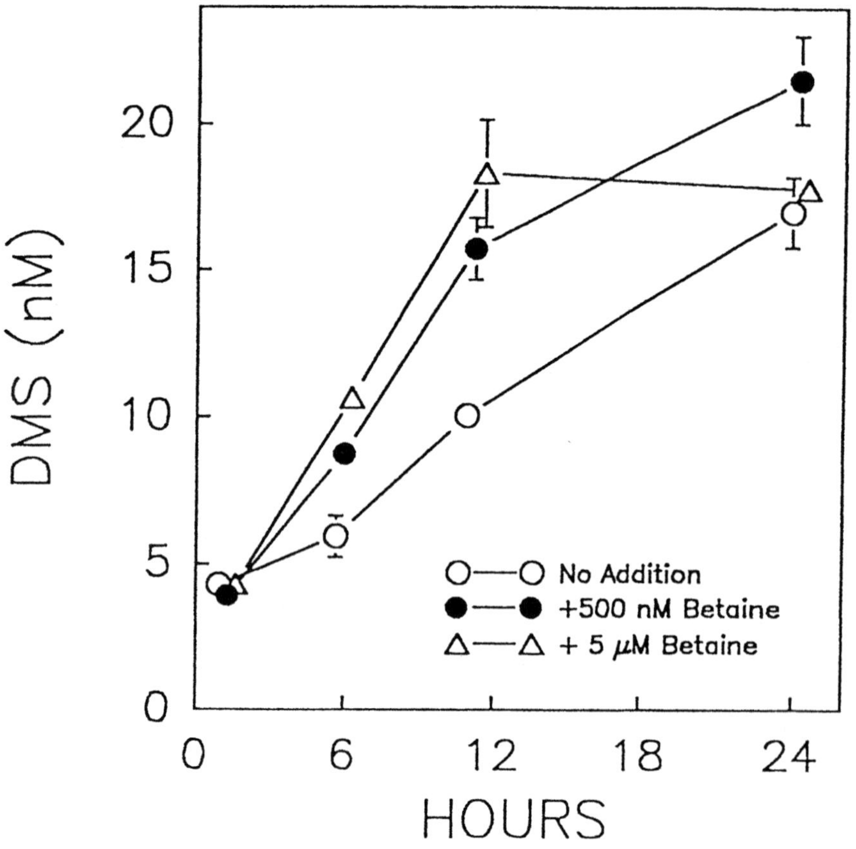

Figure 1. The effects of two different concentrations of glycine betaine (GBT) on DMS concentrations in estuarine water samples. Samples were incubated in the dark at the in situ temperature (25°C). Data represent the mean of triplicate bottles of each treatment with the bars indicating one standard deviation.

ment began to decline after 4 h and reached the level of controls (3 nM) after 12 h. During the decline in $DMSP_{diss}$, DMS concentrations increased above control levels in the 5-μM GBT treatment (Figure 8B). Only slight elevations in DMS were observed in the 500-nM and 50-μM GBT treatments. This experiment was conducted in room light and concentrations of particulate DMSP ($DMSP_{part}$) increased from 40 to 75 nM in the unamended samples (data not shown). However, this increase was not seen in any of the GBT-treated samples. A subsequent experiment with water samples incubated in a light incubator showed that

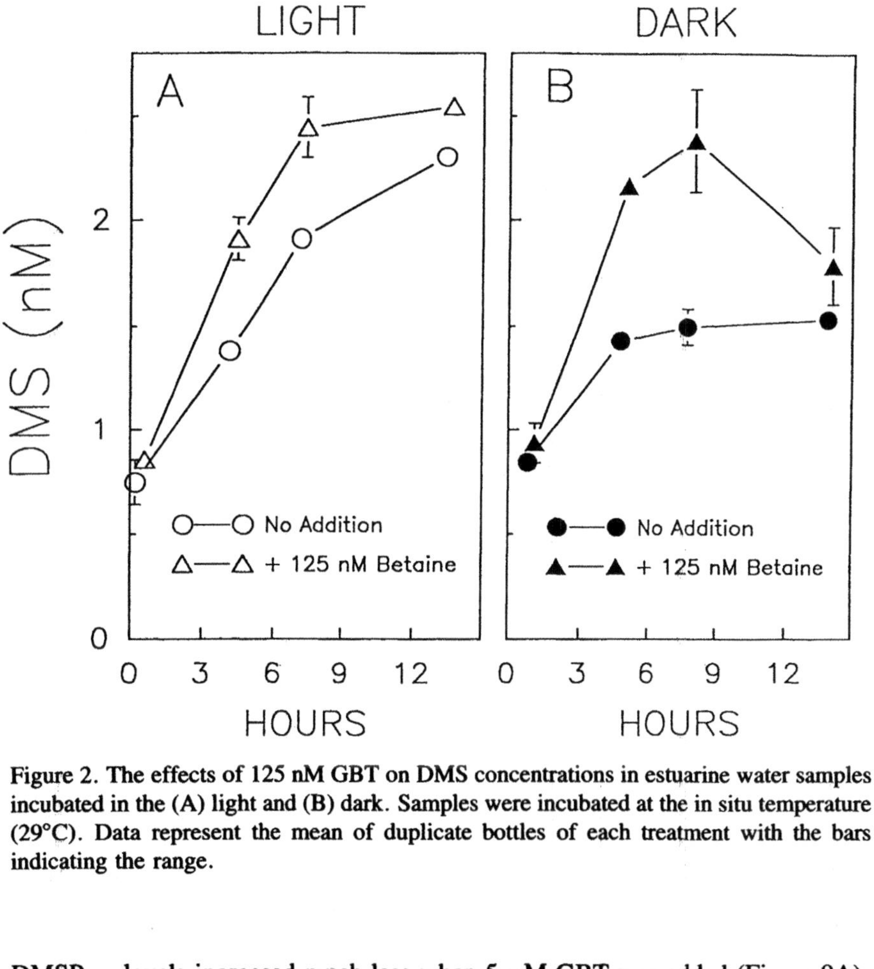

Figure 2. The effects of 125 nM GBT on DMS concentrations in estuarine water samples incubated in the (A) light and (B) dark. Samples were incubated at the in situ temperature (29°C). Data represent the mean of duplicate bottles of each treatment with the bars indicating the range.

$DMSP_{part}$ levels increased much less when 5 μM GBT was added (Figure 9A). Chlorophyll *a* also increased during the incubation and there was no difference between unamended and GBT-treated samples (Figure 9B).

Discussion

Our findings suggest that some interesting and potentially important microbial biogeochemical interactions exist between GBT and DMSP in natural systems. We observed that DMS concentrations in seawater samples increased after GBT was added over a range of concentrations from 0.125 to 5 μM and under a variety of conditions (i.e., light and dark) (Figures 1–4). In seeming contrast, we

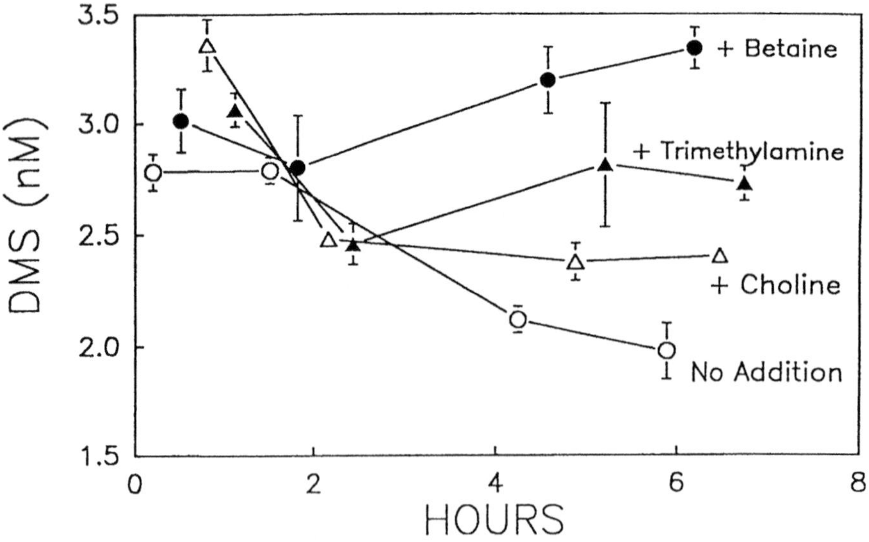

Figure 3. The effects of three different N-methyl compounds, GBT, choline and trimethylamine, each at 500 nM, on DMS concentrations in estuarine water samples. Samples were incubated in the dark at the in situ temperature (25°C). Data represent the mean of triplicate bottles of each treatment with the bars indicating one standard deviation.

found that DMS production from $DMSP_{diss}$ was inhibited by GBT (Figure 7). Why, then, did DMS concentrations generally rise in the presence of GBT, and why was the effect variable from one experiment to the next? The most likely explanation suggested by our data is that DMSP degradation was initially inhibited by GBT (Figure 7) and, as a result, $DMSP_{diss}$ concentrations increased (Figure 8A). At certain concentrations of GBT, the inhibition of $DMSP_{diss}$ metabolism was transient and the accumulated $DMSP_{diss}$ was subsequently metabolized with accelerated production of DMS (Figure 8B). The interaction between the positive (increasing $DMSP_{diss}$) and negative (inhibition of $DMSP_{diss}$ conversion to DMS) effects of GBT on DMS are complex and probably dependent on the assemblage of microorganisms present in the samples and the relative concentrations of GBT and $DMSP_{diss}$.

We observed inhibition of DMSP consumption in the presence of 500 nM GBT (Figure 7A) and this explains the lower DMS production from added DMSP observed in the presence of GBT (Figures 6 and 7B). $DMSP_{diss}$ is metabolized by at least two pathways in seawater,[26] a lyase pathway which cleaves DMSP into DMS and acrylate and also a demethylation pathway which has only recently been recognized (Refs 12, 34; Visscher, Diaz, and Taylor et al., un-

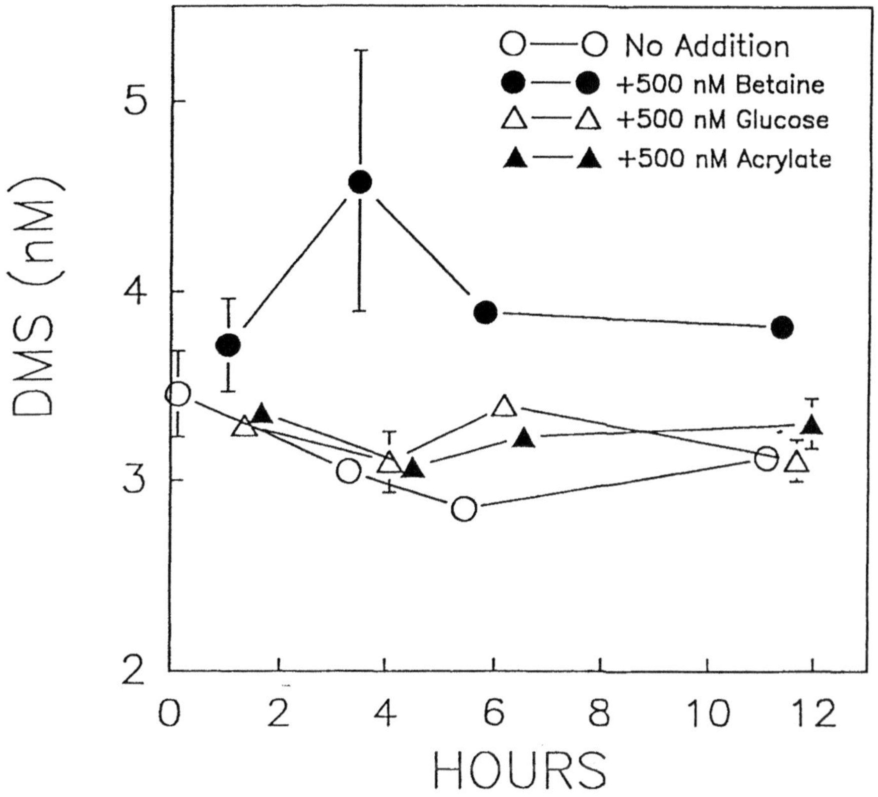

Figure 4. The effects of GBT, acrylate, glycine, and glucose, each at 500 nM on the concentration of DMS in estuarine water samples. Samples were incubated in a light incubator at the in situ temperature (26°C). Data represent the mean of duplicate bottles of each treatment with the bars indicating the range.

published data). GBT is chemically similar to DMSP and it might interfere with the direct utilization of DMSP by lyase or demethylation enzymes or with transport of DMSP into cells. Diaz et al.,[12] observed that a single marine bacterium (MD-14-50) metabolized both DMSP and GBT, with DMSP being cleaved and GBT being demethylated. Both compounds served as osmolytes in MD-14-50 and apparently could be taken up from the media.[12]

We can only speculate at the present time whether the inhibition of $DMSP_{diss}$ consumption and the accumulation of $DMSP_{diss}$ after the addition of GBT is a result of the inhibition of $DMSP_{diss}$ transport into cells or perhaps due to greater release of DMSP from particulate pools. It is not known if DMSP must be

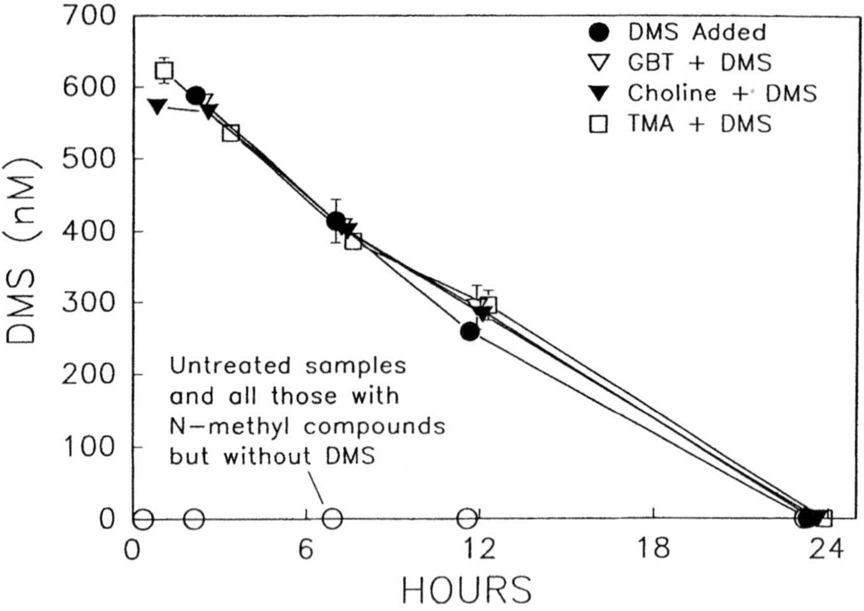

Figure 5. The effects of various N-methylated compounds (each at 5 μM) on the consumption of DMS in seawater samples. DMS was measured in the head space of the bottles. Data represent the mean of triplicate bottles of each treatment with the bars indicating one standard deviation.

transported across the cell membrane before it can be metabolized by bacteria. The degradation of DMSP to DMS has been observed in 0.2 μM filtered seawater (Ref. 35; Kiene, unpublished data) arguing that the enzymes involved are capable of extracellular activity. Evidence for the transport of DMSP into bacterial cells is given by the fact that DMSP alleviates osmotic stress in some species.[12,28] Specific membrane transport systems for DMSP have not, to our knowledge, been demonstrated, but it is well-known that certain bacteria have specific transport systems for GBT.[17,33] It is possible that a DMSP transport system might recognize GBT (and vice versa) because of functional and chemical similarities between the two. Thus, we speculate that GBT could competitively inhibit DMSP uptake. It is interesting to note that with addition of 50-μM GBT, inhibition of $DMSP_{diss}$ consumption appeared to be prolonged (Figure 8A). With added GBT levels of 5 μM or less, the inhibition of $DMSP_{diss}$ metabolism was transient and this suggests that the inhibition is reversible. GBT is undoubtedly consumed by microorganisms in aquatic systems either for os-

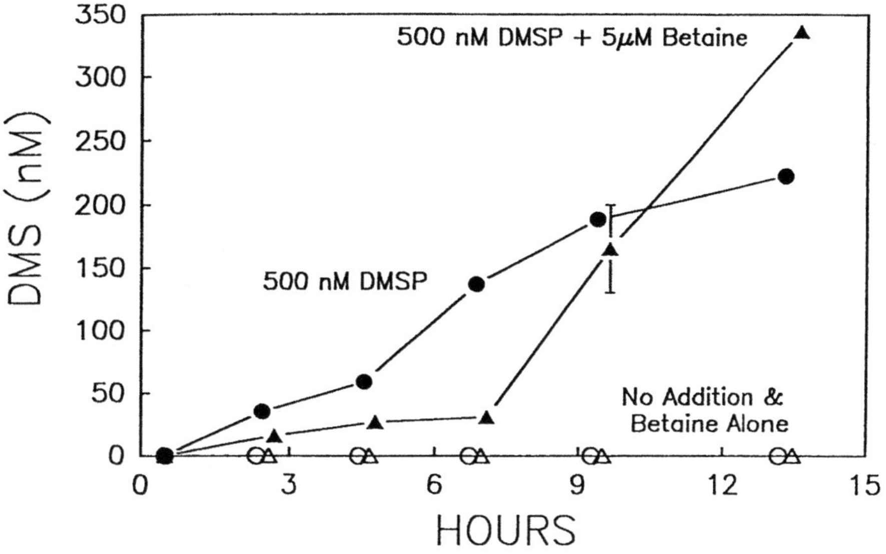

Figure 6. The effects of 5 μM GBT on DMS production from 500 nM added DMSP. Samples were incubated in the dark at the in situ temperature (25°C). DMS was measured in the head space of the bottles. Data represent the mean of four replicates of each treatment with the bars indicating one standard deviation.

motic purposes (such as by algae and bacteria?) or as a methyl substrate.[12,30] Over time, its concentration may decrease to the point where inhibition of $DMSP_{diss}$ metabolism is alleviated. Accelerated metabolism of $DMSP_{diss}$ would then result in elevated DMS concentrations as observed in several of our experiments. It does not appear that GBT caused a greater fraction of the $DMSP_{diss}$ to be degraded by the lyase pathway. The data in Figure 7 show that in the GBT treatment a slightly lower fraction (20%) of the $DMSP_{diss}$ consumed was recovered as DMS after 7 h compared to 28% in the control treatment. The difference between these two DMS yields was not statistically significant ($p > 0.14$).

The effect of GBT on DMS appears to be related to the fact that this compound contains an N-methyl moiety because additions of related compounds including choline and trimethylamine also elevated DMS levels in natural water samples (Figure 3). Their influence on DMS, however, was less than that of GBT. In contrast, nonmethylated compounds including glycine, glucose, and acrylate had little effect on DMS concentrations (Figure 4). We found no evidence that addition of 5 μM N-methyl compounds including GBT, choline, dimethylglycine, and trimethylamine inhibited the consumption of DMS (Figure 5; Wolfe and Kiene, unpublished data). It is possible that these other N-methyl com-

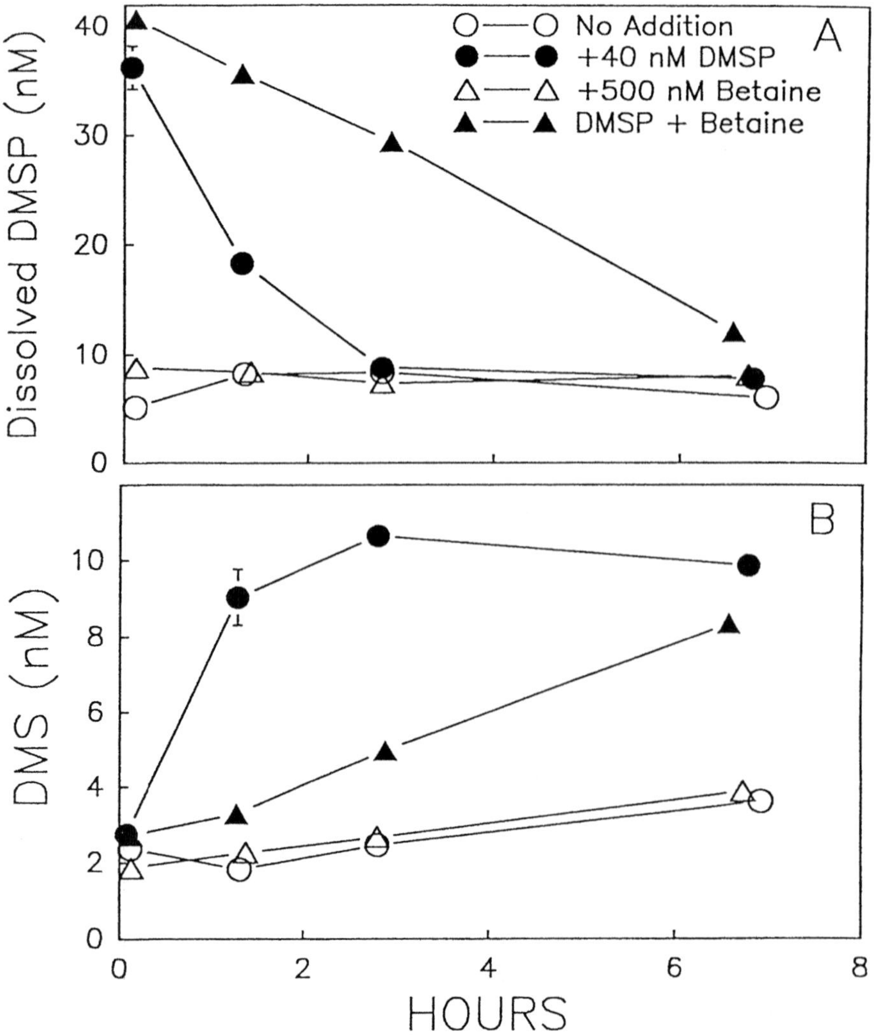

Figure 7. The effects of 500 nM GBT on (A) the decomposition of added dissolved DMSP and (B) the production of DMS from added DMSP. Samples were incubated in the dark at the in situ temperature (27°C). Data represent the mean of duplicate bottles of each treatment with the bars indicating the range. For many points the range was smaller than the symbol.

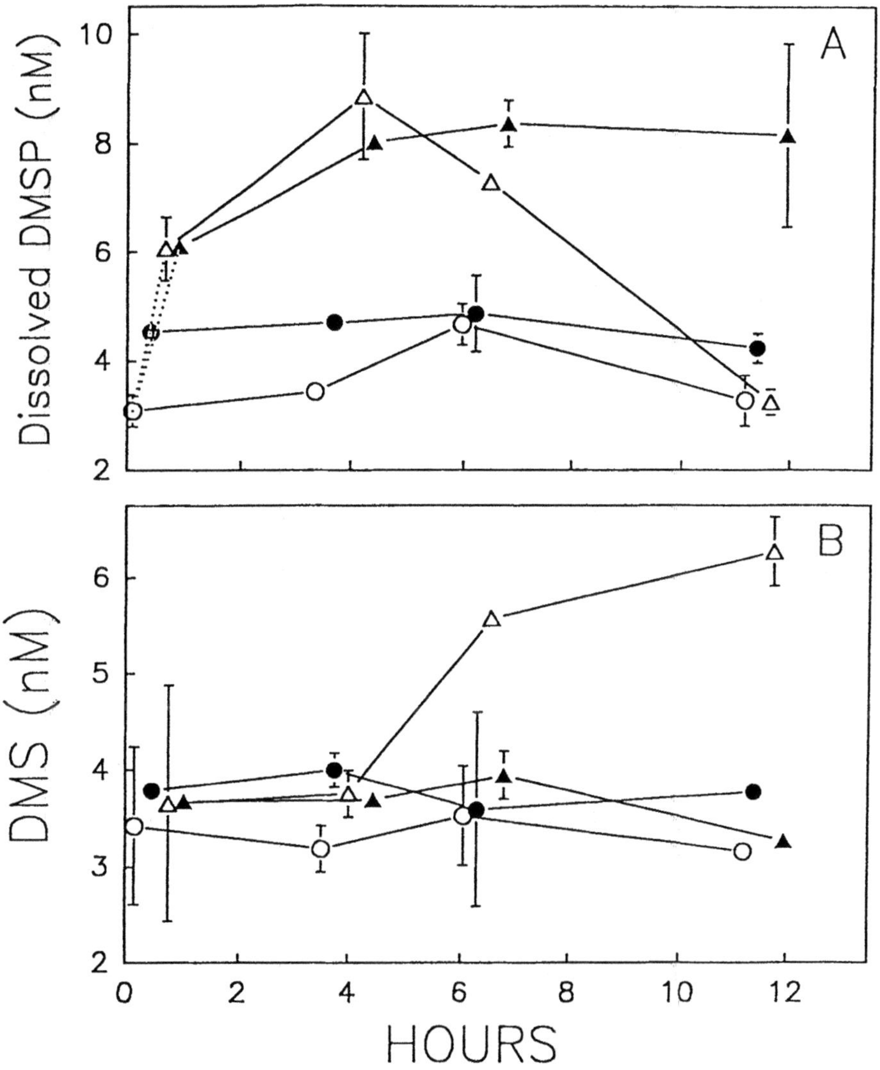

Figure 8. The effect of different concentrations of GBT on (A) dissolved DMSP concentrations and (B) DMS concentrations in estuarine water samples. Samples were incubated in the dark at the in situ temperature (25°C). Symbols: ○, no addition; ●, 0.5 μM GBT; △, 5 μM GBT; and ▲, 50 μM GBT. Data represent the mean of triplicate bottles of each treatment with the bars indicating one standard deviation.

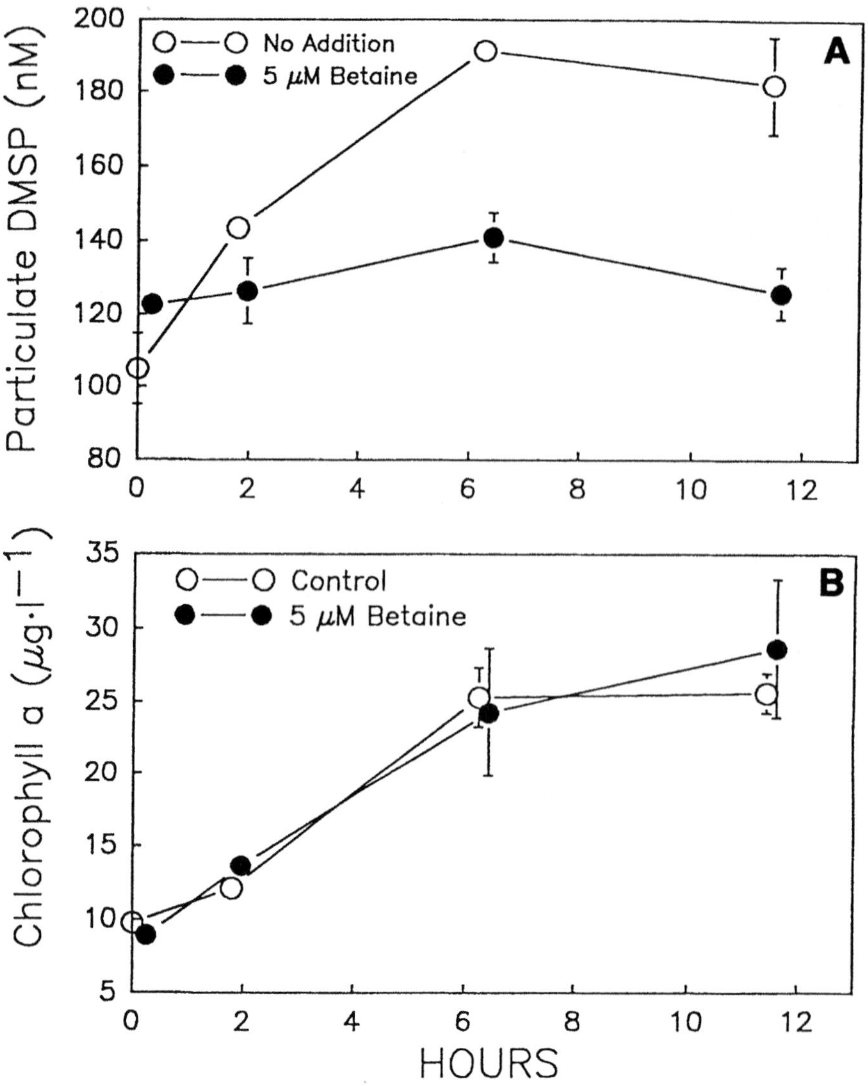

Figure 9. The effects of 5 μM GBT on (A) particulate DMSP and (B) chlorophyll *a* concentrations in estuarine water samples held in a light incubator at 26°C. Data represent the mean of duplicate bottles of each treatment with the bars indicating the range.

pounds affected $DMSP_{diss}$ metabolism in a manner similar to that observed with GBT (Figure 7). We did not test this, however.

In addition to the effects of GBT on DMS and $DMSP_{diss}$, we observed that GBT affected the production of particulate DMSP. A natural assemblage of growing phytoplankton produced less $DMSP_{part}$ in the presence of 5 μM GBT (Figure 9A), whereas growth of the phytoplankton (as evidenced by Chl *a*) was not affected by GBT (Figure 9B). This suggests that an external source of GBT can reduce the need to synthesize DMSP. It remains to be demonstrated whether GBT is taken up directly by phytoplankton cells or whether it is first degraded. Other investigators have observed that nitrogen availability plays a critical role in DMSP production in phytoplankton cultures.[18,35] Our findings with the N-compound GBT are consistent with those of Gröne and Kirst[18] who found that cultures of GBT-producing *Tetraselmis subcordiformis* grown with excess NO_3^- produced less DMSP than cultures with excess nitrogen. Similarly, Dacey et al.[11] found that DMSP levels are lower and GBT levels higher in N-fertilized *Spartina alterniflora*. Our findings suggest that the "N effect" observed by others can also be produced with GBT.

Although GBT is known to be widespread in marine macroalgae[7] and some rooted halophytes,[11,31] relatively little is known about the nitrogen-containing osmolytes of marine phytoplankton. Dickson and Kirst[13–15] reported the occurrence of osmotically significant levels of GBT in some prasinophytes, diatoms, and a rhodophyte. Some algae, particularly the prasinophytes contain both DMSP and GBT.[13–15] Several authors[1,35] have discussed the possible advantages conferred upon phytoplankton which could utilize the sulfur-containing osmolyte, DMSP, instead of nitrogen-containing compounds. Sulfur, in the form of sulfate, is never limiting in a seawater environment (20–28 mM SO_4^{2-}), whereas combined nitrogen is often severely limiting. If phytoplankton have the ability to take up GBT from the bulk pool in seawater, this could supplement their requirements for combined nitrogen in a readily utilizable form and could possibly lessen their need for compounds such as DMSP.

Acknowledgments

Support for this research was provided by the National Science Foundation (Grant OCE-8817442) and The University of Georgia Marine Institute. We thank Pieter Visscher and Gordon Wolfe for reading this manuscript. Thanks are extended to Barrie Taylor for providing access to an in-press manuscript (Diaz, Visscher, and Taylor).

References

1. Andreae, M.O. 1985. The emission of sulfur to the atmosphere: background paper. In J.N. Galloway, R.J. Charlson, M.O. Andreae, and H. Rodhe (eds.), The Bio-

geochemical Cycles of Sulfur and Nitrogen in the Remote Atmosphere, D. Reidel, Dordrecht, pp. 5–25.

2. Andreae, M.O. 1986. The ocean as a source of atmospheric sulphur compounds. In P. Buat-Menard (ed.), The Role of Air Sea Exchange in Geochemical Cycling, D. Reidel, Dordrecht, pp. 331–362.
3. Andreae, M.O. 1990. Ocean–atmosphere interactions in the global biogeochemical sulfur cycle. Marine Chem. **30**:1–29.
4. Andreae, M.O., and W.R. Barnard. 1983. Determination of trace quantities of dimethyl sulfide in aqueous solutions. Anal. Chem. **55**:608–612.
5. Bates, T.S., R.J. Charlson, and R.H. Gammon. 1987. Evidence for the climatic role of marine biogenic sulphur. Nature **329**:319–321.
6. Bates, T.S., B.K. Lamb, A. Guenther, J. Digon, and R.E. Stoiber. 1992. Sulfur emissions to the atmosphere from natural sources. J. Atmos. Chem. **14**:315–337.
7. Blunden, G., and S.M. Gordon. 1986. Betaines and their sulfonio analogues in marine algae. In F. Round and D. Chapman (eds.), Progress in Phycological Research, Biopress Ltd, Bristol, pp. 39–80.
8. Charlson, R.J., J.E. Lovelock, M.O. Andreae, and S.G. Warren. 1987. Oceanic phytoplankton, atmospheric sulfur, cloud albedo and climate. Nature **326**:655–661.
9. Dacey, J.W.H., and N. Blough. 1987. Hydroxide decomposition of DMSP to form DMS. Geophys. Res. Lett. **14**:1246–1249.
10. Dacey, J.W.H., and S.G. Wakeham. 1986. Oceanic dimethylsulfide: production during zooplankton grazing on phytoplankton. Science **233**:1314–1316.
11. Dacey, J.W.H., G. King, and S.G. Wakeham. 1987. Factors controlling the emission of dimethylsulfide from salt marshes. Nature **330**:643–645.
12. Diaz, M.R., P.T. Visscher, and B.F. Taylor. 1992. Metabolism of dimethylsulfoniopropionate and glycine betaine by a marine bacterium. FEMS Microbiol. Lett. **96**:61–66.
13. Dickson, D.M.J., and G.O. Kirst. 1986. The role of β-dimethylsulfoniopropionate, glycine betaine and homarine in the osmoacclimation of *Platymonas subcordiformis*. Planta **167**:536–543.
14. Dickson, D.M.J., and G.O. Kirst. 1987. Osmotic adjustment in marine eucaryotic algae: the role of inorganic ions, quarternary ammonium, tertiary sulfonium and carbohydrate solutes. I. Diatoms and a rhodophyte. New Phytol. **106**:645–655.
15. Dickson, D.M.J., and G.O. Kirst. 1987. Osmotic adjustment in marine eucaryotic algae: the role of inorganic ions, quarternary ammonium, tertiary sulfonium and carbohydrate solutes. II. Prasinophytes and haptophytes. New Phytol. **106**:657–666.
16. Erickson, D.J., S.J. Ghan, and J.E. Penner. 1990. Global ocean-to-atmosphere dimethyl sulfide flux. J. Geophys. Res. **95**:7543–7552.
17. Gowrishankar, J. 1988. Osmoregulation in *Enterobacteriaceae:* role of proline/betaine transport systems. Current Sci. **57**:225–234.

18. Gröne, T., and G.O. Kirst. 1992. The effect of nitrogen deficiency, methionine and inhibitors of methionine metabolism on the DMSP contents of *Tetraselmis subcordiformis* (Stein). Marine Biol. **112**:497–503.

19. Ishida, Y. 1968. Physiological studies on the evolution of dimethylsulfide. Mem. Fac. Coll. Agric. Kyoto Univ. **94**:47–82.

20. Iverson, R.L., F.L. Nearhoof, and M.O. Andreae. 1989. Production of dimethylsulfonium propionate and dimethyl sulfide by phytoplankton in estuarine and coastal waters. Limnol. Oceanogr. **34**:53–67.

21. Jumars, P.A., D.L. Penry, J.A. Baross, M.J. Perry, and B.W. Frost. 1989. Closing the microbial loop: dissolved carbon pathway to heterotrophic bacteria from incomplete ingestion, digestion and absorption in animals. Deep-Sea Res. **36**:483–495.

22. Keller, M.D., W.K. Bellows, and R.R.L. Guillard. 1989. Dimethyl sulfide production in marine phytoplankton. In E. Saltzman and W.J. Cooper (eds.), Biogenic Sulfur in the Environment, American Chemical Society, Washington, DC, pp. 167–182.

23. Kiene, R.P. 1990. Dimethyl sulfide production from dimethylsulfoniopropionate (DMSP) in coastal seawater and bacterial cultures. Appl. Environ. Microbiol. **56**: 3292–3297.

24. Kiene, R.P. 1992. Dynamics of dimethylsulfide and dimethylsulfoniopropionate in oceanic water samples. Marine Chem. **37**:29–52.

25. Kiene, R.P., and T.S. Bates. 1990. Biological removal of dimethyl sulfide from seawater. Nature **345**:702–705.

26. Kiene, R.P., and S.K. Service. 1991. Decomposition of dissolved DMSP and DMS in estuarine waters: dependence on temperature and substrate concentrations. Marine Ecol. Prog. Ser. **76**:1–11.

27. King, G.M. 1988. Distribution and metabolism of quaternary amines in marine sediments. In T. Blackburn and J. Sorensen (eds.), Nitrogen Cycling in Coastal Marine Environments, J. Wiley and Sons, New York, pp. 143–173.

28. Mason, T.G., and G. Blunden. 1989. Osmoprotectants for bacteria obtained from marine algae. Bot. Marine **32**:313–316.

29. Shaw, G.E. 1982. Bio-controlled thermostasis involving the sulfur cycle. Climate Change **5**:297–303.

30. Shieh, H.S. 1966. Further studies on the oxidation of betaine by a marine bacterium, *Achromobacter cholinophagum*. Can. J. Microbiol. **12**:299–302.

31. Storey, R., N. Ahmad, and R.G. Wyn Jones. 1977. Taxonomic and ecological aspects of the distribution of glycinebetaine and related compounds in plants. Oecologia **27**:319–322.

32. Strickland, J.D.H., and T.R. Parsons. 1972. A practical handbook of seawater analysis. Bulletin 167, Fish. Res. Bd. Can., pp. 201–203.

33. Strøm, A.R., P. Falkenberg, and B. Landfald. 1986. Genetics of osmoregulation in *Eschericia coli:* uptake and biosynthesis of organic osmolytes. FEMS Microbiol. Rev. **39**:79–86.

34. Taylor, B.F., and D.C. Gilchrist. 1992. New routes for the biodegradation of dimethylsulfoniopropionate. Appl. Environ. Microbiol. **57**:3581–3584.

35. Turner, S.M., G. Malin, and P.S. Liss. 1988. The seasonal variation of dimethyl sulfide and dimethylsulfoniopropionate concentrations in nearshore waters. Limnol. Oceanogr. **33**:364–375.

36. Vairavamurthy, A., M.O. Andreae, and R.L. Iverson. 1985. Biosynthesis of dimethylsulfide and dimethylpropiothetin by *Hymenomonas carterae* in relation to sulfur source. Limnol. Oceanogr. **30**:59–70.

37. White, R.H. 1982. Analysis of dimethyl sulfonium compounds in marine algae. J. Marine Res. **40**:529–535.

Sulfur Cycling in Laminated Marine Microbial Ecosystems

Pieter T. Visscher and Hans van Gemerden*

Department of Microbiology, University of Groningen, 9715 NN Haren, The Netherlands

ABSTRACT

Depth profiles of organic and inorganic sulfur compounds were measured in the top 50 mm of a mature marine microbial mat during three growth seasons. Similarly, the vertical distribution of sulfate-reducing bacteria, colorless sulfur bacteria, and anoxygenic phototrophic bacteria was also measured. In combination with profiles of important physico-chemical parameters, which define the mat environment, and laboratory studies, sources and sinks of volatile sulfur compounds were evaluated. Calculations show that consumption of dimethyl sulfide potentially can balance its production, minimizing the flux from the sediment to the atmosphere.

Introduction

Marine sediments in shallow waters represent highly dynamic, fluctuating environments. Baas Becking[5] described that at locations, where organic matter is abundant, oxygen might disappear entirely but also could be more than 200% air saturation. Hence, organisms living in tidal sediments are exposed to extreme fluctuations of the oxygen concentration, generated by products of dissimilatory sulfate reduction.[5]

Ecosystems, in which conversions of sulfur compounds are the predominating processes, are referred to as "sulphureta."[4] Marine microbial mats are model, sulfur-dominated ecosystems: high numbers of microorganisms can be found in these laminated ecosystems,[14,19,47,48,51] whereas sulfur bacteria play a predominant role.[14,45,51] Rates of sulfur transformations are high,[12,18,19,33,35,36] and different functional groups of sulfur bacteria are found vertically stratified according to their physiology, albeit in close distance to another.[12,38]

Typically, oxygenic photosynthesis by cyanobacteria accounts for the majority of the primary production close to the sediment surface. Directly underneath this stratum, anoxygenic phototrophic bacteria can be found, whereas chemolithoau-

*Present address: Division of Marine and Atmospheric Chemistry, Rosenstiel School of Marine and Atmospheric Science, Miami, Florida 33149-1098.

totrophic sulfur bacteria (colorless sulfur bacteria) can be found throughout the upper layers of the mat,[14,38,49] depending on their capability to utilize oxygen and/or nitrate as the terminal electron acceptor. Dissimilatory sulfate- and sulfur-reducing bacteria were believed to be restricted to the anoxic part, hence, deeper layers of the sediment.[32] However, recent reports of high numbers of sulfate-reducing bacteria[6,20,51] and the observation of sulfate reduction in the presence of oxygen[11,19,51] contradict this assumption.

Sulfide-producing processes are dominated by dissimilatory sulfate reduction, but various demethylation and demethiolation processes generate small reduced sulfur molecules as well.[27,40] However, these and other mineralization processes, which eventually yield H_2S,[22,25] are of minor quantitative importance. Biotic consumption of reduced (organic and inorganic) sulfur compounds takes place during anoxygenic photosynthesis, in which S compounds serve as electron donor, and during chemolithotrophic sulfur oxidation, in which reduced sulfur compounds are the source of energy and electrons. Also, fermentation (disproportionation) of reduced sulfur compounds by sulfate-reducing organisms can play an important role in certain environments.[20] However, in microbial mats, disproportionation presumably is not as pronounced, but, instead, consumption of reduced sulfur compounds is dominated by sulfide-oxidizing organisms, like anoxygenic phototrophs and colorless sulfur bacteria.[51] Besides biotic consumption of hydrogen sulfide, a number of abiotic reactions, as well as combined biotic/abiotic reactions, yield a variety of reduced sulfur compounds. Chemical parameters, such as the concentration of oxygen, and trace metals (Fe, Cu) play an important role in inorganic sulfur transformations,[10] whereas the presence of organic carbon mediates organic sulfur formation.[2,3,29]

Large diel fluctuations of oxygen and sulfide concentrations, which typically are encountered in microbial mats,[14,43] require adaptation of microorganisms to extreme environmental conditions. Hence, these environments select for metabolically diverse organisms. As a result of diel fluctuations, production and consumption of sulfur compounds within the mat community might be in disbalance during a part of the day, facilitating a flux from the sediment to the atmosphere. Better understanding of sulfur transformations requires a detailed characterization of the ecosystem, and the purpose of this chapter is to describe the major biotic and abiotic components of the microbial mat during the growth season. Special attention will be given to the production and consumption of volatile organic sulfur compounds.

Materials and Methods

Site Description and Sample Collection

The field site, a mature microbial mat, was located in the supralittoral zone on the Frisian Island of Texel (55°08′N 4°50′E). Usually, the mat was flooded twice monthly during spring tide, and, in addition, also during occasional northwesterly storms. *Microcoleus chthonoplastes* was the dominant cyanobacterial species and *Thiocapsa roseopersicina* was the predominant anoxygenic phototroph. In the upper layers of the mat, *Thiobacillus* spp were the principal species of colorless sulfur bacteria observed, whereas, at deeper strata, representatives from the *Thiomicrospira* genus were found as well. Because of the dense cyanobacterial layer, fluctuations of the water content over time were low, and typically ranged from 15 to 35% v/v in the top 10 mm of the mat. The highest organic matter coincides with the chlorophyll *a* (Chl*a*) peak, in the top 5 mm of the sediment, with a value of 0.31 g g^{-1} dry wt. and rapidly decreased to 0.06 with increasing depth.[51]

Sediment samples were taken with stainless-steel cores (inner diameter 2.54 cm), which were stoppered on both ends immediately, and stored on ice. Samples were sliced and processed as described previously.[49,51]

Sulfur Speciation and Analysis

Inorganic sulfur compounds were differentiated according to their individual solubility in various extraction media. In situ measurements with ion-specific needle electrodes in combination with pH microelectrodes revealed the hydrogen sulfide concentration.[14,44] Porewater was used for cyanolytical analysis: thiosulfate ($S_2O_3^{2-}$) was measured at pH 4.8 at room temperature, whereas tetrathionate and pentathionate ($S_4O_6^{2-}$, $S_5O_6^{2-}$, respectively), polysulfides (S_x^{2-}), and trithionate ($S_3O_6^{2-}$) were assayed at pH 8.7, respectively after incubation for 5 min at 0°C, 30 min at 30°C, and 60 min at 90°C.[21,34,44] After extraction of water-soluble sulfur species, acid volatile sulfur compounds (predominantly FeS) were measured with the methylene blue method.[41] Elemental sulfur (S°) was measured spectrophotometrically after extraction in methanol/hexane.[37] After extraction with distilled water, methanol/hexane, and hydrochloric acid treatment, sediment samples were oven-dried and reduced with chromium (Cr^{2+}) to hydrogen sulfide to evaluate the nonacid volatile sulfur compounds pyrite (FeS_2) and S°.[52] The sulfate concentration in the porewater was measured colorimetrically, modified after Bertolacini,[9] or with high-performance liquid chromatography.[17]

Organic sulfur compounds were analyzed in the head space of slurried sediment samples by gas chromatography.[25,26,49,50] The concentrations of dimethyl sulfide (DMS), methanethiol (MSH), and dimethyl disulfide (DMDS) could be evaluated this way. Dimethylsulfoniopropionate (DMSP) was measured as DMS after hydrolysis as described previously.[50]

Sulfur compounds were either measured in two (FeS, S°), three (FeS_2, S°, SO_4^{2-}, DMS, DMSP, MSH, DMDS), or four ($S_2O_3^{2-}$, $S_3O_6^{2-}$, $S_xO_6^{2-}$, S_x^{2-}) replicate sediment samples.

Enumeration of Sulfur Bacteria

Most probable numbers (MPNs) were evaluated after slicing, sonication, and serial dilution of sediment samples, followed by incubation in specific media.[48,51] Viable counts of dissimilatory sulfate-reducing bacteria, colorless sulfur bacteria, and anoxygenic phototrophic bacteria were assessed after 6 weeks of incubation in liquid media, and MPNs calculated after De Man.[13] Routinely, the highest positive dilutions were checked microscopically and inoculated in freshly prepared, specific medium to confirm growth.

Total cell counts were obtained using epifluorescence microscopy.[16] Aliquots were taken from the samples which were used for the serial dilutions of the MPN incubation and were stained with acridine orange as described by Bergström et al.[8]

Results

Depth profiles of organic and inorganic sulfur compounds, distribution of microorganisms, and related physico-chemical parameters were measured at 6–10-week intervals, from August 1988 until August 1990. A similar seasonal pattern was found every year, with small variations due to local climatological differences. The data shown have been collected in 1989 during the early (May), middle (August), and late (November) growth season of the microorganisms in the microbial mat.

Inorganic Sulfur Compounds

Standing concentrations of the sulfur compounds measured displayed variations in time, but, in general, no large changes in depth profiles were observed during the growth season. Profiles of hydrogen sulfide (Figure 1) measured during mid-afternoon under similar conditions (clear sky) showed a significant increase of the maximum concentration from the early growth season (178 μmol L^{-1} at 6 mm depth, measured at 15:00) to the middle of the growth season (524 μmol L^{-1} at 10 mm depth, measured at 16:00), and a decrease toward the end of the year (294 μmol L^{-1} at 9 mm depth, measured at 15:30).

Of the water-soluble sulfur compounds, thiosulfate and trithionate concentrations are expressed as atoms S per milliliter of pore water, whereas values of tetrathionates and pentathionates, and polysulfides in the porewater are given as sulfane-sulfur concentrations [measured as thiocyanate (SCN^-)] because the exact chain length of these sulfur compounds is unknown. All water-soluble sulfur species showed a decrease with increasing depth. Thiosulfate (Figure 2, top, left panel), tetrathionates and pentathionates (Figure 2, top, center panel),

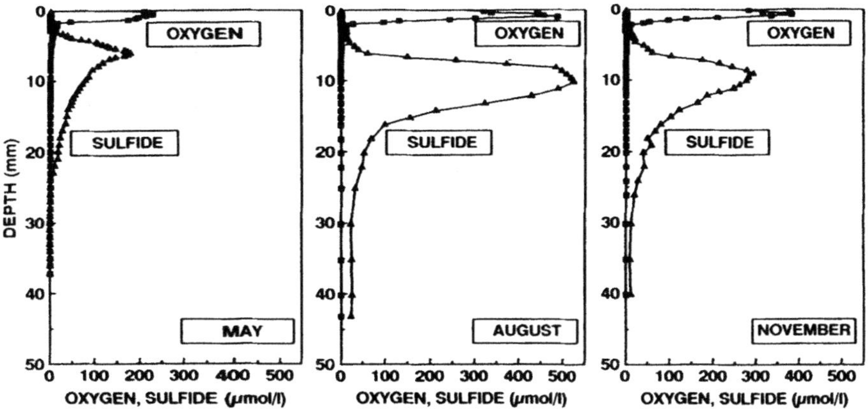

Figure 1. Depth profiles of oxygen and hydrogen sulfide measured with needle electrodes, in a mature microbial mat on the Frisian Island of Texel. Profiles shown were measured during the middle of the afternoon in May (left panel), August (center panel), and November (right panel) 1989.

and trithionate (Figure 2, top, right panel) were present in relatively low concentrations (0–100 nmol S or SCN^{-1} ml^{-1} porewater), as compared to the high concentrations of polysulfides (20–390 μmol ml^{-1}) (Figure 2, middle, center panel). The polysulfide concentration in the top 10 mm of the microbial mat decreased significantly (from 122–192 μmol ml^{-1} to 26–40 μmol ml^{-1}) after the start of the growth season, whereas at deeper strata, there was an increase observed in the polysulfide pool with time. Thiosulfate and tetra/pentathionates, especially in the top 15 mm, also displayed an increase in concentration with time.

The elemental sulfur concentration (1–9 μmol cm^{-3} sediment) displayed a sharp decrease from May to August throughout the top 50 mm of the mat, followed by an increase toward the end of the growth season (Figure 2, middle, left panel). The profile of FeS typically peaked (55–65 μmol cm^{-3} sediment) in the 5–10-mm layer and decreased with depth (Figure 2, bottom, left panel), whereas the pyrite concentration showed an increase with depth (30–40 μmol S cm^{-3} sediment; Figure 2, bottom, right panel). Iron sulfide (FeS) was found at higher concentrations at the end of the growth season, whereas the pyrite (FeS_2) profile remained unchanged.

Sulfate concentration typically ranged from 20 to 30 μmol ml^{-1} porewater and increased slightly with depth (Figure 2, middle, right panel). Because sulfate input depended on tidal flooding, fluctuations in sulfate concentration were expected to vary through the tidal cycle and was influenced by changes in water content of the sediment due to evapotranspiration.[10]

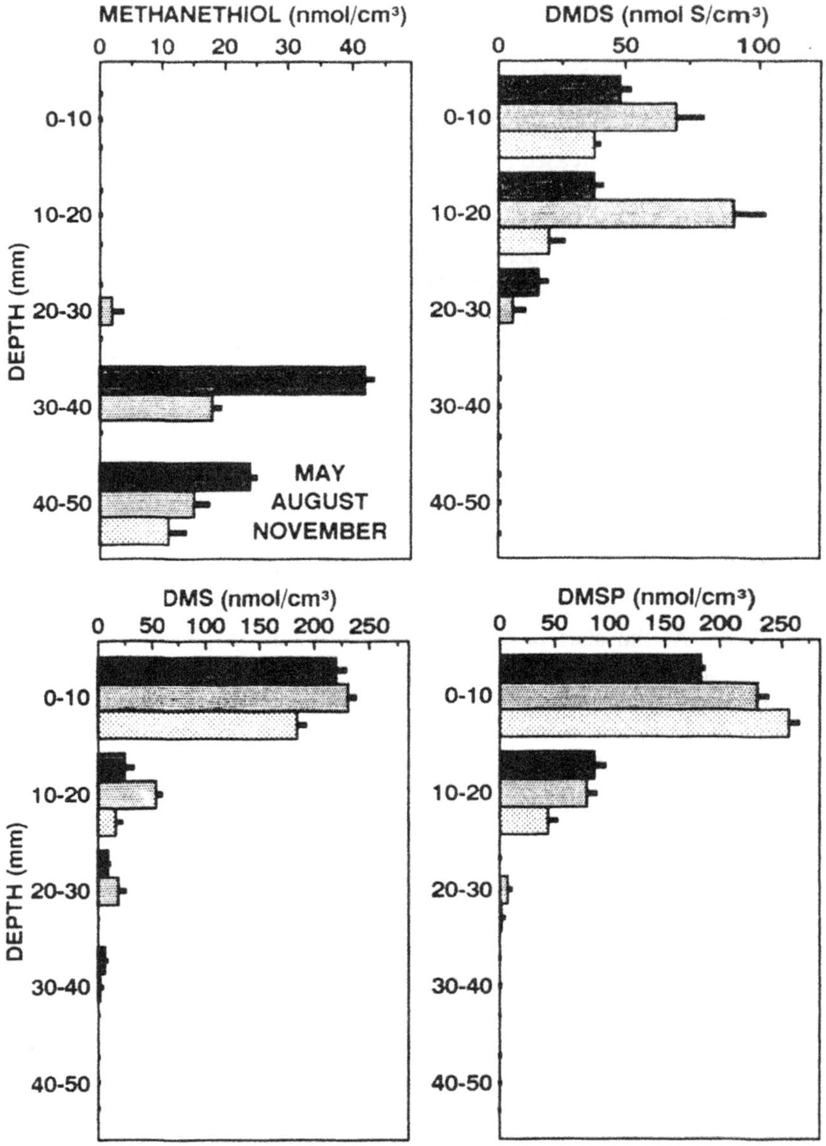

Figure 3. Depth profiles of organic sulfur compounds, measured in the top 50 mm of a mature microbial mat on the Frisian Island of Texel. Each panel displays measurements taken in May (upper bar), August (center bar), and November (lower bar) 1989. Small bars indicate standard deviation.

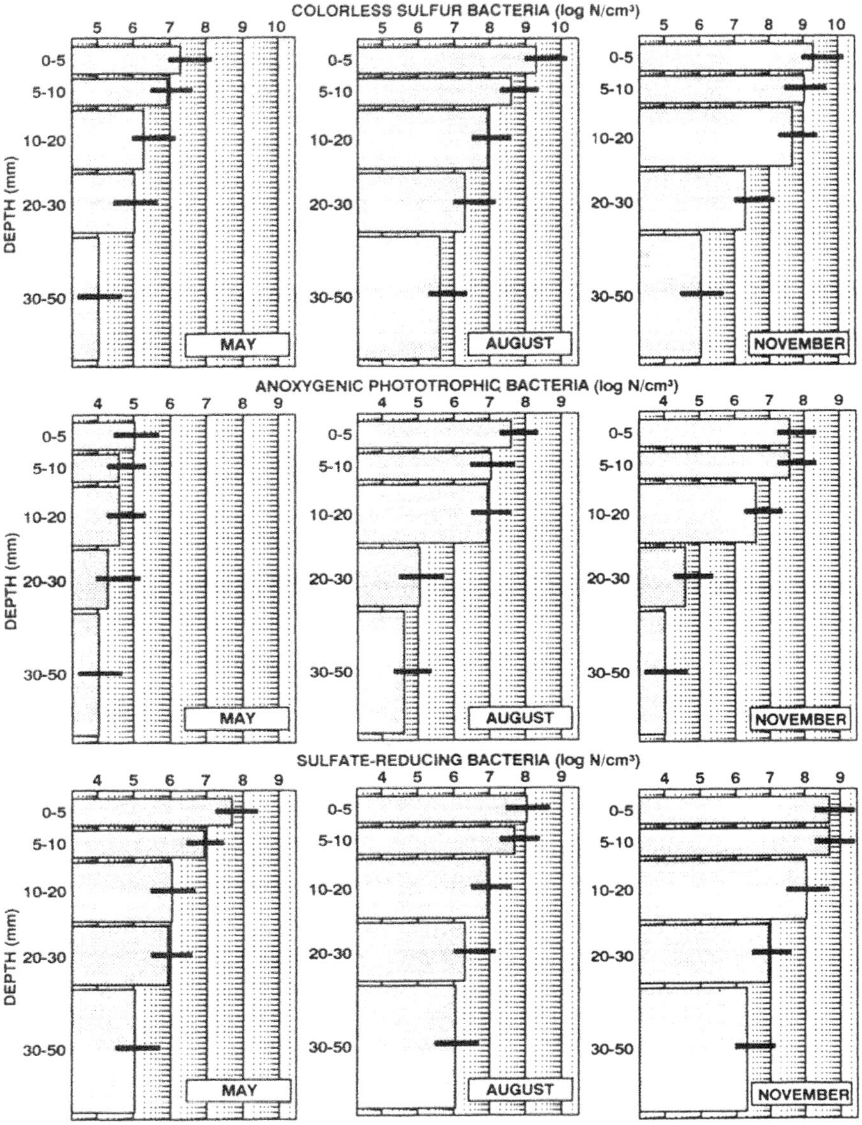

Figure 4. Depth profiles of MPNs in a microbial mat for colorless sulfur bacteria (top panels), anoxygenic phototrophic bacteria (center panels), and dissimilatory sulfate-reducing bacteria (lower panels). Enumerations shown were assessed in May (left panels), August (middle panels), and November (right panels) 1989. Error bars indicate 95% confidence interval.

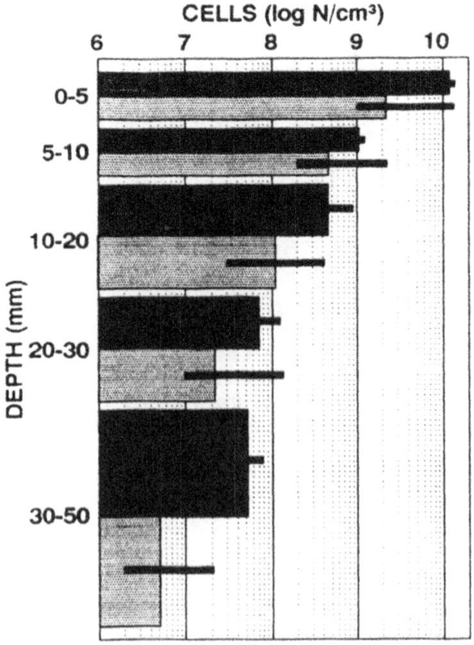

Figure 5. Depth profile of Acridine Orange Direct Counts (upper bars) and summation of the MPN Number for colorless sulfur bacteria, anoxygenic phototrophic bacteria, and sulfate-reducing bacteria (lower bars). Enumerations shown were obtained from the upper 50 mm of a mature microbial mat on the Frisian Island of Texel in August 1989.

be recovered in MPN incubations (Figure 5). This is substantially higher than the 1% recovery typically reported for comparing direct counts to viable counts.[1]

Discussion

The sulfur cycle, as all other elemental cycles, embodies the combined activity of biological, chemical, and physical processes. Measurement of different pool sizes and enumeration of bacteria is insufficient to describe the sulfur cycle, but nevertheless allows a better understanding.

Inorganic Sulfur Compounds

The depth profiles for hydrogen sulfide, thiosulfate, trithionate, tetra/pentathionate, and polysulfide typically peaked close to the sediment surface and decreased with depth, suggesting that production of these sulfur species occurred predominantly in the top layer. Because the highest cell numbers of the sulfur bacteria enumerated were also found close to the sediment surface (0–5 mm), microbial activity seems to play an important role in the establishment of these profiles. The key processes in sulfur speciation, oxygen production (by cyanobacteria), and consumption (by heterotrophs and chemolithotrophs) are largely mediated by microbes, which indirectly determine which abiotic processes can take place. Also, excretion and lysis products from cyanobacteria supply organic substrates[7] for aerobic heterotrophic consumption and anaerobic sulfate respiration.

Polysulfides

The most striking observation in the present study was the high standing concentration of polysulfides present in the microbial mat on Texel. Significant amounts of polysulfides have been reported in salt-marsh sediments[28–30] and in other microbial mats.[2,44] These compounds are known to react rapidly with organic matter,[2,3,29,30] are an intermediate product in pyrite turnover,[15,28] and might play a dominant role in the diagenesis of organic carbon in microbial mats.[2,3] In the laminated sediments of Solar Lake, where iron is limiting, a major part of the sulfide ends up in the polysulfide pool.[3,18]

The observed concentrations in the present study (<390 μmol SCN^{-1} ml^{-1}) are expressed as the amount of sulfane-sulfur present, and information on chain length is difficult to obtain. Therefore, it cannot be excluded that the values shown in Figure 2 (center, middle panel) express long-chain polysulfides: Assuming a chain length of 8 (S_x^{2-}, $x = 8$), the maximum observed concentration of 390 μmol ml^{-1} (20–25-mm-depth interval, August) would express a polysulfide concentration of 55.7 μmol ml^{-1}. Therefore, the polysulfide concentration lies between 56 and 390 mM. Long-chain polysulfides were also suggested to exist in pure cultures of the anoxygenic phototroph *Chromatium vinosum*.[39] Also, given the high organic content of the sediment (<0.31 g g^{-1} dry wt.[51]) a significant fraction of the polysulfide pool could be present as organic polysulfides. It can be expected that these organic polysulfides are included in the cyanolytical assay, but to what extent requires further study.

Hydrogen Sulfide

The hydrogen sulfide profiles, measured under comparable conditions, showed an increasing maximum value from May to August. This observation suggested a change of sulfate-reducing activity, induced by a combination of temperature, availability of organic carbon (produced by the cyanobacteria), and, consequently, growth of the sulfate reducers. Assuming that biotic and abiotic sulfide consumption were also increasing during the course of the growth season, the increase in sulfate reduction is even more pronounced than shown by the sulfide profiles (Figure 1). The number of colorless sulfur bacteria and anoxygenic phototrophs increases during the course of the growth season; hence, biotic consumption is be expected to be elevated. Similarly, the pool of intermediate sulfide oxidation products will change. In addition to elemental sulfur, polythionates are typically produced by colorless sulfur bacteria in marine environments.[42] Abiotic reaction between hydrogen sulfide and elemental sulfur yields polysulfides, as can be observed in pure cultures of colorless sulfur bacteria (P.T. Visscher and W. Stolte, unpublished results) and green sulfur bacteria.[46] Finally, chemical reactions between iron and sulfide or polysulfide generate FeS and FeS_2, which, again, can be mediated by microbial activity.[29]

Organic Sulfur Compounds

In addition to abiotic formation during reactions between polysulfides and organic carbon molecules, organic sulfur compounds are a result of degradation of amino acids and the sulfur-containing osmolyte DMSP. The similarity of the DMS and DMSP profiles (Figure 3) suggest DMS production from DMSP. The supplementary profiles of MSH and DMDS imply that DMDS, which is present at shallower depth, could arise from MSH oxidation (Figure 3). The dominant oxygenic phototroph, *Microcoleus chthonoplastes,* contained DMSP, which was released during an osmotic down-shock.[50] The relationship between Chl*a* and DMS in the mat is shown in Figure 6. As opposed to pelagic systems, where DMS concentrations are typically 2–10 times lower than the DMSP concentration,[23] the DMS and DMSP concentrations were very similar in the microbial mat. The relatively high concentrations of DMS found in the present study are not due to sampling artifacts.[48] In the microbial mat slurries, DMS could arise from MSH methylation,[26] or included DMS which is strongly absorbed to sediment particles.[23] If terminal sulfur atoms of polysulfides, which were present in high concentrations, in part are methylated, this could also represent a source of DMS in the microbial mat.

Potential Production of DMS

Sources and sinks of DMS in the microbial mat have been well-established,[48–50] which allows calculation of production (Table 1) and consumption (Table 2) of this volatile sulfur compound. The specific DMSP content of *Microcoleus chthonoplastes* was 37.1 μmol g^{-1} protein,[50] and the ratio between Chl*a* and protein in axenic cultures was reported 30 $\pm$ 5 μg mg^{-1}. The Chl*a* content in the upper 10 mm of the mat peaked at 620 μg cm^{-3} sediment, from which a protein concentration of 20.7 mg cm^{-3} sediment can be calculated. By

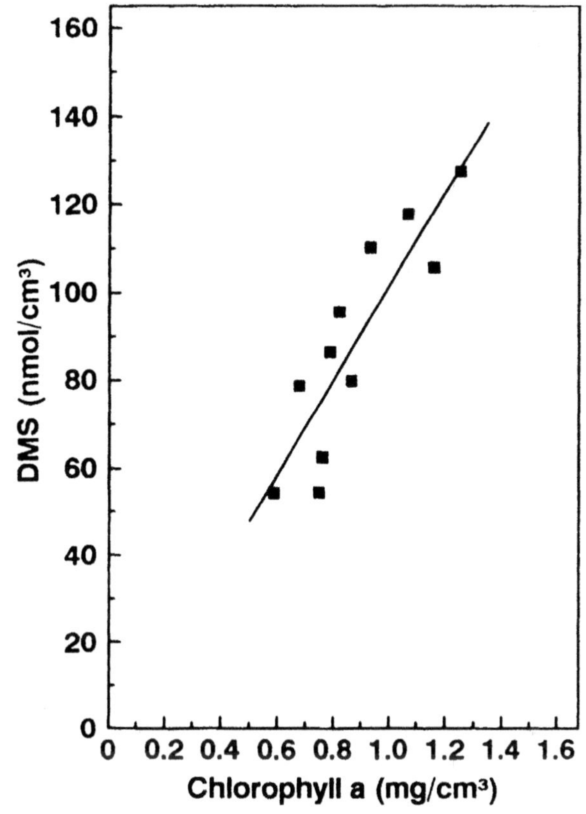

Figure 6. Relationship between the chlorophyll *a* content and the concentration of DMS in the top 10 mm of a microbial mat. Data were collected in 1989 and 1990.

extrapolation, a potential DMSP concentration of 768 nmol cm^{-3} sediment is calculated. This value agrees reasonably well with field observations of 230 nmol DMSP cm^{-3} sediment given the uncertainties.

When an axenic culture of *M. chthonoplastes,* grown at a salinity of 33 ppt, was transferred to a medium of 8 ppt, a maximum DMSP release of 12 μmol g^{-1} protein h^{-1} was observed.[50] Assuming the above-calculated protein concentration of 20.7 mg cm^{-3} sediment, a maximum release of 248 nmol DMSP cm^{-3} h^{-1} can be calculated. If this DMSP is rapidly hydrolyzed to DMS, the flux of this volatile sulfur compound to the atmosphere is determined by production and biological consumption. Preliminary studies revealed an extremely low flux of DMS from the sediment to the atmosphere. However, further mea-

Table 1. Potential production of DMSP in a *Microcoleus chthonoplastes*-dominated microbial mat. For explanation and calculations, see text (data taken from Ref. 50).

Specific DMSP content (μmol g^{-1} protein)	37.1
Specific DMSP release (μmol g^{-1} protein h^{-1})	12.0
Biomass in the mat (mg protein cm^{-3})	20.7
Potential DMSP release (μmol cm^{-3} h^{-1})	0.25

Table 2. Potential consumption of DMS by dominant mat microbes. For explanation and calculations, see text (data taken from Refs. 14 and 48–50).

	Thiocapsa	*Thiobacillus*	Homogenated Sediment	Homogenated Sediment
Metabolism	Phototrophic	Chemotrophic	—	—
Conditions	(An)oxic	Oxic	Oxic	Anoxic
μ_{max}[a] (h^{-1})	0.07	0.10	—	—
Specific DMS consumption rate (μmol mg^{-1} protein h^{-1})	55	4.5	—	—
Biomass in the mat (mg protein cm^{-3})	14.6	5	—	—
Potential DMS utilization (μmol cm^{-3} h^{-1})	803	22.65	0.24	0.08

[a]μ_{max} = maximum specific growth rate determined in chemostat cultures.

surements under better-controlled conditions are required to confirm these observations. Therefore, it seems that microbial consumption limits the flux out of the microbial mat.

Potential Consumption of DMS

There are several microorganisms present in the microbial mat capable of DMS(P) consumption, among which are colorless sulfur bacteria,[48] anoxygenic phototrophs,[49] sulfate reducers,[25,50] and methanogens.[24,31]

When DMSP was added to slurries of the upper 20 mm of the microbial mat, of which the cyanobacterial layer had been removed, rapid consumption was observed.[50] The consumption rate under anoxic conditions was 21.3

nmol ml^{-1} h^{-1}, and under oxic conditions 61.2 nmol $ml^{-1}h^{-1}$. Inhibitor studies showed that neither methanogens nor sulfate reducers stimulated DMSP consumption significantly, although methane was observed as an end product.

During organic sulfur measurements in the present study, methane was always detected (data not shown), which could imply that at least a part of the DMS is metabolized by methanogens, assuming DMS(P) stimulated CH_4 production.[24,25] Also, under anoxic conditions, pure cultures *T. roseopersicina* utilize DMS as an electron donor for photosynthesis. Simultaneous oxidation of DMS and hydrogen sulfide was found in batch cultures of this purple sulfur bacterium. The maximum rate of DMS oxidation for this organism, calculated from Visscher and Van Gemerden,[49] is approximately 55 μmol mg^{-1} protein h^{-1}. A bacteriochlorophyll *a* (BChl*a*) content of 440 μg cm^{-3} sediment has been reported for the upper 10 mm of the microbial mat,[47] whereas in pure cultures, a maximum BChl*a* : protein ratio of 30 μg mg^{-1} was found, resulting by extrapolation in a protein concentration of 14.6 mg cm^{-3} sediment in the *Thiocapsa* layer of the microbial mat. This population of anoxygenic phototrophs has the potential to oxidize 803 μmol DMS h^{-1}, which is 3000 times the maximum release by *Microcoleus*. Similarly, the population of colorless sulfur bacteria contributes to DMS consumption under oxic conditions. From data obtained in chemostat experiments,[48] a specific DMS consumption rate of 4.53 μmol mg^{-1} h^{-1} can be calculated. Protein estimates from pure cultures can be used in conversion of the MPN of colorless sulfur bacteria (2.0×10^9 cells cm^{-3}) to protein in the microbial mat (5 mg cm^{-3}), resulting in a maximum potential DMS consumption of 22.65 μmol cm^{-3} h^{-1}, which is 100 times the production calculated for *Microcoleus*.

Conclusions

From these calculations, it can be concluded that microbial biomass in the top 10 mm of the mat is capable of consuming all of the DMS produced, and that a flux to the atmosphere can only be expected during suboptimal conditions for microbial metabolism. Such conditions can be expected to occur in the mat during a variety of diel and seasonal fluctuations, e.g., in temperature, salinity, pH.

High microbial cell numbers and dynamics, as well as relatively high concentrations of sulfur compounds, appear to be in balance in the microbial mat. However, the extreme environmental conditions, which are encountered regularly, could disturb the balance and result in a flux of volatile sulfur compounds to the atmosphere. Further studies, focusing on production and consumption of these compounds under various environmental conditions, are necessary for a better understanding of exchange of sulfur gases from the microbial mat sediment to the atmosphere.

Acknowledgments

This study was subsidized by a grant from the Netherlands Organization for Scientific Research (NWO). The hospitality of Jan Vosjan and Eijsbrand Pauptit at the Netherlands Institute of Sea Research during field measurements, and of Thomas Cappenberg and Cees Hordijk at the Limnological Institute during HPLC analysis is gratefully acknowledged. Frank van den Ende and Peter Quist provided technical assistance during field work. The constructive comments made by two anonymous reviewers contributed to the manuscript in its present form.

References

1. Atlas, R.M., and R. Bartha. 1987. Microbial Ecology. The Benjamin/Cummings Publ. Co. Menlo Park, CA, pp. 203–213.
2. Aizenshtat, Z., Y. Cohen, and H. Nielsen. 1983. The geochemical sulphur enrichment of recent organic matter by polysulfides in the Solar-Lake. In M. Bjoroy, P. Albrecht, C. Cornford, K. de Groot, K. Eglington, E. Galimov, D. Leythaeuser, R. Pelet, J. Rullkotter, and G. Speers (eds.), Advances in Organic Geochemistry, J. Wiley and Sons, Chichester, pp. 279–288.
3. Aizenshtat, Z., G. Lipiner, and Y. Cohen. 1984. Biogeochemistry of carbon and sulfur cycle in microbial mats of Solar Lake (Sinai). In Y. Cohen, R.W. Castenholz, and H.O. Halvorson (eds.), Microbial Mats: Stromatolites, Alan R. Liss, Inc., New York, pp. 281–312.
4. Baas Becking, L.G.M. 1925. Studies on the sulphur bacteria. Ann. Bot. **39**:613–615.
5. Baas Becking, L.G.M. 1934. Geobiologie, W.P. van Stockum & Zoon N.V., Den Haag.
6. Bak, F., and N. Pfennig, 1991. Sulfate-reducing bacteria in littoral sediment of Lake Constance. FEMS Microbiol. Ecol. **85**:43–52.
7. Bateson, M.M., and D.M. Ward. 1988. Photoexcretion and fate of glycolate in a hot spring cyanobacterial mat. Appl. Environ. Microbiol. **54**:1738–1743.
8. Bergström, I., A. Heinaanen, and K. Salonen. 1986. Comparison of acridine orange, acriflavine, and bisbenzimide stains for enumeration of bacteria. Appl. Environ. Microbiol. **51**:664–667.
9. Bertolacini, R.J., and J.E. Barney II, 1957. Colorimetric determination of sulfate with barium chloranilate. Anal. Chem. **29**:281–283.
10. Boulegue, J., C.J. Lord III, and T.M. Church. 1982. Sulfur speciation and associated trace metals (Fe, Cu) in the pore waters of Great Marsh, Delaware. Geochim. Cosmochim. Acta **46**:453–464.
11. Canfield, D.E., and D.J. Des Marais. 1991. Aerobic sulfate reduction in microbial mats. Science **251**:1471–1473.

12. Cohen, Y., 1984. Oxygenic photosynthesis, anoxygenic photosynthesis, and sulfate reduction in cyanobacterial mats. In M.J. Klug and C.A. Reddy (ed.), Current Perspectives in Microbial Ecology, American Society for Microbiology, Washington DC, pp. 435–441.

13. De Man, J. 1975. The probability of most probable numbers. Eur. J. Appl. Microbiol. **1**:67–78.

14. De Wit, R., H.M. Jonkers, F.P. van den Ende, and H. van Gemerden. 1989. *In sitù* fluctuations of oxygen and sulphide in marine microbial sediment ecosystems. Neth. J. Sea Res. **23**:271–281.

15. Giblin, A.E. 1988. Pyrite formation in marshes during early diagenesis. Geomicrobiol. J. **6**:77–97.

16. Hobbie, J.E., R.J. Daley, and S. Jasper. 1977. Use of Nuclepore filters for counting bacteria by fluorescence microscopy. Appl. Environ. Microbiol. **33**:1225–1228.

17. Hordijk, K.A., C.P.M.M. Hagenaars, and T.E. Cappenberg. 1985. Kinetic studies of bacterial sulfate reduction in fresh water sediments by high-pressure liquid chromatography and microdistillation. Appl. Environ. Microbiol. **49**:434–440.

18. Jørgensen, B.B., and Y. Cohen. 1977. Solar Lake (Sinai). The sulfur cycle of benthic cyanobacterial mats. Limnol. Oceanogr. **22**:657–666.

19. Jørgensen, B.B., N.P. Revsbech, T.H. Blackburn, and Y. Cohen. 1979. Diurnal cycle of oxygen and sulfide microgradients and microbial photosynthesis in a cyanobacterial mat. Appl. Environ. Microbiol. **38**:46–58.

20. Jørgensen, B.B., and F. Bak. 1991. Pathways and microbiology of thiosulfate transformations and sulfate reduction in a marine sediment (Kattegat, Denmark). Appl. Environ. Microbiol. **57**:847–856.

21. Kelly, D.P., L.A. Chambers, and P.A. Trudinger. 1969. Cyanolysis and spectrometric estimation of trithionate in a mixture with thiosulfate and tetrathionate. Anal. Chem. **41**:898–901.

22. Kiene, R.P. 1988. Dimethyl sulfide metabolism in salt marsh sediments. FEMS Microbiol. Ecol. **53**:71–78.

23. Kiene, R.P. 1991. Microbial cycling of organosulfur gases in marine and freshwater environments. In D. Adams, P. Crill, and S. Seitzinger (eds.), Cycling of Reduced Gases in the Hydrosphere, E. Schweitzerbartchen Verlagsbuchhandlungen, Stuttgart.

24. Kiene, R.P., R.S. Oremland, A. Catena, L.G. Miller, and D.G. Capone. 1986. Metabolism of reduced methylated sulfur compounds in anaerobic sediments and by a pure culture of a methanogen. Appl. Environ. Microbiol. **52**:1037–1045.

25. Kiene, R.P., and P.T. Visscher. 1987. Production and fate of methylated sulfur compounds from methionine and dimethylsulfoniopropionate in anoxic salt marsh sediments. Appl. Environ. Microbiol. **53**:2426–2434.

26. Kiene, R.P., and D.G. Capone. 1988. Microbial transformations of methylated sulfur compounds in anoxic salt marsh sediments. Microbial Ecol. **15**:275–291.

27. Kiene, R.P., and B.F. Taylor. 1988. Demethylation of dimethylsulfoniopropionate

and production of thiols in anoxic marine sediments. Appl. Environ. Microbiol. **54**:2208–2212.

28. Lord, III, C.J., and T.M. Church. 1983. The geochemistry of salt marshes: sedimentary ion diffusion, sulfate reduction, and pyritization. Geochim. Cosmochim. Acta **47**:1381–1391.

29. Luther, III, G.W., T.M. Church, J.R. Scudlark, and M. Cosman. 1986. Inorganic and organic sulfur cycling in salt marsh pore waters. Science **232**:746–749.

30. Luther, III, G.W., and T.M. Church. 1988. Seasonal cycling of sulfur and iron in porewaters of a Delaware salt marsh. Marine Chem. **23**:295–309.

31. Oremland, R.S., R.P. Kiene, I. Mathrani, M.J. Whiticar, and D.R. Boone. 1989. Description of an estuarine methylotrophic methanogen which grows on dimethyl sulfide. Appl. Environ. Microbiol. **55**:994–1002.

32. Postgate, J.R. 1984. The sulphate reducing bacteria. Cambridge University Press, Cambridge.

33. Revsbech, N.P., B.B. Jørgensen, T.H. Blackburn, and Y. Cohen. 1983. Microelectrode studies of the photosynthesis and O_2, H_2S, and pH profiles in a microbial mat. Limnol. Oceanogr. **28**:1062–1074.

34. Schedel, M., 1978. PhD thesis, University of Bonn, Germany.

35. Skyring, G.W., 1984. Sulfate reduction in marine sediments associated with cyanobacterial mats in Australia. In Y. Cohen, R.W. Castenholz, and H.O. Halvorson (eds.), Microbial Mats: Stromatolites, Alan R. Liss, Inc., New York, pp. 265–275.

36. Skyring, G.W., R.M. Lynch, and G.D. Smith. 1989. Quantitative relationships between carbon, hydrogen, and sulfur metabolism in cyanobacterial mats. In Y. Cohen and E. Rosenberg (eds.), Microbial Mats, American Society for Microbiology, Washington DC, pp 170–179.

37. Stal, L.J., H. van Gemerden, and W.E. Krumbein. 1984. The simultaneous assay of chlorophyll and bacteriochlorophyll in natural microbial communities. J. Microbiol. Meth. **2**:295–306.

38. Stal, L.J., H. van Gemerden, and W.E. Krumbein. 1985. Structure and development of a benthic marine microbial mat. FEMS Microbiol. Ecol. **31**:111–1125.

39. Steudel, R., G. Holdt, P.T. Visscher, and H. van Gemerden. 1990. Search for polythionates in cultures of *Chromatium vinosum* after sulfide incubation. Arch. Microbiol. **153**:453–437.

40. Taylor, B.F., and R.P. Kiene. 1989. Microbial metabolism of dimethyl sulfide. In E.S. Saltzman and W.J. Cooper (eds.), Biogenic Sulfur in the Environment. American Chemical Society, Washington DC, pp. 202–221.

41. Trüper, H.G., and H.G. Schlegel. 1964. Sulphur metabolism in Thiorhodaceae. I. Quantitative measurements on growing cells of *Chromatium okenii*. Antonie van Leeuwenhoek J. Microbiol. Ser. **30**:225–238.

42. Tuttle, J.H., and H.W. Jannasch. 1977. Thiosulfate stimulation of microbial dark assimilation of carbon dioxide in shallow marine waters. Microb. Ecol. **4**:9–25.

43. Van Gemerden, H. (1990). Immobile anoxygenic phototrophic bacteria in tidal

areas. In J.A.M. de Bont, J. Visser, B. Mattiasson, and J. Tramper (eds.), Physiology of Immobilized Cells, Elsevier Science Publishers, Amsterdam, pp. 37–48.

44. Van Gemerden, H., C.S. Tughan, R. de Wit, and R.A. Herbert. 1989. Laminated microbial ecosystems on sheltered beaches in Scapa Flow, Orkney Islands. FEMS Microbiol. Ecol. **62**:87–102.

45. Van Gemerden, H., R. de Wit, C.S. Tughan, and R.A. Herbert. 1989. Development of mass blooms of *Thiocapsa roseopersicina* on the sheltered beaches on the Orkney Islands. FEMS Microbiol. Ecol. **62**:111–118.

46. Visscher, P.T., and H. van Gemerden. 1988. Growth of *Chlorobium limicola* f. thiosulfatophilum on polysulfides. In J.M. Olsen, J.G. Ormerod, J. Amesz, E. Stackebrandt, and H.G. Trüper (eds.), Green Sulfur Bacteria, Plenum Press, New York, pp. 287–294.

47. Visscher, P.T., J.W. Nijburg, and H. van Gemerden. 1990. Polysulfide utilization by *Thiocapsa roseopersicina.* Arch Microbiol. **155**:75–81.

48. Visscher, P.T., P. Quist, and H. van Gemerden. 1991. Methylated sulfur compounds in microbial mats: in situ concentrations and metabolism by a colorless sulfur bacterium. Appl. Environ. Microbiol. **57**:1758–1763.

49. Visscher, P.T., and H. van Gemerden. 1991. Photoautotrophic growth of Thiocapsa roseopersicina on dimethyl sulfide. FEMS Microbiol. Lett. **81**:247–250.

50. Visscher, P.T., and H. van Gemerden. 1991. Production and consumption of dimethylsulfoniopropionate in marine microbial mats. Appl. Environ. Microbiol. **57**: 3237–3242.

51. Visscher, P.T., R.A. Prins, and H. van Gemerden. 1992. Rates of sulfate reduction and thiosulfate consumption in a marine microbial mat. FEMS Microbiol. Ecol. **86**:283–294.

52. Zhabinda, N.N., and I.I. Volkov. 1978. A method of determination of various sulfur compounds in sea sediments and rocks. In W.E. Krumbein (ed.), Environmental Biogeochemistry and Geomicrobiology: Methods, Metals and Assessment. Ann Arbor Science Publishers, Ann Arbor, MI, pp. 735–745.

Biological Consumption of Dimethyl Sulfide in the Marine Euphotic Zone: Results of Radioisotope Experiments

G.V. Wolfe and T.S. Bates*

University of Washington, Seattle, Washington 98195 and NOAA/PMEL, Seattle, Washington 98115

ABSTRACT

The flux of dimethyl sulfide (DMS) to the marine atmosphere, hypothesized to have an important climatic effect via modulation of cloud albedo, depends critically on the rapidity of biological cycling of DMS in the water column, which may be much more rapid than its turnover by air–sea exchange. We developed and tested a radioisotope method to measure microbial DMS consumption using ^{14}C-labeled DMS. Results from whole-water incubations in the NE Pacific and Puget Sound, Washington show that CO_2 and particulates (cell material) are the primary products and are typically produced in ratios of about 2 : 1, suggesting methylotrophic use of DMS as a carbon source, although specific DMS consumers have not yet been isolated from ocean surface waters. Because of low specific activity and low in situ [DMS], a quasi-kinetic approach was required, and estimates of in situ DMS consumption ranged from <0.1 nM d^{-1} to 0.5 nM d^{-1}, yielding biological turnover times of 3–70 d. In the NE Pacific, rates were consistently lower than those calculated from a chloroform-inhibition method, and the radioisotope method most likely provides a lower bound on the true consumption rates. However, even the lower estimates of biological DMS consumption rates are faster than typical loss to outgassing, indicating that most DMS produced in the upper water column is consumed within the water column.

Introduction

The hypothesis that biologically produced dimethyl sulfide (DMS) may play an important climatic role via cloud modulation[1] prompted the question of what limits the flux of DMS to the atmosphere. Because this flux is highly dependent on the concentration of DMS in the upper water column, factors controlling seawater DMS concentrations may have a potential climatic impact.

The seawater concentration of DMS has been been the subject of many field investigations (see Ref. 2 for a review). It is the result of the balance between at

*Present address: Department of Microbiology, University of Washington, Seattle, WA 98195.

least two production pathways, direct and indirect enzymatic conversion of the algal metabolite dimethylsulfoniopropionate (DMSP), and several competing sinks, of which air–sea exchange is only one (Figure 1). Other potentially major sinks include photo-oxidation to dimethyl sulfoxide (DMSO) and biological consumption to cell material, CO_2, and other products.

Evidence suggests that DMSP is usually much more abundant than DMS itself. For example, in the NE Pacific Ocean off the Washington coast, data from three consecutive spring/summer cruises (Wolfe and Bates, manuscript in preparation) have shown that in the upper 20 m, [DMSP] ranged from 10 to 150 nM (particulate fraction) and from 5 to 30 nM (dissolved fraction), whereas [DMS] was typically in the range 0.5–5 nM. These values are similar to those in the North Sea and North Atlantic.[3,4] Thus, a large precursor pool exists relative to the DMS pool. However, our datasets and many others suggest that DMS surface seawater concentrations above even 10 nM are exceedingly rare, and the global average is only about 3 nM.[2]

One likely limit on the concentration of DMS is biological consumption by marine bacteria. Bacteria have been isolated from other environments (soils, sediments, and sewage wastes) which consume DMS.[5–11] DMS is one in a family of C_1 compounds present in the ocean, including methane and methanol, formaldehyde and formate, the methylamines, methyl mercaptan and dimethyl disulfide, and methylated halides and metals. Many of these compounds are known to be microbially degraded in the marine environment by a wide variety of methylotrophs and still-unidentified bacteria.

How important is the biological sink for DMS? Previous measurements of DMS consumption have been made[12,13] using chloroform to selectively inhibit C_1 catabolism. These studies indicated that DMS consumption by bacteria may, in some cases, be far faster than air–sea ventilation, making the biological sink a crucial limit on the flux of DMS to the atmosphere. The chloroform inhibition technique relies on the high sensitivity of the flame-photometric detector (FPD), allowing routine GC analysis at the picomolar to nanomolar level. However, it measures only disappearance of DMS and gives no knowledge of the biochemical fate(s), therefore yielding limited information of the mechanisms of consumption. It also requires the addition of a "specific inhibitor," a technique which invokes some risky assumptions.[14]

An obvious alternative is to use radioisotopically labeled DMS to measure DMS consumption. The isotope method has several potential advantages: It can trace the compound through metabolic products and no inhibitors are needed. This method has been used to study consumption of DMS in anaerobic sediments[15,16] but not in the marine euphotic zone. The main objectives of this study was the development of a ^{14}C–DMS method to quantify the in situ biological consumption of DMS and to identify and quantify the biochemical products.

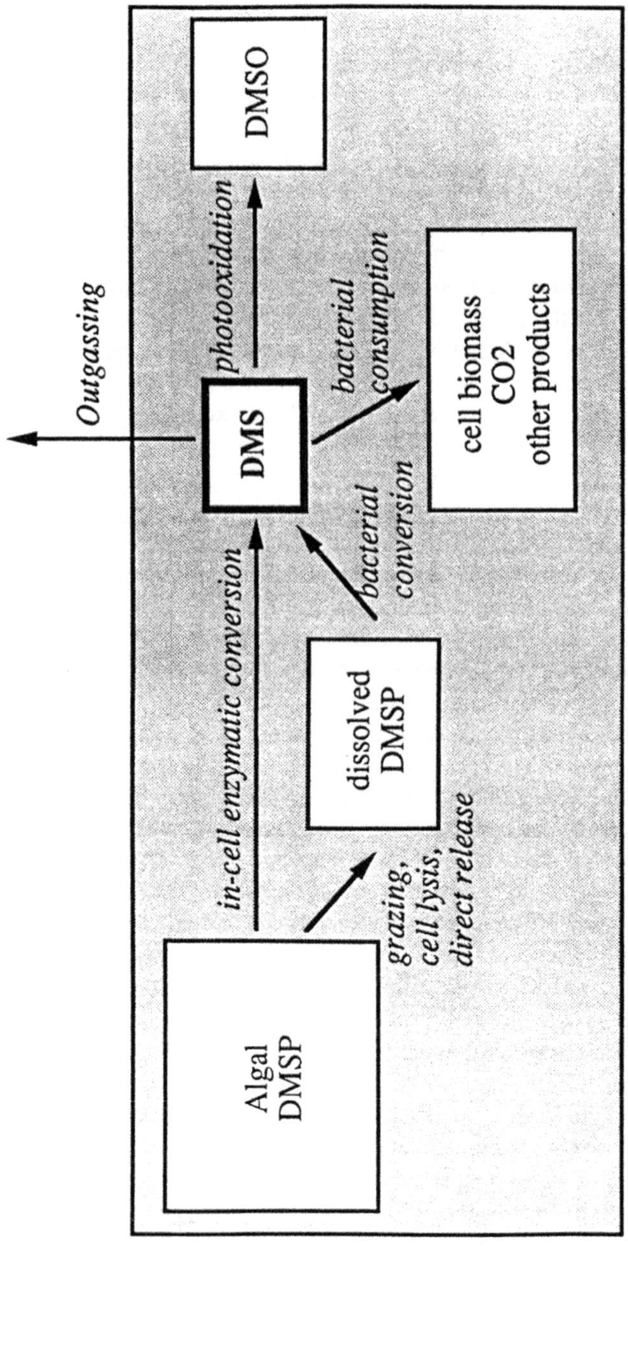

Figure 1. Box schematic of the production and destruction of DMS in the marine photic zone. Boxes denote observables and arrows denote processes.

Materials and Methods

Uniformly labeled ^{14}C–DMS was obtained from Amersham Co. (Arlington Heights, IL). The stock was quoted at 96% purity in 1985 (impurities not specified), with specific activity 22.3 mCi mmol^{-1}. This stock was dissolved into 100 ml of sterile deionized water. Subsequent dilutions were made for working stock solutions. Stock solutions were kept cold in the dark and taken out only during additions. Typical additions for experiments were 20–100 μl to 100 ml seawater, for final concentrations of 1–20 nM.

Because of the low specific activity and low concentrations required, true tracer experiments were not possible. Therefore, a quasi-kinetic approach was employed, using two to five concentration additions. Typical experiments ran for 24 h with samples taken every 8 or 12 h. Incubations were conducted in 100-ml serum bottles, capped with Teflon-lined septa. Head space was typically $< 10\%$ of the volume, so most of the DMS remained in solution. Experiments were usually conducted in duplicate with a control bottle which was fixed at time zero (t_0) but incubated for the same time period as the others. Numbers reported are the means of duplicates. Runs were also made with 0.2 μm filtered seawater.

Figure 2 outlines the incubation method. To terminate incubations, samples were killed by injection of either NaOH (final pH of 10–11) or formalin (0.5–2% final concentration). The base was used to ensure that any CO_2 produced would stay in solution during the sparging of remaining DMS and because formalin interferes with gas chromatographic analysis for DMS. However, we were concerned that the base might lyse cells and result in loss of cell retention on filters. Therefore, we also conducted experiments using formalin.

Following this, DMS and other volatiles were usually sparged from solution to avoid their contaminating the CO_2 collection. The use of phenethylamine, a standard trap for labeled CO_2, was found to also trap DMS, swamping any CO_2 signal. Increasing additions of labeled DMS into sterile artificial seawater media resulted in approximately 20% recoveries in the CO_2 trap after 16 h static degassing or 2.5 h degassing on a rotary shaker. We found that 0.2 ml of 1 N NaOH was far more selective, trapping CO_2 ($> 90\%$ recovery of $H^{14}CO_3^-$ spiked controls; results normalized to CO_2 recovery) but less than 1% of the DMS. (This amount of base was chosen as a minimum to adequately absorb all the CO_2 released from 100-ml seawater samples, but without disruption of the scintillation cocktail which occurred at higher NaOH concentrations). Therefore, sparging the remaining DMS from solution was not strictly necessary but was sometimes performed as a control. In some instances we counted the degassed volatile activity by trapping directly in Omnifluor® on dry ice ($T < -60°C$), which retained about 98% of DMS at flow rates below approximately 100 ml min^{-1}. This allowed recovery of unconverted labeled DMS, which was used as a check on the total recovery.

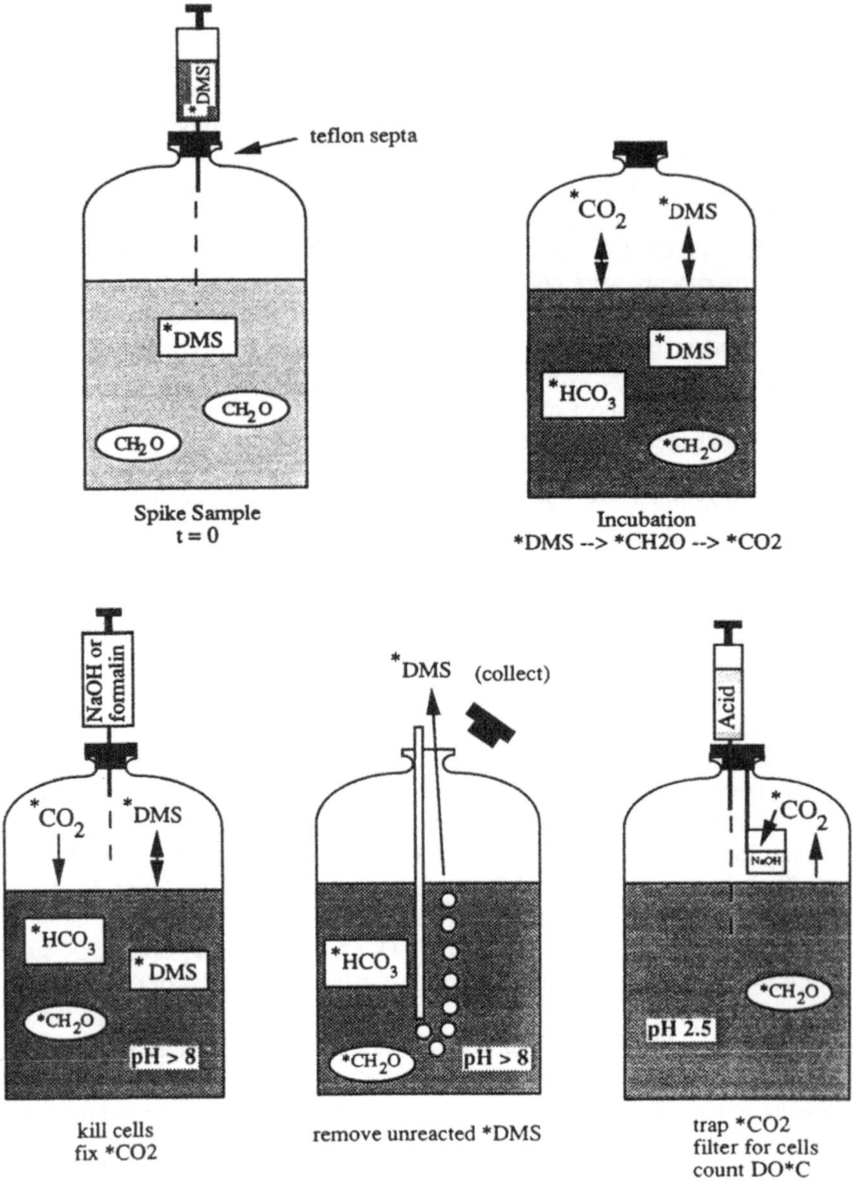

Figure 2. Schematic of ^{14}C–DMS incubation method. "CH_2O" indicates biomass.

Table 1. Fractionations performed for complete recovery of activity.

No.	Fraction	Chemical/Physical Meaning	Use
1	Solution following addition	Total activity—initial	Check activity added
2	Solution at end of incubation	Total activity—final	Check activity left in solution
3	Volatile	Leftover substrate + other volatiles	Fraction unreacted
			Other S gases (with GC)
4	Nonvolatile	(HCO_3^- + cells + DOC)	Check (3 + 4) = 2
5	Acid-volatile	CO_2	Activity respired
6	Acid-nonvolatile	(cells + DOC)	Check (5 + 6) = 4
7	Filter	Cells	Activity incorporated
8	Filtrate	DOC	Check (7 + 8) = 6

Next the serum bottle was plugged with a cap containing a cup filled with 0.2 ml of 1 N NaOH, and the solution was acidified with 5 N H_2SO_4 to about pH 2.5 to remove CO_2. CO_2 was evolved over 4–24 h on a rotary shaker, and the cup and base were counted in 5 ml of Ecolume® scintillation cocktail (ICN Biomedicals, Costa Mesa, CA).

Following CO_2 degassing, the solution was filtered for cell activity and a subsample of solution counted for nonvolatile activity. Samples were counted on a Packard model 4430 liquid scintillation counter. Counts were converted to DPM using a correction scheme based on an energy spectrum analysis applied to a calibration curve.

Laboratory experiments were also conducted to test the complete recovery of all fractions. The procedure for these was to follow activity per milliliter immediately following the addition of substrate and following termination of incubation, and in all fractions: volatile, nonvolatile, acid-volatile (CO_2), acid-nonvolatile (cells + DOM), filter material (cells), and filtrate (DOC) (Table 1). Gas chromatography was also used in a few cases to look for other volatile sulfur compounds present after incubation and to check the concentration of DMS left in solution after incubations, as well as the initial ambient DMS concentration.

Results

The main field study occurred during the April 1991 PSI-3 (Pacific Sulfur/Stratus Investigation) cruise, in the NE Pacific off the coast of Washington State. Parallel incubations were performed on the same water for both isotope and chloroform-inhibition methods (done by Dr. Ronald Kiene). Two stations were selected, a near-shore site with relatively high DMS concentrations (3–6 nM at the surface) and high biological activity, and an offshore site, approximately 200 km, with lower [DMS] (less than 1 nM) and activities.

In both instances, isotope experiments were conducted with additions of 1, 5, 10, and 20 nM labeled DMS, and were incubated 24 h in 100-ml serum bottles in a deck incubator at ambient temperature (about 9°C). The bottles were uncovered, so incubations were exposed to a day–night cycle of light intensity.

Results followed a fairly consistent pattern, with roughly 1% of the substrate recovered in the CO_2 fraction after 24 h and less than 0.5% recovered in filter fractions (> 3 μm and 0.2–3 μm). These numbers were calculated from activities normalized against t_0 activities (Figure 3). The fractions generally increased linearly with time, and 0.2-μm filtered samples showed no significant activity. The "DOC" fraction was more variable, with time trends less distinct; up to about 4% of added activity was left in solution even at t_0 due to impurities in the DMS stock (see below).

Rates were highest at the near-shore station and did not show a significant trend with substrate concentration. The in situ rate extrapolated from the ^{14}C–DMS additions was 0.6 nM d^{-1} (Table 2); in contrast, the rates calculated from the chloroform inhibition method were somewhat higher, about 2–3 nM d^{-1} (R. Kiene, personal communication).

At the offshore station, rates were very low (Table 2). Only 0.2-μm filters were used due to the low activities. Less than 0.5% of the substrate was recovered in the CO_2 fraction after 24 h, and less than 0.2% was recovered in filter fraction. These activities were close to the detection limit and sensitivity was poor. The "DOC" fraction showed no significant trend with time. In this instance, the rates of production of CO_2 and cell carbon accounted for a removal of only 0.01–0.1 nM d^{-1}, less than 10% of the rates calculated from the chloroform inhibition method (Table 2) (0.8–0.9 nM d^{-1}, R. Kiene, personal communication).

Following this field study, several laboratory experiments were run using seawater collected off of Shilshole Marina, Puget Sound, Washington in July 1991. These experiments were performed to test two possible sources of reduced activity (impurities in the labeled substrate and the possible breakup of cells from the NaOH used to kill the incubations) and to try to recover the total added activity in all fractions to check for consistency.

The substrate as purchased was found to contain approximately 10% of the activity that did not behave as DMS because it could not be removed from solution by continuous sparging. The stock was purified by sparging DMS from solution and cryotrapping it in a Teflon loop in liquid nitrogen. This was then inserted into sterile deionized water, the trap warmed, and the DMS blown into solution. Tests showed that over 98% of the purified stock's activity could be

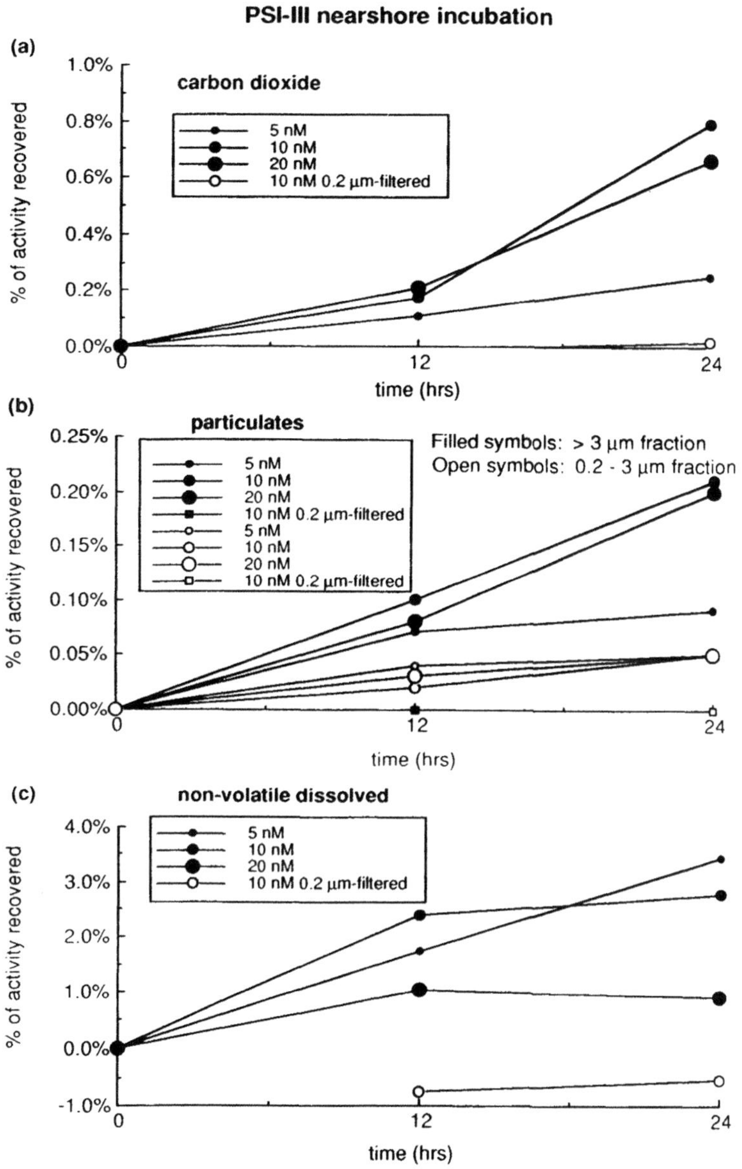

Figure 3. Incubation results from near-shore site of PSI-III study to illustrate typical incubation results. Ambient [DMS] was about 4.5 nM. Shown is relative fraction of added activity recovered as (a) CO_2 (Table 1, fraction #5), (b) > 3-µm and 0.2–3-µm particulates (Table 1, fraction #7), and (c) dissolved material (Table 1, fraction #8). Numbers are normalized to t_0.

Table 2. Comparison of DMS consumption rates during the PSI-III cruise.

Date	$[DMS]_{ambient}$ (nM)	DMS Loss (nM d^{-1})	Turnover Time (d)	Method
		Nearshore Site		
4-23-91	5.8	3.3	1.6	Chloroform[a]
4-23-91	2.0	2.2	1.0	Chloroform[a]
4-28-91	4.5	0.6	8.4	^{14}C–DMS
		Offshore Site		
4-18-91	0.9	0.9	1.0	Chloroform[a]
4-20-91	0.9	0.8	1.0	Chloroform[a]
4-28-91	0.7	<0.01	>100	^{14}C–DMS[b]

[a]Chloroform-inhibition data provided by Dr. R. Kiene.
[b]Recoveries were near limit of detection at this site and consumption rate is very uncertain.

removed after sufficient sparging. Gas chromatography showed that, as with the initial stock, the only sulfur-containing volatile present was DMS, and the specific activity was calculated at 23.8 ± 1.1 mCi mmol^{-1} (n = 6).

In the first experiment, NaOH and formalin were tested against each other as killing agents to see if the base might lyse cells and result in an undercount of cell activity. These incubations, run at 8°C in the dark for 24 h, gave very similar results. Product curves were very similar to the PSI-3 incubations. Analysis of this water by GC found nearly 10 nM ambient DMS, but DMS consumption was very low (roughly 0.1–0.2 nM d^{-1}). This gave assurance that trace impurities in the substrate and using base to terminate the incubation were not causes of low activities.

In the second experiment, the incubation was run at room temperature (21°C) for 72 h, with additions of 5, 10, and 20 nM labeled DMS. Formalin was again used to stop reactions. DMS consumption rates for this experiment were much higher than the previous experiment, consistent with increased biological rates at higher temperatures, and showed increasing rates of production of CO_2 and cell activity with time, suggesting possible adaptation of the culture to the substrate (Figure 4a). Figure 4b shows that rates of conversion of the substrate increased linearly with substrate concentration, indicating no saturation up to 30 nM total [DMS].

For this experiment, a small fraction of the solution was analyzed for complete recovery of added activity. Table 1 details the fractions recovered. Small subsample volumes (7 ml) were analyzed to allow short degassing times (10–12 min). Recoveries were good; about 90% of the added activity was recovered,

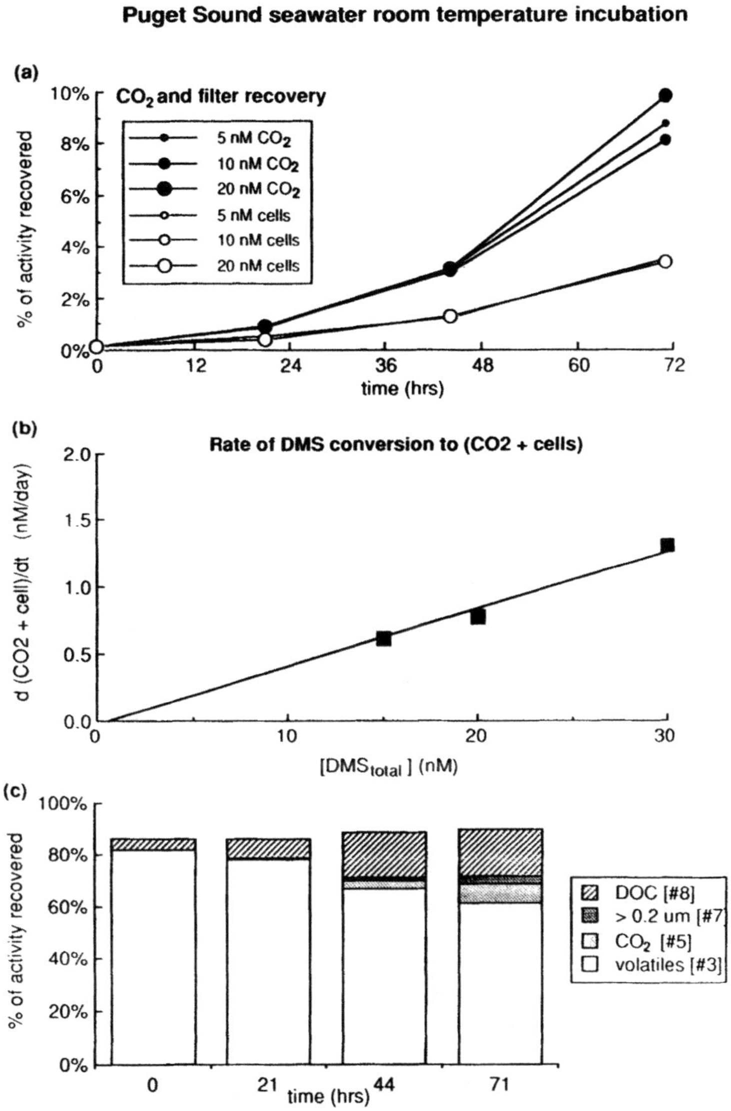

Figure 4. Incubation results from Puget Sound seawater, run at room temperature for 72 h. (a) Activities recovered as CO_2 and > 0.2 μm particulates. (b) Rates of conversion of DMS to CO_2 and > 0.2 μm particulates, averaged for a 71-h incubation period. Rates include ambient [DMS] of approximately 10 nM, plus additions of label at 5, 10, and 20 nM. Line is linear regression. (c) Fractionation results for 10 nM addition of labeled substrate. Shown are fractions recovered as DOC, particulate material > 0.2 μm, CO_2, and volatiles. Numbers in legend refer to fractions in Table 1.

with a concomitant decrease of the volatile fraction and increase in CO_2, particulate material, and DOC fractions over the 72-h time period (Figure 4c). CO_2 fractions from this method were not always identical to those recovered from the standard procedure, probably due to the low total activity of the small volumes used, but by the last time point, there was reasonable agreement.

Discussion

As with the chloroform method, consumption of labeled DMS behaves in a manner consistent with biological uptake: It is removed by 0.2-μm filtering and is inhibited with formalin or alkali. One additional experiment from the PSI-3 cruise (not shown) also demonstrated inhibition of ^{14}C–DMS consumption upon addition of chloroform (500 μM final concentration). The rate of conversion to CO_2 and cell biomass is low but increases with temperature. It appears that a significant fraction of consumed DMS may be released as nonvolatile dissolved material; whether this represents true DOC or cell material which has passed through the 0.2-μm filter is not yet known. Possible soluble compounds which might be produced from DMS include formaldehyde and dimethyl sulfoxide (DMSO). The latter compound might possibly be produced biologically, as has been demonstrated for anaerobes,[10] but could also be produced photochemically.[17] Several measurements of DMSO were taken during PSI-3 and significant concentrations, similar to DMSP, occurred in surface waters (B. Bloomquist, personal communication). Other non-sulfur-containing volatiles, such as formaldehyde or methane, might also be produced which are not currently identified by our method.

There are still several important points which need explanation:

1. Size fractionation of the particulate material results in most activity in the > 3-μm fraction, rather than in the 0.2–3-μm fraction. This is not consistent with bacterial consumption. Possible explanations include: consumption of DMS by eukaryotic organisms; consumption of DMS by bacteria, which are themselves consumed by larger grazers; consumption of DMS by bacteria which are attached to DMSP-containing phytoplankton or which reside in grazer microsites; or filtering artifacts such as flocculation of bacteria or cell breakup/flocculation which results in the bacterial material being retained on the larger-pore filters. The possibility that organisms other than marine bacteria may consume DMS cannot yet be excluded.

2. Consumption of DMS, as measured by CO_2 and cell carbon production, is lower than for the chloroform inhibition method, by as much as an order of magnitude. Because likely by-products of DMS consumption include nonvolatile organics and nonsulfur gases, which have not yet been well quantified, our results should be viewed as a lower bound on the rate of consumption. We hope

to better quantify all fractions in a future study in a direct comparison with the chloroform method.

3. Our attempts to isolate bacteria from seawater which can utilize DMS as a sole carbon or energy source have so far been unsuccessful. We do not yet know what fraction of marine bacteria can consume DMS, or what groups are involved. To our knowledge, no isolates from the marine euphotic zone have been demonstrated to grow on DMS.

4. As of now, the importance of the bacterial sink as a control of DMS outgassing is still largely unknown. It appears that consumption is widespread, but that rates vary considerably and it is likely that the magnitude of this process will be highly dependent on local biological conditions.

Acknowledgments

We wish to acknowledge the generous assistance of Dr. John Baross, University of Washington School of Oceanography. We also wish to thank Dr. James T. Staley, University of Washington Dept. of Microbiology, for his helpful discussions and ideas, and Dr. Ronald Kiene, University of Georgia, for both discussions and his incubation data from the PSI-3 cruise. This work was supported by the NASA Earth Science and Applications Division, Interdisciplinary Research Program.

References

1. Charlson, R.J., J.E. Lovelock, M.O. Andreae, and S.G. Warren. 1987. Oceanic phytoplankton, atmospheric sulphur, cloud albedo and climate. Nature **326**:655–661.
2. Cooper, W.J., and P.A. Matrai. 1989. Distribution of dimethyl sulfide in the oceans. In E.S. Saltzman and W.J. Cooper (eds.), Biogenic Sulfur in the Environment, American Chemical Society, Washington DC, pp. 140–151.
3. Turner, S.M., G. Malin, P.S. Liss, D.S. Harbour, and P.M. Holligan. 1988. The seasonal variation of dimethyl sulfide and dimethylsulfoniopropionate concentrations in nearshore waters. Limnol. Oceanogr. **33**(3):364–375.
4. Turner, S.M., G. Malin, and P.S. Liss. 1989. Dimethyl sulfide and (dimethylsulfonio) propionate in European coastal and shelf waters. In E.S. Saltzman and W.J. Cooper (eds.), Biogenic Sulfur in the Environment, American Chemical Society, Washington DC, pp. 183–200.
5. Zhang, L., I. Kuniyoshi, M. Hirai, and M. Shoda. 1991. Oxidation of dimethyl sulfide by *Pseudomonas acidovorans* DMR-11 isolated from peat biofilter. Biotechnol. Lett. **13**(3):223–228.

6. Visscher, P.T. and H.v. Gemerden. 1991. Photoautotrophic growth of *Thiocapsa roseopersicina* on dimethyl sulfide. FEMS Microbiol. Lett. **81**:247–250.

7. Ni, S., and D.R. Boone. 1991. Isolation and characterization of a dimethyl sulfide-degrading methanogen, *Methanolobus siciliae* HI350, from an oil well, characterization of *M. Siciliae* T4/M^T, and emendation of *M. siciliae*. Internat. J. System. Bacteriol. **41**(3):410–416.

8. Oremland, R.S., R.P. Kiene, I. Mathrani, M.J. Whiticar, and D.R. Boone. 1989. Description of an estuarine methylotrophic methanogen which grows on dimethyl sulfide. Appl. Environ. Microbiol. **55**(4):994–1002.

9. Suylen, G.M.H., and J.G. Kuenen. 1986. Chemostat enrichment and isolation of *Hyphomicrobium EG:* A dimethyl-sulphide oxidizing methylotroph and reevaluation of Thiobacillus MS1. Antonie van Leeuwenhoek **52**:281–293.

10. Zeyer, J., P. Eicher, S.G. Wakeham, and R.P. Schwarzenbach. 1987. Oxidation of dimethyl sulfide to dimethyl sulfoxide by phototrophic purple bacteria. Appl. Environ. Microbiol. **53**(9):2026–2032.

11. Kanagawa, T., and D.P. Kelly. 1986. Breakdown of dimethyl sulphide by mixed cultures and by *Thiobacillus thioparus*. FEMS Microbiol. Lett. **34**:13–19.

12. Kiene, R.P., and T.S. Bates. 1990. Biological removal of dimethyl sulfide from seawater. Nature **345**:702–705.

13. Kiene, R.P., and S.K. Service. 1991. Decomposition of dissolved DMSP and DMS in estuarine waters: dependence on temperature and substrate concentration. Marine Ecol. Prog. Ser. **76**:1–11.

14. Oremland, R.S., and D.G. Capone. 1988. Use of "specific" inhibitors in biogeochemistry and microbial ecology. Adv. Micro. Ecol. **10**:285–383.

15. Kiene, R.P., R.S. Oremland, A. Catena, L.G. Miller, and D.G. Capone. 1986. Metabolism of reduced methylated sulfur compounds in anaerobic sediments and by a pure culture of an estuarine methanogen. Appl. Environ. Microbiol. **52**:1037–1045.

16. Zinder, S.H., and T.D. Brock. 1978. Production of methane and carbon dioxide from methane thiol and dimethyl sulphide by anaerobic lake sediments. Nature **273**:226–228.

17. Shooter, D., and P. Brimblecombe. 1989. Dimethylsulphide oxidation in the ocean. Deep-Sea Res. **36**(4):577–585.

Aspects of the Biogeochemistry of Methane in Mono Lake and the Mono Basin of California

Ronald S. Oremland, Laurence G. Miller, Charles W. Colbertson, S.W. Robinson

U.S. Geological Survey, Menlo Park, California 94025

Richard L. Smith

U.S. Geological Survey, Denver, Colorado 80002

Derek Lovley

U.S. Geological Survey, Reston, Virginia 22092

Michael J. Whiticar

Centre for Earth & Ocean Research, University of Victoria, Victoria, B.C., Canada

Gary M. King

Darling Marine Center, University of Maine, Walpole, Maine 04573

Ronald P. Kiene

Department of Marine Science, University of Southern Alabama, Mobile, Alabama 36688

Niels Iversen

Institute of Water, Soil, and Environmental Technology, University of Aalborg, Aalborg, Denmark, DK-9000

Melinda Sargent

Department of Chemistry, University of California at Irvine, Irvine, California 92717

ABSTRACT

Above-ambient levels of methane and higher hydrocarbons were detected in the atmosphere of the Mono Basin. These gases emanated from several different sources, including natural gas seeps (thermogenic and biogenic), and methanogenic activity in sediments. Seeps were distributed over nearly 33% of the lake bottom and were also present in the exposed former lakebed. They originated from one or more natural gas deposits that underlie the basin. Seeps associated with hot springs had a thermogenic character, whereas the others had the characteristics of bacterially formed gases. The radiocarbon content of methane in all seep gases was low (5–11% modern carbon), indicating an age of greater than about 20,000 years.

Dissolved methane increased with depth in the sediments, ultimately reaching saturation levels (1–3 mM). The outward diffusive flux of methane into the anoxic bottom waters elevated concentrations to 60 μM during the 1984–1988 meromictic interval. The radiocarbon content of hypolimnion methane was equivalent to that of the dissolved inorganic carbon (~80% modern carbon), indicating derivation from current methanogenic activity. Oxidation by anaerobic bacteria was the major sink for this methane. Methane that escaped oxidation transited to the epilimnion where it attained supersaturation (0.1–3.0 μM) relative to the atmosphere. Measured outward fluxes over a 2-year period were highly variable, ranging from 0.01 to 4.82 mmoles CH_4 m^{-2} d^{-1}.

Bacterial activity was detected in the sediments of the hypolimnion. High rates of sulfate reduction (~200 μmoles L^{-1} d^{-1}) were measured in the upper 5-cm interval but decreased > 10-fold with depth. The highest rate of methanogenic activity was ~1 μmole L^{-1} d^{-1}, as measured by reduction of $^{14}CO_2$. Bacteria metabolized dimethylsulfide, trimethylamine, acetate, and glucose to primarily CO_2 and some CH_4. Turnovers of glucose ($k = 0.16$ d^{-1}) and trimethylamine ($k = 0.09$ d^{-1}) were highest in the upper 2.5-cm interval and decreased as much as 10-fold with depth. These results confirm that a low rate of methanogenic activity occurs in this extreme environment and is the source of methane to the hypolimnion.

Introduction

Concerns about global warming have focused considerable attention on methane, the most rapidly increasing,[11] and, with the exception of CO_2, the most significant radiative trace gas in the atmosphere.[17] Methane can originate from a diversity of sources, including recent or ancient bacterial methanogenesis, biomass-burning, and thermogenic natural gas emissions. Hence, interpretation of atmospheric data with regard to identifying these sources is a complicated and potentially fallible exercise. Budgeting attempts at a global scale are highly reliant on the use of stable carbon isotopic mass balances in conjunction with radiocarbon data (e.g., see Ref. 17). These investigations have determined that bacteria contribute between 70 and 90% of the total annual flux,[41,42,75] although there is great uncertainty with regard to the magnitude of any individual source. Therefore, fieldwork has focused on attaining reliable flux measurements from individual sources which have been assumed to be significant, such as rice paddies,[18,65,67] wetlands,[7,10] tundra soils,[77] coastal marine regions,[44,45,62] burning biomass,[27,39] peatlands,[21] termite nests,[66] and ruminants.[23] The biogeochemical factors which regulate the magnitude of flux and its stable carbon or hydrogen isotopic composition for any given source can be quite complex and temporally variable.[13,43,79] Hence, environments with multiple sources of methane are usually not studied because of their complexity. Although routinely used for global investigations, studies focusing on the biogeochemistry of multiple sources of methane in a single environment are rare, even though complexity is generally characteristic of our modern world.[74] In this chapter, we summarize and review our research efforts at Mono Lake, California, an environment which has several distinct sources of methane.

Site Description

Mono Lake (lake level ~1943 m) is a closed basin lake located in central eastern California at the base of the Sierran escarpment and on the western edge of the arid Great Basin (Figure 1). The Mono Basin was formed ~3×10^6 yr b.p. by subsidence, and the lake itself has an estimated age of $(0.5–1.0) \times 10^6$ yr.[16,25,64] Over this long period, the lake has experienced major climatic changes, including proximate glaciations and large expansions or contractions of the lake's size and volume in response to variations in the region's annual precipitation.[37] In the Pleistocene, when the climate of the Great Basin of the United States was much wetter than it is today, Lake Russell (the precursor of Mono Lake) covered about 5 times the area and contained 18 times the water volume of modern-day Mono Lake (Lajoie, personal communication). Lake levels have fluctuated considerably over the past 3000 yr.[73] Recent concerns about falling lake level and increasing salinity are related to the diversion since 1941 of Eastern Sierran runoff to Los Angeles. Other important geologic events include episodes of active volcanism initiated by the eruption of the Long Valley Caldera ~7×10^5 yr b.p. and continuing on to the present.[6,28,61,68,72] The last major eruption was of Panum Crater (~600 yr b.p.) located on the lake's southwestern shore. Hot springs and volcanic features are present around the Mono Basin as well as on the lake's two largest islands, Paoha and Negit. Paoha Island consists of a mass of bottom sediments uplifted by volcanic forces from the lake bottom between 200 and 400 yr b.p.[73]

In chemical terms, modern-day Mono Lake can be classified as an environmental extreme, owing to its reduced volume as well as to the influence of hydrothermal waters. General characteristics are given in Table 1. Briefly, the lakewater is highly alkaline (pH = 9.7) and hypersaline[46] (TDS = ~90 g/L) and

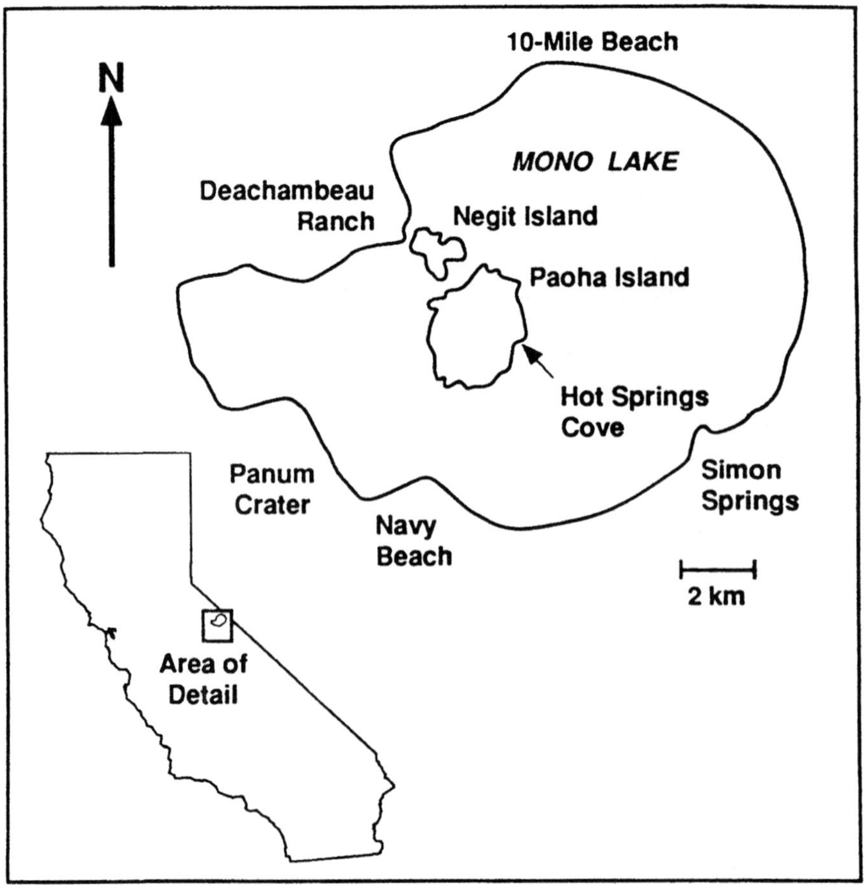

Figure 1. A map of Mono Lake showing geographic locations and major sampling sites.

contains very high concentrations of arsenic.[4] Because of its alkaline nature, the lakewater contains unusually high levels of fallout radionuclides.[5,69,70] Inorganic radiocarbon levels have increased by 60‰ since 1957.[15] Because of these extreme conditions, the food chains of the lake are primarily microbial, the only exception being the abundant macrozooplankter *Artemia* sp. and grazing waterfowl.[38] Investigations have been conducted on primary productivity,[32] sediment organic geochemistry,[60] gas exchange rates,[76] and nitrogen fixation.[51] Limnological features include a surface area of about 150 km^2, a maximum depth of ~30 m, and a total volume of ~2.64 km^3.[58] Normally, the lake is monomictic and undergoes a fall–winter overturn. However, release of freshwater into the

Table 1. Chemical properties of Mono Lake water.

Component	Concentration
Salinity	90 g/L
pH	9.8
Na^{+}	1.3 M
Cl^{-}	0.5 M
$HCO_3^{-}/CO_3^{=}$	0.4 M
$SO_4^{=}$	0.13 M
As	0.2 mM
Dissolved organic carbon	6.7 mM

NOTE: data from Mason,[46] Oremland et al,[57] Anderson and Bruland,[4] and Miller et al.[48]

lake over two consecutive wet winters (1982–1983 and 1983–1984) diluted surface salinity sufficiently to overcome increased density caused by winter cooling. Hence, the lake underwent a period of permanent annual stratification (meromixis) which persisted from 1983 to 1988.[48]

Material and Methods

Methods for extraction, quantification, and measurement of stable isotopic and radiocarbon contents of methane and other light hydrocarbons from lakewaters and sediments were as described in Reference.[57] Flux of methane from the surface waters of the lake was achieved with use of floating chambers.[47] Methane and other light hydrocarbons in the atmosphere at several locations in the Mono Basin and over Mono Lake were collected by filling evacuated 2-L stainless-steel cans.[11] The cans were allowed to fill over a time period of ~10 min while being held (1.5 m above ground) either into the prevailing wind or, in the case of calm periods, while being held in front of the walking individual collecting the sample. Hydrographic and chemical determinations of the lakewater and sediment porewaters were achieved by methods described elsewhere.[19,48,71]

Microbiological experiments employing radiotracers were conducted with anoxic lakewaters and pelagic zone sediments. Procedures for determination of sulfate reduction and methane oxidation in lakewater have been given elsewhere.[31,71] Sediments for microbiological assays were collected by subcoring extruded core sections with cutoff plastic syringes (3 cc, 5 cc, or 10 cc), and subsequently plugging the open end with stoppers or serum caps. Injections of ^{14}C-labeled acetate, trimethylamine, dimethylsulfide, and glucose were achieved with microliter syringes (see text for details on volume and specific activities). Samples were incubated at ~5°C (in situ bottom water temperature) for various time periods (up to 70 h) before being frozen to arrest activity. Analysis of $^{14}CH_4$ and $^{14}CO_2$ was achieved by extruding subsamples into test tubes or serum vials followed by capping, acidification, and quantification of the head-space gases by gas chromatography/gas proportional counting.[24] Alternatively, $^{14}CH_4$ and $^{14}CO_2$ in some samples were analyzed by placing a trapping solution downstream from a flame-ionization detector (FID) followed by liquid scintillation counting.[36] Measurement of reduction of ^{14}C–bicarbonate to methane posed a technical problem because of the high carbonate/bicarbonate pool (~0.4 M) of the lakewater (Table 1). Hence, a large quantity of radioisotope (50 μCi/9 ml sediment) was injected into cutoff glass syringes (10 cc), and samples were incubated for 60 h. Extraction of $^{14}CH_4$ from the entire sample was achieved by head-space flushing, cryogenic trapping, and furnace oxidation to $^{14}CO_2$ followed by trapping with ethanolamine and quantification by liquid scintillation spectrometry.[78] Subsamples used for determination of sulfate reduction were collected and incubated as described above, but were injected with carrier-free $^{35}SO_4^{=}$ and were analyzed for ^{35}S–sulfide by either acidification[71] or distillation with acidic Cr (II) solution.[29,34]

Results and Discussion

Light Hydrocarbons in the Atmosphere of the Mono Basin

Atmospheric samplings were collected during February and July of 1990 (Table 2). Stations from the surrounding basin were located on the northeast shoreline (10 Mile Beach), the northwest region (Dechambeau Hot Spring), the southwest corner (Panum Crater), the southeastern shore line (Simon Springs), and an out of Basin sampling at the Tioga Pass entrance station to Yosemite National Park (elevation = 3040 m). Samples were also taken from various locations on Paoha Island and at one mid-lake location about 2 km west of Paoha Island (Figure 1). Ambient methane levels ranged from 1739 to 2400 ppb (Table 2). Methane was present at significantly elevated concentrations (relative to the 1770 ppb average for this latitude[12]) at Dechambeau Hot Spring, 10 Mile Beach (0634 sample), and at all the sites on Paoha Island. In contrast, the sites on the southern shore and Tioga Pass measured at or slightly below the 1710-ppb value, the only exception being the 1830 sampling at Simon Springs (1798 ppb). Non-methane hydrocarbons (C_2 to C_4) were detected at all the sites. Elevated ethane and propane concentrations were evident at the Dechambeau and Paoha Island sites. Global average ethane concentrations for this latitude are between 0.92

Table 2. Concentrations of hydrocarbon gases in the atmosphere at various locations in the Mono Basin.

Location	Date	Time	CH_4	C_2H_4	C_2H_6	C_2H_2	C_3H_6	C_3H_8	i-C_4H_{10}	C_4H_8	n-C_4H_{10}	HR[a]
							(parts per billion by volume)					
North Shore												
10 Mile Beach	2/13	1030	1,763	0.061	1.730	0.345	0.063	0.0497	0	0.205	0.287	792
	7/25	1800	1,757	0.091	0.679	0.091	0.036	0.151	0	0.443	0.023	2117
	7/27	0630	2,400	0.096	0.748	0.096	0.115	0.229	0	0.346	0	2456
Dechambeau Hot Spring	7/25	1645	1,910	0.216	3.628	0.129	0.143	1.395	0.115	0.210	0.217	380
	7/27	0730	1,908	0.180	3.098	0.183	0.127	1.336	0.163	0.114	0.180	430
Mid-Lake	7/27	0634	1,754	0.320	0.677	0.168	0.256	0.808	0	0.239	0.024	1181
Paoha Island												
East Shore	2/15	1130	1,801	0.162	2.028	0.648	0.034	0.953	0.679	0.718	0.352	604
Crater Summit	2/15	1150	1,810	0.133	2.071	0.653	0.020	0.782	0.768	0.509	0.356	634
Crater Saddle	2/15	1200	1,813	0.117	2.087	0.584	0.027	1.013	0.106	0.624	0.356	585
Dry Fumarole	2/15	1230	2,141	0.759	707.000	0	0	250.000	52.000	0.161	58.700	2.2
Hot Spring Cove	7/27	0655	1,849	0.183	3.700	0.279	0.135	0.931	0.187	0.146	0.304	399
West Shore	7/27	0641	1,909	0.079	0.737	0.088	0.051	0.128	0	0.111	0.048	2207
South Shore												
Panum Crater	2/13	1540	1,769	0.101	1.923	0.562	0.013	0.847	0.123	0.086	0.349	639
	7/27	0615	1,739	0.168	0.617	0.094	0.081	0.135	0	0.083	0.034	2313
	7/27	1715	1,752	0.101	0.744	0.104	0.055	0.158	0	0.092	0.027	1942
Navy Beach	7/27	0645	1,739	0.160	0.637	0.122	0.153	0.156	0	0.424	0.038	2193
	7/27	1745	1,755	0.121	0.719	0.125	0.075	0.172	0.141	0.105	0.037	1970
Simon Springs	7/27	0715	1,774	0.119	0.615	0.094	0.115	0.229	0	0.346	0	2102
	7/27	1830	1,798	0.180	0.753	0.126	0.107	0.181	0	0.182	0.045	1925
Out of Basin												
Tioga Pass (3040 m)	7/29	1315	1,750	0.105	0.846	0.164	0.033	0.199	0	0.100	0.082	1675

[a]HR = Hydrocarbon ratio = $CH_4/[C_2H_6 + C_3H_8]$.

ppbv (summer) and 2.31 ppbv (winter) as a consequence of the extent of the photochemically produced OH sink.[12]

The picture is clarified somewhat by examining the $CH_4/[C_2H_6 + C_3H_8]$ ratio, a technique commonly employed in petroleum exploration geochemical surveys. In general, stations located around hot springs or fumaroles, such as Dechambeau Ranch or Paoha Island had ratio values below 650, whereas at other locations ratios were generally well above 1100. One anomalous site was a gas sample taken from a dry fumarole located in a basaltic formation 30 m above the lake level on Paoha Island. This sample had ethane, propane, and butane present at levels which were two orders of magnitude greater than that observed elsewhere, but methane was not correspondingly enriched. It appears that the primary source for the hydrocarbon enrichments observed at Paoha Island and Decambeau Ranch were from the presence of thermogenic natural gas seeps (see the following section). The other enrichments may have been caused by proximate "biogenic" seeps such as those just offshore 10 Mile Beach.

Comparisons of the absolute concentrations of these gases to other types of systems, while not really analogous (due to the varied nature of the sources), is still of interest. Crill et al.[22] reported average ambient methane levels in air taken 3 m above an Amazon floodplain lake to be 1890 ppbv, with levels ranging between 1800 and 2140 over a diurnal period. In contrast, a mid-lake early morning sample from Mono Lake had only 1754 ppbv CH_4, which indicates that the outward flux from the methane supersaturated surface waters (see below), may not always be readily detectable in the boundary layer (Table 2). Reported values for ethane and propane taken from a height of 1.2 m above a Guayana forest soil were 1.430 and 0.260 ppbv, respectively,[14] whereas samples collected 30 m above the Amazon forest canopy averaged 1.17 ppbv (ethane) and 0.370 ppbv (propane).[80] Oceanic mean values (16 stations) of these two gases were 1.803 ± 0.621 ppbv (ethane) and 0.718 ± 1.239 ppbv (propane) in marine air samples taken from stations in the tropical mid and eastern Pacific.[26] Mean tropospheric values of all three gases taken from a height of 2.86 km at Catherine Peak, New Mexico were 1590 ppbv (methane), 3.645 ppbv (ethane), and 0.358 ppbv (propane), whereas at 11.6 km, they were 1600 ppbv, 0.441 ppbv, and 0.084 ppbv, respectively.[1] Clearly, the ethane and propane levels collected at some locations around Mono Lake were significantly higher than many of the values at the above-mentioned locations. Elevated levels of methane in ambient air samples may have been associated with sources from seeps rather than the lakewater. However, more detailed and frequent samplings as well as the use of radiocarbon and stable isotopic data would be required to verify this preliminary observation.

Characterization of Light Hydrocarbons in Gas Seeps

Gas seeps were numerous (>700) and widely distributed over the surface of Mono Lake, but were particularly abundant around the margins of Paoha Island.

The seeps had measured continuous flow rates ranging from ~0.1 to 4 L min^{-1}. When their methane content was taken into account, fluxes of 2.52–305 moles CH_4 m^{-2} d^{-1} were determined.[57] A recent bathymetric survey employing reflecting sonar revealed that one-third of the lake bottom had obscured reflection patterns due to the presence of gas-charged sediments from which the seeps emanate.[58] These sediments were located in regions where seeps were observed. Flux of methane to the atmosphere from all the seeps counted was estimated to be ~2.1×10^6 moles yr^{-1}.[57]

The C_1–C_3 alkanes present in some representative seeps are given in Table 3. Methane was the major component of all the seep gases collected, as were traces of ethane and propane. The other major components included variable levels of

Table 3. Alkane content (C_1–C_3) of gas seeps from Mono Lake and the surrounding shoreline, from three locations in neighboring Long Valley, California.

Location	Methane	Ethane (percent by volume)	Propane
Mono Lake hydrocarbon seeps			
Danburg, Beach, offshore	62	0.0012	0.0007
Paoha Island, west shore	82	0.0022	0.0010
Paoha Island, hot spring	61	2.8600	1.1600
Mid-Lake, 2 km SE Paoha	76	0.0630	0.0170
Paoha–Negit Passage	89	0.0047	0.0019
Shoreline seep, NE	93	0.0033	0.0010
Onshore vent, NE basin	88	0.0010	0.0002
Other Mono Lake–Gasin seeps and entrapped gases			
Navy Beach, CO_2 seep	0.02	0.0007	0
Navy Beach, grass bubbles	13.4	0.0001	0
Dechambeau Hot Spring	2.05	0.0790	0.0450
Dechambeau Pond Bubbles	67	0.0012	0
Long Valley seeps and entrapped gases			
Hot Creek, seep 1	0.04	0.0001	0.0026
Hot Creek, seep 2	0.04	0.0006	0.0002
Hot Creek, bacterial mat	0.77	<0.0001	0
Hot Creek, mud bubbles	58.1	0.0010	0.0006
Casa Diablo, silt pond	3.74	0.0001	0.0001
Casa Diablo, bagel pond	1.16	0.0003	0.0001
Casa Diablo, steam vent	0.003	0	0
Valley, hotspring mat PI	12.1	0.0003	0
Valley, hotspring mat HI	42	0.0002	0
Valley, stream bottom	44.7	0.0001	0

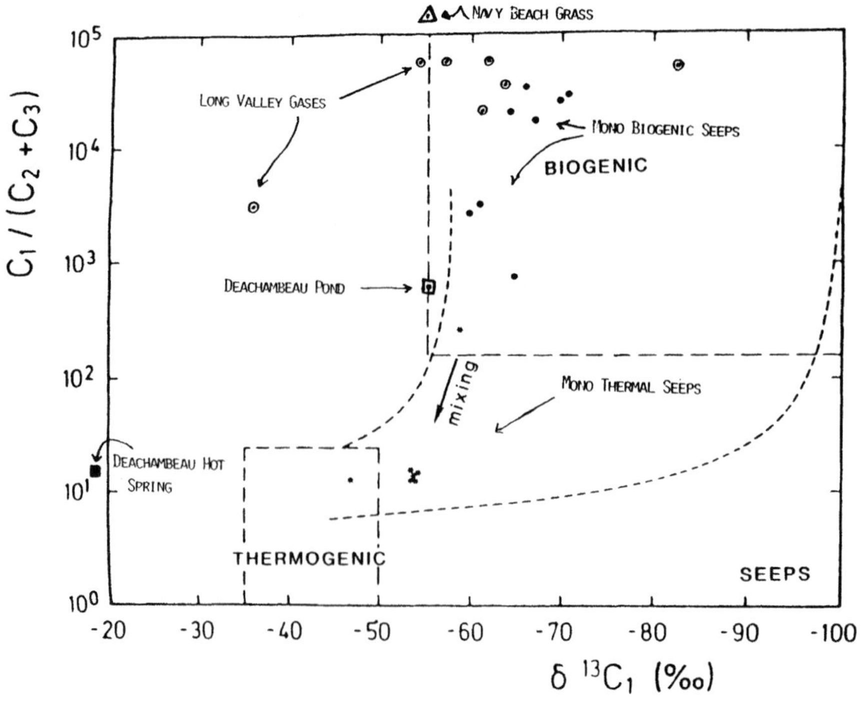

Figure 2. Ratio of $CH_4/[C_2H_6 + C_3H_8]$ versus $\delta^{13}CH_4$ for seep gases collected from various locations in Mono Lake, the Mono Basin, and from Long Valley.

nitrogen and oxygen.[57] Only the gas collected from the Paoha Island shoreline hot spring had an appreciable content of ethane and propane. When plotted as hydrocarbon ratios versus $\delta^{13}CH_4$ (Figure 2), seep gases collected from hot springs had an obvious thermogenic component, whereas those collected from "cold" seeps had the characteristics of "biogenic" (i.e., bacterially formed) gas.[63]

Seeps located onshore at Dechambeau Ranch and at Navy Beach differed in composition from those described above. Seeps located proximate to the Navy Beach shoreline were primarily composed of CO_2 (68%), contained only a trace of methane, and were of a volcanic origin.[9] The source of this methane probably stems from the neighboring tufa-entrapped bubbles. These bubbles were relatively rich in methane (13.4%), but were apparently formed by anaerobic decomposition of shoreline grass which became enveloped by a newly formed tufa crust during the lake's 1983–1984 expansion.

The gases collected from the Dechambeau hot springs were strongly thermogenic in character, based on their hydrocarbon ratios and stable isotopic composition.[63] They were enriched in ^{13}C by > 30‰ relative to the hot spring gases within Mono Lake (Figure 2). It is possible that they represent the "end-member" characteristics of the thermogenic gases formed in the region, whereas those recovered from the lake itself represent various mixtures of biogenic and thermogenic gases. Gases recovered from various geothermal sources and proximate biologically active warm sediments or mat materials from Long Valley (about 80 km south of the Mono Basin) are shown for out-of-basin comparative purposes (Table 3; Figure 2). Clearly, the Mono Lake biogenic and thermogenic seeps can be differentiated from these gases.

The radiocarbon content of methane from various sources around Mono Lake are given in Table 4. Methane dissolved in the lake's anoxic bottom water had a comparable age to that of the $HCO_3^-/CO_3^=$ pool, indicating that it is of recent biogenic origin. In contrast, methane in the thermogenic and biogenic seeps was strongly depleted in ^{14}C and yielded ages of > 20,000 years. Methane from the Navy Beach grass bubbles (52% modern carbon) was probably derived from CO_2 reduction of a mixture of atmospheric carbon dioxide with that from the proximate radiocarbon-depleted CO_2 seeps.

Water-Column-Dissolved Methane and Flux to the Atmosphere

During meromixis, the surface water of Mono Lake was ~200-fold supersaturated with respect to methane, averaging 0.49 ± 0.32 μM ($n = 14$) over a 2-yr interval (1985–1987).[47] Flux to the atmosphere from the lake's surface waters

Table 4. Radiocarbon content of dissolved inorganic carbon and methane from various sources in Mono Lake.[a]

Sources	% Modern Carbon
Dissolved inorganic carbon (1950)[b]	81
Dissolved inorganic carbon (1978)[b]	86
Dissolved CH_4, bottom water	80
Paoha Hot Spring Seep CH_4	11
Mono Lake Biogenic Seep CH_4	6
Navy Beach Seep CO_2	5
Navy Beach Dead Grass CH_4	52

[a]Data from Oremland et al.,[57] reproduced with permission of *Geochimica et Cosmochimica Acta.*
[b]Data from Peng and Broecker.[59]

was estimated to be 2.8×10^7 moles CH_4 yr^{-1} during meromixis.[57] In comparison with premeromixis conditions, surface-water-dissolved methane had declined nearly five-fold (from ~2.5 μM), whereas bottom water concentrations increased (Figure 3). At the height of meromixis (1987–1988), dissolved methane in the bottom water of the lake reached ~60 μM. Following the breakup of meromixis in 1988, the vertical profile of dissolved CH_4 reverted to its premeromixis condition. Because postmeromixis surface water has on average fivefold more dissolved CH_4, outward flux to the atmosphere should increase as well.[8] Whole-lake inventories of sulfide, methane, and ammonia were calculated during meromixis and in the years subsequent to its dissipation. After reaching maximal values in 1987, the abundance of all three reduced compounds declined markedly (Figure 4) as a consequence of oxidation and outward flux to the atmosphere.[48]

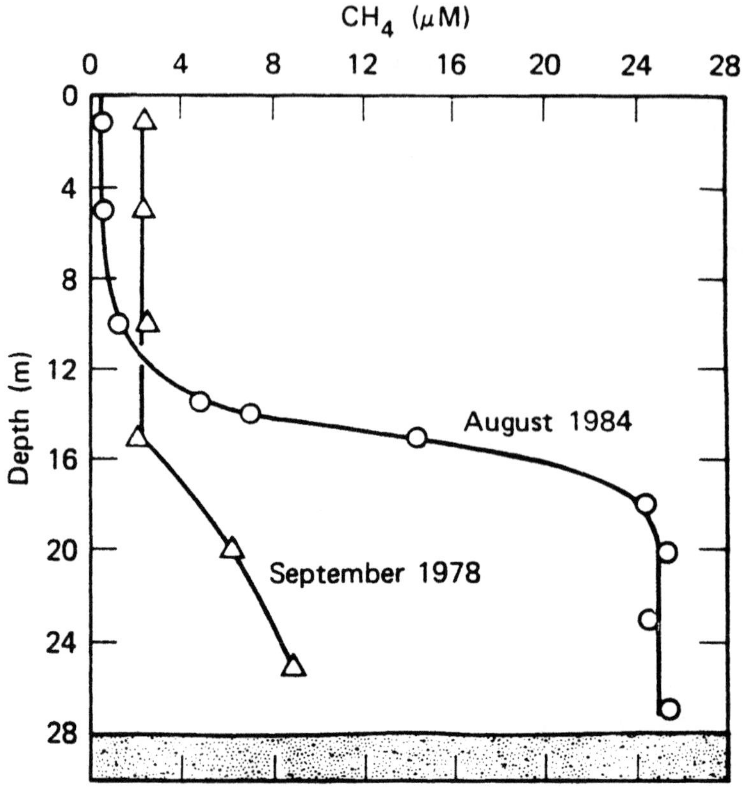

Figure 3. Profiles of dissolved methane in Mono Lake taken before meromixis (1978) and during meromixis (1984).

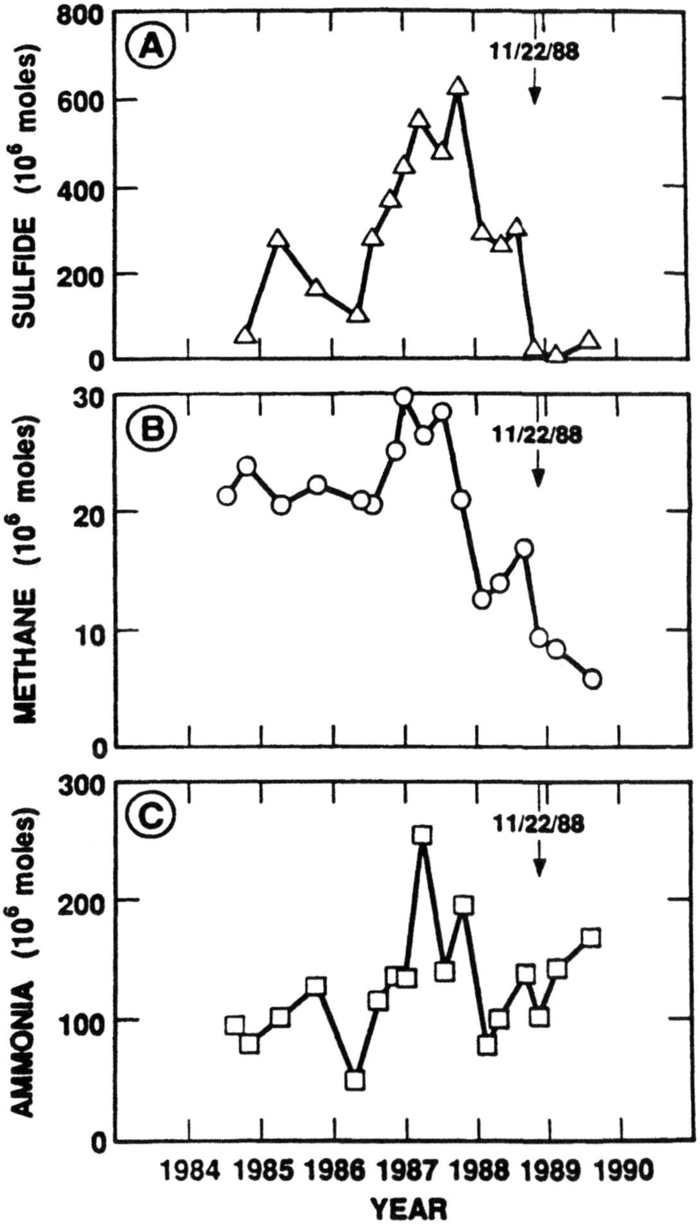

Figure 4. Whole-lake inventories of dissolved methane, ammonia, and sulfide in the Mono Lake water column during meromixis and its subsequent breakup (from Miller et al.,[48]). Arrows indicate date of turnover.

Table 5. Concentration of dissolved hydrogen in Mono Lake water and sediments during February 1988.

Depth	H_2 (nM)	O_2 (μM)	*T* (°C)
	Water Column		
1 m	ND	250	4.6
5 m	ND	266	3.0
10 m	11.4	172	1.9
14 m	14.6	141	1.9
18 m	164.4	66	1.9
22 m	12.4	16	2.6
26 m	11.6	0	4.2
	Sediments		
0–2 cm	19.6		
2–4 cm	27.4		
5–10 cm	11.8		
10–15 cm	11.8		
20–25 cm	5.2		
30–35 cm	7.9		

Dissolved hydrogen was measured in the water column during February 1988, when the lake had detectable oxygen down to a depth of about 23 m (Table 5). With the exception of an anomalously high concentration at 18 m depth, which coincided with the turbid layer, concentrations were in the range 11.4–14.6 nM. These values were high in comparison with mildly reducing freshwater sediments.[40]

Microbial Processes in the Water Column

During meromixis, a turbid microbial plate at ~16 m was evident just below the thermocline, occupying the interface between chemically reduced and oxidized conditions (Figure 5). Because this plate was in the photic zone, $^{14}CO_2$ fixation was detected at this depth. However, unlike the situation in Big Soda Lake, a similar meromictic–alkaline system,[20] a trilobate eukaryotic picoplankter (diameter ≅ 0.8–2.8 μm) occupied this niche rather than photosynthetic bacteria. This picoplankter reached maximal densities of 5×10^5 cells ml^{-1}. Annual primary productivity was estimated at 150–300 g C m^{-2}, of which this picoplankter accounted for ~40%. However, this photosynthetic plate, even though it was a striking feature, contributed < 10% of the annual productivity (C.W.

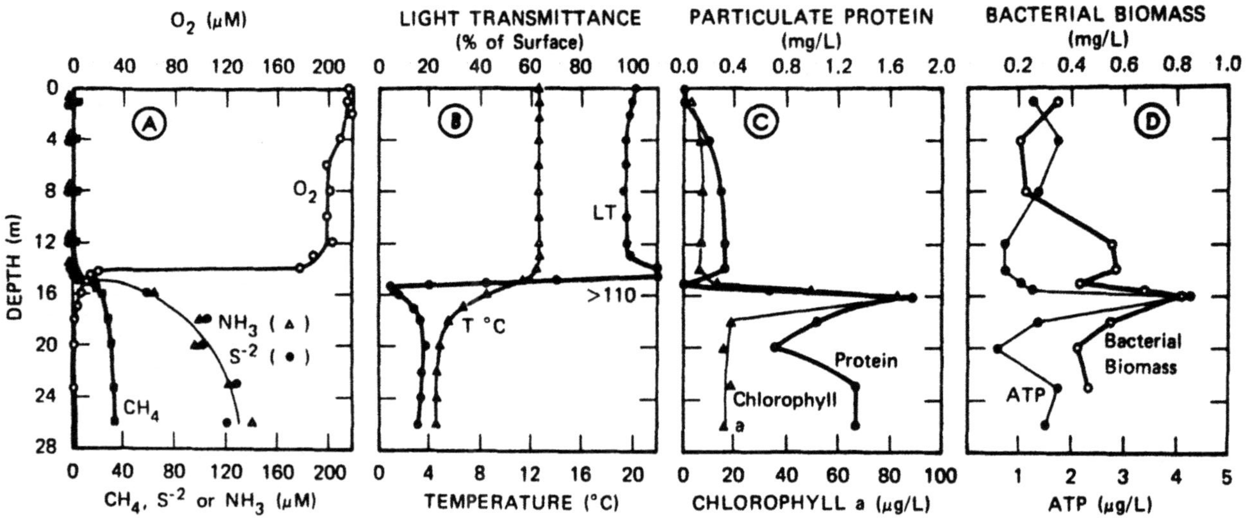

Figure 5. Water-column profiles taken during meromixis (October 1984) of (A) Dissolved oxygen, methane, sulfide, and ammonia; (B) temperature and light transmission; (C) particulate chlorophyll and protein; (D) particulate ATP and bacterial biomass.

Table 6. Estimates of sulfate reduction, methanogenesis, methane oxidation, and methane flux during meromixis in Mono Lake.[a]

Process	Water Column (mmoles m^{-2} d^{-1})	Sediments
Sulfate reduction	12.0	13.4
Methanogenesis	0	0.42–0.69
Methane oxidation	0.68	1.45
CH_4 flux from sediments	—	0.6–1.1
CH_4 flux to atmosphere[b]	0.01–4.82	—
CH_4 flux at turnover[c]	20.4	—
CH_4 flux from seeps	2,500–305,000	—

[a]Ranges indicate temporal and spatial variability observed.
[b]Range of values observed, 1986–1988.
[c]Value from Miller et al.[48]

Culbertson, unpublished data), a situation similar to that of the photosynthetic bacterial plate of Big Soda Lake.[20]

Bacterial methane oxidation was measured in the water column and sediments of Mono Lake during April 1987, while the lake was meromictic (Table 6). Activity generally followed the dissolved methane profile (Figure 6), and hence the anoxic waters below 18 m accounted for about 93% of the water-column activity. Integrated areal rates for the oxic and anoxic portions of the water column were 0.04 and 0.64 mmol CH_4 m^{-2} d^{-1}, respectively. Kinetic experiments with anoxic samples indicated a K_m of 197 μM and a V_{max} of 0.65 μmol L^{-1} d^{-1} (N. Iversen, unpublished data). Boiling the samples eliminated activity in both the anaerobic (Figure 7) and aerobic (not shown) regions, thereby underscoring the bacterial nature of the process.

Preliminary studies with inhibitors of methane oxidation yielded contrasting results for aerobic samples from 12 m versus anaerobic samples from 21 m depth (Figure 8). Acetylene caused total inhibition of aerobic oxidation, whereas dimethylether was partially effective (43–70%), and no significant inhibition was observed for methyl fluoride (Figure 8A). It is not clear why neither methyl fluoride nor dimethyl ether were ineffective at this depth because they both inhibit aerobic CH_4 oxidation in soils.[49,50] In contrast, no inhibition was observed by acetylene in the 21-m samples, and only a partial inhibition (~33%) was observed in the 24–48-h period for methyl fluoride and dimethylether (Figure 8B). Acetylene, dimethylether, and methyl fluoride all inhibit the methane monooxygenase of aerobic methanotrophs.[53] Assuming that there were no problems with inhibitor concentrations or with cell density effects,[49,50] these

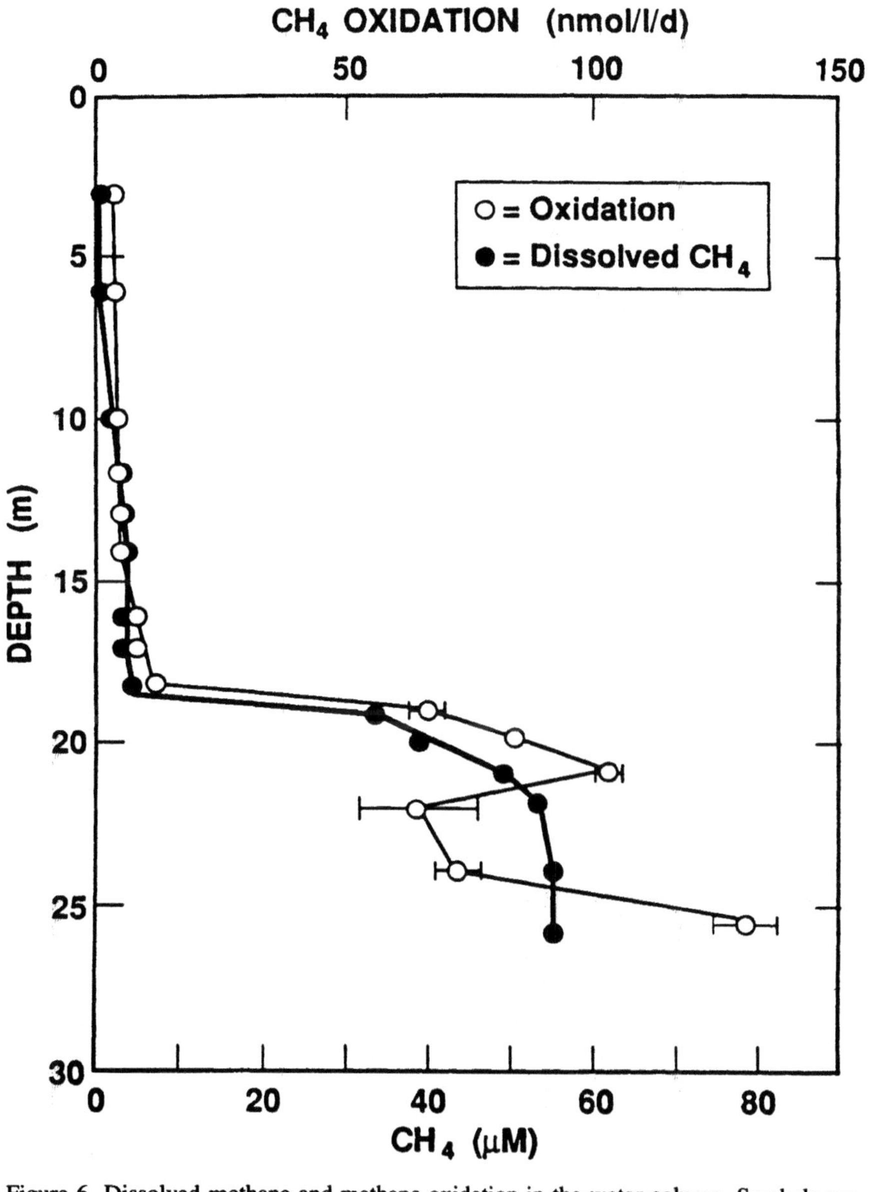

Figure 6. Dissolved methane and methane oxidation in the water column. Symbols represent the mean of three samples and bars indicate ± 1 SD.

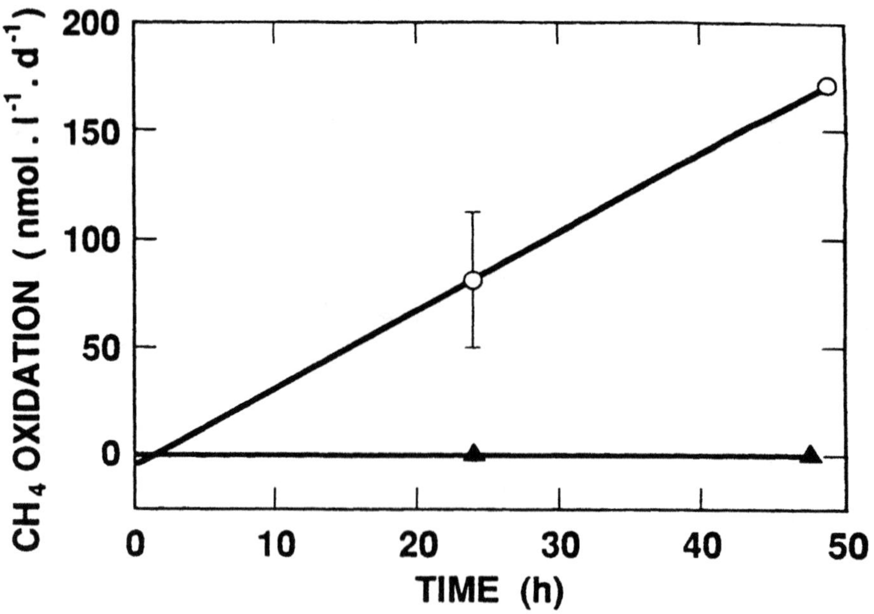

Figure 7. Time course of anaerobic methane oxidation in water taken from 21 m. Experimental (○), boiled controls (▲).

results suggest that the flora at 12 m were probably conventional aerobic methanotrophs, whereas a different population was responsible for the activity observed in the anoxic zone.

Unfortunately, the above results are insufficient to characterize the type(s) of bacteria responsible for the observed anaerobic methane oxidation. Hence, little can be said about the organism(s), mechanisms, or redox coupling of the reaction that substantially goes beyond what has already been observed.[2,3,30,31] However, bacterial consumption of dissolved methane in the water column and sediments (see below) of Mono Lake had a profound influence upon the $\delta^{13}CH_4$ observed in the water column. Values of $\delta^{13}CH_4$ in the epilimnion were as heavy as −20‰, whereas the anoxic monimolimnion averaged −74‰, and the sediments about −86‰.[57] Hence, methane oxidation (primarily anaerobic) was responsible for a ^{13}C enrichment of up to 66‰ relative to its source in the sediments. This phenomenon was also observed in Big Soda Lake[54] and to a lesser extent in marine sediments.[2] Therefore, outward flux of methane to the atmosphere from the lake's supersaturated surface waters was strongly enriched in ^{13}C, being as much as 27‰ heavier than the atmospheric value of −47‰.[57]

Sulfate-reduction was detected in the anoxic waters of Mono Lake in August

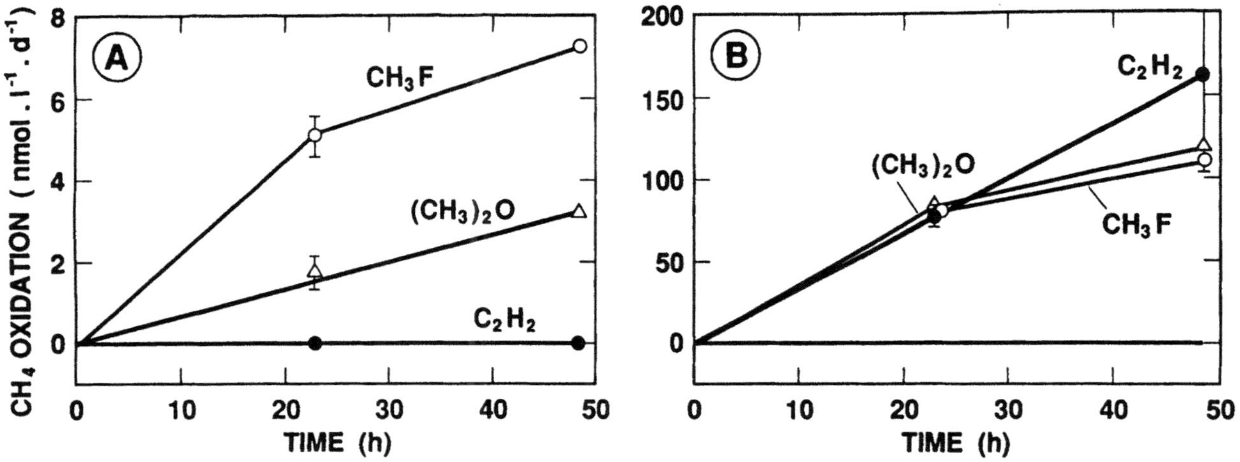

Figure 8. Effect of inhibitors of methane oxidation in water column samples from (A) 12 m and (B) 21 m depth. Concentrations employed were 5% v/v for C_2H_2 and 10% v/v for MF or DME.

of 1986, and rates increased with depth below the 16 m chemocline (Table 6). Integrated areal rates of sulfate-reduction for the water column were about 12 mmol m^{-2} d^{-1}. In experiments conducted in February, 1988, no apparent methanogenic activity was detected during 5 day incubations of anoxic lake water from 18, 22, and 26 m (data not shown).

Gases, Dissolved Ions, and Microbial Activities in the Sediments

Methane and higher hydrocarbons (ethane and propane) all increased with depth in the pelagic sediments (Figure 9) and displayed $CH_4/[C_2H_6 + C_3H_8]$ ratios of 600–750, which are generally indicative of a bacterial origin for the gas.[63] The $\delta^{13}CH_4$ of the sediment methane was strongly enriched in ^{12}C, and values ranged between −80 and −86‰.[57] This reinforced the interpretation of a bacterial origin of the gas. Porewater sulfate decreased with depth in the sediment, but substantial quantities (~30 mM) were still evident at a depth of 110 cm. Free sulfide and ammonia also increased with sediment depth and reached maximal values of 15 mM and 3 mM, respectively.[48]

Considerable variability was observed with respect to the shape and magnitude of the methane or sulfate versus sediment depth profiles collected from the pelagic sediments (Figure 10). In some cases, methane increased in a gradual, linear fashion with depth and reached saturation values (~1–2 mM) at ~70–80 cm, whereas in another core, methane reached saturation at ~20 cm and bubbles were evident at that depth. Although sulfate always decreased steeply, the absolute quantity found at the bottom of the core could be as low as 1–2 mM or ≥ 30 mM (Figures 9A and 10A).

In contrast, measurements of the rate of sulfate reduction in the sediment were very consistent (Figure 10B). Activity was highest in the top 5 cm (100–225 μmoles L^{-1} d^{-1}) and declined by 100-fold below 10 cm depth. Sulfate reduction in these sediments was inhibited with tungstate; however, the inhibition was only partial. The addition of 50, 100, and 200 mM tungstate to sediments resulted in 81, 89, and 91% inhibition, respectively (R.L. Smith, unpublished data). Integrated areal rates for the sediments were 13.4 mmoles m^{-2} d^{-1} (Table 6), which when combined with the value of 12.0 for the anoxic water column (during meromixis), yielded a value of 25.4 mmoles $SO_4^{=}$ m^{-2} d^{-1}. Assuming that 2 moles of carbon are oxidized for every mole of sulfate reduced,[71] this extrapolates to a value of 18.5 moles C mineralized m^{-2} yr^{-1}. Productivity of Mono Lake has been estimated to be 28–45 moles C m^{-2} yr^{-1}.[32] Hence, sulfate reduction can potentially mineralize 41–66% of the lake's primary productivity occurring during meromixis.

Dissolved hydrogen displayed a generally decreasing trend with depth (Table 5). Absolute concentrations ranged between 5.2 and 27.4 nM. In contrast, freshwater and brackish sulfate-reducing sediments had reported values of 1–2 nM,

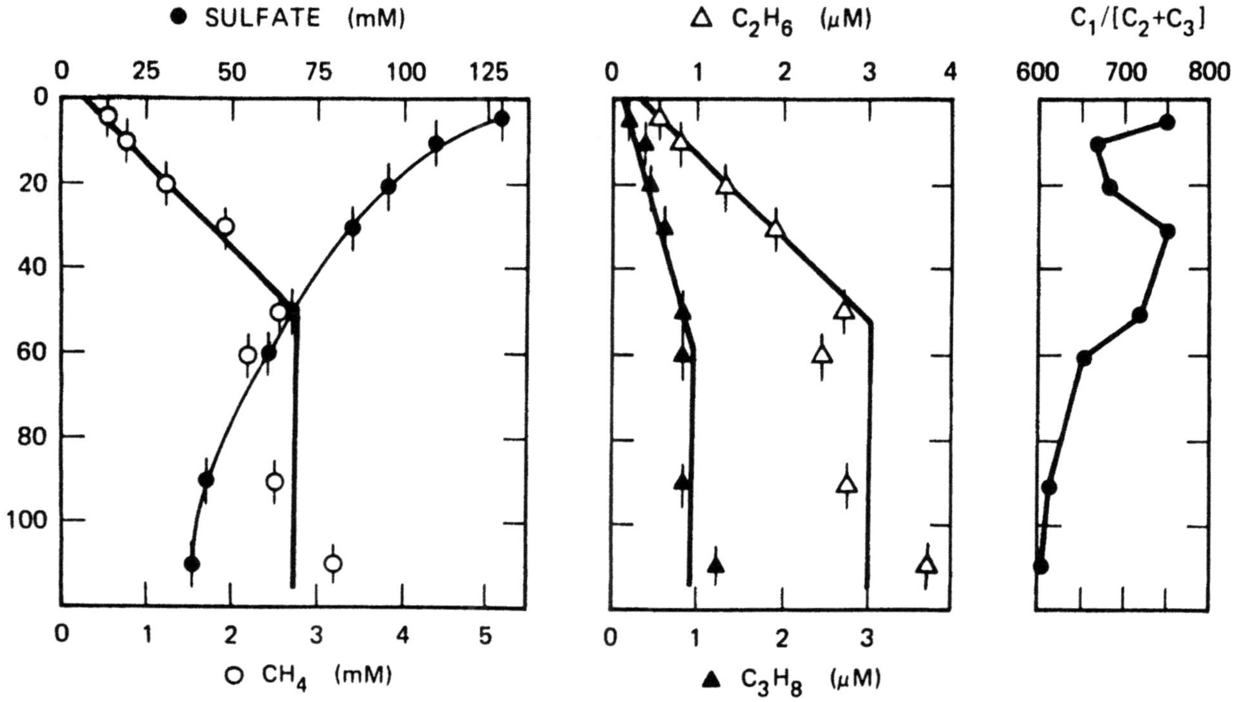

Figure 9. Sediment profiles of (A) dissolved methane and sulfate; B) dissolved ethane and propane; C) ratio of methane/(ethane + propane). Reproduced from Oremland et al. (57) with permission of Geochimica et Cosmochimica Acta.

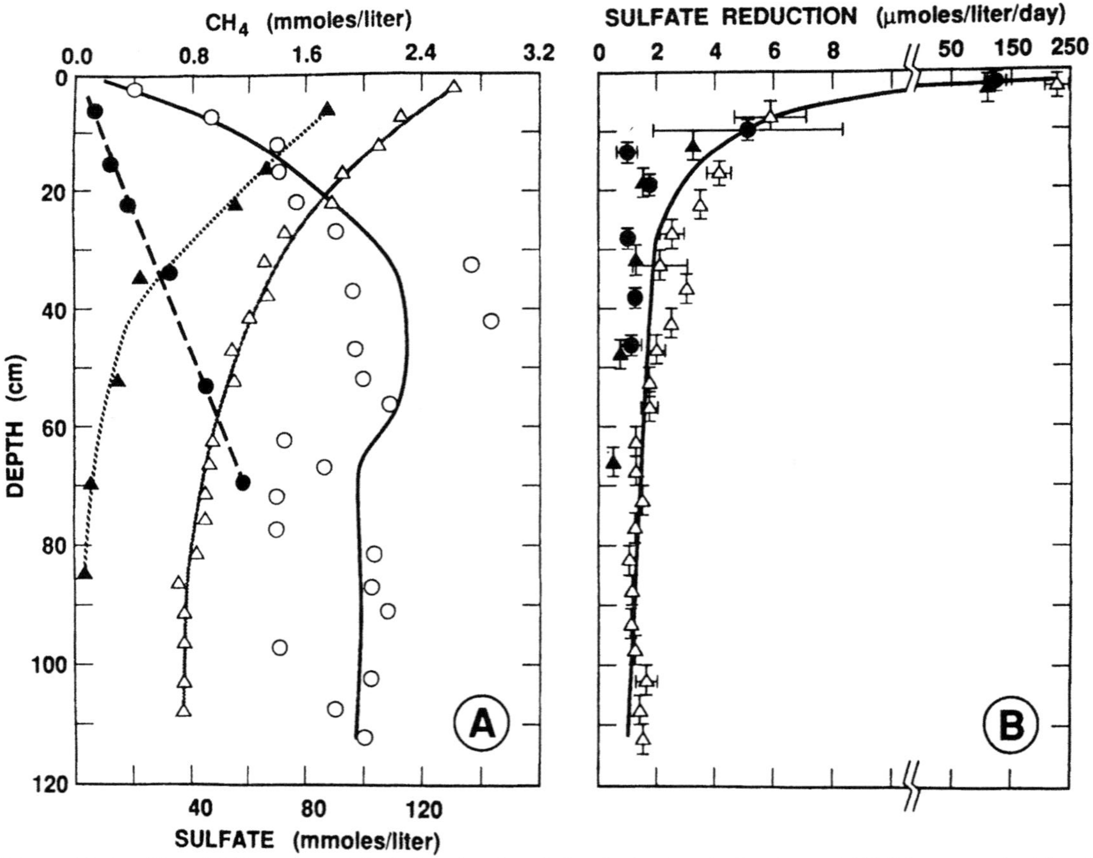

Figure 10. Profiles of (A) methane (circles) and sulfate (triangles) taken during August 1986 (open symbols) and August 1988 (closed symbols); (B) rates of sulfate reduction during August 1986 (△), August 1988 (●), and August 1989 (△). Horizontal bars indicate ± 1 SD from the mean (n = 3).

whereas methanogenic systems were 5–10 nM.[40] The comparison of absolute values of dissolved hydrogen in Mono Lake sediments to anoxic freshwater/brackish systems to ascertain the predominant terminal electron accepting process does not appear to be viable because it would indicate that Mono Lake is a system dominated by methanogenesis rather than sulfate reduction. However, sulfate reduction in Mono Lake is clearly the most important anaerobic mineralization process (see above), whereas methanogenesis is of only minor significance (see below). The results suggest that hydrogen is not as efficiently scavenged by the anaerobic terminal electron acceptors in highly alkaline/hypersaline systems as it is in conventional freshwater/brackish systems. The reason(s) for this observation, as well as for the precipitous decrease in sulfate-reduction rates over a 10-cm sediment interval are unknown, but may be related.

Incubated sediment slurries from Mono Lake demonstrated enhanced methanogenic activity when amended with methylated sulfur compounds[33] or methanol[52] and inhibition could be achieved with 2-bromoethanesulfonic acid.[33,52] Hydrogen only slightly enhanced methanogenesis, whereas, inexplicably, 10 mM trimethylamine (TMA) was strongly inhibitory.[52] Addition of ethylated sulfur compounds resulted in enhanced production of ethane.[56] These results indicated that methanogens were present and potentially active in the sediments, but they shed little light on the overall significance of this activity or the importance of these individual substrates as precursors of methane under in situ conditions.

To pursue this question, ^{14}C-tracer experiments were performed with intact subcores taken from extruded sections of 1-m cores, and gaseous end products were quantified by gas-proportional counting.[24] In a 70-h time course experiments with ^{14}C-labeled acetate and dimethylsulfide (DMS), $^{14}CO_2$ was the major product of DMS metabolism, and $^{14}CO_2$ was the only observed product of acetate metabolism (Figure 11). Production of these gases was linear for the length of the incubation. In a core profile of activity, overall metabolism of acetate, DMS, and TMA was greatest in the upper 12 cm of the core profile (Figure 12). Percent utilization per day for acetate, DMS, and TMA in the top 5-cm section were 36, 4, and 6, respectively. Because sulfate reduction was most extensive in this region (Figure 10) and because $^{14}CO_2$ was the major or sole end product, it appears that sulfate-reducing bacteria, rather than methanogens, were responsible for most of the observed activity. Actual pool sizes for acetate and TMA were ~1 μM, whereas DMS was < 0.1 μM.[52] Hydrolyzable dimethylsulfoniopropionate, the precursor of DMS, was ~112 μM at 0–5 cm, 1.7 μM at 10–15 cm, and undetectable below 20 cm depth (R.P. Kiene, unpublished data from August 1986).

In a subsequent series of experiments, turnover of labeled glucose, acetate, and TMA were followed in a 50-cm-core profile. End-product gases were quantified by oxidation with an FID, followed by trapping and liquid scintillation

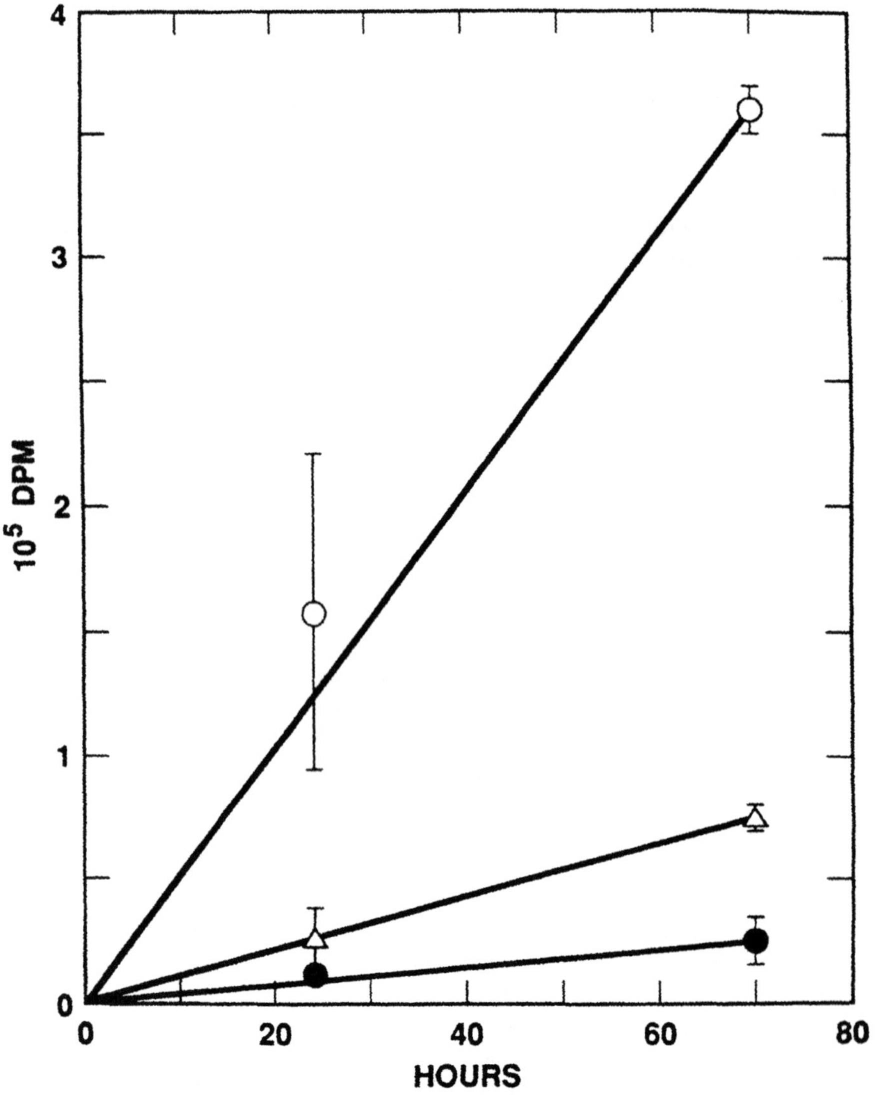

Figure 11. Time course of metabolism of ^{14}C-DMS (1 μCi/4 ml subcore) and 2-^{14}C-acetate (0.5 μCi/4 ml subcore) in experiments conducted during August 1986. DMS samples were from 0 to 5 cm and acetate samples from 60 to 65 cm. Formation of $^{14}CO_2$ (○) and $^{14}CH_4$ (●) from ^{14}C-DMS. Formation of $^{14}CO_2$ from ^{14}C-acetate (△). Symbols represent the mean of three samples and bars indicate ± 1 SD.

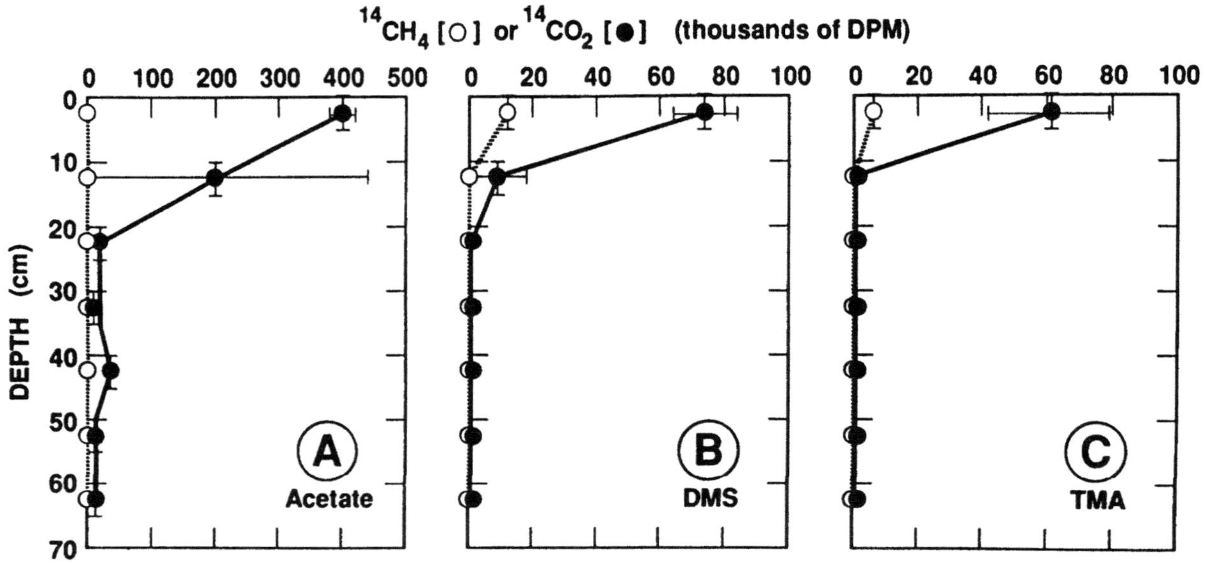

Figure 12. Production of $^{14}CH_4$ (○) and $^{14}CO_2$ (●) in sediment core subsections collected during August 1986 and incubated for 24 h at 5°C. (A) Labeled with 0.5 μCi of 2–^{14}C–acetate; (B) 1 μCi of ^{14}C–DMS; (C) 0.5 μCi of U–^{14}C–TMA. Symbols indicate the means of three samples and horizontal bars indicate ± 1 SD.

counting, a technique which is more sensitive than gas chromatography/gas proportional counting. As was observed in the previous experiment, turnover was most rapid in the upper few centimeters of the sediment profile, with glucose having 16.5% d^{-1} and TMA having 9.2% d^{-1} (Figure 13). For glucose and acetate, methane was a minor product, implying that sulfate reducers were more important in mineralizing these substrates over the length of the core (Figure 14). However, the $^{14}CH_4/^{14}CO_2$ ratios for acetate and glucose increased with depth (Figure 14A), whereas the ratio for TMA was highest and methane was a significant product only at the sediment surface (Figure 14B). These results indicate that in the case of "competitive" substrates like acetate or intermediates of glucose metabolism, methanogens assume a greater relative importance with depth in the sediment. In the case of "noncompetitive" substrates like DMS and TMA,[33,55] methylotrophic methanogens were most active at the sediment surface, although they were "out competed" for these compounds by the sulfate reducers at greater depths. That sulfate reducers can mineralize significant amounts of "noncompetitive" methylated substrates appears to be a substrate–concentration effect. Hence, when micromolar amounts are employed, as in radiotracer experiments, sulfate reducers can predominate over methanogens.[35]

Evaluating the importance of carbonate/bicarbonate reduction as a mechanism for methanogenesis posed technical difficulties because of the very large pool size (decreased steadily from 0.23 M at 0–10 cm to 0.16 M at 40–60 cm) of these anions in the sediment porewaters. To overcome this, high levels of ^{14}C–$NaHCO_3$ were employed (50 μCi injected/9 ml sediment subcore) and samples were incubated for longer time periods (60 h) and the entire $^{14}CH_4$ sample was oxidized to $^{14}CO_2$ in a copper furnace, trapped, and counted by liquid scintillation.[78] A core profile of activity for experiments conducted in August 1988 is shown in Figure 15. Highest activity was detected at 20 cm subbottom depth, which accounted for a daily turnover of at most 2.25×10^{-3}% of the carbonate/bicarbonate pool. This value corresponds to a maximum rate of methanogenesis of 1.7 μmoles L^{-1} d^{-1}, which is in reasonable agreement with experiments conducted in August 1986.[52] Because pool sizes of TMA, DMS, and acetate are quite low, it is suggested that "CO_2 reduction" is the major pathway for methanogenesis in these sediments. The integrated areal rate of methanogenesis from CO_2 reduction occurring in the upper 60 cm was 0.34 mmoles m^{-2} d^{-1} for the August 1988 profile (Table 6). Applying the interstitial carbonate + bicarbonate pools observed in August 1988 to the experiments of August 1986, an areal rate of 0.50 mmoles m^{-2} d^{-1} was calculated. Hence, the two calculations (1986 and 1988) are in reasonable agreement. Mean diffusive flux of methane out of the sediments (upper 10 cm) obtained from 13 core profiles was 0.6 ± 0.5 mmoles m^{-2} d^{-1},[48] which also agrees with the calculated rates of methanogenesis. Anaerobic methane oxidation activity increased with depth and paralleled sediment methane concentrations until a depth of 10 cm, below which rates

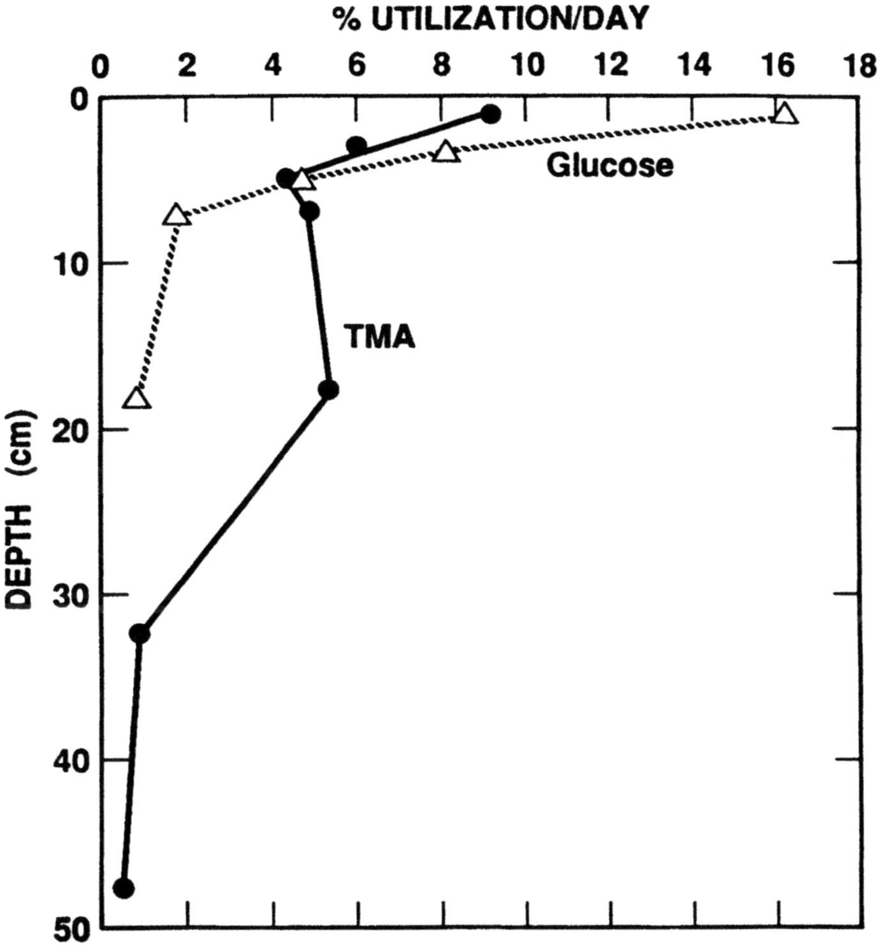

Figure 13. Percent daily utilization of U–^{14}C–glucose (1.5 μCi/2.4 ml sediment) and U–^{14}C–TMA (0.1–0.2 μCi/2.4 ml sediment) with depth in a core collected in November 1986 and incubated at 5°C for 24 h. Results represent the mean of three samples.

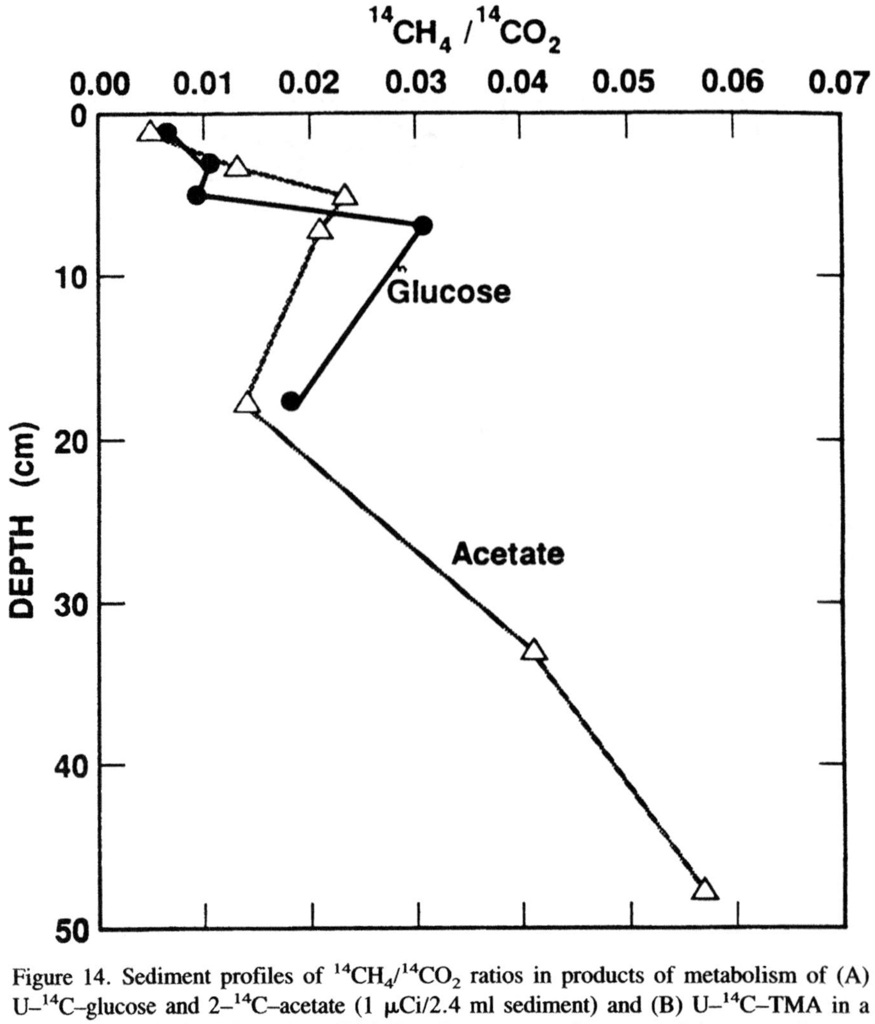

Figure 14. Sediment profiles of $^{14}CH_4/^{14}CO_2$ ratios in products of metabolism of (A) U–^{14}C–glucose and 2–^{14}C–acetate (1 μCi/2.4 ml sediment) and (B) U–^{14}C–TMA in a core collected during November 1986. Results represent the mean of three samples.

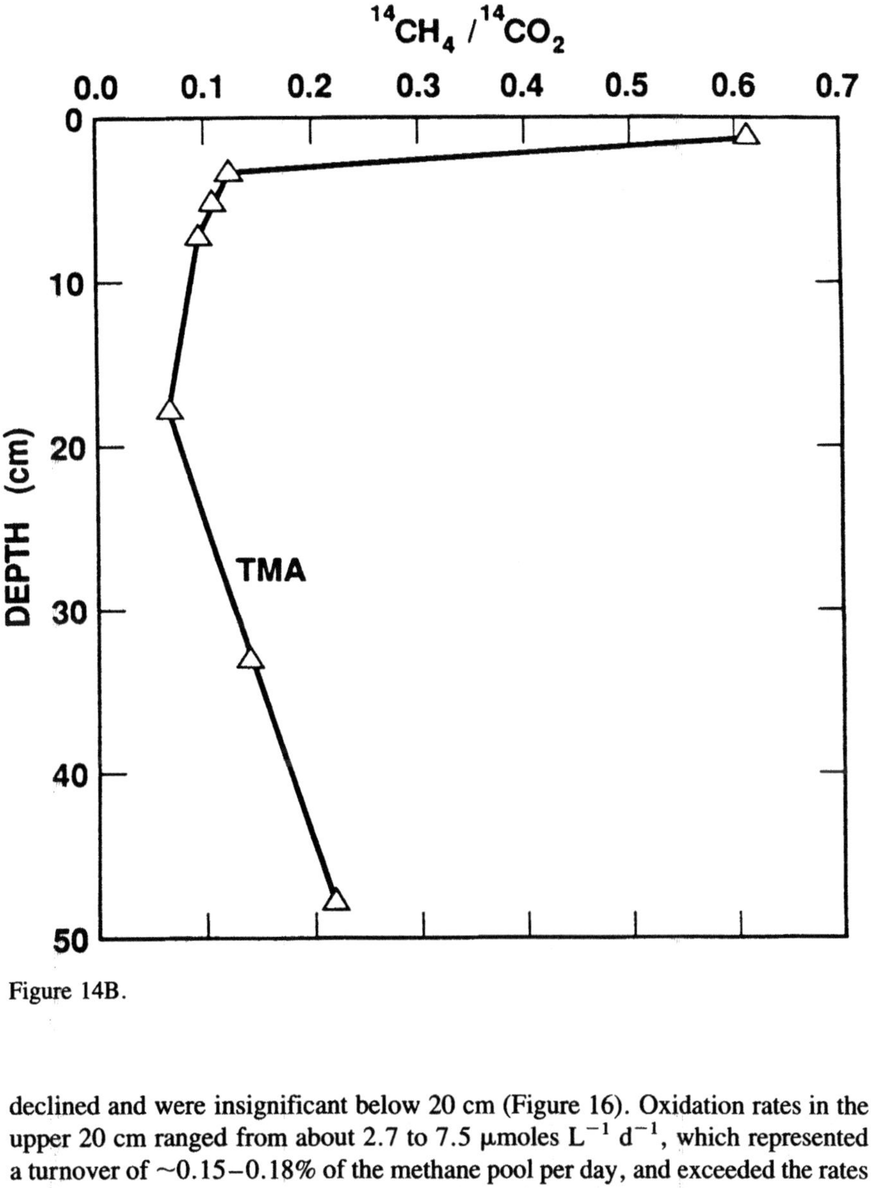

Figure 14B.

declined and were insignificant below 20 cm (Figure 16). Oxidation rates in the upper 20 cm ranged from about 2.7 to 7.5 μmoles L^{-1} d^{-1}, which represented a turnover of ~0.15–0.18% of the methane pool per day, and exceeded the rates of methanogenesis. On an areal basis, oxidation consumed 1.45 mmol CH_4 m^{-2} d^{-1}, a value that was 2.1–3.5 times greater than that calculated for methane formation from carbonate/bicarbonate reduction (Table 6). This value for methane oxidation implies that a net consumption occurred in the upper 20 cm, a facet which complicates the methane balance. Since the mean outward flux of

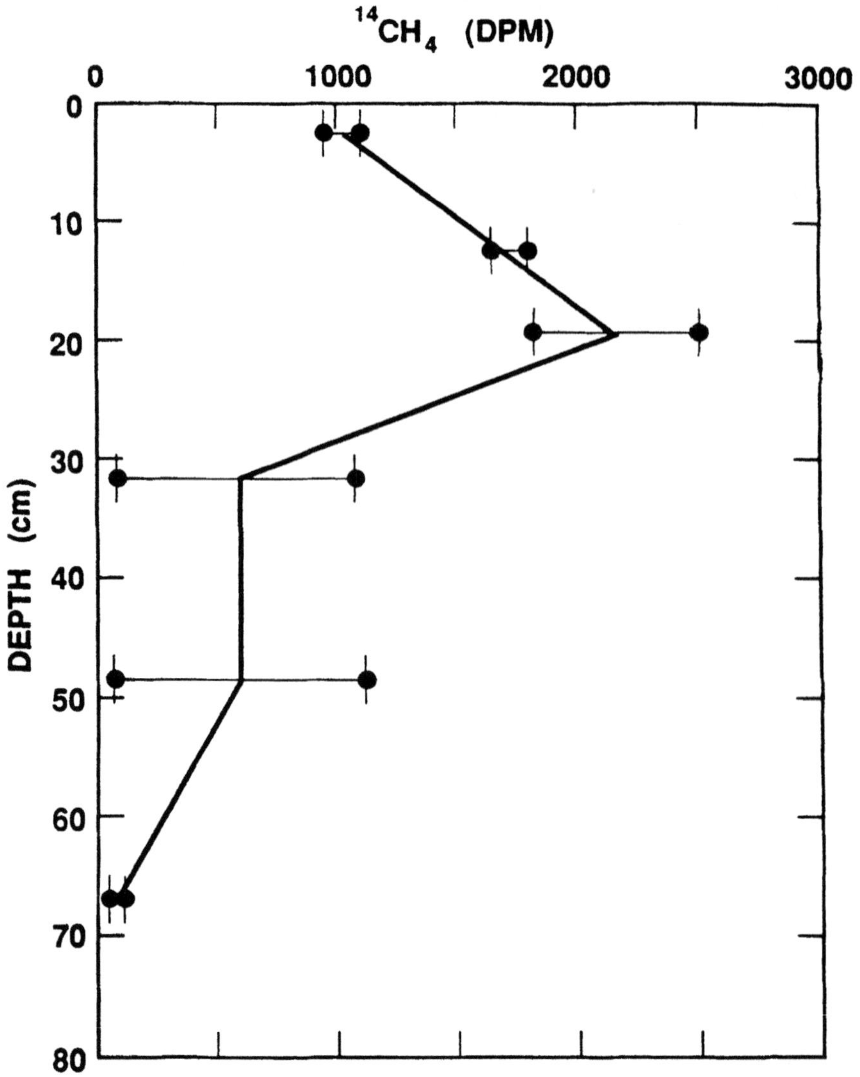

Figure 15. Profile of sediment reduction of ^{14}C–bicarbonate (50 μCi/8 ml sediment) in a core collected in August 1988. Symbols indicate activity in duplicate sets of samples.

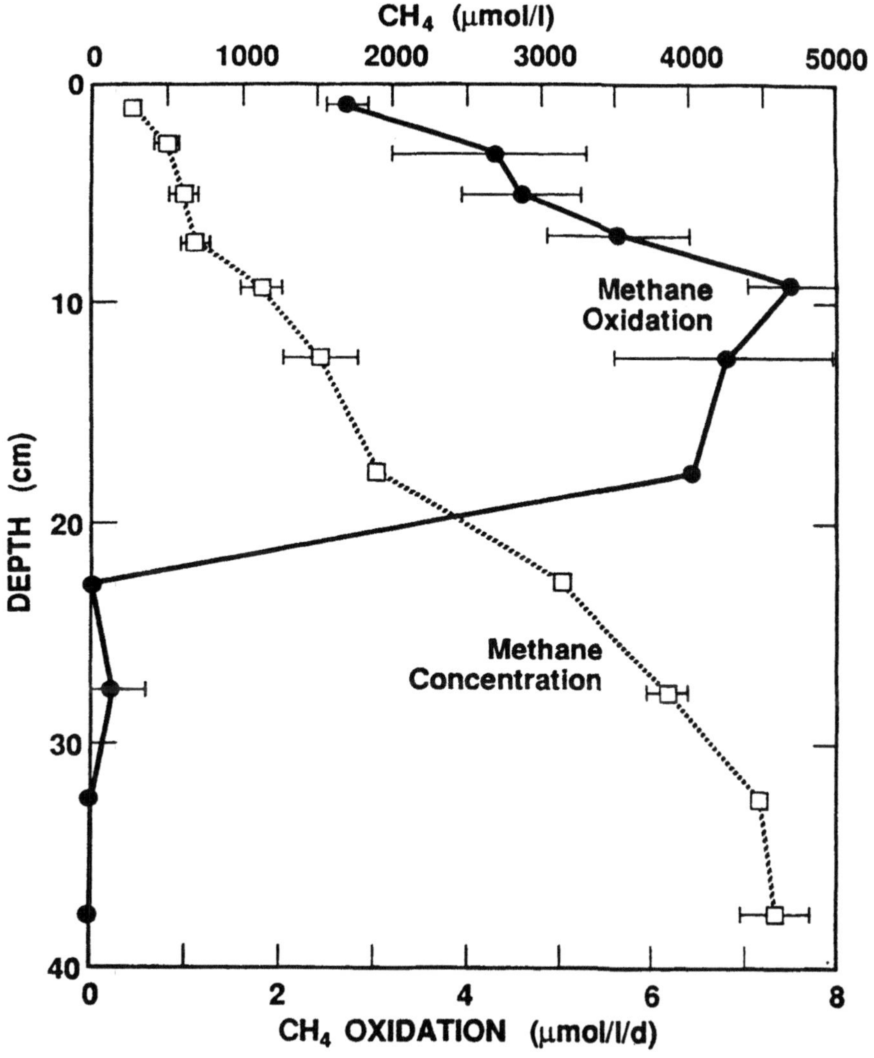

Figure 16. Anaerobic methane oxidation activity in sediment samples collected from a core taken in April 1987. Sediment subsamples (5 ml) were injected with an aqueous solution of 0.5 $\mu Ci^{14}CH_4$ and incubated for 24 h at 5°C. Results represent the mean ± 1 SD of three samples.

methane from the sediments was 0.6 mmol m^{-2} d^{-1}, either the methane consumption term must be balanced by an equivalent flux from below or the rates of methane oxidation measured in this single core were too high. Methane mass balance is given by

$$J\ CH_4\ \text{in} = J\ CH_4\ \text{out} + CH_4\ \text{oxidation} - CH_4\ \text{production},$$

where J out = methane flux out of the sediments, and J in = flux from below into the upper 25 cm of the sediment column. Applying the value of 1.5 mmoles m^{-2} d^{-1} for oxidation, 0.5 mmole m^{-2} d^{-1} for production, and a minimum to maximum range for outward flux (0.6–1.1 mmoles m^{-2} d^{-1}) yields calculated J in values of 1.6–2.1 mmoles m^{-2} d^{-1}. Measures of diffusive flux from below are difficult to obtain because, in many cases, this depth coincides with the region of saturation and bubble formation (Figures 9 and 10). Alternatively, rates of methane oxidation are probably too high because the interstitial concentrations of methane in this core were generally higher than those otherwise observed (Figure 16). It is suggested that additional measures of methane oxidation would yield mean values significantly lower than those calculated from this particular core profile. Because of the variabilities encountered, additional measurements of the rates of methane oxidation and production are warranted in future investigations before a methane balance can be calculated for these sediments.

Conclusions

1. Several sources of methane emanate from Mono Lake and its surrounding basin to the atmosphere. These include natural gas seeps of thermogenic and bacterial origin, as well as methane formed by current methanogenesis in the lake's sediments. Use of gaseous hydrocarbon abundance ratios, stable C isotopes, radiocarbon data, and a consideration of microbial activities (e.g., methane oxidation) permits identification of these sources.
2. Elevated methane and nonmethane hydrocarbons have been detected in ambient air samples at various locations around Mono Lake. Preliminary observations suggest that outgassing from natural gas seeps is the main source for these enrichments rather than flux from the lake's surface water as estimated by Oremland et al.[57] However, more detailed and frequent samplings as well as use of radiocarbon and stable isotopes will be required to discern between various local sources and the characteristic "background" signals for methane in the atmosphere from outside the basin.

3. Microbial activities detected in the anoxic water column include sulfate reduction and methane oxidation.
4. Microbial activities detected in the sediments include sulfate reduction, methanogenesis, and anaerobic methane oxidation. Sulfate reduction is the major process for the mineralization of organic matter in the lake. The methane budget for the lake has not been balanced.

Acknowledgments

We are grateful to K. Kvenvolden, D. Herbst, and D. Des Marais for their helpful review of this manuscript.

References

1. Aikin, A.C., J.R. Herman, E.J. Maier, and G.J. McQuillan. 1982. Atmospheric chemistry of ethane and ethylene, J. Geophys. Res. **87**:3105–3118.
2. Alperin, M.J., W.S. Reeburgh, and M.J. Whiticar. 1988. Carbon and hydrogen isotope fractionation resulting from anaerobic methane oxidation, Global Biogeochem. Cycles **2**:279–288.
3. Alperin, M.J., and W.S. Reeburgh. 1985. Inhibition experiments on anaerobic methane oxidation, Appl. Environ. Microbiol. **50**:940–945.
4. Anderson, L.C.D., and K.W. Bruland 1991. Biogeochemistry of arsenic in natural waters: The importance of methylated species. Environ. Sci. Technol. **25**:420–427.
5. Anderson, R.F., M.P. Bacon, and P.G. Brewer. 1982. Elevated concentrations of actinides in Mono Lake. Science **216**:514–516.
6. Bailey, R.A., G.B. Dalyrymple, and M.A. Lanphere. 1976. Volcanism, structure, and geochronology of Long Valley caldera, Mono County, California. J. Geophys. Res. **81**:725–744.
7. Baker-Blocker, A., T.M. Donahue, and K.H. Mancy. 1977. Methane flux from wetland areas, Tellus **29**:245–250.
8. Barber, T.R., R.A. Burke, Jr., and W.M. Sackett. 1988. Diffusive flux of methane from warm wetlands. Global Biogeochem. Cycles **2**:411–426.
9. Barnes, I., R.W. Kistler, R.H. Mariner, and T.S. Presser. 1981. Geochemical evidence of the basement rocks of the Sierra Nevada, California. USGS Water Supply Paper 2181, U.S. Government Printing Office, Washington, DC.
10. Bartlett, K.B., P.M. Crill, D.I. Sebacher, R.C. Harriss, J.O. Wilson, and J.M. Melack. 1988. Methane flux from the central Amazonian floodplain. J. Geophys. Res. **91**:1571–1582.

11. Blake, D.R., and F.S. Rowland. 1988. Continuing worldwide increase in tropospheric methane, 1978 to 1987. Science **239**:1129–1131.

12. Blake, D.R., and F.S. Rowland. 1986. Global atmospheric concentrations and source strength of ethane. Nature (London) **321**:231–233.

13. Blair, N.E., C.S. Martens, and D.J. Des Marais. 1987. Natural abundances of carbon isotopes in acetate from a coastal marine sediment. Science **236**:66–68.

14. Bonsang, B., M. Kanakidou, and G. Lambert. 1987. Non-methane hydrocarbons chemistry in the atmosphere of an equatorial forest: A case of indirect photochemical production of OH radicals. Geophys. Res. Lett. **14**:1250–1253.

15. Broecker, W.S., R. Wanninkhof, G. Mathieu, T.H. Peng, S. Stine, S. Robinson, A. Herczeg, and M. Stuiver. 1991. The radiocarbon budget for Mono Lake: an unsolved mystery. Earth Planet. Sci. Lett. **88**:16–26.

16. Christensen, M.N., C.M. Gilbert, K.R. Lajoie, and Y. Al Rawi. 1969. Geological–geophysical interpretation of Mono Basin, California–Nevada. J. Geophys. Res. **74**:5221–5239.

17. Cicerone, R.J., and R.S. Oremland 1988. Biogeochemical aspects of atmospheric methane. Global Biogeochem. Cycles **2**:299–327.

18. Cicerone, R.J., J.D. Shetter, and C.C. Delwiche. 1983. Seasonal variation of methane flux from a California rice paddy. J. Geophys. Res. **88**:11022–11024.

19. Cloern, J.E., B.E. Cole, and R.S. Oremland. 1983. Seasonal changes in the chemistry and biology of a meromictic lake (Big Soda Lake, Nevada, U.S.A.). Hydrobiologia **105**:195–206.

20. Cloern, J.E., B.E. Cole, and R.S. Oremland. 1983. Autotrophic processes in meromictic Big Soda Lake, Nevada. Limnol. Oceanogr. **28**:1049–1061.

21. Crill, P.M., K.B. Bartlett, R.C. Harriss, E. Gorham, E.S. Verry, D.I. Sebacher, L. Madzar, and W. Sanner. 1988. Methane flux from Minnesota peatlands. Global Biogeochem. Cycles **2**:371–384.

22. Crill, P.M., K.B. Bartlett, J.O. Wilson, D.I. Sebacher, R.C. Harriss, J.M. Melack, S. MacIntyre, L. Lesack, and L. Smith-Morrill. 1988. Tropospheric methane from an Amazonian floodplain lake. J. Geophys. Res. **93**:1564–1570.

23. Crutzen, P.J., I. Aselman, and W. Seiler. 1986. Methane production by domestic animals, wild ruminants, other herbivorous fauna, and humans. Tellus. **38B**:271–284.

24. Culbertson, C.W., A.J.B. Zehnder, and R.S. Oremland. 1981. Anaerobic oxidation of acetylene by estuarine sediments and enrichment cultures. Appl. Environ. Microbiol. **41**:396–403.

25. Gilbert, C.M., N.M. Christensen, Y. Al-Rawi, and K.R. Lajoie. 1968. Structural and volcanic history of the Mono Basin, California. In R.R. Coats, R.L. Hay, and Anderson (eds.), Studies in Volcanology, Geol. Soc. Amer. Mem. 116, Geological Society of America, Washington DC.

26. Greenberg, J.P., and P.R. Zimmerman. 1984. Non-methane hydrocarbons in remote tropical, continental, and marine atmospheres. J. Geophys. Res. **89**:4767–4778.

27. Greenberg, J.P., P.R. Zimmerman, L. Heidt, and W. Pollock. 1984. Hydrocarbon and carbon monoxide emissions from biomass burning in Brazil. J. Geophys. Res. **89**:1350–1354.

28. Hill, D.P., R.A. Bailey, and A.S. Ryall. 1985. Active tectonic and magmatic processes beneath Long Valley caldera, eastern California: An overview. J. Geophys. Res. **90**:11111–11120.

29. Howes, B.L., J.W. Dacey, and G.M. King. 1984. Carbon flow through oxygen and sulfate reduction pathways in salt marsh sediments. Limnol. Oceanogr. **29**:1037–1051.

30. Iversen, N.I., R.S. Oremland, and M. Klug. 1987. Big Soda Lake (Nevada). 2. Pelagic methanogenesis and anaerobic methane oxidation. Limnol. Oceanogr. **32**: 804–814.

31. Iversen, N.I., and T.H. Blackburn. 1981. Seasonal rates of methane oxidation in anoxic marine sediments. Appl Environ. Microbiol. **41**:1295–1300.

32. Jellison, R., and J.M. Melack. 1988. Photosynthetic activity of phytoplankton and its relation to environmental factors in hypersaline Mono Lake, California. Hydrobiologia **158**:69–88.

33. Kiene, R.P., R.S. Oremland, A. Catena, L.G. Miller, and D.G. Capone. 1986. Metabolism of reduced methylated sulfur compounds by anaerobic sediments and a pure culture of an estuarine methanogen. Appl. Environ. Microbiol. **52**:1037–1045.

34. King, G.M., B.L. Howes, and J.W.H. Dacey. 1985. Sulfate reduction in Georgia salt marsh soils: An evaluation of pyrite formation using ^{35}S and ^{55}Fe tracers. Limnol. Oceanogr. **28**:987–995.

35. King, G.M. 1984. Utilization of hydrogen, acetate, and "non-competitive" substrates by methanogenic bacteria in marine sediments. Geomicrobiol. J. **3**:275–306.

36. King, G.M., M.J. Klug, and D.R. Lovley. 1981. Metabolism of acetate, methanol, and methylated amines in intertidal sediments of Lowes Cove, Maine. Appl. Environ. Microbiol. **45**:1848–1853.

37. Lajoie, K.R. 1968. Quaternary stratigraphy and geologic history of Mono Basin, eastern California, PhD dissertation, University of California, Berkeley.

38. Lenz, P.H., S.D. Cooper, J.M. Melack, and D.M. Winkler. 1986. Spatial and temporal distribution patterns of three trophic levels in a saline lake. J. Plankton Res. **8**:1051–1064.

39. Levine, J.S., W.R. Cofer III, D.I. Sebacher, R.P. Rhinehart, E.L. Winstead, S. Sebacher, C.R. Hinkle, P.A. Schmalzer, and A. Koller, Jr. 1990. The effect of fire on biogenic emissions of methane and nitric oxide from wetlands. J. Geophys. Res. **95**:1853–1864.

40. Lovley, D.R., and S. Goodwin. 1988. Hydrogen concentrations as an indicator of the predominant terminal electron-accepting reactions in aquatic sediments. Geochim. Cosmochim. Acta **52**:2993–3003.

41. Lowe, D.C., C.A.M. Brenninkmeijer, M.R. Manning, R. Sparks, and G. Wallace. 1988. Radiocarbon determination of atmospheric methane at Baring Head, New Zealand. Nature (London) **332**:522–525.

42. Manning, M.R., D.C. Lowe, W.H. Melhuish, R.J. Sparks, G. Wallace, C.A.M. Brenninkmeijer, and R.C. McGill. 1990. The use of radiocarbon measurements in atmospheric studies. Radiocarbon **32**:37–58.

43. Martens, C.S., N.E. BLair, C.D. Green, and D.J. Des Marais. 1986. Seasonal variations in the stable carbon isotopic signature of biogenic methane in a coastal sediment. Science **233**:1300–1303.

44. Martens, C.S., and J.V. Klump. 1984. Biogeochemical cycling in an organic-rich coastal marine basin. 4. An organic carbon budget for sediments dominated by sulfate reduction and methanogenesis. Geochim. Cosmochim. Acta **48**:1987–2004.

45. Martens, C.S., G.W. Kipphut, and J.V. Klump. 1980. Sediment–water chemical exchange in the coastal zone traced by *in situ* Radon-222 flux measurements. Science **208**:285–288.

46. Mason, D.T. 1967. Limnology of Mono Lake, California, PhD. dissertation, University California, Berkeley.

47. Miller, L.G., and R.S. Oremland. 1988. Methane efflux from the pelagic regions of four lakes. Global Biogeochem. Cycles. **2**:269–277.

48. Miller, L.G., R. Jellison, R.S. Oremland, and C.W. Culbertson. 1993. Meromixis in hypersaline Mono Lake, California III: Biogeochemical response to stratification and overturn. Limnol. Oceanog. (in press).

49. Oremland, R.S., and C.W. Culberson. 1992. Importance of methane-oxidizing bacteria in the methane budget as revealed by the use of a specific inhibitor. Nature (London) **356**:421–423.

50. Oremland, R.S., and C.W. Culberson. 1992. Evaluation of methyl fluoride and dimethyl ether as inhibitors of aerobic methane oxidation. Appl. Environ. Microbiol. **58**:2983–2992.

51. Oremland, R.S. 1990. Nitrogen fixation dynamics of two diazotrophic communities in Mono Lake, California. Appl. Environ. Microbiol. **56**:614–622.

52. Oremland, R.S., and G.M. King. 1989. Methanogenesis in hypersaline environments. In Y. Cohen and E. Rosenberg (eds.), Microbial Mats: Physiological Ecology of Benthic Microbial Communities. American Society of Microbiology, Washington DC.

53. Oremland, R.S., and D.G. Capone. 1988. Use of "specific" inhibitors in biogeochemistry and microbial ecology. Adv. Microbial Ecology. **10**:285–383.

54. Oremland, R.S., and D.J. Des Marais. 1983. Distribution, abundance, and carbon isotopic composition of gaseous hydrocarbons in Big Soda Lake, Nevada: An alkaline, meromictic lake. Geochim. Cosmochim. Acta **47**:2107–2114.

55. Oremland, R.S., and S. Polcin. 1982. Methanogenesis and sulfate-reduction: Competitive and non-competitive substrates in estuarine sediments. Appl. Environ. Microbiol. **44**:1270–1276.

56. Oremland, R.S., M.J. Whiticar, F.E. Strohmaier, and R.P. Kiene. 1988. Bacterial ethane formation from reduced, ethylated sulfur compounds in anoxic sediments. Geochim. Cosmochim. Acta **52**:1895–1904.

57. Oremland, R.S., L.G. Miller, and M.J. Whiticar. 1987. Sources and flux of natural gases from Mono Lake, California. Geochim. Cosmochim. Acta **51**:2915–2929.

58. Pelagos. 1987. A bathymetric survey of Mono Lake, California, for the Los Angeles Department of Water and Power, Pelagos Corporation, San Diego.

59. Peng, T-H., and W. Broecker. 1980. Gas exchange rates for three closed-basin lakes. Limnol. Oceanogr. **25**:789–796.

60. Reed, W.E. 1977. Biogeochemistry of Mono Lake, California. Geochim. Cosmochim. Acta **41**:1231–1245.

61. Russell, I.C. 1889. Quaternary history of Mono Valley, California. U.S. Geological Survey 8th Annual Report, pp. 261–394.

62. Sansone, F.J., and C.S. Martens. 1981. Methane production from acetate and associated methane fluxes from anoxic coastal sediments. Science **211**:707–709.

63. Schoell, M. 1983. Genetic characterization of natural gases. Amer. Assoc. Petrol. Geol. Bull. **67**:2225–2238.

64. Scholl, D.W., R. Von Huene, P. St. Amand, and J.B. Ridlon. 1967. Age and origin of topography beneath Mono Lake, a remnant Pleistocene lake, California. Bull. Geol. Soc. Amer. **78**:583–600.

65. Schutz, H., A. Holzapfel-Pschorn, R. Conrad, H. Rennenberg, and W. Seiler. 1989. A 3-year continuous record on the influence of daytime, season, and fertilizer treatment on methane emission rates from an Italian rice paddy. J. Geophys. Res. **94**:16405–16416.

66. Seiler, W., R. Conrad, and D. Scharfee. 1984. Field studies of methane emission from termite nests into the atmosphere and measurements of methane uptake by tropical soils. J. Atmos. Chem. **1**:171–186.

67. Seiler, W., A. Holzapfel-Pschorn, R. Conrad, and D. Scharffe. 1984. Methane emission from rice paddies. J. Atmos. Chem. **1**:241–268.

68. Sieh, K., and M. Bursik. 1986. Most recent eruption of the Mono Craters, eastern central California. Geophys. Res. **91**:12539–12571.

69. Simpson, H.J., R.M. Trier, J.R. Toggweiler, G. Mathieu, B.L. Deck, C.R. Olsen, D.E. Hammond, C. Fuller, and T.L. Ku. 1982. Radionuclides in Mono Lake, California. Science **216**:512–514.

70. Simpson, H.J., R.M. Trier, C.R. Olsen, D.E. Hammond, A. Ege, L. Miller, and J.M. Melack. 1980. Fallout plutonium in an alkaline, saline lake. Science **207**: 1071–1073.

71. Smith, R.L., and R.S. Oremland. 1987. Big Soda Lake (Nevada). 3. Pelagic sulfate reduction. Limnol. Oceanogr. **32**:794–803.

72. Sorey, M.L. 1985. Eruption and present state of the hydrothermal system in Long Valley caldera. J. Geophys. Res. **90**:11219–11228.

73. Stine, S. 1990. Late Holocene fluctuations of Mono Lake, eastern California. Paleogeogr. Paleoclimatol. Paleoecol. **78**:333–381.

74. Tyler, S.C., P.R. Zimmerman, C. Cumberbatch, J.P. Greenberg, G. Westberg, and

J.P.E.C. Darlington. 1988. Measurements and interpretation of ^{13}C data from termites, rice paddies, and wetlands in Kenya. Global Biogeochem. Cycles **2**:341–356.

75. Wahlen, M., N. Tanaka, R. Henry, B. Deck, J. Zeglin, J.S. Vogel, J. Southon, A. Shemesh, R. Fairbanks, and W. Broecker. 1989. Carbon-14 in methane sources and in atmospheric methane: The contribution of fossil carbon. Science **245**:286–290.

76. Wanninkof, R., J.R. Ledwell, and W.S. Broecker. 1987. Gas exchange on Mono Lake and Crowly lake, California. J. Geophys. Res. **92**:14567–14580.

77. Whalen, S.C., and W.S. Reeburgh. 1988. A methane flux time series for tundra environments. Global Biogeochem. Cycles **2**:399–409.

78. Whelan, J.K., R.S. Oremland, M. Tarafa, R.L. Smith, R. Howarth, and C. Lee. 1986. Evidence for sulfate reducing and methane producing microorganisms from sites 618, 619, and 622. In A. Buoma et al. (eds.), Initial Reports of the Deep Sea Drilling Project Vol 96, U.S. Government Printing Office, Washington DC, pp. 767–775.

79. Whiticar, M.J., E. Faber, and M. Schoell. 1986. Biogenic methane formation in marine and freshwater environments: CO_2 reduction vs. acetate fermentation–isotope evidence. Geochim. Cosmochim. Acta **49**:2069–2080.

80. Zimmerman, P.R., J.P. Greenberg, and C.E. Westberg. 1988. Measurements of atmospheric hydrocarbons and biogenic emission fluxes in the Amazon Boundary layer. J. Geophys. Res. **93**:1407–1416.

Biogenesis and Catabolism

Bacterial Transformations of Organic Sulfur Compounds in Marine Environments

Barrie F. Taylor

Rosenstiel School of Marine and Atmospheric Science, University of Miami, Miami, Florida 33149–1098

ABSTRACT

Inputs of biogenic sulfur into the atmosphere from marine environments are largely organic, as dimethyl sulfide (DMS), in contrast to mainly inorganic contributions from terrestrial regions. DMS is photochemically oxidized in the troposphere to sulfate and methane sulfonic acid (MSA) which serve as nuclei for cloud formation, thereby affecting climate. DMS derives from the osmolyte dimethylsulfoniopropionate (DMSP) of some marine phytoplankton, macroalgae, and marsh grasses. Lesser amounts of DMS originate in transmethylation reactions with sulfide and methanethiol using methoxylated aromatic compounds, from lignin, as methyl donors. Marine plants directly liberate some DMS, but most is produced after release of DMSP into the environment by senescence, lysis, and zooplankton feeding and digestion. Bacteria that cleave DMS from DMSP function in oxic and anoxic environments. Net entry of DMS to the marine atmosphere is influenced by bacteria which demethylate DMSP or oxidize DMS. DMSP is sequentially demethylated to 3-methiolpropionate (MMPA) and 3-mercaptopropionate (MPA) in oxic and anoxic environments; bacteria also aerobically demethiolate MMPA to give methanethiol. DMS is aerobically oxidized by methylotrophic hyphomicrobia and autotrophic thiobacilli although isolates from marine environments have received little attention. In anoxic marine sediments, DMS is oxidized to CO_2 during sulfate reduction and is also metabolized by methylotrophic methanogens. Understanding the environmental factors that regulate the fluxes of volatile organic sulfur compounds, most notably DMS, to the marine atmosphere requires more knowledge on the ecology and physiology of microorganisms that synthesize, transform, and degrade DMSP and DMS.

Introduction

Upon examining bacterial transformations of organic sulfur compounds in natural environments, it becomes apparent that major differences between marine and freshwater or terrestrial regions arise from the presence of dimethylsulfoniopropionate (DMSP) in marine plants. DMSP is an osmolyte which is synthesized by some species of marine phytoplankton, macroalgae, and marsh plants.[22,57,101,141] Cytoplasmic concentrations of DMSP may reach molar levels and its release into the environment has a quantitative impact on the marine and global biogeochemistry of sulfur. Biodegradation of DMSP generates dimethyl sulfide (DMS). Although there are many uncertainties in global estimates,[3] it is generally agreed that DMS, not H_2S, is the most quantitatively important biogenic sulfur gas entering the atmosphere.[2,60] However, there is still a need for much work to understand mechanisms which control the release of H_2S from natural environments.[2] Current estimates are that DMS provides over 50% of the volatile biogenic sulfur entering the atmosphere, with about 95% of it issuing from marine environments.[60] Figure 1 outlines the main transformations of the sulfur cycle and connections between organosulfur compounds and sulfur gases with biological sources.

Possible links between DMS and climate[16] have stimulated research into microbial processes that produce and consume DMS. DMS molecules persist in the troposphere for only days,[66] being rapidly oxidized to sulfuric acid and methane sulfonic acid (MSA).[2] Sulfuric acid and MSA, as strong acids, attract water and promote cloud formation. The albedo or reflectivity of clouds lowers global temperatures and inverse correlations between MSA concentrations in ice cores and the Earth's temperature have been observed.[81]

In addition to H_2S, the inorganic sulfur gases carbon disulfide (CS_2) and carbonyl sulfide (COS) are biogenic components of the sulfur trace gas fraction of the atmosphere. COS occurs at about 500 pptv in the atmosphere as compared to 10–100 pptv and 1–10 pptv for DMS and H_2S, respectively.[22] COS has a residence time of 1 year[2] and penetrates into the stratosphere where it is oxidized to a sulfate aerosol layer that reflects solar radiation.[127]

Dominant bacterial processes in the sulfur cycle are those of recycling and reworking inorganic and organic components within sediments or the water column. This situation is well-illustrated for the transformations of inorganic sulfur compounds in marine sediments by the studies of Jorgensen and his colleagues.[49–52] The marine environment, with respect to bacterial transformations of organic sulfur compounds, can be considered as pelagic and benthic compartments. Clearly, materials pass between these major compartments, and volatile components escape into the atmosphere. However, net fluxes to the marine atmosphere are only a small fraction of the material turning over within the water column or sediment; concealed is a complexity of competing and interacting bacterial transformations of organic sulfur compounds.[21,62,64,134] Kiene[62] could not detect DMS in the head space over sediment cores from a salt marsh until metabolic inhibitors were added. DMS turnover within the surface layer of the ocean exceeds that in the atmosphere by a factor of about 10,[66] and the rate of DMS consumption by microorganisms within a stratified salt pond was approximately seven-fold higher than losses to the atmosphere.[21]

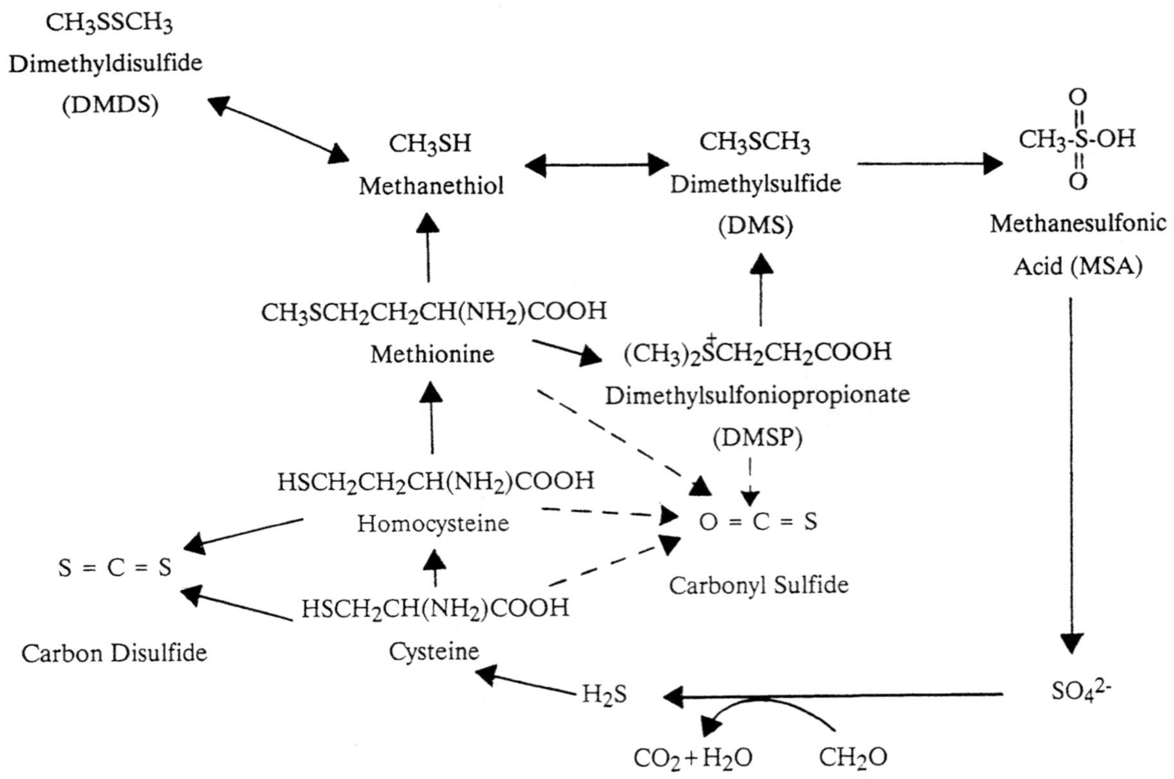

Figure 1. Principal transformations of biogeochemically important organosulfur compounds. Dotted lines show photochemical conversions in surface ocean waters. DMS is oxidized to methanesulfonic acid (MSA) in the atmosphere. CH_2O refers to organic matter which is oxidized anaerobically to CO_2 by sulfate-reducing bacteria in marine sediments to form H_2S, some of which is assimilated by plants and microorganisms to make cysteine and, thus, homocysteine, methionine, and DMSP.

Production of C_1–Organosulfur Compounds

Kelly and Baker[59] drew attention to the importance of C_1–sulfur compounds in the biogeochemical cycling of sulfur. This group of molecules includes DMS, methanethiol, dimethyl disulfide (DMDS), and their oxidized products MSA, dimethyl sulfoxide (DMSO), and dimethylsulfone. The bacterial formation and metabolism of methylated sulfides has been studied in coastal marine and salt marsh sediments, and estuarine and oceanic waters. The principal transformations are summarized in Figure 2. Quantitatively important precursors are DMSP and methionine but 3-methiolpropionate (MMPA), S-methylcysteine, and S-methylmethionine may also participate. DMSP is the main precursor of DMS in marine environments being subject to enzymatic and, in some microhabitats, to chemical cleavage to yield DMS (Figure 3). Methionine, MMPA, and S-methylcysteine initially yield methanethiol, whereas S-methylmethionine generates DMS. Enzymes which generate methylated sulfides from organosulfur compounds are described in Table 1.

Dimethylsulfoniopropionate as a Precursor

DMSP is biosynthesized by some species of microalgae, macroalgae, and higher plants which grow in saline environments. DMSP occurs in several species of green and red seaweeds, and the salt marsh grass *Spartina alterniflora* but not in the red mangrove *Rhizophora mangle*.[22] DMSP was reported in the temperate seagrass *Zostera*,[141] but we have detected DMSP only in association with epiphytized leaves of the subtropical–tropical seagrass *Thalassia testudinum*.[29] DMSP-producing microorganisms growing on marine plants and sediment surfaces[132] undoubtedly contribute to DMSP entering the sediments and the water column. Kiene[62] recorded levels of DMSP up to 190 μmol/L of sediment in the top 5 cm of sediment from a *Spartina* salt marsh and about 50 μmol/L of sediment at 25 cm depth. Although nearly all of the DMS emanating to the atmosphere from salt marshes comes directly from foliage,[22] DMSP will enter the sediments through the plant roots and from microorganisms growing on the sediment surface.

Radioisotopic labeling experiments with the green macroalga *Ulva lactuca* identified the methyl group and sulfur atom of methionine as precursors of the DMS group of DMSP, with the C_2 of methionine forming the C_1 of DMSP.[43] The biosynthesis of DMSP by bacteria has not been reported. Bacteria may not synthesize DMSP, but they probably acquire exogenous DMSP for use as an osmolyte in saline environments. DMSP substitutes for glycine betaine in *E. coli* exposed to high salinity.[15] Apparently, the uptake system for glycine betaine accepts structurally similar molecules, such as DMSP and acetothetin ($(CH_3)_2S^+CH_2COO^-$),[15] and this may be characteristic of other bacteria.

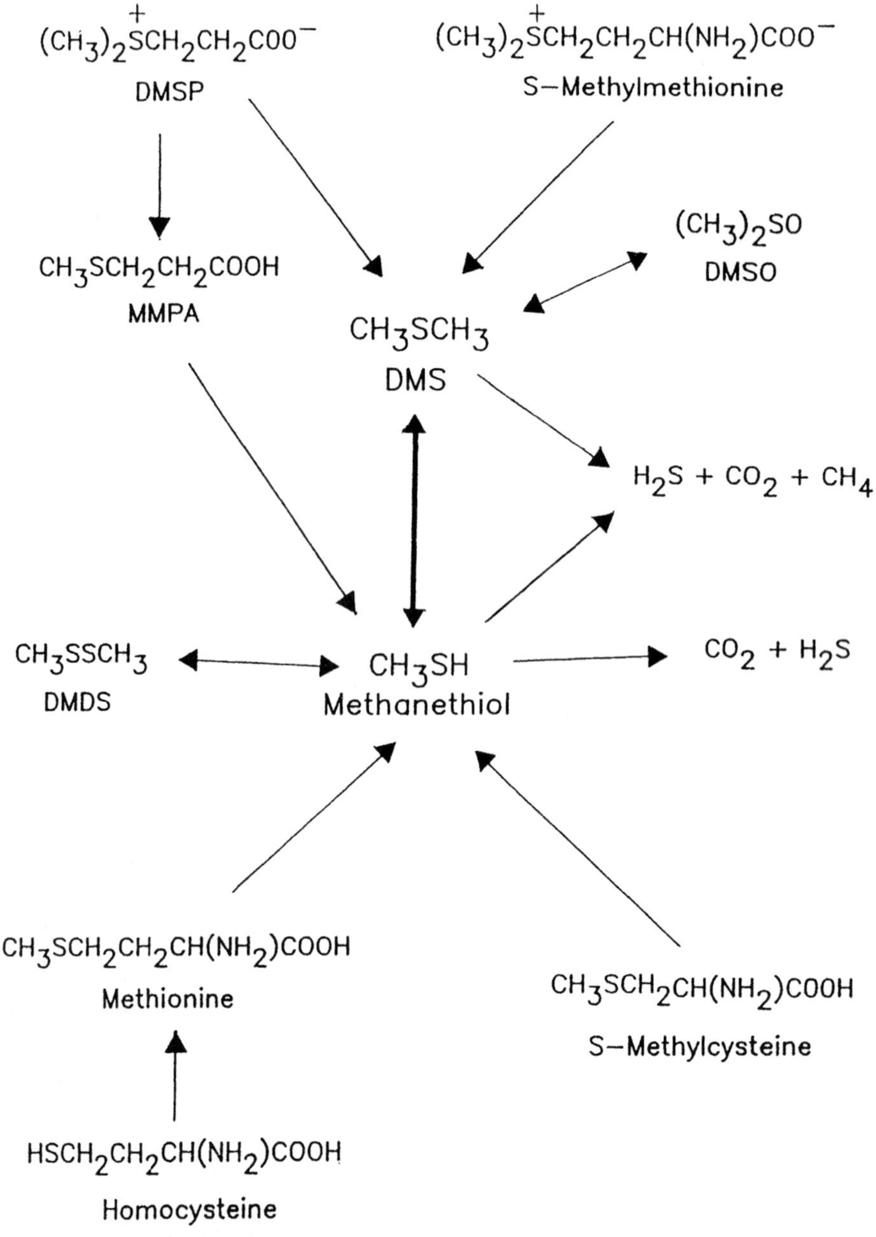

Figure 2. Principal routes for bacterial formation of methylated sulfur compounds from organosulfur precursors, and the interconversions and degradation of the methylated sulfides.

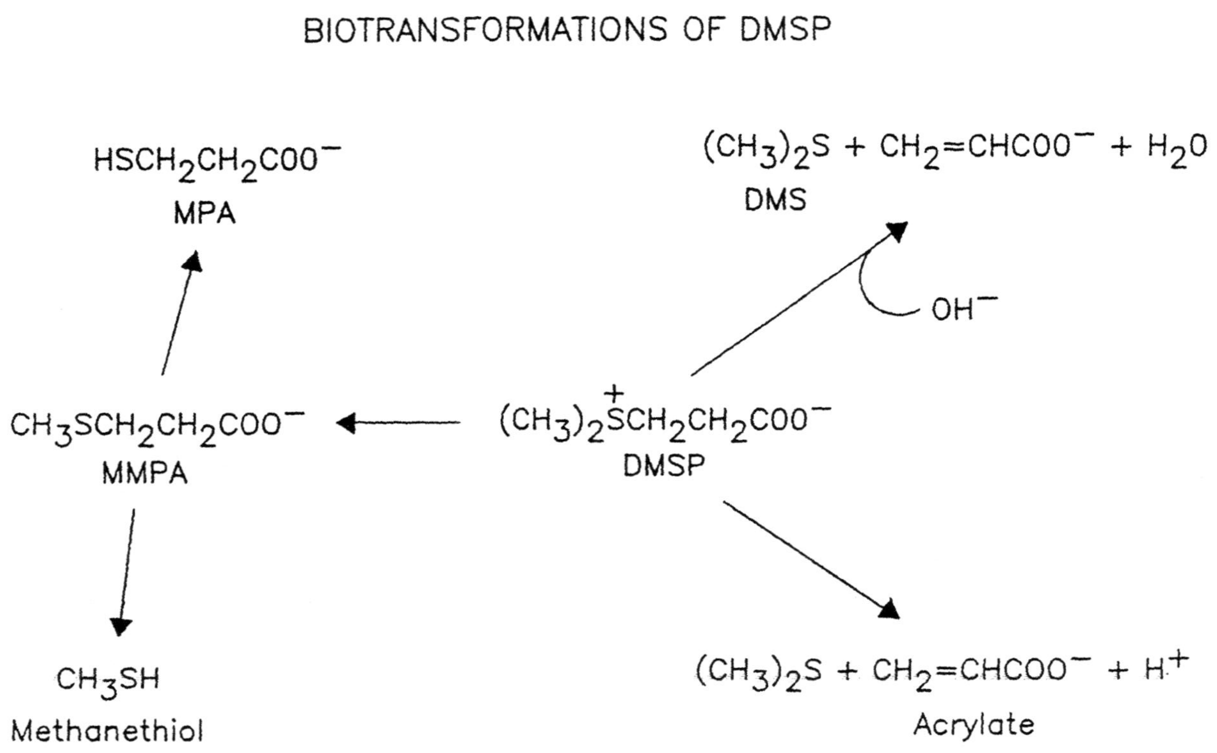

Figure 3. Biotransformations of DMSP. (1) Chemical alkaline decomposition which may occur in environments with high rates of photosynthesis; (2) DMSP lyase giving cleavage to DMS, acrylate, and a proton; (3) Enzymatic demethylation of DMSP to MMPA; (4) enzymatic demethylation of MMPA to MPA; (5) aerobic demethiolation to methanethiol.

Table 1. Enzymes producing methylated sulfides from organosulfur compounds.

Enzyme	Substrate(s)	Products	Inhibitor/ Coenzyme	Organisms
DMSP Lyase	DMSP	DMS	PCMB	Marine bacteria and algae
Sulfonium Lyase	S-methylmethionine	DMS	—	*Corynebacterium*–p.
Trimethylsulfonium–THFA methyltransferase	$(CH_3)_3S^+$	DMS	—	*Pseudomonas* MS
Methionine γ-lyase	Methionine	CH_3SH	PCMB,	Aerobic and anaerobic
	S-methylcysteine	CH_3SH	PALP	bacteria
S-alkylcysteinase	S-methylcysteine	CH_3SH	PALP	*Pseudomonas cruciviae*
Thiol S-methyltransferase	H_2S	CH_3SH	—	Ubiquitous
	CH_3SH	DMS		
MMPA-ase	MMPA	CH_3SH	—	Aerobic
	S-methylcysteine	CH_3SH		marine bacteria

NOTE: PCMB = *p*-chloromercuribenzoate; PALP = pyridoxal phosphate; MMPA = 3-Methiolpropionate.

DMS is produced from DMSP, after its release from marine plants, mainly by the activities of bacteria even though only algal enzymes which catalyze its decomposition have only been studied from *Polysiphonia lanosa* and *Gyrodinium cohnii.*[13,55] DMS is enzymatically cleaved from DMSP with the accompanying production of acrylate and a proton:

$$(CH_3)_2S^+CH_2CH_2COO^- = (CH_3)_2S + CH_2{=}CHCOO^- + H^+.$$

Aerobic bacteria effecting this transformation have been isolated from coastal and open-ocean seawater and marine sediments.[20,28,63,121] There is one report of a strict anaerobe, a *Clostridium* species from river mud, which anaerobically produced DMS from DMSP.[137] DMS was cleaved from DMSP and not further metabolized, but acrylate was fermented to a mixture of acetate and propionate. The use of acrylate but not DMS for growth is characteristic of bacteria with DMSP lyase(s). Acrylate is an inexpensive but selective substrate for the enrichment and isolation of marine bacteria that cleave DMSP.[63,123] DMSP is the only natural source of acrylate in marine environments and its metabolism may be genetically linked to DMSP lyase(s) in marine bacteria.

DMSP, in addition to its pivotal role in the sulfur cycle, may be an important precursor of volatile fatty acids in marine sediments. Acrylate derived from DMSP probably makes a significant contribution to the acetate pool in some marine environments. Using anoxic slurries of coastal marine sediments, Kiene and Taylor[71] observed the fermentation of acrylate to a 1 : 2 molar mixture of acetate and propionate when molybdate[96,124] was added to prevent acetate and propionate consumption by sulfate-reducing bacteria. Propionate may also contribute to the acetate pool, probably being metabolized by *Desulfobulbus* species which are the dominant sulfate reducers in estuarine and marine sediments.[78,142] Enzymatic mechanisms for acrylate fermentation have not been investigated in marine bacteria, but possible pathways are outlined in Figure 4. A well-recognized route involves acrylyl-CoA conversion to acetate via lactyl CoA and pyruvate.[42] A novel pathway implicates β-hydroxypropionate as an intermediate with its oxidation to malonate and subsequent decarboxylation to acetate.[107] The decarboxylation of malonate to acetate was recently reported for strict anaerobes from anoxic marine and freshwater muds and previously from termites.[10,25,26]

The anaerobic metabolism of DMSP gave mainly DMS and small amounts of methanethiol in anoxic slurries of salt-marsh sediments.[67,72] DMS production from DMSP was blocked by autoclaving or glutaraldehyde but was generally insensitive to metabolic inhibitors [molybdate, 2-bromoethanesulfonate (BES), chloramphenicol, cycloheximide] and it was suggested that DMSP cleavage was probably by an extracellular enzyme.[72] The continued production of DMS from DMSP in ocean water after filtration (0.2-μm pores) also indicated an extracellular enzyme or cell rupture during filtration.[129] Similar treatment of estuarine water slowed DMS production from DMSP for about 12 h, after which the original rate resumed.[63] Location of DMSP lyase(s) in the periplasm, for organisms that degrade DMSP, would allow capture of energy released during the elimination reaction as a proton gradient.[123] Such organisms, unlike those using DMSP as an osmolyte, would lack a transport system for DMSP. Some bacteria use exogenous DMSP for growth and osmotic purposes,[28] perhaps possessing an extracytoplasmic DMSP lyase and an uptake system for DMSP which only functions under osmotic stress.

In addition to possible extracellular functioning of DMSP lyase there is an indication of genetic heterogeneity, at least between algal and bacterial enzymes. Kiene and Service[68] could not inhibit DMSP cleavage, presumably due to bacteria, in estuarine water samples with *p*-chloromercuribenzoate (pCMB), iodoacetamide, or azide. In contrast, these inhibitors were effective with the enzyme in cell-free extracts of a dinoflagellate, *Gyrodinium cohnii*.[55]

Methanethiol, not DMS, is generated in an alternative route for DMSP catabolism; DMSP is initially demethylated to MMPA which is then demethiolated to liberate methanethiol[121] (Figure 3). Bacteria that use this pathway were isolated from a variety of marine environments and from cultures of marine phytoplankton. Enrichments with DMSP are the sole source of carbon and energy selected for organisms that cleaved DMS from DMSP, whereas MMPA enrichments secured bacteria that grew on MMPA or DMSP with the formation of methanethiol.[136] The isolates also demethylated MMPA because low levels of

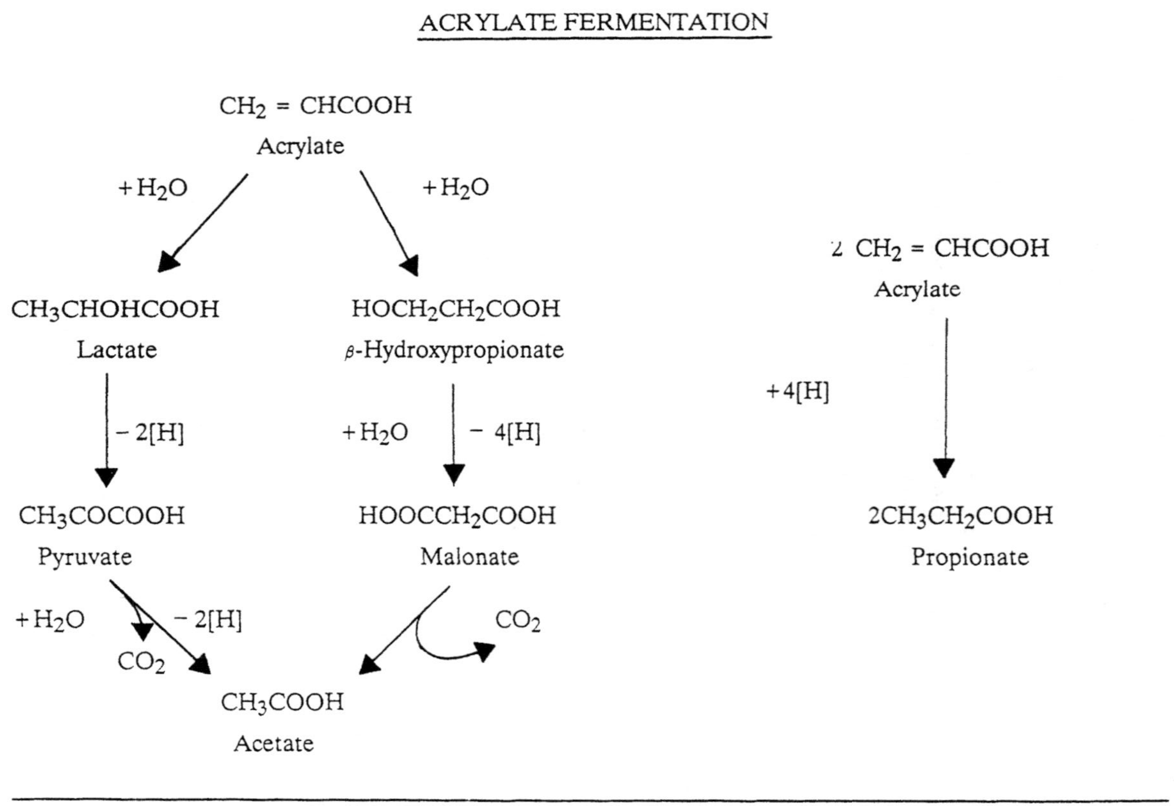

$$3\,CH_2 = CHCOOH + 2H_2O \longrightarrow 2CH_3CH_2COOH + CH_3COOH$$

Figure 4. Possible pathways for fermentation of acrylate.

3-mercaptopropionate (MPA) were detected during growth on MMPA. Demethiolation of MMPA only proceeded aerobically and activity was weak with methionine, but S-methylcysteine was attacked at about 50% of the rate of MMPA.[122] Demethylation of DMSP to MMPA was previously observed with extracts of *Gyrodinium cohnii* using homocysteine as the methyl acceptor to produce methionine.[55] The cofactor involved in the transmethylation was not identified, but a folate is a likely candidate as in the demethylation of trimethylsulfonium chloride.[138] MMPA was previously identified in plants and as a metabolite of methionine in rat liver.[115,116] In rat liver tissues, methionine was converted to its keto acid, 2-keto-4-methiolbutyrate which was then oxidatively decarboxylated to MMPA; MMPA was further metabolized to sulfate and CO_2, and methanethiol and H_2S were detected as intermediates.[115,116] The demethiolation mechanism for MMPA was not determined, but in aerobic marine bacteria the transformation could occur either reductively or oxidatively:

$$CH_3SCH_2CH_2COOH + 2[H] = CH_3SH + CH_3CH_2COOH,$$
$$CH_3SCH_2CH_2COOH + \tfrac{1}{2}O_2 = CH_3SH + CH_3COCOOH + H_2O$$

The bacterial transformations of DMSP are outlined and summarized in Figure 3. The alkaline hydrolysis of DMSP [20] may occur in some natural environments such as marine microbial mats,[132] where intense rates of photosynthesis generate microhabitats with high pH. Chemical decomposition of DMSP probably involves an E2 elimination mechanism with the abstraction of a proton from the α-carbon atom; first-order kinetics for DMS production with respect to DMSP and OH^- support this mechanism.[20] The enzymatic reaction probably also proceeds with proton abstraction because diethylsulfoniopropionate (DESP), but not dimethylsulfonio-2-methylpropionate, is subject to both chemical and microbial cleavage.[122]

Biogenesis from Methionine

Methionine is mainly converted to methanethiol, with some DMS and DMDS, in salt-marsh sediments,[67,72,74] as was previously shown for anaerobic metabolism in soil,[38] lake sediments,[150] and the rumen.[105] In all environments, methionine is demethiolated to give methanethiol and there is no evidence for demethylation. In sediments from a freshwater lake, methionine was degraded, even at a low concentration (2 μM), to both methane and CO_2.[150] In salt-marsh sediments, in contrast, $^{14}CH_3$–methionine at similar levels is recovered as $^{14}CO_2$ with little formation of $^{14}CH_4$.[7,72,143]

Studies have concentrated on anaerobic transformations of methionine in marine sediments even though aerobic demethiolation is common in soil bacteria and fungi.[104,106] Organisms involved in methanethiol production from methionine in marine environments have not been isolated. Anaerobic demethiolation of methionine in salt-marsh sediment slurries, in contrast to DMSP cleavage, was sensitive to molybdate and chloramphenicol but not the methanogenic inhibitor BES.[45,72] Molybdate only partially inhibited methanethiol formation and presumably other anaerobes (but not methanogens), as well as sulfate-reducing bacteria, participated in demethiolation.[72]

Microbial processes for methanethiol formation from methionine are summarized in Table 2. Four different mechanisms have been detected, one of which is solely aerobic. Methionine γ-lyase catalyzes the simultaneous αγ-elimination of NH_3 and methanethiol from methionine (Table 2, reaction 1). Methionine γ-lyase has been purified from several species of nonmarine bacteria, including aerobes and anaerobes,[34] but neither the enzyme nor the transformation has been described for marine bacteria. There is, however, evidence for a pathway with deamination of methionine prior to reductive demethiolation, a mechanism previously described for soil bacteria and fungi (Table 2, reaction 2).[104,106] 2-Keto-4-methiolbutyrate stimulated methanethiol formation at similar rates to methionine in anoxic slurries of coastal marine sediments suggesting deamination prior to demethiolation.[73] A peculiar aspect of 2-keto-4-methiolbutyrate metabolism is its aerobic conversion to ethylene by a variety of bacteria.[85] The transformation requires oxygen and generates ethylene and methanethiol:

$$CH_3SCH_2CH_2COCOOH + \tfrac{1}{2}O_2 = CH_3SH + CH_2{=}CH_2 + 2CO_2.$$

There have been reports of 2-aminobutyrate and methanethiol formation from methionine, indicating reductive demethiolation without the elimination of ammonia (Table 2, reaction 3).[104] Neither the aerobic demethiolation of 2-keto-4-methiolbutyrate with ethylene formation nor methionine degradation to aminobutyrate have been reported for marine bacteria or environments.

Other Organosulfur Precursors

Other organosulfur compounds which generate C_1–organosulfur compounds in marine environments include S-methylcysteine, S-methylmethionine, and MMPA. S-methylmethionine and S-methylcysteine are characteristic of terrestrial plants,[41,46] but their metabolism by marine bacteria suggests that terriginous inputs of these compounds may contribute to the production of sulfur gases in coastal marine regions. MMPA is derived from DMSP by bacterial activities in aerobic and anaerobic marine environments.[70,121]

Methanethiol was formed from S-methyl cysteine in slurries of salt-marsh sediments.[67] Little is known about the metabolism of S-methylcysteine by ma-

Table 2. Microbial processes for methanethiol formation from methionine.

1.	$CH_3SCH_2CH_2CH(NH_2)COOH + H_2O$	=	$CH_3SH + NH_3 + CH_3CH_2COCOOH$
			2-Ketobutyrate
2.	$CH_3SCH_2CH_2CH(NH_2)COOH + H_2O$	=	$CH_3SCH_2CH_2COCOOH + NH_3 + 2[H]$
			2-Keto-4-methiolbutyrate
	$CH_3SCH_2CH_2COCOOH + 2[H]$	=	$CH_3SH + CH_3CH_2COCOOH$
			2-Ketobutyrate
3.	$CH_3SCH_2CH_2CH(NH_2)COOH + 2[H]$	=	$CH_3SH + CH_3CH_2CH(NH_2)COOH$
			2-Aminobutyrate
4.	$CH_3SCH_2CH_2COCOOH + 1/2\ O_2$	=	$CH_3SH + CH_2 = CH_2 + 2CO_2$
			Ethylene

rine bacteria, but an enzyme which demethiolates S-methylcysteine was partially purified from an aerobic soil bacterium, *Pseudomonas cruciviae* (Table 1).[92] Mechanistically, the enzyme resembles methionine γ-lyase in requiring pyridoxal phosphate as a cofactor and catalyzing the elimination of methanethiol and ammonia:

$$CH_3SCH_2CH(NH_2)COOH + H_2O = CH_3SH + CH_3COCOOH + NH_3.$$

The enzyme was active with a range of S-alkylcysteines but methionine was not a substrate. S-Methylcysteine may also be demethiolated in oxic marine environments by bacteria which produce methanethiol from MMPA.[122]

S-Methylmethionine stimulated methanethiol production, after a lag of 2–3 d, in anoxic slurries of coastal sediments.[73] Methanethiol was probably formed via DMS cleaved from S-methylmethionine; a sulfonium lyase that liberates DMS from S-methylmethionine has been described for a soil isolate,[86] *Corynebacterium* sp. (Table 1):

$$(CH_3)_2{}^+SCH_2CH_2CH(NH_2)COOH + H_2O =$$
$$(CH_3)_2S + HOCH_2CH_2CH(NH_2)COOH + H^+.$$

The 2–3-d lag before S-methylmethionine was utilized suggests that organisms with DMSP lyases did not attack S-methylmethionine and degradation could not occur before induction and growth in the slurries of microbial populations with the appropriate enzyme system(s). This agrees with the specificities of DMSP lyase, which accepts DMSP but not S-methylmethionine, and sulfonium lyase which attacks S-methylmethionine but not DMSP (Table 1). S-methylmethionine was not degraded via DMSP because MPA was not produced, only methanethiol.

Interconversions and Biodegradation of C_1–Organosulfur Compounds

There is considerable information on the anaerobic transformations of methylated sulfides in marine environments, but much less is known about their aerobic consumption. Bacteria which grow aerobically on MSA have been isolated from soil but not yet from marine habitats. The anaerobic degradation of MSA and the microbial transformations of dimethylsulfone have not been observed.

Anaerobic Metabolism

The biodegradation of methylated sulfides in marine sediments yields H_2S and CO_2. In both oxic and anoxic regions of the sediments, H_2S is further oxidized to thiosulfate, elemental sulfur, and sulfate. At high concentrations of methylated sulfides, some methane is formed in anoxic zones. Transmethylation reactions convert sulfide sequentially to methanethiol and then DMS,[32,33,37] and DMS is demethylated to methanethiol. Side reactions involve the reversible oxidative-reductive conversions of DMS and DMSO, and methanethiol and DMDS. The degradation and interconversions of methylated sulfides (Figure 2) have been studied in salt-marsh and coastal marine sediments.[7,62,67,72,74,135,143] The emphasis has been on anaerobic transformations, but Kiene[62] showed that both aerobic and anaerobic metabolism of DMS occurs. Porewaters from sediments of a *Spartina* salt marsh contained from about 0.1 μM DMS, but slurrying released 1–10 μm DMS into the aqueous phase, which was consumed at similar rates, either aerobically or anaerobically. Methanethiol and methane were produced during anaerobic metabolism but not during aerobic incubations. Inhibitor experiments, with molybdate to block sulfate reduction and BES to inhibit methanogenesis, showed DMS was anaerobically metabolized primarily by sulfate-reducing bacteria. About 20% of the DMS was converted to methane and the rest was used by sulfate reducers. Also sulfate-reducing bacteria accounted for more than 90% of ^{14}C–DMS use, at about a 5-μM level, in anoxic slurries of sediment from a salt marsh in San Francisco Bay.[74] At DMS concentrations above 200 μM, more DMS was used by methanogens than sulfate reducers.[74] Even at low concentrations (1–10 μM), DMS accounted for a major portion (about 30%) of the methanogenic rate, whereas DMS-dependent sulfate reduction amounted to 1% or less of the total sulfate-reduction rates.[62] Sulfate reducers were mainly responsible for the demethylation of DMS to methanethiol and subsequent consumption of methanethiol because BES only slightly retarded methanethiol utilization rates. DMDS was not detected as a product of anaerobic DMS or methanethiol metabolism perhaps due to low analytical sensitivity.[62]

Zinder and Brock[149] first showed the conversion of $^{14}CH_3SH$ and ^{14}C–DMS to $^{14}CH_4$ by sediments from a freshwater lake and sewage sludge. They noted the similarity of methanethiol degradation to that for methanogenesis from methanol by *Methanosarcina* but could not show the conversion of labeled methanethiol or DMS to methane and CO_2 by pure cultures (including *M. barkeri*) nor secure enrichment cultures which metabolized the methylated sulfides. Oremland and his colleagues isolated an obligately methylotrophic methanogen (strain GS-16) from an estuarine sediment using DMS as the sole substrate.[74,98] The organism was marine, with optimal growth on DMS at pH 7.0–8.7 in media containing 0.2–0.4M Na^+ and 0.04M Mg^{2+}. Strain GS-16 also grew on trimethylamine (TMA) or methanol but not on methanethiol, DMDS, methionine, formate, acetate, or H_2 plus CO_2. DMS-grown cells simultaneously metabolized TMA and DMS, but TMA-grown cells did not use DMS, indicating a need for additional enzymes to use DMS as compared to TMA. No growth occurred on methanethiol, but it was a transient intermediate during growth on DMS.[74]

However, several methylotrophic methanogens, including strain GS-16, have recently been grown on methanethiol at concentrations up to 30–40 mM.[6,91]

Methylotrophic methanogens occupy an ecological niche in marine sediments by using trimethylamine derived from the marine osmolytes glycine betaine and trimethylamine N-oxide (TMAO). TMA is not metabolized by sulfate reducers, and so methylotrophic methanogens avoid competing for hydrogen and acetate which are more effectively consumed by sulfate reducers.[39,76,97] The ecological advantage for obligate methylotrophic methanogens to metabolize DMS, and even methanol, is uncertain since, unlike TMA, these compounds are used preferentially by sulfate reducers at low concentrations.[74,76,95] Perhaps fluctuating levels of DMS and methanol frequently occur in marine sediments, with surplus consumption by methylotrophic methanogens over the low levels which are monopolized by sulfate reducers.

Proteolytic clostridia form methane from methionine and methanethiol,[103] but they are unlikely to significantly contribute to methanogenesis and methylated sulfide conversions in marine sediments. Clostridia also produce traces of DMDS, dimethyltrisulfide, and S-methylthioacetate[103] which may explain an earlier report of S-methylthioacetate formation in soils incubated anaerobically with methionine and glucose.[38]

Successive methylations produce methanethiol and then DMS from sulfide in marine sediments.[37,67,72] These transformations may be carried out by organisms possessing thiol S-methyltransferase catalyzing the following reactions:

$$H_2S + \text{S-adenosylmethionine} = CH_3SH + \text{S-adenosylhomocysteine},$$
$$CH_3SH + \text{S-adenosylmethionine} = CH_3SCH_3 + \text{S-adenosylhomocysteine}.$$

The enzyme was first purified from rat liver[140] but has since been detected in a variety of microorganisms and plants.[31–33] Finster et al.[37] recently identified lignin-derived methoxylated aromatics as sources of methyl groups for enzymatic transmethylations of sulfide and methanethiol in marine sediments. Syringate (3,5-dimethoxy-4-hydroxybenzoate) and 3,4,5-trimethoxybenzoate stimulated methanethiol, DMS, and methane formation in slurries of marine and estuarine sediments. It was suggested that acetogenic bacteria were responsible for the transmethylation of sulfide and methanethiol. The high concentrations of the methylated sulfides (about 100 μM) generated from 0.1-mM syringate probably allowed their use by methanogens, even in the presence of sulfate-reducing bacteria.[76,95] Slurries of a freshwater sediment, in contrast, showed methanogenesis independently of syringate addition, presumably driven by endogenous precursors in the sediments. Syringate, however, stimulated DMS production but without the detectable formation of methanethiol, perhaps because low sulfide levels in the freshwater system allowed rapid sequential methylation to DMS without accumulation of methanethiol.[37] Transmethylation processes may account for occasional reports of high rates of DMS production in some freshwater environments.[37,93]

DMSO concentrations were reported to equal or even exceed those of DMS in surface ocean waters.[1] However, the analytical assay may have included DMSP[139] and the estimates for DMSO levels were probably too high. Many facultative and strict anaerobes, including photosynthetic nonsulfur bacteria but not sulfate reducers, reduce DMSO to DMS.[102,144,147,148] DMSO is chemically reduced to DMS by sulfide,[148] but only biological reduction was observed in slurries of salt-marsh sediments.[67] DMSO was metabolized to DMS and some methanethiol and then eventually methane, and these transformations were blocked by glutaraldehyde or chloramphenicol. DMS production from DMSO was stimulated by molybdate presumably because DMS consumption by sulfate reducers was prevented. DMSO reduction also occurs during aerobic growth of hyphomicrobia on DMSO as a sole carbon and energy source (see below).

The microbial oxidation of DMS to DMSO in oxic marine environments has not been shown. However, DMS to DMSO conversion occurs in anoxic marine habitats. Zeyer et al.[146] obtained enrichment cultures from the H_2S-containing hypolimnion of Salt Pond, Massachusetts, which grew anaerobically on DMS with stoichiometric production of DMSO. DMS use was light-dependent and proceeded in parallel with methanogenesis; the two processes were independent because BES blocked methane production without effect on DMS consumption. A pure culture, a *Thiocystis* species, was isolated which grew phototrophically on DMS and produced DMSO. Efficient transformation of DMS to DMSO, for both enrichments and pure culture, required acetate and H_2S. Visscher and van Gemerden[133] recently isolated a strain of *Thiocapsa rosepersicina* from an intertidal microbial mat, which grew on DMS as its sole electron donor with DMSO formation.

Biological reduction of DMDS to methanethiol has been observed with anoxic slurries from salt-marsh sediments[67] and with cultures of thiobacilli which grow aerobically on DMDS as their sole source of carbon and energy.[109]

Aerobic Transformations

So far, pure cultures which grow aerobically on methylated sulfides as sole sources of carbon and energy are either autotrophic thiobacilli or methylotrophic hyphomicrobia.[60] Only thiobacilli have been isolated from marine environments. A common biochemical pathway operates for the oxidation of DMS and methanethiol in both groups of organisms. Key enzymes in the catabolic sequence are oxygenases catalyzing the incorporation of molecular oxygen into the products. DMS monooxygenase initially yields methanethiol and formaldehyde:

$$(CH_3)_2S + O_2 + NAD(P)H_2 = CH_3SH + HCHO + H_2O + NAD(P).$$

Methanethiol oxidase then catalyzes the oxidation of methanethiol to formaldehyde:

$$CH_3SH + O_2 + H_2O = HCHO + H_2S + H_2O_2.$$

Formaldehyde is oxidized to formate and then CO_2 in reactions catalyzed by pyridine nucleotide-dependent formaldehyde and formate dehydrogenases. Peroxide generated during methanethiol oxidation[120] is removed by catalase and the addition of 3-amino-1,2,4-triazole, a catalase inhibitor, blocks growth on DMS or DMDS.[109,119]

Sulfide is oxidized to sulfate, via tetrathionate and thiosulfate in the thiobacilli,[109,110] with energy conservation in both groups of organisms. The hyphomicrobia are strict methylotrophs, requiring methylated substrates to supply cell carbon, but chemostat experiments with *Hyphomicrobium* EG showed improved growth yields from the oxidation of thiosulfate or sulfide.[119] High levels of hydroxypyruvate reductase and low levels of ribulose bisphosphate carboxylase in the hyphomicrobia indicated carbon assimilation as formaldehyde via the serine pathway.[24,119] The thiobacilli, in contrast, fixed $^{14}CO_2$ during growth on methylated sulfides.[56,109] *Thiobacillus thioparus* strain E6 grown on DMDS possessed high levels of ribulosebisphosphate carboxylase, but hydroxypyruvate reductase was undetectable, in accordance with autotrophic growth via the Calvin cycle.[110] Chemostat yields showed that *T. thioparus* strain E6 benefited from the oxidation of the methyl groups of DMDS, with energy probably obtained during the oxidation of formaldehyde and formate.[110] However, unlike *T. versutus*,[61] *T. thioparus* strain E6 did not grow autotrophically on formaldehyde or formate.

Information on the aerobic metabolism and transformation of methylated sulfides by marine bacteria is limited. Visscher et al.[135] recently isolated *Thiobacillus thioparus* strain 5 from an intertidal microbial mat, which grew on DMS, methanethiol, DMDS, and DMSO. Growth also occurred on thiosulfate, H_2S, polysulfides, elemental sulfur, and thiocyanate. Smith and Kelly[109] isolated thiobacilli that grew aerobically on DMS and DMDS from several environments. *Thiobacillus* sp. strain E5 was isolated from marine mud enriched with DMS as sole source of carbon and energy. The isolate grew on methylated sulfides (methanethiol, DMS, DMDS) and inorganic sulfur compounds (sulfide, thiosulfate, tetrathionate, thiocyanate, COS). No growth occurred on DMSO, methane, methanol, methylamines, formaldehyde, or formate or heterotrophic substrates.

The isolation of aerobic bacteria which form colonies on media containing methylated sulfides has often proved difficult.[56,109,118] Kanagawa and Kelly[56] could only obtain DMS utilizers by plating, from sludge enrichments degrading DMS, onto thiosulfate-agar. Smith and Kelly[109] were unable to isolate pure cultures with DMS or DMDS, after initially successful enrichments on DMS or DMDS from a variety of environments. Pure cultures which grew on DMS and DMDS were obtained by plating enrichments on thiosulfate–mineral–salts agar; even plating from the DMS or DMDS enrichments onto methylotrophic (methylamine) medium failed to secure DMS or DMDS utilizers.[109] *Hyphomicrobium* S was isolated from soil by batch culture enrichment and plating on DMSO media.[24] *Hyphomicrobium* EG was derived from a fixed bed reactor used to remove H_2S and methylated sulfides from waste water by chemostat enrichment on DMSO, and then plating on methylamines because colony formation on DMSO plates was very slow.[118] *Hyphomicrobium* S only grew on DMSO or DMS, whereas *Hyphomicrobium* EG also developed on methylamine, dimethylamine, or TMAO.

DMSO utilization by aerobic bacteria depends on the presence of pyridine nucleotide reductases[24,118]:

$$(CH_3)_2SO + NAD(P)H + H^+ = (CH_3)_2S + NAD(P)^+ + H_2O.$$

As noted above, the aerobic oxidation by microorganisms of DMS to DMSO has not been shown.

DMDS, like DMSO, feeds into the metabolic sequence for aerobic breakdown of methylated sulfides by an initial enzymatic reduction[110]:

$$CH_3SSCH_3 + NADH + H^+ = 2CH_3SH + NAD^+.$$

Thiobacillus thioparus strain E6 (from pond mud) showed DMDS metabolism via methanethiol with NADPH-stimulated reduction in cells permeabilized with cetyltrimethylammonium bromide.[110]

Although methanethiol appears to be oxidized to DMDS in a variety of environments, marine microorganisms which carry out this transformation have not been isolated. DMDS occurs in similar amounts to methanethiol as a product of methionine breakdown in soils and sediments.[9,150] The oxidation probably occurs mainly in oxic zones and may occur abiotically, at least to some extent. Segal and Starkey[106] recorded variation in the relative rates of methanethiol and DMDS production during the growth of different species of soil bacterial on methionine, which was demethiolated to methanethiol. Presumably some species catalyzed the oxidation of methanethiol to DMDS, and so produced more DMDS relative to methanethiol. However, definitive investigations are needed to substantiate and differentiate between chemical and biochemical mechanisms for the oxidation of methanethiol to DMDS in marine environments.

Trimethylsulfonium chloride (TMS), $(CH_3)_3S^+Cl^-$, is aerobically metabolized by enzymatic demethylation in which a methyl group is transferred to tetrahydrofolate to form DMS and 5-methyltetrahydrofolate.[77,138] The enzyme was partially purified from a soil bacterium, *Pseudomonas* MS, which was isolated on TMS as its sole source of carbon. The enzyme lacked vitamin B_{12} and neither DMSP nor S-adenosylmethionine substituted for TMS in the methyl-transfer reaction. The carbon-assimilation pathway was not determined but is probably methylotrophic. Unlike the other aerobic methylated sulfide users so far described, *Pseudomonas* MS displayed nutritional versatility, growing on a variety of sugars and amino acids. Noteworthy in this respect is a report of diverse bacteria (Gram-negative and Gram-positive) isolated from a variety of habitats, which were enriched on DMS, DMDS, and diethylsulfide (DES).[60] One isolate when grown on DES completely oxidized the following compounds to sulfate and CO_2: DMS, DES, DMDS, diethyldisulfide (DEDS), ethanethiol, and CS_2. Dipropylsulfide and butanethiol were partially oxidized.

Several soil isolates which grew aerobically on MSA were secured from soil by chemostat enrichment on a mixture of MSA and methylammonium.[5] Nine cultures, three Gram-negative and six Gram-positive, were obtained from a single soil sample. The diversity of organisms suggests MSA degradation may be common. A Gram-negative isolate (strain M2) grew aerobically on 20 mM MSA with a doubling time of 7–8 h. It also grew on methylamines, trimethylsulfonium chloride, methanol, formaldehyde, formate, and heterotrophic substrates such as glucose and glutamate. No growth occurred on methane, thiosulfate, DMS, DMDS, or methane phosphonate. The organism was a methylotroph lacking ribulosebisphosphate carboxylase and hexulose phosphate synthase but possessing hydroxypyruvate reductase indicating formaldehyde and CO_2 assimilation via the serine pathway.[18]

MSA degradation may involve a hydrolase or a monooxygenase[5,59]:

$$CH_3SO_3H + H_2O = CH_3OH + H_2SO_3,$$
$$CH_3SO_3H + O_2 + 2[H] = HCHO + H_2SO_3 + H_2O,$$

but hydrolytic cleavage to yield methane, by analogy to methane phosphonate degradation,[23] was precluded. Another suggested alternative involved a methyl–tetrahydrofolate transfer as for *Pseudomonas* MS.[77,138] Hydrolytic cleavage may be difficult considering the chemical stability of MSA,[60] but oxygenative attack has been observed for C_4–C_{12} *n*-alkane-1-sulfonates.[125,126] Hydrolytic cleavage render would MSA available as methanol which is used by various anaerobes, including nitrate reducers (especially hyphomicrobia), methanogens, and sulfate reducers.[88,95,114] MSA would then provide stoichiometric amounts of both methanol and sulfite for sulfate reducers:

$$CH_3SO_3H + H_2O = CH_3OH + H_2SO_3,$$
$$CH_3OH + H_2SO_3 = CO_2 + H_2S + 2H_2O.$$

The possible direct use of MSA by methanogens and/or sulfate reducers is intriguing[59]:

$$CH_3SO_3H + 4H_2 = CH_4 + H_2S + 3H_2O.$$

This could be accomplished by methanogens or allow methanogenesis by sulfate reducers.

Novel Routes for Biodegradation of Organosulfur Compounds

In addition to DMS cleavage, DMSP is anaerobically degraded in marine sediments with sequential demethylations to produce MMPA and MPA (Figure 3). This catabolic route was proposed to explain the high levels of MPA in porewaters of sediments from Biscayne Bay, Florida.[87] MPA occurred at concentrations up to 1 μM and beneath the maximal horizon for glutathione[69] which functions only in aerobes.[35] DMSP, homocysteine and methionine were suggested as possible precursors of MPA.[87]

MPA and/or methanethiol production in anoxic slurries of coastal sediment was stimulated by DMSP, MMPA, homocysteine, 2-keto-4-methiolbutyrate, and S-methylmethionine.[69,70,73] The transformations were inhibited by antibiotics (chloramphenicol plus tetracycline). DMSP produced MPA and methanethiol at about equal rates, whereas MMPA, the immediate precursor of MPA, gave primarily MPA. Methanethiol probably arose from DMSP via cleavage to acrylate and DMS, with subsequent demethylation of the DMS; added DMS was converted to methanethiol in slurries. Acrylate was detected as an intermediate during metabolism of high levels of DMSP. Acrylate, at the concentrations detected, did not contribute to MPA formation by chemical addition reactions with sulfide or polysulfides in the slurries[130,131]; instead, it was fermented to acetate and propionate which were oxidized by sulfate reducers.[71] High DMSP levels favored the cleavage rather than demethylation pathway. Methionine and 2-keto-4-methiolbutyrate were transformed at similar rates to methanethiol and MPA, with a preponderance of methanethiol production. Presumably demethiolation of methionine occurred after its deamination or transamination. Homocysteine produced both methanethiol and MPA but mostly the latter compound with probable formation via its keto acid. Methanethiol formation from homocysteine probably proceeded by methylation to methionine. Glycine betaine functions in the enzymatic transmethylation of homocysteine[19,44] and endogenous marine osmolytes (glycine betaine and/or DMSP) in the sediment slurries may act as extracytoplasmic methylating agents.

MPA metabolism in marine sediments is poorly understood, but its utilization in anoxic sediment slurries was inhibited by tungstate or molybdate, thereby implicating sulfate reducers.[64,71] MPA presents metabolic problems because it is a potent inhibitor of fatty acid oxidation, mimicking CoA by virtue of its —SH group. Aerobic degradation may occur during which MPA may be chemically oxidized to its disulfide form or even undergo chemical oxidation of the sulfhydryl group. Enzymatic mechanisms for dealing with sulfoxy derivatives of MPA probably resemble those for cysteine and its sulfoxy acids. The disulfide molecule may be reduced back to MPA but attack from one end of the carbon chain could circumvent the metabolic problems presented by the free sulfhydryl group. MPA may be metabolized after H_2S elimination, similar to cysteine desulfhydrase, to give acrylate which is readily metabolized aerobically or anaerobically.

Novel routes for the biotransformation of cysteine were recently detected in anoxic coastal marine sediments.[73] Glutathione gave rise to cysteine which was further transformed to 3-mercaptopyruvate, and then to mercaptoethanol and mercaptoacetate. Mercaptoethanol was oxidized to mercaptoacetate. All transformations, including the consumption of mercaptoacetate, were inhibited by antibiotics (chloramphenicol plus tetracycline). Detection of mercaptopyruvate as an environmental product of cysteine is novel, although mercaptopyruvate is produced from cysteine in mammalian tissues by cysteine aminotransferase.[44] Further metabolism in mammals involves 3-mercaptopyruvate transsulfurylase which displaces H_2S and generates pyruvate.[44] Clearly, novel catabolic routes for cysteine are operating in marine sediments which do not, at least immediately, displace the sulfur atom. The transformations of mercaptopyruvate resemble those for ethanol and acetate production from pyruvate, and the organisms and biochemical mechanisms need investigating.

Metabolism of Sediment-Bound Thiols

The most common organosulfur compounds in the porewaters of coastal sediments are DMS, glutathione, methanethiol, MPA, and cysteine.[87,108,113] Organosulfur compounds interact with the particulate phase in marine sediments and the sediment-bound fraction usually constitutes the bulk of organosulfur present in the sediment.[64,87] Bound molecules may be released by physically disturbing the sediment when preparing a slurry in seawater.[62,64] Slurrying may also release substrates, for example, DMSP or sulfur-containing amino acids which are rapidly metabolized to methanethiol or DMS.[54] Chemical displacement or exchange mechanisms also liberate organosulfur compounds into the aqueous phase. Zinder and Brock[150] observed the displacement of $^{14}CH_3SH$ by cold methanethiol in sediment from a freshwater lake. Sulfide displaces sediment-bound thiols[64] as does the reductive cleavage of disulfide bonds with tributylphosphine (TBP)[48,87]:

$$R'SSR + (C_4H_9)_3P + H_2O = R'SH + RSH + (C_4H_9)_3PO.$$

A wide range of thiols were detected in the porewaters of coastal marine sediments and their concentrations increased about 20-fold upon treatment with TBP.[87] Thus, >95% of the thiols were present attached to sediment particles, probably via disulfide bonds. Other mechanisms such as hydrophobic interactions may operate for some organosulfur molecules, e.g., alkyl sulfides. Similar studies by Kiene[64] showed up to 100- to 200-fold increases in porewater levels of methanethiol and MPA on adding TBP to a coastal marine sediment. Methanethiol and MPA were commonly present at peak levels of 300 nM in the top 20 cm of sediment and glutathione up to 600 nM, although it was restricted to the top 5 cm of sediment. TBP did not increase glutathione levels in the upper 5 cm fraction, but it liberated, from the lower anoxic zone, a compound that was tentatively identified as glutathione.[64]

At least some of the sediment-bound organosulfur compounds are readily exchangeable with the porewaters and then consumed by microorganisms.[62,64] Slurries of coastal marine sediment showed release of methanethiol and MPA into the aqueous phase up to levels of about 100–300 nM. After 24 h incubation, the levels decreased and inhibitor addition at this time identified sulfate reducers but not methanogens as important consumers of methanethiol and MPA.[64]

In general, it seems that organosulfur compounds bind to sediment particles, but the bound fraction is usually in equilibrium with the dissolved component which is readily available for microbial transformation and consumption.

Production and Metabolism of COS and CS_2

CS_2 was first detected in the marine environment by Lovelock,[83] with average values of about 7 and 14 pM, respectively, in pelagic and inshore waters of the Atlantic Ocean. Up to 1 nM CS_2 occurred in porewaters of anoxic marine muds. Low levels of CS_2 in surface oceanic waters have been confirmed in recent studies in the North Atlantic with an average value of 8 pM, and about a twofold increase in coastal samples.[2,75] Turner and Liss[128] detected up to 150 pM CS_2 in English coastal waters. Higher coastal levels of CS_2 agree with a source in coastal sediments.[83] Wakeham et al.[139] found CS_2 up to about 10 nM levels in the anoxic, H_2S-containing waters below the midwater oxic–anoxic interface of a stratified salt pond. CS_2 levels in the anoxic hypolimnion increased toward the sediment surface and its vertical profile paralleled that of methane, indicating a sedimentary source.[139] CS_2 and COS were also detected in the bottom waters of an Antarctic lake.[27] COS levels in the oxic, surface, mixed layer of the oceans range from 0.03 to 1.0 nM and show distinct diurnal variations, with daytime maxima.[2] COS is produced by the photochemical degradation of biologically generated organic sulfur molecules such as DMSP, methionine, cysteine, and glutathione.[36] The photochemical formation of COS in seawater is greatly stimulated by natural photosensitizers such as humic and fulvic acids.[145]

The highest production rates of CS_2 and COS for natural systems are from salt marshes.[60,117] Plants may generate and consume COS and CS_2,[60] but anaerobic microbial activities in sediments of coastal marsh and marine environments undoubtedly also produce CS_2 and perhaps COS.

Bremner and Steele[9] during studies on the biodegradation of plant residues and animal manures in soil, identified cysteine, cystine, and homocysteine precursors of CS_2. Thiocyanate and isothiocyanates stimulated COS formation and both COS and CS_2 were generated from lanthionine (present in wool hydrolysates) and djenkolic acid from the djenkole bean (Figure 5). All the transformations occurred in either aerobic or waterlogged soils. Dicotyledonous plants, especially the members of the Cruciferae, contain glucosinates from which are derived organic thiocyanates.[30] CS_2 is generated from organic isothiocyanates by hydrolysis[14] and acid hydrolysis of thiocyanate produces COS:

$$HSCN + H_2O = COS + NH_3.$$

CS_2 is an assimilatory sulfur source for some methanogens,[100] but metabolism of CS_2 and/or COS as analogues of CO_2 by methanogens has not been reported. COS as a structural analogue of CO_2 is exemplified by its ability to substitute for CO_2 in enzymatic reactions involving CO_2. COS inhibits inorganic carbon uptake in cyanobacteria[94] and is a substrate, but not activator, of ribulosebisphosphate carboxylase.[82] Chemolithotrophic growth on CS_2 was first reported for *Thiobacillus thio-oxidans*.[12] More recently, Smith and Kelly[111] showed autotrophic growth of *T. thioparus* strain TK-m on CS_2 and COS. *T. thioparus* strain TK-m also grew on thiocyanate,[109] another structural analogue of CS_2, COS, and CO_2. COS was detected as an intermediate during growth on CS_2 and sulfate production from CS_2 was stoichiometric. Cells grown on CS_2 and incubated anaerobically sequentially produced COS and H_2S. Smith and Kelly[111] postulated hydrolytic cleavage of CS_2 and COS to yield H_2S for lithotrophic growth:

$$CS_2 + H_2O = COS + H_2S,$$
$$COS + H_2O = CO_2 + H_2S.$$

COS is a substrate for carbonic anhydrase which catalyzes its hydrolysis to H_2S and CO_2, and this may occur during microbial catabolism of COS. Even though

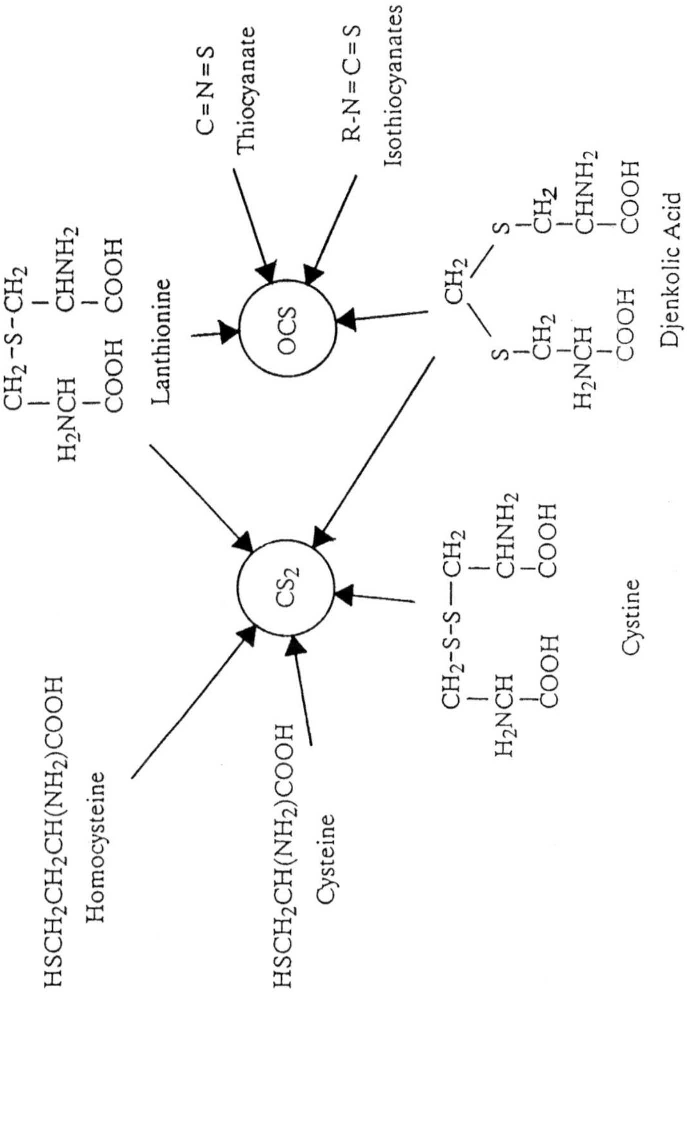

Figure 5. Sources of COS and CS_2 in aerobic and waterlogged soils (from Ref. 9).

Smith and Kelly[111] failed to secure enrichments on CS_2, other organisms with diverse mechanisms and physiologies which use this compound and COS undoubtedly exist in natural environments. Kelly and Smith[60] drew attention to the oxidative degradation of CS_2 by a monooxygenase, from rat liver, which involves NADPH and cytochrome P450.[17] Hydrolytic cleavage of CS_2 and COS, perhaps promoted by carbonic anhydrase or more specialized enzymes, to yield H_2S clearly provides a lithotrophic substrate for photosynthetic sulfur bacteria in anoxic environments. In anoxic situations, methanogens may exist which utilize the carbon of CS_2 or COS, rather than just assimilating the sulfur atom:

$$CS_2 + 4H_2 = CH_4 + 2H_2S,$$
$$COS + 4H_2 = CH_4 + H_2S + H_2O.$$

Zinder and Brock[150] found increased endogenous production of methanethiol, DMDS, propanethiol, and CS_2 when $CHCl_3$ was added to anaerobic sediments of a freshwater lake. Inhibition by $CHCl_3$ implicates methanogens in CS_2 production,[8] but further work with a more specific inhibitor such as BES is needed. Methionine was identified as a precursor of the methylated sulfides, but cysteine, unlike the studies of Bremner and Steele,[9] did not stimulate CS_2 production. Anaerobic mechanisms in marine and other environments for the microbial production of CS_2 and COS await investigation.

Discussion

DMSP concentrations in marine waters often exceed those for DMS by several-fold.[65,80,129] There is a correlation between DMS and DMSP which is better than that for phytoplankton biomass (chlorophyll *a*), and DMS emissions to the atmosphere ultimately relate back to microalgal species that synthesis DMSP.[57] DMSP release from phytoplankton occurs during grazing, senescence, and by cell lysis,[21,90] but DMS levels in the water, and consequently net atmospheric fluxes, are principally the outcome of the interacting and competing activities of different groups of bacteria. In open-ocean and estuarine environments less than 30% of DMSP is converted to DMS.[65,68] Furthermore, turnover rates of DMS in the water greatly exceed those for DMS in the atmosphere.[66] Little is known about the web of organisms which transforms and consumes DMSP and DMS. Distinct groups of bacteria metabolize DMSP and DMS in marine waters,[63] and it is conceivable that flagellates and ciliates[63] participate in DMSP cleavage since the enzyme occurs in eukaryotes.[13,55] Transformations of DMSP in zooplankton guts and fecal pellets by microflora or host enzymes need examination.

Alternative routes for DMSP metabolism operate in marine waters since less than 30% appears as DMS. Aerobic demethylation and subsequent metabolism of MMPA by demethylation and demethiolation are distinct possibilities. DMSP transmethylation reactions may supply methyl groups for producing methanethiol (from sulfide), methyl halides (from halide ions), and DMS from methanethiol. Transfer may be catalyzed by thiol S-methyltransferases, with folates as C_1 carriers which would also allow oxidation of the methyl groups.[138] Methanethiol has been detected in oceanic water samples[79] and may arise from the bacterial catabolism of DMS, DMSP, MMPA, or MPA. The turnover of methanethiol may be difficult to quantify because of rapid chemical and biochemical transformations.

DMS, MMPA, methanethiol, and MPA are all potential precursors of the H_2S detected in the surface layers of the ocean.[84] In addition, DMS and methanethiol catabolism may contribute to H_2O_2 levels in oceanic waters.[99] The catalase inhibitor, 1,2,4-azatriozole, may be a useful experimental tool for inhibiting DMS and methanethiol consumption in oxic environments.

Bacteria which degrade DMS have not been isolated from the ocean environment. Organisms similar to autotrophic thiobacilli and methylotrophic hyphomicrobia of terrestrial environments probably consume DMS, but novel organisms may participate. Isolation of organisms that catabolize DMS from nonmarine regions has not been simple, and capture of open-ocean bacteria that use DMS at nanomolar levels may require chemostats. Continuous culture will also allow exploration of the interactions between bacteria that compete for DMS in the ocean. Predominant DMS utilizers may be methylotrophs rather than autotrophs because the assimilation of HCHO rather than CO_2 would give more efficient reprocessing of primary production through a microbial loop.[4]

A neglected aspect of DMS metabolism in oxic marine waters is its oxidation to DMSO. Naturally occurring photosensitizers catalyze the photo-oxidation of DMS to DMSO at rates those of DMS fluxes to the atmosphere.[11] Peroxide from the catabolism of DMS and methanethiol may participate in photo-oxidations.[112] Direct biochemical contributions to DMSO formation in oxic environments are unknown but certainly possible. Enzymes catalyze the production of sulfoxides and sulfones from organic sulfides in reactions involving O_2.[47] Monooxygenase systems carry out the following general reaction:

$$RSR + NADPH + H^+ = RS(O)R + NAD^+ + H_2O$$

using specifically either cytochrome P450 or flavin adenine dinucleotide (FAD) to activate the O_2. Such enzymes are widely distributed in microorganisms, plants, and animals. DMS is a substrate for the pig liver enzyme and DMSO is further oxidized to dimethylsulfone. Another enzyme catalyzing the oxidation of DMSO to dimethylsulfone is a haloperoxidase from the fungus *Caldariomyces fumago.*[40] Haloperoxidases abound in marine organisms[89] and biochemical production of dimethylsulfone in oceanic waters is feasible. More speculative is the oxidative metabolism of dimethylsulfone to yield MSA. Monooxygenative attack of a methyl group initially,

$$(CH_3)_2SO_2 + O_2 + 2[H] = CH_3SO_2CH_2OH + H_2O,$$

and oxidation gives an acid that is susceptible to decarboxylation,

$$CH_3SO_2COO^- = CH_3SO_2^- + CO_2$$

with subsequent oxidation to MSA.

Several novel biotransformations of organosulfur compounds in sediments need further investigation. DMSP degradation via MPA raises questions about anaerobic demethylation. *Desulfobacterium* strains, isolated from marine sediments, grow on glycine betaine with the oxidation of one methyl group to CO_2 and the formation of dimethylglycine.[53] Analogous transformations may proceed for DMSP. MPA may also be produced from homocysteine by the activities of sulfate reducers which incompletely oxidize organic compounds:

$$4HSCH_2CH_2CH(NH_2)COOH + 4H_2O = 4HSCH_2CH_2COOH + 4NH_3 + 4CO_2 + 8[H],$$
$$8[H] + SO_4^{2-} = 4H_2O + S^{2-}.$$

Methionine could be channeled into this route by demethylation.

MPA catabolism, as suggested earlier, may occur with the elimination of H_2S to generate acrylate which is aerobically or anaerobically degraded. MPA may also be assimilated by marine microorganisms and used biosynthetically or remethylated to DMSP for osmotic functions.

New transformations of mercaptopyruvate to mercaptoethanol and mercaptoacetate in marine sediments were previously discussed. Organisms possessing novel enzymes with potential in biotechnology may be involved in the conversions of mercaptopyruvate, mercaptoethanol, and mercaptoacetate. Mercaptoacetate may be oxidized:

$$HSCH_2COOH + 2H_2O = H_2S + 2CO_2 + 6[H];$$

sulfate reducers may accomplish the transformation. Another completely unexplored aspect in marine sediments is the bacterial generation of CS_2 and COS in marine sediments. In all, the transformations of organosulfur compounds in marine environments hold much interest for microbiologists, biochemists, and biotechnologists. Research should provide new insights into biogeochemical cycles of sulfur, most notably with respect to gas fluxes of trace sulfur gases to the atmosphere.

Acknowledgments

Thanks to Ron Kiene and Pieter Visscher for providing unpublished manuscripts and for discussions. Thanks also to Bob Gauley, Tom Mende, Eric Saltzman, A. Vairavamurthy, Darrin Gilchrist, and Mara Diaz for their input. Financial support was provided by NSF grant OCE-9012157.

References

1. Andreae, M.O. 1980. Dimethylsulfoxide in marine and freshwaters. Limnol. Oceanogr. **25**:1054–1063.
2. Andreae, M.O. 1990. Ocean–atmosphere interactions in the global biogeochemical sulfur cycle. Marine Chem. **30**:1–29.
3. Aneja, V.P. 1990. Natural sulfur emissions into the atmosphere. J. Air Waste Manag. Assoc. **40**:469–476.
4. Azam, F., T. Fenchel, J.G. Field, J.S. Gray, L.A. Meyer-Reil, and F. Thingstad. 1983. The ecological role of water-column microbes in the sea. Marine Ecol. Progr. Ser. **10**:257–263.
5. Baker, S.C., D.P. Kelly, and J.C. Murrell. 1991. Microbial degradation of methanesulfonic acid: a missing link in the biogeochemical cycle. Nature **350**:627–628.
6. Bak, F., and K. Finster. 1993. Formation of dimethyl sulfide and methanethiol from methoxylated aromatic compounds and inorganic sulfide by newly-isolated, anaerobic bacteria. In R.S. Oremland (ed.), Biogeochemistry of Global Change: Radiatively Active Trace Gases, Chapman and Hall, New York.
7. Banat, I.M., D.B. Nedwell, and M.T. Balba. 1983. Stimulation of methanogenesis by slurries of salt marsh sediment after the addition of molybdate to inhibit sulfate reducing bacteria. J. Gen. Microbiol. **129**:123–129.
8. Bauchop, T. 1967. Inhibition of rumen methanogenesis by methane analogs. J. Bacteriol. **94**:171–175.
9. Bremner, J.M., and C.G. Steele. 1978. Role of microorganisms in the atmospheric sulfur cycle. Adv. Microbial Ecol. **2**:155–201.
10. Breznak, J.A., J.M. Switzer, and H-J. Seitz. 1988. *Sporomusa termitida* sp. nov., an H_2/CO_2-utilizing acetogen isolated from termites. Arch. Microbiol. **150**:282–288.
11. Brimblecombe, P., and D. Shooter. 1986. Photo-oxidation of dimethylsulfide in aqueous solution. Marine Chem. **19**:343–353.

12. Butler, R.G., B. Rothschild, and J.R. Keller. 1969. Metabolism of carbon disulfide by *Thiobacillus thiooxidans*. Abstracts of 70th Annual Meeting, p. 64. American Society for Microbiology, Washington DC.

13. Cantoni, G.L., and D.G. Anderson. 1956. Enzymatic cleavage of dimethylpropiothetin by *Polysiphonia lanosa*. J. Biol. Chem. **222**:171–177.

14. Challenger, F. 1959. Aspects of the Organic Chemistry of Sulfur, Butterworth, London.

15. Chambers, S.T., C.M. Kunin, D. Miller, and A. Hamada. 1987. Dimethylthetin can substitute for glycine betaine as an osmoprotectant molecule for *Escherichia coli*. Appl. Environ. Microbiol. **169**:4845–4847.

16. Charlson, R.J., J.E. Lovelock, M.O. Andreae, and S.G. Warren. 1987. Oceanic phytoplankton, atmospheric sulfur, cloud albedo and climate. Nature **326**:655–661.

17. Chengelis, C.P., and R.A. Neal. 1987. Oxidative metabolism of carbon disulfide by isolated rat liver hepatocytes. Biochem. Pharmacol. **36**:363–368.

18. Colby, J., D.I. Stirling, and H. Dalton. 1977. The soluble methane mono-oxygenase of *Methylococcus capsulatus* (Bath). Its ability to oxygenate *n*-alkanes, *n*-alkenes, ethers, and alicyclic, aromatic and heterocyclic compounds. Biochem. J. **165**:395–402.

19. Cooper, A.J.L. 1983. Biochemistry of sulfur-containing amino acids. Ann. Rev. Biochem. **52**:187–222.

20. Dacey, J.W.H., and N.V. Blough. 1987. Hydroxide decomposition of dimethylsulfoniopropionate to form dimethylsulfide. Geophys. Res. Lett. **14**:1246–1249.

21. Dacey, J.W.H., and S.G. Wakeham. 1986. Oceanic dimethylsulfide: production during zooplankton grazing on phytoplankton. Science **233**:1314–1316.

22. Dacey, J.W.H., G.M. King, and S.G. Wakeham. 1987. Factors controlling emission of dimethylsulfide from salt marshes. Nature **330**:643–645.

23. Daughton, C.G., A.M. Cook, and M. Alexander. 1979. Biodegradation of phosphonate toxicants yields methane or ethane on cleavage of the C–P bond. FEMS Microbiol. Lett. **5**:91–93.

24. De Bont, J.A.M., J.P. van Dijken, and W. Harder. 1981. Dimethyl sulfoxide and dimethyl sulfide as a carbon, sulfur and energy sources for growth of *Hyphomicrobium* S. J. Gen. Microbiol. **127**:315–323.

25. Dehning, I., and B. Schink. 1989. *Malonomonas rubra* gen. nov. sp. nov., a microaerotolerant anaerobic bacterium growing by decarboxylation of malonate. Arch. Microbiol. **151**:427–433.

26. Dehning, I., M. Stieb, and B. Schink. 1989. *Sporomusa malonica* sp. nov., a homoacetogenic bacterium growing by decarboxylation of malonate or succinate. Arch. Microbiol. **151**:421–426.

27. Deprez, P.P., P.D. Franzman, and H.R. Burton. 1986. Determination of reduced sulfur gases in Antarctic lakes and seawater by gas chromatography after solid adsorbent preconcentration. J. Chromatogr. **362**:9–21.

28. Diaz, M, and B.F. Taylor. 1992. Metabolism of dimethylsulfoniopropionate

(DMSP) and glycine betaine by a marine bacterium. Gen. Meet. Amer. Soc. Microbiol. Abstract K144, p. 280.

29. Diaz, M, and B.F. Taylor. Unpublished data.

30. Drobnica, L., P. Kristian, and J. Augustin. 1977. The chemistry of the —NCS group. In S. Patai (ed.), The Chemistry of Cyanates and their Derivatives, Wiley, New York, pp. 1003–1221.

31. Drotar, A.M., and R. Fall. 1985. Methylation of xenobiotic thiols by *Euglena gracilis:* characterization of a cytoplasmic thiol methyltransferase. Plant Cell Physiol. **26**:847–854.

32. Drotar, A.M., G.A. Burton, Jr., J.E. Tavernier, and R. Fall. 1987. Widespread occurrence of bacterial thiol methyltransferases and the biogenic emission of methylated sulfur gases. Appl. Environ. Microbiol. **53**:1626–1631.

33. Drotar, A.M., L.R. Fall, E.A. Mishalanie, J.E. Tavernier, Jr., and R. Fall. 1987. Enzymatic methylation of sulfide, selenide, and organic thiols by *Tetrahymena thermophila*. Appl. Environ. Microbiol. **53**:2111–2118.

34. Esaki, N., and K. Soda. 1987. L-Methionine γ-lyase from *Pseudomonas putida* and *Aeromonas*. In W.B. Jakoby and O.W. Griffith (eds.), Methods in Enzymology, Vol 143, Academic Press, New York, pp. 459–465.

35. Fahey, R.C., and G.L. Newton. 1983. Occurrence of low molecular weight thiols in biological systems. In A. Larrson, S. Orrenius, A. Holmgren, and B. Mannervick (eds.), Functions of Glutathione: Biochemical, Physiological, Toxicological, and Clinical Aspects, Raven Press, New York, pp. 251–260.

36. Ferek, R.J., and M.O. Andreae. 1984. Photochemical production of carbonyl sulfide in marine surface waters. Nature **307**:148–150.

37. Finster, K., G.M. King, and F. Bak. 1990. Formation of methyl mercaptan and dimethyl sulfide from methoxylated aromatic compounds in anoxic marine and sediments. FEMS Microbial Ecol. **74**:295–302.

38. Francis, A.J., J.M. Duxbury, and M. Alexander. 1975. Formation of volatile organic products in soils under anaerobiosis. II. Metabolism of amino acids. Soil Biol. Biochem. **7**:51–56.

39. Franklin, M.J., W.J. Wiebe, and W.B. Whitman. 1988. Populations of methanogenic bacteria in a Georgia salt marsh. Appl. Environ. Microbiol. **54**:1151–1157.

40. Geigert, J., S.K. De Witt, S.L. Neidleman, G. Lee, D.J. Dalientos and M. Moreland. 1983. DMSO is a substrate for chloroperoxidase. Biochem. Biophys. Res. Commun. **116**:82–85.

41. Giovanelli, J. 1987. Sulfur amino acids of plants: an overview. In W.B. Jakoby and O.W. Griffith (eds.), Methods in Enzymology, Vol 143, Academic Press, New York, pp. 419–426.

42. Gottschalk, G. 1986. Bacterial Metabolism, Springer-Verlag, New York.

43. Greene, R.C. 1962. Biosynthesis of dimethyl-β-propiothetin. J. Biol. Chem. **237**: 2251–2254.

44. Griffith, O.W. 1987. Mammalian sulfur amino acid metabolism: an overview. In

W.B. Jacoby and O.W. Griffith (eds.), Sulfur and Sulfur Amino Acids, Methods in Enzymology, Vol 143, Academic Press, New York, pp. 366–376.

45. Gunsalus, R.P., J.A. Romesser, and R.S. Wolfe. 1978. Preparation of coenzyme M analogs and their activity in the methyl coenzyme M reductase system of *Methanobacterium thermoautotrophicum*. Biochemistry **17**:2374–2377.

46. Hatula, T., and B. Granroth. 1974. Formation of dimethyl sulfide from S-methylmethionine in onion seedlings (*Allium cepa*). J. Sci. Food Agric. **25**:1517–1521.

47. Holland, H.L. 1988. Chiral sulfoxidation by biotransformation of organic sulfides. Chem. Rev. **88**:473–485.

48. Humphrey, R.E., and J.L. Potter. 1965. Reduction of disulfides with tributylphosphine. Anal. Chem. **37**:164–165.

49. Jorgensen, B.B. 1977. The sulfur cycle of a coastal marine sediment (Limfjorden, Denmark). Limnol. Oceanogr. **22**:814–822.

50. Jorgensen, B.B. 1988. Ecology of the sulfur cycle: oxidative pathways in sediments. Symp. Soc. Gen. Microbiol. **42**:31–63.

51. Jorgensen, B.B. 1990. A thiosulfate shunt in the sulfur cycle of marine sediments. Science **249**:152–154.

52. Jorgensen, B.B., and F. Bak. 1991. Pathways and microbiology of thiosulfate transformations and sulfate reduction in a marine sediment (Kattegat, Denmark). Appl. Environ. Microbiol. **57**:847–856.

53. Heijthuijsen, J.H.F.G., and T.A. Hansen. 1989. Anaerobic degradation of betaine by marine *Desulfobacterium* strains. Arch. Microbiol. **152**:393–396.

54. Howes, B.L., J.W.H. Dacey, and S.G. Wakeham. 1985. Effects of sampling technique on measurements of pore water constituents in salt marsh sediments. Limnol. Oceanogr. **30**:221–227.

55. Kadota, H., and Y. Ishida. 1972. Production of volatile sulfur compounds by microorganisms. Ann. Rev. Microbiol. **26**:127–138.

56. Kanagawa, T., and D.P. Kelly. 1986. Breakdown of dimethyl sulfide by mixed cultures and by *Thiobacillus thioparus*. FEMS Microbiol. Lett. **34**:13–19.

57. Keller, M.D., W.K. Bellows, and R.R.L. Guillard. 1989. Dimethyl sulfide production in marine phytoplankton. In E.S. Saltzman and W.J. Cooper (eds.), Biogenic Sulfur in the Environment, American Chemical Society, Washington DC., pp. 167–200.

58. Kelly, D.P. 1988. Oxidation of sulfur compounds. Symp. Soc. Gen. Microbiol. **42**:65–98.

59. Kelly, D.P., and S.C. Baker. 1990. The organosulfur cycle: aerobic and anaerobic processes leading to the turnover of C1-sulfur compounds. FEMS Microbiol. Rev. **87**:241–246.

60. Kelly, D.P., and N.A. Smith. 1990. Organic sulfur compounds in the environment: Biogeochemistry, microbiology, and ecological aspects. Adv. Microbial Ecol. **11**:345–385.

61. Kelly, D.P., and A.P. Wood. 1984. Potential for methylotrophic autotrophy in *Thiobacillus versutus* (*Thiobacillus* sp. strain A2). In R.L. Crawford and R.S. Hanson (eds.), Microbial Growth on C_1 Compounds, American Society for Microbiology, Washington DC, pp. 324–329.

62. Kiene, R.P. 1988. Dimethyl sulfide metabolism in salt marsh sediments. FEMS Microbiol. Ecol. **53**:71–78.

63. Kiene, R.P. 1990. Dimethyl sulfide production from dimethylsulfoniopropionate in coastal seawater and bacterial cultures. Appl. Environ. Microbiol. **56**:3292–3297.

64. Kiene, R.P. 1991. Evidence for the biological turnover of thiols in marine sediments. Biogeochemistry **13**:117–135.

65. Kiene, R.P. 1992. Dynamics of dimethyl sulfide and dimethylsulfoniopropionate in oceanic water samples. Marine Chem. **37**:29–52.

66. Kiene, R.P., and T.S. Bates. 1990. Biological removal of dimethyl sulfide from sea water. Nature **345**:702–705.

67. Kiene, R.P., and D.G. Capone. 1988. Microbial transformations of methylated sulfur compounds in anoxic saltmarsh sediments. Microbial Ecol. **15**:275–291.

68. Kiene, R.P., and S.K. Service. 1991. Decomposition of dissolved DMSP and DMS in estuarine waters: dependence on temperature and substrate concentration. Marine Ecol. Progr. Ser. **76**:1–11.

69. Kiene, R.P., and B.F. Taylor. 1988. Biotransformations of organosulfur compounds in sediments via 3-mercaptopropionate. Nature **332**:148–150.

70. Kiene, R.P., and B.F. Taylor. 1988. Demethylation of dimethylsulfoniopropionate and production of thiols in anoxic marine sediments. Appl. Environ. Microbiol. **54**:2208–2212.

71. Kiene, R.P., and B.F. Taylor. 1989. Metabolism of acrylate and 3-mercaptopropionate, decomposition products of dimethylsulfoniopropionate, in anoxic marine sediments. In E.S. Saltzman and W. Cooper (eds.), Biogenic Sulfur in the Environment, American Chemical Society, Washington DC., pp. 222–230.

72. Kiene, R.P., and P.T. Visscher. 1987. Production and fate of methylated sulfur compounds from methionine and dimethylsulfoniopropionate in anoxic salt marsh sediments. Appl. Environ. Microbiol. **53**:2426–2434.

73. Kiene, R.P., K.D. Malloy, and B.F. Taylor. 1990. Sulfur-containing amino acids as precursors of thiols in anoxic coastal sediments. Appl. Environ. Microbiol. **56**:156–161.

74. Kiene, R.P., R.S. Oremland, A. Catena, L.G. Miller, and D.G. Capone. 1986. Metabolism of reduced methylated sulfur compounds in anaerobic sediments and by a pure culture of an estuarine methanogen. Appl. Environ. Microbiol. **52**: 1037–1045.

75. Kim, K.-H., and M.O. Andreae. 1987. Carbon disulfide in seawater and the marine atmosphere over the North Atlantic. J. Geophys. Res. **92**:14733–14738.

76. King, G.M. 1984. Utilization of hydrogen, acetate, and "noncompetitive" sub-

strates by methanogenic bacteria in marine sediments. Geomicrobiol. J. **3**:275–306.

77. Kung, H.F., and C. Wagner. 1970. Oxidation of C1 compounds by *Pseudomonas* sp. MS. Biochem. J. **116**:357–365.
78. Laanbroek, H.J., and N. Pfennig. 1981. Oxidation of short-chain fatty acids by sulfate-reducing bacteria in freshwater and marine sediments. Arch. Microbiol. **128**:330–335.
79. Leck, C., U. Larsson, L.E. Bagender, S. Johansson, and S. Hajdu. 1990. Dimethyl sulfide in the Baltic Sea: annual variability in relation to biological activity. J. Geophys. Res. **95**:3353–3364.
80. Lee, C., and S.G. Wakeham. 1988. Organic matter in seawater: Biogeochemical processes. In J. Riley (ed.), Chemical Oceanography, Vol. 8. Academic Press, New York, pp. 1–51.
81. Legrand, M., C. Feniet-Saigne, E.S. Saltzman, C. Germain, N.I. Barkov, and V.N. Petrov. 1991. Ice-core record of oceanic emissions of dimethylsulfide during the last climate cycle. Nature **350**:144–146.
82. Lorimer, G.H., and J. Pierce. 1990. Carbonyl sulfide: an alternative substrate for but not an activator of ribulose-1,5-bisphosphate carboxylase. J. Biol. Chem. **264**:2764–2772.
83. Lovelock, J.E. 1974. CS_2 and the natural sulfur cycle. Nature **248**:625–626.
84. Luther, G.W., III, and E. Tsamakis. 1989. Concentration and form of dissolved sulfide in the oxic water column. Marine Chem. **27**:165–177.
85. Mansouri, S., and A.W. Bunch. 1989. Bacterial ethylene synthesis from 2-oxo-4-thiobutyric acid and from methionine. J. Gen. Microbiol. **135**:2819–2827.
86. Mazelis, M., B. Levin, and N. Mallinson. 1965. Decomposition of methyl methionine sulfonium salts by a bacterial enzyme. Biochim. Biophys. Acta **105**:106–114.
87. Mopper, K., and B.F. Taylor. 1986. Biogeochemical cycling of sulfur: thiols in marine sediments. In M. Sohn (ed.), Organic Marine Geochemistry. American Chemical Society Symposium Series. No. 305, American Chemical Society, Washington DC, pp. 324–339.
88. Nanninga, H.J., and J.C. Gottschal. 1986. Isolation of a sulfate-reducing bacterium growing with methanol. FEMS Microbiol Ecol. **38**:125–130.
89. Neidleman, S.L., and J. Geigert. 1986. Biodehalogenation: Principles, Basic Roles and Applications, Wiley, New York.
90. Nguyen, B.C., S. Belviso, N. Mihalopoulos, J. Gostan, and P. Nival. 1988. Dimethyl sulfide production during natural phytoplanktonic blooms. Marine Chem. **24**:133–141.
91. Ni, S., and D.R. Boone. 1992. Catabolism of dimethyl sulfide and methanethiol by methylotrophic methanogens. In R.S. Oremland (ed.), Biogeochemistry of Global Change: Radiatively Active Trace Gases, Chapman and Hall, New York.
92. Nomura, J., Y. Nishizuka, and O. Hayaishi. 1963. S-Alkylcysteinase: Enzymatic cleavage of S-methyl-L-cysteine and its sulfoxide. J. Biol. Chem. **238**:1441–1446.

93. Nriagu, J.O., D.A. Holdway, and R.D. Coker. 1987. Biogenic sulfur and the acidity of rainfall in remote areas of Canada. Science **237**:1189–1192.

94. Ogawa, T., and R.K. Togasaki. 1988. Carbonyl sulfide: An inhibitor of inorganic carbon transport in cyanobacteria. Plant Physiol. **88**:800–804.

95. Oremland, R.S. 1988. Biogeochemistry of methanogenic bacteria. In A.J.B. Zehnder (ed.), Biology of Anaerobic Microorganisms, Wiley, New York, pp. 641–705.

96. Oremland, R.S., and B.F. Taylor. 1978. Sulfate reduction and methanogenesis in marine sediments. Geochim. Cosmochim. Acta **42**:209–214.

97. Oremland, R.S., L.M. Marsh, and S. Polcin. 1982. Methane production and simultaneous sulfate reduction in anoxic salt marsh sediments. Nature **296**:143–145.

98. Oremland, R.S., R.P. Kiene, I. Mathrani, M.J. Whiticar, and D.R. Boone. 1989. Description of an estuarine methanogen which grows on dimethyl sulfide. Appl. Environ. Microbiol. **55**:994–1002.

99. Petasne, R.G., and R.G. Zika. 1987. Fate of superoxide in coastal seawater. Nature **325**:516–518.

100. Rajagopal, B.S., and L. Daniels. 1986. Investigation of mercaptans, organic sulfides, and inorganic sulfur compounds as sulfur sources for the growth of methanogenic bacteria, Curr. Microbiol. **14**:137–144.

101. Reed, R.H. 1983. Measurement and osmotic significance of β-dimethylsulfoniopropionate in marine macroalgae. Marine Biol. Lett. **34**:173–181.

102. Richardson, D.J., G.F. King, D.J. Kelly, A.G. McEwan, S.J. Ferguson, and J.B. Jackson. 1988. The role of auxiliary oxidants in maintaining redox balance during growth of *Rhodobacter capsulatus* on propionate or butyrate. Arch. Microbiol. **150**:131–137.

103. Rimbault, A., P. Niel, H. Virelizier, J.C. Darbord, and G. Leluan. 1988. L-Methionine, a precursor of trace methane in some proteolytic clostridia. Appl. Environ. Microbiol. **54**:1581–1586.

104. Ruiz-Herrera, J., and R.L. Starkey. 1969. Dissimilation of methionine by fungi. J. Bacteriol. **99**:544–551.

105. Salsbury, R.L., and D.L. Merricks. 1975. Production of methanethiol and dimethyl sulfide from methionine and identity of the resulting sulfur products. Plant Soil **43**:191–209.

106. Segal, W., and R.L. Starkey. 1969. Microbial decomposition of methionine and identification of the resulting sulfur products. J. Bacteriol. **98**:908–913.

107. Shanker, R., C. Ramakrishna, and P.K. Seth. 1990. Microbial degradation of acrylamide polymer. Arch. Microbiol. **154**:192–198.

108. Shea, D., and W.A. MacCrehan. 1988. Determination of hydrophilic thiols in sediment porewater using ion-pair liquid chromatography coupled to electrochemical detection. Anal. Chem. **60**:1449–1454.

109. Smith, N.A., and D.P. Kelly. 1988. Isolation and physiological characterization of

autotrophic sulfur bacteria oxidizing dimethyl disulfide as sole source of energy. J. Gen. Microbiol. **134**:1407–1417.

110. Smith, N.A., and D.P. Kelly. 1988. Mechanism of oxidation of dimethyl disulfide by *Thiobacillus thioparus* strain E6. J. Gen. Microbiol. **134**:3031–3039.

111. Smith, N.A., and D.P. Kelly. 1988. Oxidation of carbon disulfide as the sole source of energy for the autotrophic growth of *Thiobacillus thioparus* strain TK-m. J. Gen. Microbiol. **134**:3041–3048.

112. Snow, J.T., J.W. Finlay, and G.O. Kohler. 1976. Fate and metabolism of DMSO in agricultural crops. Ann. New York Acad. Sci. **243**:228–236.

113. Sorensen, J. 1988. Dimethylsulfide and methanethiol in sediment porewater of a Danish estuary. Biogeochem. **6**:201–210.

114. Sperl, G.T., and D.S. Hoare. 1971. Denitrification with methanol: Selective enrichment for *Hyphomicrobium* species. J. Bacteriol. **108**:733–736.

115. Steele, R.D., and N.J. Benevenga. 1979. Identification of 3-methylpropionic acid as an intermediate in mammalian methionine metabolism in vitro. J. Biol. Chem. **253**:7844–7850.

116. Steele, R.D., and N.J. Benevenga. 1979. The metabolism of 3-methylpropionic acid in rat liver homogenates. J. Biol. Chem. **253**:8885–8890.

117. Steudler, P.A., and B.J. Peterson. 1984. Contribution of gaseous sulfur from salt marshes to the global sulfur cycle. Nature **311**:455–457.

118. Suylen, G.M.H., and J.G. Kuenen. 1986. Chemostat enrichment and Isolation of *Hyphomicrobium* EG, a dimethylsulfide oxidizing methylotroph and reevaluation of *Thiobacillus* MS1. Antonie van Leeuwenhoek **52**:281–293.

119. Suylen, G.M.H., G.C. Stefess, and J.G. Kuenen. 1986. Chemolithotrophic potential of a *Hyphomicrobium* species, capable of growth on methylated Sulfur Compounds. Arch. Microbiol. **146**:192–198.

120. Suylen, G.M.H., P.J. Large, J.P. van Dijken, and J.G. Kuenen. 1987. Methyl mercaptan oxidase, a key enzyme in the metabolism of methylated sulfur compounds by *Hyphomicrobium* EG. J. Gen. Microbiol. **133**:2989–2997.

121. Taylor, B.F., and D.C. Gilchrist. 1991. New routes for aerobic biodegradation of dimethylsulfoniopropionate. Appl. Environ. Microbiol. **57**:3581–3584.

122. Taylor, B.F., and D.C. Gilchrist. Unpublished data.

123. Taylor, B.F., and R.P. Kiene. 1989. Microbial metabolism of dimethyl sulfide. In E.S. Saltzman and W.J. Cooper (eds.), Biogenic Sulfur in the Environment, American Chemical Society Symposium Series No. 393, American Chemical Society, Washington DC, pp. 202–221.

124. Taylor, B.F., and R.S. Oremland. 1979. Depletion of adenosine triphosphate in *Desulfovibrio* by oxyanions of Group VI elements. Curr. Microbiol. **3**:101–103.

125. Thysse, G.J.E., and T.H. Wanders. 1972. Degradation of *n*-alkane-1-sulfonates by *Pseudomonas*. Antonie van Leeuwenhoek J. Microbiol. Ser. **38**:53–63.

126. Thysse, G.J.E., and T.H. Wanders. 1974. Initial steps in the degradation of

n-1-alkane-sulfonates by *Pseudomonas*. Antonie van Leeuwenhoek J. Microbiol. Ser. **40**:25–37.

127. Turco, R.P., R.C. Whitten, O.B. Toon, J.B. Pollack, and P. Hamill. 1980. OCS, stratospheric aerosols and climate. Nature **283**:283–286.
128. Turner, S.M., and P.S. Liss. 1985. Measurements of various sulfur gases in a coastal marine environment. J. Atmos. Chem. **2**:223–232.
129. Turner, S.M., G. Malin, P.S. Liss, D.S. Harbour, and P.M. Holligan. 1988. The seasonal variation of dimethylsulfide and dimethylsulfoniopropionate concentrations in nearshore waters. Limnol. Oceanogr. **33**:364–375.
130. Vairavamurthy, A., and K. Mopper. 1987. Geochemical formation of organosulphur compounds (thiols) by addition of H_2S to sedimentary organic matter. Nature **329**:623–625.
131. Vairavamurthy, A., and K. Mopper. 1989. Mechanistic studies of organosulfur (thiol) formation in coastal marine sediments. In E.S. Saltzman and W.J. Cooper (eds), Biogenic Sulfur in the Environment, American Chemical Society Symposium Series No. 393., American Chemical Society, Washington, DC, pp. 231–242.
132. Visscher, P.T., and H. van Gemerden. 1991. Production and consumption of dimethylsulfoniopropionate in marine microbial mats. Appl. Environ. Microbiol. **57**:3237–3242.
133. Visscher, P.T., and H. van Gemerden. 1991. Photoautotrophic growth of *Thiocapsa roseopersicina* on dimethyl sulfide. FEMS Microbiol. Lett. **81**:247–250.
134. Visscher, P.T., and H. van Gemerden. 1992. Sulfur cycling in marine microbial mats: distribution of compounds and functional groups of sulfur bacteria. In R.S. Oremland (ed.), Biogeochemistry of Global Change: Radiatively Active Trace Gases, Chapman and Hall, New York.
135. Visscher, P.T., P. Quist, and H. van Gemerden. 1991. Methylated sulfur compounds in microbial mats: in situ concentrations and metabolism by a colorless sulfur bacterium. Appl. Environ. Microbiol. **57**:1758–1763.
136. Visscher, P.T., M.R. Diaz, and B.F. Taylor. 1992. Enumeration of bacteria which cleave or demethylate dimethylsulfoniopropionate in the Caribbean Sea. Marine Ecol. Progr. Ser. (in press).
137. Wagner, C., and E.R. Stadtman. 1962. Bacterial fermentation of dimethyl-β-propiothetin. Arch. Biochem. Biophys. **98**:331–336.
138. Wagner, C., S.M. Lusty, Jr., H-F. Kung, and N.L. Rogers. 1967. Preparation and properties of trimethylsulfonium-tetrahydrofolate methyltransferase. J. Biol. Chem. **242**:1287–1293.
139. Wakeham, S.G., B.L. Howes, and J.W.H. Dacey. 1987. Biogeochemistry of dimethyl sulfide in a seasonally stratified coastal salt pond. Geochim. Cosmochim. Acta **51**:1675–1684.
140. Weisinger, R.A., and W.B. Jakoby. 1980. S-methylation: Thiol S-methyltransferase. In W.B. Jakoby (ed.), Enzymatic Basis of Detoxification, Vol. 1, Academic Press, New York, pp. 131–140.

141. White, R.H. 1982. Analysis of dimethylsulfonium compounds in marine algae. J. Marine Res. **40**:529–536.

142. Widdel, F., and N. Pfennig. 1982. Studies on dissimilatory sulfate-reducing bacteria that decompose fatty acids II. Incomplete oxidation of propionate by *Desulfobulbus propionicus* gen. nov., sp. nov. Arch. Microbiol. **131**:360–365.

143. Winfrey, M.R., and D.M. Ward. 1983. Substrates for sulfate reduction and methane production in intertidal sediments. Appl. Environ. Microbiol. **45**:193–199.

144. Yen, H.C., and B. Marrs. 1977. Growth of *Rhodopseudomonas capsulata* under dark anaerobic conditions with dimethyl sulfoxide. Arch. Biochem. Biophys. **181**: 411–418.

145. Zepp, R.G., and M.O. Andreae. 1989. Factors affecting the photochemical production of carbonyl sulfide in seawater. EOS **70**:1023.

146. Zeyer, J., P. Eicher, S.G. Wakeham, and R.P. Schwarzenbach. 1987. Oxidation of dimethylsulfide to dimethylsulfoxide by phototrophic purple bacteria. Appl. Environ. Microbiol. **53**:2026–2032.

147. Zinder, S.H., and T.D. Brock. 1978. Dimethyl sulfoxide reduction by microorganisms. J. Gen. Microbiol. **105**:335–342.

148. Zinder, S.H., and T.D. Brock. 1978. Dimethyl sulfoxide as an electron acceptor for anaerobic growth. Arch. Microbiol. **116**:35–40.

149. Zinder, S.H., and T.D. Brock. 1978. Production of methane and carbon dioxide from methane thiol and dimethyl sulfide by anaerobic lake sediments. Nature **273**:226–228.

150. Zinder, S.H., and T.D. Brock. 1978. Methane, carbon dioxide, and hydrogen sulfide production from the terminal methiol group of methionine by anaerobic lake sediments. Appl. Environ. Microbiol. **35**:344–352.

Formation of Dimethylsulfide and Methanethiol from Methoxylated Aromatic Compounds and Inorganic Sulfide by Newly Isolated Anaerobic Bacteria

*Friedhelm Bak**

Max-Planck-Institut für Terrestrische Mikrobiologie
Karl-von-Frisch Strasse, D-3550 Marburg, Germany

Kai Finster

Institute of Biological Sciences, Department of Microbial Ecology, University of Aarhus, Ny Munkegade, DK-8000 Aarhus C, Denmark

ABSTRACT

Anaerobic enrichment cultures supplemented with methoxylated aromatic acids (3,4,5-trimethoxybenzoic acid or syringic acid) showed intense gas production and smell of methylated sulfur compounds. Via direct dilution of mud samples, two strictly anaerobic rod-shaped bacteria were isolated which produced dimethylsulfide and methanethiol during growth on methoxylated aromatic compounds. The volatile organic sulfur compounds were derived from the methoxyl carbon and the sulfide added as a reducing agent to the medium. Both isolates also grew with gallate, phloroglucinol, or pyrogallol without producing dimethylsulfide or methanethiol. The aromatic substrates were degraded to acetate. Determination of the stoichiometries of substrate degradation and sulfur compound formation indicated a stepwise methylation of sulfide to dimethylsulfide with methanethiol as an intermediate. The principle biochemical pathways leading to formation of methylated sulfur compounds by the new isolates have not been elucidated so far.

Further experiments established that a variety of anoxic sediment samples had a relatively high capacity for dimethylsulfide and methanethiol production via sulfide-linked demethoxylation. Dimethylsulfide was the dominant sulfur compound in freshwater sediments, in contrast to marine sediments where methanethiol predominated. Autoclaved sediment samples did not produce organosulfur compounds. The release of dimethylsulfide and methanethiol during demethoxylation reactions by the acetogenic bacteria represents a new mode by which methylated sulfur compounds can be formed in natural anoxic habitats.

*Deceased.

Introduction

Investigations during the last decade have demonstrated the importance of volatile methylated sulfur compounds in the global cycling of sulfur (e.g., Refs 1, 2, 7, and 19). The list of relevant sulfur species includes dimethylsulfide (DMS), methanethiol (MT), dimethyl disulfide, dimethyl sulfoxide, and a few other alkyl sulfides of higher molecular weight (e.g., Refs 17, 22, and 23). Evidence is increasing that much of the atmospheric sulfur, originating from natural marine or terrestrial habitats, enters the atmosphere in the form of DMS (e.g., Refs. 1, 6, 25). It is, thus, not surprising that numerous recent reports on the biogeochemistry of sulfur have dealt with the production and fate of DMS and related compounds.

A variety of biogenic sources for DMS or MT have been identified. On a global basis, the major source for DMS is probably the marine environment. There, DMS is released under oxic conditions from marine algae during enzymatic cleavage of the osmolyte dimethylsulfonium propionate.[9,10,24] Additional sources for DMS include sulfonium compounds like S-methyl methionine produced by plants (for review, see Ref. 17). DMS may also be released from soils on addition of various sulfur-containing fertilizers.[8] However, the mechanisms by which DMS is produced in soils are largely unknown. Under anoxic conditions, e.g., in anoxic sediments, some bacteria may utilize DMSO as respiratory electron acceptor and produce DMS.[19,28,29] DMS production by rumen microorganisms has also been observed.[26]

MT is produced by numerous microorganisms during degradation of the amino acid methionine.[16] Considering the ubiquitous occurrence of methionine in living organisms, it is not surprising that MT generation has been observed in various oxic and anoxic environments.[9,15,20] Recently, Drotar et al.[11] demonstrated that MT can also be formed enzymatically from sulfide by many heterotrophic bacteria. The reaction is catalyzed by a methyl-transferase, where S-adenosylmethionine serves as a methyl donor. In anoxic sediments, MT formation may also be the result of a demethylation of DMS.[20]

During enrichment studies, originally designed to obtain sulfate reducers degrading aromatic compounds, we discovered a new mode by which DMS and MT can be formed from inorganic sulfide and methoxylated aromatic compounds under strictly anoxic conditions. The following sections summarize the most important findings from pure culture and sediment slurry experiments concerning this novel mechanism of methylated sulfur-compound formation.[4,12]

Discovery and Isolation of DMS- and MT-Producing Acetogenic Bacteria

Freshly inoculated enrichment cultures in defined sulfide-reduced media that were supplemented with either trimethoxybenzoate or syringate (Figure 1)

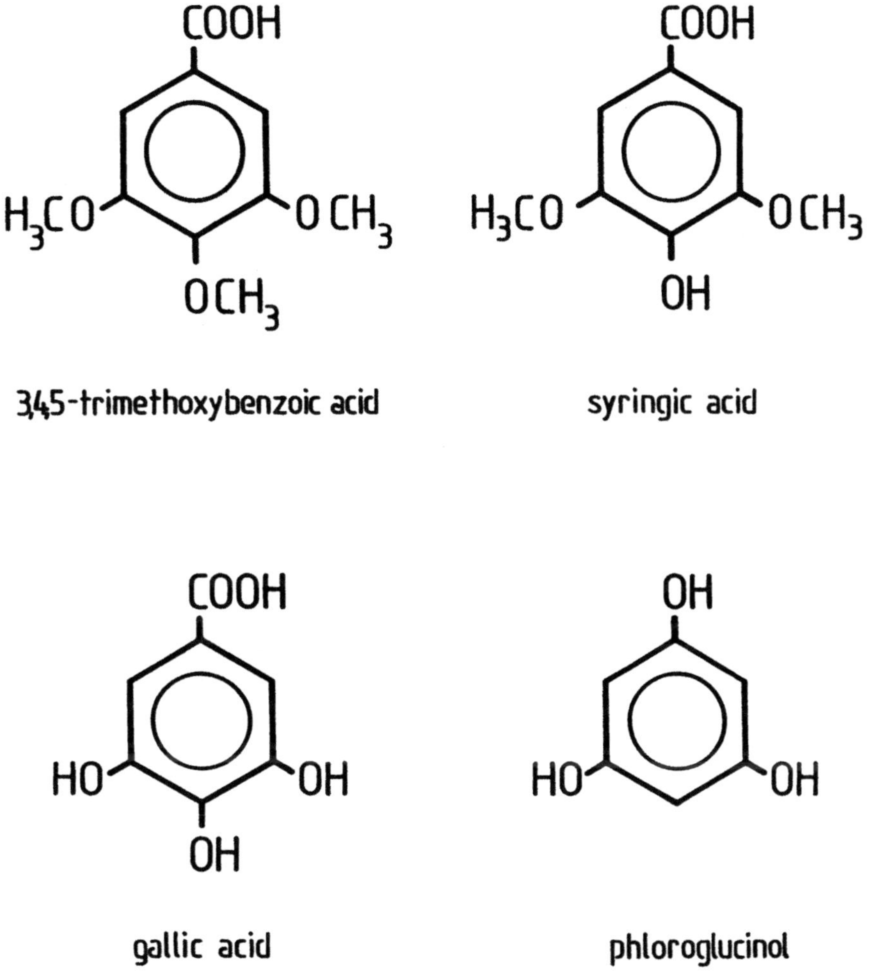

Figure 1. Illustration of some methoxylated and hydroxylated aromatic compounds used in the present study.

showed intense formation of gas and they smelled strongly of organic sulfur compounds after 3–4 d of incubation. On repeated transfer, the gas formation and the offensive smell disappeared and growth of rod-shaped *Acetobacterium*-like bacteria occurred. The development of *Acetobacterium* species was consistent with earlier findings by Bache and Pfennig[3] who demonstrated that anoxic media with methoxylated aromatic compounds may be selective for these bacteria. However, *Acetobacterium* have, so far, not been reported to produce gas or organic sulfur compounds during growth on methoxylated aromatic compounds. Apparently, some other type of bacteria must have been the cause for the gas formation in the described enrichment cultures.

Indeed, via direct dilution of anoxic freshwater mud samples, a tiny rod-shaped bacterium was obtained that degraded the aromatic acids with concomitant formation of methylated sulfur compounds. This organism developed only in higher dilutions where, obviously, it was not outcompeted by the numerically less abundant but more efficiently growing *Acetobacterium* species. When transferred into fresh medium, the tiny rods continued to produce volatile sulfur compounds. By applying deep agar dilutions, we succeeded in isolating the organism (strain TMBS4, DSM No. 6591). Furthermore, using similar dilution techniques with syringate as substrate, a marine strain was isolated which catalyzed the same process of DMS and MT formation.

The discovery and isolation of these two strains demonstrates well the shortcomings of batch enrichments. Such enrichments usually select for those microorganisms which are best adapted and which grow most efficiently under the given conditions (here *Acetobacterium*). However, the fast-growing strains are not necessarily the numerically predominant organisms in the natural habitat. Others, such as the DMS- and MT-producing acetogenic bacteria described herein, may be overgrown even though they occur in higher numbers. It is likely that many ecologically relevant organisms escape detection in batch enrichment.

Stoichiometries of Sulfur-Compound Formation

The two isolates produced the DMS and MT from the methoxyl groups of the aromatic ring and the hydrogen sulfide which was initially added as reducing agent to the medium. Formation of methylated sulfur compounds was not observed when the strains were cultivated with aliphatic substrates or aromatic substrates lacking methoxyl groups (e.g., pyruvate, phloroglucinol, or gallic acid). In addition, during growth in sulfide-free dithionite-reduced medium, DMS and MT were not produced. In the presence of excess aromatic substrate (e.g., syringic acid), all sulfide added to the medium was completely consumed and could be recovered in the thiol pool. This indicates a very efficient methyl transfer onto sulfide. The major organosulfur compound produced by the cultures was DMS. MT occurred only transiently and in significantly lower concentrations (Figure 2). The appearance of MT only as an intermediate indicates a stepwise methylation of hydrogen sulfide with MT as a first reaction product: (R = aromatic residue):

1. methylation of sulfide:

$$R\text{—}OCH_3 + H_2S \rightarrow R\text{—}OH + CH_3SH;$$

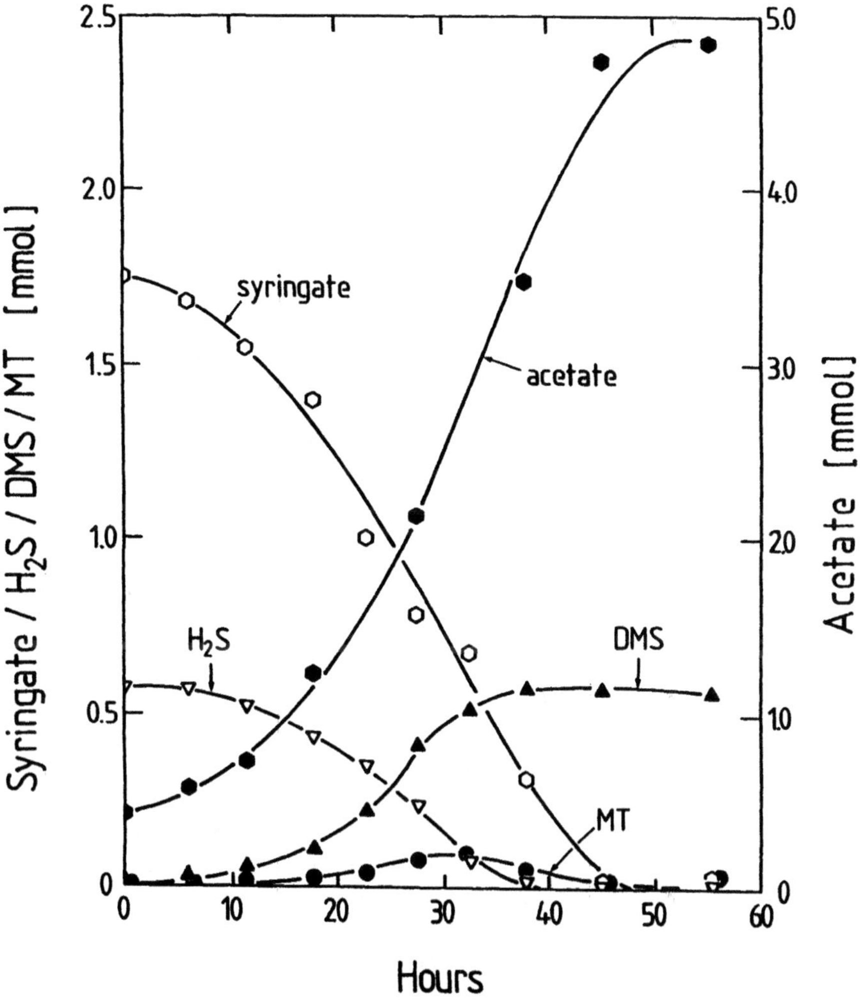

Figure 2. Time course of syringate degradation with concomitant production of dimethylsulfide (DMS) and methanethiol (MT) by the newly isolated acetogenic freshwater strain. The experiment was conducted using stopper-sealed 1-L bottles with 800 ml culture volume and 200 ml anoxic (N_2/CO_2) head space (redrawn from Ref. 4).

2. methylation of methanethiol:

$$R\text{—}OCH_3 + CH_3SH \rightarrow R\text{—}OH + (CH_3)_2S;$$

Sum:

$$2\,R\text{—}OCH_3 + H_2S \rightarrow 2\,R\text{—}OH + (CH_3)_2S.$$

Indirect evidence for such a stepwise mechanism of DMS formation was recently obtained in experiments with cell suspensions of the freshwater isolate, where it was shown that the strain could utilize a variety of organic sulfides as methyl-group acceptors (Kreft and Schink, in preparation).

The biochemical pathways and the enzymes involved in these methylation reactions are, at present, entirely unknown but may be partly similar to the recently described MT-yielding sulfide methylations in heterotrophic bacteria.[11] It remains unknown whether the new isolates derive any benefit from these reactions or if they are purely fortuitous. The process is probably not a detoxification of sulfide since both isolates grew in sulfide-reduced media with non-methoxylated substrates without being adversely affected. In fact, it was impossible to cultivate the marine strain in a sulfide-free medium.

Some Further Characteristics of the Newly Isolated Acetogenic Bacteria

The freshwater and the marine isolate were mesophilic rod-shaped, strictly anaerobic bacteria. Although clearly different in size and salt requirements, the isolates had a rather similar physiology. Of a great variety of aliphatic and aromatic compounds tested, only a few methoxylated and trihydroxylated aromatic compounds were utilized as growth substrates. In addition, the freshwater strain could grow slowly on pyruvate. The major metabolic end product of the substrate degradation was, excluding the thiols, acetate (Figure 2). The marine strain also formed minor amounts of butyrate.

The two isolates are the first anaerobes in pure culture which are able to carry out both the demethoxylation of the aromatic ring and the subsequent cleavage of the ring. The overall anaerobic degradation of methoxylated aromatic compounds to acetate was previously believed to be catalyzed only by the interaction of a demethoxylating acetogen (e.g., *Acetobacterium* or *Clostridium* species) and a ring-degrading organism like *Pelobacter acidigallici*.[3,27] The new isolates, thus, combine the metabolic capabilities of these two groups of anaerobic bacteria. The taxonomic status of the freshwater strain is presently under investigation.

An interesting feature, especially of the marine strain, was the fact that even well-grown cultures survived only relatively short storage periods. Even after storage of only 3–4 weeks at 4°C in the dark, it was difficult to activate the marine culture again. Storage of the strain at room temperature caused death within 4–5 d. The freshwater isolate was somewhat more resistant. Reasons for this may be the toxicity of the formed DMS or an increased sensitivity toward trace contaminations of oxygen because of lack of the reductant sulfide, which had been converted into DMS during growth.

Demonstration of the Process of DMS- and MT-Formation in Anoxic Sediment Slurries

The fact that the DMS- and MT-producing acetogens could only be obtained from higher dilutions (10^{-5}–10^{-6}), where they were not outcompeted by *Acetobacterium,* indicated the presence of a relatively large population of these organisms in anoxic habitats. Hence, the described process of DMS- and MT-release may, in fact, be an important property of anoxic sediments. To investigate this possibility, we have conducted slurry experiments with anoxic reduced sediment samples of different origin. The samples were from a salt-marsh area at the Danish west coast, from a freshwater lake, and from a fjord near the town of Aarhus (Denmark). The experiments were performed using airtight sealed serum bottles filled with a homogenized and diluted (1 : 5 ratio; vol sediment : vol water) sediment from the upper 0–2 cm. After preincubation, experiments were started by injection of small amounts of methoxylated aromatic substrates (e.g., syringic acid or trimethoxybenzoate). The concentrations of DMS, MT, and methane were analyzed periodically.

Two typical time courses are depicted in Figures 3a and 3b. All three sediments had a relatively high potential for DMS- and MT-production from methoxylated aromatic compounds. Autoclaved slurries did not produce methylated sulfur compounds, demonstrating that the process is biologically catalyzed. Also on addition of gallate (Figure 1), release of DMS and MT was not observed. This indicated that the methyl carbon of DMS and MT is derived from the methoxy-group carbon and that the process of organosulfur compound formation observed in sediment slurries is similar to the one established in the preceding pure culture studies. In slurries of marine origin, the formation of the volatile sulfur only started after a lag phase of about 2–3 d. The reason for the delay may be that the enzymes involved in methylation of sulfide first have to be induced. Toxicity of the employed aromatic compounds may be a further explanation, although added amounts were relatively small (0.5 mM final concentration).

Interestingly, the saline slurries produced predominantly MT, whereas in the freshwater samples, DMS was the major sulfur compound. This may be due to

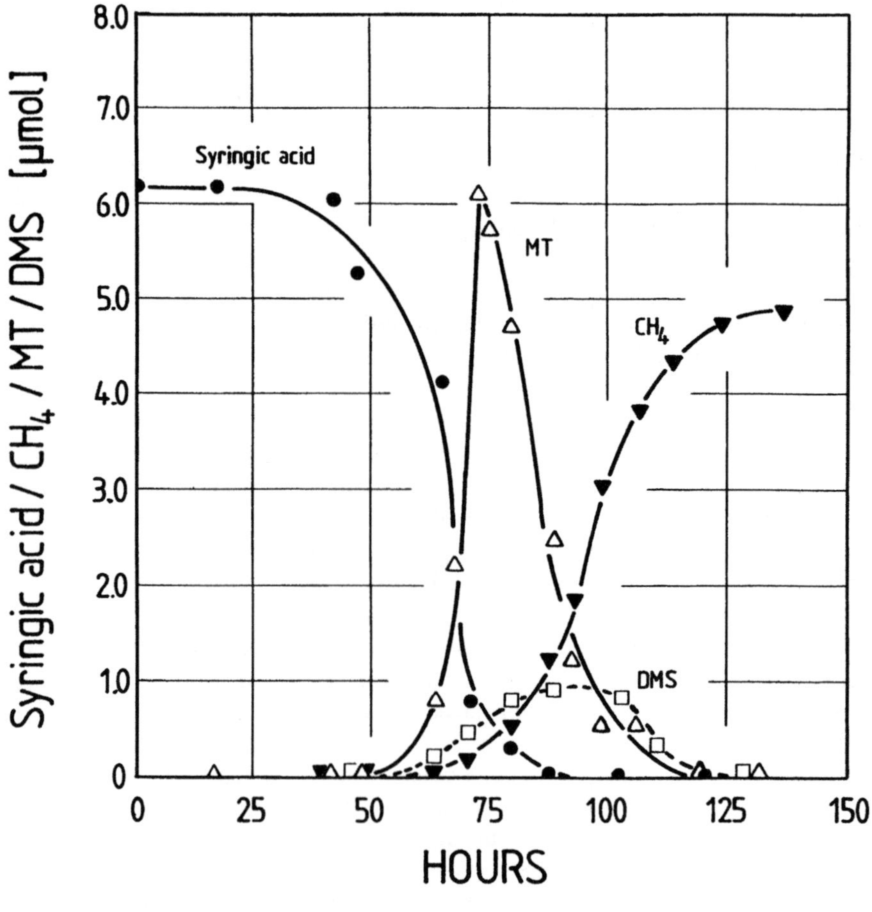

Figure 3. Formation of dimethylsulfide (DMS) and methanethiol (MT) in anoxic sediment slurries after addition of syringic acid. The experiment was conducted using stopper-sealed 120-ml septum bottles with 60 ml slurry and anoxic (N_2) head space (redrawn from Ref. 12). (a) Marine sediment from Kysing Fjord (Denmark); (b) sediment from Vilhelmsborg Lake (Denmark).

the differences in the available sulfur pools in both types of sediments. Marine sediments are characterized by high sulfate concentrations and corresponding high rates of sulfate reduction which sometimes results in substantial accumulation of free hydrogen sulfide (e.g., H_2S concentrations in Kysing Fjord may exceed of 10 mM; K. Finster, unpublished data). In contrast, H_2S concentrations in freshwater sediments are often undetectably low (e.g., Ref. 5). Assuming that the observed DMS production is based on a stepwise methylation of sulfide (as

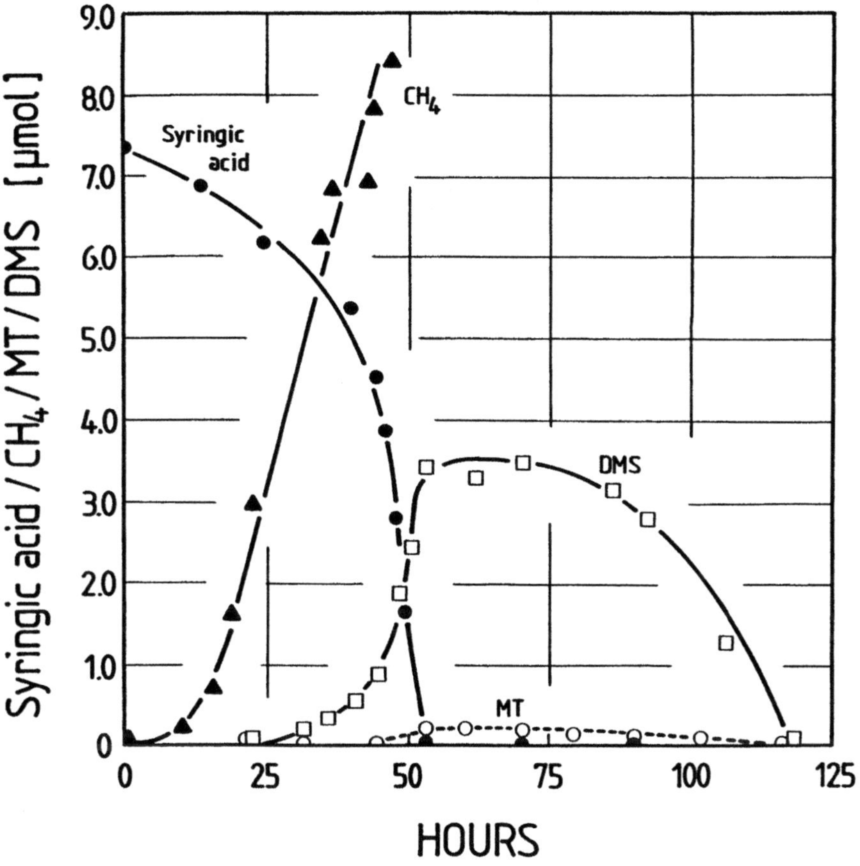

Figure 3b.

suggested above), it becomes understandable why marine sediments mainly produced MT and freshwater sediments produced DMS. In the marine samples, the methyl groups will be preferentially captured by H_2S that is present in excess. Only occasionally will some of the formed MT react with the methyl-group donor to form DMS. In freshwater samples, on the other hand, the majority of the formed MT can be methylated to DMS because H_2S is limiting.

The transient release of DMS and MT from sediment slurries was in all cases followed by an intense formation of methane, indicating an involvement of methanogenic bacteria in the breakdown of the methylated sulfur compounds. This confirms previous studies with anoxic sediment samples[18,20] where it was demonstrated that methanogens contribute to the degradation of these substrates. Like methanol or methylamine, methylated sulfur compounds may be regarded as so-called ''noncompetitive substrates'' which allow methanogens to survive in habitats dominated by the activity of sulfate-reducing bacteria (e.g., Ref. 21). It is not surprising, therefore, that enrichments with DMS were selective for methanogenic bacteria.[18] By employing similar dilution techniques as described above, we recently succeeded in isolating a methanogenic bacterium in pure culture which is able to grow with MT as sole source of energy and carbon.[13]

Ecological Considerations

The production of DMS and MT by acetogenic bacteria provides a additional and previously unknown source of DMS or MT in natural habitats. A similar biological pathway leading to formation of methylated sulfur compounds has, to our knowledge, not been demonstrated. The overall reactions are the same as those suggested for the chemical formation of methylated sulfur species during paper manufacturing.[14] Interestingly, the source of the methyl group is the same in both cases, namely, the methyl carbon from methoxylated aromatic acids released during lignin decomposition.

The quantitative role of the biological process catalyzed by our newly isolated acetogens has not been established yet. The estimation of the ecological relevance of this process requires detailed information about the in situ concentrations and turnover of methoxylated aromatic acids. Such measurements have, to our knowledge, not been undertaken so far. However, considering the widespread occurrence of lignin, it is very likely that the process contributes to the natural DMS or MT turnover. This is further indicated by the fact that all sediments tested had a relatively high potential for production of DMS and MT from methoxylated aromatic compounds.

Due to the strictly anaerobic nature of the new isolates, the described formation of DMS and MT is probably restricted to anoxic environments. In shallow habitats where the anoxic zones are closely located to the water–air interface, some of the methylated sulfur formed may escape to the atmosphere before it undergoes microbial degradation. Significant emissions of organic sulfur compounds of so far unknown origin have been observed in freshwater wetlands of Canada.[25] The formation of methylated sulfur compounds by sulfide-linked demethoxylation may be an important mechanism explaining these emissions.

The interaction of the described acetogenic bacteria with DMS- or MT-degrading methanogens within a natural ecosystem demonstrates an interesting carbon flow from more complex substrates (methoxylated aromatic compounds) via methylated sulfur compounds into methane and carbon dioxide. By this methyl transfer to an external inorganic acceptor (sulfide), the carbon ends up in

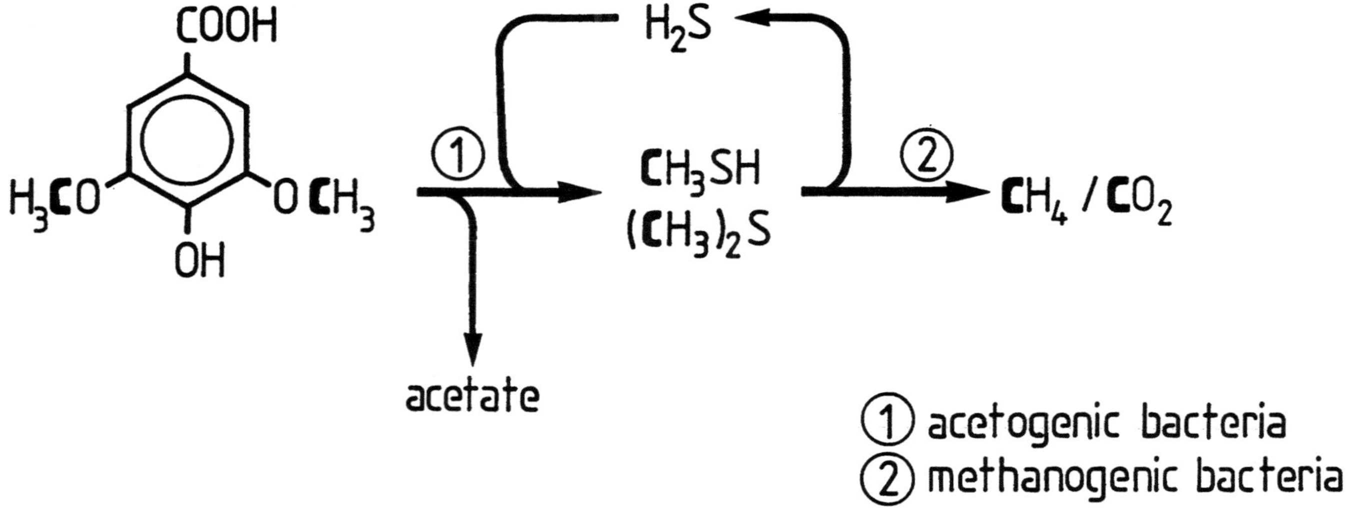

Figure 4. Possible direct microbial transfer of methylgroups from aromatic compounds via methylated sulfur compounds into methane and carbon dioxide involving sulfide as mediating compound (redrawn from Ref. 12).

a substrate pool where it may become available to bacteria which otherwise would not be involved in the degradation. Sulfur, in the form of sulfide, functions in this process as a mediating element which catalyzes the transfer of the methylgroup from one bacterial group to the other (Figure 4).

Acknowledgment

The authors wish to thank Dr. R. Conrad for helpful comments on the manuscript.

References

1. Andreae, M.O. 1986. The ocean as a source of atmospheric sulfur compounds. In P. Buat-Menard (ed.), The Role of Air–Sea Exchange in Geochemical Cycling, E. Reidel, Hingham, MA, pp. 331–362.
2. Andreae, M.O., and T.W. Andreae. 1988. The cycle of biogenic sulfur compounds over the amazon basin. J. Geophys. Res. **93**:1487–1497.
3. Bache, R., and N. Pfennig. 1981. Selective isolation of *Acetobacterium woodii* on methoxylated aromatic acids and determination of growth yields. Arch. Microbiol. **130**:255–261.
4. Bak, F., K. Finster, and F. Rothfuß. 1992. Formation of dimethylsulfide and methanethiol from methoxylated aromatic compounds and inorganic sulfide by newly isolated anaerobic bacteria. Arch. Microbiol. **157**:529–534.
5. Bak, F., and N. Pfennig. 1991. Microbial sulfate reduction in littoral sediment of Lake Constance. FEMS Microbiol. Ecol. **85**:31–42.
6. Banard, W.R., M.O. Andreae, W.R. Watkins, H. Bingemer, and H.W. Georgii. 1982. The flux of dimethyl sulfide from the ocean to the atmosphere. J. Geophys. Res. **87**:8787–8793.
7. Bates, T.S., R.J. Charlson, and R.H. Gammon. 1987. Evidence for the climatic role of marine biogenic sulphur. Nature **329**:319–321.
8. Bremner, J.M., and G.C. Steele. 1978. Role of microorganisms in the atmospheric sulfur cycle. In M. Alexander (ed.), Advances in Microbial Ecology, Vol. 2, Plenum, New York, p. 155.
9. Challenger, F., R. Bywood, D. Thomas, and B.J. Hayward. 1957. The natural occurrence and chemical reactions of some thetins. Arch. Biochem. Biophys. **69**: 514–523.
10. Dacey, J.W.D., and S.G. Wakeham. 1986. Oceanic dimethylsulfide: Production during zooplankton grazing on phytoplankton. Science. **233**:1314–1316.
11. Drotar, A., G.A. Burton, J.E. Travenier, and F. Fall. 1987. Widespread occurrence

of bacterial thiol methyltransferase and the biogenic emission of methylated sulfur gases. Appl. Environ. Microbiol. **53**:1626–1631.

12. Finster, K., King, G.M., and F. Bak. 1990. Formation of methyl mercaptan and dimethylsulfide from methoxylated aromatic compounds in anoxic marine and freshwater sediments. FEMS Microbiol. Ecol. **74**:295–302.
13. Finster, K., Y. Tanimoto, and F. Bak. 1992. Fermentation of methanethiol and dimethylsulfide by a newly isolated methanogenic bacterium. Arch. Microbiol. **157**: 425–430.
14. Hyninnen, P. 1971. On the production of methyl mercaptane and its oxidation products in air or steam stripping of sulfate mill condensates. Paper och trä. Specialnummer 4a, pp. 159–169.
15. Headley, J.V. 1987. GC/MS identification of organosulfur compounds in environmental samples. Biomed. Environ. Mass Spectrom. **14**:275.
16. Kadota, J., and Y. Ishida. 1972. Production of volatile sulfur compounds by microorganisms. Annu. Rev. Microbiol. **26**:127–143.
17. Kelly, D.P., and N.A. Smith. 1990. Organic sulfur compounds in the environment. In K.C. Marshall (ed.), Advances in Microbial Ecology, Vol 11, Plenum, New York, pp. 345–385.
18. Kiene, R.P., R.S. Oremland, A. Catena, L.G. Miller, and D.G. Capone. 1986. Metabolism of reduced methylated sulfur compounds by anaerobic sediments and a pure culture of an estuarine methanogen. Appl. Environ. Microbiol. **56**:1037–1045.
19. Kiene, R.P., and D.G. Capone. 1988. Microbial transformations of methylated sulfur compounds in anoxic salt marsh sediments. Microb. Ecol. **15**:275–291.
20. Kiene, R.P., and B.F. Taylor. 1988. Demethylation of dimethylsulfoniopropionate and production of thiols in anoxic marine sediment. Appl. Environ. Microbiol. **54**:2208–2212.
21. King G.M. 1984. Utilization of hydrogen, acetate, and "noncompetitive" substrates by methanogenic bacteria in marine sediments. Geomicrobiol. J. **3**:275–306.
22. Kjaer, A. 1977. Low molecular weight sulphur-containing compounds in nature: A survey. Pure Appl. Chem. **49**:137–152.
23. Krouse, H.R., and R.G.L. McCready. 1979. Biogeochemical cycling of sulfur. In P.A. Trudinger (ed.), Cycling of the Mineral Forming Elements, Elsevier, Amsterdam, pp. 401–430.
24. Lovelock, J.E., R.J. Maggs, and R.A. Rasmussen. 1972. Atmospheric dimethylsulfide and the natural sulfur cycle. Nature **237**:452–453.
25. Nriagu, J.O., D.A. Holdway, and R.D. Coker. 1987. Biogenic sulfur and the acidity of rainfall in remote areas of Canada. Science **237**:1189–1192.
26. Salsbury, R.L., and D.L. Merricks. 1975. Production of methanethiol and dimethyl sulfide by rumen microorganisms. Plant and Soil **43**:191–209.
27. Schink, B., and N. Pfennig. 1982. Fermentation of trihydroxybenzenes by *Pelobacter acidigallici* gen. nov. sp. nov., a new strictly anaerobic, non-sporeforming bacterium. Arch. Microbiol. **133**:195–201.

28. Yen, H.C., and B. Marrs. 1977. Growth of *Rhodopseudomonas capsulatus* under anaerobic dark conditions with dimethylsulfoxide. Arch. Biochem. Biophys. **181**: 411.

29. Zinder, S.H., and T.D. Brock. 1978. Dimethyl sulfoxide reduction by microorganisms. J. Gen. Microbiol. **105**:335–342.

Catabolism of Dimethylsulfide and Methane Thiol by Methylotrophic Methanogens

Shuisong Ni

Department of Chemical and Biological Sciences, Oregon Graduate Institute, Beaverton, Oregon 97006-1999

David R. Boone

Department of Environmental Science and Engineering, and Department of Chemical and Biological Sciences, Oregon Graduate Institute, Beaverton, Oregon 97006-1999

ABSTRACT

We demonstrated the ability of several methylotrophic methanogens to degrade dimethylsulfide and methane thiol to hydrogen sulfide, methane, and carbon dioxide. This is the first report of the growth of a pure culture of methanogen on methane thiol. *Methanolobus siciliae* HI350, a methylotrophic methanogen isolated from an oil well, was grown routinely on trimethylamine. When this culture was inoculated into a medium with 5 mM dimethyl sulfide, it began producing methane and hydrogen sulfide after a lag of several weeks. Methane production was slow, with an apparent microbial growth rate of 0.0033 h^{-1} (about 3% as fast as growth on trimethylamine or methanol). The lag was shorter when 2 mM methane thiol was substrate. When a culture of *M. siciliae* HI350 growing on dimethylsulfide was subcultured on dimethylsulfide, the lag disappeared and growth rate was higher (0.087 h^{-1}). Dimethylsulfide-grown cultures also had no lag when transferred to media with methane thiol. Studies of cell-free extracts suggested that enzymes for the degradation of dimethylsulfide and methane thiol were inducible, whereas those for the degradation of methanol and trimethylamine were constitutive. Degradation of dimethylsulfide or methane thiol was complete, and stoichiometric quantities of methane and hydrogen sulfide were formed. Most surprisingly, this strain could be adapted to grow with high concentrations of dimethylsulfide or methane thiol, as high as 30 mM. Other methanogens which have been reported to catabolize dimethylsulfide are *Methanolobus siciliae* T4/M, *Methanohalophilus zhilinaeae* WeN5, *Methanohalophilus oregonensis* WAL1, and strain GS-16. The *M. siciliae* strains and strain GS-16 were also able to use methane thiol as their catabolic substrate (other strains were not tested). We tested other methylotrophic methanogens for their ability to use dimethylsulfide at concentrations which did not inhibit their ability to degrade trimethylamine and found the following cultures unable to catabolize dimethylsulfide: *Methanococcoides methylutens* TMA-10, *Methanolobus tindarius* Tindari 3, *Methanolobus vulcani* PL-12/M, *Methanohalophilus mahii* SLP, *Methanohalophilus halophilus* Z-7982, *Methanosarcina mazeii* LYC, and *Methanosarcina mazeii* C 16.

Introduction

Dimethylsulfide (DMS) and methane thiol (MT) are important compounds of the sulfur cycle in a wide variety of terrestrial and aquatic environments.[1–5,10–13,20,21,24,25] These methylated sulfides are formed as catabolic products during the degradation of sulfur-containing organic matter, especially of the compounds dimethylsulfoniopropionate and methionine. Perhaps more importantly, DMS and MT are produced by transfer of methyl groups to sulfide during biodegradation of organic matter.[4] DMS and MT are degraded to hydrogen sulfide (and sometimes methane) in various anoxic environments, but little is known about the pathways by which these compounds are degraded or about the organisms responsible. Recently, several strains of obligately methyltrophic methanogens were shown to use DMS as their substrate: strain GS-16,[19] *Methanohalophilus zhilinaeae* WeN5,[16] *Methanohalophilus oregonensis* WAL1.[14]

Methanolobus siciliae HI350, a DMS-degrading strain from an oil well in the Gulf of Mexico, and the type strain (strain T4/M) can degrade DMS.[18] These two strains also grow on trimethylamine (TMA) and methanol. Cultures grown either on TMA or methanol produce methane from 5 mM DMS after a lag of 2 weeks, and stoichiometric quantities of methane and sulfide are formed according to

$$(1)\quad 2CH_3SCH_3 + 2H_2O \rightarrow CO_2 + 3CH_4 + 2H_2S, \qquad \Delta G^{\circ\prime} = -73.8 \text{ kJ/mol.}$$

These cultures also degrade MT, although growth on this substrate has not been established:

$$(2)\quad 4CH_3SH + 2H_2O \rightarrow 4H_2S + CO_2 + 3CH_4, \qquad \Delta G^{\circ\prime} = -36.6 \text{ kJ/mol.}$$

Whole-cell protein analysis by denaturing gel electrophoresis indicated that major changes in intracellular proteins of *M. siciliae* HI350 occur when its growth substrate is switched between DMS and TMA; this is consistent with the long lags we observed.

In this chapter we show that DMS can be used by only a few species of obligately methylotrophic methanogens. Also, *M. siciliae* HI350 grew on MT as well as DMS, and the products from either of these were methane, hydrogen sulfide, and (presumably) carbon dioxide.

Materials and Methods

Source of Organisms

All the organisms were obtained from the Oregon Collection of Methanogens (OCM): *Methanolobus siciliae* HI350 (OCM 210), *Methanolobus siciliae* T4/M (OCM 156), *Methanolobus tindarius* Tindari 3 (OCM 150), *Methanolobus vulcani* PL-12/M (OCM 157), *Methanococcoides methylutens* TMA-10 (OCM 158), strain GS-16 (OCM 122), *Methanosarcina mahii* LYC (OCM 34), *Methanosarcina mazeii* C 16 (OCM 98), *Methanohalophilus halophilus* Z-7982 (OCM 160) and *Methanohalophilus zhilinaeae* WeN5 (OCM 62).

Media and Culture Techniques

Modified anaerobic techniques of Hungate[7,18] were used, and growth rates were calculated from specific methane production rates during the exponential phase of growth.[18] The MSHCO2 medium used throughout this study contained (per liter): 2.0 g of Trypticase peptone (BBL Microbiology Systems, Cockeysville, MD), 2.0 g of yeast extract, 0.5 g of 2-mercaptoethanesulfonic acid, 29.2 g of NaCl, 4.0 g of NaOH, 2.7 g of $MgCl_2 \cdot 6H_2O$, 0.45 g of KCl, 1.0 g of NH_4Cl, 0.4 g of $K_2PO_4 \cdot 3H_2O$, 0.25 g of $Na_2S \cdot 9H_2O$, 1.0 mg of resazurin, 5.0 mg of Na–EDTA$\cdot 2H_2O$, 1.5 mg of $CoCl_2 \cdot 6H_2O$, 1.0 mg of $MnCl_2 \cdot 4H_2O$, 1.0 mg of $FeSO_4 \cdot 7H_2O$, 1.0 mg of $ZnCl_2$, 0.4 mg of $AlCl_3 \cdot 6H_2O$, 0.3 mg of $Na_2WO_4 \cdot 2H_2O$, 0.2 mg of $CuCl_2 \cdot 2H_2O$, 0.2 mg of $NiSO_4 \cdot 6H_2O$, 0.1 mg of H_2SeO_3, 0.1 mg of H_3BO_3, 0.1 mg of $NaMoO_4 \cdot 2H_2O$. The medium was prepared by dissolving NaOH in O_2-free water and equilibrating the preparation with CO_2; final pH was 6.8. For growth of alkaliphilic strains (stain GS-16, *M. oregonensis* and *M. zhilinaeae*), the gas phase was then switched to 100% N_2, and the medium was called MSHA; its pH was 8.3. The other constituents (except sulfide) were then added from concentrated stock solutions. The medium was dispensed into serum bottles which were sealed and autoclaved (121°C, 20 min). Cultures with DMS and MT were routinely grown in 300 ml of MSHCO2 or MSHA medium contained in a 600-ml bottle. Large culture bottles minimize the effects of loss of those substrates by the absorption of butyl-rubber stoppers. Although we did not measure the pH after the addition of catabolic substrate, it is unlikely that the addition of DMS or MT changed the pH of the medium; DMS is not known to ionize, and MT does not strongly ionize at the pH of the media (pK_a = 10.75).[3] Sulfide and most substrates were added from sterile anoxic stock solutions approximately 1 h prior to inoculation. DMS was added as concentrated liquid, and MT was added as a gas. Inoculation was 10% in all the experiments. Cultures were tested in duplicate and incubated at 40°C.

Preparation of Crude Cell-Free Extracts

M. siciliae HI350 was grown in 13 L of MSHCO2 medium with 50 mM methanol, 50 mM TMA, or 30 mM DMS as catabolic substrate. Cells were harvested in the late exponential phase by continuous-flow centrifugation at 10,000 *g*. A suspension of about 40% wet cells was made in O_2-free buffer (20 mM $KHCO_3$, 10 mM $MgCl_2$, 2 mM titanous chloride,[23] and 4 mM sodium citrate, adjusted to pH 7.2) and cells were broken by ultrasonification at 35 W for 5–10 min. After centrifugation at 20,000 *g* for 30 min, the supernatant was used for cell-free enzymatic assay. The assay mixture (final volume, 1.0 ml) contained 10 mM $MgCl_2$, 10 mM coenzyme M, 2 mM ATP, 4 mM titanous chloride, and 8 mM sodium citrate. The enzyme reaction was started by adding 0.4 ml of cell-free extract (about 20 mg of total protein per ml) to the mixture and incubating at 40°C.

Analytical Methods

Methane was determined by gas chromatography.[15] DMS, MT, and H_2S in the gas phase were determined by combined gas chromatography–mass spectrometry with sulfur hexaflouride (SF_6) as internal standard.[5,6] These compounds were separated on a SPB-1 SULFUR fused capillary column 30 m long with 0.32 mm internal diameter, with 3-μm film (Supelco) at 40°C with helium as carrier gas (linear velocity, 30 cm/s). Injection volume was 10 μl with a split ratio to the detector of 35 : 1. The ionizer of the mass detector was operated at 233°C. Standard gas mixtures were prepared by transferring gaseous DMS or MT in a pressure-lock syringe at room temperature to a sealed serum bottle previously flushed with N_2. Concentrations of DMS, MT, and H_2S were measured only in the gas phase. We measured the Henry's constant of those compounds in our culture medium and used these values to calculate the total quantity present in culture vessels (gaseous plus aqueous amounts) from the gaseous concentrations. Methanol was also measured by gas chromatography of liquid samples.

Results and Discussion

Henry's Constants for DMS and MT in MS Medium

We measured the Henry's constants of DMS in MSH medium by adding 3.4 mmol of DMS to 300 ml of culture medium in a 600-ml sealed vessel, equilibrating at 40°C, and measuring the concentration of DMS in the head space. Analysis indicated a Henry's constant for DMS of 90.3 mmol per liter of gas per molar concentration in the liquid. The Henry's constant of MT measured in the same way was 118 (mmol/L) M^{-1}. Thus, at 40°C, culture vessels with equal volumes of gas and liquid had 15% of the DMS and 20% of the MT in the gas phase.

Growth of M. siciliae HI350 on DMS and MT

In addition to methanol, TMA and DMS, strain HI350 grew on MT as substrate. As shown in Figure 1A, TMA-grown cultures began to produce methane after a lag of only 3 days when transferred to 2 mM MT, but the lag was substantially longer with DMS. The lag before growth on DMS was unaffected by DMS concentrations between 2 and 10 mM. Similar lag was also observed when methanol-grown cells were first transferred to DMS. Methane production from either DMS or MT was initially slow, with an estimated specific growth rate of 0.003 h^{-1} (about 3% as fast as growth on TMA or methanol). Although our initial studies[18] found that the lag continued to decline over several transfers, more recent data (Figure 1A) indicated that the lag disappeared after one transfer on DMS when cells were transferred in late exponential phase. Growth rate of subsequent cultures increased to a maximum of 0.087 h^{-1} and 0.068 h^{-1} on 5 mM DMS and MT, respectively. When cultures growing on 5 mM DMS were transferred (10% vol/vol) to medium with 50 mM TMA as substrate, growth occurred without a lag (data not shown). When this culture was transferred in late exponential phase to medium with 5 mM DMS, DMS was catabolized without a lag (Figure 1B). However, if the culture was transferred three or more times in medium with TMA, a long lag preceded DMS utilization (Figure 1B). This suggested that enzymes necessary for toleration or catabolism of DMS are not produced constitutively.

Most surprisingly, this organism could be adapted to high concentrations of DMS or MT. An unadapted (TMA-grown) culture showed no growth on 5 mM MT, but after gradual adaptation, it grew on 25 mM MT and 40 mM DMS. Although cultures grew rapidly on 40 mM DMS, they ceased producing methane before the stoichiometrically expected methane was formed. It is possible that in the latter stages of growth these cultures were inhibited by accumulating sulfide.

Catabolism of DMS and MT by M. siciliae HI350

We measured H_2S, CH_4, DMS, and MT during growth of *M. siciliae* HI350 on DMS or MT. We found that DMS or MT were completely degraded, with expected amounts of H_2S and CH_4 formed (Figures 2 and 3). In addition, small

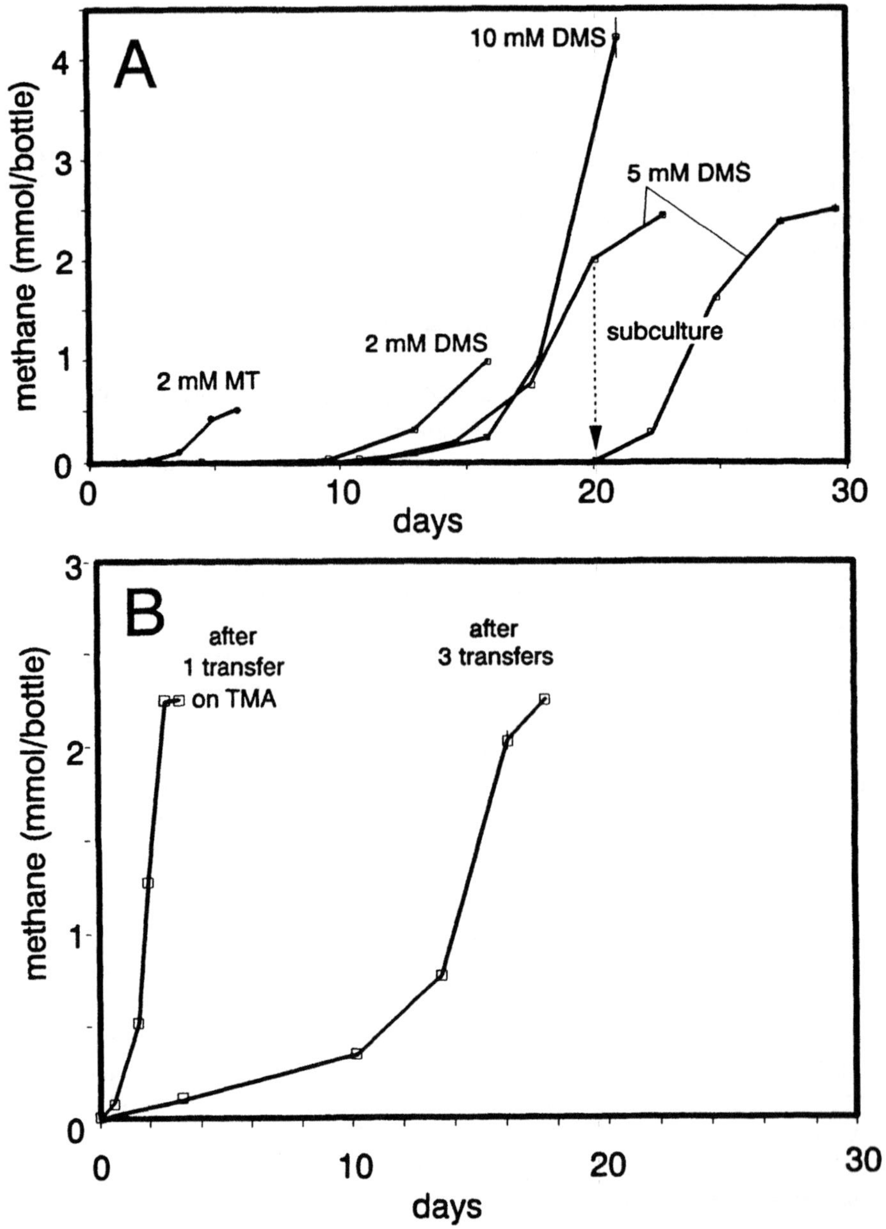

Figure 1. Growth of TMA-grown *M. siciliae* HI350 on DMS and MT. (A) Cultures grown on 20 mM TMA and not previously exposed to DMS or MT were inoculated into media with DMS or MT. After 20 d incubation, one of the replicate cultures in 5 mM DMS (in late exponential phase) was transferred to fresh medium with 5 mM DMS (indicated by dotted line and arrow). (B) DMS-grown cells were transferred and grown once or three times (10% vol/vol inoculum) in medium with 20 mM TMA and then transferred back to medium with 5 mM DMS.

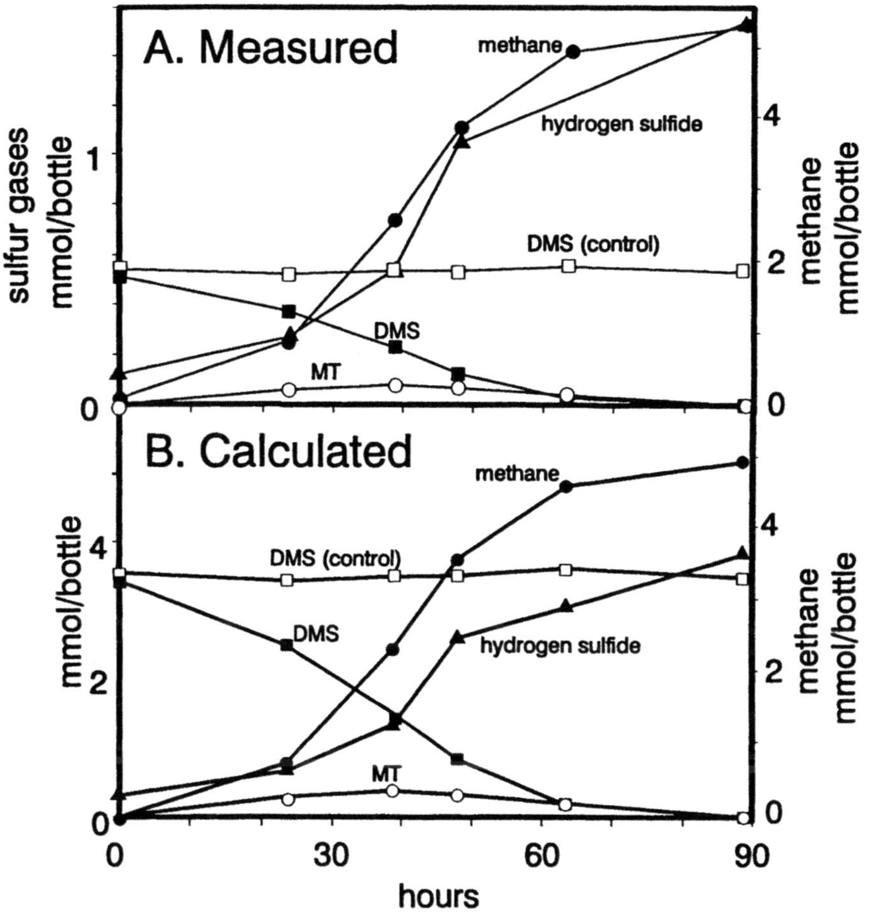

Figure 2. Growth of DMS-adapted cultures. Total gases measured (A) in the head space of cultures growing on DMS, or DMS in an uninoculated control and (B) in the entire bottle (with dissolved gases calculated from the measured gas-phase concentration from Henry's constants of each constituent, which we measured in identical culture medium).

amounts of MT formed during DMS degradation (Figure 2), and even smaller amounts of DMS formed during MT degradation (Figure 3). Kiene and Capone also found these compounds transiently during DMS and MT degradation in salt-marsh sediments.[11] It is possible that one of these compounds may be a free intermediate in the degradation of the other, but other explanations are also plausible, including the presence of methyltransferases which catalyze reversible reactions.

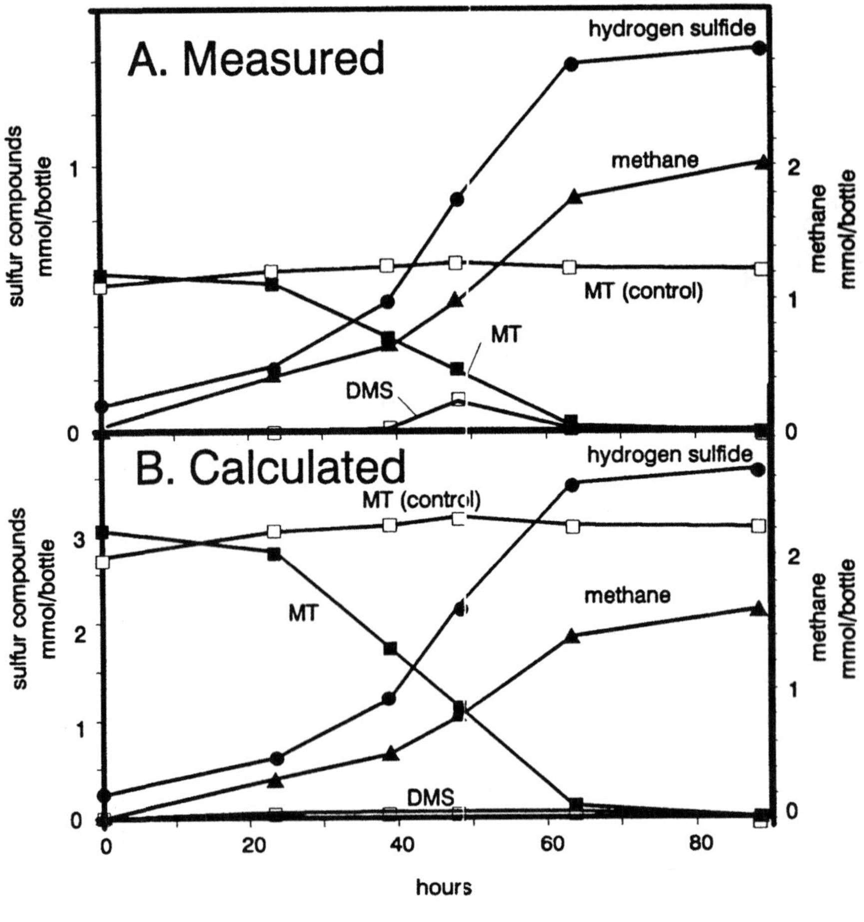

Figure 3. Growth of MT-adapted cultures. Total gases measured (A) in the head space of cultures growing on MT, or MT in an uninoculated control, and (B) in the entire bottle (with dissolved gases calculated from the measured gas-phase concentration).

Catabolic Activities in Crude Cell-Free Extracts of Strain HI350

The notion that enzymes of strain HI350 for the degradation of DMS and MT are inducible was also supported by the results of cell-free enzymatic assays. Only the cell-free extract prepared from DMS- or MT-grown cells showed enzymatic activities for DMS and MT degradation and produced methane from DMS and MT (Figures 4 and 5). Cell-free extracts of either methanol- or TMA-

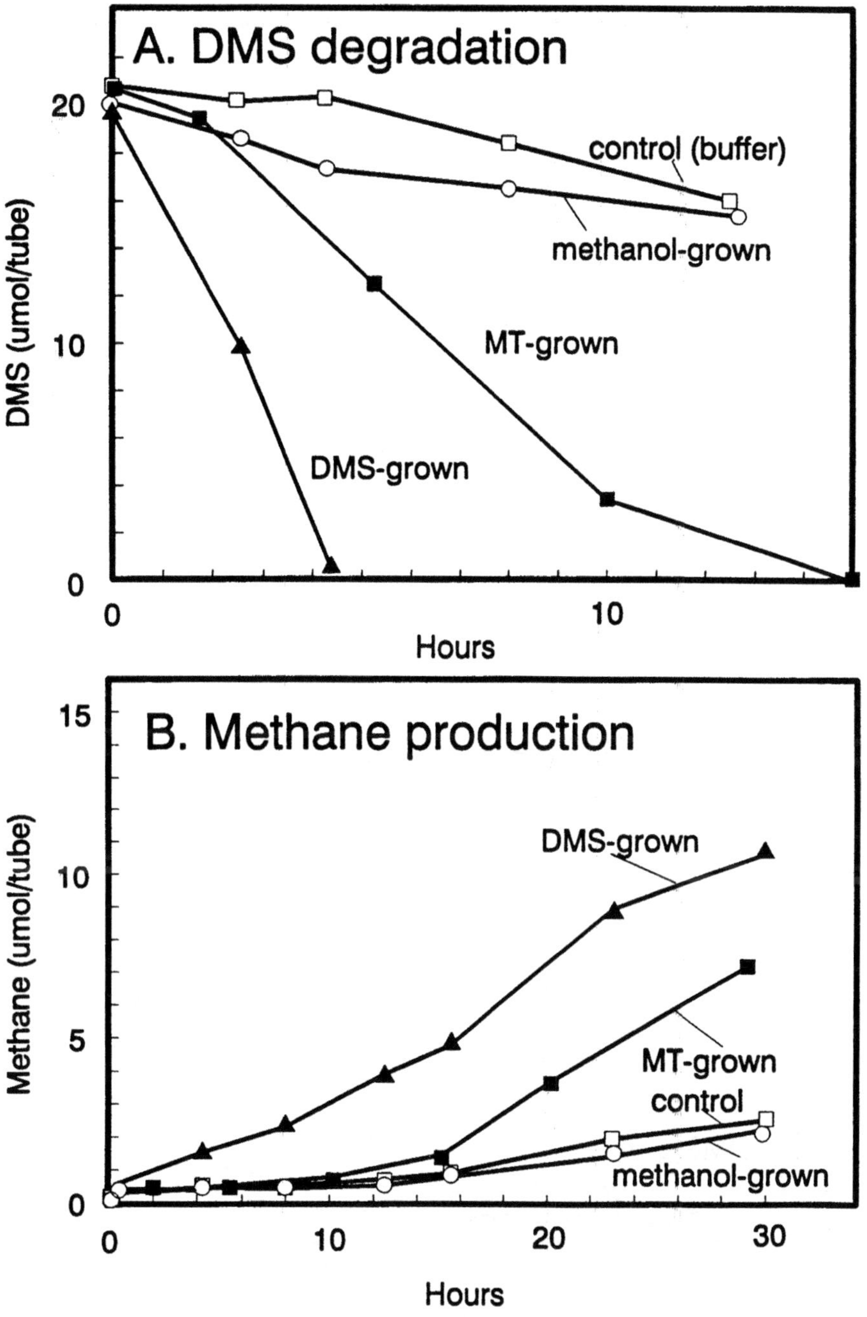

Figure 4. DMS degradation (A) and CH4 production (B) by cell-free extracts prepared from cells grown on methanol, DMS, and MT. DMS degradation and methane formation by TMA-grown cells (not shown) was similar to that by methanol-grown cell-free extracts. The control in (A) has cell-free extract replaced by buffer and the control in (B) is DMS-grown cell-free extract without added substrate.

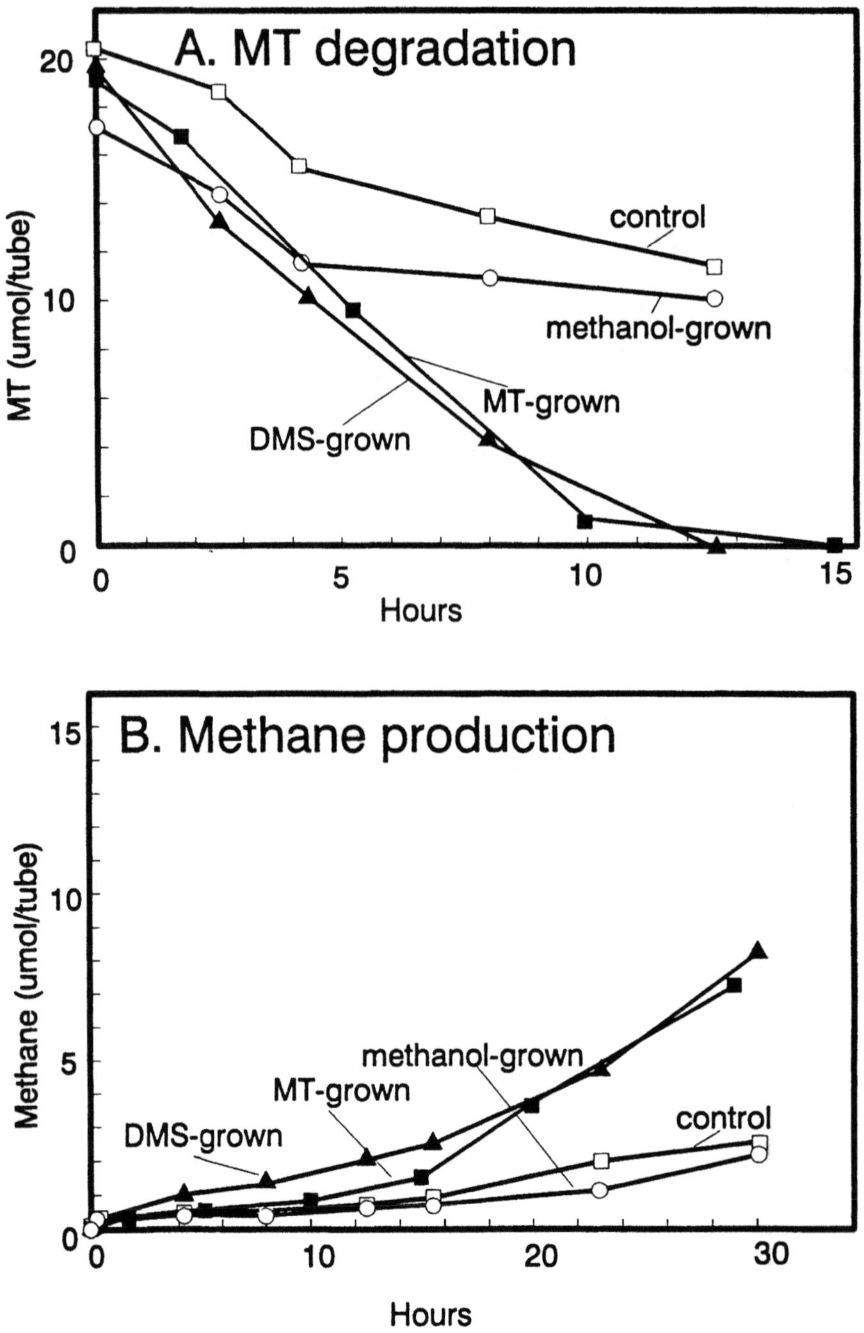

Figure 5. MT degradation (A) and CH4 production (B) by cell-free extracts of methanol-, DMS-, and MT-grown cells. MT degradation and methane formation by TMA-grown cells (not shown) was similar to that by methanol-grown cell-free extracts.

grown cells showed no enzymatic activity with either of DMS or MT. However, activities for methanol degradation were present in cell-free extracts of all the substrates (Figure 6). Similarly, cell-free extracts from TMA, methanol, DMS, and MT produced methane from TMA (data not shown). These results suggested that the enzymes for the utilization of methanol and TMA were constitutively expressed by this organism,[17] which was consistent with our earlier observation that DMS-adapted cultures could resume growth on TMA or methanol without a lag.

In cell-free extracts which degraded DMS or MT (Figures 4 and 5), substrate depletion was more rapid than methane formation, suggesting that the methane-liberating reaction (methylreductase) was the rate-limiting reaction, and that the transfer of methyl groups from DMS and MT (methyltransferase) could occur more rapidly, resulting in the transient accumulation of intermediates. We did not quantify these intermediates, which may have been methylated corrinoids or coenzyme M.[8,9,22]

Growth of Strain GS-16 on MT

Strain GS-16 grows on DMS, but it was reported to be unable to grow on MT.[19] After we found that *M. siciliae* HI350 could grow on MT, we rechecked strain GS-16 by inoculating cultures grown on DMS into medium with 5 mM MT. The cultures grew and produced methane in amounts expected by the stoichiometry of Eq. (2). When these cultures were subcultured into medium with 5 mM MT, exponential growth occurred. Differences between these results and earlier reports may be due to differences in media or incubation temperature.

DMS Utilization Among other Methanogens

The following methanogens were selected from different methylotrophic genera and tested: *Methanococcoides methylutens* TMA-10, *Methanolobus tindarius* Tindari 3, *Methanolobus vulcani* PL-12/M, *Methanohalophilus mahii* SLP, *Methanohalophilus halophilus* Z-7982, *Methanosarcina mazei* LYC, and *Methanosarcina mazeii* C 16. TMA-grown cultures were inoculated into medium with 5 mM DMS, as well as three control media with the following catabolic substrates: (i) none; (ii) 5 mM TMA; and (iii) 5 mM TMA plus 5 mM DMS. For all strains listed, cultures with trimethylamine plus DMS produced similar amounts of methane as cultures with TMA alone; no extra quantities of methane was formed in cultures with DMS only as compared to those without substrate. Thus, none of those strains were able to catabolize DMS.

Thus, only a few obligately methyltrophic halotolerant species were capable of catabolizing DMS. As suggested by Finster et al.,[4] this capability would allow DMS (and MT) to be interspecies methyl-group carriers in anoxic systems, in a

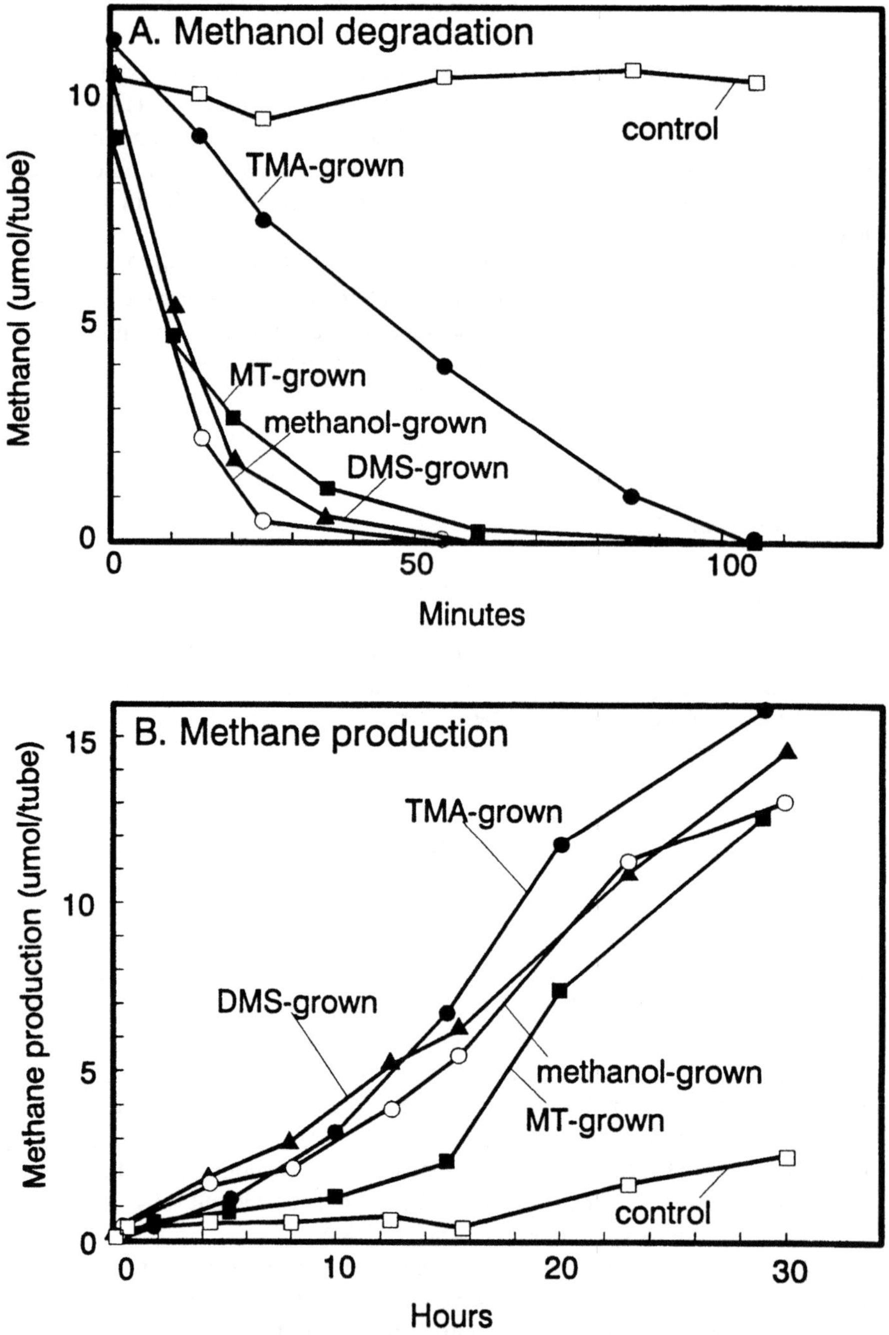

Figure 6. Methanol degradation (A) and CH4 production (B) by cell-free extracts of HI350 cells grown on methanol, TMA, DMS, and MT.

manner analogous to H_2 and formate being interspecies electron carriers. These methylated sulfides supported the growth of *M. siciliae* HI350 and T4/M and of strain GS-16 and were degraded to methane, hydrogen, sulfide, and (presumably) carbon dioxide according to Eqs. (1) and (2).

Acknowledgments

We thank Lorne Isabelle and Bruce Tiffany for help in the analysis of DMS and MT, and we thank José Menaia and Yitai Liu for helpful discussions.

This work was supported by grants from the U.S. Department of the Interior, U.S. Geological Survey (grants 14-08-0001-G1636 and 14-08-0001-G2098). However, this chapter does not necessarily represent the policy of that agency, and you should not assume endorsement by the Federal Government.

References

1. Andreae, M.O., and H. Raemdonck. 1983. Dimethyl sulfide in the surface ocean and the marine atmosphere: a global view. Science **221**:744–747.
2. Charlson, R.J., J.E. Lovelock, M.O. Andreae, and S.G. Warren. 1987. Oceanic phytoplankton, atmospheric sulfur, cloud albedo and climate. Nature (London) **326**:655–661.
3. Dean, J.A. 1987. Handbook of Organic Chemistry, McGraw-Hill, New York, Section 8, p. 37.
4. Finster, K., F. Bak, and G.M. King. 1991. Formation of methylmercaptan and dimethyl sulfide from methyoxylated aromatic compounds in anoxic marine and freshwater sediments. 10th International Symposium on Environmental Biogeochemistry, San Francisco.
5. Henatsch, J.J., and F. Jüttner. 1990. Occurrence and distribution of methane thiol and other volatile organic sulphur compounds in a stratified lake with anoxic hypolimnion. Arch. Hydrobiol. **119**:314–323.
6. Hutte, R.S., N.G. Johansen, and M.F. Legier. 1990. Column selection and optimization for sulfur compound analyses by gas chromatography. J. High Resolution Chromatogr. **13**:421–426.
7. Hungate, R.E. 1969. A roll tube method for cultivation of strict anaerobes. In R. Norris and D.W. Ribbons (eds.), Methods in Microbiology, Vol. 3B. Academic Press, New York, pp. 117–132.
8. Jablonski, P.E., and J.G. Ferry. 1991. Purification and properties of methyl coenzyme M methylreductase from acetate-grown *Methanosarcina thermophila*. J. Bacteriol. **173**:2481–2487.
9. Kengen, S.W.M., J.J. Mosterd, R.L.H. Nelissen, J.T. Keltjens, C. van der Drift,

and G.D. Vogels. 1988. Reductive activation of the methyl-tetrahydromethanopterin: coenzyme M methyltransferase from *Methanobacterium thermoautotrophicum* strain ΔH. Arch. Microbiol. **150**:405–412.

10. Kiene, R.P. 1988. Dimethyl sulfide metabolism in salt marsh sediments. FEMS Microb. Ecol. **53**:71–78.

11. Kiene, R.P., and D.G. Capone. 1988. Microbial transformations of methylated sulfur compounds in anoxic salt marsh sediments. Microb. Ecol. **15**:275–291.

12. Kiene, R.P., R.S. Oremland, A. Catena, L.G. Miller, and D.G. Capone. 1986. Metabolism of reduced methylated sulfur compounds in anaerobic sediments and by a pure culture of an estuarine methanogen. Appl. Environ. Microbiol. **52**:1037–1045.

13. Kiene, R.P., and P.T. Visscher. 1987. Production and fate of methylated sulfur compounds from methionine and dimethylsulfoniopropionate in anoxic salt marsh sediments. Appl. Environ. Microbiol. **53**:2426–2434.

14. Liu, Y., D.R. Boone, and C. Choy. 1990. *Methanohalophilus oregonense* sp. nov., a methylotrophic methanogen from an alkaline, saline aquifer. Int. J. Syst. Bacteriol. **40**:111–116.

15. Maestrojuán, G.M., and D.R. Boone. Characterization of *Methanosarcina barkeri* strains MS^T and 227, *Methanosarcina mazei* $S\text{-}6^T$, and *Methanosarcina vacuolata* $Z\text{-}761^T$ Int. J. Syst. Bacteriol. **41**:267–274.

16. Mathrani, I.M., D.R. Boone, R.A. Mah, G.E. Fox, and P.P. Lau. 1988. *Methanohalophilus zhilinae* sp. nov., an alkaliphilic, methyltrophic methanogen. Int. J. Syst. Bacteriol. **38**:139–142.

17. Naumann, E.K. Fahlbusch, and G. Gottschalk. 1984. Presence of a trimethylamine: HS-coenzyme M methyltransferase in *Methanosarcina barkeri*. Arch. Microbiol. **138**:79–83.

18. Ni, S., and D.R. Boone. 1991. Isolation and characterization of a dimethyl sulfide-degrading methanogen, *Methanolobus siciliae* HI350, from an oil well, characterization of *M. siciliae* $T4/M^T$, and emendation of *M. siciliae*. Int. J. Syst. Bacteriol. **41**:410–416.

19. Oremland, R.S., R.P. Kiene, M.J. Whiticar, and D.R. Boone. 1989. Description of an estuarine methylotrophic methanogen which grows on dimethylsulfide. Appl. Environ. Microbiol. **55**:944–1022.

20. Salsbury, R.L., and D.L. Merricks. 1975. Production of methane thiol and dimethyl sulfide by rumen microorganisms. Plant and Soil **43**:191–209.

21. Taylor, B.F., and R.P. Kiene. 1987. Microbial metabolism of dimethyl sulfide. In E.S. Saltzman and W.J. Cooper (eds.), Biogenic Sulfur in the Environment, American Chemical Society, Washington, DC, pp. 202–221.

22. Van der Meijden, P., B.W.J. te Brömmelstroet, C.M. Poirot, C. van der Drift, and G.D. Vogels. 1984. Purification and properties of methanol:5-hydroxybenzimidazolylcobamide methyltransferase from *Methanosarcina barkeri*. J. Bacteriol. **160**: 629–635.

23. Zehnder, A.J.B., and K. Wuhrmann. 1976. Titanium (III) citrate as a nontoxic oxidation-reduction buffering system for the culture of obligate anaerobes. Nature **194**:1165–1166.

24. Zinder, S.H., and T.D. Brock. 1978. Production of methane and carbon dioxide from methane thiol and dimethylsulfide by anaerobic lake sediments. Nature **273**: 226–228.

25. Zinder, S.H., W.N. Doemel, and T.D. Brock. 1977. Production of volatile sulfur compounds during the decomposition of algal mats. Appl. Environ. Microbiol. **34**:859–860.

Bromoperoxidases: Their Role in the Formation of HOBr and Bromoform by Seaweeds

R. Wever, M.G.M. Tromp, J.W.P.M. van Schijndel, and E. Vollenbroek

E.C. Slater Institute for Biochemical Research and Biotechnological Centre, Plantage Muidergracht 12, 1018 TV Amsterdam. The Netherlands

R.L. Olsen

Norwegian Institute of Fisheries and Aquaculture N-9001 Tromsø, Norway

E. Fogelqvist

Swedish Meteorological and Hydrological Institute, Oceanographic Laboratory, S403 14 Göteborg, Sweden

ABSTRACT

The biogenic release of halogenated volatile organics from the ocean by natural sources and the biochemical pathways which lead to the production of these substances are not well understood. In this chapter, some of the marine sources are discussed and results are presented and reviewed on the role of vanadium bromoperoxidases from seaweeds in the formation of halomethanes. Further, it will be shown that most of the predominant seaweed species from the North Atlantic, which are also present in the Arctic Ocean, contain bromoperoxidases. In contrast, in members of the Desmarestiales collected in the Weddell Sea (Antarctic) which provide the bulk of the biomass of benthic seaweeds in the Antarctic waters, no bromoperoxidase activity could be detected. The distribution of seaweeds with bromoperoxidase activity is correlated with the seawater concentration of bromoform.

Introduction

There is circumstantial evidence that macro-algae release large amounts of volatile halogenated compounds into coastal seawater.[6,9,15,24,25,36,37,41] This natural source of halogenated organic substances is especially of interest in light of the atmospheric processes such as their reaction with ozone in the troposphere[1,46] and stratosphere[7] and their potential greenhouse effect.[47] Such natural sources have to be quantified to assess the impact of man-made halogenated compounds.

Important halogenated volatiles in terms of production on a global scale are methyl iodide, methyl bromide and methyl chloride.[65] The source and biosynthetic pathways of these naturally produced methyl halides are poorly understood,[65,67] except that it is not very likely that peroxidases are involved in their production. Much more is known about the formation and biosynthesis of polybromomethanes by macro-algae[17,18,19,62] and the discussion will be focused on them. As has been demonstrated,[24,25] some marine macro-algae (*Ascophyllum nodosum, Fucus vesiculosis*) release large amounts of volatile halogenated organic compounds like bromoform, dibromomethane, and dibromochloromethane into the seawater. For example, the brown seaweed *Ascophyllum nodosum* produces 4.5 μg of organobromides per gram of dry algae per day.[24] Assuming that all seaweeds, corresponding to a biomass of 10^{13} g algae,[58] release 1–10 μg bromine in the form of organobromide per gram per day, it is possible to estimate an annual global input of 10^{10} g/yr^{-1}.[24] Also Class et al.[9] reported that macro-algae (*Fucus sargassum* and *Laminaria* species) collected near the Bermuda islands and Cape of Good Hope, respectively, emit dibromomethane, bromoform, and dibromochloromethane in large amounts.

There is substantial evidence that bromoperoxidases are involved in the biosynthesis of the halogenated gases by seaweeds. These enzymes catalyze bromination reactions[12,13,63] by the formation of HOBr [Eq. (1)] which in a subsequent reaction with a nucleophilic acceptor (AH) gives rise to the formation of the brominated product ABr [Eq. (2)].

$$H^+ + Br^- + H_2O_2 \rightarrow HOBr + H_2O, \tag{1}$$

$$HOBr + AH \rightarrow ABr + H_2O. \tag{2}$$

For the red seaweed *Bonnemaisonia hamifera*, Theiler et al.[54] have shown that the nucleophilic acceptors are ketones present in the alga which, upon bromination, decay to give rise to formation of brominated methanes. An example of this is the bromination of 3-oxo-octanoic acid[54] which yields a number of intermediates according to

$$HOBr + R\text{—}C{=}O\text{—}CH_2\text{—}COOH \rightarrow R\text{—}C{=}O\text{—}CH_2Br + CO_2 + H_2O, \tag{3}$$

$$HOBr + R\text{—}C{=}O\text{—}CH_2Br \rightarrow R\text{—}C{=}O\text{—}CHBr_2 + H_2O, \tag{4}$$

(5) $$H_2O + R\text{—}C{=}O\text{—}CBr_2 \rightarrow R\text{—}C{=}O\text{—}CHBr_3 + H_2O,$$

(6) $$H_2O + R\text{—}C{=}O\text{—}CBr_3 \rightarrow R\text{—}C{=}O\text{—}OH + CHBr_3.$$

The tri- and dibrominated compounds are unstable and decay to bromoform [Eq. (6)] and dibromomethane, respectively (not shown).

An alternative pathway for the formation of bromoform has been proposed by Wever et al.[64] They showed that in some macro-algae, bromoperoxidases are located on the surfaces of the plants[33] and are able to release HOBr in seawater. HOBr will react rapidly with organic matter which also gives rise to formation of bromoform.[27,29] However, the intermediates involved in this reaction have not been identified yet. Whatever the pathway of formation of bromoform may be, the bromoperoxidases play a crucial role and it is of importance to assess which seaweed species are responsible for the formation of these halogenated compounds and whether there is a link between the concentration of bromoform in seawater, the occurrence of macro-algae, and the presence of these enzymes. Therefore, a number of seaweeds coming from various parts of the globe were screened for the presence of these enzymes. In particular, our efforts were concentrated on the species coming from the Arctic Ocean and North Atlantic since it has been reported[15, 22] that the bromoform which is present in these oceans is, in part, produced by algal belts. In some areas of the Arctic Ocean, these algal belts, which are mainly consisting of Laminariales, are luxuriously developed.[30] For comparison a group of algae from the Antarctic were also tested for the presence of these enzymes. The Antarctic possesses the only cold-water flora without Laminariales and the dominating algae are the Desmarestiales.[40] The concentration of bromoform in Antarctic seawater has been reported to be very low, less than 2 ng/L (M. Kryssel, personal communication) which is considerably lower than in the Arctic Ocean, 5–20 ng/L in the photic layer,[22,34] and this would suggest that in these species no bromoperoxidase activity is present. Indeed, in the Antarctic species tested, no bromoperoxidase activity could be detected by us. The hypothesis that bromoform in the Arctic is, in part, due to bromoperoxidase activity in seaweeds would be in line with this observation.

Materials and Methods

Table 1 summarizes the seaweed species which were screened for the presence of bromoperoxidase activity and the sites of collection. The brown seaweeds from the Antarctic were collected by divers as part of the British Antarctic Survey in the Weddell Sea near Signy Island. All seaweeds were frozen and stored at −20°C. No effect was observed of freezing and thawing on the activity of the isolated bromoperoxidase or on the enzyme in the intact plants in which peroxidase is present.

Table 1. Seaweed species investigated and site of collection

Species	Site of Collection	Depth/Zone	Period
Laminaria hyperborea	Ramsö, Southern part of the Kosterrchipelago (border Norway and Sweden)	5 m	Summer 1989
Laminaria saccharina	Tromsø, Norway	Subtidal zone	Summer 1989
	Oosterschelde, The Netherlands	Subtidal zone	Spring 1990
	Iceland	Subtidal zone	January 1989
Laminaria digitata	Oban, Scotland	Subtidal zone	Spring 1990
Pelvetia canaliculata	Oban, Scotland	Littoral zone	Spring 1990
Fucus vesiculosis	Oosterschelde, The Netherlands	Littoral zone	March 1990
Chondrus crispus	Oosterschelde, The Netherlands	Littoral zone	March 1990
Fucus distichus	Iceland	Littoral zone	January 1989
Alaria esculenta	Iceland	Subtidal zone	January 1989
Ascophyllum nodosum	Iceland	Littoral zone	January 1989
	Tromsø, Norway	Littoral zone	Spring 1989
	Kornwerderzand, The Netherlands	Littoral zone	Spring 1989
Lessonia flavicans (Lessoniaceae)	Weddell Sea, Signy Island, Billie Rocks	4 m under ice	December 1988
Desmaresta anceps (Desmarestiaceae)	Weddell Sea, Signy Island, Billie Rocks	4 m under ice	December 1988
D. chordalis (Desmarestiaceae)	Weddell Sea, Signy Island, Factory cove	1 m under ice	December 1988
Himantothallus grandifolius (Desmarestiaceae)	Weddell Sea, Signy Island, Billy Rocks	4 m under ice	December 1988

To determine the activity of bromoperoxidase in seaweeds, the plants were chopped in a meat mill or in a Waring blendor in 0.2 M Tris-SO_4 (pH 8.3). Alternatively small amounts of plant material were frozen in liquid nitrogen and ground in a mortar. The material was stirred for 30 min in 0.2 M Tris-SO_4 (pH 8.3). After centrifugation, the supernatants were tested for bromoperoxidase activity in an assay system containing 100 mM KBr, 40 μM phenol red, and 100 mM potassium phosphate (pH 6.5). If bromoperoxidase activity is present, bromination of the dye will occur and this results in formation of bromophenol blue. This tetrabrominated compound has a typical absorption spectrum with an absorption band at 592 nm.[14,64] The bromoperoxidase-catalyzed reaction of phenol red has been studied in some detail and is used as a convenient and specific assay to detect bromoperoxidases (cf. Refs. 14, 42, and 64). Optical absorption spectra were obtained with a Hewlett-Packard 8452A spectrophotometer.

Bromoperoxidase from *Laminaria hyperborea* was extracted and purified according to Ref. 11. To test whether the enzyme from *Laminaria hyperborea* did contain vanadium as a prosthetic group, the enzyme was inactivated by dialysis against 0.1 M potassium phosphate (pH 3.8) containing 1 mM EDTA.[11] Reconstitution of enzymic activity by orthovanadate was carried out at neutral pH. The enzyme was completely reactivated by VO_4^{3-}. This reconstitution by vanadate[56] is a property of all vanadium-containing bromoperoxidases and makes it likely that this enzyme is also a vanadium bromoperoxidase.[63]

Results and Discussion

The presence of bromoperoxidases in the marine environment is well-established. A survey has been carried out along the Carribean Coast of Central America which showed that most of the seaweeds investigated (55 out of 72) were positive for the presence of bromoperoxidase activity.[26] A number of these bromoperoxidases are hemoproteins,[39] whereas, as has been discovered recently, others contain vanadium as a prosthetic group.[62] Table 2 summarizes the various seaweed species which up to now have been shown to contain the novel vanadium bromoperoxidases and the areas where they have been collected. As may be seen, the enzymes are found in seaweeds from all parts of the world such as the Great Barrier Reef in Australia, the Japanese Sea, the East Pacific, and the North Atlantic Ocean.

As has been shown by Dyrssen and Fogelqvist,[15] bromoform is found near algal belts in the Arctic Ocean, and from data[21] obtained near and in the Isfjorden, Spitsbergen, it is possible to conclude that this bromoform is produced directly or indirectly by seaweeds. Further, direct measurements have shown that the seaweeds *Ascophyllum nodosum, Fucus vesiculosis,* and *Laminaria laminaria* release substantial amounts of bromoform[9,24,25] in seawater. In particular,

Table 2. Vanadium bromoperoxidases in seaweeds.

	Collection Site	Ref.
Red seaweeds (Rhodophytae)		
Ceramium rubrum	Oosterschelde, The Netherlands	31
Corallina pilulifera	Shirahama, Takahama, Japan	28, 32
Corallina vancouverensis	California, USA	16
Corallina officinalis	Woods Hole, MA, USA	68
Plocamium hamatum	Great Barrier Reef, Australia	64
Brown seaweeds (Phaeophytae)		
Macrocystis pyrifera	California, USA	51
Chorda filum	Vlieland, The Netherlands	11
Ascophyllum nodosum	Kornwerderzand, The Netherlands; Iceland; Roscoff, France; Tromsø, Norway	10, 57, 59, 61
Laminaria saccharina	Tromsø, Norway; Iceland Oosterschelde, The Netherlands	11, this work
Laminaria digitata	Oban, Scotland; Roscoff, France	64
Laminaria hyperborea	Ramsö, Sweden	This work
Fucus vesiculosis	Oosterschelde, The Netherlands	11, 64
Fucus distichus	California, USA; Iceland	51, 61
Alaria esculenta	Iceland	61

the marine algal vegetation along the coast of Spitsbergen is known in detail[53] and accurate data on bromoform concentration are available for these waters. For exposed localities at the outerpart of Isfjorden, the most important sublittoral species are *Alaria grandfolia, Laminaria digitata,* and *Laminaria saccharina* (Table 3) which cover considerable areas and are luxuriously developed to a depth of approximately 15 m. Phaeophyceae other than Laminariales are also present. At depths greater than 20 m and extending to 55 m, various Rhodophyceae are found. It is of some interest to note that some red seaweeds, e.g., *Lithothamnion,* may be abundant even at 88 m depth and have been recorded to be present at a depth of 120 m in East Greenland.

The vegetation in unexposed localities differs in the sense that less different species are found. Table 3 summarizes the most abundant seaweeds at approximately 2–5 m depth. Below 5 m depth, the following five species are predominant: *Chaetopteris plumosa, Desmarestia aculeata, D. viridis, Laminaria saccharina,* and *Phyllophora brodiaei.* A comparison of Table 2 with the species present at exposed and unexposed localities at Spitsbergen (Table 3) clearly shows that in many of the reported species vanadium bromoperoxidases are present.

The distribution pattern of bromoform in the Idefjorden on the border of

Table 3. Sublittoral vegetation in the Isfjorden (Spitsbergen) according to Ref. 53.

Unexposed Localities (2–5 m depth)	
Cladophora spp	*Elachista fucicola*
Monostroma spp.[a]	*Fucus distichus*[a]
Alaria grandifolia	*Laminaria digitata*[a]
Chaetopteris plumosa[a]	*Laminaria saccharina*[a]
Chorda filum	*Phylllaria dermatodea*
Chorda tomentosa[a]	*Ceramium* sp.[a]
Chordaria flagelliformis[a]	*Halosaccion ramentaceum*[a]
Desmarestia aculeata	*Rhodomela lycopodioides*[a]
Dictyosiphon chordaria	*Rhodymenia palmata*[a]
Ectocarpus spp	
Most Important Sublittoral Species at Exposed Localities (Ref. 53)	
Alaria grandfolia	
Laminaria digitata	
Laminaria saccharina	

[a]The most abundant species.

Norway and Sweden and the Skagerrak[23] confirms that bromoform has a natural source and is derived from algal belts outside the fjord at a depth of about 5–30 m. Although the species present have not been identified, it is likely again that Laminariales are present. In fact, the *Laminaria hyperborea* analyzed in this work was collected about 30 km south of the fjord mouth and, as shown in Table 2, this seaweed also contains a vanadium bromoperoxidase. These various observations demonstrate a link between bromoform in seawater and seaweeds and suggest strongly that vanadium bromoperoxidases are involved in the biosynthesis of this halogenated compound. It should be noted that there is strong evidence that pelagic micro-algae also are able to produce bromoform and these algae form a substantial amount of the biomass present[48] in the Arctic Ocean. Studies on bromoform depth profiles carried out by Krysell[34] in the Nansen Bassin indicate the involvement of these algae.

The bromoform concentration in seawater in the Antarctic is considerably lower than that reported for the Arctic. All over the Weddell Sea and at all depths sampled (> 500 samples) during the Swedish Antarctic expedition SWEDARP 88/89, the bromoform concentration was less than 2 ng/L (M. Krysell, personal communication). If the suggestion is correct that there is a direct relation between the presence of vanadium bromoperoxidases in seaweeds and the concentration of bromoform in seawater, these species should lack such enzymes. The dominating algae in the Antarctic are members of the Desmarestiales. Further, they are mainly indigenous and their presence is restricted to these cold waters.[40] Indeed, as expected, the following brown seaweeds which were collected in the Weddell Sea did not contain bromoperoxidase activity at all: *Lessonica flavicans, Desmarestia anceps, Desmarestia chordalis,* and *Himantothallus grandifolius.*

A question that must be answered is via which pathways the bromoform is synthesized and to assess the role of vanadium bromoperoxidase in this process. There may be at least two pathways present and the pathway used may depend on the seaweed species. For the tropical red seaweed *Bonnemaisonia hamifera* from which a variety of halogenated ketones[50] and volatile bromomethanes have been extracted, it has been shown[54] that the purified bromoperoxidase is able to brominate ketones found in this seaweed. The brominated ketones formed are unstable and decay via the haloform reaction to volatile brominated hydrocarbons.[54] It is interesting to note that thalli of members of the family Bonnemaisoniaceae contain so-called vesicle cells in which bromine is present,[66] probably in the form of brominated compounds. Similarly, it has been reported[44] that bromine in the form of organic compounds is also present as granular deposits in the mucilage of the thallus medulla in *Thysanochladia densa* (Gigertinales, Rhodophytae) and in the cuticle of *Polysiphonia nigrescens* (Rhodophytae).[45]

Another, and in light of the above, more likely pathway for the formation of haloforms has been suggested by Wever et al.[64] This suggestion is based on the observation that the brown seaweeds *Laminaria digitata, Laminaria saccharina, Ascophyllum nodosum,* and *Fucus vesiculosis* and the red seaweeds *Chondrus crispus, Plocamium hamatum,* and *Antithamnion plumula*[49] are able to brominate exogeneous organic compounds when these are added to the intact plants in seawater in the presence of hydrogen peroxide.[43] So, for these plants there is evidence that the bromoperoxidases are present extracellularly on the thallus surface of these plants and release HOBr in seawater. Also, in terrestrial plants peroxidases are found which are located extracellularly.[2] In line with the proposed location of the enzyme in some seaweeds, Kylin[35] reported the presence of an iodoperoxidase in the exterior wall of the surface cells of *Laminaria digitata.* This iodoperoxidase is probably the same as the bromoperoxidase since bromoperoxidases are able to oxidize iodide. When the algae produce HOBr on their surfaces, this will result in bromination of dissolved organic matter to form unstable brominated compounds which decay to give rise to bromoform.[27,29] As demonstrated by Jaworske and Helz,[29] HOBr will react extremely rapidly with organic matter in contaminated coastal water to give rise to formation of $CHBr_3$. Evidence for such a process occurring at or near the surface of the seaweed *Ascophyllum nodosum* was also obtained by Gschwend et al.[24] From a comparison of the amount of volatile halogenated organic compounds present within algal tissue with that released to seawater, they concluded that bromine-containing compounds are either produced and transported to the outside of the thallus extremely rapidly, or alternatively are synthesized near the plant surface.

This latter possibility would be in excellent agreement with our finding that some algae release HOBr in seawater which may react with dissolved organic matter to form bromoform.

There is growing interest in the formation of biogenic halogenated volatiles, in particular since it has been reported that there is a strong negative correlation between the ozone concentration in the troposphere and the amounts of bromine containing particles in aerosols collected on filters at Polar Sunrise in the Arctic.[1,5] There is strong evidence that the bromine in these particles is of biogenic origin[3,4] and may come from photolysis of bromoform. There is, however, some dispute concerning the source of bromine, e.g., it has been suggested to result from the interaction of NO_x compounds with NaBr from sea salt in aerosols.[20] Still it is obvious that the Arctic Ocean acts as a source of bromoform which is produced biologically by seaweed belts[15] and possibly phytoplankton.[22,34] Bromoform concentration has been reported[8] to peak in December–January in the Arctic atmosphere and an annual sharp maximum is also observed in the bromine content of Arctic aerosols at Point Barrow (Alaska), Alert (Canadian Arctic), and Spitsbergen (N. Alesund) between February and May at concentrations that are the highest in the world.[3,4,52]

It has been speculated that the presence of bromoform in the Arctic atmosphere is due to the enzymic activity of vanadium bromoperoxidases present in brown seaweeds that grow in the Arctic Ocean.[60] However, data concerning the seasonal release of bromoform by seaweeds from the Arctic Ocean are not available. For some seaweed species, seasonal variations in the release of bromoform have been reported. Experiments carried out in the laboratory on intertidal macro-algae *Ascophyllum nodosum* and *Fucus vesiculosis* collected near Cape Cod, Massachusetts show maximal release in October and November.[24] In contrast, Cape Cod Canal seawater collected at high tide, i.e., when both seaweed species are submerged, showed no seasonal pattern in the formation of polybromomethanes.[25] Wever[60] and Vilter et al.[55] also observed a seasonal pattern in bromoperoxidase activity extracted from the seaweed *Ascophyllum nodosum*. The highest activity was found in winter and spring and low activity in summer. This may be related to a decrease in biomass due to the shedding of the so-called fruiting bodies. Due to major changes in growth rate and reproduction physiology, a clear picture has yet to emerge. Further, other factors like light intensity and temperature may also affect the release of brominated methanes by the seaweeds. As discussed, bromoform may not be produced directly, but may be the product of a secondary reaction between HOBr and dissolved organic matter. Therefore, the nature of the organic matter and its concentration may also greatly affect the nature and yield of the brominated compounds being formed.

The surprising discovery[64] was also made that *Ascophyllum nodosum* is able to brominate organic compounds directly when the plant is exposed to sunlight. This phenomenon was not observed with the other seaweeds tested. The amount of HOBr generated by this seaweed is substantial. From the rate of HOBr formation induced by sunlight (68 nmol g^{-1} h^{-1}) and a biomass of 2×10^8 g of *Ascophyllum nodosum* along a dike in The Netherlands, an annual production of 1.8×10^6 g of HOBr was estimated at this site. This seaweed grows in large quantities along the shores of the North Atlantic and western Russian polar seas[38] and its biomass may well be of the order of 10^{11} g. Assuming that the rate of HOBr formation is the same as that under laboratory conditions under sunlight, an annual global input of 10^9 g of HOBr can be calculated. Most of the HOBr will react with organic matter in seawater but some fraction of HOBr may also escape directly to the air since this seaweed is not always submerged. Therefore, it may well be that the annual sharp maximum observed in bromine content of Arctic aerosol, which peaks every year just after Arctic dawn[3,4,52] is related to photosynthetic activity of seaweeds present in the Arctic Ocean.

It is obvious that a number of questions remain to be answered. What is clearly needed is a study of the bromoform concentration in seawater in the vicinity of algal belts and in open seas as a function of the season. Further, the seaweed species able to produce HOBr and bromoform should be identified and the factors should be studied which induce activity of the vanadium bromoperoxidases. Considering the biomass of micro-algae in the Arctic Ocean, their blooming in spring, and their ability to produce bromoform, it would be most interesting to determine whether these algae also contain vanadium bromoperoxidases.

Acknowledgments

We thank the Dutch State Mines, Geleen for their support and Gunnar Olafsson for collecting the seaweeds on Iceland. We greatly acknowledge Dr. L. Lindner (University of Utrecht) and Anita Buma (NIOZ) for their help in organizing the collection and transportation of seaweeds from the Antarctic to The Netherlands. We especially thank Dr. I. Lovegrove, commander of the base of the British Antarctic Survey at Signy Island as well as the divers and other members of his team, who under harsh conditions were involved in collecting the seaweeds and identifying them. This work was partly supported by the Netherlands Foundation for Chemical Research.

References

1. Barrie, L.A., J.W. Bottenheim, R.C. Schnell, P.J. Crutzen, and R.A. Rasmussen. 1988. Ozone destruction and photochemical reactions at polar sunrise in the lower Arctic atmosphere. Nature **334**:138–141.
2. Barcelo, A.R., R. Munoz, and Sabater F. 1989. Subcellular location of basic and acidic soluble isoperoxidases in lupinus. Plant Sci. **63**:31–38.

3. Berg, W.W., P.D. Sperry, K.A. Rahn, and E.S. Gladney. 1983. Atmospheric bromine in the Arctic. J. Geophys. Res. **88**:6719–6736.
4. Berg, W.W., L.E. Heidt, W. Pollock, P.D. Sperry, and R.J. Cicerone. 1984. Brominated organic species in the Arctic atmosphere. Geophys. Res. Lett. **11**: 429–432.
5. Bottenheim, J.W., L.A. Barrie, E. Atlas, L.E. Heidt, H. Niki, R.A. Rasmussen, and P.B. Shepson. 1990. Depletation of lower tropospheric ozone during Arctic Spring: The Polar Sunrise experiment 1988. J. Geophys. Res. **95**(D11):18555–18568.
6. Burreson, J.A., R.E. Moore, and P.P. Rohler. 1976. Volatile halogen compounds in the alga *Asparagopsis taxiformis* rhodophyta). J. Agric. Food Chem. **24**:856–861.
7. Cicerone, R.J. 1981. Halogens in the atmosphere. Rev. Geophys. Space Phys. **19**:123–139.
8. Cicerone, R.J., L.E. Heidt, and W.H. Pollock. 1988. Measurements of atmospheric methyl bromide and bromoform. J. Geophys. Res. **93**:3745–3749.
9. Class, Th., R. Kohnle, and K. Ballschmiter. 1986. Chemistry of organic traces in air VII: bromo- and bromochloromethanes in air over the Atlantic Ocean. Chemosphere **4**:429–436.
10. De Boer, E., Y. Van Kooyk, M.G.M. Tromp, H. Plat, and R. Wever. 1986. Bromoperoxidase from *Ascophyllum nodosum:* A novel class of enzymes containing vanadium as a prosthetic group. Biochim. Biophys. Acta **869**:48–53.
11. De Boer, E., M.G.M. Tromp, H. Plat, G.E. Krenn, and R. Wever. 1986. Vanadium (V) as an essential element for haloperoxidase activity in marine brown algae: purification and characterization of a vanadium (V) containing bromoperoxidase from *Laminaria saccharina*. Biochim. Biophys. Acta **872**:104–115.
12. De Boer, E., and R. Wever. 1987. Some structural and kinetic aspects of vanadium bromoperoxidases from the marine brown alga *Ascophyllum nodosum*. Receuil. Trav. Chim. Pays-Bas **106**:409.
13. De Boer, E., and R. Wever. 1988. The reaction mechanism of the novel vanadium bromoperoxidase: a steady-state kinetic analysis. J. Biol. Chem. **263**:12326–12332.
14. De Boer, E., H. Plat, M.G.M. Tromp, M.C.R. Franssen, H.C. Van Der Plas, E.M. Meijer, H.E. Schoemaker, and R. Wever. 1987. Vanadium-containing bromoperoxidase, an example of an oxido-reductase with high operational stability in aqueous and organic media. Biotechnol Bioeng. **30**:607–610.
15. Dyrssen, D., and E. Fogelqvist. 1981. Bromoform concentrations of the Arctic Ocean in the Svalbard area. Oceanol. Acta **4**:313–317.
16. Everett, R.R., J.R. Kanofsky, and A. Butler. 1990. Mechanistic investigations of the novel non-heme vanadium bromoperoxidases: evidence for singlet oxygen formation J. Biol. Chem. **265**:4908–4914.
17. Faulkner, D.J. 1984. Marine natural products: metabolites of marine algae and herbivorious marine molluscs. Natural Prod. Repts. **1**:251–280.

18. Fenical, W. 1979. Molecular aspects of halogen-based biosynthesis of marine natural products. Recent Adv. Phytochem. **13**:219–239.

19. Fenical, W. 1982. Natural products chemistry in the marine environment. Science **215**:923–928.

20. Finlayson-Pitts, B.L., F.E. Livingstone, and H.N. Berko. 1990. Ozone destruction and bromine photochemistry at ground level in the Arctic spring. Nature **343**: 622–625.

21. Fogelqvist, E., B. Josefsson, and C. Roos. 1982. Halocarbons as tracer substances in studies of the distribution pattern of chlorinated waters in coastal areas. Environ. Sci. Technol. **16**:479–482.

22. Fogelqvist, E. 1985. Carbon tetrachloride, tetrachloroethane, 1,1,1,-trichloroethane and bromoform in Arctic sea water. J. Geophys. Res. **90**:9181–9193.

23. Fogelqvist, E., and M. Kryssel. 1988. The antropogenic and biogenic origin of low molecular weight halocarbons in a polluted fjord, the Idefjorden. Marine Pollut. Bull. **17**:378–382.

24. Gschwend, P.M., J.K. MacFarlane, and K.A. Newman. 1985. Volatile halogenated organic compounds released to sea water from temperate marine macroalgae. Science **227**:1033–1036.

25. Gschwend, P.M., and J.K. MacFarlane. 1986. Polybromomethanes, a year round study of their release to seawater from Ascophyllum nodosum and Fucus vesiculosis. In M.L. Sohn (ed.), Organic Marine Geochemistry, ACS Symposium Series 305, American Chemical Society, Washington DC, pp. 314–322.

26. Hewson, W.D., and L.P. Hager. 1980. Bromoperoxidases and halogenated lipids in marine algae. J. Phycol. **16**:340–345.

27. Helz, G.R., and R.Y. Hsu. 1970. Volatile chloro- and bromocarbons in coastal waters. Limnol. Oceanogr. **23**:859–869.

28. Itoh, N., Y. Izumi, and H. Yamada. 1986. Characterization of nonheme type bromoperoxidase in *Corallina pilulifera*. J. Biol. Chem. **261**:5194–5200.

29. Jaworske, D.A., and G.R. Helz. 1985. Rapid consumption of bromine oxidants in river and estuarine waters. Environ. Sci. Technol. **19**:1188–1191.

30. Kjellman, F.R. 1883. The algae of the Arctic Sea. Kongl. Svenska Vetenskaps Akademiens Handlingar **20**:1–61. Reprinted (1971) by Otto Koeltz Antiquariat Koenigstein-Taunus, B.R.D.

31. Krenn, B.E., H. Plat, and R. Wever. 1987. The bromoperoxidase from the red alga *Ceramium rubrum* also contains vanadium as a prosthetic group. Biochim. Biophys. Acta **912**:287–291.

32. Krenn, B.E., Y. Izumi, H. Yamada, and R. Wever. 1989. A comparison of different (vanadium) bromoperoxidases: the bromoperoxidase from *Corallina pilulifera* is also a vanadium enzyme. Biochim. Biophys. Acta **998**:63–68.

33. Krenn, B.E., M.G.M. Tromp, and R. Wever. 1989. The brown alga *Ascophyllum nodosum* contains two different vanadium bromoperoxidases. J. Biol. Chem. **264**: 19287–19292.

34. Krysell, M. 1991. Bromoform in the Nansen Basin in the Arctic Ocean. Marine Chem. **33**:187–197.

35. Kylin, H. 1929. Über das Vorkommen vom Jodiden, Bromiden and Jodidoxydasen bei den Meeresalgen. Hoppe Seyler's Zeit. Physiol. Chemie **186**:50–84.

36. Lovelock, J.E., R.J. Maggs, and R.J. Wade. 1973. Halogenated hydrocarbons in and over the Atlantic. Nature **241**:194–196.

37. Lovelock, J.E. 1975. Natural halocarbons in the air and in the sea. Nature **256**: 193–194.

38. Lüning, K. 1990. Seaweeds: Their Environment, Biogeography, and Ecophysiology, J. Wiley and Sons, New York, pp. 193–194.

39. Manthey, J.A., and L.P. Hager. 1989. Characterization of the catalytic properties of bromoperoxidase. Biochemistry **28**:3052–3057.

40. Moe, R.L., and P.C. Silva. 1977. Antarctic marine flora: Uniquely devoid of kelps. Science **196**:1206–1208.

41. Moore, R.E. 1977. Volatile compounds from marine algae. Acc. Chem. Res. **10**:40–47.

42. Neidleman, S.L., and J. Geigert. 1986. Biohalogenation: Principles, Basic Roles and Applications, Ellis Horwood Ltd., Chichester.

43. Palinek, B., O.C. Zafiriou, and F.M.M. Morel. 1981. Hydrogen peroxide production by a marine phytoplankter. Limnol. Oceanogr. **33**:1365–1369.

44. Pallaghy, C.K., J. Minchinton, G.T. Kraft, and R. Wetherby. 1983. Presence and distribution of bromine in Thysanochladioa densa (Solieriaceae, Gigertinales), a marine red alga from the Great Barrier Reef. J. Phycol. **19**:204–208.

45. Pedersén, M.E.E., G. Roomans, and A. v. Hofsten. 1981. Bromine in the cuticle of Polysiphonia nigrescens: Localization and content. J. Phycol. **17**:105–108.

46. Penkett, S.A., B.M.R. Jones, M.J. Rycrofft, and D.A. Simons. 1985. An interhemispheric comparison of the concentrations of bromine compounds in the atmosphere. Nature **318**:550–553.

47. Ramanathan, V., R.J. Cicerone, H.B. Singh, and J.T. Kiehl. 1985. Trace gas trends and their potential role in climate changes. J. Geophys. Res. **90**:5547–5566.

48. Sakshang, E., and H.R. Skjoldal. 1989. Life at the ice edge. AMBIO **1**:60–67.

49. Sauvageau, C. 1926. Zur quelques algues floridées renformant du brome a l'état libre. Bull. Stat. Arch. **23**:5–23.

50. Siuda, J.F., G.R. van Blaricom, P.D. Shaw, R.D. Johnson, H. White, L.P. Hager, and R.L. Rinehart. 1975. 1-iodo-3,3—dibromo—2-heptanone, 1,1,3,3,tetrabromo-2 heptanone and related compounds from the red algae *Bonnemaisonia hamifera*. J. Amer. Chem. Soc. **97**:937–939.

51. Soedjak, H.S., and A. Butler. 1990. Characterization of vanadium bromoperoxidase from *Macrocystis* and *Fucus:* Reactivity of vanadium bromoperoxidase towards acyl and alkyl peroxides and bromination of amines. Biochemistry **29**:7974–7981.

52. Sturges, W.T., and L.A. Barrie. 1988. Chlorine, bromine and iodine in Arctic aerosols. Atmos. Environ. **22**:1179–1194.

53. Svendson, P. 1959. The algal vegetation of Spitsbergen. Norsk Polar Institutt Skrifter **116**:3–52.

54. Theiler, R., J.C. Cook, L.P. Hager, and J.F. Siuda. 1978. Halohydrocarbon synthesis by bromoperoxidase. Science **202**:1094–1096.

55. Vilter, H., K.-W. Glombitza, and A. Grawe. 1983. Peroxidases from Phaeophyceae I: extraction and detection of the peroxidases. Bot. Marine **26**:331–340.

56. Vilter, H. 1984. Peroxidases from Phaeophyceae: a vanadium (V) dependent peroxidase from *Ascophyllum nodosum*. Phytochemistry **23**:1387–1390.

57. Vilter, H. 1983. Peroxidases from Phaeophyceae IV. Fractionation and location of peroxidase isoenzymes in *Ascophyllum nodosum* (L.) Le Jol. Bot. Marine **26**: 451–455.

58. Waaland, R.J. 1981. Commercial utilization. In C.S. Lobban and M.J. Wynne (eds.), The Biology of Seaweeds, University of California Press, Berkeley and Los Angeles, pp. 726–741.

59. Wever, R., H. Plat, and E. De Boer. 1985. Isolation procedure and some properties of the bromoperoxidase from the seaweed *Ascophyllum nodosum*. Biochim. Biophys. Acta **830**:181–186.

60. Wever, R. 1988. Ozone destruction by algae in the Arctic atmosphere. Nature **335**:501.

61. Wever, R., G. Olafsson, B.E. Krenn, and M.G.M. Tromp. 1988. Ozone destruction and bromoform production in the Arctic: pieces of a puzzle. Abstr. 32nd IUPAC Congress, No. 210, Stockholm, Sweden.

62. Wever, R., and B.E. Krenn. 1990. Vanadium haloperoxidases. In N.D. Chasteen (ed.), Vanadium in Biological Systems, Kluwer Academic Publishers, Amsterdam, pp. 81–97.

63. Wever, R., and K. Kustin. 1990. Vanadium: A biologically relevant element. Adv. Inorg. Chem. **35**:81–115.

64. Wever, R., M.G.M. Tromp, B.E. Krenn, A. Marjani, and M. Van Tol. 1991. Brominating activity of the seaweed *A. nodosum:* Impact on the biosphere. Environ. Sci. Technol. **25**:446–449.

65. Wever, R. 1991. Formation of halogenated gases by natural sources. In J.E. Rogers and W.B. Whitman (eds.), Microbial Production and Consumption of Greenhouse Gases: Methane, Nitrogen Oxides, and Halomethanes, American Society for Microbiology, Washington DC, pp. 277–285.

66. Wolk, C.P. 1968. Role of bromine in the formation of refractile inclusions of the vesicle cells of the Bonnemaisoniaceae (Rhodophyta). Planta **78**:371–378.

67. Wuosmaa, A.M., and L.P. Hager. 1990. Methyl chloride transferase: a carbocation route for biosynthesis of halometabolites. Science **249**:160–162.

68. Yu, H., and J.W. Whittaker. 1989. Vanadate activation of bromoperoxidase from *Corallina officinalis*. Biochem. Biophys. Res. Commun. **160**:87–92.

Microbiological Transformations of Low-Molecular-Weight Carbon Compounds in the Deep Subsurface

Sergei S. Belyaev and Igor A. Borzenkov

Institute of Microbiology, Russian Academy of Sciences, Moscow, 117811, Russia

ABSTRACT

Methanogenesis, sulfate reduction, and acetogenesis were studied in the deep subsurface waters of two exploited oil fields. Higher rates of methanogenesis (mean value of 2.16 μg CH_4 L^{-1} d^{-1}) were found in the faintly mineralized waters of the Mychpayskoe oil field, which contained low sulfate concentrations. In the waters of the Bondyuzhskoe oil field (salinity = 15.8–274.5 g/L) methanogenesis was by one to two orders of magnitude lower, with high rates of sulfate reduction (0.14 μg H_2S L^{-1} d^{-1}). The main precursor of methane in waters of the Mychpayskoe oil field was acetate, whereas in the Bondyuzhskoe oil field, it was bicarbonate. Acetogenesis from CO_2 and H_2 was detected at rates varying from 6 to 196 μg L^{-1} d^{-1}. However, the acetate destruction rate under these conditions was more than five times higher.

Introduction

Modern microbiological processes in natural and man-made ecosystems contribute to global elemental cycles.[10,13] Studies on the microbial geochemical activity within the deep subsurface depend on the investigation of their role in generation of the chemical composition of underground waters as well as in the formation, transformation, and degradation of organic and inorganic compounds of oil, gas, and other deposits.[3,16,21,23] During the past decade, these works have received a new stimulus because of fuel- and energy-supply problems in the world economy. Therefore, microbiological investigations have emphasized the development of new biotechnologies for enhanced oil recovery.[6,15]

In addition, microbiological investigations have also focused on bioremediation of chemically contaminated aquifers.[16,19] The aim of this chapter was to study methanogenesis, sulfate reduction, and acetogenesis in the stratal waters of two exploited oil fields with different geophysical and hydrochemical characteristics.

Materials and Methods

Objectives of Field Investigations

Field investigations were carried out in the Bondyuzhskoe and Mychpayskoe oil fields. The Bondyuzhskoe oil field is located on the right bank of the Kama river (Tartar Republic). The oil-bearing rocks are sandstones and siltstones of the Upper Devonian located at a depth of 1500–1700 m and having an in situ temperature of 30–40°C. The mean porosity of the oil-bearing rocks is 22% and permeability of about 500 millidarcy. The oil from this field is paraffinic, highly bituminous, and sulfurous, with a specific gravity of 0.871 g cm^{-3}. The associated gas consists mainly of methane, the content of which reaches up to 42%.[2]

The Mychpayskoe oil field is located in the Nizhnevartovsky district (Western Siberia). The productive layers of Cretaceous deposits are located at a depth of 1700–1850 m, and the in situ temperature was 60–70°C. The oil-bearing rocks are heterogeneous by their lithology and characterized by alteration of sandstones, siltstones, and argillites. Their mean porosity was ~22% and their permeability was ~200 millidarcy. The oil of this field is light, low sulfur content, of the naphtheno-methanous type, and contains significant amounts of light dissolved hydrocarbons. Its specific gravity was 0.779 g cm^{-3}. The associated gas contains 70% of methane, 3.14% of N_2, and 0.33% and CO_2.[5]

Chemical Determinations

Contents of HCO_3^-, O_2, H_2S, and NH_4^+ were determined with the help of an "Aqua Merck" kit of Merck (Germany). SO_4^{2-} and Ca^{2+} were determined by the ion chromatography on a "Biotronic 1C-1000" chromatograph. For SO_4^{2-} determination, a BT11AN column and 1 mM Na_2CO_3 and 1.2 mM $NaHCO_3$ were used as eluents with a flow rate of 1.5 ml/min. For Ca^{2+}, a BT IV KA column and 1 mM HNO_3 + 1.2 mM phenylene-diammonium dichloride were used with a flow rate of 1.5 ml/min. The pH and redox potential values were measured by the OP-104 "Radelkis" (Hungary) conductometer and pH-150 millivoltmeter (USSR), respectively.

Before acetate determination, the sample desalting was carried out by vacuum distillation procedures[27] from 5 ml aliquots. The distillate volume was 1 ml. The distillate was neutralized with 10% $NaHCO_3$ solution until pH 7.0 and evaporated to dryness at 60°C. The dried material was dissolved in 200 μl of 5% H_3PO_4 directly before analysis. The acetate content was determined by gas chromatography on a Pye Unicam-4550 gas chromatograph equipped with the flame-ionization detector using a glass column of Porapak Q at 180°C with an argon carrier. The methylamine content was determined according to the method of Lovley and Klug[20] on a 3700 gas chromatograph (USSR) equipped with a flame-ionization detector. Methane was analyzed on the same gas chromatograph employing a glass column of Porapak Q at 60°C with N_2 as a carrier gas.

Radioisotopes

Radioisotopes of the firm "Isotope" (the Leningrad branch) were used throughout this work. The specific activities were for $NaH^{14}CO_3$—806 MBK/mmole, $^{14}CH_3COONa$—1640 MBK/mmole, $^{14}CH_3\ ^{14}COONa$—1530 MBK/mmole, $^{14}CH_3NH_2$—1960 MBK/mmole, and $Na_2{}^{35}SO_4$—7200 MBK/mmole.

Radioisotopic Investigations

The stratal water samples were put into 50-ml sterile flasks preliminarily filled with argon and closed with a rubber stopper. Two hundred microliters of a radioisotope solution was added to each flask with a syringe and incubation proceeded for 3 d at the stratal temperature. In the case of the bacterial methanogenesis rate determinations, the total activity added to the flask was 2.9 MBK for $NaH^{14}CO_3$, 0.5 MBK for $^{14}CH_3COONa$, and 0.31 MBK for $^{14}CH_3NH_2$. After incubation, the probes were inactivated. For this purpose, we used 0.5 ml of 40% NaOH for probes with hydrocarbonate and acetate and 0.5 ml of concentrated phosphoric acid for those with monomethylamine. The activity of methane formed from hydrocarbonate and acetate was determined by the earlier described method.[18] For this purpose, the methane combustion procedure was used converting methane to CO_2 followed by trapping in a mixture of the scintillation liquid "Lipoluma" with β-phenylethylamine and methanol at the ratio of 3 :1 : 1. Methane from the probes with labeled methylamine before combustion was further purified by bubbling through a trap with 10% NaOH.[28]

For the bacterial sulfate-reduction-rate determination, $Na_2{}^{35}SO_4$ with the total activity in each probe of 1.6 MBK was used. After incubation, probes were inactivated with 0.5 ml of 40% NaOH. In the laboratory, 0.5 mM of Na_2S was added to each sample as a carrier and then the samples were acidified with 10% H_3PO_4 to pH 3.0 and hydrogen sulfide was distilled into the trap containing scintillation liquid with β-phenylethylamine and methanol (3 : 1 : 1).

To determine the fraction of total acetate production derived from the reduction of carbon dioxide, $NaH^{14}CO_3$ was added to each sample (activity 2.9 MBK). After incubation, samples were inactivated with 0.5 ml of 40% NaOH. In the laboratory, samples (30 ml) were acidified with 10% H_3PO_4 to pH 4.0 and, for HCO_3^- removal, they were boiled in the flask with return condenser for 45 mins. Subsequently, 5 ml aliquot 0.1 μM CH_3COOH was added to each as a carrier and acidified with 10% H_3PO_4 to pH 2.0 followed by vacuum distillation.[27] The distillate volume was 1.0 ml. Acetate was isolated on a BT 111-OS column using a Biotronic IC-1000 ion chromatograph with 10 N HNO_3

as eluent and a flow rate of 0.8 ml/min. The acetate peak was collected into scintillation vials with Lipoluma scintillation liquid.

The acetate incorporation rate into bacterial cells was determined with $^{14}CH_3{}^{14}COONa$ (0.31 MBK per sample). After incubation, 5-ml samples were filtered through membrane filters (Millipore; 0.22 μm). The filter washed with 25 ml of 0.05 N HCl and 50 ml of distilled water and air-dried. Activity of the biomass was determined in the Lipoluma scintillation liquid.

Acetate oxidation of CO_2 was determined with $^{14}CH_3{}^{14}COONa$ (0.31 MBK per sample). After incubation, samples were inactivated with 0.5 ml of 40% NaOH. In the laboratory, samples were acidified with 10% H_3PO_4 to pH 5.4 and $^{14}CO_2$ was forced out with an air flow to a trap containing scintillation liquid and β-phenylethylamine and methanol (3 : 1 : 1). All radioisotopic determinations were carried out in the liquid scintillation counter Rack Betta (LKB, Finland).

Mass-Spectrometric Determination

Mass-spectrometric determinations were carried out with an isotopic mass spectrometer MI-1201. Dissolved carbonates were precipitated with the saturated $Ba(OH_2)$ solution. Methane rectification from homologues was done by gas chromatography on a "Chromo-80" chromatograph (USSR) on a Porapak Q containing a preparative column with the catarometer as detector and argon as the carrier gas. Further treatment of samples for mass-spectrometric analysis was carried out as published previously.[9] Obtained results are given on the basis of PDB of the carbonate standard. Error was $\leq 0.2‰$ for CO_2 and ≤ 0.5 ‰ for CH_4.

Microbial Biomass

Enumeration of microorganisms was done by performing 10-fold dilutions on the appropriate liquid elective media. Incubations were carried out at 30°C. All chemical compounds used for nutrient media were reagent grade. Liquid media and supplements were prepared and sterilized in strictly anaerobic conditions in the N_2, $N_2{\cdot}CO_2$ (95 : 5), and H_2CO_2 (80 : 20) as given previously.[29] For methanogens, a mineral medium was used.[1] As energy substrates, a mixture of $CO_2 : H_2$ (20 : 80) in the gas phase or acetate (0.5%) were used. Development of methanogens was followed by the methane content increase in the gas phase. The number of sulfate-reducing bacteria was determined in the Postgate's B medium with lactate.[24] The number of acetogens was estimated with nutrient medium[7] modified by us.[4] To inhibit development of sulfate reducer and methanogens, $NaMoO_4$ (1 mM) and the sodium salt of bromoethanesulfonic acid (10 mM) were added to the medium immediately before inoculation. Development of acetogens was followed by acetate production.

Results

The Bondyuzhskoe Oil Field

Exploitation of the field has been carried out for over 30 years. For the maintenance of pressure, flooding the oil-bearing stratum with Kama river waters is used. Intrastratal freshwater injections result in changes in the chemical composition of the Devonian relic brines. As the salinity of the stratal waters decrease, the concentration of Ca^{2+} cations decreases, whereas bicarbonate and pH increase (Table 1). The sulfate content in the waters is not higher than 69 mg/l. Oxygen and hydrogen sulfide were not found. The E_h value of stratal waters was consistent with the faintly reduced environmental conditions. On the whole, freshwaters flooding of the field results in the establishment of a wide spectrum of ecological niches with the salinity ranging from 1 to 292 g L^{-1}. Another important property of the injection of fresh waters is the fact that they are a source of different microorganisms, including the anaerobic bacteria (Table 2). Sulfate-reducing and acetogenic bacteria and methanogenic bacteria were found in these waters.

Table 1. Chemical and physico-chemical characteristics of the Bondyuzhskoe oil field waters from production wells.

Well No.	Salinity (g L^{-1})	pH	E_h(mV)	Content (mg L^{-1}) HCO_3^-	SO_4^{2-}	CH_3COO^-	NH_4^+	Ca^{2+}
12140[a]	0.7	6.6	+70	140	48	1.0	1.2	n.d.[b]
304	15.8	6.7	+145	353	63	5.3	15	1,100
303	63.0	6.8	+135	219	17	10.6	20	4,400
323	83.0	6.4	+95	177	69	5.1	20	5,800
351	131.0	6.5	+140	195	42	9.7	30	9,200
15	152.0	5.9	+145	152	69	6.1	30	10,600
365	162.0	5.9	+175	152	48	7.4	30	11,400
352	175.0	6.1	+170	146	17	13.6	30	12,400
283	180.1	6.1	+105	171	24	3.8	35	12,600
367	237.0	5.4	+170	116	21	8.8	40	16,600
353	274.5	5.3	+165	30	32	4.6	55	19,200
272	292.0	5.6	+200	37	29	6.3	60	20,400
Average for production wells	160.5	6.1	+150	150.6	39.2	7.4	33.2	9,200

[a]Injection well.
[b]No data.

Table 2. Anaerobic microflora of injected waters of the Bondyuzhskoe oil field.

Well No.	Methanogens (cells ml^{-1}) $H_2 + CO_2$	Methanogens (cells ml^{-1}) Acetate	Sulfate reducers (cells ml^{-1}) Lactate	Acetogens (cells ml^{-1}) $H_2 + CO_2$
428	25,000	6,000	4,000	n.d.[a]
422	2,500	2,500	2,000	n.d.
28	1,300	6,000	6,000	n.d.
366	10,000	25,000	10,000	600
12140	100,000	60,000	100,000	1000

[a]No data.

Table 3 shows data about the development of anaerobic microbiological processes in the stratal waters of the Bondyuzhskoe oil field taken from the injection (N° 12140) and production wells. The bacterial methanogenesis was shown to proceed extensively along the stratum in waters with salinity up to 274.5 g L^{-1}. Its maximum rate of 4.95 μg L^{-1} d^{-1} was found in the desalinized near-bottom zone of the injection well. The rate of methanogenesis decreased as salinity increased from the tenths to the hundredth parts of a microgram per liter per day (average ≅ 0.19 μg CH_4 L^{-1} d^{-1}). The sulfate-reduction rate was of the same order, but higher values were found in waters with salinity from 131 to 175 g L^{-1}. If the salinity was from 15.8 to 83.0 and from 180.1 to 292.0 g L^{-1}, the sulfate reduction rate was one order lower and was equal to several hundredths of a microgram H_2S L^{-1} d^{-1}. The maximal rate of H_2S production was observed in the near-bottom zone of the injection well. The production of acetate proceeded in waters with the salinity up to 237 g L^{-1} at a mean rate of 17 μg L^{-1} d^{-1}, which is by two orders higher than the methanogenesis and sulfate reduction rates. Higher acetogenesis rates were observed in the low salinity stratal waters.

The main methanogenesis substrates in the Bondyuzhskoe oil field stratal waters were hydrocarbonate and molecular hydrogen, the mean contribution of which was 83.2% (Table 4). The contribution of the acetate methyl groups to methane production was between 4 and 26%. A small part of methane originated from monomethylamine, the ratio of which in methanogenesis increased slightly with salinity.

The data in Table 4 are also given on the carbon isotopic composition of methane and hydrocarbonates dissolved in the Bondyuzhskoe oil field stratal waters. The carbon isotopic composition of methane varied from −42.1 to −55.9‰. No distinct correlation was found between $\delta^{13}C$ of methane and the

Table 3. Rate of anaerobic bacterial processes in the Bondyuzhskoe oil field waters.

Well No.	Salinity ($g\ L^{-1}$)	Production rate ($\mu L^{-1}\ d^{-1}$)		
		CH_4	CH_3COOH	H_2S
12140[a]	0.7	4.95	12.0	2.66
304	15.8	0.47	44.3	0.04
303	63.0	0.27	21.7	0.02
323	83.0	0.38	16.0	0.03
351	131.0	0.34	37.1	0.61
15	152.0	0.16	15.4	0.16
365	162.0	0.06	11.0	0.28
352	175.0	0.18	13.2	0.25
283	180.1	0.14	10.1	0.09
367	237.0	0.05	6.0	0.03
353	274.5	0.03	0	0.02
273	292.0	0	0	0.01
Average for production wells	160.5	0.19	17.0	0.14

[b]Injection well.

bacterial methanogenesis, although there was a slight tendency toward lighter carbon in methane associated with increased methanogenesic activity. Perhaps this may be explained by a low ratio of the contemporary biogenic methane in the total volume of relict gaseous hydrocarbons. The carbon isotopic composition of dissolved carbonates became much heavier along with the water decreased salinity and increased methanogenesis.

There was a significant level of acetogenesis observed without any noticeable acetate accumulation in the subsurface waters (Tables 1 and 3). Less than 1% of the acetate was used for methanogenesis (Table 5). On an average, more than 25% of the acetate carbon was incorporated into bacterial cells. However, the bulk of the acetate carbon was oxidized to CO_2. The rate of this path was almost six times higher than that of acetogenesis from H_2 and CO_2. In this ecosystem, there must be other mechanisms for acetogenesis. A partial acetate input from the aerobic zone with the injected waters is possible. Among microbiological processes responsible for acetate oxidation to CO_2, sulfate reduction is undoubtedly important.

The Mychpayskoe Oil Field

This oil field has been exploited for 15 years, using contour flooding. Initially, fresh water was injected into the stratum, but recently water collected during the

Table 4. Contribution of different substrates to the bacterial methanogenesis and carbon isotopic composition of methane and carbonates in the Bondyuzhskoe oil field waters.

Well No.	Methanogenesis Rate (μg CH_4 L^{-1} d^{-1})	Contribution to Methanogenesis (%)			$\delta^{13}C$ (‰)	
		HCO_3^-	CH_3COO^-	CH_3NH_2	CH_4	ΣHCO_3^- + CO_2 + CO_3^{2-}
304	0.47	86	14	0	−54.2	−4.6
303	0.27	72	26	2	−53.1	−10.7
323	0.38	92	6	2	−54.5	−15.5
351	0.34	92	7	1	−49.8	−10.3
15	0.16	87	7	6	−55.9	−13.4
365	0.06	83	11	6	−46.8	−15.8
352	0.18	77	23	0	−52.6	−15.6
283	0.14	88	6	6	−42.1	−14.9
367	0.05	79	4	17	−51.4	−16.3
353	0.03	76	4	20	−49.6	−18.1
273	0	0	0	0	−45.7	−18.6
Average	0.19	83.2	10.8	6.0	−50.5	−14.0

oil field pumping is used for reinjection. This water is similar to the stratal waters in chemical composition. Table 6 indicates that the total mineralization of injected and stratal waters changes in a narrow interval of concentrations—salinity from 15.3 to 28.4 g L^{-1} and Ca^{2+} from 0.74 to 1.52 g L^{-1}. A very low sulfate concentration and high acetate content are notable.

In the stratal waters of the Mychpayskoe oil field, active microbiological processes take place (Table 7). The methanogenesis rate varied from 0.48 to 12.62 μg CH_4 L^{-1} d^{-1} without any dependence on the water's salinity. Maximum methanogenesis rate, like that of other anaerobic processes (acetogenesis and sulfate reduction), has been observed in the near-bottom zone of injection well 409. The acetogenesis rate was approximately 10-fold greater than that of methanogenesis. Sulfate reduction was detectable, but its intensity was 100-fold lower than that of methanogenesis. Comparison of bicarbonate and acetate contribution to the methane yield indicates that the acetate was responsible for most of the methanogenesis from the Mychpayskoe oil field (Table 8). On average, 70.4% of CH_4 originated from acetate and only 29.7% was due to CO_2 reduction by molecular hydrogen.

The carbon isotope composition of methane taken from the Mychpayskoe oil field production wells was characterized by close $\delta^{13}C$ values, from −50.1 to −52.5‰. The $\delta^{13}C$ values of dissolved carbonates varied from −2.1 to

Table 5. Acetate formation and utilization during microbiological processes in the Bondyuzhskoe oil field waters.

		Rate of Acetate Carbon Uptake (μg C L^{-1} d^{-1})			
Well No.	Acetogenesis Rate (μg C L^{-1} d^{-1})	$CH_4(R_2)$	Bacterial Cells (R_3)	$CO_2(R_4)$	$R_1 - (R_2 + R_3 + R_4)$
304	17.91	0.05	12.43	96.2	−90.77
303	8.77	0.05	1.13	17.3	−9.71
323	6.51	0.02	0.94	15.4	−9.85
351	15.09	0.02	0.07	16.2	−1.20
15	6.26	0.01	3.39	39.1	−36.14
365	4.47	0.01	0.06	12.6	−8.20
352	5.37	0.03	0.71	67.4	−62.77
283	4.11	0.01	0.87	74.0	−70.77
367	2.44	0.005	0.48	61.6	−59.64
353	0	0.005	0.13	11.8	−11.93
273	0	0	1.01	0	−0.01
Average	6.45	0.02	1.82	37.42	−32.81

−10.9‰. Both values did not correlate with variability of the water salinity and rate of bacterial methanogenesis.

Discussion

The Bondyuzhskoe and Mychpayskoe oil fields differed from each other by some physical and hydrochemical characteristics. For the Mychpayskoe oil field, high stratal temperatures (60–70°C) were prevalent. Stratal waters of this oil field differed from those of the Bondyuzhskoe oil field by their low salinity, extremely low sulfate concentrations, and a significant content of low molecular organic compounds (Tables 1 and 6).

The results obtained in the present work demonstrate rather convincingly that active anaerobic microbiological processes take place in the stratal waters of oil fields exploited with flooding by surface waters. In addition to previously demonstrated methanogenesis and sulfate reduction,[2–4,25] acetogenesis due to CO_2 reduction by hydrogen is also active in the subsurface waters.

Acetogenesis predomination over all the other anaerobic processes studied. It was found in waters of both oil fields with the salinity up to 237 g L^{-1} and its rate was equal to 6–196 μg L^{-1} d^{-1} (Tables 3 and 7). However, the rate of the

Table 6. Chemical and physico-chemical characteristics of the Mychpayskoe oil field waters from production wells.

Well No.	Salinity ($g L^{-1}$)	pH	E_h (mV)	Content ($mg L^{-1}$) HCO_3^-	SO_4^{2-}	CH_3COO^-	Ca^{2+}
409[a]	19.7	6.7	+40	317	3.9	69.0	n.d.[b]
366	15.3	7.1	+25	181	0.1	21.0	1120
505	17.6	7.8	+90	183	1.4	40.0	1180
308	18.2	7.9	+80	153	0.1	23.1	1158
314	18.9	7.3	+180	176	5.1	27.9	1119
352	19.2	7.8	+110	165	0.7	37.7	741
336	19.8	7.9	+110	133	0.7	49.9	874
251	20.1	8.0	+100	195	0.6	18.0	817
374	22.8	7.3	+65	122	2.4	56.9	1083
367	23.9	7.1	+65	98	2.0	74.0	1064
345	28.4	7.2	+130	18	0.1	277.0	1520
Average for production wells	20.4	7.5	+100	142.4	1.2	62.6	1058

[a]Injection well.
[b]No data.

anaerobic acetate destruction to CO_2 exceeded the above values more than by five times (Table 5), thus allowing us to suggest the existence of other sources of acetate production in these ecosystems.

Sulfate reduction was found in all studied stratal water samples. At the same time, its rate in the Mychpayskoe oil field waters was 10-fold lower than in the Bondyuzhskoe oil field, evidently due to the former's low sulfate content. The mean daily rate of sulfate reduction in stratal waters of the Bondyuzhskoe oil field was equal to 0.14 $\mu g\ H_2S\ L^{-1}\ d^{-1}$ and was by 10–100-fold lower than that of waters with a high sulfate content of the Apsheron oil field in Azerbaijan or of freshwater lakes.[8,25]

Current methanogenesis was also found in the waters of both oil fields. In the Mychpayskoe oil field, about 70% of the methane was derived from acetate, whereas in waters of the Bondyuzhskoe oil field, the bulk of methane was formed by CO_2 reduction by hydrogen. In addition, some methane in the Bondyuzhskoe oil field was partially formed from methylamine—an important methane precursor in ecosystems of extremely high salinity.[22] The methanogenesis rate was 10–100-fold higher in the low salinity waters of the Mychpayskoe oil field as compared with the Bondyuzhskoe field.

Table 7. Rate of anaerobic bacterial processes in the Mychpayskoe oil field waters.

Well No.	Salinity (g L^{-1})	Production Rate (μg L^{-1} d^{-1})		
		CH_4	CH_3COOH	H_2S
409*	19.7	12.62	196	0.070
366	15.3	1.23	16	0.004
505	17.6	3.51	36	0.003
308	18.2	0.66	6	0.010
314	18.9	0.48	6	0.010
352	19.2	1.44	38	0.004
336	19.8	1.47	27	0.003
251	20.1	0.86	48	0.001
374	22.8	1.45	52	0.009
367	23.9	3.77	68	0.014
345	28.4	6.76	74	0.008
Average for production wells	20.4	2.16	37.1	0.014

[a]Injection well.

Table 8. Different substrates contribution to the bacterial methanogenesis and methane and carbonate carbon isotopic composition in the Mychpayskoe oil field waters.

Well No.	Methanogenesis Rate (μg CH_4 L^{-1} d^{-1})	Contributions to Methanogenesis (%)		$\delta^{13}C$ (‰)	
		HCO_3^-	CH_3CHOO^-	CH_4	$\Sigma HCO_3^- + CO_2 + CO_3^{2-}$
366	1.23	58	42	−51.2	−4.1
505	3.51	24	76	−50.1	−10.9
308	0.66	19	81	−51.9	−6.6
314	0.48	53	47	−51.4	−6.4
352	1.44	24	76	−52.5	−4.2
336	1.47	19	81	−51.5	−5.6
251	0.86	57	43	−52.1	−2.1
374	1.45	21	79	−51.4	−7.4
367	3.77	11	89	−50.3	−5.5
345	6.76	11	89	−50.4	−7.2
Average	2.16	29.7	70.3	−51.3	−6.0

In the Bondyuzhskoe oil field, a definite correlation was observed between some lightening of the methane carbon isotopic composition and current methanogenesis rate increases (Table 4). Separation of stable carbon isotopes during bacterial methanogenesis resulting in the lightening of the methane carbon isotopic composition and increased ^{13}C of the residual substrate have been known from a series of microbiological[12,14,17] and geochemical[11,26] works.

Our results indicate that the contribution of modern biogenic methane becomes noticeable in the course of field exploitation, and activity increases with decreased salinity associated with the injection of fresh water into the stratal fluids.

References

1. Balch, W.E., G.E. Fox, C.R. Woese, and R.S. Wolfe. 1979. Methanogens: Reevaluation of a unique biological group. Microbiol. Rev. **43**:260–296.
2. Belyaev, S.S., R. Wolkin, W.R. Kenealy, M.J. DeNiro, S. Epstein, and J.G. Zeikus. 1983. Methanogenic bacterial from Bondyuzhskoe oil field: General characterization and analysis of stable-carbon isotopic fractionation. Appl. Environ. Microbiol. **128**:288–293.
3. Belyaev, S.S., and M.V. Ivanov. 1983. Bacterial methanogenesis in underground waters. Ecol. Bull. **35**:273–280.
4. Belyaev, S.S., I.A. Borzenkov, E.I. Milekhina, I.A. Charakhchyan, and M.V. Ivanov. 1990. Microbiological processes in oil-bearing layers of the Romashkino oil field. Microbiology (USSR) **59**:1118–1126.
5. Belyaev, S.S., E.P. Rozanova, I.A. Borzenkov, I.A. Charakhchyan, Yu. M. Miller, M. Yu. Sokolov, and M.V. Ivanov. 1990. Specificity of the microbiological processes in a water-flooded oil field located in the middle Ob region. Microbiology (USSR) **59**:1075–1082.
6. Belyaev, S.S., I.A. Charakhchyan, and V.G. Kuznetsova. 1991. Strict anaerobic bacteria and their possible contribution to the enhancement of oil recovery. In E.C. Donaldson (ed.), Microbial Enhancement of Oil Recovery—Recent Advances, Elsevier, Amsterdam, pp. 163–172.
7. Braun, M., F. Mayer, and G. Gottschalk. 1981. *Clostridium aceticum* (Wieringa), a microorganism producing acetic acid from molecular hydrogen and carbon dioxide. Arch. Microbiol. **128**:288–293.
8. Chebotarev, E.N. 1974. Microbiological H_2S production in freshwater lakes. Microbiology (USSR) **43**:1105–1110.
9. Craig, H. 1957. Isotopic standard for carbon and oxygen and correction factor for mass-spectrometric analysis of carbon dioxide. Geochim. Cosmochim. Acta **12**: 133–149.
10. Frayer, H.D. 1979. Atmospheric cycles of trace gases containing carbon. In B.

Bolin, E.T. Degens, S. Kempe, and P. Ketner (eds.), The Global Carbon Cycle, Wiley and Sons, New York, pp. 101–128.

11. Fuex, A.N. 1977. The use of stable carbon isotopes in hydrocarbon exploration. J. Geophys. Explor. **7**:155–188.
12. Games, L.M., J.M. Hayes, and R.P. Gunsalus. 1978. Methane-producing bacteria: Natural fractionations of the stable isotopes. Geochim. Cosmochim. Acta **42**:1295–1297.
13. Ivanov, M.V., and J.R. Freney (eds.). 1983. Global Biogeochemical Sulfur Cycle and Influence on It of Human Activity, Nauka, Moscow, p. 421.
14. Ivanov, M.V., S.S. Belyaev, V.A. Bondar, O.V. Shipin, and K.S. Laurinavichus. 1983. Fractionation of stable carbon isotopes by methanogenic bacteria. In: "Biological Processes Related to Petroleum Recovery," W.E. Hainer (ed.), Seattle, Vol. 28, N 3, American Chemical Society, Washington D.C., p. 802–804.
15. Ivanov, M.V., and S.S. Belyaev. 1991. Biotechnology of enhancement of oil recovery based on the geochemical activity of microorganisms (field experiments). In E.C. Donaldson (ed.), Microbial Enhancement of Oil Recovery—Recent Advances, Elsevier, Amsterdam, pp. 421–432.
16. Jones, R.E., R.E. Beeman, and J.M. Suflita. 1989. Anaerobic metabolic processes in the deep terrestrial subsurface. Geomicrobiol. J. **7**:117–130.
17. Krzychi, J.A., W.R. Kenealy, M.J. DeNiro, and J.G. Zeikus. 1987. Stable carbon isotope fractionation by *Methanosarcina barkeri* during methanogenesis from acetate, methanol or carbon dioxide-hydrogen. Appl. Environ. Microbiol. **53**:2597–2599.
18. Laurinavichus, K.S., and S.S. Belyaev. 1978. The rate of microbial methane production determined by the radioisotopic technique. Microbiology (USSR) **47**:1115–1117.
19. Lee, M.D., J.M. Thomas, R.C. Borden, P.B. Bedient, C.H. Ward, and J.T. Wilson. 1988. Biorestoration of aquifers contaminated with organic compounds. CRC Crit. Rev. Environ. Contr. **18**:29–89.
20. Lovley, D.R., and M.J. Klug. 1983. Methanogenesis from methanol and methylamines and acetogenesis from hydrogen and carbon dioxide in the sediments of a eutrophic lake. Appl. Environ. Microbiol. **45**:1310–1315.
21. Nissenbaum, A., B.J. Presley, and I.R. Kaplan. 1972. Early diagenesis in a reducing fiord, Saanich Inlet, British Columbia, 1. Chemical and isotopic changes in major components of interstitial water. Geochim. Cosmochim. Acta **42**:209–214.
22. Oremland, R.S., L. Marsh, and D.J. DesMarais. 1982. Methanogenesis in Big Soda Lake, Nevada: an alkaline, moderately hypersaline desert lake. Appl. Environ. Microbiol. **43**:462–468.
23. Atlas, R.M. (ed.). 1984. Petroleum Microbiology, Macmillan, New York, p. 692.
24. Postgate, J.R. 1965. Media for sulfur bacteria. Lab. Practice **15**:1214–1244.
25. Rosenfeld, W.D., and S.R. Silverman. 1959. Carbon isotope fractionation in bacterial production of methane. Science **130**:1658–1659.

26. Rozanova, E.P., and S.I. Kuznetsov. 1974. Microflora of Oil Deposits, Nauka, Moscow, p. 196.

27. Sorensen, J., D. Christensen, and B.B. Jorgensen. 1981. Volatile fatty acids and hydrogen as substrates for sulfate-reducing bacteria in anaerobic marine sediments. Appl. Environ. Microbiol. **36**:330–340.

28. Walther, R., K. Flebig, K. Fahlbusch, D. Caspari, H. Hippe, and G. Gottschalk. 1981. Growth of methanogens on methylamines. In H. Dalton (ed.), Microbial Growth on C-1 Compounds, Heyden and Son. London, pp. 146–151.

29. Zeikus, J.G., A. Ben-Bassat, and P.W. Hegge. 1980. Microbiology of methanogenesis in thermal, volcanic environments. J. Bacteriol. **143**:432–440.

Biological and Chemical Transformations of Halogenated Aliphatic Compounds in Aquatic and Terrestrial Environments

Perry L. McCarty and Martin Reinhard

Environmental Engineering and Science, Department of Civil Engineering, Stanford University, Stanford, California 94305-4020

ABSTRACT

Halogenated C_1 and C_2 aliphatic compounds (HACs) including chlorofluorocarbons (CFCs) are among the most important trace gases affecting stratospheric ozone levels and are contributors to the greenhouse effect. HACs are among the most frequently found chemicals emanating from hazardous waste sites into the atmosphere. Although once believed to be resistant to transformation by aquatic and terrestrial processes, such transformations are now known to occur. Abiotic and microbial metabolic processes can transform in parallel or sequentially a wide range of HACs including CFC-11, CFC-12, CFC-113, 1,1,1-trichloroethane, carbon tetrachloride, and trichloroethylene. Rates and pathways depend on chemical structure and physico-chemical and biogeochemical conditions. The significance of such rates in terms of effect on global distributions is not known.

Introduction

Halogenated aliphatic compounds (HACs) including chlorofluorocarbons (CFCs) are widely used as solvents, refrigerants, pesticides, and precursors for production of a wide variety of industrial products.[38] In general, HACs are sparingly soluble in water and highly volatile. Halogenation of aliphatic compounds makes them less susceptible to oxidative biological attack so that they tend to be persistent in the environment. Several are known human carcinogens [vinyl chloride (VC)] or potential human carcinogens [chloroform (CF), carbon tetrachloride (CT), trichloroethylene (TCE), and tetrachloroethylene (PCE)]. These as well as others [1,1,1-trichloroethane (TCA), *cis*- and *trans*-1,2-dichloroethylene (*c*-DCE, *t*-DCE), dibromoethane (EDB), and dibromochloropropane (DBCP)] are acutely toxic to humans at relatively low doses. Some [1,1,2-trichloro-1,2,2-difluoroethane (CFC-113), dichlorodifluoromethane (CFC-12), trichlorofluoroethane (CFC-11), CT, and TCA] are relatively resistant to photolysis and persist in the atmosphere for decades so that a significant portion reaches the stratosphere.

Here, chlorine release from solar UV photolysis leads to ozone depletion. In addition, the CFCs are of concern as they are among the gases that can lead to global warming.[33] Thus, as a class, the halogenated aliphatic compounds are useful to society, but also present severe environmental problems.

Although HACs are relatively resistant to both chemical and biological transformations in the environment, especially under aerobic conditions, many can be transformed under certain environmental conditions. Such transformations are sufficiently rapid in some aquatic or terrestrial systems that they should be considered when making global HAC mass balances. This chapter presents a summary of abiotic and biotic transformations that have been found of common significance for the HACs in general, but, more specifically, with HACs identified as being ozone-depleting or greenhouse gases (TCA, CT, CFC-11, CFC-12, and CFC-113). This discussion does not include photochemical reactions which are of major significance in the atmosphere.

Greenhouse and Ozone-Depleting Gases

Rowland[33] summarized the photochemically related reactions and chemicals that are believed to be responsible for depletion of ozone in the Antarctic stratosphere in recent years. Briefly, ozone is formed by the reaction of an oxygen atom O with molecular oxygen O_2 ($O + O_2 \rightarrow O_3$). Photolytic cleavage of a chlorinated compound such as CCl_2F_2, produces two radicals, a chlorine atom Cl and the trichloromethyl radical $CClF_2$:

(1) $$CCl_2F_2 + UV\ (<220\ nm) \rightarrow Cl + CClF_2.$$

Cl then initiates a chain reaction in which Cl reacts with O_3 to form chlorine monoxide ClO. ClO, in turn, reacts with an oxygen atom O to form O_2 and Cl, thereby consuming the UV absorbing ozone:

(2) $$Cl + O_3 \rightarrow ClO + O_2,$$

(3) $$ClO + O \rightarrow Cl\cdot + O_2.$$

O_3 depletion via this chain reaction is highly effective because both the reactant and the product of the principle O_3-forming reaction is consumed.

A summary of changes in the atmospheric concentrations of these chemicals over the past 30 years is contained in Table 1. In addition to CFCs, other HACs [TCA, CT, and methyl chloride (MC)] have a significant impact on stratospheric ozone. The concentration of MC has not changed greatly over the past 30 years, its primary source being natural biological formation. However, the other HACs were almost nonexistent in the atmosphere 30 years ago.

Table 1. Ozone-depleting and greenhouse chemicals [after Rowland (Ref. 33)].

Chemical	Formula	Atmospheric Concentration (ppbv)		Half-Life Years
		1960	1990	
CFC-113	CCl_2FCClF_2	0.00	0.07	100
CFC-12	CCl_2F_2	0.00	0.5	120
CFC-11	CCl_3F	0.00	0.26	75
TCA	CH_3CCl_3	0.00	0.17	
CT	CCl_4	0.05	0.15	
MC	CH_3Cl	0.6	0.6	
Methane	CH_4		1,720	
Carbon dioxide	CO_2		353,000	

Methane and carbon dioxide, which are included in Table 1 to give an indication of relative concentrations, are among the significant gases affecting global climate. They are not ozone-depleting gases, however. Both HACs and CFCs are trace gases that are also implicated to increase global temperature through the greenhouse effect.[32, 33] The significant recent increase in atmospheric CFC concentration is estimated to have had a greater impact on global climate over the past 10 years than has methane increases (Table 2), accounting for about one-quarter of the change. In contrast to the rapid recent rate of increase in atmospheric HAC concentration (about 5% organic chlorine increase per year), carbon dioxide concentration has increased only about 0.3% per year and methane about 1% per year over the past 30 years.[33] Thus, the HACs listed in Table 1 are highly important contributors to serious global environmental problems. A better understanding of their movement and fate in aquatic and terrestrial systems, as well as in the atmosphere, is, thus, of importance.

Abiotic and Biotic HAC Reactions

Abiotic transformations of HACs depend on structural features of the substrate (e.g., degree of unsaturation, halogenation, and branching), the presence of different nucleophiles, sorbing or catalyzing surfaces, as well as on environmental factors such as pH, redox conditions, and temperature.[38] Biotic transformations depend on many factors including the presence of organisms, nutrient and redox conditions, the type and concentration of primary substrates, and the environmental factors listed above for abiotic reactions. In aquatic and terrestrial systems, biotic and abiotic transformations may be linked in ways which are

Table 2. Contribution of human-made greenhouse gases to changes in radiative forcing from 1980 to 1990 [from Rosswall (Ref. 32)]

Greenhouse Gas	Change Caused (%)
CFC-11, & CFC-12	17
Other CFCs	7
Nitrous oxide	6
Methane	15
Carbon dioxide	55
Total	100

poorly understood. Because of this linkage, both abiotic and biotic factors and their mutual interactions must be considered. A summary of chemical and biological reactions of importance to HAC fate in terrestrial and aquatic systems is given in Table 3.

In chemical systems, *R–X* bond cleavage can occur in one of two distinct reactions: heterolytic cleavage if the two bond-forming electrons are shifted to one fragment leading to ions, and homolytic cleavage if the two electrons are split between the two fragments leading to one or two radicals. Heterolytic reactions of significance include substitution and elimination (dehydrohalogenation).

An example for homolytic cleavage is dissociative electron transfer in which cleavage of *R–X* bonds is induced by transfer of a single electron from a donor.[12]

The rate of reduction induced by single electron transfer reactions tends to increase with the degree of halogenation of the substrate. In contrast to the nonhalogenated hydrocarbons which are used as fuels, halogenated hydrocarbons are inherently resistant toward oxidants and some are used as flame retardants. The resistance toward oxidation tends to increase with degree of halogenation. Thus, rates of halogenated hydrocarbon oxidations and reductions tend to follow opposite trends.[38] Recently, biological oxidations of HACs, especially chlorinated ethylenes, has also received considerable attention. However, much more remains to be learned before quantitative predictions based on system characteristics and structure are possible. Reactions typical of various HACs are discussed in the following.

Trichloroethane

TCA undergoes significant abiotic and biotic transformations in aquatic systems and in soils. Abiotic reactions of importance include both hydrolysis[5,7,18,42] and dehydrohalogenation[7,18,30,38]:

Table 3. Typical terrestrial and aquatic abiotic and biotic reactions of halogenated aliphatic compounds.

Name	Reaction
1. Substitution	
(a) Solvolysis, hydrolysis	$RX + H_2O \rightarrow ROH + HX$
(b) Nucleophilic substitution	$RX + N^- \rightarrow RN + X^-$
2. Dehydrohalogenation (elimination)	$RCXCHR' \rightarrow RC{=}CR' + HX$
3. Reduction	
(a) Hydrogenolysis	$RX + H^+ + 2e^- \rightarrow RH + X^-$
(b) Dihalo-elimination	$RCXCXR' + 2e^- \rightarrow RC{=}CR' + 2X^-$
4. Oxidation	
(a) α-Hydroxylation	$RCHX + H_2O \rightarrow RCOHX + 2H^+ + 2e^-$
(b) Epoxidation	$RC{=}CXR' + H_2O \rightarrow RCOCXR' + 2H^+ + 2e^-$

(4) $CH_3CCl_3 + 2\,H_2O \rightarrow CH_3COOH + 3H^+ + 3Cl^-$ (hydrolysis),

(5) $CH_3CCl_3 \rightarrow CH_2{=}CCl_2 + H^+ + Cl^-$ (dehydrohalogenation).

The rate of the first reaction is about four times that of the second, leading to half-lives in water at neutral pH and 20°C of about 0.8 yr.[7] Although the second reaction is slower, it is the one of primary concern as it leads to the formation of 1,1-dichloroethylene, a potential carcinogen with a U.S. EPA Maximum Concentration Limit (MCL) in drinking water of 7 mg L^{-1}. TCA, in contrast, is regulated primarily because of toxicity with an MCL of 200 mg L^{-1}.

Biological reactions associated with TCA are reductions and oxidations. TCA can undergo hydrogenolysis under anaerobic conditions, and most rapidly under reducing methanogenic conditions.[38–40] The dominant reaction here generally is as follows:

(6) $CH_3CCl_3 + H^+ + 2e^- \rightarrow CH_3CHCl_2 + Cl^-$ (hydrogenolysis).

It appears that TCA reduction is associated with co-metabolism in which the cell receives no energy or growth benefit from the fortuitous transformation. TCA can also be transformed as above under less-reducing conditions associated with fermentation of amino acids.[15,16] Here, using a pure culture of *Clostridium*, only about 30–40% of the TCA was found as 1,1-dichloroethane (1,1-DCA), the remainder appeared in a variety of products that included acetic acid (7%). The other products were not readily identified.

Criddle and McCarty[10] used electrolytic reduction as a model system for simulating the biological process. Here, reduction led to some 1,1-DCA formation, but this accounted for only about 7% of the TCA loss. The remainder of the products were not identified. It was suggested that TCA may first undergo a one-electron reduction to give an ethyldichloro radical and a chloride ion. Depending on environmental conditions, further transformation of the radical may proceed through many possible routes, one of which leads to 1,1-DCA formation (reaction 3a, Table 3), another to acetic acid, and others to the unidentified products noted.

As with TCA, 1,1-DCA can undergo hydrogenolysis, leading to the formation of chloroethane[39] (CA):

$$\text{(7)} \qquad CH_3CHCl_2 + H^+ + 2e^- \rightarrow CH_3CH_2Cl + Cl^- \quad \text{(hydrolysis).}$$

The rate of this transformation is much slower than that of TCA reduction to 1,1-DCA. This is consistent with the general observation that more highly halogenated compounds are reduced more readily than less-halogenated compounds.[38] The subsequent biological reduction of CA tends to be very slow. However, CA is readily hydrolyzed chemically, leading to the formation of ethanol and HCl:

$$\text{(8)} \qquad CH_3CH_2Cl + H_2O \rightarrow CH_3CH_2OH + H^+ + Cl^- \quad \text{(hydrolysis).}$$

The half-life of this reaction at 25°C has been indicated to be about 30 d based on extrapolation from higher-temperature results,[26] although analyses of CA in the biological systems indicated above[38] suggest a half-life of nearly 2 years. This difference needs to be resolved.

Oxidation of TCA is also possible biologically This has been noted for methanotrophic bacteria.[36] The oxidation of simple halogenated aliphatic compounds by methanotrophic bacteria has been indicated,[29] and this has been extended to trichloroethylene, dichloroethylene, and several other halogenated compounds as well.[19,43] Less-halogenated compounds are oxidized more readily than more highly oxidized compounds. The oxidation of HACs by methanotrophs is brought about by methane monooxygenase (MMO), which is used by the organisms in the initial step in the oxidation of methane to methanol. Several other microorganisms that contain monooxygenases or dioxygenases have also been shown to oxidize HACs through cometabolism. Examples include bacteria that oxidize toluene,[28] propane,[41] ethylene,[31] and ammonia.[1]

TCA oxidation probably involves hydroxylation as the initial step:

$$CH_3CCl_3 + H_2O \rightarrow HOCH_2CCl_3 + 2H^+ + 2e^- \quad \text{(hydroxylation).} \qquad (9)$$

Although this reaction may occur naturally, its relatively slow rate and requirement of a primary substrate and aerobic conditions suggests that this is not a significant transformation mechanism in the natural environment.

The above discussion with TCA indicates the variety of biological and chemical processes that can lead to its transformation in aquatic and terrestrial environments. On a global scale, chemical transformations are probably of greatest significance [Eqs. (4) and (5)] as they are relatively fast. They can occur independent of bacteria and under essentially any environmental conditions. Thus, transformation following rain-out and absorptive transfer into aquatic environments such as lakes, oceans, and groundwater, with residence times of several years, as well as in rain drops and clouds, may serve as significant sinks for atmospheric TCA.

Tetrachloroethylene and Trichloroethylene

Tetrachloroethylene (PCE) and trichloroethylene (TCE) are not generally mentioned as ozone-depleting or greenhouse gases, but they tend to be among the most frequently found HACs in groundwater and waste-disposal sites. Thus, they deserve brief mentioning. Hydrolysis of PCE and TCE is very slow (1.3×10^6 and 9.9×10^8 yr, respectively) and are thus not of environmental significance.[20] Most natural transformations are the biological reductions by hydrogenolysis[2,14,39]:

$$\text{(10a)}\quad CCl_2{=}CCl_2 + H^+ + 2e^- \rightarrow CHCl{=}CCl_2 + Cl^-,$$

$$\text{(10b)}\quad CHCl{=}CCl_2 + H^+ + 2e^- \rightarrow CHCl{=}CHCl\ (\textit{cis}\ \text{or}\ \textit{trans}) + Cl^-,$$

$$\text{(10c)}\quad CHCl{=}CHCl + H^+ + 2e^- \rightarrow CH_2{=}CHCl + Cl^-,$$

$$\text{(10d)}\quad CH_2{=}CHCl + H^+ + 2e^- \rightarrow CH_2{=}CH_2 + Cl^-.$$

Thus, vinyl chloride and three dichloroethylene isomers can be formed from such anaerobic co-metabolic processes. Of some concern is the production of vinyl chloride [Eq. (10c)], which is a known human carcinogen with a U.S. EPA drinking water MCL of 2 $\mu g\ L^{-1}$. Its reduction to ethene [Eq. (10d)] has been demonstrated,[14] although this reaction tends to be quite slow.

Oxidation of the four halogenated ethenes represented in Eqs. (10a)–(10d) also occurs by co-metabolism associated with methane, propane, toluene, or ammonia oxygenases as noted previously. The rates are much faster with the less-halogenated compounds. PCE oxidation either does not occur or is very slow. Attempts to use oxidative co-metabolism for remediation of contaminated soils and groundwaters are underway, but naturally occurring oxidations of these compounds are probably not of global significance.

Carbon Tetrachloride

The carbon in CT is in its most oxidized state and so CT can be transformed only by substitution or reduction. Known pathways for abiotic and biotic CT transformation have been summarized.[10] CT transformation is generally the one-electron reduction of CT to give a trichloromethyl radical and a chloride ion. Depending on environmental conditions, transformations of the radical then proceed by one of five principal routes: (1) dimerization to hexachloroethane, (2) sequential reduction to chloroform, (3) addition to molecular oxygen, (4) formation of a carbenoid or chloromethyl complex, and (5) covalent binding to other organics such as cellular material. The products and intermediates of these routes may be hydrolyzed or oxidized either chemically or biologically, thus leading to a variety of possible products.

The second route of hydrogenolysis to chloroform occurs in a number of microbial systems.[3,4,6,8,9,13,15] Although it occurs most rapidly under the reducing conditions of methanogenesis, it can occur with denitrification as well.[4] The transformation under denitrification conditions has been carried out in the field using acetate as a primary electron acceptor.[35] No evidence for CT transformation under aerobic conditions is available.

Although biological transformation of CT to chloroform is often the dominant anaerobic biological process, conversion to other compounds through alternative pathways also occurs. Chloroform itself can be reduced further to dichloromethane,[15] which, in turn, can be transformed further.[17] However, at least a portion of the other four pathways may also be operative. Transformation to other products has been observed with pure cultures[8,9,13,15] and mixed cultures.[35] With at least one denitrifying Pseudomonad,[8] CT transformation led to formation of chloroform only as a minor product of transformation while other products dominated. Thus, anaerobic biotransformation of CT can lead to a variety of products depending on environmental conditions and the microorganisms involved.

Chemically, CT is very resistant towards attack by nucleophiles. Hydrolysis of CT is very slow [$t_{1/2}$ = 40 yr (Ref. 20)] and probably slower than for most other environmental removal mechanisms. HS^- was found to accelerate CT transformation relative to the rate observed in water, presumably by a reductive mechanism.[23]

Solid surfaces were found to promote CT transformation significantly. In HS^- solutions containing biotite and vermiculite, CT degraded within 4.5 and 2.9 d, respectively, approximately two orders of magnitude faster than in water. Iron sulfides (pyrite and marcasite) which contain both reduced iron and sulfide were shown to increase the rate of CT degradation[23] above the rate observed in sulfide or ferrous sulfide solutions. The reaction was presumed to occur at the particle surface.

Hexachloroethane

Hexachloroethane (HCA) is a compound that has been used as a probe to examine reducing properties of environmental reductants including reduced humic acid.[11] HCA was found to transform nearly quantitatively into PCE via dihaloelimination in the presence of sterile aquifer material, and humic acid was found to be the most reactive component. HCA was also found to undergo dihaloelimination in the presence of Fe(II) containing sheet silicate such as vermiculite and biotite, presumably by oxidizing the structural Fe(II), but the details of the reaction remain to be elucidated.[23] As with CT, the process was much faster when HS^- was added to the system.

Chlorofluorocarbons

CFCs are very resistant to most chemical and biological reactions, especially oxidations. However, because of their highly halogenated characteristic, they are susceptible to reductions. In general, fluorine–carbon bonds are more stable than chlorine–carbon bonds. Thus, reductions are more likely to result in replacement of chlorine with hydrogen through hydrogenolysis, or else the removal of chlorine atoms from adjacent carbon atoms in 2-carbon CFCs to produce halogenated alkenes.

Wolf et al.[44] demonstrated that trichlorofluoromethane (CFC-11) is reductively changed to dichlorofluoromethane in rat liver. Since then, field studies of CFC loss in soils[21] and rice fields[22] have been reported. Evidence for microbial transformations of CFCs has been reported as well.[24,25,34,35,37] Recent observation of CFC-11 decrease along with reduction of CT to chloroform in a field study where denitrification of acetate was induced[35] suggests such transformations can occur under proper environmental conditions and that the transformation is biologically induced. In addition, the disappearance of 1,1,2-dichloro-1,2,2-difluoroethane (CFC-113) was found following the stimulation of denitrification conditions. Decrease in the concentrations of CFC-11 and CFC-12 in methanogenic aquatic sediments also tends to support a biological mechanism.[25] End products of these transformations were not identified. Lesage et al.[24] noted, in groundwater leachate from a municipal landfill, the presence of 1,2-dichloro-1,2,2-trifluoroethane (HCFC-123a) and also chlorotrifluoroethene (CFC-1113). Because of the significant concentration of CFC-113 present, they speculated that these compounds were formed under the reductive conditions present. HCFC-123a formation could have occurred through hydrogenolysis, similar to that found with TCA conversion to 1,1-DCA and hexachloroethane conversion to pentachloroethane. CFC-1113 could have been formed through dihaloelimination, similar to that for hexachloroethane reduction to PCE.

The reductions of CFC-11 and CFC-113 would be expected to be slower than of their completely chlorinated counterparts because of the resistance to conversion at the fluorine–carbon bond. Nevertheless, reactions tend to be similar.

Field Examples of HAC conversion

There are numerous examples where products of abiotic and biotic conversions of HACs have occurred. An example of the biological conversion of trichloroethylene into dichloroethylene and vinyl chloride is at the St. Joseph, Michigan site where the aquifer was contaminated more than 10 years ago with trichloroethylene.[28] Reduction has resulted in a groundwater plume containing in the mg L^{-1} range of TCE, *cis*- and *trans* 1,2-DCE, and VC. A possibility of stimulating methanotrophic bacteria to oxidize these compounds through the addition of methane and oxygen to the aquifer is under evaluation.

A good example of the reductive transformations that can take place under highly active anaerobic conditions is given by analyses of gases from municipal landfills. A summary of such data by Wolfe[45] is contained in Table 4. Similar

Table 4. Composition of gaseous emissions (ppbv) from eight municipal refuse sites [(after W.H. Wolfe as reported by Charnley et al. (Ref. 45)].

Chemical	Formula	Average	Range[a]
PCE	$CCl_2=CCl_2$	7150	1400–12000
TCE	$CHCl=CCl_2$	5090	nd–16000
c-1,2-DCE	—	—	—
t-1,2-DCE	$CHCl=CHCl$	240	nd–1300
VC	$CH_2=CHCl$	5600	2600–12000
TCA	CH_3CCl_3	170	nd–770
1,1-DCE	$CH_2=CCl_2$	100	nd–490
1,1-DCA	CH_3CHCl_2	2500	200–6200
CA	CH_3CH_2Cl	370	nd–950
CFC-11	CCl_3F	1370	130–7600
HCFC-21	$CHCl_2F$	1430	nd–4600
CFC-12	CCl_2F_2	95	nd–500
HCFC-22	$CHClF_2$	1280	400–4800
Methane	CH_4	$473(10^6)$	$370(10^6)$–$510(10^6)$

[a]nd = not detected.

results for analysis of landfill gaseous emanations were given by Wood and Porter.[46] These data suggest that significant reductions of HACs occur in landfills. This is evidenced in Table 4 by the high concentration of VC relative to PCE and TCE, of 1,1-DCA relative to TCA, of HCFC-21 relative to CFC-11, and of HCFC-22 relative to CFC-12. It, thus, appears that although landfills produce much methane, which is a greenhouse gas, they do reduce the concentration of ozone-depleting gases (TCA, CFC-11, CFC-12, and probably CFC-113). In addition, CFC-11 and CFC-12 are about 500 times as efficient in IR absorption as methane,[33] and thus the reductions noted in the CFCs can offset somewhat the greenhouse effect of the methane produced, although not to a great extent because of the much greater quantity of methane that comes from landfills.

Acknowledgments

This work was supported by the U.S. Environmental Protection Agency sponsored Western Region Hazardous Substance Research Center and through Grants R-815738, CR-814823, and CR-815816. It has not been subjected to EPA's peer and administrative review and, therefore, should not be assumed to reflect the views or the endorsement of that agency.

References

1. Arciero, D., T. Vannelli, M. Logan, and A. B. Hooper. 1989. Degradation of trichloroethylene by the ammonia-oxidizing bacterium *Nitrosomonas europaea.* Biochem. Biophys. Res. Commun. **159**:640–643.
2. Barrio-Lage, G., F.Z. Parsons, R.S. Nassar, and P.A. Lorenzo. 1986. Sequential Dehalogenation of Chlorinated Ethenes. Environ. Sci. Technol. **20**:96–99.
3. Bouwer, E.J., and P.L. McCarty. 1983. Transformations of halogenated organic compounds under denitrification conditions. Appl. Environ. Microbiol. **45**:1295–1299.
4. Bouwer, E.J., and J.P. Wright. 1988. Transformations of trace halogenated aliphatics in anoxic biofilm columns. J. Contam. Hydrol. **2**:155–169.
5. Britton, E.C., and W.R. Reed. 1932. Hydrolysis of 1,1,1-trichloroethane. Chem. Abstr. **26**:5578.
6. Castro, C.E., R.S. Wade, and N.O. Belser. 1985. Biodehalogenation: reactions of cytochrome P-450 with polyhalomethanes. Biochemistry **24**:204–210.
7. Cline, P.V., and J.J. Delfino. 1989. Transformation Kinetics of 1,1,1-trichloroethane to the stable product 1,1-dichloroethene. In R.A. Larson (ed.), Biohazards of Drinking Water Treatment, Lewis Publishers, Inc., Chelsa, MI, pp. 47–56.
8. Criddle, C.S., J.T. DeWitt, D. Grbic-Galic, and P.L. McCarty. 1990. Transfor-

mation of carbon tetrachloride by *Pseudomonas* sp. strain KC under denitrification conditions. Appl. Environ. Microbiol. **56**:3240–3246.

9. Criddle, C.S., J.T. DeWitt, and P.L. McCarty. 1990. Reductive dehalogenation of carbon tetrachloride by *Escherichia coli* K-12. Appl. Environ. Microbiol. **56**:3247–3254.

10. Criddle, C.S., and P.L. McCarty. 1991. Electrolytic model system for reductive dehalogenation in aqueous environments. Environ. Sci. Technol. **25**:973–978.

11. Curtis, G.P. 1991. Ph.D. dissertation, Stanford University, Stanford, CA.

12. Eberson, L. 1987. Electron Transfer Reactions in Organic Chemistry, Springer-Verlag, Berlin.

13. Egli, C., T. Tschan, R. Scholtz, A.M. Cook, and T. Leisinger 1988. Transformation of tetrachloromethane to dichloromethane and carbon dioxide by *Acetobacterium woodii.* Appl. Environ. Microbiol. **54**:2819–2824.

14. Freedman, D.L., and J.M. Gossett. 1989. Biological reductive dechlorination of tetrachloroethylene and trichloroethylene to ethylene under methanogenic conditions. Appl. Environ. Microbiol. **55**:2144–2151.

15. Gälli, R., and P.L. McCarty. 1989. Biotransformations of 1,1,1-trichloroethane, trichloromethane, and tetrachloromethane by a *Clostridium* sp. Appl. Environ. Microbiol. **55**:837–844.

16. Gälli, R., and P.L. McCarty. 1989. Kinetics of biotransformation of 1,1,1-trichloroethane by clostridium sp. Strain TCA IIB. Appl. Environ. Microbiol. **55**:845–851.

17. Gossett, J.M., and D.L. Freedman. 1988. Biodegradation of dichloromethane under methanogenic conditions. In NATO/CCMS Pilot Study on Demonstration of Remedial Action Technologies for Contaminated Land and Groundwater. Bilthoven, the Netherlands, National Institute of Public Health and Environmental Protection.

18. Haag, W.R., and T. Mill. 1988. Effect of Subsurface Sediment on Hydrolysis of Haloalkanes and Epoxides. Environ. Sci. Technol. **22**:658–633.

19. Henson, J.M., M.V. Yates, J.W. Cochran, and D.L. Shackleford. 1988. Microbial removal of halogenated methanes, ethanes, and ethylenes in an aerobic soil exposed to methane. FEMS Microbiol. Ecol. **53**:193–201.

20. Jeffers, P.M., L.M. Ward, L.M. Woytowitch, and N.L. Wolfe. 1989. Homogenous hydrolysis rate constants for selected chlorinated methanes, Ethanes and propanes. Environ. Sci. Technol. **23**:965–969.

21. Khalil, M.A.K., and R.A. Rasmussen. 1989. The Potential of soils as a sink of chlorofluorocarbons and other manmade chlorocarbons. Geophys. Res. Lett. **16**: 679–682.

22. Khalil, M.A.K., R.A. Rasmussen, M.X. Wang, and L. Ren. 1990. Emissions of trace gases from Chinese rice fields and biogas generators: CH_4, N_2O, CO, CO_2, chlorocarbons, and hydrocarbons. Chemosphere **20**:207–226.

23. Kriegman, M.R., and M. Reinhard. 1991. Reduction of hexachloroethane and

carbon tetrachloride at surfaces of biotite, vermiculite, pyrite and marcasite. In R. Baker (ed.), Organic Substances and Sediments in Water, Vol. 2, Processes and Analytical, Lewis Publishers, Chelsea, MI. pp. 349–364.

24. Lesage, S., R.E. Jackson, M. W. Priddle, and P.G. Riemann. 1990. Occurrence and fate of organic solvent residues in anoxic groundwater at the Gloucester landfill, Canada. Environ. Sci. Technol. **24**:559–566.

25. Lovley, D.R., and J.C. Woodward. 1990. Consumption of Freons F-11 and F-12 in methane-producing aquatic sediments. EOS **71**:1236.

26. Mabey, W., and T. Mill. 1978. Critical review of hydrolysis of organic compounds in water under environmental conditions, J. Phys. Chem. Ref. Data **7**:383–415.

27. McCarty, P.L., L. Semprini, M.E. Dolan, T.C. Harmon, C. Tiedeman, and S.M. Gorelick. 1991. In-Situ methanotrophic bioremediation for contaminated groundwater at St. Joseph, Michigan. In R.E. Hinchee and R.F. Olfenbuttel (eds.), On-site Bioreclamation, Butterworth-Heinemann, Boston, MA, pp. 16–40.

28. Nelson, M.J.K., S.O. Montgomery, and P.H. Pritchard. 1988. Trichloroethylene metabolism by microorganisms that degrade aromatic compounds. Appl. Environ. Microbiol. **54**:604–606.

29. Patel, R.N., C.T. Hou, A.I. Laskin, and A. Felix. 1982. Microbial oxidation of hydrocarbons: Properties of a soluble methane monooxygenase from a facultative methane-utilizing organism, Methylobacterium sp. Strain CRL-26. Appl. Environ Microbiol. **44**:1130–1137.

30. Pearson, C.R., and G. McConnell. 1975. Chlorinated C_1 and C_2 hydrocarbons in the marine environment. Proc. Royal Soc. London, Series B **189**:305–332.

31. Roberts, P.V., L. Semprini, G.D. Hopkins, P.L. McCarty, and M. Reinhard. 1989. In-situ aquifer restoration of chlorinated aliphatics by methanotrophic bacteria. Report No. EPA/600/2-89/033. U.S. EPA Center for Environmental Information, Cincinnati, OH.

32. Rosswall, T. 1991. Greenhouse gases and global change: International collaboration. Environ. Sci. Technol. **25**:567–573.

33. Rowland, E.S. 1991. Stratospheric ozone in the 21st century. Environ. Sci. Technol. **25**:622–628.

34. Semprini, L., G.D. Hopkins, P.V. Roberts, and P.L. McCarty. 1990. In-situ biotransformation of carbon tetrachloride, 1,1,1-trichloroethane, freon-11, and freon-113 under anoxic conditions. EOS **71**:1324.

35. Semprini, L., G.D. Hopkins, P.V. Roberts, and P.L. McCarty. 1992. In-situ Transformation of Carbon Tetrachloride and other Halogenated Compounds Resulting From Biostimulation Under Anoxic Conditions, Environ. Sci. Technol. **26**:2454–2461.

36. Strand, S.E., M.D. Bjelland, and H.D. Stensel. 1990. Kinetics of chlorinated hydrocarbon degradation by suspended cultures of methane-oxidizing bacteria. Res. J. Water Pollut. Cont. Fed. **62**:124–129.

37. Strand, S.E., and B.A. Denovan. 1991. Anaerobic biodegradation of chlorofluo-

rocarbons. Abstract Q-57, 91st General Meeting of the American Society for Microbiology, Dallas, p. 286.

38. Vogel, T.M., C.S. Criddle, and P.L. McCarty. 1987. Transformations of halogenated aliphatic compounds. Environ. Sci. Technol. **21**:722–736.

39. Vogel, T.M., and P.L. McCarty. 1985. Biotransformation of tetrachloroethylene to trichloroethylene, dichloroethylene, vinyl chloride, and carbon dioxide under methanogenic conditions. Appl. Environ. Microbiol. **49**:1080–1083.

40. Vogel, T.M., and P.L. McCarty. 1987. Abiotic and biotic transformations of 1,1,1-trichloroethane under methanogenic conditions. Environ. Sci. Technol. **21**:1208–1213.

41. Wackett, L.P., G.A. Brusseau, S.R. Householder, and R.S. Hanson. 1989. Survey of microbial oxygenases: Trichloroethlene degradation by propane-oxidizing bacteria. Appl. Environ. Microbiol. **55**:2960–2964.

42. Walraevens, R., P. Trouillet, and A. Devos. 1974. Basic elimination of HCl from chlorinated ethanes. Int. J. Chem. Kinet. **6**:777–786.

43. Wilson, J.T., and B.H. Wilson. 1985. Biotransformation of trichloroethylene. Appl. Environ. Microbiol. **49**:242–243.

44. Wolf, C.R., L.K. King, and D.V. Parke. 1978. The anaerobic dechlorination of trichlorofluoromethane by rat liver preparations in vitro. Chem. Biol. Interact. **21**:277–288.

45. Wolfe, W.H., as referenced by G. Charnley, E.A.C. Crouch, L.C. Green, and T.L. Lash. 1988. Municipal Solid Waste Landfilling: A Review of Environmental Effects, Meta Systems, Inc., Cambridge, MA.

46. Wood, J.A., and M.L. Porter. 1987. Hazardous pollutants in Class II landfills. Hazardous Waste Manag. **37**:609–615.

*Index**

*Page numbers printed in boldface type refer to tables or figures.